Principles of Analysis

Principles of Analysis

Measure, Integration, Functional Analysis, and Applications

Hugo D. Junghenn

CRC Press
Taylor & Francis Group
Boca Raton London New York

CRC Press is an imprint of the
Taylor & Francis Group, an **informa** business

CRC Press
Taylor & Francis Group
6000 Broken Sound Parkway NW, Suite 300
Boca Raton, FL 33487-2742

First issued in paperback 2022

Version Date: 20180324

ISBN 13: 978-1-03-247621-6 (pbk)
ISBN 13: 978-1-4987-7328-7 (hbk)

DOI: 10.1201/9781315151601

Library of Congress Cataloging-in-Publication Data

Names: Junghenn, Hugo D. (Hugo Dietrich), 1939- author.
Title: Principles of real analysis : measure, integration, functional analysis, and applications / Hugo D. Junghenn.
Description: Boca Raton : CRC Press, Taylor & Francis Group, 2018. | Includes bibliographical references and index.
Identifiers: LCCN 2017061660 | ISBN 9781498773287
Subjects: LCSH: Functions of real variables--Textbooks. | Mathematical analysis--Textbooks.
Classification: LCC QA331.5 .J86 2018 | DDC 515/.8--dc23
LC record available at https://lccn.loc.gov/2017061660

Visit the e-resources at: https://www.crcpress.com/9781498773287

Visit the Taylor & Francis Web site at
http://www.taylorandfrancis.com

and the CRC Press Web site at
http://www.crcpress.com

TO MY CHILDREN

Katie and Patrick

AND TO MY WIFE

Mary

AS ALWAYS

Contents

Preface

The purpose of this book is to provide a rigorous and detailed treatment of the essentials of measure, integration, and functional analysis at the graduate level. It is assumed that the reader has an undergraduate background in what is now traditionally called real analysis, including elementary set theory and a rigorous treatment of limits, continuity, differentiation, Riemann integration, and uniform convergence. An acquaintance with complex function theory, in particular the complex exponential function e^z and Cauchy's integral equation, is needed for a few applications. A knowledge of basic linear algebra, at least the notions of subspace, basis, and linear transformation, is also assumed. Metric spaces and general topology are developed in detail in Chapter 0. The former topic will be needed for the treatment of L^p spaces and the latter for the chapters on Radon measures and weak topologies.

The book has four parts. Part I consists of Chapters 1–7 and develops the general theory of Lebesgue integration. A course in the subject could consist of Chapters 1–5 with Chapters 6 or 7 optional.

Part II is organized as a course in functional analysis. Chapters 8–12 could form the core of such a course, with Chapter 13 optional. Some of the applications and examples in Part II rely on the measure and integration developed in Part I. The reader with a background in this subject could safely omit Part I. Chapter 14 consists of deeper theorems in functional analysis as well as applications. Some of the applications in the remainder of the book rely on results of this chapter.

Part III consists of a variety of topics and applications that depend on, and indeed are meant to illustrate the power of, topics developed in the first two parts. The chapters here are largely independent, with the exception of Chapter 17, which depends on some results in Chapter 16. The goal of these chapters is to provide a relatively quick overview of the essentials of the subjects treated therein. The approach to these is sufficiently detailed so that the reader can follow the development with relative ease. It is hoped that the treatment here will inspire the reader to consult some of the many fine texts that specialize in these subjects, some of which are listed in the bibliography.

Part IV consists of two appendices with proofs of the change of variables theorem and a theorem on separate and joint continuity. The reader may safely omit the proofs without disturbing the flow of the text.

The book contains nearly 700 exercises. Hints and/or a framework of intermediate steps are given for the more difficult exercises. Many of these are extensions of material in the text or are of special independent interest. Exercises related in a critical way to material elsewhere in the text are marked with either an upward arrow, referring to earlier results, or a downward arrow, referring to later material. Instructors with suitable bona fides may obtain complete solutions to the exercises from the publisher.

A word about numbering: Proclamations (theorems, lemmas, examples, etc.) are numbered consecutively in each section. Thus 1.2.3 refers to the third proclamation in Section 2 of

Chapter 1. Important equations are numbered consecutively in each chapter. Thus (4.5) refers to the fifth such equation of Chapter 4. Equations within a proof that are only locally relevant are referenced by symbols such as (†), (α), etc. Exercises are numbered consecutively within each chapter. Thus Ex. 6.7 refers to the seventh exercise of Chapter 6.

The book is an outgrowth of courses in analysis taught at The George Washington University. Specific notes for the book have been tested in classes over the last three years and have benefitted greatly from comments, questions, and corrections from students; for these I am grateful. It goes without saying that the book has also benefitted from several excellent texts in analysis that have served as valuable resources—several of these are listed in the bibliography. Finally, I wish to express my gratitude to my teacher C.T. Taam who first exposed me to much of the mathematics that appears in this book.

Hugo D. Junghenn
Washington, D.C.

Chapter 0

Preliminaries

In this chapter we assemble the basic material needed for the topics treated in the book. The reader may wish to simply skim the chapter at first, returning to specific topics as the need arises.

0.1 Sets

The terms *set, collection*, and *family* are synonymous, although in some contexts one term may be preferred over another, as in a *collection of sets* or a *family of functions*. Sets are usually denoted by capital letters in various styles, and members of sets by small letters. As usual, the notation $x \in A$ denotes membership of x in A.

A concrete set may be described either by (perhaps only partially) listing its members or by *set-builder notation*. The latter is of the form $\{x : P(x)\}$, which is read "the set of all x such that $P(x)$," where $P(x)$ is a well-defined property that x must possess to belong to the set. For example, the set of all odd integers may be described as

$$\{\pm 1, \pm 3, \pm 5, \ldots\} = \{n : n = 2m - 1 \text{ for some integer } m\}.$$

Set Operations

If A is a subset of B, we write $A \subseteq B$. If all sets in a particular discussion are subsets of a set X, then X is called a **universal set**. The **power set** of a set X is the collection $\mathcal{P}(X)$ of all subsets of X. If $A, B \subseteq X$, then $A \cup B$, $A \cap B$, and $A \setminus B$ denote the union, intersection and relative difference of A and B, respectively, and A^c denotes the complement of A in X. If $\mathcal{A} \subseteq \mathcal{P}(X)$ and $B \in \mathcal{P}(X)$, we define the **trace of $\mathcal{A}$ on** B by

$$\mathcal{A} \cap B = \{A \cap B : A \in \mathcal{A}\}.$$

The union and intersection of an indexed family $\mathcal{A} = \{A_i : i \in \mathfrak{I}\}$ of sets are denoted, respectively, by

$$\bigcup \mathcal{A} = \bigcup_{i \in \mathfrak{I}} A_i \quad \text{and} \quad \bigcap \mathcal{A} = \bigcap_{i \in \mathfrak{I}} A_i.$$

If the index set in these operations is $\{1, 2 \ldots, n\}$ or $\{1, 2 \ldots\}$, we write instead

$$\bigcup_{j=1}^{n} A_j = A_1 \cup \cdots \cup A_n, \quad \bigcap_{j=1}^{n} A_j = A_1 \cap \cdots \cap A_n, \quad \bigcup_{j=1}^{\infty} A_j = A_1 \cup A_2 \cup \ldots, \quad \bigcap_{j=1}^{\infty} A_j = A_1 \cap A_2 \cap \ldots$$

0.1.1 Proposition. *Union and intersection have the following properties:*

(a) $\left(\bigcup_{i\in\mathfrak{I}} A_i\right)^c = \bigcap_{i\in\mathfrak{I}} A_i^c.$ **(b)** $\left(\bigcap_{i\in\mathfrak{I}} A_i\right)^c = \bigcup_{i\in\mathfrak{I}} A_i^c.$

(c) $A \cup \left(\bigcap_{i\in\mathfrak{I}} A_i\right) = \bigcap_{i\in\mathfrak{I}} A \cup A_i.$ **(d)** $A \cap \left(\bigcup_{i\in\mathfrak{I}} A_i\right) = \bigcup_{i\in\mathfrak{I}} A \cap A_i.$

Parts (a) and (b) of the proposition are known as *DeMorgan's laws*, and parts (c) and (d) are called *distributive laws.*

A family $\{A_i : i \in \mathfrak{I}\}$ of sets is **disjoint** if $A_i \cap A_j = \emptyset$ whenever $i \neq j$. In this case, the union $\bigcup_{i\in\mathfrak{I}} A_i$ is said to be **disjoint**. A **partition** of a set X is a collection of nonempty, disjoint sets whose union is X.

A sequence of sets A_n is said to be **increasing** if $A_1 \subseteq A_2 \subseteq \cdots$, in which case we write $A_n \uparrow$. Similarly, the sequence is **decreasing** if $A_1 \supseteq A_2 \supseteq \cdots$, written $A_n \downarrow$. In the first case we also write $A_n \uparrow A$, where $A = A_1 \cup A_2 \cup \cdots$, and in the second $A_n \downarrow A$, where $A = A_1 \cap A_2 \cap \cdots$.

Cartesian products of finite or infinite sequences of sets $A_1, A_2, \ldots$ are denoted, respectively, by

$$\prod_{n=1}^{d} A_n = A_1 \times \cdots \times A_d \quad \text{and} \quad \prod_{n=1}^{\infty} A_n = A_1 \times A_2 \times \ldots.$$

In the case $A_1 = A_2 = \cdots = A$, we write instead A^d or A^∞.

0.1.2 Proposition. *Cartesian products have the following properties:*

(a) $A \times (A_1 \cup A_2 \cup \cdots) = (A \times A_1) \cup (A \times A_2) \cup \cdots.$

(b) $A \times (A_1 \cap A_2 \cap \cdots) = (A \times A_1) \cap (A \times A_2) \cap \cdots.$

(c) $(A_1 \cap A_2 \cap \cdots) \times (B_1 \cap B_2 \cap \cdots) = (A_1 \times B_1) \cap (A_2 \times B_2) \cap \cdots.$

The Cartesian product $X := \prod_{i\in\mathfrak{I}} X_i$ of a family of nonempty sets X_i is defined as the collection of all functions $f : \mathfrak{I} \to \bigcup_{i\in\mathfrak{I}} X_i$ such that $f(i) \in X_i$ for each i. (The axiom of choice asserts that the Cartesian product is nonempty.) The value $f(i)$ is called the $\boldsymbol{i}$**th coordinate of** $\boldsymbol{f}$. Note that if we identify a finite index set $\mathfrak{I}$ with the set $\{1, \ldots, n\}$, then a function f in X is completely described by the n-tuple $(f(1), \ldots, f(n))$. Thus the general definition of Cartesian product reduces in this case to the "concrete" definition given above. A similar remark applies if $\mathfrak{I}$ is countably infinite.

Number Systems

The following notation is used for the standard number systems:

$\mathbb{N} :=$ the set of **positive integers.**

$\mathbb{Z} :=$ the set of **integers.**

$\mathbb{Q} :=$ the set of **rational numbers.**

$\mathbb{R} :=$ the set of **real numbers.**

$\mathbb{C} :=$ the set of **complex numbers.**

Two subsets of $\mathbb{C}$ are of particular importance:

$\mathbb{D} := \{z \in \mathbb{C} : |z| < 1\}$ the **open unit disk** and $\mathbb{T} := \{z \in \mathbb{C} : |z| = 1\}$ the **circle group.**

The symbol $\mathbb{K}$ serves double-duty:

$$\mathbb{K} := \mathbb{R} \text{ or } \mathbb{C}.$$

This will be convenient, for example, when considering vector spaces where the scalar field may be either $\mathbb{R}$ or $\mathbb{C}$. If $A \subseteq \mathbb{R}$ we write A^+ for the set $A \cap [0, +\infty.)$. For example, the set of nonnegative integers is

$$\mathbb{Z}^+ = \{n \in \mathbb{Z} : n \geq 0\}.$$

If $A \subseteq \mathbb{C}$ we denote by A_* the set of nonzero members of A.

Real and complex d-dimensional Euclidean space are defined, respectively, by

$$\mathbb{R}^d := \mathbb{R} \times \cdots \times \mathbb{R} \quad \text{and} \quad \mathbb{C}^d := \mathbb{C} \times \cdots \times \mathbb{C} \quad (d \text{ factors}).$$

We also use the notation

$$\mathbb{K}^d := \mathbb{K} \times \cdots \times \mathbb{K}$$

where appropriate. A **d-dimensional interval** in $\mathbb{R}^d$ is indicated with boldface, as in

$$(\boldsymbol{a}, \boldsymbol{b}] := (a_1, b_1] \times \cdots \times (a_d, b_d], \quad \text{where } \boldsymbol{a} := (a_1, \ldots, a_d) \text{ and } \boldsymbol{b} := (b_1, \ldots, b_d),$$

The **Euclidean norm** on $\mathbb{K}^d$ is denoted by

$$|\boldsymbol{z}| := \sqrt{|z_1|^2 + \cdots + |z_d|^2}.$$

The **extended real number system** is the set

$$\overline{\mathbb{R}} := \mathbb{R} \cup \{\pm\infty\} = [-\infty, +\infty]$$

with the following order structure and operations:

$$\begin{gathered}
-\infty < x < +\infty \quad \text{for all } x \in \mathbb{R}, \\
x + \infty = \infty \text{ if } -\infty < x \leq +\infty, \quad x - \infty = -\infty \text{ if } -\infty \leq x < +\infty, \\
x \cdot (+\infty) = +\infty \text{ if } 0 < x \leq +\infty, \quad x \cdot (+\infty) = -\infty \text{ if } -\infty < x < 0, \\
x \cdot (-\infty) = -\infty \text{ if } 0 < x < +\infty, \quad x \cdot (-\infty) = +\infty \text{ if } -\infty \leq x < 0, \\
\frac{x}{+\infty} = \frac{x}{-\infty} = 0 \text{ if } -\infty < x < +\infty, \\
0 \cdot (\pm\infty) = 0.
\end{gathered}$$

Hereafter, we shall use the customary abbreviation ∞ for $+\infty$. The members of $[0, \infty] = \overline{\mathbb{R}}^+$ are called **extended nonnegative real numbers**. The symbol $\mathbb{K}$ is modified as follows to apply to the extended setting:

$$\overline{\mathbb{K}} := \overline{\mathbb{R}} \text{ or } \mathbb{C}.$$

Relations

A **relation** on a nonempty set X is a nonempty set $\sim$ of ordered pairs from X. It is customary to write $x \sim y$ rather than the prolix $(x, y) \in \sim$. A relation is said to be

(a) **reflexive** if $x \sim x$ for every $x \in X$;

(b) **symmetric** if $x \sim y \Rightarrow y \sim x$;

(c) **transitive** if $x \sim y$ and $y \sim z \Rightarrow x \sim z$;

(d) **antisymmetric** if $x \sim y$ and $y \sim x \Rightarrow x = y$.

A relation $\sim$ on X that is reflexive, symmetric, and transitive is called an **equivalence relation.** The **equivalence class of** $x \in X$ is the set

$$[x] := \{y \in X : x \sim y\}.$$

The collection $X/\sim$ of distinct equivalence classes is a partition of X. Conversely, given a partition of X, the relation defined by $x \sim y$ iff x and y are in the same partition member is an equivalence relation on X whose equivalence classes are precisely the members of the partition. Thus equivalence relations and partitions are two versions of the same concept. An example of an equivalence relation on $\mathbb{R}$ is given by the definition $x \sim y$ iff $x - y \in \mathbb{Q}$. Here, the distinct equivalence classes are the sets $x + \mathbb{Q}$, where x is irrational, and these classes partition $\mathbb{R}$. Additional examples of equivalence relations appear throughout the text.

A relation that is reflexive, antisymmetric, and transitive is called a **partial order**. For partial orders we use the more suggestive notation $x \preceq y$ (equivalently, $y \succeq x$) rather than $x \sim y$. A set with a partial order is called a **partially ordered set**. An **upper (lower) bound** of a subset Y of a partially ordered set X is a member $x \in X$ such that $y \preceq x$ ($x \preceq y$) for all $y \in Y$. The **supremum (infimum)** of Y is an upper (lower) bound x_0 of Y such that $x_0 \preceq x$ ($x_0 \succeq x$) for all upper (lower) bounds x of Y. For example, if the power set $\mathcal{X} = \mathcal{P}(X)$ is ordered upward by inclusion, that is, $A \preceq B$ iff $A \subseteq B$, and if $\mathcal{Y} \subseteq \mathcal{X}$, then $\bigcap \mathcal{Y}$ and $\bigcup \mathcal{Y}$ are, respectively, the infimum and supremum of $\mathcal{Y}$. The terms **least upper bound** and **greatest lower bound** are synonymous with supremum and infimum, respectively. A member x of X is **maximal** if $y \in X$ and $x \preceq y \Rightarrow y = x$, and is **minimal** if $y \in X$ and $x \succeq y \Rightarrow y = x$.

A nonempty subset Y of a partially ordered set X is said to be **totally ordered** or a **chain** if for all $y_1, y_2 \in Y$ either $y_1 \preceq y_2$ or $y_1 \succeq y_2$. We shall make frequent and significant use of the following important result. (For a proof based on the axiom of choice see, for example, [23] or [30].)

0.1.3 Zorn's Lemma. *Let X be a partially ordered set such that every chain has an upper bound in X. Then X has a maximal element.*

Functions

The terms **mapping, transformation**, and **function** are synonymous. A function f with domain $\operatorname{dom} f = X$ and range $\operatorname{ran} f \subseteq Y$ is symbolized by $f : X \to Y$. We also occasionally write $x \mapsto f(x)$, as in $x \mapsto \sqrt{x}$, to describe a function. The collection of all functions from X to Y is denoted by Y^X.

The **image of** $A \subseteq X$ and the **preimage of** $B \subseteq Y$ under a function $f : X \to Y$ are defined, respectively, by

$$f(A) = \{f(x) : x \in A\} \quad \text{and} \quad f^{-1}(B) = \{x : f(x) \in B\}.$$

A function $f : X \to Y$ is **surjective** or **onto** Y if $f(X) = Y$, and **injective**, or **one-to-one** (1-1), if $x_1 \neq x_2 \Rightarrow f(x_1) \neq f(x_2)$. A **surjection (injection)** is a map that is surjective (injective). A mapping that is both surjective and injective is said to be **bijective** and is called a **bijection** or a **one-to-one correspondence**.

An important example of a surjective function is the **quotient map** $Q : X \to X/\sim$ associated with an equivalence relation $\sim$ on a nonempty set X. Here $Q(x) := [x]$, the equivalence class containing x. The preimage of a subset B of $X/\sim$ under Q is the union of all equivalence classes $[x]$ in B.

The basic properties of images and preimages are summarized in the following proposition.

0.1.4 Proposition. *Let* $\{X_i : i \in \mathfrak{I}\}$ *be a collection of subsets of* X *and* $\{Y_j : j \in \mathfrak{J}\}$ *a collection of subsets of* Y. *Then*

(a) $f^{-1}\left(\bigcup_{j\in\mathfrak{J}} Y_j\right) = \bigcup_{j\in\mathfrak{J}} f^{-1}(Y_j)$.
(b) $f^{-1}\left(\bigcap_{j\in\mathfrak{J}} Y_j\right) = \bigcap_{j\in\mathfrak{J}} f^{-1}(Y_j)$.
(c) $f\left(\bigcup_{i\in\mathfrak{I}} X_i\right) = \bigcup_{i\in\mathfrak{I}} f(X_i)$.
(d) $f\left(\bigcap_{i\in\mathfrak{I}} X_i\right) \subseteq \bigcap_{i\in\mathfrak{I}} f(X_i)$.
(e) $f^{-1}(Y_j^c) = \left[f^{-1}(Y_j)\right]^c$.
(f) $f(X_i^c) \subseteq \left[f(X_i)\right]^c$.
(g) $f(f^{-1}(Y_j)) \subseteq Y_j$.
(h) $X_i \subseteq f^{-1}(f(X_i))$.

Equality holds in (d) and (h) if f *is injective. Equality holds in (f) and (g) if* f *is surjective.*

For $f : X \to Y$, $\mathcal{A} \subseteq \mathcal{P}(X)$ and $\mathcal{B} \subseteq \mathcal{P}(Y)$, we define the collections

$$f(\mathcal{A}) = \{f(A) : A \in \mathcal{A}\} \subseteq \mathcal{P}(Y) \quad \text{and} \quad f^{-1}(\mathcal{B}) = \{f^{-1}(B) : B \in \mathcal{B}\} \subseteq \mathcal{P}(X).$$

If $f : X \to Y$ and $g : Y \to Z$ are functions with $f(X) \subseteq Y$, then $g \circ f : X \to Z$ denotes the **composition of** g **and** f:

$$(g \circ f)(x) = g\big(f(x)\big), \quad x \in X.$$

The following relation holds for subsets $A \subseteq Z$:

$$(g \circ f)^{-1}(A) = f^{-1}\big(g^{-1}(A)\big).$$

The **identity function** id_X on a set X is defined by $\mathrm{id}_X(x) = x$ for all $x \in X$. If $A \subseteq X$, then the restriction of id_X to A is called the **inclusion map** and is frequently denoted by $\iota_A : A \hookrightarrow X$.

If $f : X \to Y$ is bijective, then the **inverse** $f^{-1} : Y \to X$ of f is defined by the rule $x = f^{-1}(y)$ iff $y = f(x)$. One then has

$$f^{-1} \circ f = \mathrm{id}_X \quad \text{and} \quad f \circ f^{-1} = \mathrm{id}_Y.$$

If X is a universal set and $A \subseteq X$, then the **indicator function of** A is defined by

$$\mathbf{1}_A(x) = \begin{cases} 1 & \text{if } x \in A, \\ 0 & \text{if } x \in A^c. \end{cases}$$

Clearly, indicator functions satisfy the relations

$$\mathbf{1}_{AB} = \mathbf{1}_A \mathbf{1}_B, \quad \mathbf{1}_{A\cup B} = \mathbf{1}_A + \mathbf{1}_B - \mathbf{1}_A \mathbf{1}_B, \quad \text{and} \quad \mathbf{1}_{A^c} = 1 - \mathbf{1}_A.$$

A special case of an indicator function is the function δ_{xy} defined by

$$\delta_{xy} = \begin{cases} 1 & \text{if } x = y, \\ 0 & \text{if } x \neq y. \end{cases}$$

Indeed, if $A = \{(x, x) : x \in X\}$, then $\delta_{xy} = \mathbf{1}_A(x, y)$.

The functions x^+ and x^- on $\mathbb{R}$ are defined by

$$x^+ = \max\{x, 0\} \quad \text{and} \quad x^- = \max\{-x, 0\}.$$

The following relations are easily established:

$$x = x^+ - x^- \quad \text{and} \quad |x| = x^+ + x^-.$$

The real and imaginary parts of a complex number z are denoted, respectively, by $\operatorname{Re} z$ and $\operatorname{Im} z$, the conjugate by $\overline{z}$, and the modulus by $|z|$. Thus

$$z = \operatorname{Re} z + i \operatorname{Im} z, \quad \overline{z} = \operatorname{Re} z - i \operatorname{Im} z, \quad \text{and} \quad |z| = \sqrt{(\operatorname{Re} z)^2 + (\operatorname{Im} z)^2}.$$

The **signum** $\operatorname{sgn}(z)$ of a complex variable z is defined by

$$\operatorname{sgn}(z) = \begin{cases} \dfrac{|z|}{z} & \text{if } z \neq 0, \\ 0 & \text{otherwise.} \end{cases}$$

Thus $|z| = z \operatorname{sgn}(z)$ for all $z \in \mathbb{C}$ and $|\operatorname{sgn}(z)| = 1$ for all $z \neq 0$.

If $F \subseteq Y^X$, then the **evaluation functional at** $x \in X$ is the function $\widehat{x} : F \to Y$ defined by

$$\widehat{x}(f) = f(x), \quad f \in F.$$

The notation δ_x is sometimes used for $\widehat{x}$. The **adjoint** or **dual** of a map $\phi : Z \to X$ (with respect to F) is the function $\phi^* : F \to Y^Z$ defined by

$$\phi^*(f) = f \circ \phi, \quad f \in F.$$

The notation ϕ' is also used.

The following notation for real-valued functions on a set X will be needed in subsequent chapters:

$$\begin{aligned} f^+(x) &:= \max\{f(x), 0\} & f^- &:= (-f)^+ \\ (f_1 \vee \cdots \vee f_n)(x) &:= \max_{1 \le k \le n} f_k(x) & (f_1 \wedge \cdots \wedge f_n)(x) &:= \min_{1 \le k \le n} f_k(x) \\ (\sup_n f_n)(x) &:= \sup_n f_n(x) & (\inf_n f_n)(x) &:= \inf_n f_n(x) \\ (\overline{\lim_n} f_n)(x) &:= \overline{\lim_n} f_n(x) & (\underline{\lim_n} f_n)(x) &:= \underline{\lim_n} f_n(x). \end{aligned}$$

For complex-valued functions we define $\operatorname{Re} f$, $\operatorname{Im} f$, $\overline{f}$ and $|f|$ by

$$(\operatorname{Re} f)(x) = \operatorname{Re} f(x), \quad (\operatorname{Im} f)(x) = \operatorname{Im} f(x), \quad \overline{f}(x) = \overline{f(x)}, \quad \text{and} \quad |f|(x) = |f(x)|.$$

Cardinality

Two sets A and B are said to **have the same cardinality** if there exists a bijection from A to B. A set A is **finite** if either A is the empty set or A has the same cardinality as $\{1, 2, \ldots, n\}$ for some positive integer n. In the latter case, the members of A may be labeled with the numbers $1, 2, \ldots, n$ so that A may be written $\{a_1, a_2, \ldots, a_n\}$. A set A is **countably infinite** if it has the same cardinality as the set $\mathbb{N}$ of positive integers, in which case we may write $A = \{a_1, a_2, \ldots\}$. A set is **countable** if it is either finite or countably infinite; otherwise, it is said to be **uncountable**. The set of all integers is countably infinite, as is the set of rational numbers. The set of all real numbers is uncountable, as is any (nondegenerate) interval of real numbers. The **cardinality** of $\mathbb{R}$ is denoted by $\mathfrak{c}$ and that of $\mathbb{N}$ by $\aleph_0$. For a detailed discussion of cardinality, the reader is referred to [23].

0.2 Algebraic Structures

Semigroups and Groups

A **semigroup** is a nonempty set G together with an **associative binary operation** $(s,t) \to st : G \times G \to G$, that is, $r(st) = (rs)t$ for all $r, s, t \in G$. A semigroup G is **commutative** or **abelian** if $st = ts$ for all $s, t \in G$. An **identity element** for G is a member e such that $se = es = s$ for all $s \in G$. Identities, if they exist, are unique. Indeed, if also $se' = e's = s$ for all s, then in particular $e' = ee' = e$. An element $s \in G$ has an **inverse** t if $st = ts = e$. The inverse of s is unique: if also $sr = rs = e$ then $r = re = rst = et = t$. The inverse of s, if it exists, is denoted by s^{-1}. A **group** is a semigroup with identity such that every element has an inverse. Semigroups and groups may be written additively, so that $s + t$ replaces st, 0 replaces e, and $-s$ replaces s^{-1}.

A **subsemigroup** of a semigroup G is a nonempty subset H that is closed under multiplication. A **subgroup** of a group G is a subsemigroup that contains the identity of G and is closed under the inverse operation (and hence is a group in its own right).

If G is a semigroup and A, $B \subseteq G$, $t \in G$, we set

$$At = \{at : a \in A\}, \; tA = \{ta : a \in A\}, \; AB = \{ab : a \in A, \, b \in B\}, \; A^{-1} = \{a^{-1} : a \in A\},$$

the last notation assuming that G is a group. The notation is modified in the obvious way if multiplication is written additively.

The sets $\mathbb{R}$ and $\mathbb{C}$ are groups under addition and are semigroups under multiplication. Removing zero in each case yields a group under multiplication. The interval $[1, \infty)$ is a semigroup under both addition and multiplication. The disk $\mathbb{D}$ and its closure are subsemigroups of $\mathbb{C}$ under multiplication, and $\mathbb{T}$ is a group. These examples are obviously all commutative. The collection of nonsingular $n \times n$ matrices over $\mathbb{K}$ $(n \geq 2)$ is a noncommutative group under matrix multiplication. The subset of matrices with determinant 1 is a subgroup.

It G and G' are semigroups, then a function $\varphi : G \to G'$ satisfying

$$\varphi(st) = \varphi(s)\varphi(t), \quad s, t \in G,$$

is called a **homomorphism**. A bijective homomorphism of semigroups is called an **isomorphism**. Note that in this case, the inverse map is automatically a homomorphism. A homomorphism of groups necessarily maps the identity onto the identity and inverses onto inverses. For example, the determinant function is a homomorphism from the semigroup of $n \times n$ matrices over $\mathbb{K}$ under multiplication to the multiplicative semigroup $\mathbb{K}$. The mapping $x \to e^x$ is an isomorphism from the additive group of real numbers onto the multiplicative group of positive reals.

If G is a group and H is a subgroup, then $x \sim y$ iff $x^{-1}y \in H$ defines an equivalence relation on G with equivalence classes xH, **the left coset of H with respect to x**. The collection G/H of all left cosets is called the **left coset space**. If H is **normal**, that is, $xH = Hx$ for all x, then G/H is a group under the operation $xH \cdot yH = xyH$, and the quotient map $Q : G \to G/H$ is a homomorphism. Conversely, if $\varphi : G \to G'$ is a surjective homomorphism, then the set

$$\ker \varphi := \{x \in G : \varphi(x) = e'\},$$

called **kernel** of φ, is a normal subgroup of G and the mapping $Q(x) \mapsto \varphi(x)$ is an isomorphism of $G/\ker \varphi$ onto G'.

Linear Spaces

A **linear space** (or **vector space**) **over** $\mathbb{K}$ is an additively written abelian group $\mathscr{V}$ with identity $\mathbf{0}$ and an operation **scalar multiplication** $\mathbb{K} \times \mathscr{V} \to \mathscr{V}$, $(s, \boldsymbol{v}) \to s\boldsymbol{v}$, satisfying

$$(st)\boldsymbol{v} = s(t\boldsymbol{v}),\ 1\,\boldsymbol{v} = \boldsymbol{v},\ s(\boldsymbol{v} + \boldsymbol{w}) = s\boldsymbol{v} + s\boldsymbol{w},\ \text{and}\ \ (s+t)\boldsymbol{v} = s\boldsymbol{v} + t\boldsymbol{v}$$

for all $s, t \in \mathbb{K}$ and $\boldsymbol{v}, \boldsymbol{w} \in \mathscr{V}$. It follows that $0\,\boldsymbol{v} = \mathbf{0}$ for all $\boldsymbol{v} \in \mathscr{V}$. Linear spaces are always taken over $\mathbb{K}$, whether or not explicitly mentioned. Euclidean space is a familiar example of a linear space. Numerous additional examples appear throughout the text. It is assumed that the reader has some familiarity with the basic theory of finite dimensional vector spaces.

A **subspace** of a linear space $\mathscr{V}$ is a nonempty subset $\mathscr{W}$ that is closed under the operations of addition and scalar multiplication. If $A \subseteq \mathscr{V}$, then the **span of** A is the subspace of $\mathscr{V}$ consisting of all **linear combinations** of members of A:

$$\operatorname{span} A := \left\{ \sum_{j=1}^{m} c_j \boldsymbol{a}_j : \boldsymbol{a}_j \in A,\ c_j \in \mathbb{K},\ m \in \mathbb{N} \right\}.$$

If A and B are nonempty subsets of $\mathscr{V}$ and $c \in \mathbb{K}$, we define

$$A + B := \{\boldsymbol{x} + \boldsymbol{y} : \boldsymbol{x} \in A,\ \ \boldsymbol{y} \in B\}\ \text{ and }\ cA := \{c\boldsymbol{x} : \boldsymbol{x} \in A\}.$$

A subset C of $\mathscr{V}$ is said to be **convex** if

$$tC + (1-t)C \subseteq C\ \text{ for all }\ 0 \le t \le 1,$$

and **balanced** if

$$cC \subseteq C,\quad \text{for all } c \in \mathbb{K} \text{ with } |c| \le 1.$$

A subspace of a linear space is obviously convex and balanced. The **line segment from** $\boldsymbol{a}$ **to** $\boldsymbol{b}$, defined by

$$[\boldsymbol{a} : \boldsymbol{b}] = \{(1-t)\boldsymbol{a} + t\boldsymbol{b} : 0 \le t \le 1\},$$

is convex but generally not balanced. The disk $\mathbb{D}$ is both convex and balanced in the real linear space $\mathbb{R}^2$, while $\mathbb{T}$ is neither balanced nor convex.

The **convex hull** $\operatorname{co} A$ of a subset A of a linear space $\mathscr{V}$ is the intersection of all convex subsets of $\mathscr{V}$ containing A. It is the smallest convex set (in the sense of containment) containing A. Similarly, the **convex balanced hull** $\operatorname{cobal} A$ of A is the intersection of all convex balanced subsets of $\mathscr{V}$ containing A. Here are important alternate descriptions of these sets.

0.2.1 Proposition. *Let A be a subset of a linear space $\mathscr{V}$. Then*

(a) $\operatorname{co} A$ *consists of all sums of the form* $\sum_{j=1}^{n} t_j \boldsymbol{x}_j$, *where* $n \in \mathbb{N}$, $\boldsymbol{x}_j \in A$, $t_j \ge 0$, *and* $\sum_{j=1}^{n} t_j = 1$.

(b) $\operatorname{cobal} A$ *consists of all sums of the form* $\sum_{j=1}^{n} c_j \boldsymbol{x}_j$, *where* $n \in \mathbb{N}$, $\boldsymbol{x}_j \in A$, $c_j \in \mathbb{K}$, *and* $\sum_{j=1}^{n} |c_j| \le 1$.

Proof. Let C denote the collection of all sums in (a). One easily checks that C is convex. Since $C \supseteq A$, we have $C \supseteq \operatorname{co} A$. For the reverse inclusion, let D be any convex set containing A. By induction, $D \supseteq C$. Since $\operatorname{co} A$ is the intersection of all such sets D, $\operatorname{co} A \supseteq C$. This proves (a). The proof of (b) is similar. $\square$

The sum in part (a) of the proposition is called a **convex combination** and the sum in (b) an **absolutely convex combination**.

0.2.2 Theorem. *Every linearly independent set A in a vector space may be extended to a basis. Thus every (nontrivial) vector space has a basis.*

Proof. Partially order the collection of linearly independent sets containing A by inclusion and note that the union of a chain of such sets is linearly independent. By Zorn's lemma, there exists a maximal linearly independent set, which is necessarily a basis. $\square$

A basis for $\mathscr{V}$ is also called a **Hamel basis** to distinguish it from other types of bases, for example Schauder bases.

Linear Transformations

Let $\mathscr{V}$ and $\mathscr{W}$ be linear spaces over $\mathbb{K}$. A **linear transformation** from $\mathscr{V}$ into $\mathscr{W}$ is a function $T : \mathscr{V} \to \mathscr{W}$ such that

$$T(\boldsymbol{x} + \boldsymbol{y}) = T\boldsymbol{x} + T\boldsymbol{y} \text{ and } T(c\boldsymbol{x}) = cT\boldsymbol{x} \text{ for all } \boldsymbol{x}, \boldsymbol{y} \in \mathscr{V} \text{ and } c \in \mathbb{K}.$$ [1]

The collection of all linear transformations from $\mathscr{V}$ to $\mathscr{W}$ is a linear space under pointwise addition and scalar multiplication

$$(S + T)(\boldsymbol{x}) := S\boldsymbol{x} + T\boldsymbol{x}, \quad (cT)(\boldsymbol{x}) = c(T\boldsymbol{x}), \quad \boldsymbol{x} \in \mathscr{V}, \ c \in \mathbb{K}.$$

The **kernel** of a linear transformation $T : \mathscr{V} \to \mathscr{W}$ is the subspace

$$\ker T = \{\boldsymbol{x} \in \mathscr{V} : T\boldsymbol{x} = 0\}.$$

By linearity, T is injective iff $\ker T = \{0\}$. If T is a bijection, then $T^{-1} : \mathscr{W} \to \mathscr{V}$ is easily seen to be linear. Such a mapping is called an **isomorphism** of linear spaces.

A linear transformation $f : \mathscr{V} \to \mathbb{K}$ is called a **linear functional**. The following proposition will be useful later.

0.2.3 Proposition. *Let $f, f_1, \ldots, f_n$ linear functionals on a linear space $\mathscr{V}$. Then f is a linear combination of $f_1, \ldots, f_n$ iff $\bigcap_{j=1}^{n} \ker f_j \subseteq \ker f$.*

Proof. The necessity is clear. For the sufficiency, define $F(\boldsymbol{v}) = (f_1(\boldsymbol{v}), \ldots, f_n(\boldsymbol{v}))$, $\boldsymbol{v} \in \mathscr{V}$. Then F maps $\mathscr{V}$ onto a subspace $\mathscr{X}$ of $\mathbb{K}^n$. Define a linear functional g on $\mathscr{X}$ so that $g \circ F = f$. The hypothesis implies that g is well-defined. Extend g linearly in a natural way to $\mathbb{K}^n$ by expanding a basis for $\mathscr{X}$ to a basis for $\mathbb{K}^n$. Then g must be of the form $g(x) = \alpha \cdot x$, hence $f(\boldsymbol{v}) = \alpha \cdot F(\boldsymbol{v})$, which is a linear combination of $f_1, \ldots, f_n$. $\square$

Let C be a convex subset $\mathscr{V}$. A function $f : C \to \mathscr{W}$ is said to be **affine** if

$$f\big((1 - t)\boldsymbol{x} + t\boldsymbol{y}\big) = (1 - t)f(\boldsymbol{x}) + tf(\boldsymbol{y}), \quad \forall \ \boldsymbol{x}, \boldsymbol{y} \in C \text{ and } 0 < t < 1.$$

For example, the restriction of a linear transformation to a convex set is affine. The function $x \mapsto \boldsymbol{a} \cdot x + \boldsymbol{b}$ on a convex subset of $\mathbb{R}^d$ is affine.

---

[1] The notation $T\boldsymbol{x}$ for $T(\boldsymbol{x})$ is standard for linear transformations.

Quotient Linear Spaces

The notion of cosets in a group applies to linear spaces as follows: Let $\mathscr{U}$ be a subspace of a linear space $\mathscr{V}$. The relation $\boldsymbol{x} \sim \boldsymbol{y}$ iff $\boldsymbol{x} - \boldsymbol{y} \in \mathscr{U}$ is an equivalence relation on $\mathscr{U}$. The **quotient space** $\mathscr{V}/\mathscr{U}$ is the vector space of all sets of the form $\boldsymbol{x} + \mathscr{U}$, $\boldsymbol{x} \in \mathscr{V}$, where addition and scalar multiplication are defined by

$$(\boldsymbol{x} + \mathscr{U}) + (\boldsymbol{y} + \mathscr{U}) = (\boldsymbol{x} + \boldsymbol{y}) + \mathscr{U} \text{ and } c(\boldsymbol{x} + \mathscr{U}) = c\boldsymbol{x} + \mathscr{U}.$$

The quotient map $\boldsymbol{x} \mapsto \boldsymbol{x} + \mathscr{U}$ is easily seen to be a linear transformation. If $T : \mathscr{V} \to \mathscr{W}$ is a linear transformation, then the mapping $T\boldsymbol{v} \mapsto \boldsymbol{v} + \ker T$ is an isomorphism from $\operatorname{ran} T$ onto $\mathscr{V} / \ker T$.

Algebras

An **algebra** (over $\mathbb{K}$) is a linear space $\mathscr{A}$ with an associative multiplication $(\boldsymbol{x}, \boldsymbol{y}) \to \boldsymbol{x}\boldsymbol{y}$ that satisfies

$$\boldsymbol{z}(\boldsymbol{x} + \boldsymbol{y}) = \boldsymbol{z}\boldsymbol{x} + \boldsymbol{z}\boldsymbol{y}, \quad (\boldsymbol{x} + \boldsymbol{y})\boldsymbol{z} = \boldsymbol{x}\boldsymbol{z} + \boldsymbol{y}\boldsymbol{z}, \text{ and } c(\boldsymbol{x}\boldsymbol{y}) = (c\boldsymbol{x})\boldsymbol{y} = \boldsymbol{x}(c\boldsymbol{y})$$

for all $\boldsymbol{x}, \boldsymbol{y}, \boldsymbol{z} \in \mathscr{A}$ and $c \in \mathbb{K}$. An algebra $\mathscr{A}$ is **commutative** if $\boldsymbol{x}\boldsymbol{y} = \boldsymbol{y}\boldsymbol{x}$ for all $\boldsymbol{x}, \boldsymbol{y} \in \mathscr{A}$. An **identity** of $\mathscr{A}$ is an element $\boldsymbol{e}$ such that $\boldsymbol{e}\boldsymbol{x} = \boldsymbol{x}\boldsymbol{e} = \boldsymbol{x}$ for all $\boldsymbol{x} \in \mathscr{A}$. An identity, if it exists, is unique. An algebra with identity is said to be **unital**. For example, the vector space of $n \times n$ matrices over $\mathbb{K}$ is an algebra with respect to matrix multiplication. More generally, the space of linear transformations from a vector space into itself is an algebra with respect to function composition. Additional examples appear throughout the text.

A **homomorphism** of algebras $\mathscr{A}$ and $\mathscr{B}$ is a linear map $\varphi : \mathscr{A} \to \mathscr{B}$ such that $\varphi(\boldsymbol{x}\boldsymbol{y}) = \varphi(\boldsymbol{x})\varphi(\boldsymbol{y})$. A bijective homomorphism is called an algebra **isomorphism**. A **subalgebra** of an algebra $\mathscr{A}$ is a linear subspace of $\mathscr{A}$ that is closed under multiplication. An **ideal** $\mathscr{I}$ of $\mathscr{A}$ is a subalgebra with the stronger property $\boldsymbol{x} \in \mathscr{A}$ and $\boldsymbol{y} \in \mathscr{I} \Rightarrow \boldsymbol{x}\boldsymbol{y}, \boldsymbol{y}\boldsymbol{x} \in \mathscr{A}$. It then follows that the quotient linear space $\mathscr{A}/\mathscr{I}$ is an algebra under multiplication $(\boldsymbol{x}+\mathscr{I})(\boldsymbol{y}+\mathscr{I}) = \boldsymbol{x}\boldsymbol{y}+\mathscr{I}$, and the quotient map is an algebra homomorphism. For example, in the algebra $\mathscr{A}$ of all functions $f : \mathbb{R} \to \mathbb{R}$ under pointwise addition and multiplication, the subset $\{f \in \mathscr{A} : f(x) = 0 \ \forall\, x \in [0, 1]\}$ is an ideal, and the map $f + \mathscr{I} \to f\big|_{[0,1]}$ is an algebra isomorphism from $\mathscr{A}/\mathscr{I}$ onto the algebra of all real functions on $[0, 1]$.

0.3 Metric Spaces

A **metric** on a nonempty set X is a function $d : X \times X \to \mathbb{R}$ with the following properties:

(a) $d(x, y) \geq 0$ (nonnegativity).

(b) $d(x, y) = 0$ iff $x = y$ (coincidence).

(c) $d(x, y) = d(y, x)$ (symmetry).

(d) $d(x, y) \leq d(x, z) + d(y, z)$ (triangle inequality).

The ordered pair (X, d), as well as the set X, is called a **metric space**. A nonempty subset Y of X with the metric $d\big|_{Y \times Y}$ is called a **subspace of** X. A metric has the property

$$|d(x, y) - d(u, v)| \leq d(x, u) + d(y, v) \tag{0.1}$$

as may be seen from the triangle inequality $d(x,y) \le d(x,u) + d(u,v) + d(v,y)$ and its counterpart.

The real number system $\mathbb{R}$ is a metric space under the **usual metric** $d(x,y) = |x-y|$. More generally, the set $\mathbb{R}^d$ is a metric space under the **Euclidean metric**

$$d(\boldsymbol{x},\boldsymbol{y}) = |\boldsymbol{x}-\boldsymbol{y}| = \left(\sum_{j=1}^{d}(x_j - y_j)^2\right)^{1/2}.$$

For another example, let X be a nonempty set and define $d(x,y) = 1$ if $x \ne y$ and $d(x,x) = 0$. Then d is a metric, called the **discrete metric** on X.

Open and Closed Sets

Let (X,d) be a metric space. For $x \in X$ and $r > 0$, the sets

$$B_r(x) = B(x,r) := \{y \in X : d(x,y) < r\} \quad \text{and} \quad C_r(x) = C(x,r) := \{y \in X : d(x,y) \le r\}$$

are called, respectively, the **open and closed balls with center x and radius r**. The set

$$S_r(x) = S(x,r) := \{y \in X : d(x,y) = r\} = C_r(x) \setminus B_r(x)$$

is called the **sphere with center x and radius r**. For example, the open (closed) balls in $\mathbb{R}$ with the usual metric are the bounded open (closed) intervals. The open (closed) balls in Euclidean space $\mathbb{R}^2$ are open (closed) disks and the spheres are circles. The open and closed balls in a discrete metric space X are the sets X and $\{x\}$; the spheres are $X \setminus \{x\}$ and the empty set.

A subset U of X is said to be **open** if either $U = \emptyset$ or else for each $x \in U$ there exists an $r > 0$ such that $B_r(x) \subseteq U$. A subset of X is **closed** if its complement is open. An application of the triangle inequality shows that an open ball is open. Indeed, if $y \in B_\varepsilon(x)$, then $B_\delta(y) \subseteq B_\varepsilon(x)$ for $\delta = \varepsilon - d(x,y)$, which shows that $B_\varepsilon(x)$ is a union of open balls $B_\delta(y)$. A similar argument shows that a closed ball is closed.

A **neighborhood** of a point a in X is any set containing an open set containing a. As we shall see, certain concepts such as continuity and convergence are conveniently phrased in terms of neighborhoods.

0.3.1 Proposition. *Open and closed sets have the following properties:*

(a) *If $\mathfrak{U}$ is a collection of open sets, then $\bigcup \mathfrak{U}$ is open.*

(b) *If $V_1, \ldots, V_n$ are open, then $V := V_1 \cap \cdots \cap V_n$ is open.*

(c) *If $\mathfrak{C}$ is a family of closed sets, then $\bigcap \mathfrak{C}$ is closed.*

(d) *If $C_1, \ldots, C_n$ are closed, then $C_1 \cup \cdots \cup C_n$ is closed.*

Proof. (a) Let $x \in \bigcup \mathfrak{U}$. Then $x \in U$ for some $U \in \mathfrak{U}$, and since U is open there exists $r > 0$ such that $B_r(x) \subseteq U \subseteq \bigcup \mathfrak{U}$.

(b) Let $x \in V$. For each j there exists $r_j > 0$ such that $B_{r_j}(x) \subseteq V_j$. Then $B_r(x) \subseteq V$, where $r := \min\{r_1, \ldots, r_n\}$. Therefore, V is a union of open balls and so is open.

Parts (c) and (d) follow from (a) and (b) and DeMorgan's laws. □

Interior, Closure, and Boundary

Let E be a subset of a metric space X. The **interior** $\text{int}(E) = \text{int}_X(E)$ of E is the union of all open subsets of X contained in E. The **closure** $\text{cl}(E) = \text{cl}_X(E)$ of E in X is the intersection of all closed subsets of X containing E. The **boundary** $\text{bd}(E) = \text{bd}_X(E)$ of E is the set $\text{cl}(E) \setminus \text{int}(E)$. Thus the interior of E is the largest open set contained in E and the closure of E is the smallest closed set containing E. A point $x \in X$ is in the boundary of E iff every neighborhood of x meets both E and E^c.

If $\text{cl}\,E = X$, then E said to be **dense in** X. A metric space with a countable dense subset is said to be **separable**. For example, $\mathbb{R}^d$ is separable with respect to the Euclidean metric, as may be seen by considering $\mathbb{Q}^d$.

Sequential Convergence. Completeness

A sequence (x_n) in a metric space (X, d) is said to **converge** to $x \in X$ if $\lim_n d(x_n, x) = 0$. In this case we write $x_n \to x$ or $\lim_n x_n = x$. In terms of neighborhoods, $x_n \to x$ iff x_n is **eventually** in each neighborhood N of x, that is, $x_n \in N$ for all sufficiently large n. The triangle inequality shows that the limit of a sequence (x_n) in X, if it exists, is unique.

A **cluster point** of a sequence in X is the limit of a convergent subsequence. Thus x is a cluster point of (x_n) iff (x_n) is **frequently** in every neighborhood N of x, that is, $x_n \in N$ for infinitely many n.

A sequence (x_n) is said to be **Cauchy** if $\lim_{m,n\to\infty} d(x_m, x_n) = 0$. If every Cauchy sequence in (X, d) converges to a member of X, then (X, d) is said to be **complete**. For example, Euclidean space is complete. It may be shown that every metric space X has a **completion**, that is, a complete metric space that contains a dense copy of X. (See [2]).

It follows from (0.1) that $x_n \to x$ and $y_n \to y \Rightarrow d(x_n, y_n) \to d(x, y)$, which expresses the *continuity of the metric.*

The following result is sometimes useful in establishing completeness of a metric space.

0.3.2 Proposition. *A Cauchy sequence with a convergent subsequence converges.*

Proof. Let (x_n) be a Cauchy sequence and $x_{n_k} \to x$. Given $\varepsilon > 0$, by the Cauchy property the inequality $d(x_n, x_{n_k}) < \varepsilon$ holds for all sufficiently large n and k. Fixing such an n and letting $k \to \infty$ yields $d(x_n, x) \le \varepsilon$, by continuity of the metric. Therefore, $x_n \to x$. □

0.3.3 Proposition. *A subset C of X is closed iff C contains the limit of each convergent sequence in C.*

Proof. Assume that C is closed and let (x_n) be a sequence in C with $x_n \to x$. If $x \notin C$, then, because C^c is open, there exists $\varepsilon > 0$ such that $B_\varepsilon(x) \cap C = \emptyset$. But this is impossible, since x_n is eventually in $B_\varepsilon(x) \subseteq C^c$. Therefore, $x \in C$.

Now suppose C is not closed. Then C^c is not open, hence there exists $x \in C^c$ such that $B_{1/n}(x) \cap C \neq \emptyset$, for every $n \in \mathbb{N}$. Choosing a point x_n in this intersection we obtain a sequence (x_n) in C that converges to a member of C^c. □

0.3.4 Corollary. *Let Y be a subspace of X.*

(a) *If X is complete and Y is closed, then Y is complete.*

(b) *If Y is complete, then Y is closed.*

Proof. (a) Let (y_n) be a Cauchy sequence in Y. Since X is complete, there exists $x \in X$ such that $y_n \to x$. Since Y is closed, $x \in Y$. Therefore, Y is complete.

(b) Let (y_n) be a sequence in Y such that $y_n \to x \in X$. Then (y_n) is Cauchy and hence converges to some $y \in Y$. Since limits are unique, $x = y \in Y$. Therefore, Y is closed. □

0.3.5 Proposition. *Let $A \subseteq X$. Then $x \in \text{cl}(A)$ iff there exists a sequence (a_n) in A such that $a_n \to x$.*

Proof. Let C be the set of all limits of convergent sequences in A including constant sequences, so $A \subseteq C \subseteq \text{cl}(A)$, the second inclusion by 0.3.3. We show that C is closed, proving the assertion.

Suppose C is not closed. Then C^c is not open, hence there exists $y \in C^c$ and for each n a point $y_n \in B_{1/n}(y) \cap C$. Since each y_n is the limit of a sequence in A, there exists $a_n \in A$ such that $d(y_n, a_n) < 1/n$. By the triangle inequality, $d(a_n, y) < 2/n$, hence $a_n \to y$. But then $y \in C$, a contradiction. □

Continuity

Let (X, d) and (Y, ρ) be metric spaces. A function $f : X \to Y$ is said to be **continuous at** $a \in X$ if for each $\varepsilon > 0$ there exists a $\delta > 0$ such that $d(x, a) < \delta \Rightarrow \rho\big(f(x), f(a)\big) < \varepsilon$. In terms of open balls,

$$f\big(B_\delta(a)\big) \subseteq B_\varepsilon\big(f(a)\big). \tag{0.2}$$

If $E \subseteq X$ and f is continuous at each point of E, then f is said to be **continuous on** E. If f is continuous at each member of X, then f is said to be **continuous**. A **homeomorphism** from X to Y is a bijection $f : X \to Y$ such that both f and f^{-1} are continuous.

The following proposition describes a useful characterization of continuity in terms of neighborhoods. It will have implications later in the formulation of the definition of continuity in the more general setting of topological spaces.

0.3.6 Proposition. *A function $f : X \to Y$ is continuous at a iff for each neighborhood M of $f(a)$ there exists a neighborhood N of a such that $f(N) \subseteq M$.*

Proof. Let f be continuous at a and let M be a neighborhood of $f(a)$. Choose $\varepsilon > 0$ such that $B_\varepsilon(f(a)) \subseteq M$ and choose $\delta > 0$ as in 0.2. Then $B_\delta(a)$ is the required neighborhood N.

Conversely, assume the neighborhood property holds and let $\varepsilon > 0$. Choose a neighborhood N of a such that $f(N) \subseteq B_\varepsilon(f(a))$ and choose δ so that $B_\delta(a) \subseteq N$. Then (0.2) holds. □

It is clear from the proof that the neighborhoods M and N in 0.3.6 may be taken to be open.

0.3.7 Proposition. *Let $f : (X, d) \to (Y \rho)$ and $a \in X$. Then f is continuous at a iff $f(a_n) \to f(a)$ for any sequence (a_n) in X with $a_n \to a$.*

Proof. If f is continuous at a, then for any neighborhood M of $f(a)$ there exists a neighborhood N of a such that $f(N) \subseteq M$. If $a_n \to a$, then $a_n \in N$ for all sufficiently large n, and for such n, $f(a_n) \in M$. Therefore, $f(a_n) \to f(a)$.

Conversely, if f is not continuous at a, then for some $\varepsilon > 0$ and each $n \in \mathbb{N}$ there exists an $a_n \in B_{1/n}(a)$ with $f(a_n) \notin B_\varepsilon\big(f(a)\big)$. Thus the sequential property fails. □

0.3.8 Theorem. *Let $f : (X, d) \to (Y \rho)$. The following statements are equivalent:*

(a) *f is continuous.*

(b) *$f^{-1}(V)$ is open in X for each open subset V of Y.*

(c) *$f^{-1}(C)$ is closed in X for each closed subset C of Y.*

(d) *$f(\text{cl}\, A) \subseteq \text{cl}(f(A))$ for each subset A of X.*

Proof. (a) $\Rightarrow$ (b): Let $V \subseteq Y$ be open. If $x \in f^{-1}(V)$, then V is a neighborhood of $f(x)$, hence there exists an open neighborhood N of x such that $f(N) \subseteq V$ and so $N \subseteq f^{-1}(V)$. Therefore, $f^{-1}(V)$ is a union of open sets hence is open.

(b) $\Rightarrow$ (c): This follows from $\left(f^{-1}(C)\right)^c = f^{-1}(C^c)$.

(c) $\Rightarrow$ (d): Let $C = \mathrm{cl}\, f(A)$. Then $f^{-1}(C)$ is closed and $f^{-1}(C) \supseteq f^{-1}\big(f(A)\big) \supseteq A$, hence $f^{-1}(C) \supseteq \mathrm{cl}\, A$ and so $C \supseteq f(f^{-1}(C)) \supseteq f(\mathrm{cl}\, A)$.

(d) $\Rightarrow$ (a): If f is not continuous at x, then there exists $\varepsilon > 0$ and a sequence (x_n) in X such that $x_n \to x$ and $d\big(f(x_n), f(x)\big) \geq \varepsilon$ for all n. But this is impossible, since $f(x) \in f\big(\mathrm{cl}\{x_1, x_2, \ldots\}\big) \subseteq \mathrm{cl}\, f\big(\{x_1, x_2, \ldots\}\big)$. □

A function $f : (X, d) \to (Y, \rho)$ is said to be **uniformly continuous on a set** $E \subseteq X$ if, given $\varepsilon > 0$, there exists $\delta > 0$ such that $u, v \in E$ and $d(u, v) < \delta \Rightarrow \rho(f(u), f(v)) < \varepsilon$. The proof of the following theorem is similar to that of 0.3.7

0.3.9 Proposition. *A function $f : (X, d) \to (Y, \rho)$ is uniformly continuous on $E \subseteq X$ iff $\rho\big(f(u_n), f(v_n)\big) \to 0$ for any pair of sequences (u_n) and (v_n) in E with $d(u_n, v_n) \to 0$.*

Let A and B nonempty subsets of X. The **distance between** A **and** B is defined as

$$d(A, B) := \inf\{d(x, y) : x \in A,\ y \in B\}.$$

To simplify notation we set

$$d(x, A) := d(\{x\}, A).$$

It follows easily from the triangle inequality that $|d(x, A) - d(y, A)| \leq d(x, y)$, which shows that $d(x, A)$ is continuous, indeed uniformly continuous. A simple consequence of this is Urysohn's lemma for metric spaces:

0.3.10 Lemma. *Let (X, d) be a metric space. If A and B are disjoint, nonempty, closed sets in X, then the function*

$$f(x) = \frac{d(x, A)}{d(x, A) + d(x, B)}$$

is well-defined and continuous, $0 \leq f \leq 1$, $A = \{x : f(x) = 0\}$, and $B = \{x : f(x) = 1\}$.

Category

The **diameter** of a nonempty subset E of a metric space (X, d) is defined by

$$d(E) = \sup\{d(x, y) : x, y \in E\}.$$

Note that the continuity of the metric implies that $d(E) = d(\mathrm{cl}(E))$.

Here is an important characterization of completeness of a metric space in terms of diameters.

0.3.11 Cantor Intersection Theorem. *A metric space X is complete iff the intersection of any decreasing sequence of nonempty closed sets C_n in X with $d(C_n) \to 0$ consists of a single point.*

Proof. Assume X is complete. For each n choose $x_n \in C_n$. Since $C_n \downarrow$ and $d(C_n) \to 0$, (x_n) is Cauchy. Let $x_n \to x$. Since $x_m \in C_n$ for all $m \geq n$ and C_n is closed, letting $m \to \infty$ we see that $x \in C_n$ for all n, that is, $x \in C := \bigcap_n C_n$. Since $d(C) \leq d(C_n) \to 0$, $C = \{x\}$.

Conversely, let X have the stated intersection property and let (x_n) be a Cauchy sequence in X. Set $C_n := \mathrm{cl}\{x_k : k \geq n\}$. By the Cauchy property, $d(C_n) \to 0$. Since $C_n \downarrow$, by our hypothesis $\bigcap_n C_n$ contains a point x. It follows easily that $x_{n_k} \to x$ for some subsequence (x_{n_k}). By 0.3.2, $x_n \to x$. □

The following consequence of Cantor's theorem is a key step in the proofs of several important results in analysis. In §0.12 we give a version of the theorem for locally compact spaces.

0.3.12 Baire Category Theorem. *Let X be a complete metric space. If (X_n) a sequence of closed sets with union X, then $\operatorname{int} X_n \neq \emptyset$ for some n.*

Proof. Suppose for a contradiction that $\operatorname{int} X_n = \emptyset$ for all n. Choose an open ball $B(x_0, r_0)$ with $r_0 = 1$. Since $\operatorname{int} X_1 = \emptyset$, there exists $x_1 \in B(x_0, r_0) \setminus X_1$, and since $B(x_0, r_0) \setminus X_1$ is open, there exists $r_1 \in (0, 1/2)$ such that $C(x_1, r_1) \subseteq B(x_0, r_0) \setminus X_1$. Since $\operatorname{int} X_2 = \emptyset$ and $B(x_1, r_1) \setminus X_2$ is open there exists $x_2 \in B(x_1, r_1) \setminus X_2$ and $r_2 \in (0, 1/3)$ such that $C(x_2, r_2) \subseteq B(x_1, r_1) \setminus X_2$. In this way we construct sequences (x_n) in X and (r_n) in $\mathbb{R}$ such that

$$C(x_n, r_n) \subseteq B(x_{n-1}, r_{n-1}) \setminus X_n, \quad 0 < r_{n-1} \leq 1/n, \quad n \geq 1.$$

Since the closed balls are decreasing and the diameters are tending to zero, their intersection C is nonempty (0.3.11). But this is impossible because $C \cap X_n = \emptyset$ for all n. □

0.4 Normed Linear Spaces

Norms and Seminorms

A **norm** on a linear space $\mathscr{X}$ over $\mathbb{K}$ is a function $\|\cdot\| : \mathscr{X} \to \mathbb{R}$ with the following properties:

(a) $\boldsymbol{x} \neq 0$ implies $\|\boldsymbol{x}\| \neq 0$ (positivity).

(b) $\|c\boldsymbol{x}\| = |c| \, \|\boldsymbol{x}\|$ (absolute homogeneity).

(c) $\|\boldsymbol{x} + \boldsymbol{y}\| \leq \|\boldsymbol{x}\| + \|\boldsymbol{y}\|$ (triangle inequality).

If $\|\cdot\|$ satisfies (b) and (c) but not necessarily (a), then $\|\cdot\|$ is called a **seminorm**. Note that by (b), $\|0\| = 0$ and $\|\boldsymbol{x}\| = \|-\boldsymbol{x}\|$, hence by (c), $0 = \|\boldsymbol{x} - \boldsymbol{x}\| \leq \|\boldsymbol{x}\| + \|-\boldsymbol{x}\| = 2\,\|\boldsymbol{x}\|$. Therefore, seminorms and norms are nonnegative.

Property (d) has the following useful extensions:

$$\left\| \sum_{j=1}^{n} \boldsymbol{x}_j \right\| \leq \sum_{j=1}^{n} \|\boldsymbol{x}_j\| \quad \text{and} \quad \big|\, \|\boldsymbol{x}\| - \|\boldsymbol{y}\| \,\big| \leq \|\boldsymbol{x} - \boldsymbol{y}\|. \tag{0.3}$$

Indeed, the first inequality may be established by a simple induction argument, and the second by applying the triangle inequality to $\|\boldsymbol{x}\| = \|\boldsymbol{x} - \boldsymbol{y} + \boldsymbol{y}\|$ and $\|\boldsymbol{y}\| = \|\boldsymbol{y} - \boldsymbol{x} + \boldsymbol{x}\|$.

If $\|\cdot\|$ is a norm on $\mathscr{X}$, then the pair $(\mathscr{X}, \|\cdot\|)$ is called a **normed space**. It is easy to check that the mapping $(\boldsymbol{x}, \boldsymbol{y}) \mapsto \|\boldsymbol{x} - \boldsymbol{y}\|$ is a metric on $\mathscr{X}$, making the entire machinery of metric spaces available. Unless stated otherwise, convergence and continuity in a normed space are taken relative to this metric.

Banach Spaces

A normed space $(\mathscr{X}, \|\cdot\|)$ that is complete in the metric $(\boldsymbol{x}, \boldsymbol{y}) \to \|\boldsymbol{x} - \boldsymbol{y}\|$ is called a **Banach space.** A familiar example is Euclidean space $\mathbb{K}^d$. Many other examples appear throughout the text. For now we content ourselves with the following.

0.4.1 Example. (*The space of bounded functions*). Let X be a nonempty set and let $B(X)$ denote the vector space (under pointwise addition and scalar multiplication) of all bounded functions $f : X \to \mathbb{K}$. The **supremum norm** or **uniform norm** on $B(X)$ is defined by

$$\|f\|_\infty = \sup\{|f(x)| : x \in X\}.$$

That $\|\cdot\|_\infty$ is a norm is easily established using familiar properties of absolute value. For example, the triangle inequality follows by taking the supremum over X in

$$|(f+g)(x)| \le |f(x)| + |g(x)| \le \|f\|_\infty + \|g\|_\infty.$$

To verify completeness, let (f_n) be a Cauchy sequence in $B(X)$ and $\varepsilon > 0$. Choose N such that $\|f_n - f_m\| < \varepsilon$ for all m, $n \ge N$. For such indices and each $x \in X$ we then have

$$|f_n(x) - f_m(x)| < \varepsilon, \tag{†}$$

which shows that $(f_n(x))$ is a Cauchy sequence in $\mathbb{K}$. Since $\mathbb{K}$ is complete, $f_n(x) \to f(x) \in \mathbb{K}$. Fixing $n \ge N$ in (†) and letting $m \to \infty$ yields $|f_n(x) - f(x)| \le \varepsilon$ for all $x \in X$ and $n \ge N$. Therefore, $f = f - f_n + f_n \in B(X)$ and $\|f_n - f\|_\infty \le \varepsilon$. ◊

Completion of a Normed Space

The following theorem asserts that every normed space may be realized as a dense subspace of a Banach space. A proof may be given at this point based on the fact that $\mathcal{X}$ has a metric space completion; however, we prefer to wait until Chapter 8 when the machinery for simpler proof will be available. (See 8.5.9.)

0.4.2 Theorem. *Let $(\mathcal{X}, \|\cdot\|)$ be a normed space. Then there exists a Banach space $\overline{\mathcal{X}}$ and a linear transformation from $\mathcal{X}$ onto a dense subspace of $\overline{\mathcal{X}}$ that preserves norm.*

Infinite Series in Normed Spaces

An infinite series $\sum_{n=1}^\infty \boldsymbol{x}_n$ of members of a normed space $\mathcal{X}$ is said to **converge** if the sequence of **partial sums** $\sum_{k=1}^n \boldsymbol{x}_k$ converges in $\mathcal{X}$. By analogy with numerical series, $\sum_{n=1}^\infty \boldsymbol{x}_n$ is said to **converge absolutely** if $\sum_{n=1}^\infty \|\boldsymbol{x}_n\| < \infty$. The following result is sometimes useful in establishing completeness of a normed space.

0.4.3 Proposition. *Let $\mathcal{X}$ be a normed vector space. Then $\mathcal{X}$ is complete (and hence is a Banach space) iff every absolutely convergent series converges.*

Proof. Assume that $\mathcal{X}$ is complete and $\sum_{n=1}^\infty \|\boldsymbol{x}_n\| < \infty$. Then for $m > n \ge 1$,

$$\left\|\sum_{j=1}^m \boldsymbol{x}_j - \sum_{j=1}^n \boldsymbol{x}_j\right\| \le \sum_{j=n+1}^m \|\boldsymbol{x}_j\| \le \sum_{j=n+1}^\infty \|\boldsymbol{x}_j\|.$$

Since the right side tends to 0 as $n \to \infty$, the sequence of partial sums $\sum_{j=1}^n \boldsymbol{x}_j$ is Cauchy and so converges, that is, the series $\sum_{n=1}^\infty \boldsymbol{x}_n$ converges in $\mathcal{X}$.

Conversely, assume that every absolutely convergent series converges. Let $(\boldsymbol{x}_n)$ be a Cauchy sequence in $\mathcal{X}$. One may then obtain a strictly increasing sequence (n_k) in $\mathbb{N}$ such that $\|\boldsymbol{x}_m - \boldsymbol{x}_n\| < 2^{-k}$ for all $m, n \ge n_k$. In particular, $\|\boldsymbol{x}_{n_{k+1}} - \boldsymbol{x}_{n_k}\| < 2^{-k}$, which implies that the series $\sum_k \|\boldsymbol{x}_{n_{k+1}} - \boldsymbol{x}_{n_k}\|$ converges. By hypothesis, the sequence of partial sums $\sum_{j=1}^k (\boldsymbol{x}_{n_{j+1}} - \boldsymbol{x}_{n_j})$ converges. But these sums collapse to $\boldsymbol{x}_{n_{k+1}} - \boldsymbol{x}_{n_1}$. Therefore, $(\boldsymbol{x}_{n_k})$ is a convergent subsequence of $(\boldsymbol{x}_n)$ and so the latter must converge (0.3.2). □

Unordered Sums in Normed Spaces

Let $\{\boldsymbol{x}_i : i \in \mathfrak{I}\}$ be a family of vectors in a normed vector space $\mathcal{X}$ and let $\boldsymbol{x} \in \mathcal{X}$. We say that $\{\boldsymbol{x}_i : i \in \mathfrak{I}\}$ **converges unconditionally to** $\boldsymbol{x}$ if for each $\varepsilon > 0$ there exists a finite set $F_\varepsilon \subseteq \mathfrak{I}$ such that

$$\Big\| \sum_{i \in F} \boldsymbol{x}_i - \boldsymbol{x} \Big\| < \varepsilon \ \text{ for all finite } F \subseteq \mathfrak{I} \text{ with } F \supseteq F_\varepsilon. \tag{0.4}$$

We then call $\boldsymbol{x}$ the **unordered sum of** $\{\boldsymbol{x}_i : i \in \mathfrak{I}\}$ and write $\boldsymbol{x} = \sum_{i\in\mathfrak{I}} \boldsymbol{x}_i$. In this case, we also say that $\{\boldsymbol{x}_i : i \in \mathfrak{I}\}$ is **summable to** $\boldsymbol{x}$.

0.4.4 Proposition. *If the unordered sums on the right in the following equality exist, then the unordered sum on the left exists and*

$$\sum_{i\in\mathfrak{I}} (a\boldsymbol{x}_i + b\boldsymbol{y}_i) = a\sum_{i\in\mathfrak{I}} \boldsymbol{x}_i + b\sum_{i\in\mathfrak{I}} \boldsymbol{y}_i.$$

Proof. Let $\boldsymbol{x} = \sum_{i\in\mathfrak{I}} \boldsymbol{x}_i$ and $\boldsymbol{y} = \sum_{i\in\mathfrak{I}} \boldsymbol{y}_i$. Given $\varepsilon > 0$, choose finite F_ε, $G_\varepsilon \subseteq \mathfrak{I}$ such that

$$\Big\| \sum_{i\in F} \boldsymbol{x}_i - \boldsymbol{x} \Big\| < \varepsilon/2 \text{ and } \Big\| \sum_{i\in G} \boldsymbol{x}_i - \boldsymbol{x} \Big\| < \varepsilon/2$$

for all finite $F \supseteq F_\varepsilon$ and $G \supseteq G_\varepsilon$. Then for finite $F \supseteq F_\varepsilon \cup G_\varepsilon$, by the extended triangle inequality (0.3) we have

$$\Big\| \sum_{i\in F} (\boldsymbol{x}_i + \boldsymbol{y}_i) - (\boldsymbol{x} + \boldsymbol{y}) \Big\| \le \sum_{i\in F} \|\boldsymbol{x}_i - \boldsymbol{x}\| + \sum_{i\in F} \|\boldsymbol{y}_i - \boldsymbol{y}\| < \varepsilon.$$

This shows that $\sum_{i\in\mathfrak{I}} (\boldsymbol{x}_i + \boldsymbol{y}_i) = \sum_{i\in\mathfrak{I}} \boldsymbol{x}_i + \sum_{i\in\mathfrak{I}} \boldsymbol{y}_i$. A even simpler argument shows that $\sum_{i\in\mathfrak{I}} a\boldsymbol{x}_i = a\sum_{i\in\mathfrak{I}} \boldsymbol{x}_i$. □

0.4.5 Proposition. *If the family $\{\boldsymbol{x}_i : i \in \mathfrak{I}\}$ is unconditionally convergent to $\boldsymbol{x}$, then all but countably many members of the family are zero. If the nonzero members are arranged in any order, say $\boldsymbol{x}_{i_1}, \boldsymbol{x}_{i_2}, \ldots$, then $\boldsymbol{x} = \sum_{n=1}^\infty \boldsymbol{x}_{i_n}$.*

Proof. For each $n \in \mathbb{N}$, let F_n be a finite set of indices such that

$$\Big\| \sum_{i\in F} \boldsymbol{x}_i - \boldsymbol{x} \Big\| < 1/n \ \text{ for all finite } F \supseteq F_n.$$

Thus if $j \notin F_n$ and $F = \{j\} \cup F_n$, then

$$\|\boldsymbol{x}_j\| = \Big\| \sum_{i\in F} \boldsymbol{x}_i - \sum_{i\in F_n} \boldsymbol{x}_i \Big\| \le \Big\| \sum_{i\in F} \boldsymbol{x}_i - \boldsymbol{x} \Big\| + \Big\| \sum_{i\in F_n} \boldsymbol{x}_i - \boldsymbol{x} \Big\| < 2/n.$$

Thus if $j \notin \bigcup_n F_n$, then the preceding inequality holds for all n and so $\boldsymbol{x}_j = \boldsymbol{0}$.

For the last assertion of the proposition, choose a finite set F_ε such that (0.4) holds. We may assume that F_ε contains no index j for which $\boldsymbol{x}_j = \boldsymbol{0}$. Then, for all sufficiently large n, $F_\varepsilon \subseteq \{i_1, \ldots, i_n\}$, hence $\big\| \sum_{j=1}^n \boldsymbol{x}_{i_j} - \boldsymbol{x} \big\| < \varepsilon$. □

The next result follows easily from the approximation property of suprema.

0.4.6 Proposition. *A family $\{t_i : i \in \mathfrak{I}\} \subseteq [0, \infty)$ is unconditionally convergent to $\sup_F \sum_{i\in F} t_i$, provided that the supremum, which is taken over all finite $F \subseteq \mathfrak{I}$, is finite.*

0.4.7 Theorem. *Let $\mathscr{X}$ be a Banach space and let $\{\boldsymbol{x}_i : i \in \mathfrak{I}\} \subseteq \mathscr{X}$ such that $s := \sup_F \sum_{i\in F} \|\boldsymbol{x}_i\| < \infty$, where the supremum is taken over all finite $F \subseteq \mathfrak{I}$. Then the family $\{\|\boldsymbol{x}_i\| : i \in \mathfrak{I}\}$ converges unconditionally to s, $\{\boldsymbol{x}_i : i \in \mathfrak{I}\}$ converges unconditionally to some $\boldsymbol{x} \in \mathscr{X}$, and $\|\boldsymbol{x}\| \le s$, that is,*

$$\left\| \sum_{i\in\mathfrak{I}} \boldsymbol{x}_i \right\| \le \sum_{i\in\mathfrak{I}} \|\boldsymbol{x}_i\| .$$

Proof. By 0.4.5 and 0.4.6, $s = \sum_{i\in\mathfrak{I}} \|\boldsymbol{x}_i\| = \sum_{k=1}^{\infty} \|\boldsymbol{x}_{i_k}\|$. Thus the series $\sum_{k=1}^{\infty} \boldsymbol{x}_{i_k}$ is absolutely convergent, so converges to some $\boldsymbol{x} \in \mathscr{X}$ (0.4.3). Given $\varepsilon > 0$, choose m such that

$$\left\| \sum_{k=1}^{n} \boldsymbol{x}_{i_k} - \boldsymbol{x} \right\| < \varepsilon \quad \text{for all } n \ge m.$$

then $F_\varepsilon := \{i_1, \dots, i_m\}$ satisfies 0.4, hence the convergence is unconditional. □

Bounded Linear Transformations

0.4.8 Proposition. *Let $\mathscr{X}$ and $\mathscr{Y}$ be normed spaces and $T : \mathscr{X} \to \mathscr{Y}$ linear. Then T is continuous iff there exists a constant $M \ge 0$ such that*

$$\|T\boldsymbol{x}\| \le M\|\boldsymbol{x}\| \quad \textit{for all } \boldsymbol{x} \in \mathscr{X}.$$

Proof. If the condition holds, then continuity (in fact uniform continuity) follows from the inequality $\|T\boldsymbol{x} - T\boldsymbol{y}\| = \|T(\boldsymbol{x} - \boldsymbol{y})\| \le M\|\boldsymbol{x} - \boldsymbol{y}\|$.

Conversely, if T is continuous there exists a $\delta > 0$ such that $\|\boldsymbol{y}\| \le \delta \Rightarrow \|T\boldsymbol{y}\| \le 1$. Thus for any $\boldsymbol{x} \ne 0$,

$$\frac{\delta}{\|\boldsymbol{x}\|}\|T\boldsymbol{x}\| = \left\| T\left(\frac{\delta}{\|\boldsymbol{x}\|}\boldsymbol{x}\right)\right\| \le 1$$

and so $\|T\boldsymbol{x}\| \le (1/\delta)\|\boldsymbol{x}\|$. □

The proposition implies that a linear transformation T is continuous iff it is bounded on bounded sets, that is, on sets of the form $\{\boldsymbol{x} \in \mathscr{X} : \|\boldsymbol{x}\| \le r\}$. For this reason, a continuous linear transformation is said to be **bounded**, reflecting the fact that T maps bounded sets onto bounded sets. Note that by the above proof, T is bounded iff it is continuous at zero.

A **topological isomorphism** from a normed space $\mathscr{X}$ onto a normed space $\mathscr{Y}$ is a linear bijection $T : \mathscr{X} \to \mathscr{Y}$ that is also a homeomorphism. A linear map $T : \mathscr{X} \to \mathscr{Y}$ with the property $\|T\boldsymbol{x}\| = \|\boldsymbol{x}\|$ for all $\boldsymbol{x} \in \mathscr{X}$ is called an **isometry**. If also $T(\mathscr{X}) = \mathscr{Y}$, then T is called an **isometric isomorphism onto** $\mathscr{Y}$. For example, if X and Y are sets and $\varphi : X \to Y$ is any function, then the adjoint map $\varphi^* : B(Y) \to B(X)$ is a bounded linear transformation, an isometry if φ is surjective, and an isometric isomorphism if φ is a bijection.

Banach Algebras

A **normed algebra** is an algebra $\mathscr{A}$ over $\mathbb{C}$ with a norm that satisfies

$$\|\boldsymbol{x}\boldsymbol{y}\| \le \|\boldsymbol{x}\|\,\|\boldsymbol{y}\|, \quad \boldsymbol{x}, \boldsymbol{y} \in \mathscr{A}.$$

A complete normed algebra is called a **Banach algebra**. These structures occur in many important settings, particularly in the theory of operators on Hilbert spaces. The Banach space $B(X)$ of all bounded functions under pointwise multiplication is a simple example of a commutative unital Banach algebra. Other examples appear throughout the text. General commutative Banach algebras are discussed in detail in Chapter 13.

0.5 Topological Spaces

Open and Closed Sets

A **topology** on a set X is a collection $\mathcal{T}$ of subsets of X with the following properties:

$$\begin{aligned} &\text{(a)}\ X, \emptyset \in \mathcal{T}, \\ &\text{(b)}\ \mathcal{U} \subseteq \mathcal{T} \Rightarrow \bigcup \mathcal{U} \in \mathcal{T}, \\ &\text{(c)}\ U, V \in \mathcal{T} \Rightarrow U \cap V \in \mathcal{T}. \end{aligned} \tag{0.5}$$

A member of $\mathcal{T}$ is said to be **$\mathcal{T}$-open**, or simply **open** if there is no possibility of confusion. A set X with a topology $\mathcal{T}$ is called a **topological space** and is denoted by $X_{\mathcal{T}}$. In the absence of ambiguity we omit the subscript $\mathcal{T}$. We shall occasionally also use the notation $\mathcal{O}$ to denote the collection of open sets of a topology.

A metric space is an important example of a topological space. The open sets here are unions of open balls. The topology of a metric space X called the **metric topology** of X. In particular, $\mathbb{R}^d$ with the Euclidean metric is a topological space, its topology called the **usual topology of** $\mathbb{R}^d$. Other examples of topological spaces appear throughout the text.

As in the case of metric spaces, a subset of a topological space is said to be **closed** if it is the complement of an open set. Proposition 0.3.1 then clearly holds for topological spaces.

If $\mathcal{T}_1$ and $\mathcal{T}_2$ are topologies and $\mathcal{T}_1 \subseteq \mathcal{T}_2$, then $\mathcal{T}_1$ is said to be **weaker than** $\mathcal{T}_2$ and $\mathcal{T}_2$ **stronger than** $\mathcal{T}_1$. In this case we also write $\mathcal{T}_1 \leq \mathcal{T}_2$. Every nonempty set X has a weakest topology, the **indiscrete topology**, whose only open sets are $\emptyset$ and X, and a strongest topology, the **discrete topology**, for which *every* subset is open. The latter is the metric topology generated by the discrete metric.

If X is a set and $\{\mathcal{T}_i : i \in \mathfrak{I}\}$ is a family of topologies on X, then the intersection $\bigcap\{\mathcal{T}_i : i \in \mathfrak{I}\}$ is easily seen to be a topology on X. In particular, if $\mathcal{S}$ is a family of subsets of X, then intersection $\mathcal{T}(\mathcal{S})$ of all topologies containing $\mathcal{S}$ (the discrete topology being one such) is well-defined and is the weakest topology relative to which all members of $\mathcal{S}$ are open. $\mathcal{T}(\mathcal{S})$ is called the **topology generated by** $\mathcal{S}$. For example, the usual topology of $\mathbb{R}$ is generated by the finite open intervals with rational endpoints.

A topological space $X_{\mathcal{T}}$ is said to be **Hausdorff** if distinct points in X can be **separated by open sets**, that is, if for each pair of distinct points $x, y \in X$, there exist disjoint open sets U and V with $x \in U$ and $y \in V$. For example, a metric space is a Hausdorff space, since x and y may be separated by the open balls $B_r(x)$ and $B_r(y)$, where $0 < r \leq d(x,y)/2$.

A **base** for a topology is a collection of open sets $\mathcal{U}$ such that every open set is a union of members of $\mathcal{U}$. A topological space is said to be **second countable** if it has a countable base. For example, $\mathbb{R}^d$ is second countable, as may be seen by considering the open balls $B_{1/n}(x)$ where x has rational coordinates.

The **interior**, **closure**, and **boundary** of a subset of a topological space are defined exactly as in the case of a metric space, as are the notions of **dense subspace** and **separability**. [2]

0.5.1 Proposition. *A second countable topological space X is separable. A separable metric space is second countable.*

[2]Generally speaking, a metric concept phrased entirely in terms of open sets typically has a valid topological analog.

Proof. Let (U_n) be a countable base and $x_n \in U_n$. For any open neighborhood U of x, there exists n such that $U_n \subseteq U$, hence $x_n \in U$. Therefore, (x_n) is dense in X.

Now let X be a metric space with countable dense set $\{x_1, x_2, \ldots\}$. The collection $\mathcal{B} := \{B_{1/m}(x_n) : m, n \in \mathbb{N}, \}$ is then countable. We show that every nonempty open set U is a union of members of $\mathcal{B}$. Let $x \in U$ and choose m such that $B_{2/m}(x) \subseteq U$. Next, choose $x_n \in B_{1/m}(x)$. Then $x \in B_{1/m}(x_n) \subseteq B_{2/m}(x) \subseteq U$. Therefore, U is a union of the balls $B_{1/m}(x_n)$. □

Neighborhood Systems

The notion of **neighborhood of a point** x in a topological space X is defined as in the case of a metric space, namely as a superset of a open set containing x. The collection of all neighborhoods of x is called the **neighborhood system at** x and is denoted by $\mathcal{N}(x)$. Neighborhood systems clearly have the following properties:

(a) $X \in \mathcal{N}(x)\ \forall\ x \in X$.
(b) $N \in \mathcal{N}(x) \Rightarrow x \in N$.
(c) $N \in \mathcal{N}(x)$ and $M \supseteq N \Rightarrow M \in \mathcal{N}(x)$. (0.6)
(d) $N_1, N_2 \in \mathcal{N}(x) \Rightarrow N_1 \cap N_2 \in \mathcal{N}(x)$.
(e) $N \in \mathcal{N}(x) \Rightarrow$ there exists $U \in \mathcal{N}(x)$ with $U \subseteq N$ such that $U \in \mathcal{N}(y)\ \forall\ y \in U$.

The following is a converse. It allows a topological space X to be defined "locally", that is, by specifying a neighborhood system at each point of X.

0.5.2 Proposition. *Let X be a nonempty set and for each $x \in X$ let $\mathcal{N}(x)$ be a collection of subsets of X for which properties* (a) – (e) *of* (0.6) *hold. Then there exists a unique topology $\mathcal{T}$ on X such that* (i) *$\mathcal{N}(x)$ is the $\mathcal{T}$-neighborhood system at x and* (ii) $x \in U \in \mathcal{T} \Rightarrow U \in \mathcal{N}(x)$.

Proof. Let $\mathcal{T}$ be the collection of all sets U such that either $U = \emptyset$ or $U \in \mathcal{N}(x)$ for each $x \in U$. By (a), $X \in \mathcal{T}$, and, by (c) and (d), $\mathcal{T}$ is closed under arbitrary unions and finite intersections. Therefore, $\mathcal{T}$ is a topology for X satisfying (ii).

Now let $\{\mathcal{N}_{\mathcal{T}}(x) : x \in X\}$ be the $\mathcal{T}$-neighborhood system. If $M \in \mathcal{N}_{\mathcal{T}}(x)$ and U is open with $x \in U \subseteq M$ then, by definition of $\mathcal{T}$, $U \in \mathcal{N}(x)$, hence, by (c), $M \in \mathcal{N}(x)$. Conversely, if $N \in \mathcal{N}(x)$, then the set U in (e) is in $\mathcal{T}$, hence $N \in \mathcal{N}_{\mathcal{T}}(x)$. Therefore, $\mathcal{N}(x) = \mathcal{N}_{\mathcal{T}}(x)$.

To prove uniqueness, let $\mathcal{T}'$ be a topology satisfying (i) and (ii). If $x \in V \in \mathcal{T}'$, then $V \in \mathcal{N}(x)$ by (ii), hence there exists $U \in \mathcal{T}$ such that $x \in U \subseteq V$. Therefore, V is a union of $\mathcal{T}$-open sets and so is $\mathcal{T}$-open. This shows that $\mathcal{T}' \subseteq \mathcal{T}$. Similarly, $\mathcal{T} \subseteq \mathcal{T}'$. □

Neighborhood Bases

Let X be a topological space. A **neighborhood base** at $x \in X$ is a subset $\mathcal{B}(x)$ of $\mathcal{N}(x)$ such that every member of $\mathcal{N}(x)$ contains a member of $\mathcal{B}(x)$. For example, the collection of open neighborhoods of x is clearly a neighborhood base at x.

If each $x \in X$ has a neighborhood base $\mathcal{B}(x)$, then the resulting system $\{\mathcal{B}(x) : x \in X\}$ has the following properties, derived from those of $\mathcal{N}(x)$:

(a) $B \in \mathcal{B}(x) \Rightarrow x \in B$.
(b) $B_1, B_2 \in \mathcal{B}(x) \Rightarrow$ there exists $B_3 \in \mathcal{B}(x)$ with $B_3 \subseteq B_1 \cap B_2$. (0.7)
(c) $B \in \mathcal{B}(x) \Rightarrow$ there exists $U \in \mathcal{B}(x)$ with $U \subseteq B$ such that U contains a member of $\mathcal{B}(y)$ for each $y \in U$.

Here is a converse based on 0.5.2:

0.5.3 Proposition. *Let X be a nonempty set and for each $x \in X$ let $\mathcal{B}(x)$ be a collection of subsets of X with properties* (a) – (c) *of* (0.7). *Then there exists a unique topology $\mathcal{T}$ on X such that* (i) *$\mathcal{B}(x)$ is a neighborhood base at x and* (ii) *every open set is a neighborhood of each of its points.*

Proof. Let $\mathcal{N}(x)$ be the collection of all supersets of members of $\mathcal{B}(x)$. Then $\mathcal{N}(x)$ satisfies the conditions (a) – (e) of (0.6), and the assertions follow from 0.5.2. □

A topological space is said to be **first countable** if each point x has a countable neighborhood base. A metric space is first countable; for example, the collection of open (or closed) balls at x with radii $1/n$ ($n \in \mathbb{N}$) is a countable neighborhood base.

Relative Topology

If Y is a subset of a topological space $X_{\mathcal{T}}$, then the trace $\mathcal{T} \cap Y$ is a topology called the **relative topology of** Y. The collection of closed sets in Y is easily seen to be the trace of the collection of closed sets in X. Open (closed) sets of Y are frequently referred to as **relatively open (closed)**. The neighborhood system of $y \in Y$ is the trace on Y of the $\mathcal{T}$-neighborhood system of y. If $\mathcal{B}(y)$ is a $\mathcal{T}$-neighborhood base at $y \in Y$, then $\mathcal{B}(y) \cap Y$ is a neighborhood base at y. For example, the collection of intervals $[0, 1/n)$ ($n \in \mathbb{N}$) is a neighborhood base at 0 in the relative topology of $[0, 1]$.

Nets

A **directed set** is a nonempty set A together with a relation $\preceq$ that is reflexive, transitive, and has the property that every pair of elements has an upper bound. For example, the neighborhood system of a point x in a topological space is directed by **reverse inclusion**, that is,

$$N_x \preceq M_x \text{ iff } M_x \subseteq N_x.$$

The collection of all partitions of an interval $[a, b]$ is directed by **inclusion**:

$$\mathcal{P} \preceq \mathcal{Q} \text{ iff } \mathcal{Q} \text{ is a refinement of } \mathcal{P}.$$

The Cartesian product $A \times B$ of directed sets A and B is directed by the **product ordering**

$$(a_1, b_1) \preceq (a_2, b_2) \text{ iff } a_1 \preceq a_2 \text{ and } b_1 \preceq b_2.$$

A **net** is a function whose domain A is a directed set. We shall use notation such as $(x_\alpha)_A$, or simply (x_α), for a net with values x_α. A net (x_α) in a set X is said to be **eventually** in a subset E of X if there exists a β such that $x_\alpha \in E$ for all $\alpha \succeq \beta$. The net is **frequently** in E if for each $\alpha \in A$ there exists an $\beta \succeq \alpha$ such that $x_\beta \in E$. A net (x_α) in a topological space $X_{\mathcal{T}}$ is said to **$\mathcal{T}$-converge to** $x \in X$ if (x_α) is eventually in every neighborhood of x. In this case we write $\mathcal{T}\text{-}\lim_\alpha x_\alpha = x$ or $x_\alpha \xrightarrow{\mathcal{T}} x$. In the absence of ambiguity, we usually drop the symbol $\mathcal{T}$ from this notation. The reader may easily verify that in a Hausdorff space limits are unique.

An infinite sequence in a metric space X is a net directed by the set $\mathbb{N}$ with the usual order $<$. Net convergence in this case is simply sequential convergence. If (x_n) is a Cauchy sequence in X, then under the product ordering of $\mathbb{N} \times \mathbb{N}$ the distances $d(x_m, x_n)$ form a net which converges to zero. For another example, consider the set of all tagged partitions $(\mathcal{P}, \xi)$ of an interval $[a, b]$ directed by

$$(\mathcal{Q}, \zeta) \preceq (\mathcal{P}, \xi) \text{ iff } \|\mathcal{P}\| \leq \|\mathcal{Q}\|.$$

If $f : [a,b] \to \mathbb{R}$ is Riemann integrable, then the Riemann sums $S(f, \mathcal{P}, \xi)$ form a net such that $\lim_{(\mathcal{P},\xi)} S(f, \mathcal{P}, \xi) = \int_a^b f(x)\,dx$. (See §3.3.)

Many properties of sequential convergence in a metric space carry over to nets in general topological spaces. In fact, the notion of net was introduced to describe convergence in topological spaces that are not first countable. Here is the net analog of 0.3.5.

0.5.4 Proposition. *Let X be a topological space and $E \subseteq X$. Then $x \in \text{cl}(E)$ iff there exists a net (x_α) in E converging to x.*

Proof. If $x \in \text{cl}(E)$, then $N \cap E \neq \emptyset$ for each neighborhood N of x. Choosing $x_N \in N \cap E$ and directing the neighborhood system at x by reverse inclusion, we obtain a net in E converging to x. Conversely, if $x \notin \text{cl}(E)$ then every net converging to x is eventually in the open set $\text{cl}(E)^c \subseteq E^c$, hence no net in E can converge to x. □

The notion of subsequence has the following net counterpart: A net $(y_\beta)_B$ is a **subnet** of a net $(x_\alpha)_A$ if there exists a function $\beta \mapsto \alpha_\beta : B \to A$ such that

(i) $y_\beta = x_{\alpha_\beta}$.

(ii) For each $\alpha_0 \in A$ there exists $\beta_0 \in B$ such that $\alpha_\beta \succeq \alpha_0$ for all $\beta \succeq \beta_0$.

While this generalizes the notion of subsequence, it should be noted that a subnet of a sequence need not be a subsequence. (Consider a subnet (x_{n_β}) of a sequence (x_n) with $\beta(1) = \beta(2) = 1$, $\beta(3) = \beta(4) = 2$, etc.)

A point x in a topological space X is a said to be **cluster point** of a net (x_α) if (x_α) is frequently in every neighborhood of x. The connection between cluster points and subnets is analogous to the connection between cluster points and subsequences in a metric space:

0.5.5 Proposition. *A net $(x_\alpha)_A$ in a topological space X has a cluster point x iff (x_α) has a subnet converging to x*

Proof. Let (x_{α_β}) be a subnet of (x_α) converging to x. If N is a neighborhood of x, then there exists β_1 such that $x_{\alpha_\beta} \in N$ for all $\beta \succeq \beta_1$. This implies that x_α is frequently in N. Indeed, by definition of subnet, for each $\alpha_0 \in A$ there exists $\beta_0 \in B$ such that $\beta \succeq \beta_0 \Rightarrow \alpha_\beta \succeq \alpha_0$. Thus if $\beta \succeq \beta_0$ and $\beta \succeq \beta_1$, then $\alpha_\beta \succeq \alpha_0$ and $x_{\alpha_\beta} \in N$.

Conversely, let x be a cluster point of (x_α) and let $\mathcal{N}(x)$ be the neighborhood system at x directed by reverse inclusion. Direct the pairs $(\alpha, N) \in A \times \mathcal{N}(x)$ by the product ordering. For each (γ, N), the net (x_α) is frequently in N, hence there exists $\alpha(\gamma, N) \succeq \gamma$ such that $x_{\alpha(\gamma,N)} \in N$. Then $(x_{\alpha(\gamma,N)})_{A\times\mathcal{N}(x)}$ is the required subnet. Indeed, for any $\alpha_0 \in A$ and $N_0 \in \mathcal{N}(x)$, $(\gamma, N) \succeq (\alpha_0, N_0) \Rightarrow \alpha(\gamma, N) \succeq \gamma \succeq \alpha_0 \Rightarrow x_{\alpha(\gamma,N)} \in N \subseteq N_0$. □

0.5.6 Corollary. *Let X have topologies $\mathcal{T}$ and $\mathcal{T}'$. Then $\mathcal{T}' \leq \mathcal{T}$ iff every net (x_α) that $\mathcal{T}$-converges to some member x of X also $\mathcal{T}'$-converges to x.*

Proof. The necessity is clear, since every $\mathcal{T}'$-neighborhood of x is a $\mathcal{T}$-neighborhood of x. For the sufficiency, suppose for a contradiction that C is $\mathcal{T}'$-closed but not $\mathcal{T}$-closed. Take any $x \in \text{cl}_{\mathcal{T}}(C) \setminus C$ and for each $\mathcal{T}$-neighborhood N of x choose $x_N \in C \cap N$. Then $x_N \to x$ in $X_{\mathcal{T}}$, hence also in $X_{\mathcal{T}'}$. But then $x \in C$. □

0.6 Continuity in Topological Spaces

Definition and General Properties

Let X and Y be topological spaces. Motivated by 0.3.6, we say that a function $f : X \to Y$ is **continuous at a point** $x \in X$ if for each neighborhood $N_{f(x)}$ of $f(x)$ there exists a neighborhood N_x of x such that $f(N_x) \subseteq N_{f(x)}$. If f is continuous at each point of a subset E of X, then f is said to be **continuous on** E. Note that this is not the same as the assertion that $f\big|_E$ is continuous. For example, the indicator function $\mathbf{1}_{\mathbb{Z}} : \mathbb{R} \to \{0, 1\}$ is not continuous at any integer, yet its restriction to $\mathbb{Z}$ is continuous in the relative topology. If f is continuous on X, we say that f is **continuous**. A continuous function $f : X \to Y$ with a continuous inverse $f^{-1} : Y \to X$ is called a **homeomorphism**.

A function $f : X \to Y$ is said to be **open** if $f(U)$ is open in Y for each open subset U of X. An open map need not be continuous (take Y to be discrete for suitable X), and a continuous map need not be open (take X to be discrete). Obviously, a continuous bijection is a homeomorphism iff it is open.

0.6.1 Proposition. *A function $f : X \to Y$ is continuous at x iff $f(x_\alpha) \to f(x)$ for each net (x_α) in X that converges to x.*

Proof. Suppose that f is continuous at x and let $x_\alpha \to x$. Given $N_{f(x)}$ choose N_x such that $f(N_x) \subseteq N_{f(x)}$. Next, choose α_0 so that $x_\alpha \in N_x$ for all $\alpha \succeq \alpha_0$. For such α, $f(x_\alpha) \in N_{f(x)}$. Therefore, $f(x_\alpha) \to f(x)$.

Conversely, if f is not continuous at x, then there exists a neighborhood $N_{f(x)}$ such that $f(N) \not\subseteq N_{f(x)}$ for all $N \in \mathcal{N}(x)$. For each N choose $x_N \in N$ so that $f(x_N) \notin N_{f(x)}$. Then the net (x_N) converges to x, but $f(x_N)$ is never in $N_{f(x)}$. □

The next result is proved exactly as in the metric case (0.3.8), except that in the proof of (d) $\Rightarrow$ (a) one must use nets instead of sequences. We leave the details to the reader.

0.6.2 Theorem. *The following statements are equivalent:*

(a) *f is continuous.*

(b) *$f^{-1}(V)$ is open in X for each open subset V of Y.*

(c) *$f^{-1}(C)$ is closed in X for each closed subset C of Y.*

(d) *$f(\operatorname{cl} A) \subseteq \operatorname{cl}(f(A))$ for each subset A of X.*

0.6.3 Corollary. *Let X and Y be topological spaces and let the topology of Y be generated by a collection $\mathcal{S}$ of subsets of Y. Then a function $f : X \to Y$ is continuous iff $f^{-1}(U)$ is open in X for each $U \in \mathcal{S}$.*

Proof. The necessity is clear. For the sufficiency, assume $f^{-1}(U)$ is open in X for each $U \in \mathcal{S}$. Then the collection of all sets $V \subseteq Y$ for which $f^{-1}(V)$ is open in X is a topology containing $\mathcal{S}$ and so contains $\mathcal{T}(\mathcal{S})$. □

For example, a real-valued function f on a topological space is continuous iff $f^{-1}\big((-\infty, a)\big)$ and $f^{-1}\big((a, \infty)\big)$ are open for all rational numbers a.

Initial Topologies

Let X be a set, Y a topological space, and $\mathscr{F}$ a family of maps $f : X \to Y$. The topology $\mathfrak{T}$ on X generated by the sets $f^{-1}(U)$, where $f \in \mathscr{F}$ and U is open in Y, is called the **initial topology on X with respect to $\mathscr{F}$.**

0.6.4 Proposition. *The initial topology $\mathfrak{T}$ has the following properties:*

(a) *$\mathfrak{T}$ is the weakest topology on X relative to which every member of $\mathscr{F}$ is continuous.*

(b) *For each $y \in Y$, let $\mathfrak{B}_y$ be an open neighborhood base at y. For $x \in X$, let $\mathfrak{B}_x$ denote the collection of all sets of the form $f_1^{-1}(U_1) \cap \cdots \cap f_n^{-1}(U_n)$ containing x, where $f_j \in \mathscr{F}$ and $U_j \in \mathfrak{B}_{f_j(x)}$. Then $\mathfrak{B}_x$ is a neighborhood base at x for the initial topology.*

(c) *A net (x_α) in X $\mathfrak{T}$-converges to x iff $f(x_\alpha) \to f(x)$ for every $f \in \mathscr{F}$.*

(d) *If Z is a topological space, a mapping $\varphi : Z \to X$ is $\mathfrak{T}$-continuous iff $f \circ \varphi : Z \to Y$ is continuous for each $f \in \mathscr{F}$.*

Proof. (a) Let $\mathfrak{T}'$ be a topology relative to which every member of $\mathscr{F}$ is continuous. Then $f^{-1}(U) \in \mathfrak{T}'$ for all open $U \subseteq Y$, hence $\mathfrak{T} \leq \mathfrak{T}'$.

(b) The system $\{\mathfrak{B}_x : x \in X\}$ satisfies (0.7) and hence defines a topology $\mathfrak{T}'$ on X relative to which $\mathfrak{B}(x)$ is a neighborhood base at x (0.5.3). Since every $\mathfrak{T}'$-open set is a union of these basic neighborhoods and since the latter are $\mathfrak{T}$-open, $\mathfrak{T}' \leq \mathfrak{T}$. On the other hand, every member of $\mathscr{F}$ is obviously $\mathfrak{T}'$-continuous, hence $\mathfrak{T} \leq \mathfrak{T}'$.

(c) The necessity is clear. For the sufficiency, let (x_α) be a net in X such that $f(x_\alpha) \to f(x)$ for every $f \in \mathscr{F}$. By (b), (x_α) is eventually in any basic neighborhood of x.

(d) For the sufficiency, let $f \circ \varphi : Z \to Y$ be continuous for each $f \in \mathscr{F}$. Then for any open set $U \subseteq Y$ and $f \in \mathscr{F}$, $\varphi^{-1}(V)$ is open in Z, where $V := f^{-1}(U)$. Since the sets V generate the topology of X, φ is continuous by 0.6.3. □

Product Topology

Let $\{X_i : i \in \mathfrak{I}\}$ be a family of topological spaces and set $X := \prod_{i \in \mathfrak{I}} X_i$. The **product topology on** X is the initial topology with respect to the family of projection mappings $\pi_i : X \to X_i$. By 0.6.4(c), a net (f_α) in X converges to f in this topology iff $f_\alpha(i) = \pi_i(f_\alpha) \to \pi_i(f) = f(i)$ for each $i \in \mathfrak{I}$. For this reason the product topology is also called the **topology of pointwise convergence on** $\mathfrak{I}$. Note that the product topology on $\mathbb{R}^d$ is simply the topology defined by the Euclidean metric.

Final Topologies

Let X be a topological space, Y a nonempty set and $\mathscr{F}$ a family of maps $f : X \to Y$. The collection $\mathfrak{T}$ of all subsets V of Y such that $f^{-1}(V)$ is open in X for each $f \in \mathscr{F}$ is a topology called the **final topology on Y with respect to $\mathscr{F}$.**

0.6.5 Proposition. *The final topology $\mathfrak{T}$ has the following properties:*

(a) *$\mathfrak{T}$ is the strongest topology on Y relative to which every member of $\mathscr{F}$ is continuous.*

(b) *If Z is a topological space, then a function $\varphi : Y \to Z$ is continuous iff $\varphi \circ f : X \to Z$ is continuous for every $f \in \mathscr{F}$.*

Proof. (a) This follows immediately from the definition of $\mathfrak{T}$.

(b) For the sufficiency, let U be open in Z. Then $f^{-1}\big(\varphi^{-1}(U)\big) = (\varphi \circ f)^{-1}(U)$ is open in X for every $f \in \mathscr{F}$, hence $\varphi^{-1}(U)$ is open in Y. □

Quotient Topology

Let X be a topological space and $\sim$ an equivalence relation on X. The final topology on $X/\sim$ with respect to the quotient map $Q : X \to X/\sim$ is called the **quotient topology**. The open sets of $X/\sim$ are precisely those collections V of equivalence classes $[x]$ such that $Q^{-1}(V) = \bigcup_{[x]\in V}[x]$ is open in X. Quotient topologies play an important role in the theory of normed linear spaces (see §8.4).

The Space of Continuous Functions

Let X and Y be topological spaces. The set of continuous functions $f : X \to Y$ is denoted by $C(X, Y)$. If $Y = \mathbb{C}$, we use the simpler notation $C(X)$. Thus

$$C(X) := \text{the space of all continuous functions } f : X \to \mathbb{C}.$$

Clearly, $C(X)$ is closed under addition and multiplication and so is an algebra. We define the related space of all bounded, continuous functions $f : X \to \mathbb{C}$ by

$$C_b(X) := C(X) \cap B(X).$$

As the uniform limit of a sequence of continuous functions is continuous, $C_b(X)$ is a closed subalgebra of the Banach algebra $B(X)$ under the supremum norm. Thus $C_b(X)$ is a Banach algebra.

Notation. For spaces such as $C\big((0, 1)\big)$, etc. we usually omit the outer parentheses and write instead $C(0, 1)$. This convention holds for other function spaces as well.

F-sigma and G-delta Sets

A countable union of closed sets in a topological space X is called an **F_σ-set**. Dually, a countable intersection of open sets is called a **G_δ-set**. For example, the half-open interval $(a, b]$ is both an F_σ-set and a G_δ-set, the set of rationals is an F_σ set, and the set of irrationals is a G_δ set. By DeMorgan's laws, a set is F_σ iff its complement is G_δ. In a metric space every closed set C is a G_δ set and every open set U is an F_σ set, as may be seen from

$$C = \bigcap_n \{x \in X : d(x, C) < 1/n\} \quad \text{and} \quad U = \bigcup_n \{x \in X : d(x, U^c) \ge 1/n\}.$$

Proposition 0.6.7 below characterizes the set of continuity points of a function as a G_δ set. For the proof we need

0.6.6 Lemma. *Let X be a topological space and (Y, d) a metric space. For each $x \in X$, define*

$$F(x) = \inf_U \sup\{d(f(x'), f(x'')) : x'\, x'' \in U\},$$

where the infimum is taken over all open neighborhoods of x. Then for each $r > 0$ the set $W_r := \{x \in X : F(x) < r\}$ is open. Thus $W_0 := \{x \in X : F(x) = 0\}$ is a G_δ set.

Proof. Let $x_0 \in W_r$ and choose an open neighborhood U of x_0 such that

$$\sup\{d(f(x'), f(x'')) : x'\, x'' \in U\} < r.$$

Then $F(x) < r$ for all $x \in U$, that is, $U \subseteq W_r$. Therefore, W_r is open and so $W_0 = \bigcap_{n=1}^\infty W_{1/n}$ is a G_δ. □

0.6.7 Proposition. *Let X be a topological space and (Y, d) a metric space. Then the set of points where a function $f : X \to Y$ is continuous is a G_δ set.*

Proof. We claim that W_0 is the set of continuity points of f. Indeed, f is continuous at x iff for each $\varepsilon > 0$ there exists a neighborhood U of x such that $d(x', x) < \varepsilon$ for all $x' \in U$ iff for each $\varepsilon > 0$ there exists a neighborhood U of x such that $d(x', x'') < \varepsilon$ for all x', $x'' \in U$ iff $F(x) < \varepsilon$ for all ε iff $F(x) = 0$. □

From the proposition we see that no function $f : \mathbb{R} \to \mathbb{R}$ can be continuous precisely at the rationals. The reader may easily find examples of functions that are continuous precisely at the irrationals. (See Ex. 2.20.)

0.7 Normal Topological Spaces

A Hausdorff topological space is said to be **normal** if every pair of disjoint closed subsets A and B may be **separated** by open sets U and V, that is, $A \subseteq U$, $B \subseteq V$, and $U \cap V = \emptyset$. In this section we describe the two most important properties of normal spaces. The first shows that an ostensibly stronger separation property holds. For its proof we need the following lemma.

0.7.1 Lemma. *Let X be normal, C closed, and U open with $C \subseteq U$. Then there exists an open set V such that $C \subseteq V \subseteq \mathrm{cl}(V) \subseteq U$.*

Proof. Since C and U^c are disjoint closed sets, by normality there exist disjoint open sets $V \supseteq C$ and $W \supseteq U^c$. If $x \in \mathrm{cl}(V) \cap U^c$, then W, as a neighborhood of x, meets V, a contradiction. Therefore, $\mathrm{cl}(V) \subseteq U$. □

Urysohn's Lemma

0.7.2 Theorem (Urysohn). *If X is a normal topological space and A and B are disjoint closed subsets, then there exists a continuous function $f : X \to [a, b]$ such that $f = a$ on A and $f = b$ on B.*

Proof. We may assume $a = 0$ and $b = 1$ (otherwise, replace f by $(f - a)/(b - a)$). Let $D := \{r = k2^{-n} : n \in \mathbb{N},\ 0 < k < 2^n\}$, the set of dyadic rational numbers in $(0, 1)$. We show by induction on n that there exists a family of open sets U_r indexed by members r of D such that

$$A \subseteq U_r \subseteq \mathrm{cl}\, U_r \subseteq U_s \subseteq D^c \quad \text{for all } r, s \in D \text{ with } r < s. \tag{†}$$

By 0.7.1, there exists an open set $U_{1/2}$ such that $A \subseteq U_{1/2} \subseteq \mathrm{cl}\, U_{1/2} \subseteq B^c$. This defines U_r for the case $k = n = 1$. Now assume that sets U_r have been constructed for $r = k/2^n$ $(0 < k < 2^n)$. Since $k/2^n = 2k/2^{n+1}$, it remains to construct U_r for $r = (2k+1)/2^{n+1}$. But since $\mathrm{cl}\, U_{k2^{-n}} \subseteq U_{(k+1)2^{-n}}$, there exists by 0.7.1 an open set U_r such that

$$A \subseteq \mathrm{cl}\, U_{k2^{-n}} \subseteq U_r \subseteq \mathrm{cl}\, U_r \subseteq U_{(k+1)2^{-n}} \subseteq B^c,$$

establishing (†).

Now set $U_1 = X$ and define f on X by $f(x) = \inf\{r \in D : x \in U_r\}$. Obviously, $0 \le f \le 1$. Also, since no member of B is in U_r for $r < 1$, $f(B) = 1$. Moreover, since $A \subseteq U_r$ for all r, $f(A) = 0$. To see that f is continuous, let $0 < t < 1$ and note that $f(x) < t$ iff $x \in U_r$ for some $r < t$, and $f(x) > t$ iff $x \notin \mathrm{cl}(U_r)$ for some $r > t$. Thus we have open sets

$$\{f < t\} = \bigcup_{r<t} U_r \quad \text{and} \quad \{f > t\} = \bigcup_{r>t} (\mathrm{cl}\, U_r)^c.$$

Since the intervals $(-\infty, t)$, (t, ∞) generate the topology of $\mathbb{R}$, f is continuous by 0.6.3. □

Tietze Extension Theorem

Here is the one of the main applications of Urysohn's lemma. A variation of the theorem is given in the section on locally compact spaces.

0.7.3 Theorem (Tietze). *If X is a normal topological space and $Y \subseteq X$ is closed, then for each continuous $f : Y \to [a, b]$ there exists a continuous $F : X \to [a, b]$ such $F\big|_Y = f$.*

Proof. We may assume $a = 0$ and $b = 1$. To construct F we first construct inductively a sequence of continuous functions $g_n : X \to [0, 1]$ such that

$$\text{(i)}\ \ 0 \le g_n \le 2^{n-1}/3^n \text{ on } X \quad \text{and} \quad \text{(ii)}\ \ 0 \le f - \sum_{j=1}^{n} g_j \le (2/3)^n \text{ on } Y. \qquad (\dagger)$$

To obtain g_1, note that the disjoint subsets $f^{-1}([0, 1/3])$ and $f^{-1}([2/3, 1])$ of Y are closed in X, hence by Urysohn's lemma there exists a continuous function $g_1 : X \to [0, 1/3]$ with $g_1 = 0$ on $f^{-1}([0, 1/3])$ and $g_1 = 1/3$ on $f^{-1}([2/3, 1])$. Thus ($\dagger$) holds for $n = 1$. Now assume that $g_1, \ldots, g_n$ satisfy ($\dagger$). By Urysohn's lemma again, there exists a continuous function $g_{n+1} : X \to [0, 2^n/3^{n+1}]$ such that $g_{n+1} = 0$ whenever $f - \sum_{j=1}^{n} g_j \le 2^n/3^{n+1}$, and $g_{n+1} = 2^n/3^{n+1}$ whenever $f - \sum_{j=1}^{n} g_j \ge (2/3)^{n+1}$, completing the construction. Now set $F = \sum_{n=1}^{\infty} g_n$. By (i), the convergence is uniform, so F is continuous. By (ii), $F = f$ on Y. □

A Hausdorff topological space is **completely regular** if for each $x \in X$ and closed set $Y \subseteq X$ there exists a continuous function $f : X \to [0, 1]$ such that $f = 0$ on Y and $f(x) = 1$. It follows from Urysohn's lemma that every normal space is completely regular. The notion of complete regularity finds special importance in the Stone-Čech compactification theorem (§13.4).

0.8 Compact Topological Spaces

A collection $\mathfrak{U}$ of open subsets of a topological space X such that $\bigcup \mathfrak{U} = X$ is called an **open cover of** X. A subcollection of $\mathfrak{U}$ that is a cover of X is called a **subcover**. A space X is said to be **compact** if each open cover of X has a finite subcover. A subset Y of X is **compact** if it is compact in the relative topology, that is, if for each collection $\mathfrak{U}$ of open sets in X there exists a finite subcollection $\mathfrak{U}_0$ such that $Y \subseteq \bigcup \mathfrak{U}_0$. A subset Y of X is **relatively compact** if its closure is compact.

Finite subsets of a topological space are obviously compact. In a discrete topological space X these are the only compact sets, since any nonempty set has an open cover of singletons.

A family $\mathcal{A}$ of subsets of a set X is said to have the **finite intersection property** (f.i.p.) if every finite subcollection of $\mathcal{A}$ has a nonempty intersection. The following is a useful characterization of compactness in terms of this notion.

0.8.1 Proposition. *A topological space X is compact iff $\bigcap_{A \in \mathcal{A}} \mathrm{cl}(A) \ne \emptyset$ for every collection $\mathcal{A}$ of subsets of X with the* f.i.p.

Proof. Let X be compact and $\mathcal{A}$ a collection of subsets of X. If $\bigcap_{A \in \mathcal{A}} \mathrm{cl}(A) = \emptyset$, then the collection of complements $\mathrm{cl}(A)^c$ is an open cover of X and so has a finite subcover $\{\mathrm{cl}(A_1)^c, \ldots, \mathrm{cl}(A_n)^c\}$. Then $\bigcap_{j=1}^{n} A_j \subseteq \bigcap_{j=1}^{n} \mathrm{cl}(A_j) = \emptyset$, hence $\mathcal{A}$ does not have the f.i.p.

Conversely, if X is not compact, then there exists an open cover $\mathfrak{U}$ with no finite subcover.

Taking $\mathcal{C}$ to be the collection of complements of members of $\mathcal{U}$, we see that $\mathcal{C}$ has the finite intersection property but has empty intersection. □

0.8.2 Proposition. *A compact subset of a Hausdorff space X is closed.*

Proof. Let $A \subseteq X$ be compact. We show that A^c is open. Let $b \in A^c$. For each $x \in A$, let M_x and N_x be disjoint open neighborhoods of x and b, respectively. Then $\{M_x : x \in A\}$ is an open cover of A, hence there exists a finite subset A_0 of A such that $U_b := \bigcup_{x \in A_0} M_x \supseteq A$. Set $V_b := \bigcap_{x \in A_0} N_x$. Then V_b is a neighborhood of b, and since $V_b \cap M_x = \emptyset$ for every $x \in A_0$, $V_b \subseteq A^c$. Therefore A^c is open. □

0.8.3 Proposition. *A compact subset Y of a metric space (X, d) is bounded.*

Proof. Fix $y \in Y$. The collection of open balls $B_n(y)$ with center $y \in Y$ and radius $n \in N$ is an open cover of Y and so has a finite subcover. Therefore, $Y \subseteq B_n(y)$ for some n. □

0.8.4 Proposition. *A closed subset of a compact space X is compact.*

Proof. Let $Y \subseteq X$ be closed. If $\mathcal{U}$ is a cover of Y by open sets of X, then enlarging $\mathcal{U}$ by including the open set $X \setminus Y$ results in an open cover of X. Since X is compact, there exist $U_1, \ldots, U_n \in \mathcal{U}$ such that $X = (X \setminus Y) \cup \bigcup_j U_j$. Then $Y \subseteq \bigcup_j U_j$. □

0.8.5 Corollary. *Let X have topologies $\mathcal{T}_1 \leq \mathcal{T}_2$ such that $(X, \mathcal{T}_2)$ is compact and $(X, \mathcal{T}_1)$ is Hausdorff. Then $\mathcal{T}_1 = \mathcal{T}_2$.*

Proof. Let C be $\mathcal{T}_2$-closed. By 0.8.4, C is $\mathcal{T}_2$-compact hence $\mathcal{T}_1$-compact. By 0.8.2, C is $\mathcal{T}_1$-closed. Therefore, $\mathcal{T}_1$ and $\mathcal{T}_2$ have the same closed sets and so are equal. □

The following proposition asserts that disjoint compact sets in a Hausdorff space may be separated by open sets.

0.8.6 Proposition. *Let A and B be disjoint compact subsets of a Hausdorff space X. Then there exist disjoint open sets U and V with $A \subseteq U$ and $B \subseteq V$.*

Proof. By the proof of 0.8.2, for each $b \in B$ there exist disjoint open sets $U_b \supseteq A$ and $V_b \ni b$. Then $\{V_b : b \in B\}$ is an open cover of B, so by compactness there exists a finite set $B_0 \subseteq B$ such that $B \subseteq V := \bigcup_{b \in B_0} V_b$. Now set $U := \bigcap_{b \in B_0} U_b$. □

From 0.8.4 and 0.8.6 we have the following:

0.8.7 Corollary. *A compact Hausdorff space is normal.*

Convergence in Compact Spaces

Here is an important characterization of compactness in terms of nets.

0.8.8 Theorem. *A topological space X is compact iff each net in X has a convergent subnet.*

Proof. Let X be compact and let $(x_\alpha)_A$ be a net in X. By 0.5.5 it suffices to show that (x_α) has a cluster point in X. For $\alpha \in A$, define $E_\alpha := \{x_\beta : \beta \geq \alpha\}$. Since every finite subset of A has an upper bound, the collection $\{E_\alpha : \alpha \in A\}$ has the f.i.p. By compactness, there exists an $x \in \bigcap \operatorname{cl} E_\alpha$. Thus if N is a neighborhood of x then $N \cap E_\alpha \neq \emptyset$ for every α. By definition of E_α, (x_α) is frequently in N. Therefore, x is the required cluster point.

Conversely, if X is not compact, then there exists an open cover $\{U_i : i \in \mathfrak{I}\}$ of X with no finite subcover. Direct the finite subsets α of $\mathfrak{I}$ upward by inclusion. For each α choose a point $x_\alpha \in X \setminus \bigcup_{i \in \alpha} U_i$. Then (x_α) is a net in X with no cluster points. Indeed, if $x \in X$, then $x \in U_j$ for some $j \in \mathfrak{I}$, but $x_\alpha \notin U_j$ for all finite sets α containing j (that is, $\alpha \succeq \{j\}$), hence x_α is not frequently in U_j. □

We shall see in the next section that in a metric space the nets in the last theorem may be replaced by sequences.

Compactness of Cartesian Products

The following theorem is among the most powerful theorems in topology. It will have important analytical consequences in a variety of contexts later.

0.8.9 Tychonoff's Theorem. *If $\{X_i : i \in \mathfrak{I}\}$ is a family of compact topological spaces, then the product space $X := \prod_{i\in\mathfrak{I}} X_i$ is compact in the product topology.*

Proof. We show that $\bigcap \mathrm{cl}(A)_{A\in\mathcal{A}_0} \neq \emptyset$ for any collection $\mathcal{A}_0 \subseteq \mathcal{P}(X)$ with the f.i.p. Order upward by inclusion the family $\mathfrak{A}$ of all collections $\mathcal{A} \supseteq \mathcal{A}_0$ with the f.i.p. It is easy to check that every chain in $\mathfrak{A}$ has an upper bound, namely the union of all the collections in the chain. By Zorn's lemma, $\mathfrak{A}$ has a maximal element $\mathcal{A}$. We show that $\bigcap \mathrm{cl}(A)_{A\in\mathcal{A}} \neq \emptyset$. Since $\mathcal{A}_0 \subseteq \mathcal{A}$, it will follow that $\bigcap \mathrm{cl}(A)_{A\in\mathcal{A}_0} \neq \emptyset$.

Observe that

$$\text{(i) } A,\, B \in \mathcal{A} \Rightarrow A \cap B \in \mathcal{A} \quad \text{and} \quad \text{(ii) } A \cap B \neq \emptyset \;\; \forall\, B \in \mathcal{A} \Rightarrow A \in \mathcal{A},$$

since the negation of either would allow $\mathcal{A}$ to be enlarged while still retaining the f.i.p., contradicting the maximality of $\mathcal{A}$. Let $\pi_i : X \to X_i$ denote the projection map. Since $\pi_i(\mathcal{A})$ has the f.i.p., by compactness of X_i there exists an $x_i \in \bigcap_{A\in\mathcal{A}} \mathrm{cl}\big(\pi_i(A)\big)$. Thus if U_i is an open neighborhood of x_i, then for every $A \in \mathcal{A}$, $U_i \cap \pi_i(A) \neq \emptyset$ and so $\pi_i^{-1}(U_i) \cap A \neq \emptyset$. By (ii), $\pi_i^{-1}(U_i) \in \mathcal{A}$, hence by (i) $U_F := \bigcap_{i\in F} \pi_i^{-1}(U_i) \in \mathcal{A}$ for any finite $F \subseteq \mathfrak{I}$. Since $\mathcal{A}$ has the f.i.p., $U_F \cap A \neq \emptyset$ for all $A \in \mathcal{A}$. Now set $f(i) := x_i$, $i \in \mathfrak{I}$. Since the sets U_F form a neighborhood base at f, $f \in \bigcap_{A\in\mathcal{A}} \mathrm{cl}(A)$. □

Continuity and Compactness

0.8.10 Theorem. *Let X and Y be topological spaces with X compact, and let $f : X \to Y$ be continuous. Then the following hold:*

(a) *$f(X)$ is compact.*

(b) *If f is a bijection and Y is Hausdorff, then f is a homeomorphism.*

(c) *If X and Y are metric spaces, then $f : X \to Y$ is uniformly continuous.*

Proof. (a) Let $\{V_i : i \in \mathfrak{I}\}$ be an open cover of $f(X)$ in Y. Then $\{f^{-1}(V_i) : i \in \mathfrak{I}\}$ is an open cover of X, hence there exists a finite subset $\mathfrak{I}_0$ of $\mathfrak{I}$ such that $\{f^{-1}(V_i) : i \in \mathfrak{I}_0\}$ is a cover of X. It follows that $\{V_i : i \in \mathfrak{I}_0\}$ is a finite cover of $f(X)$.

(b) Let $g = f^{-1}$ and let C be a closed subset of X. Then C is compact, hence $g^{-1}(C) = f(C)$ is compact and therefore closed. By 0.6.2, g is continuous.

(c) Let X and Y have metrics d and ρ, respectively. Let $\varepsilon > 0$. For each $x \in X$ choose $\gamma_x > 0$ such that

$$f\big(B_{\gamma_x}(x)\big) \subseteq B_{\varepsilon/2}\big(f(x)\big). \tag{$\dagger$}$$

Set $\delta_x = \gamma_x/2$. The collection $\{B_{\delta_x}(x) : x \in X\}$ is an open cover of X, hence there exists a finite set $F \subseteq X$ such that $\{B_{\delta_x}(x) : x \in F\}$ covers X. Let $\delta := \min_{x\in F} \delta_x$ and let $a,\, b \in X$ with $d(a,b) < \delta$. Choose $x \in F$ such that $a \in B_{\delta_x}(x)$. Then

$$d(x,a) < \delta_x < \gamma_x \quad \text{and} \quad d(x,b) \le d(a,b) + d(x,a) < \delta_x + \delta_x = \gamma_x,$$

so $a,\, b \in B_{\gamma_x}(x)$. By ($\dagger$),

$$\rho\big(f(a), f(b)\big) \le \rho\big(f(a), f(x)\big) + \rho\big(f(x), f(b)\big) < \varepsilon/2 + \varepsilon/2 = \varepsilon.$$

Therefore, f is uniformly continuous. □

0.8.11 Corollary. *If $f : X \to \mathbb{R}$ is continuous and X is compact, then there exist points x_m and x_M in X such that*

$$f(x_m) \le f(x) \le f(x_M) \quad \textit{for all} \quad x \in X.$$

Proof. By 0.8.10(a), $f(X)$ is compact, hence closed and bounded in $\mathbb{R}$. Thus $f(X)$ must contain its supremum and infimum. □

0.9 Totally Bounded Metric Spaces

Let (X, d) be a metric space. In this section we give two alternate characterizations of compactness of X.

A subset E of X is said to be *totally bounded* if for each $\varepsilon > 0$ there exist points $x_1, \ldots, x_n \in X$ such that $E \subseteq \bigcup_j B_\varepsilon(x_j)$. Since a finite union of open balls is bounded, every totally bounded set is bounded. The converse is false. For example, in a discrete metric space all sets are bounded, but no infinite set can be totally bounded.

A subset E of X is said to be *sequentially compact* if every sequence in E has a cluster point in E.

0.9.1 Theorem. *The following statements are equivalent:*

(a) *X is compact.*

(b) *X is sequentially compact.*

(c) *X is complete and totally bounded.*

Proof. (a) $\Rightarrow$ (b): Let (a_n) be a sequence in X with no cluster point. Then for each $x \in X$ there must exist an open ball $B(x)$ with center x that contains only finitely many terms of (a_n). This implies that every finite subcover of the open cover $\{B(x) : x \in X\}$ of X contains only finitely many terms of the sequence and so cannot cover X. Therefore, X is not compact.

(b) $\Rightarrow$ (c): Let X be sequentially compact. That X is complete follows from 0.3.2. Suppose X is not totally bounded. Then there exists $\varepsilon > 0$ such that no finite collection of open balls of radius ε covers X. Choose any $a_1 \in X$. Since $B_\varepsilon(a_1)$ does not cover X, there exists $a_2 \in X \setminus B_\varepsilon(a_1)$. Since $B_\varepsilon(a_1) \cup B_\varepsilon(a_2)$ does not cover X, there exists $a_3 \in X \setminus \big(B_\varepsilon(a_1) \cup B_\varepsilon(a_2)\big)$. Continuing in this manner we obtain a sequence (a_n) in X with

$$a_n \in X \setminus \big[B_\varepsilon(a_1) \cup B_\varepsilon(a_2) \cup \cdots \cup B_\varepsilon(a_{n-1})\big].$$

It follows that $d(a_n, a_m) \ge \varepsilon$ for all $m \ne n$. But then no subsequence of $\{a_n\}$ can converge. Therefore, X must be totally bounded.

(c) $\Rightarrow$ (a): Assume that X is complete and totally bounded but not compact. Then X has an open cover $\mathfrak{U} = \{U_i : i \in \mathfrak{I}\}$ with no finite subcover. For each k let F_k be a finite set of points in X such that $\{B_{1/k}(x) : x \in F_k\}$ is a cover of X. Consider the case $k = 1$. If for each $x \in F_1$ the ball $B_1(x)$ could be covered by finitely many members of $\mathfrak{U}$, then X itself would have a finite cover, contradicting our assumption. Thus there exists $x_1 \in F_1$ such that $E_1 := B_1(x_1)$ cannot be covered by finitely many members of $\mathfrak{U}$. Since $\{B_{1/2}(x) : x \in F_2\}$ covers X, $\{E_1 \cap B_{1/2}(x) : x \in F_2\}$ covers E_1, so by similar reasoning applied to E_1 there exists $x_2 \in F_2$ such that $E_2 := E_1 \cap B_{1/2}(x_2)$ cannot be covered by finitely many members of $\mathfrak{U}$. In this way we obtain a sequence (x_n) in X and decreasing sets

$$E_n = B_1(x_1) \cap B_{1/2}(x_2) \cap \cdots \cap B_{1/n}(x_n) = E_{n-1} \cap B_{1/n}(x_n) \qquad (\dagger)$$

that cannot be covered by finitely many members of $\mathfrak{U}$. In particular, $E_n \neq \emptyset$. For each n, choose a point $y_n \in E_n$. If $n > m$, then $y_n \in E_m$, hence from (†)

$$d(x_m, x_n) \leq d(x_m, y_n) + d(y_n, x_n) < 1/m + 1/n,$$

from which it follows that (x_n) is a Cauchy sequence. Since X is complete, $x_n \to x$ for some $x \in X$. Choose $i \in \mathfrak{I}$ such that $x \in U_i$. Since U_i is open, there exists $r > 0$ such that $B_r(x) \subseteq U_i$. Taking $n > 2/r$ so that $d(x_n, x) < r/2$ we then have $E_n \subseteq B_{1/n}(x_n) \subseteq B_r(x) \subseteq U_i$, contradicting the non-covering property of E_n. Therefore, X must be compact. □

The following result is known as the *Heine-Borel theorem.*

0.9.2 Corollary. *A subset of $\mathbb{R}^d$ is compact iff it is closed and bounded.*

Proof. We have already proved the necessity. For the sufficiency, let $C \subseteq \mathbb{R}^d$ be closed and bounded. Since $\mathbb{R}^d$ is complete, C is complete (0.3.4). Since C is bounded, it is totally bounded as may be seen by enclosing C in bounded d-dimensional interval I and then subdividing I into finitely many congruent subintervals of arbitrarily small diameter. Therefore, C is compact. □

0.10 Equicontinuity

We have seen that every closed ball in $\mathbb{R}^d$ is compact. By contrast, closed balls in the space $C[0,1]$ with the supremum norm are not compact, as may be inferred from the fact that the sequence of functions $f_n(x) = x^n$ has no convergent subsequence in $C[0,1]$. The additional property of *equicontinuity* is needed to characterize compact subsets of such spaces.

Let X be a topological space. A family $\mathscr{F}$ of functions in $C(X)$ is said to be **equicontinuous at a point** $a \in X$ if, for each $\varepsilon > 0$, there exists a neighborhood N of a such that $|f(x) - f(a)| < \varepsilon$ for all $x \in N$ and all $f \in \mathscr{F}$. If $\mathscr{F}$ is equicontinuous at each point of X, then $\mathscr{F}$ is said to be **equicontinuous**. The distinguishing feature of equicontinuity is that, while the neighborhood N may vary with the point a, the same N works for all $f \in \mathscr{F}$.

Here is the main result regarding equicontinuity.

0.10.1 Theorem (Arzelá–Ascoli). *Let X be a compact Hausdorff space. A subset $\mathscr{F}$ of $C(X)$ is relatively compact in the uniform norm topology iff it is equicontinuous and pointwise bounded, that is,*

$$\sup\left\{|f(x)| : f \in \mathscr{F}\right\} < \infty \quad \textit{for all } x \in X.$$

Proof. Suppose $\mathscr{F}$ is relatively compact in $C(X)$. Then $\mathscr{F}$ is bounded, hence certainly pointwise bounded. If $\mathscr{F}$ is not equicontinuous at some $a \in X$, then there exists an $\varepsilon > 0$ and for every $N \in \mathfrak{N}_a$ a point $x_N \in N$ and a function $f_N \in \mathscr{F}$ such that $|f_N(x_N) - f_N(a)| \geq \varepsilon$. By relative compactness of $\mathscr{F}$, there exists a subnet (f_{N_α}) and $f \in C(X)$ such that

$\|f_{N_\alpha} - f\|_\infty \to 0$. Furthermore, since X is compact there exists a subnet (x_{N_β}) of (x_{N_α}) and a point $a \in X$ such that $x_{N_\beta} \to a$. But then

$$\begin{aligned}\varepsilon &\le |f_{N_\beta}(x_{N_\beta}) - f_{N_\beta}(a)| \le |f_{N_\beta}(x_{N_\beta}) - f(x_{N_\beta})| + |f(x_{N_\beta}) - f(a)| + |f(a) - f_{N_\beta}(a)| \\ &\le 2\left\|f_{N_\beta} - f\right\|_\infty + |f(x_{N_\beta}) - f(a)|, \qquad (\dagger)\end{aligned}$$

impossible since the expressions in (†) tend to zero.

Conversely, assume that $\mathscr{F}$ is equicontinuous. Since $C(X)$ is complete, to show that $\mathscr{F}$ is relatively compact it suffices by 0.9.1 to show that the closure of $\mathscr{F}$ in $C(X)$ is totally bounded. Since the closure of a totally bounded set is totally bounded, it is enough to show that $\mathscr{F}$ is totally bounded.

Let $\varepsilon > 0$. By equicontinuity, for each $x \in X$ we may choose an open neighborhood N_x of x such that

$$|f(y) - f(x)| < \varepsilon/4 \text{ for all } y \in N_x \text{ and } f \in \mathscr{F}. \qquad (\ddagger)$$

By compactness, there exists a finite set $F \subseteq X$ such that $X = \bigcup_{x\in F} N_x$. Since $\mathscr{F}$ is pointwise bounded, the set $\{f(x) : x \in F, f \in \mathscr{F}\}$ is bounded in $\mathbb{C}$ and hence is totally bounded. Thus we may choose a finite set $C \subseteq \mathbb{C}$ such that for each $f \in \mathscr{F}$ and $x \in F$ there exists $\varphi_f(x) \in C$ such that $|f(x) - \varphi_f(x)| < \varepsilon/4$. For each function $\varphi : F \to C$ set

$$\mathscr{F}_\varphi = \{f \in \mathscr{F} : |f(x) - \varphi(x)| < \varepsilon/4 \ \forall\, x \in F\}.$$

As there are only finitely many functions φ, there are only finitely many sets $\mathscr{F}_\varphi$. Moreover, $f \in \mathscr{F}_{\varphi_f}$ for each $f \in \mathscr{F}$. Let $\mathscr{F}_j = \mathscr{F}_{\varphi_j}$ $(j = 1, \ldots, m)$ denote the nonempty $\mathscr{F}_\varphi$, so that $\mathscr{F} = \bigcup_j \mathscr{F}_j$. Choosing $f_j \in \mathscr{F}_j$ we then have

$$\begin{aligned} f \in \mathscr{F} &\Rightarrow f \in \mathscr{F}_j \text{ for some } j \\ &\Rightarrow |f(x) - f_j(x)| \le |f(x) - \varphi_j(x)| + |\varphi_j(x) - f_j(x)| < \varepsilon/2 \ \ \forall x \in F \\ &\Rightarrow |f(y) - f_j(y)| \le |f(y) - f(x)| + |f(x) - f_j(x)| + |f_j(x) - f_j(y)| < \varepsilon \\ &\qquad \forall\, x \in F \text{ and } y \in N_x \qquad \text{(by } (\ddagger)) \\ &\Rightarrow \|f - f_j\|_\infty < \varepsilon \qquad (\text{since } X = \textstyle\bigcup_{x\in\mathscr{F}} N_x)\end{aligned}$$

Thus $\mathscr{F} \subseteq \bigcup_{j=1}^m B_\varepsilon(f_j)$, proving that $\mathscr{F}$ is totally bounded. □

0.11 The Stone-Weierstrass Theorem

Weierstrass's classical approximation theorem asserts that any function in $C[a, b]$ may be uniformly approximated by polynomials. Stone's generalization of Weierstrass's theorem replaces $[a, b]$ by an arbitrary compact Hausdorff topological space and replaces the set of polynomials by a more general class of functions.

For the statement of the theorem, the following terminology will be needed. A collection $\mathscr{A}$ of complex-valued functions on a set X is said to **separate points of** X if for each pair of distinct points x and y in X there exists $f \in \mathscr{A}$ such that $f(x) \ne f(y)$. For example, the algebra of all polynomials on $[a, b]$ separates points. The set $\{\sin x, \cos x\}$ separates the points of $[\varepsilon, 2\pi]$ but not of $[0, 2\pi]$.

Here is the statement of the theorem. Rather than giving Stone's original proof, we prefer to wait until Chapter 14 when a shorter proof using the Krein-Milman theorem is available.

0.11.1 Stone-Weierstrass Theorem. *Let X be a compact Hausdorff space and $\mathscr{A}$ a subalgebra of $C(X)$ that contains the constant functions, separates points of X, and is closed under complex conjugation. Then $\mathscr{A}$ is dense in $C(X)$ in the uniform norm.*

0.12 Locally Compact Topological Spaces

General Properties

A topological space is said to be **locally compact** if each member of the space has a compact neighborhood. For example, discrete spaces and Euclidean spaces are locally compact. We shall see in Chapter 8 that no infinite dimensional normed space can be locally compact.

The following proposition is immediate from the definition of relative topology.

0.12.1 Proposition. *If X is locally compact and Y is an open or closed subset of X, then Y is locally compact.*

The next proposition gives a key property of locally compact spaces that underlies the utility and importance of these spaces.

0.12.2 Proposition. *If X is a locally compact Hausdorff space, then for each $x \in X$ the collection of compact neighborhoods of x is a neighborhood base.*

Proof. Let N be an open neighborhood of x. We may assume that $\mathrm{cl}(N)$ is compact, otherwise replace N by the smaller open neighborhood $\mathrm{int}(M) \cap N$, where M is a compact neighborhood of x. By 0.8.6, there exist disjoint open sets U and V with $x \in U$ and $\mathrm{cl}(N) \setminus N \subseteq V$. If $y \in \mathrm{cl}(U \cap N) \setminus N$, then $y \in V$, hence $V \cap (U \cap N) \neq \emptyset$, which is impossible. Therefore, $x \in U \cap N \subseteq \mathrm{cl}(U \cap N) \subseteq N$, so $\mathrm{cl}(U \cap N)$ is the desired compact neighborhood contained in N. □

The following version of 0.7.1 will be needed below.

0.12.3 Proposition. *Let X be locally compact and Hausdorff. If $K \subseteq U \subseteq X$ with U open and K compact, then there exists an open set V with compact closure such that $K \subseteq V \subseteq \mathrm{cl}(V) \subseteq U$.*

Proof. By 0.12.2, for each $x \in K$ there exists an open neighborhood V_x of x with compact closure contained in U. By compactness of K, there exists a finite set $F \subseteq K$ such that $V := \bigcup_{x \in F} V_x \supseteq K$. Then $\mathrm{cl}(V) \subseteq \bigcup_{x \in F} \mathrm{cl}(V_x)$, hence $\mathrm{cl}(V)$ is compact and $\subseteq U$. □

Baire Spaces

0.12.4 Proposition. *Let X be a topological space. The following statements are equivalent:*

(a) *If U_n is open and dense in X for each n, then $\bigcap_{n=1}^{\infty} U_n$ is dense in X.*

(b) *If C_n is closed and $\bigcup_{n=1}^{\infty} C_n$ has an interior point, then some C_n has an interior point.*

Proof. The equivalence follows from De Morgan's laws and the fact that an open set is dense in X iff its complement has empty interior. □

A **Baire space** is a topological space X with the equivalent properties in the proposition. For example, a complete metric space is a Baire space (0.3.12). Here is another important example.

0.12.5 Theorem. *A locally compact Hausdorff space X is a Baire space.*

Proof. We show that (a) of 0.12.4 holds. Set $D := \bigcap_{n=1}^{\infty} U_n$ and let U be any nonempty open set in X. We show that $D \cap U \neq \emptyset$. Since $U \cap U_1$ is open and nonempty, there exists a nonempty open set V_1 such that $\mathrm{cl}\, V_1$ is compact and contained in $U \cap U_1$ (0.12.3). Since $V_1 \cap U_2$ is open and nonempty, there exists a nonempty open set V_2 such that $\mathrm{cl}\, V_2$ is compact and contained in $V_1 \cap U_2$ and hence is contained in $U \cap U_1 \cap U_2 \cap V_1$. Proceeding in this manner we construct a sequence of nonempty open sets V_n with compact closure contained in $U \cap U_1 \cap \cdots \cap U_n \cap V_{n-1}$. Since the compact sets $\mathrm{cl}\, V_n$ are decreasing, their intersection is nonempty. Any point in this intersection is a member of $D \cap U$ □

Functions with Compact Support

Let X be a topological space. The **support** of a function $f : X \to \mathbb{C}$ is defined as

$$\mathrm{supp}(f) := \mathrm{cl}\,\{x \in X : f(x) \neq 0\}.$$

Thus $\mathrm{supp}(f)$ is the smallest closed set on whose complement $f = 0$. The collection of all functions $f \in C(X)$ with compact support is denoted by $C_c(X)$:

$$C_c(X) := \{f \in C(X) : \mathrm{supp}(f) \text{ is compact in } X\}.$$

Clearly, $C_c(X) \subseteq C_b(X)$. Moreover, the relations

$$\mathrm{supp}(f+g) \subseteq \mathrm{supp}(f) \cup \mathrm{supp}(g), \;\; \mathrm{supp}(cf) = c\,\mathrm{supp}(f) \;\, (c \neq 0), \;\text{ and } \mathrm{supp}(fg) \subseteq \mathrm{supp}(f)$$

show that $C_c(X)$ is an ideal in the algebra $C_b(X)$.

The next two theorems are versions of Urysohn's lemma and Tietze's extension theorem for locally compact Hausdorff spaces. They imply that such spaces have a rich supply of continuous functions, a property crucial in the development of integration on locally compact spaces.

0.12.6 Theorem. *Let X be locally compact and Hausdorff. If $K \subseteq U \subseteq X$ with K compact and U open, then there exists a continuous function $h : X \to [0,1]$ with compact support such that $h = 1$ on K and $h = 0$ on U^c, that is, $\mathbf{1}_K \leq h \leq \mathbf{1}_U$.*

Proof. Let V be as in 0.12.3. Since $\mathrm{cl}(V)$ is compact it is normal, hence, by Urysohn's lemma, there exists a continuous function $h : \mathrm{cl}(V) \to [0,1]$ such that $h = 1$ on K and $h = 0$ on $\mathrm{cl}(V) \setminus V$. Extend h to X by setting $h = 0$ on $\mathrm{cl}(V)^c$. Then h is clearly continuous on the open set $V \cup \mathrm{cl}(V)^c$. A simple argument shows that h is continuous at each point of $\mathrm{cl}(V) \setminus V$ as well. □

The proof of the following result is the same as that of Tietze's extension theorem for normal spaces, except that one uses the preceding theorem instead of Urysohn's lemma.

0.12.7 Theorem. *Let X be locally compact and Hausdorff. If $K \subseteq X$ is compact and $f : K \to [a,b]$ is continuous, then there exists a continuous function $F : X \to [a,b]$ such that $f = F$ on K.*

0.12.8 Corollary. *Let X be locally compact and Hausdorff and $K \subseteq U \subseteq X$ with U open and K compact. If $g : K \to [a,b]$ is continuous, then there exists a continuous function $f : X \to \mathbb{R}$ with compact support contained in U such that $f = g$ on K.*

Proof. Let $G : X \to [a,b]$ be a continuous extension of g and let h be as in 0.12.6. Then $f := Gh$ satisfies the requirements. □

Functions That Vanish at Infinity

Let X be a topological space. A function $f : X \to \mathbb{C}$ is said to **vanish at infinity** if for each $\varepsilon > 0$ the set

$$K(f,\varepsilon) := \{x \in X : |f(x)| \geq \varepsilon\}$$

is compact. In $\mathbb{R}^d$ this is simply the assertion that $\lim_{|x|\to\infty} f(x) = 0$. The collection of all functions in $C(X)$ vanishing at infinity is denoted by $C_0(X)$:

$$C_0(X) := \{f \in C(X) : K(f,\varepsilon) \text{ is compact in } X \;\forall\; \varepsilon > 0\}.$$

Note that if $f \in C_0(X)$, then $|f(x)| \leq \varepsilon + \sup\big|f\big(K(f,\varepsilon)\big)\big|$ for all $x \in X$, hence $f \in C_b(X)$. Thus we have the inclusions $C_c(X) \subseteq C_0(X) \subseteq C_b(X)$, with equality holding throughout if X is compact. The relations

$$K(f+g,\varepsilon) \subseteq K(f,\varepsilon) \cup K(g,\varepsilon), \quad K(cf,\varepsilon) = K(f,\varepsilon/|c|) \text{ and } K(fg,\varepsilon) \subseteq K(f,\varepsilon/\left\|g\right\|_\infty)$$

imply that $C_0(X)$ is an ideal of $C_b(X)$. More can be said:

0.12.9 Proposition. *$C_0(X)$ is a Banach space under the uniform norm.*

Proof. By 0.3.4(a), it suffices to show that $C_0(X)$ is closed in $C_b(X)$. Let $f_n \in C_0(X)$ and $f_n \to f \in C_b(X)$. To show that $f \in C_0(X)$, given $\varepsilon > 0$, choose n so that $\|f_n - f\| < \varepsilon/2$. To see that $K(f,\varepsilon)$ is compact, let (x_α) be a net in $K(f,\varepsilon)$. Then for all α,

$$|f_n(x_\alpha)| = |f(x_\alpha) + f_n(x_\alpha) - f(x_\alpha)| \geq |f(x_\alpha)| - |f_n(x_\alpha) - f(x_\alpha)| > \varepsilon/2,$$

hence, $x_\alpha \in K(f_n,\varepsilon/2)$. Since $K(f_n,\varepsilon/2)$ is compact, there exists a subnet (x_β) that converges to some $x \in K(f_n,\varepsilon/2)$. Since $|f(x_\alpha)| \geq \varepsilon$ for all α and f is continuous, $|f(x)| \geq \varepsilon$. Therefore, $x \in K(f,\varepsilon)$. By 0.8.8, $K(f,\varepsilon)$ is compact. □

0.12.10 Proposition. *If X is locally compact and Hausdorff, then $C_c(X)$ is dense in $C_0(X)$.*

Proof. Let $f \in C_0(X)$ and $\varepsilon > 0$. Since $K(f,\varepsilon) \subseteq U := \{x \in X : |f(x)| > \varepsilon/2\}$, there exists a function $g : X \to [0,1]$ in $C_c(X)$ such that $g = 1$ on $K(f,\varepsilon)$ and $g = 0$ on U^c (0.12.6). Then $fg \in C_c(X)$ and $\|fg - f\|_\infty \leq \varepsilon$. □

The One-Point Compactification

Let X be a noncompact locally compact Hausdorff space with topology $\mathfrak{T}$ and let ∞ be a point not in X. Define $X_\infty := X \cup \{\infty\}$ and let $\mathfrak{T}_\infty$ consist of the members of $\mathfrak{T}$ together with all sets of the form $X_\infty \setminus K$, where $K \subseteq X$ is compact. It is straightforward to check that $\mathfrak{T}_\infty$ is closed under finite intersections and arbitrary unions and hence is a topology on X_∞. The pair $(X_\infty, \mathfrak{T}_\infty)$ is called the **one-point compactification** of X. This construction is useful in extending results from a compact setting to a locally compact one, as, for example, in 0.12.13, below.

The following proposition makes the basic connection between $C_0(X)$ and $C(X_\infty)$ and justifies the terminology "vanishing at infinity" for functions $f \in C_0(X)$.

0.12.11 Proposition. *Let $f \in C_b(X)$. Then $f \in C_0(X)$ iff $f(x_\alpha) \to 0$ for any net (x_α) in X with $x_\alpha \to \infty$.*

Proof. The necessity is clear, since if $x_\alpha \to \infty$ then x_α is eventually in $K(f,\varepsilon)^c$ for any $\varepsilon > 0$. For the sufficiency, we show that the convergence hypothesis implies that $K(f,\varepsilon)$ is compact. Let (x_α) be a net in $K(f,\varepsilon)$ with no cluster point in X. Then for each compact

$K \subseteq X$ and each α there exists an $\alpha_K \succeq \alpha$ with $x_{\alpha_K} \notin K$. Direct the collection of compact subsets upward by inclusion. Then (x_{α_K}) is a subnet of (x_α), and for any compact K_0, $x_{\alpha_K} \notin K_0$ for all $K \supseteq K_0$. Therefore, $x_{\alpha_K} \to \infty$ and so by hypothesis $f(x_{\alpha_K}) \to 0$. But this is impossible, since $|f(x_\alpha| \geq \varepsilon$ for all α. This shows that (x_α) must have a cluster point in $K(f, \varepsilon)$ and so $K(f, \varepsilon)$ is compact. □

We may now prove the following *extension property* of one-point compactifications:

0.12.12 Proposition. *Let X and Y be noncompact, locally compact Hausdorff spaces and let $\varphi : X \to Y$ be continuous such that $g \circ \varphi \in C_0(X)$ for all $g \in C_0(Y)$. Then the extension $\varphi_\infty : X_\infty \to Y_\infty$ of φ defined by $\varphi_\infty(\infty) = \infty$ is continuous.*

Proof. Let $x_\alpha \in X$ and $x_\alpha \to \infty$. Let $K \subseteq Y$ be compact and choose $g \in C_c(Y)$ such that $g = 1$ on K. By hypothesis, $f := g \circ \varphi \in C_0(X)$, hence $f(x_\alpha) \to 0$. Thus the net $(\varphi(x_\alpha))$ is eventually in K^c, hence $\varphi(x_\alpha) \to \infty$. This establishes continuity of φ at ∞. □

We conclude this subsection with a locally compact version of the Stone-Weierstrass theorem. It is derived from the compact version via the one-point compactification.

0.12.13 Stone-Weierstrass Theorem. *Let X be a locally compact noncompact Hausdorff topological space and let $\mathscr{A}$ be a conjugate closed subalgebra of $C_0(X)$ that separates points of X and with the property that $\bigcap_{f \in \mathscr{A}} \{x \in X : f(x) = 0\} = \emptyset$. Then $\mathscr{A}$ is dense in $C(X)$ in the uniform norm.*

Proof. Identify $C_0(X)$ with the closed subspace of $C(X_\infty)$ consisting of all f with $f(\infty) = 0$. Let $\mathscr{A}_1$ denote the subalgebra of $C_\infty(X)$ generated by $\mathscr{A}$ and the constant function 1. Then $\mathscr{A}_1$ trivially separates points of X_∞, hence is dense in $C(X_\infty)$. Moreover, every member g of $\mathscr{A}_1$ may be written uniquely as $g = g_0 + g(\infty)$, where $g_0 \in C_0(X)$. Now let $f \in C_0(X)$ and $\varepsilon > 0$, and choose $g \in \mathscr{A}_1$ such that $\|g - f\|_\infty < \varepsilon$. In particular, $|g(\infty)| < \varepsilon$, hence setting $g_0 = g - g(\infty)$ we have $g_0 \in \mathscr{A}$ and for all $x \in X$

$$|f(x) - g_0(x)| = |f(x) - g(x) + g(\infty)| \leq |g(x) - f(x)| + |g(\infty)| < 2\varepsilon. \qquad \square$$

0.13 Spaces of Differentiable Functions

In this section we define several spaces of differentiable functions on open sets $U \subseteq \mathbb{R}^d$ that will appear in later chapters. We shall need the following terminology and notation.

A **multi-index** is a d-tuple $\alpha = (\alpha_1, \ldots, \alpha_d)$ of nonnegative integers. We set

$$|\alpha| = \alpha_1 + \cdots + \alpha_d.$$

While this conflicts with the notation for the Euclidean norm on $\mathbb{R}^d$, context will make clear which notion is being referenced. The **partial differential operator of order** $|\alpha|$ is defined by

$$\partial^\alpha = \partial^\alpha_x = \left(\frac{\partial}{\partial x_1}\right)^{\alpha_1} \cdots \left(\frac{\partial}{\partial x_d}\right)^{\alpha_d}.$$

If $\alpha = (0, \ldots, 0)$, then ∂^α is the identity operator. The following spaces of differentiable functions figure prominently in the study of Fourier analysis and distributions on $\mathbb{R}^d$. (See Chapters 6 and 15.)

$$C^k(U) = \{f : \partial^\alpha f \in C(U) \text{ for all } |\alpha| \le k\}, \quad C^\infty(U) := \bigcap_{k=1}^\infty C^k(U).$$
$$C_c^k(U) = \{f : \partial^\alpha f \in C_c(U) \text{ for all } |\alpha| \le k\}, \quad C_c^\infty(U) := \bigcap_{k=1}^\infty C_c^k(U).$$

By the standard rules of differentiation, these spaces are closed under addition, multiplication, and scalar multiplication and so are algebras. Moreover, the C^∞ spaces satisfy $\partial^\alpha C^\infty \subseteq C^\infty$ for all α.

0.14 Partitions of Unity

In this section we prove two related results that are useful for piecing together local data to form a global construct such as a surface integral. The first result occurs in the general setting of locally compact spaces; the second is a C^∞ version of the first in the context of $\mathbb{R}^d$. In each case, the functions ϕ_i in the statement of the theorem are said to form a **partition of unity subordinate to the open sets** U_i.

0.14.1 Theorem. *Let K be a compact subset of a locally compact Hausdorff space X and let $\{U_i : i \in \mathfrak{I}\}$ be an open cover of K. Then there exists a finite subcover $\{U_1, \ldots, U_p\}$ of K and nonnegative $\phi_i \in C_c(X)$ such that* $\operatorname{supp}(\phi_j) \subseteq U_j$ *and* $\sum_{j=1}^p \phi_j = 1$ *on K.*

Proof. For each $x \in K$, let $j(x)$ be an index such that $x \in U_{j(x)}$. Choose an open neighborhood V_x of x with compact closure such that $\operatorname{cl}\left(V_x\right) \subseteq U_{j(x)}$. Since K is compact, finitely many of the sets V_x cover K. Denote these by $V_1, \ldots, V_p$ and denote the corresponding sets $U_{j(x)}$ by $U_1, \ldots, U_p$. Since $V_j \subseteq K_j := \operatorname{cl}(V_j) \subseteq U_j$, there exists by 0.12.6 a continuous function $\psi_j : X \to [0, 1]$ with compact support such that $\psi_j = 1$ on K_j and $\operatorname{supp}(\psi_j) \subseteq U_j$. Now set

$$\phi_1 := \psi_1 \text{ and } \phi_j := (1 - \psi_1)(1 - \psi_2) \cdots (1 - \psi_{j-1})\psi_j, \quad j > 1.$$

Then $\phi_j \in C_c(X)$, $0 \le \phi_j \le 1$, and $\operatorname{supp}(\phi_j) \subseteq \operatorname{supp}(\psi_j) \subseteq U_j$. Finally, let

$$\eta_j := (1 - \psi_1)(1 - \psi_2) \cdots (1 - \psi_j).$$

For $j > 1$, $\eta_{j-1} - \eta_j = (1 - \psi_1)(1 - \psi_2) \cdots (1 - \psi_{j-1})\big[1 - (1 - \psi_j)\big] = \phi_j$, hence

$$\sum_{j=1}^p \phi_j = \phi_1 + \sum_{j=2}^p (\eta_{j-1} - \eta_j) = \phi_1 + \eta_1 - \eta_p = 1 - \eta_p.$$

Since $K \subseteq \bigcup_j V_j \subseteq \bigcup_j K_j$ and $\phi_j = 1$ on K_j, $\eta_p = 0$ on K, hence $\sum_{j=1}^p \phi_j = 1$ on K, completing the proof. □

For the C^∞ version of 0.14.1, we need the following lemmas.

0.14.2 Lemma. *Let $a < b$. Then there exists a C^∞ function $h : \mathbb{R} \to [0, +\infty)$ such that $h > 0$ on (a, b), and $h = 0$ on $(a, b)^c$.*

Proof. Define h by

$$h(x) = \begin{cases} \exp\left[(x - a)^{-1}(x - b)^{-1}\right] & \text{if } a < x < b, \\ 0 & \text{otherwise.} \end{cases}$$

Clearly, $h^{(m)} = 0$ on $[a,b]^c$ for all $m \geq 0$. Moreover, if $x \in (a,b)$, then $h^{(m)}(x)$ is a sum of terms of the form

$$\frac{\pm h(x)}{(x-a)^p(x-b)^q}, \quad p, q \in \mathbb{Z}^+.$$

Since the exponent $(x-a)^{-1}(x-b)^{-1}$ in $h(x)$ is negative on (a,b), l'Hospital's rule is applicable and yields

$$\lim_{x \to a^+} \frac{h(x)}{(x-a)^p(x-b)^q} = 0, \quad a < x < b.$$

Therefore, $\lim_{x\to a} h^{(m)}(x) = 0$. An induction argument then shows that $h^{(m)}(a) = 0$ for all m. A similar argument holds for b. Thus h is C^∞ on $\mathbb{R}$. □

0.14.3 Lemma. *Let $a < b$. Then there exists a C^∞ function $g : \mathbb{R} \to \mathbb{R}$ such that $0 \leq g \leq 1$, $g = 0$ on $(-\infty, a]$, and $g = 1$ on $[b, +\infty)$.*

Proof. Take $g(x) := \left(\int_a^b h\right)^{-1} \int_a^x h$, where h is the function in 0.14.2. □

The following consequence of 0.14.2 is useful for obtaining smooth approximations of functions on $\mathbb{R}^d$.

0.14.4 Lemma. *Let $I = (a_1, b_1)\times\cdots\times(a_n, b_n)$. Then there exists a C^∞ function $f : \mathbb{R}^d \to \mathbb{R}$ such that $f > 0$ on I and $f = 0$ on I^c.*

Proof. For each j, let $h_j : \mathbb{R} \to [0, +\infty)$ be a C^∞ function such that $h_j > 0$ on (a_j, b_j) and $h_j = 0$ on $(a_j, b_j)^c$. Now set $f(x_1, \ldots, x_n) := h_1(x_1)\cdots h_n(x_n)$. □

We may now prove the following C^∞ version of Urysohn's lemma:

0.14.5 Theorem. *Let $K \subseteq U \subseteq \mathbb{R}^d$, where K is compact and U is open. Then there exists a C^∞ function $\psi : \mathbb{R}^d \to [0,1]$ such that $\operatorname{supp}(\psi) \subseteq U$ and $\psi = 1$ on K.*

Proof. For each $x \in K$, let V_x be an open cube with center x and edge $2r$:

$$V_x := \{y \in \mathbb{R}^d : x_j - r < y_j < x_j + r, \ j = 1, \ldots, n\},$$

where r is chosen so that $\operatorname{cl}\left(V_x\right) \subseteq U$. Let $W_x \subseteq V_x$ denote the concentric open cube with center x and edge r. Since K is compact, there exist finitely many cubes W_x whose union contains K. Denote these by $W_1, \ldots, W_m$ and denote the corresponding cubes V_x by $V_1, \ldots, V_m$. By 0.14.4, for each i there exists a C^∞ function $f_i : \mathbb{R}^d \to \mathbb{R}$ such that $f_i > 0$ on W_i and $f_i = 0$ on W_i^c. Set

$$f := \sum_{i=1}^m f_i, \quad V := \bigcup_{i=1}^m V_i, \quad \text{and} \quad W := \bigcup_{i=1}^m W_i.$$

Then f is nonnegative and C^∞ on $\mathbb{R}^d$, $f > 0$ on $W \supseteq K$, and $\operatorname{supp}(f) \subseteq \operatorname{cl}(V) \subseteq U$. Set $a := \min_{x\in K} f(x)$. Since $a > 0$, by 0.14.3 there exists a C^∞ function $g : \mathbb{R} \to [0,1]$ such that $g = 0$ on $(-\infty, 0]$ and $g = 1$ on $[a, +\infty)$. Now take $\psi := g \circ f$. □

The following theorem is proved as in 0.14.1 but using the above C^∞ version of Urysohn's lemma.

0.14.6 Theorem. *Let K be a compact subset of $\mathbb{R}^d$ and let $\{U_i : i \in \mathfrak{I}\}$ be an open cover of K. Then there exists a finite subcover $\{U_1, \ldots, U_p\}$ of K and nonnegative functions $\phi_i \in C_c^\infty(\mathbb{R}^d)$ such that $\operatorname{supp}(\phi_i) \subseteq U_i$ and $\sum_{i=1}^p \phi_i = 1$ on K.*

0.15 Connectedness

A pair of open sets U, V in a topological space X is said to **separate** X if

$$X = U \cup V, \quad U \neq \emptyset, \quad V \neq \emptyset, \quad \text{and } U \cap V = \emptyset.$$

The pair (U, V) is then called a **separation of** X. The space X is said to be **disconnected** if it has a separation, and **connected** if no separation exists. A subset E of X is **disconnected** (**connected**) if it is disconnected (connected) as a subspace of X. Thus if E is disconnected, then there exist sets U, V open in X such that $(E \cap U, E \cap V)$ is a separation of E.

In any topological space, the singletons $\{x\}$ are trivially connected. In a discrete space the only connected sets are the singletons. The set $\mathbb{Q}$ is not connected in $\mathbb{R}$, since the open sets $(-\infty, \sqrt{2}) \cap \mathbb{Q}$ and $(\sqrt{2}, +\infty) \cap \mathbb{Q}$ separate $\mathbb{Q}$.

0.15.1 Theorem. *A topological space X is disconnected iff there exists a continuous function from X onto $\{0, 1\}$. Equivalently, X is connected iff every continuous function from X into $\{0, 1\}$ is constant.*

Proof. Assume that X is disconnected and let (U, V) separate X. The function

$$g(x) = \begin{cases} 0 & \text{if } x \in U, \\ 1 & \text{if } x \in V. \end{cases}$$

from X onto $\{0, 1\}$ is easily seen to be continuous. Conversely, if $h : X \to \{0, 1\}$ is continuous and surjective, then the open sets $h^{-1}((-1, 1/2))$ and $h^{-1}((1/2, 2))$ separate X. □

0.15.2 Corollary. *The nonempty, connected subsets of $\mathbb{R}$ are the intervals.*

Proof. By the intermediate value theorem, no continuous function from an interval into $\{0, 1\}$ can be surjective. Therefore, intervals must be connected. On the other hand, if E is a nonempty subset of $\mathbb{R}$ that is not an interval, then there exist real numbers $a < c < b$ with $a, b \in E$ and $c \notin E$. The sets $(-\infty, c)$ and $(c, +\infty)$ then separate E. □

0.15.3 Corollary. *If $f : X \to Y$ is continuous and X is connected, then $f(X)$ is connected.*

Proof. If $g : f(X) \to \{0, 1\}$ is continuous, then $g \circ f : X \to \{0, 1\}$ is continuous and so is constant. Therefore, g must be constant. □

0.15.4 Corollary. *If $A \subseteq X$ is connected and $A \subseteq B \subseteq \operatorname{cl}(A)$, then B is connected. In particular, the closure of a connected set is connected.*

Proof. Let $g : B \to \{0, 1\}$ be continuous. Then $g|_A$ is continuous and hence is constant. Since $B \subseteq \operatorname{cl} A$, g is constant. Therefore, B is connected. □

0.15.5 Corollary. *Let $\{E_i : i \in \mathfrak{I}\}$ be a family of connected subsets of X. If $\bigcap_i E_i \neq \emptyset$, then $E := \bigcup_i E_i$ is connected.*

Proof. Let $g : E \to \{0, 1\}$ be continuous. Then $g\big|_{E_i}$ is constant for each i. Since there is a point common to all E_i, the constant is the same for all i. Therefore, g is constant. □

The **component** C_x of a member x of a topological space X is the union of all connected subsets of X containing x. By 0.15.5, C_x is connected and is therefore the largest connected set containing x. For example, the components in a discrete space are the singletons. The following theorem summarizes the main properties of components.

0.15.6 Theorem. *Let X be a topological space.*

(a) *C_x is not properly contained in a connected set.*

(b) *The set of all distinct components of X is a partition of X.*

(c) *C_x is closed in X.*

(d) *In a normed space $\mathscr{X}$, the components of an open set are open.*

Proof. Part (a) follows directly from the definition of component. Part (b) follows from (a) and 0.15.5. Part (c) follows from 0.15.4 and (a) by considering the closure of C_x. For (d) let $U \subseteq \mathscr{X}$ be open and C a component of U. If $\boldsymbol{x} \in C$ and r is chosen so that $B_r(\boldsymbol{x}) \subseteq U$, then, since $B_r(\boldsymbol{x})$ is connected, $C \cup B_r(\boldsymbol{x})$ is connected (0.15.5), hence $B_r(\boldsymbol{x}) \subseteq C$ by (a). □

Part I

Measure and Integration

Chapter 1

Measurable Sets

1.1 Introduction

This chapter begins the development of Lebesgue integration, which constitutes Part I of the text. The theory may be seen as arising from the need to overcome some of the shortcomings of the Riemann integral, which is restrictive in both the kind of function that may be integrated and the space over which the integration takes place. These shortcomings make the Riemann integral unsuitable for certain applications, for example those involving random parameters. A further complication with the Riemann theory concerns the integration of a pointwise limit of a sequence of Riemann integrable functions, such limits sometimes failing to be Riemann integrable. The removal of these limitations may be seen as a reason for the wide applicability of the Lebesgue theory.

Nevertheless, the Riemann integral still occupies an important position in analysis. Indeed, as we shall see, the set of Lebesgue integrable functions on $[a, b]$ is the completion in a precise sense of the set of Riemann integrable functions, much as the real number system is the completion of the rational number system.

It is illuminating to compare the construction of the two integrals in terms of how the domain $[a, b]$ of an integrand f is partitioned. In the case of the Riemann integral, $[a, b]$ is partitioned into subintervals $[x_{i-1}, x_i]$ and a point x_i^* is chosen in each. A suitable limit of the corresponding Riemann sums $\sum_i f(x_i^*)\Delta x_i$ then produces the Riemann integral of f. By contrast, in the Lebesgue theory it is the *range* of the function that is partitioned into subintervals, these inducing, via preimages under f, a partition of $[a, b]$. This partition will in general *not* consist of intervals. However, the Lebesgue theory provides a way of "measuring" the members of the partition. The Lebesgue integral is then constructed by multiplying these measured values by (approximate) function values, summing, and taking limits.

The preceding discussion suggests (correctly) that a fundamental feature of the Lebesgue theory is the notion of "measure" of a set. Such measures are constructed by starting with a collection $\mathcal{A}$ of elementary sets, such as intervals in $\mathbb{R}$ or rectangles in $\mathbb{R}^2$, and a *set function* that assigns a natural "size" to each member of $\mathcal{A}$, for example length in the case of intervals and area in the case of rectangles. The collection $\mathcal{A}$ is then enlarged to a richer class of sets that can still be "measured," the so-called *σ-field of measurable sets.* Unlike $\mathcal{A}$, this collection is closed under standard set-theoretic operations, including countable unions and intersections, a feature eventually resulting in limit theorems of a sort unavailable in Riemann integration, these theorems underlying much of modern analysis. The first step then in the construction of the Lebesgue integral is to develop the notion of measurable set and measure, which is the goal of this chapter.

1.2 Measurable Spaces

For a robust theory of integration that admits the standard combinatorial and limit operations, one requires that the collections of measurable sets on which the integration is based be closed under the usual set-theoretic operations. In this section we discuss the most common of such collections.

Fields and Sigma Fields

Let X be a nonempty set. A **field** on X is a family $\mathcal{F}$ of subsets of X satisfying (a)–(c) of the following. If $\mathcal{F}$ also satisfies (d), then $\mathcal{F}$ is called a **σ-field**:

(a) $X \in \mathcal{F}$.

(b) $A \in \mathcal{F} \Rightarrow A^c \in \mathcal{F}$.

(c) $A, B \in \mathcal{F} \Rightarrow A \cup B \in \mathcal{F}$.

(d) $A_1, A_2, \dots \in \mathcal{F} \Rightarrow \bigcup_{n=1}^{\infty} A_n \in \mathcal{F}$.

Note that (a) and (b) imply that $\emptyset \in \mathcal{F}$. An induction argument using (c) shows that a field $\mathcal{F}$ is closed under finite unions, that is,

$$A_1, \dots, A_n \in \mathcal{F} \Rightarrow A_1 \cup \dots \cup A_n \in \mathcal{F}.$$

Of course, every field with only finitely many members is a σ-field, since in this case countable unions reduce to finite unions. De Morgan's law

$$A_1 \cap A_2 \cap \dots \cap A_n = \left(A_1^c \cup A_2^c \cup \dots \cup A_n^c\right)^c$$

together with (b) shows that a field is closed under finite intersections and thus, for example, under the operation of **symmetric difference** defined by

$$A \triangle B := (A \cup B) \setminus (A \cap B) = (A \setminus B) \cup (B \setminus A).$$

Furthermore, every finite union of members of a field may be expressed as a *disjoint* union of members of the field via the construction

$$\bigcup_{k=1}^{n} A_k = A_1 \cup (A_2 \cap A_1^c) \cup \dots \cup (A_n \cap A_1^c \cap \dots \cap A_{n-1}^c). \tag{1.1}$$

Similar remarks apply to σ-fields: Part (d) of the above definition asserts that a σ-field is closed under countable unions, and an application of De Morgan's law shows that a σ-field is closed under countable intersections as well. As a consequence, a σ-field $\mathcal{F}$ is closed under the operations of **limit infimum** and **limit supremum** defined, respectively, by

$$\underline{\lim}_n A_n := \bigcup_{n=1}^{\infty} \bigcap_{k=n}^{\infty} A_k \quad \text{and} \quad \overline{\lim}_n A_n := \bigcap_{n=1}^{\infty} \bigcup_{k=n}^{\infty} A_k.$$

Moreover, every countable union of members of $\mathcal{F}$ may be expressed as a countable *disjoint* union of members of $\mathcal{F}$ in the manner of (1.1):

$$\bigcup_{n=1}^{\infty} A_n = A_1 \cup (A_2 \cap A_1^c) \cup \dots \cup (A_n \cap A_1^c \cap \dots \cap A_{n-1}^c) \cup \cdots. \tag{1.2}$$

Members of a σ-field $\mathcal{F}$ on X are called **$\mathcal{F}$-measurable sets**. The qualifier $\mathcal{F}$ is usually dropped if the σ-field is understood and there is no possibility of confusion. The pair $(X, \mathcal{F})$ is called a **measurable space**. A finite or countably infinite sequence of disjoint measurable sets with union A is called a **measurable partition** of A.

1.2.1 Examples.

(a) The power set $\mathcal{P}(X)$ is obviously a σ-field, as is the collection $\{\emptyset, X\}$. A field clearly cannot have exactly three members. All fields with exactly four members are of the form $\{\emptyset, X, A, A^c\}$.

(b) A subset A of X is said to be **cofinite** if A^c is finite. The collection $\mathcal{F}$ of all sets that are either finite or cofinite is a field. If X is infinite, then $\mathcal{F}$ is not a σ-field (Ex. 1.9).

(c) A subset A of X is said to be **cocountable** if A^c is countable. The collection $\mathcal{F}$ of all sets that are either countable or cocountable is a σ-field. For example, to see that $\mathcal{F}$ is closed under countable unions $A = \bigcup_{n=1}^{\infty} A_n$, note that if each A_n is countable, then A is countable and if some A_n is cocountable then A is cocountable. In either case, $A \in \mathcal{F}$.

(d) If $\mathcal{F}$ is a field (σ-field) on X, then the trace

$$\mathcal{F} \cap E = \{A \cap E : A \in \mathcal{F}\}$$

is a field (σ-field) on E. For example, if A, $B \in \mathcal{F}$, then the relations

$$(A \cup B) \cap E = (A \cap E) \cup (B \cap E) \quad \text{and} \quad (A \setminus B) \cap E = (A \cap E) \setminus (B \cap E)$$

show that $A \cup B$, $A \setminus B \in \mathcal{F}$. Note that $\mathcal{F} \cap E \subseteq \mathcal{F}$ iff $E \in \mathcal{F}$, in which case $\mathcal{F} \cap E$ is simply the collection of all sets $A \in \mathcal{F}$ with $A \subseteq E$. ◊

Generated Sigma Fields

The intersection of a nonempty family of σ-fields on a nonempty set X is easily seen to be a σ-field. In particular, if $\mathcal{A}$ is an arbitrary nonempty collection of subsets of X, then the intersection $\sigma(\mathcal{A})$ of all σ-fields on X containing $\mathcal{A}$ is a σ-field, called **σ-field generated by $\mathcal{A}$**. Note that there is at least one σ-field containing $\mathcal{A}$, namely, $\mathcal{P}(X)$, hence $\sigma(\mathcal{A})$ is well-defined. Generated σ-fields have the important **minimality property**:

$$\mathcal{F} \textit{ a } \sigma\textit{-field and } \mathcal{A} \subseteq \mathcal{F} \Rightarrow \sigma(\mathcal{A}) \subseteq \mathcal{F}.$$

The **field generated by $\mathcal{A}$**, denoted by $\varphi(\mathcal{A})$, is defined in a similar manner and enjoys the analogous minimality property.

1.2.2 Example. Let $\mathcal{A} = \{A_1, A_2, \ldots\}$ be a countable partition of X. Then $\sigma(\mathcal{A})$ consists of all unions $\bigcup_{n \in S} A_n$, where $S \subseteq \mathbb{N}$. (If $S = \emptyset$, then the union is defined to be $\emptyset$.)

To see this, note first that the collection $\mathcal{F}$ of all such unions is a σ-field. Indeed, $\mathcal{F}$ is obviously closed under countable unions, and by disjointness

$$\left(\bigcup_{n \in S} A_n\right)^c = \bigcup_{n \in S^c} A_n,$$

hence $\mathcal{F}$ is closed under complements as well. Since $\mathcal{A} \subseteq \mathcal{F} \subseteq \sigma(\mathcal{A})$, the minimality property implies that $\sigma(\mathcal{A}) = \mathcal{F}$. The analogous assertions hold for finite partitions of X. ◊

Borel Sets

Let X be a topological space. The σ-field generated by the collection of all open subsets of X is called the **Borel σ-field on X** and is denoted by $\mathcal{B}(X)$. A member of $\mathcal{B}(X)$ is called a **Borel set**. The minimality property of $\mathcal{B}(X)$ takes the following form:

If a σ-field $\mathcal{F}$ contains all open sets, then it contains all Borel sets.

Borel σ-fields provide a bridge between topology and measure theory, allowing, for example, the entry of continuous functions into integration theory.

Since closed sets are complements of open sets, $\mathcal{B}(X)$ is also generated by the collection of closed sets. For Euclidean space $\mathbb{R}^d$, more can be said:

1.2.3 Proposition. *The σ-field $\mathcal{B}(\mathbb{R}^d)$ is generated by the collection*

(a) $\mathcal{O}_I$ *of all bounded, open d-dimensional intervals* $(a_1, b_1) \times \cdots \times (a_d, b_d)$.

(b) $\mathcal{C}_I$ *of all bounded, closed d-dimensional intervals* $[a_1, b_1] \times \cdots \times [a_d, b_d]$.

(c) $\mathcal{H}_I$ *of all bounded, left-open d-dimensional intervals* $(a_1, b_1] \times \cdots \times (a_d, b_d]$.

Proof. For ease of notation we prove the proposition for $d = 1$; the proof for the general case is entirely similar.

(a) Let $\mathcal{O}$ denote the collection of all open sets in $\mathbb{R}$. Since $\mathcal{O}_I \subseteq \mathcal{O}$, by minimality we have $\sigma(\mathcal{O}_I) \subseteq \sigma(\mathcal{O}) = \mathcal{B}(\mathbb{R})$. On the other hand, every member of $\mathcal{O}$ is a countable union of sets in $\mathcal{O}_I$, hence $\mathcal{O} \subseteq \sigma(\mathcal{O}_I)$ and so $\mathcal{B}(\mathbb{R}) \subseteq \sigma(\mathcal{O}_I)$.

(b) Let $\mathcal{C}$ denote the collection of all closed sets in $\mathbb{R}$. As in part (a), $\sigma(\mathcal{C}_I) \subseteq \sigma(\mathcal{C}) = \mathcal{B}(\mathbb{R})$. Moreover, every bounded open interval (a, b) may be expressed as $\bigcup_n [a + 1/n, b - 1/n]$, hence $\mathcal{O}_I \subseteq \sigma(\mathcal{C}_I)$. By part (a) and minimality, $\mathcal{B}(\mathbb{R}) = \sigma(\mathcal{O}_I) \subseteq \sigma(\mathcal{C}_I)$.

(c) From the representations $(a, b) = \bigcup_n (a, b - 1/n]$ and $(c, d] = \bigcap_n (c, d + 1/n)$, we see that $\mathcal{O}_I \subseteq \sigma(\mathcal{H}_I)$ and $\mathcal{H}_I \subseteq \sigma(\mathcal{O}_I)$. By minimality, $\sigma(\mathcal{O}_I) \subseteq \sigma(\mathcal{H}_I)$ and $\sigma(\mathcal{H}_I) \subseteq \sigma(\mathcal{O}_I)$. An application of (a) completes the argument. □

The collection $\mathcal{H}_I$ will figure prominently in the development of the Lebesgue integral on Euclidean space $\mathbb{R}^d$.

Extended Borel Sets

To deal with functions that take values in $\overline{\mathbb{R}}$, we need to augment $\mathcal{B}(\mathbb{R})$ with the sets

$$B \cup \{-\infty\}, \quad B \cup \{\infty\}, \quad B \cup \{-\infty, \infty\}, \quad B \in \mathcal{B}(\mathbb{R}).$$

The collection of all such sets, together with the Borel subsets of $\mathbb{R}^d$, is called the **extended Borel σ-field** and is denoted by $\mathcal{B}(\overline{\mathbb{R}})$. One easily checks that $\mathcal{B}(\overline{\mathbb{R}})$ is indeed a σ-field with trace $\mathcal{B}(\mathbb{R})$ on $\mathbb{R}$. It may be shown that $\overline{\mathbb{R}}$ has a natural topology whose open sets generate $\mathcal{B}(\overline{\mathbb{R}})$ (Exercise 2.30).

Product Sigma Fields

Let $X_1, \ldots, X_d$ be nonempty sets and set $X := X_1 \times \cdots \times X_d$. For arbitrary nonempty collections $\mathcal{A}_j \subseteq \mathcal{P}(X_j)$ define

$$\mathcal{A}_1 \times \cdots \times \mathcal{A}_d = \{A_1 \times \cdots \times A_d : A_j \in \mathcal{A}_j,\ j = 1, \ldots, d\}.$$

If $\mathcal{F}_j$ is a σ-field on X_j, then the σ-field on X generated by $\mathcal{F}_1 \times \cdots \times \mathcal{F}_d$ is called the **product σ-field** and is denoted by $\mathcal{F}_1 \otimes \cdots \otimes \mathcal{F}_d$. Thus

$$\mathcal{F}_1 \otimes \cdots \otimes \mathcal{F}_d := \sigma(\mathcal{F}_1 \times \cdots \times \mathcal{F}_d).$$

Members of $\mathcal{F}_1 \times \cdots \times \mathcal{F}_d$ are called **measurable rectangles.**

1.2.4 Theorem. *If $\mathcal{A}_j \subseteq \mathcal{P}(X_j)$, then*

$$\sigma(\mathcal{A}_1) \otimes \cdots \otimes \sigma(\mathcal{A}_d) = \sigma(\mathcal{A}_1 \times \cdots \times \mathcal{A}_d). \tag{1.3}$$

Proof. The inclusion $\supseteq$ follows from $\sigma(\mathcal{A}_1) \otimes \cdots \otimes \sigma(\mathcal{A}_d) \supseteq \mathcal{A}_1 \times \cdots \times \mathcal{A}_d$ and minimality. For the reverse inclusion, let $A_j \in \mathcal{A}_j$, $j = 2, \ldots, d$. Then

$$\sigma(\mathcal{A}_1) \times \{A_2\} \times \cdots \times \{A_d\} \subseteq \sigma\left(\mathcal{A}_1 \times \cdots \times \mathcal{A}_d\right). \tag{†}$$

Indeed, the collection $\mathcal{F}_1$ of all $B_1 \in \sigma(\mathcal{A}_1)$ for which $B_1 \times A_2 \times \cdots \times A_d \in \sigma(\mathcal{A}_1 \times \cdots \times \mathcal{A}_d)$ is easily seen to be a σ-field containing $\mathcal{A}_1$ and so by minimality $\mathcal{F}_1 = \sigma(\mathcal{A}_1)$.

Next, let $B_1 \in \sigma(\mathcal{A}_1)$ and $A_j \in \mathcal{A}_j$, $j = 3, \ldots, d$. By (†)

$$\{B_1\} \times \mathcal{A}_2 \times \{A_3\} \times \cdots \times \{A_d\} \subseteq \sigma(\mathcal{A}_1 \times \cdots \times \mathcal{A}_d).$$

Arguing as before, this time on the second coordinate, we see that

$$\{B_1\} \times \sigma(\mathcal{A}_2) \times \{A_3\} \cdots \times \{A_d\} \subseteq \sigma(\mathcal{A}_1 \times \cdots \times \mathcal{A}_d).$$

We have now shown that

$$\sigma(\mathcal{A}_1) \times \sigma(\mathcal{A}_2) \times \mathcal{A}_3 \cdots \times \mathcal{A}_d \subseteq \sigma(\mathcal{A}_1 \times \cdots \times \mathcal{A}_d).$$

Continuing in this manner we eventually obtain the inclusion $\subseteq$ in (1.3). □

1.2.5 Corollary. *Let $d = d_1 + \cdots + d_k$, where $d_j \in \mathbb{N}$. Then*

$$\mathcal{B}(\mathbb{R}^d) = \mathcal{B}(\mathbb{R}^{d_1}) \otimes \cdots \otimes \mathcal{B}(\mathbb{R}^{d_k}). \tag{1.4}$$

In particular,

$$\mathcal{B}(\mathbb{R}^d) = \mathcal{B}(\mathbb{R}) \otimes \cdots \otimes \mathcal{B}(\mathbb{R}) \quad (d \text{ factors}).$$

Proof. By definition, $\mathcal{B}(\mathbb{R}^{d_j}) = \sigma(\mathcal{O}_j)$ and $\mathcal{B}(\mathbb{R}^d) = \sigma(\mathcal{O})$, where $\mathcal{O}_j$ is the collection of all open subsets of $\mathbb{R}^{d_j}$ and $\mathcal{O}$ is the collection of all open subsets of $\mathbb{R}^d$. By the theorem,

$$\sigma(\mathcal{O}_1 \times \cdots \times \mathcal{O}_k) = \sigma(\mathcal{O}_1) \times \cdots \times \sigma(\mathcal{O}_k) = \mathcal{B}(\mathbb{R}^{d_1}) \otimes \cdots \otimes \mathcal{B}(\mathbb{R}^{d_k}).$$

It therefore suffices to show that

$$\mathcal{O}_1 \times \cdots \times \mathcal{O}_k \subseteq \mathcal{O} \subseteq \mathcal{B}(\mathbb{R}^{d_1}) \otimes \cdots \otimes \mathcal{B}(\mathbb{R}^{d_k}); \tag{†}$$

the desired equality (1.4) will then follow by minimality. The first inclusion in (†) follows from the definition of the product topology of $\mathbb{R}^{d_1} \times \cdots \times \mathbb{R}^{d_k}$ (the latter identified with $\mathbb{R}^d$). For the second inclusion, recall that each $U \in \mathcal{O}$ is a countable union of open intervals $I = (a_1, b_1) \times \cdots \times (a_d, b_d)$. Since each such interval may be written as $I_{d_1} \times \cdots \times I_{d_k}$, where I_{d_j} is a d_j-dimensional open interval, $U \in \mathcal{B}(\mathbb{R}^d_1) \otimes \cdots \otimes \mathcal{B}(\mathbb{R}^{d_k})$. Therefore, (†) holds, completing the proof. □

Pi-Systems and Lambda-Systems

A collection $\mathcal{P}$ of subsets a set X is called a **π-system** if it is closed under finite intersections. Clearly, every field is a π-system, as is the collection of all open (or closed) intervals of $\mathbb{R}$.

A collection $\mathcal{L}$ of subsets a set X is called **λ-system** if it has the following properties:

(a) $X \in \mathcal{L}$.

(b) $A, B \in \mathcal{L}$ and $A \subseteq B \Rightarrow B \setminus A \in \mathcal{L}$. (1.5)

(c) $A_n \in \mathcal{L}$ and $A_n \uparrow A \Rightarrow A \in \mathcal{L}$.

Note that (a) and (b) imply that a λ-system is closed under complements and contains the empty set. The importance of λ-systems is that they provide an indirect method for establishing various properties of certain collections of sets. (See, for example, 1.6.8.) The method is based on Dynkin's π-λ theorem, which makes a connection between π-systems, λ-systems, and σ-fields.

1.2.6 Theorem (Dynkin). *Let $\mathcal{L}$ be a λ-system and $\mathcal{P} \subseteq \mathcal{L}$ a π-system. Then $\sigma(\mathcal{P}) \subseteq \mathcal{L}$.*

Proof. Let $\ell(\mathcal{P})$ denote the intersection of all λ-systems containing $\mathcal{P}$. Then $\ell(\mathcal{P})$ is a λ-system, as is easily verified, and $\ell(\mathcal{P}) \subseteq \sigma(\mathcal{P})$. If we show that $\ell(\mathcal{P})$ is a σ-field, it will then follow by minimality that $\sigma(\mathcal{P}) = \ell(\mathcal{P}) \subseteq \mathcal{L}$, establishing the theorem.

To show that $\ell(\mathcal{P})$ is closed under finite intersections, let $A \in \ell(\mathcal{P})$ and define

$$\mathcal{L}_A := \{B \in \ell(\mathcal{P}) : A \cap B \in \ell(\mathcal{P})\}.$$

One easily checks that $\mathcal{L}_A$ is a λ-system. Furthermore, if $A \in \mathcal{P}$, then $\mathcal{P} \subseteq \mathcal{L}_A$, so by minimality $\ell(\mathcal{P}) \subseteq \mathcal{L}_A$. Thus $A \cap B \in \ell(\mathcal{P})$ for all $A \in \mathcal{P}$ and $B \in \ell(\mathcal{P})$. Fixing such a B we have $\mathcal{P} \subseteq \mathcal{L}_B$, hence by minimality $\ell(\mathcal{P}) \subseteq \mathcal{L}_B$. Thus $A,\ B \in \ell(\mathcal{P}) \Rightarrow A \cap B \in \ell(\mathcal{P})$.

Now let (E_n) be a sequence in $\ell(\mathcal{P})$. By the preceding result and induction,

$$A_n := \bigcup_{k=1}^{n} E_k = \left(\bigcap_{k=1}^{n} E_k^c\right)^c \in \ell(\mathcal{P}).$$

By (c) of (1.5), $\bigcup_{k=1}^\infty E_k = \bigcup_{n=1}^\infty A_n \in \ell(\mathcal{P})$. Therefore, $\ell(\mathcal{P})$ is a σ-field, completing the proof. □

Exercises

1.1 Let $A,\ B,\ C,\ A_n,\ B_n \subseteq X$. Verify the following:

(a) $\mathbf{1}_{A \triangle B} = |\mathbf{1}_A - \mathbf{1}_B|$.

(b) $(A \triangle B)^c = A^c \triangle B = A \triangle B^c$.

(c) $A^c \triangle B^c = A \triangle B$.

(d) $(A \triangle B) \cap C = (A \cap C) \triangle (B \cap C)$.

(e) $\left(\bigcup_{n=1}^\infty A_n\right) \triangle \left(\bigcup_{n=1}^\infty B_n\right) \subseteq \bigcup_{n=1}^\infty A_n \triangle B_n$.

1.2 Let $A_n,\ B_n \subseteq X$. Verify the following:

(a) $x \in \underline{\lim}_n A_n$ iff $x \in A_n$ for all sufficiently large n.

(b) $x \in \overline{\lim}_n A_n$ iff $x \in A_n$ for infinitely many n.

(c) $\underline{\lim}_n A_n \subseteq \overline{\lim}_n A_n$.

(d) $\left(\underline{\lim}_n A_n\right)^c = \overline{\lim}_n A_n^c$.

(e) $\left(\overline{\lim}_n A_n\right)^c = \underline{\lim}_n A_n^c$.

(f) $\overline{\lim}_n (A_n \cap B_n) \subseteq \overline{\lim}_n A_n \cap \overline{\lim}_n B_n$.

(g) $\overline{\lim}_n (A_n \cup B_n) = \overline{\lim}_n A_n \cup \overline{\lim}_n B_n$.

(h) $\underline{\lim}_n (A_n \cap B_n) = \underline{\lim}_n A_n \cap \underline{\lim}_n B_n$.

(i) $\underline{\lim}_n (A_n \cup B_n) \supseteq \underline{\lim}_n A_n \cup \underline{\lim}_n B_n$.

Show that the inclusions in (c), (f), and (i) may be strict.

1.3 For $A_n \subseteq X$, write $A_n \to A$ if $\overline{\lim}_n A_n = \underline{\lim}_n A_n = A$. Let $A_n \to A$ and $B_n \to B$. Show that
(a) $A_n \cup B_n \to A \cup B$. (b) $A_n \cap B_n \to A \cap B$. (c) $A_n^c \to A^c$. (d) $A_n \triangle B_n \to A \triangle B$.

1.4 Let $A_n, A \subseteq X$ and set $B = \underline{\lim}_n A_n$ and $C = \overline{\lim}_n A_n$. Prove that
(a) $\mathbf{1}_B = \underline{\lim}_n \mathbf{1}_{A_n}$. (b) $\mathbf{1}_C = \overline{\lim}_n \mathbf{1}_{A_n}$ (c) $A_n \to A$ iff $\mathbf{1}_{A_n} \to \mathbf{1}_A$.

1.5 Let $\{a_n\}$ be a sequence in $\mathbb{R}$ and set $A_n = (-\infty, a_n)$ and $B_n = (a_n, \infty)$. Prove:

(a) $x \in \overline{\lim}_n A_n \Rightarrow x \leq \overline{\lim}_n a_n$.

(b) $x < \overline{\lim}_n a_n \Rightarrow x \in \overline{\lim}_n A_n$.

(c) $x \in \underline{\lim}_n A_n \Rightarrow x \leq \underline{\lim}_n a_n$.

(d) $x < \underline{\lim}_n a_n \Rightarrow x \in \underline{\lim}_n A_n$.

(e) $x \in \overline{\lim}_n B_n \Rightarrow \underline{\lim}_n a_n \leq x$.

1.6 Determine all sets in the field on $X = \{1, 2, 3, 4, 5, 6\}$ generated by the sets

(a) $\{1, 2\}$, $\{2, 3\}$, $\{3, 4\}$, $\{4, 5\}$.

(b) $\{1, 2, 3\}$, $\{2, 3, 4\}$, $\{3, 4, 5\}$.

(c) $\{1, 2, 3, 4\}$, $\{2, 3, 4, 5\}$, $\{3, 4, 5, 6\}$.

1.7 Let $\mathcal{F}$ be a σ-field on X and $E \subseteq X$. Show that $\sigma\big(\mathcal{F} \cup \{E\}\big)$ consists of all sets of the form $(A \cap E) \cup (B \cap E^c)$, A, $B \in \mathcal{F}$.

1.8 Let $\mathcal{F} \subseteq \mathcal{P}(X)$ such that $X \in \mathcal{F}$ and $A \setminus B \in \mathcal{F}$ whenever A, $B \in \mathcal{F}$. Show that $\mathcal{F}$ is a field.

1.9 Show that if X is infinite, then the field consisting of all finite or cofinite sets is not a σ-field.

1.10 Let $\mathcal{F}_1$, $\mathcal{F}_2, \ldots$ be a sequence of σ-fields on X such that $\mathcal{F}_1 \subseteq \mathcal{F}_2 \subseteq \cdots$. Show that $\mathcal{F} := \bigcup_{n=1}^{\infty} \mathcal{F}_n$ is a field. Show by example that $\mathcal{F}$ need not be a σ-field.

1.11 Find examples of fields $\mathcal{F}$ and $\mathcal{G}$ on $X = \{1, 2, 3\}$ such that $\mathcal{F} \cup \mathcal{G}$ is not a field.

1.12 Describe the σ-field $\mathcal{F}$ on $(0, 1)$ generated by all singletons $\{x\}$, $x \in (0, 1)$. Show that $\mathcal{F}$ is contained in $\mathcal{B}(0, 1)$ and contains no proper open subinterval of $(0, 1)$.

1.13 Let $\mathcal{F}$ be the collection of all finite disjoint unions of intervals $[a, b) \subseteq [0, 1)$. Show that $\mathcal{F}$ is a field on $[0, 1)$ but not a σ-field.

1.14 Let $\mathcal{A} \subseteq \mathcal{P}(X)$. Show that $\sigma\big(\varphi(\mathcal{A})\big) = \sigma(\mathcal{A})$.

1.15 Let $\mathcal{F}_f$ denote the field consisting of the subsets of X that are either finite or cofinite. Show that $\sigma(\mathcal{F}_f)$ is the σ-field $\mathcal{F}_c$ consisting of the countable or cocountable subsets of X.

1.16 Show that $\mathcal{B}(\mathbb{R}^d)$ is generated by the collection

(a) $\mathcal{K}$ of all compact sets.

(b) $\mathcal{I}_r$ of all intervals $(a_1, \infty) \times \cdots \times (a_d, \infty)$, $a_j \in \mathbb{Q}$.

1.17 Let $\mathcal{F}$ be a field. Prove that the following are equivalent:

(a) $\mathcal{F}$ is a σ-field.

(b) $\bigcup_{n=1}^{\infty} A_n \in \mathcal{F}$ for every sequence of disjoint sets $A_n \in \mathcal{F}$.

(c) $\bigcup_{n=1}^{\infty} B_n \in \mathcal{F}$ for every increasing sequence of sets $B_n \in \mathcal{F}$.

1.18 Let $\mathcal{A} \subseteq \mathcal{P}(X)$ and $E \subseteq X$. Prove that $\sigma(\mathcal{A} \cap E) = \sigma(\mathcal{A}) \cap E$.

1.19 Let X be a topological space and let $E \subseteq X$ have the relative topology. Prove that $\mathcal{B}(X) \cap E = \mathcal{B}(E)$.

1.20 [$\downarrow$2.30] Let a, $b \in \mathbb{R}$ and let $[a, b]$ and (a, b) have the relative topology from $\mathbb{R}$. Show that $\mathcal{B}([a, b])$ consists of the sets B, $B \cup \{a\}$, $B \cup \{b\}$, and $B \cup \{a, b\}$ where $B \in \mathcal{B}\big((a, b)\big)$.

1.21 For $j = 1, \ldots, d$, let $\mathcal{A}_j \subseteq \mathcal{P}(X_j)$ and $E_j \subset \mathcal{P}(X_j)$. Set $E := E_1 \times \cdots \times E_d$. Show that $\sigma(\mathcal{A}_1 \cap E_1) \otimes \cdots \otimes \sigma(\mathcal{A}_d \cap E_d) = \sigma(\mathcal{A}_1 \times \cdots \times \mathcal{A}_d) \cap E$.

1.22 Let $B \in \mathcal{B}(\mathbb{R}^d)$, $x \in \mathbb{R}^d$, and $r \in \mathbb{R}$. Prove that $B + x := \{b + x : b \in B\}$ and $rB := \{rb : b \in B\}$ are Borel sets.

1.23 Let $\mathcal{A} \subseteq \mathcal{P}(X)$ and let $\mathcal{F}$ be the union of all σ-fields $\sigma(\mathcal{C})$, where $\mathcal{C}$ is a countable subfamily of $\mathcal{A}$. Prove that $\mathcal{F} = \sigma(\mathcal{A})$.

1.24 Let $\mathcal{F} = \{B_1, \ldots, B_m\}$ be a finite field on X. Show that there exists a finite partition $\mathcal{A}$ of X by sets in $\mathcal{F}$ such that every member of $\mathcal{F}$ is a union of members of $\mathcal{A}$. [[Consider $C_1 \cap \cdots \cap C_m$, where $C_j = B_j$ or B_j^c.]]

1.25 Show that every infinite σ-field $\mathcal{F}$ has an infinite sequence of disjoint nonempty sets. Conclude that $\mathcal{F}$ has cardinality at least that of the continuum. Conclude that no σ-field can have cardinality $\aleph_0$. Find a *field* that has cardinality $\aleph_0$.

1.26 A nonempty collection $\mathcal{M}$ of subsets of X is a **monotone class** if for any sequence $\{A_n\}$ in $\mathcal{M}$, $A_n \uparrow A$ or $A_n \downarrow A \Rightarrow A \in \mathcal{M}$. Carry out steps (a)–(f) below to prove the **monotone class theorem**, due to Halmos: *If $\mathcal{F}$ is a field, $\mathcal{M}$ is a monotone class, and $\mathcal{F} \subseteq \mathcal{M}$, then $\sigma(\mathcal{F}) \subseteq \mathcal{M}$.*

(a) Show that a monotone class that is closed under finite unions (intersections) is closed under countable unions (intersections).

(b) Let $m(\mathcal{F})$ denote the intersection of all monotone classes containing $\mathcal{F}$. Show that $m(\mathcal{F})$ is a monotone class.

(c) Show that $\mathcal{A} := \{A \in m(\mathcal{F}) : A^c \in m(\mathcal{F})\}$ is monotone and $m(\mathcal{F}) = \mathcal{A}$. Conclude that $m(\mathcal{F})$ is closed under complements.

(d) Let $\mathcal{B} = \{B \in m(\mathcal{F}) : A \cup B \in m(\mathcal{F})$ for all $A \in \mathcal{F}\}$. Show that $\mathcal{B}$ is a monotone class and $\mathcal{B} = m(\mathcal{F})$. Conclude that $A \cup B \in m(\mathcal{F})$ for all $B \in m(\mathcal{F})$ and all $A \in \mathcal{F}$.

(e) Let $\mathcal{C} = \{C \in m(\mathcal{F}) : C \cup B \in m(\mathcal{F})$ for all $B \in m(\mathcal{F})\}$. Show that $\mathcal{C}$ is monotone and $\mathcal{C} = m(\mathcal{F})$. Conclude that $m(\mathcal{F})$ is closed under finite unions.

(f) Show that $m(\mathcal{F})$ is closed under countable unions. Conclude that $\sigma(\mathcal{F}) \subseteq m(\mathcal{F}) \subseteq \mathcal{M}$.

1.3 Measures

Set Functions

Let X be a nonempty set. A collection of subsets of X containing the empty set is called a **paving of** X. A function μ on a paving $\mathcal{A}$ of X that takes values in $\overline{\mathbb{R}}$ is called a **set function on** $\mathcal{A}$. Until Chapter 5, we consider only **nonnegative set functions**, that is, those taking values in $[0, \infty]$. An important example is the function that assigns the length $b - a$ to intervals $[a, b]$. This set function and its d-dimensional generalization will be examined in detail in §1.7.

Let μ be a nonnegative set function on a paving $\mathcal{A}$ and let $A_1, A_2, \ldots \in \mathcal{A}$. Then μ is said to be

- **monotone** if $A_1 \subseteq A_2$ implies $\mu(A_1) \leq \mu(A_2)$.
- **finitely additive** if $A := \bigcup_{k=1}^n A_k$ disjoint and $A \in \mathcal{A}$ implies $\mu(A) = \sum_{k=1}^n \mu(A_k)$.
- **finitely subadditive** if $A := \bigcup_{k=1}^n A_k \in \mathcal{A}$ implies $\mu(A) \leq \sum_{k=1}^n \mu(A_k)$.
- **countably additive** if $A := \bigcup_{n=1}^\infty A_n$ disjoint and $A \in \mathcal{A}$ implies $\mu(A) = \sum_{n=1}^\infty \mu(A_n)$.
- **countably subadditive** if $A := \bigcup_{n=1}^\infty A_n \in \mathcal{A}$ implies $\mu(A) \leq \sum_{n=1}^\infty \mu(A_n)$.
- **finite** if $\mu(A) < \infty$ for every $A \in \mathcal{A}$.
- **σ-finite** if there exist pairwise disjoint $X_1, X_2, \ldots \in \mathcal{A}$ with union X and $\mu(X_n) < \infty$.
- **a measure on** $\mathcal{A}$ if μ is countably additive and $\mu(\emptyset) = 0$.

If μ is a measure on a σ-field $\mathcal{F}$, then the triple $(X, \mathcal{F}, \mu)$ is called a **measure space**. A member E of $\mathcal{F}$ that is a countable union of sets of finite measure is called a **σ-finite set**. If $\mu(X) = 1$, then μ is said to be a **probability measure**. Note that a measure on a field is finitely additive: simply apply countable additivity to the sequence $A_1, \ldots, A_n, \emptyset, \emptyset, \ldots$.

Notation. In the sequel, if μ is a set function defined on intervals we write $\mu(a, b)$ for $\mu((a, b))$, $\mu[a, b]$ for $\mu([a, b])$, etc. No confusion should arise from these abbreviations, as context will make clear the intended meaning.

Properties and Examples of Measures

1.3.1 Proposition. *A measure μ on a σ-field $\mathcal{F}$ is monotone and countably subadditive. Moreover, for $A_n \in \mathcal{F}$ the following hold:*

(a) *(Continuity at A from below). $A_n \uparrow A$ implies $\mu(A_n) \uparrow \mu(A)$.*

(b) *(Continuity at A from above). $A_n \downarrow A$ and $\mu(A_1) < \infty$ implies $\mu(A_n) \downarrow \mu(A)$.*

Proof. If $A_1 \subseteq A_2$ then $\mu(A_2) = \mu(A_2 \setminus A_1) + \mu(A_1) \geq \mu(A_1)$, hence μ is monotone. For subadditivity use (1.2), countable additivity, and monotonicity:

$$\mu\Big(\bigcup_{k=1}^{\infty} A_k\Big) = \mu(A_1) + \mu(A_2 \cap A_1^c) + \mu(A_3 \cap A_1^c \cap A_2^c) + \cdots \leq \sum_{k=1}^{\infty} \mu(A_k).$$

Part (a) is clear if some A_k has infinite measure, so assume $\mu(A_k) < \infty$ for all k. Set $A_0 = \emptyset$ and $E_k = A_k \setminus A_{k-1}$. Then A is the disjoint union $\bigcup_{k=1}^{\infty} E_k$, hence

$$\mu(A) = \sum_{k=1}^{\infty} \mu(E_k) = \lim_n \sum_{k=1}^{n} \big[\mu(A_k) - \mu(A_{k-1})\big] = \lim_n \mu(A_n).$$

For (b), note that $A_1 \setminus A_n \uparrow A_1 \setminus A$, hence, by (a),

$$\mu(A_1) - \mu(A) = \mu(A_1 \setminus A) = \lim_n \mu(A_1 \setminus A_n) = \mu(A_1) - \lim_n \mu(A_n). \qquad \square$$

The preceding proposition has a converse:

1.3.2 Proposition. *Let μ be a finitely additive, nonnegative set function on a field $\mathcal{F}$.*

(a) *If μ is continuous from below, then μ is a measure.*

(b) *If $\mu(X) < \infty$ and μ is continuous at $\emptyset$ from above, then μ is a measure.*

Proof. For (a), let $\{A_n\}$ be a sequence of disjoint sets in $\mathcal{F}$ with union $A \in \mathcal{F}$ and set $B_n := \bigcup_{k=1}^{n} A_k$. Then $B_n \in \mathcal{F}$ and $B_n \uparrow A$. By finite additivity and continuity from below,

$$\sum_{k=1}^{\infty} \mu(A_k) = \lim_n \sum_{k=1}^{n} \mu(A_k) = \lim_n \mu(B_n) = \mu(A).$$

The proof of (b) is left as an exercise (1.39). $\square$

1.3.3 Examples.

(a) Set $\mu(\emptyset) = 0$ and $\mu(A) = \infty$ if $A \neq \emptyset$. Then μ is a measure on $\mathcal{P}(X)$.

(b) Let X be an infinite set and define $\mu(A) = 0$ if A is countable and $\mu(A) = \infty$ otherwise. Then μ is a measure on $\mathcal{P}(X)$.

(c) Let X be uncountable and $\mathcal{F}$ the σ-field of countable or cocountable subsets of X (see 1.2.1(c)). Define $\mu(A) = 0$ if A is countable and $\mu(A) = 1$ if A is cocountable. Then μ is a probability measure on $\mathcal{F}$.

(d) *Dirac measure.* Let $(X, \mathcal{F})$ be a measurable space. For $x \in X$ and $A \in \mathcal{F}$ define $\delta_x(A) = \mathbf{1}_A(x)$. Then δ_x is a probability measure on $\mathcal{F}$.

(e) If μ_j are measures on a σ-field $\mathcal{F}$ and $a_j \geq 0$, then $\sum_{j=1}^n a_j\mu_j$ is a measure on $\mathcal{F}$. In particular, a nonnegative linear combination of Dirac measures is a measure.

(f) If $(X, \mathcal{F}, \mu)$ is a measure space and $E \in \mathcal{F}$, then $\mu_E(A) := \mu(A \cap E)$ defines a measure on $\mathcal{F}$. Note that μ_E agrees with μ on the trace $\mathcal{F} \cap E$.

(g) *Counting measure.* Let X be a nonempty set. For $A \subseteq X$ let $\mu(A)$ be the number of elements in A if A is finite and $\mu(A) = \infty$ otherwise. Then μ is clearly finitely additive on $\mathcal{P}(X)$. To show that μ is a measure, let $A_n \uparrow A$. If there exists an m such that $A_m = A$, then $A_n = A$ for all $n \geq m$ and so, trivially, $\mu(A_n) \uparrow \mu(A)$. On the other hand, if no such m exists, then A must be infinite and $A_{n_{k-1}} \subsetneqq A_{n_k}$ for some sequence of indices. Since $\mu(A_{n_k}) \geq \mu(A_{n_{k-1}}) + 1$,

$$\lim_n \mu(A_n) = \lim_k \mu(A_{n_k}) = \infty = \mu(A).$$

By 1.3.2, μ measure on $\mathcal{P}(X)$.

(h) *Infinite series measure.* For an arbitrary sequence (p_n) in $[0, \infty)$, define

$$\mu(E) = \sum_{k \in E} p_k, \quad E \subseteq \mathbb{N},$$

where the sum may be infinite. (By convention, the sum over the empty set is zero.) The rearrangement theorem for nonnegative series implies that μ is well-defined and finitely additive. Let $A_n \uparrow A$. If A is finite, then eventually $A_n = A$, so obviously $\mu(A_n) \uparrow \mu(A)$. If A is infinite, then $\mu(A)$ may be written as an infinite series $\mu(A) = \sum_{k=1}^{\infty} p_{n_k}$. Let $r < \mu(A)$, choose k such that $\sum_{i=1}^{k} p_{n_i} > r$, and choose m so that A_m contains the indices $n_1, \ldots, n_k$. Then $\mu(A_n) \geq \mu(A_m) > r$ for all $n \geq m$. Since r was arbitrary, $\mu(A_n) \to \mu(A)$. By 1.3.2, μ is a measure on $\mathcal{P}(\mathbb{N})$. Note that if $p_k \equiv 1$, then μ is simply counting measure on $\mathbb{N}$. ◊

Exercises

1.27 Let $\mathcal{A} \subseteq \mathcal{P}(X)$ and $\emptyset \in \mathcal{A}$. Show that if μ is a countably additive, finite set function on $\mathcal{A}$, then $\mu(\emptyset) = 0$.

1.28 Verify that the set functions defined in 1.3.3 (c) and (d) are measures.

1.29 Give an example of a measure μ on a σ-field $\mathcal{F}$ and a sequence of sets $A_n \in \mathcal{F}$ decreasing to A such that $\lim_n \mu(A_n) \neq \mu(A)$.

1.30 [↑ 1.2.1] Let $\mathcal{F}$ be the field of finite or cofinite subsets of X and define $\mu(A) = 0$ if A is finite and $\mu(A) = 1$ if A is cofinite. (a) Show that μ is finitely additive but in general is not countably additive. (b) Show that μ is countably additive if X is uncountable.

1.31 Let μ be a finitely additive, nonnegative set function on a field $\mathcal{F}$. Prove that if $\mu(A)$ and $\mu(B)$ are finite, then $|\mu(A) - \mu(B)| \leq \mu(A \triangle B)$.

1.32 (Inclusion-exclusion I). Let μ be a finitely additive nonnegative set function on a field $\mathcal{F}$. Prove that $\mu(A) + \mu(B) = \mu(A \cup B) + \mu(A \cap B)$.

1.33 Let μ be a finitely additive, nonnegative set function on a field $\mathcal{F}$ and let $A, B \in \mathcal{F}$ with $\mu(B) = 0$. Show that $\mu(A \cup B) = \mu(A \setminus B) = \mu(A)$.

1.34 (Inclusion-exclusion II). Let μ be a finitely additive, nonnegative set function on a field $\mathcal{F}$ and let $A_1, \ldots, A_n \in \mathcal{F}$ with union A such that $\mu(A) < \infty$. Prove that for $n \geq 2$

$$\mu(A) = \sum_{i=1}^{n} \mu(A_i) - \sum_{1 \leq i < j \leq n}^{n} \mu(A_i \cap A_j) + \sum_{1 \leq i < j < k \leq n}^{n} \mu(A_i \cap A_j \cap A_k) - \cdots + (-1)^{n-1} \mu(A_1 \cap \cdots \cap A_n).$$

1.35 (Inclusion-exclusion III). Let μ be a finitely additive, nonnegative set function on a field $\mathcal{F}$ with $\mu(X) < \infty$ and let $B_1, \ldots, B_n \in \mathcal{F}$ with intersection B. Prove that for $n \geq 2$,

$$\mu(B) = \sum_{i=1}^{n} \mu(B_i) - \sum_{1 \leq i < j \leq n}^{n} \mu(B_i \cup B_j) + \sum_{1 \leq i < j < k \leq n}^{n} \mu(B_i \cup B_j \cup B_k) - \cdots + (-1)^{n-1} \mu(B_1 \cup \cdots \cup B_n).$$

1.36 Let $(X, \mathcal{F}, \mu)$ be a measure space and let $A_n \in \mathcal{F}$ such that $\mu(A_m \cap A_n) = 0$ for $m \neq n$. Prove that $\mu\left(\bigcup_{n=1}^{\infty} A_n\right) = \sum_{n=1}^{\infty} \mu(A_n)$.

1.37 [↓5.3.2] Let $(X, \mathcal{F}, \mu)$ be a measure space and $A_n \in \mathcal{F}$. Prove:

(a) $\mu\left(\underline{\lim}_n A_n\right) \leq \underline{\lim}_n \mu(A_n)$.

(b) $\mu\left(\overline{\lim}_n A_n\right) \geq \overline{\lim}_n \mu(A_n)$ if $\mu\left(\bigcup_n A_n\right) < \infty$.

(c) $\mu\left(\overline{\lim}_n A_n\right) = 0$ if $\sum_n \mu(A_n) < \infty$.

1.38 Let $(X, \mathcal{F})$ be a measurable space and let $x_1, x_2 \in X$. For $A \in \mathcal{P}(X)$, define $\mu(A) = 1$ if $\{x_1, x_2\} \subseteq A$ and $\mu(A) = 0$ otherwise. Prove that μ is continuous from below. Is μ a measure?

1.39 Prove 1.3.2(b).

1.40 [↓Ex. 3.3] Let μ_n be a sequence of measures on a σ-field $\mathcal{F}$ on X such that $\mu_n(A) \leq \mu_{n+1}(A)$ for all $A \in \mathcal{F}$. Define the set function μ on $\mathcal{F}$ by $\mu(A) = \lim_n \mu_n(A)$. Prove that μ is a measure.

1.41 Let μ_n be a sequence of measures on a σ-field $\mathcal{F}$ on X and define μ on $\mathcal{F}$ by $\mu(A) = \sum_n \mu_n(A)$. Prove that μ is a measure.

1.42 Let $(X, \mathcal{F}, \mu)$ be a finite measure space. Show that there can be at most countably many pairwise disjoint sets of positive measure.

1.43 Let $(X, \mathcal{F}, \mu)$ be a σ-finite measure space and $\mathcal{E}$ a collection of pairwise disjoint members of $\mathcal{F}$. Show that for any $A \in \mathcal{F}$, $\mu(A \cap E) > 0$ for at most countably many members of $\mathcal{E}$.

1.44 Let $(X, \mathcal{F}, \mu)$ be a measure space and for $A \in \mathcal{F}$ define

$$\mu_0(A) = \sup\{\mu(B) : B \in \mathcal{F},\ B \subseteq A \text{ and } \mu(B) < \infty\}.$$

Show that μ_0 is a measure on $\mathcal{F}$. Show also that $\mu_0 = \mu$ iff the following condition holds: For each $A \in \mathcal{F}$ with $\mu(A) = \infty$ there exists $B \in \mathcal{F}$ such that $B \subseteq A$ and $0 < \mu(B) < \infty$.

1.45 Let $(X, \mathcal{F}, \mu)$ be a measure space and $\{E_k\}$ be a sequence in $\mathcal{F}$. For fixed $m \in \mathbb{N}$, let A denote the set of all x such that $x \in E_k$ for exactly m values of k; B the set of all x such that $x \in E_k$ for finitely many and at least m values of k; and C the set of all x such that $x \in E_k$ for at most m values of k. Prove that $A, B, C \in \mathcal{F}$. If $s(D) := \sum_{k=1}^{\infty} \mu(D \cap E_k)$, prove that

(a) $\mu(A) = s(A)/m$. (b) $\mu(B) \geq s(B)/m$. (c) $\mu(C) \leq s(C)/m$.

1.4 Complete Measure Spaces

A measure space $(X, \mathcal{F}, \mu)$ is said to be **complete** if

$$M \in \mathcal{F}, \ \mu(M) = 0, \ \text{and } N \subseteq M \Rightarrow N \in \mathcal{F}.$$

Examples (a)–(c), (g), and (h) of 1.3.3 are complete measure spaces. In this section we show that any measure space $(X, \mathcal{F}, \mu)$ may be enlarged in a minimal way to produce a complete measure space. The following simple example illustrates the basic idea behind the construction.

1.4.1 Example. Let $X = \{1, 2, 3\}$ and $\mathcal{F} = \{\emptyset, \{1\}, \{2, 3\}, X\}$. The measure μ defined by $\mu\{1\} = 1$ and $\mu\{2, 3\} = 0$ is not complete. However, by enlarging $\mathcal{F}$ to include $\{2\}$, $\{3\}$ and defining a new measure $\overline{\mu}$ on the augmented σ-field so that $\overline{\mu}\{1\} = 1$ and $\overline{\mu}\{2\} = \overline{\mu}\{3\} = 0$, we obtain an extension of $(X, \mathcal{F}, \mu)$ that is complete. ◊

Completion Theorem

Here is the general technique for completing a measure space. Part (a) of the theorem gives the construction and part (b) describes a minimality property of a completion.

1.4.2 Theorem. *Let $(X, \mathcal{F}, \mu)$ be a measure space. Define*

$$\mathcal{F}_\mu := \{A \cup N : A \in \mathcal{F}, \ N \subseteq M \in \mathcal{F}, \ \mu(M) = 0\} \ \text{ and } \ \overline{\mu}(A \cup N) := \mu(A). \qquad (1.6)$$

(a) *$\mathcal{F}_\mu$ is a σ-field containing $\mathcal{F}$ and $\overline{\mu}$ is a measure on $\mathcal{F}_\mu$ that extends μ such that $(X, \mathcal{F}_\mu, \overline{\mu})$ is complete.*

(b) *If $(X, \mathcal{G}, \nu)$ is a complete measure space such that $\mathcal{F} \subseteq \mathcal{G}$ and ν is an extension of μ, then $\mathcal{F}_\mu \subseteq \mathcal{G}$ and the restriction of ν to $\mathcal{F}_\mu$ is $\overline{\mu}$.*

Proof. (a) To see that $\overline{\mu}$ is well-defined, let $A_1 \cup N_1 = A_2 \cup N_2$, where $N_j \subseteq M_j$, A_j, $M_j \in \mathcal{F}$ and $\mu(M_j) = 0$. Then $A_1 \subseteq A_2 \cup M_2$ and $A_2 \subseteq A_1 \cup M_1$, hence $\mu(A_2) = \mu(A_1)$.

Clearly, $\mathcal{F} \subseteq \mathcal{F}_\mu$. To see that $\mathcal{F}_\mu$ is closed under complements note that in the notation of (1.6)

$$(A \cup N)^c = (A^c \cap M^c) \cup (A^c \cap N^c \cap M), \quad A^c \cap M^c \in \mathcal{F} \text{ and } A^c \cap N^c \cap M \subseteq M.$$

For closure under countable unions, let $B_n := A_n \cup N_n \in \mathcal{F}_\mu$ and $B := \bigcup_n B_n$, where $N_n \subseteq M_n$, A_n, $M_n \in \mathcal{F}$, and $\mu(M_n) = 0$. Then

$$B = A \cup N, \ \text{ where } A := \bigcup_{n=1}^{\infty} A_n \text{ and } N := \bigcup_{n=1}^{\infty} N_n \subseteq M := \bigcup_{n=1}^{\infty} M_n.$$

Since $\mu(M) = 0$, $B \in \mathcal{F}_\mu$. Moreover, if the sets B_n are disjoint, then

$$\overline{\mu}(B) = \mu(A) = \sum_n \mu(A_n) = \sum_n \overline{\mu}(B_n).$$

Therefore, $\mathcal{F}_\mu$ is a σ-field and $\overline{\mu}$ is a measure on $\mathcal{F}_\mu$. Clearly, $(X, \mathcal{F}_\mu, \overline{\mu})$ is complete.

(b) Let A, N and M be as in (1.6). Then $\nu(M) = \mu(M) = 0$, hence, since $(X, \mathcal{G}, \nu)$ is complete, $N \in \mathcal{G}$. Therefore, $A \cup N \in \mathcal{G}$ and $\overline{\mu}(A \cup N) = \mu(A) = \nu(A) = \nu(A \cup N)$, so ν is an extension of $\overline{\mu}$. □

Note that the completion theorem produces nothing new if $(X, \mathcal{F}, \mu)$ is already complete, since then the sets N in the above construction are already in $\mathcal{F}$.

Null Sets

The sets N in the completion theorem, namely the subsets of $\mathcal{F}$-measurable sets M with measure zero, are called **μ-null sets**. Such sets appear throughout measure theory, frequently in the following context:

A property $P(x)$ of points $x \in X$ is said to hold **μ-almost everywhere**, abbreviated **μ-a.e.**, if the set of all x for which $P(x)$ is false is a μ-null set, that is,

$$\overline{\mu}\{x \in X : P(x) \text{ is false}\} = 0.$$

In this case we also say that the property $P(x)$ holds for **μ-almost all** x, abbreviated **μ-a.a.** x. If the measure is clear from context we drop the qualifier μ and simply write a.e. or a.a. For example, if a function f in 1.4.1 is defined by $f(j) = j$, then $f = 1$ a.e. For an example with far reaching implications, consider functions $f_n, f : X \to \mathbb{C}$. The notation $f_n \to f$ a.e. then means that

$$\overline{\mu}\{x \in X : \lim_n f_n(x) \neq f(x)\} = 0.$$

This type of convergence will be examined in Chapter 2.

Exercises

1.46 [$\uparrow$ 1.3.3(d).] Let $(X, \mathcal{F})$ be a measurable space, E a finite subset of X, and $\mu := \sum_{x \in E} \delta_x$. Describe the completion of $(X, \mathcal{F}, \mu)$.

1.47 Show that if $\mathcal{G} \subseteq \mathcal{F}$ are sigma fields, μ is a measure on $\mathcal{F}$, and $\nu = \mu\big|_{\mathcal{G}}$, then $\mathcal{G}_\nu \subseteq \mathcal{F}_\mu$ and $\overline{\nu} = \overline{\mu}\big|_{\mathcal{G}_\nu}$.

1.48 [$\uparrow$ 1.44] Prove that $\overline{\mu_0} = \overline{\mu}_0$.

1.49 Let $\{\mathcal{F}^i : i \in \mathfrak{I}\}$ be a collection of σ-fields on X and μ a measure on $\mathcal{G} := \sigma\big(\bigcup_i \mathcal{F}^i\big)$. For each i let μ_i denote the restriction of μ to $\mathcal{F}^i$. Show that $\mathcal{G}_\mu = \mathcal{H}_\mu$, where $\mathcal{H} := \sigma\big(\bigcup_i \mathcal{F}^i_{\mu_i}\big)$.

1.50 Let ν and η be measures on a σ-field $\mathcal{F}$ and set $\mu := \nu + \eta$. Show that $\mathcal{F}_\mu \subseteq \mathcal{F}_\nu \cap \mathcal{F}_\eta$ and $\overline{\mu} := \overline{\nu} + \overline{\eta}$ on $\mathcal{F}_\mu$.

1.51 [$\uparrow$ 1.3.3(f)] Let $E \in \mathcal{F}$. Prove that $\mathcal{F}_{\mu_E} \cap E = \mathcal{F}_\mu \cap E$ and $\overline{\mu_E} = \overline{\mu}_E$ on $\mathcal{F}_{\mu_E}$.

1.52 Let $(X, \mathcal{F}, \mu)$ be a finite measure space. For $E \subseteq X$ define

$$\mu_*(E) = \sup\{\mu(A) : A \in \mathcal{F},\ A \subseteq E\} \text{ and } \mu^*(E) = \inf\{\mu(B) : B \in \mathcal{F},\ B \supseteq E\}.$$

Show that $\mathcal{F}_\mu = \{E \subseteq X : \mu_*(E) = \mu^*(E)\}$.

1.5 Outer Measure and Measurability

As mentioned in the introduction to the chapter, the construction of a measure generally begins with a collection $\mathcal{A}$ of "elementary" subsets of X and a set function μ on $\mathcal{A}$, and culminates with an extension of μ to a measure on a σ-field containing $\mathcal{A}$. Of course, there may be several σ fields containing $\mathcal{A}$, $\mathcal{P}(X)$ being an obvious one. However, in many cases it is impossible to extend μ to $\mathcal{P}(X)$. For example, in §1.7 it is shown that the length set-function on the collection bounded intervals of $\mathbb{R}$ cannot be extended to a measure on $\mathcal{P}(\mathbb{R})$. In general, the best one can hope for is an extension of μ to the completion of the σ-field generated by $\mathcal{A}$. This is accomplished by first constructing a related set function on $\mathcal{P}(X)$, called *outer measure*, and then restricting this function to the class of so-called *measurable sets*. The details follow.

Construction of an Outer Measure

An **outer measure** on a nonempty set X is a nonnegative, monotone, countably subadditive set function μ^* on $\mathcal{P}(X)$ such that $\mu^*(\emptyset) = 0$. Clearly, every measure on $\mathcal{P}(X)$ is an outer measure. In particular, the set function that assigns 0 to the empty set and ∞ to every nonempty set is an outer measure. By contrast, the set function that assigns 0 to the empty set and 1 to every nonempty set is an outer measure that is not a measure.

The following proposition describes a general class of outer measures which are typically not measures. The outer measure μ^* defined in (1.7) is said to be **generated by the pair** $(\mathcal{A}, \mu)$. The sequences (A_n) in (1.7) are said to **cover** E.

1.5.1 Proposition. *Let $\mathcal{A}$ be a paving of X and let μ be a nonnegative set function on $\mathcal{A}$ such that $\mu(\emptyset) = 0$. Define a set function μ^* on $\mathcal{P}(X)$ by*

$$\mu^*(E) = \inf\left\{\sum_{n=1}^{\infty} \mu(A_n) : \ A_n \in \mathcal{A} \text{ and } E \subseteq \bigcup_{n=1}^{\infty} A_n\right\}, \tag{1.7}$$

where $\inf \emptyset := \infty$. *Then μ^* is an outer measure.*

Proof. That $\mu^*(\emptyset) = 0$ can be seen by taking as a cover the sequence $A_1 = A_2 = \cdots = \emptyset$. Monotonicity of μ^* follows from the observation that if $A \subseteq B$, then every cover of B is a cover of A. For countable subadditivity, let $E_n \in \mathcal{P}(X)$ and $E := \bigcup_{n=1}^{\infty} E_n$. We may assume that $\sum_{n=1}^{\infty} \mu^*(E_n) < \infty$. Given $\varepsilon > 0$, for each n choose a cover $\{A_{n,j}\}_j$ of E_n in $\mathcal{A}$ such that

$$\sum_{j=1}^{\infty} \mu(A_{n,j}) < \mu^*(E_n) + \varepsilon/2^n.$$

Since $\{A_{n,j}\}_{n,j}$ is a cover of E,

$$\mu^*(E) \le \sum_{n,j} \mu(A_{n,j}) < \sum_n \mu^*(E_n) + \varepsilon.$$

Thus $\mu^*(E) \le \sum_n \mu^*(E_n)$, as required. □

Carathéodory's Theorem

Let μ^* be any outer measure on X. A subset E of X is said to be **μ^*-measurable** if

$$\mu^*(C) = \mu^*(C \cap E) + \mu^*(C \cap E^c) \ \text{ for all } C \subseteq X. \tag{1.8}$$

The definition asserts that E "splits" the outer measure of each subset C of X, a property that may be seen as a precursor to finite additivity. Note that by subadditivity the inequality $\le$ in (1.8) always holds. Thus the measurability criterion singles out precisely those sets E for which the inequality $\ge$ in (1.8) is satisfied. The collection of all μ^*-measurable subsets of X is denoted by $\mathcal{M}(\mu^*)$. Here is the main result regarding outer measure.

1.5.2 Theorem (Carathéodory). *Let μ^* be an outer measure on X. Then $\mathcal{M} := \mathcal{M}(\mu^*)$ is a σ-field and the restriction $\overline{\mu} := \mu^*\big|_{\mathcal{M}}$ is a complete measure.*

Proof. Clearly, $\emptyset$, $X \in \mathcal{M}$, and since E and E^c appear symmetrically in (1.8), $E^c \in \mathcal{M}$ iff $E \in \mathcal{M}$. Furthermore, if $\mu^*(E) = 0$, then, by monotonicity,

$$\mu^*(C \cap E) + \mu^*(C \cap E^c) \le \mu^*(E) + \mu^*(C \cap E^c) = \mu^*(C \cap E^c) \le \mu^*(C),$$

hence $E \in \mathcal{M}$. Thus $\mathcal{M}$ contains all sets of outer measure zero.

It remains to show that for any sequence (E_n) in $\mathcal{M}$,

$$\text{(a)} \quad \bigcup_{n=1}^{\infty} E_n \in \mathcal{M} \quad \text{and} \quad \text{(b)} \quad \mu^*\left(\bigcup_{n=1}^{\infty} E_n\right) = \sum_{n=1}^{\infty} \mu^*(E_n) \quad \text{if the union is disjoint.}$$

The verifications of (a) and (b) are carried out in the following steps. For convenience, call a set C for which the equality in (1.8) holds a **test set for** E.

(1) *$\mathcal{M}$ closed under finite unions and hence is a field.*

⟦Let E, $F \in \mathcal{M}$. Take any set C as a test set for E and take $C \cap E^c$ as a test set for F. This gives

$$\begin{aligned} \mu^*(C) &= \mu^*(C \cap E) + \mu^*(C \cap E^c) \quad \text{and} \\ \mu^*(C \cap E^c) &= \mu^*(C \cap E^c \cap F) + \mu^*(C \cap E^c \cap F^c). \end{aligned}$$

Combining these we have

$$\begin{aligned} \mu^*(C) &= \mu^*(C \cap E) + \mu^*(C \cap E^c \cap F) + \mu^*(C \cap E^c \cap F^c) \\ &\geq \mu^*\left[(C \cap E) \cup (C \cap E^c \cap F)\right] + \mu^*(C \cap E^c \cap F^c) \qquad \text{(by subadditivity)} \\ &= \mu^*\left[C \cap (E \cup F)\right] + \mu^*\left[C \cap (E \cup F)^c\right]. \end{aligned}$$

Therefore, $E \cup F \in \mathcal{M}$.⟧

(2) *$C \subseteq X$, E, $F \in \mathcal{M}$ and $E \cap F = \emptyset \Rightarrow \mu^*\big(C \cap (E \cup F)\big) = \mu^*(C \cap E) + \mu^*(C \cap F)$.*

⟦Using $C \cap (E \cup F)$ as a test set for E we have

$$\begin{aligned} \mu^*\left[C \cap (E \cup F)\right] &= \mu^*\left[C \cap (E \cup F) \cap E\right] + \mu^*\left[C \cap (E \cup F) \cap E^c\right] \\ &= \mu^*(C \cap E) + \mu^*(C \cap F).⟧ \end{aligned}$$

(3) *If the sets E_n are disjoint, then $F := \bigcup_{n=1}^{\infty} E_n \in \mathcal{M}$ and $\mu(F) = \sum_{n=1}^{\infty} \mu(E_n)$.*

⟦Let $F_n := \bigcup_{k=1}^{n} E_k$ and $C \subseteq X$. By steps (1) and (2) and induction, $F_n \in \mathcal{M}$ and $\mu^*(C \cap F_n) = \sum_{k=1}^{n} \mu^*(C \cap E_k)$. Therefore, by monotonicity,

$$\mu^*(C) = \mu^*(C \cap F_n) + \mu^*(C \cap F_n^c) \geq \sum_{k=1}^{n} \mu^*(C \cap E_k) + \mu^*(C \cap F^c)$$

for all n and so

$$\mu^*(C) \geq \sum_{k=1}^{\infty} \mu^*(C \cap E_k) + \mu^*(C \cap F^c) \geq \mu^*(C \cap F) + \mu^*(C \cap F^c) \geq \mu^*(C).$$

This shows that $F \in \mathcal{M}$. Taking $C = F$ verifies countable additivity.⟧

(4) *If $E_n \in \mathcal{M}$, then $\bigcup_n E_n \in \mathcal{M}$.*

⟦Use (1), (3) and (1.2).⟧ □

Exercises

1.53 Define an outer measure μ^* on $\mathcal{P}(X)$ by $\mu^*(\emptyset) = 0$ and $\mu^*(E) = 1$ if $E \neq \emptyset$. Find $\mathcal{M}(\mu^*)$.

1.54 Let $\mathcal{O}_I$ denote the collection of all bounded open subintervals of $\mathbb{R}$ and let $\mu := \delta_0$ be the Dirac measure at 0 on $\mathcal{O}_I$. Show that the outer measure μ^* generated by $(\mathcal{O}_I, \mu)$ is the Dirac measure at 0 on $\mathcal{P}(\mathbb{R})$. Find $\mathcal{M}(\mu^*)$.

1.55 Let X be an uncountable set and define $\mu^*(E) = 0$ if $E = \emptyset$ and $\mu^*(E) = 1$ otherwise. Show that $\mu^*(E) = 0$ or 1 according as E is countable or uncountable. Show also that $\mathcal{M}(\mu^*)$ is the σ-field of sets that are countable or cocountable.

1.56 [$\uparrow$ 1.3.3(f)] Let μ be a monotone set function on a field $\mathcal{F}$. For $E \in \mathcal{F}$, let μ_E denote the set function on $\mathcal{F}$ defined by $\mu_E(A) = \mu(E \cap A)$ and let $(\mu_E)^*$ be the outer measure generated by $(\mathcal{F}, \mu_E)$. Prove that $(\mu^*)_E = (\mu_E)^*$.

1.57 [$\downarrow$ 1.8.1.] Let $\mathcal{A}$ and $\mathcal{B}$ be pavings of X such that each contains sequence with union X. Let μ be a measure on $\mathcal{A} \cup \mathcal{B}$ and let μ_a^* and μ_b^* be the outer measures generated by $(\mathcal{A}, \mu)$ and $(\mathcal{B}, \mu)$, respectively. Suppose that

$$\mu_a^*(E) = \mu_b^*(E) = \mu(E) \ \forall \ E \in \mathcal{A} \cup \mathcal{B}. \tag{†}$$

Prove that $\mu_a^* = \mu_b^*$. Show that assertion fails if the condition in (†) is not assumed.

1.58 Let μ^* be an outer measure on X, $E \subseteq X$, and $A \in \mathcal{M}(\mu^*)$ with $E \cap A = \emptyset$. Show that $\mu^*(E \cup A) = \mu^*(E) + \mu(A)$.

1.59 Let μ^* be an outer measure on X, $E \subseteq X$, and $A, B \in \mathcal{M}(\mu^*)$ with $A \cap B = \emptyset$. Show that $\mu^*\big(E \cap (A \cup B)\big) = \mu^*(E \cap A) + \mu(E \cap B)$. Show that the conclusion holds for countable disjoint unions as well.

1.60 Let μ a nonnegative set function on a paving $\mathcal{A}$ of X with $\mu(\emptyset) = 0$, and let μ^* be the outer measure generated by $(\mathcal{A}, \mu)$. Prove that $E \in \mathcal{M}(\mu^*)$ for any $E \subseteq X$ satisfying

$$\mu^*(A) = \mu^*(A \cap E) + \mu^*(A \cap E^c) \ \text{ for all } A \in \mathcal{A}.$$

1.6 Extension of a Measure

We have seen that a suitably defined pair $(\mathcal{A}, \mu)$ generates an outer measure μ^* and that the restriction of μ^* to the σ-field $\mathcal{M}(\mu^*)$ of measurable sets is a complete measure. A more intimate connection between μ and μ^* is possible if certain additional conditions are imposed on $(\mathcal{A}, \mu)$. For this we need the following definitions.

A nonempty collection $\mathcal{A}$ of subsets X is called a

- **semiring** if $\mathcal{A}$ is a π-system and for any $A, B \in \mathcal{A}$, there exist finitely many disjoint members C_j of $\mathcal{A}$ with $A \setminus B = \bigcup_{j=1}^n C_j$.

- **ring** if and $A, B \in \mathcal{A}$ implies $A \cup B$, $A \setminus B \in \mathcal{A}$.

Every ring is a π system and hence a semiring, since $A \cap B = A \setminus (A \setminus B)$ The collection of all bounded intervals on $\mathbb{R}$ is a semiring that is not a ring. A ring that contains X is closed under complements and hence is a field. If $(X, \mathcal{F}, \mu)$ is a measure space, then the collection of all members of $\mathcal{F}$ with finite measure is a ring that obviously need not be a field.

In this section we show that a measure μ on a semiring $\mathcal{A}$ may be extended to a measure on $\sigma(\mathcal{A})$ and that under suitable conditions the extension is unique and possesses certain approximation and completeness properties.

The Measure Extension Theorem

Let $\mathcal{A}$ be a semiring on a set X, μ a measure on $\mathcal{A}$, and μ^* the outer measure generated by $(\mathcal{A}, \mu)$. The proof of the measure extension theorem is based on the following lemmas.

1.6.1 Lemma. *The set $\mathcal{A}_{\mathrm{u}}$ of all finite disjoint unions of members of $\mathcal{A}$ is a ring.*

Proof. Let A, $B \in \mathcal{A}_{\mathrm{u}}$, say

$$A = \bigcup_{j=1}^{m} A_j, \ A_j \in \mathcal{A}, \ \text{ and } \ B = \bigcup_{k=1}^{n} B_k, \ B_k \in \mathcal{A} \ \text{ (disjoint unions)}.$$

To see that $A \setminus B \in \mathcal{A}_{\mathrm{u}}$, for each j and k choose finitely many disjoint sets $C_{ijk} \in \mathcal{A}$ such that $A_j \setminus B_k = \bigcup_i C_{ijk}$. Then $A_j \setminus B_k \in \mathcal{A}_{\mathrm{u}}$ and

$$A \setminus B = \bigcup_{j=1}^{m} A_j \cap B^c = \bigcup_{j=1}^{m} \bigcap_{k=1}^{n} A_j \setminus B_k = \bigcup_{j=1}^{m} \bigcap_{k=1}^{n} \bigcup_{i} C_{ijk}.$$

Since this is a disjoint union of members of $\mathcal{A}$, $A \setminus B \in \mathcal{A}_{\mathrm{u}}$.

To show that $A \cup B \in \mathcal{A}_{\mathrm{u}}$, write $A \cup B$ as the disjoint union $(A \setminus B) \cup (B \setminus A) \cup (A \cap B)$ and note that $A \cap B$ is the disjoint union $\bigcup_{j,\,k} A_j \cap B_k$ of members of $\mathcal{A}$. □

1.6.2 Lemma. *Define a set function μ_{u} on $\mathcal{A}_{\mathrm{u}}$ by*

$$\mu_{\mathrm{u}}\left(\bigcup_{j=1}^{m} A_j\right) = \sum_{j=1}^{m} \mu(A_j), \quad A_j \in \mathcal{A} \ \text{ (disjoint union)}.$$

Then μ_{u} is a well-defined measure on $\mathcal{A}_{\mathrm{u}}$ and $\mu_{\mathrm{u}}\big|_{\mathcal{A}} = \mu$.

Proof. To show that μ_{u} is well-defined, let $\bigcup_{j=1}^{m} A_j = \bigcup_{k=1}^{n} B_k$ be disjoint unions of members of $\mathcal{A}$. Then $A_j = \bigcup_{k=1}^{n} A_j \cap B_k$ and $B_k = \bigcup_{j=1}^{m} A_j \cap B_k$, hence

$$\mu(A_j) = \sum_{k=1}^{n} \mu(A_j \cap B_k) \ \text{ and } \ \mu(B_k) = \sum_{j=1}^{m} \mu(A_j \cap B_k).$$

Summing, we obtain

$$\sum_{j=1}^{m} \mu(A_j) = \sum_{j=1}^{m} \sum_{k=1}^{n} \mu(A_j \cap B_k) = \sum_{k=1}^{n} \mu(B_k).$$

To show countable additivity, let $E_1, E_2 \ldots \in \mathcal{A}_{\mathrm{u}}$ be disjoint with union $E \in \mathcal{A}_{\mathrm{u}}$. Choose disjoint sets $A_1, \ldots, A_m \in \mathcal{A}$ such that $E = \bigcup_{i=1}^{m} A_i$, and for each k choose disjoint sets $B_{k,1}, \ldots, \in B_{k,m_k} \in \mathcal{A}$ such that $E_k = \bigcup_{j=1}^{m_k} B_{k,j}$. Then

$$E_k = E \cap E_k = \bigcup_{j=1}^{m} A_i \cap E_k = \bigcup_{i=1}^{m} \bigcup_{j=1}^{m_k} A_i \cap B_{k,j}, \ \text{ (disjoint unions)}.$$

By definition of μ_{u},

$$\mu_{\mathrm{u}}(E) = \sum_{i=1}^{m} \mu(A_i) \ \text{ and } \ \mu_{\mathrm{u}}(E_k) = \sum_{i=1}^{m} \sum_{j=1}^{m_k} \mu(A_i \cap B_{k,j}). \tag{α}$$

Also, for each i,

$$A_i = A_i \cap E = \bigcup_{k=1}^{\infty} A_i \cap E_k = \bigcup_{k=1}^{\infty} \bigcup_{j=1}^{m_k} A_i \cap B_{k,j} \quad \text{(disjoint unions)},$$

hence, by the countable additivity of μ,

$$\mu(A_i) = \sum_{k=1}^{\infty} \sum_{j=1}^{m_k} \mu(A_i \cap B_{k,j}). \tag{β}$$

By (α) and (β) and a rearrangement,

$$\mu_{\mathrm{u}}(E) = \sum_{i=1}^{m} \sum_{k=1}^{\infty} \sum_{j=1}^{m_k} \mu(A_i \cap B_{k,j}) = \sum_{k=1}^{\infty} \sum_{i=1}^{m} \sum_{j=1}^{m_k} \mu(A_i \cap B_{k,j}) = \sum_{k=1}^{\infty} \mu_{\mathrm{u}}(E_k). \qquad \square$$

1.6.3 Lemma. *The outer measures generated by* $(\mathcal{A}, \mu)$ *and* $(\mathcal{A}_{\mathrm{u}}, \mu_{\mathrm{u}})$ *are the same.*

Proof. Let $E \subseteq X$. Typical sums in the definitions of $\mu(E)$ and $\mu_{\mathrm{u}}(E)$ are, respectively,

$$s = \sum_{n=1}^{\infty} \mu(A_n),\ A_n \in \mathcal{A},\ E \subseteq \bigcup_{n=1}^{\infty} A_n, \ \text{and}\ t = \sum_{n=1}^{\infty} \mu_{\mathrm{u}}(B_n),\ B_n \in \mathcal{A}_{\mathrm{u}},\ E \subseteq \bigcup_{n=1}^{\infty} B_n.$$

Since $\mathcal{A} \subseteq \mathcal{A}_{\mathrm{u}}$, every sum s is also a sum t. On the other hand, since each B_n is a finite disjoint union of members of $\mathcal{A}$ and μ_{u} is additive, every t may be decomposed and written as an s. The infima over these sums are therefore the same. $\square$

We may now prove

1.6.4 Theorem. *Let* $\mathcal{A}$ *be a semiring on a set* X, μ *a measure on* $\mathcal{A}$, μ^* *the outer measure generated by* $(\mathcal{A}, \mu)$, *and* $\mathcal{M} = \mathcal{M}(\mu^*)$ *the* σ*-field of* μ^**-measurable sets. Then* $\sigma(\mathcal{A}) \subseteq \mathcal{M}$ *and the measure* $\mu^*\big|_{\mathcal{M}}$ *is an extension of* μ.[1]

Proof. By the last lemma, we may assume that $\mathcal{A}$ is a ring. To show that $\mathcal{A} \subseteq \mathcal{M}(\mu^*)$, let $A \in \mathcal{A}$ and $C \subseteq X$. We show that

$$\mu^*(C \cap A) + \mu^*(C \cap A^c) \le \mu^*(C). \tag{$\dagger$}$$

Let $C_n \in \mathcal{A}$ such that $C \subseteq \bigcup_{n=1}^{\infty} C_n$. Since $\mathcal{A}$ is a ring, $C_n \cap A$, $C_n \cap A^c \in \mathcal{A}$. Moreover, $C \cap A \subseteq \bigcup_{n=1}^{\infty}(C_n \cap A)$ and $C \cap A^c \subseteq \bigcup_{n=1}^{\infty}(C_n \cap A^c)$, so

$$\mu^*(C \cap A) \le \sum_{n=1}^{\infty} \mu(C_n \cap A) \quad \text{and} \quad \mu^*(C \cap A^c) \le \sum_{n=1}^{\infty} \mu(C_n \cap A^c).$$

Adding we have

$$\mu^*(C \cap A) + \mu^*(C \cap A^c) \le \sum_{n=1}^{\infty} \mu(C_n \cap A) + \sum_{n=1}^{\infty} \mu(C_n \cap A^c) = \sum_{n=1}^{\infty} \mu(C_n).$$

Since the cover (C_n) of C was arbitrary, ($\dagger$) holds.

To show that $\mu^*\big|_{\mathcal{A}} = \mu$, let A, $A_n \in \mathcal{A}$ with $A \subseteq \bigcup_{n=1}^{\infty} A_n$. Then

$$\mu(A) \le \sum_{n=1}^{\infty} \mu(A \cap A_n) \le \sum_{n=1}^{\infty} \mu(A_n).$$

Taking infima over all such sequences $\{A_n\}$ yields $\mu(A) \le \mu^*(A)$. On the other hand, the sequence $A, \emptyset, \emptyset, \ldots$ is a cover of A by members of $\mathcal{A}$, hence $\mu^*(A) \le \mu(A)$. Therefore, $\mu^*\big|_{\mathcal{A}} = \mu$, completing the proof of the theorem. $\square$

[1] We frequently denote this extension also by μ, depending on context.

Approximation Property of the Extension

1.6.5 Theorem. *Let $E \in \sigma(\mathcal{A})$ with $\mu(E) < \infty$. Then for each $\varepsilon > 0$ there exist disjoint sets $A_1, \ldots, A_n \in \mathcal{A}$ such that*

$$\mu\Big(E \triangle \bigcup\nolimits_{j=1}^{n} A_j\Big) < \varepsilon.$$

Proof. Choose a cover $\{B_n\}$ of E in $\mathcal{A}$ such that $\sum_n \mu(B_n) < \mu(E) + \varepsilon/2$. Define

$$E_1 = B_1 \quad\text{and}\quad E_n = B_n \cap B_1^c \cdots \cap B_{n-1}^c = (B_n \setminus B_1) \cap \cdots \cap (B_n \setminus B_{n-1}), \quad n \geq 2.$$

The sets E_n are disjoint and cover E. Choose n so large that $\sum_{j=n+1}^{\infty} \mu(B_j) < \varepsilon/2$. From the inclusion

$$E \triangle \bigcup_{j=1}^{n} E_j = \left[\Big(\bigcup_{j=1}^{n} E_j\Big) \setminus E\right] \cup \left[E \setminus \bigcup_{j=1}^{n} E_j\right] \subseteq \left[E^c \cap \bigcup_{j=1}^{\infty} E_j\right] \cup \bigcup_{j=n+1}^{\infty} E_j$$

we have

$$\begin{aligned}\mu\Big(E \triangle \bigcup_{j=1}^{n} E_j\Big) &\leq \mu\Big(E^c \cap \bigcup_{j=1}^{\infty} E_j\Big) + \mu\Big(\bigcup_{j=n+1}^{\infty} E_j\Big) \leq \sum_{j=1}^{\infty} \mu(B_j) - \mu(E) + \sum_{j=n+1}^{\infty} \mu(E_j)\\ &< \varepsilon/2 + \varepsilon/2 = \varepsilon.\end{aligned}$$

Noting that each E_j is a disjoint union of members of $\mathcal{A}$ (because $\mathcal{A}$ is a semiring), we obtain the desired approximation. □

Completeness of the Extension

1.6.6 Theorem. *If $(\mathcal{A}, \mu)$ is σ-finite, then $\mathcal{M}(\mu^*)$ is the completion of $\big(\sigma(\mathcal{A}), \mu\big)$.*

Proof. Let $\mathcal{F} = \sigma(\mathcal{A})$. Since $\mathcal{M}(\mu^*)$ is complete, by minimality $\mathcal{F}_\mu \subseteq \mathcal{M}(\mu^*)$. For the reverse inclusion, assume first that $\mu(X) < \infty$. Let $E \in \mathcal{M}(\mu^*)$ and for each n choose sequences $\{A_{n,j}\}_{j=1}^{\infty}$ and $\{B_{n,j}\}_{j=1}^{\infty}$ in $\mathcal{A}$ such that

$$\mu^*(E^c) \leq \mu(A_n) \leq \sum_{j=1}^{\infty} \mu(A_{n,j}) \leq \mu^*(E^c) + 1/n, \quad \text{where } A_n := \bigcup_{j=1}^{\infty} A_{n,j} \supseteq E^c, \quad \text{and}$$

$$\mu^*(E) \leq \mu(B_n) \leq \sum_{j=1}^{\infty} \mu(B_{n,j}) \leq \mu^*(E) + 1/n, \quad \text{where } B_n := \bigcup_{j=1}^{\infty} B_{n,j} \supseteq E.$$

Then B_n, $A_n^c \in \sigma(\mathcal{A})$, $A_n^c \subseteq E \subseteq B_n$, and

$$\mu(B_n) \to \mu^*(E), \quad \mu(A_n^c) = \mu(X) - \mu(A_n) \to \mu(X) - \mu^*(E^c) = \mu^*(E). \tag{†}$$

Next, let

$$A = \bigcup_{n=1}^{\infty} A_n^c \quad\text{and}\quad B = \bigcap_{n=1}^{\infty} B_n.$$

Then $A_n^c \subseteq A \subseteq E \subseteq B \subseteq B_n$, hence from (†), $\mu(B \setminus A) = 0$. Setting $M = B \setminus A$ and $N = E \setminus A$ we have $E = A \cup N$, $N \subseteq M$, A, $M \in \mathcal{F}$, and $\mu(M) = 0$ and so $E \in \mathcal{F}_\mu$.

In the general case, let $X = \bigcup_{n=1}^{\infty} X_n$, where $X_n \in \mathcal{A}$ and $\mu(X_n) < \infty$. Set $\mathcal{A}_n = \mathcal{A} \cap X_n$ and $\mu_n = \mu\big|_{\mathcal{A}_n}$. Then $\mathcal{A}_n$ is a semiring on X_n (Ex. 1.62) and μ_n is a measure on $\mathcal{A}_n$, so the outer measure μ_n^* generated by $(\mathcal{A}_n, \mu_n)$ is a measure on $\mathcal{F}_n := \sigma(\mathcal{A}_n) = \mathcal{F} \cap X_n$ with completion $\mathcal{M}(\mu_n^*)$. By Ex. 1.62 again, $\mathcal{M}(\mu_n^*) = \mathcal{M}(\mu^*) \cap X_n$ and μ_n^* is the restriction of μ^* to $\mathcal{M}(\mu_n^*)$. Now let $E \in \mathcal{M}(\mu^*)$. By the preceding paragraph, for each n there exist M_n, $A_n \in \mathcal{F}_n$ with $\mu_n(M_n) = 0$ and $N_n \subseteq M_n$ such that $E \cap X_n = A_n \cup N_n$. Setting

$$A = \bigcup_{n=1}^{\infty} A_n, \quad M = \bigcup_{n=1}^{\infty} M_n, \quad \text{and} \quad N = \bigcup_{n=1}^{\infty} N_n,$$

we have $E = A \cup N$, $N \subseteq M$, M, $A \in \mathcal{F}$ and $\mu(M) = 0$, hence $E \in \mathcal{F}_\mu$. □

1.6.7 Remark. The σ-finite hypothesis in the completeness theorem cannot be removed. For example, let $\mathcal{A} = \{\emptyset, \mathbb{R}\}$ with $\mu(\emptyset) = 0$ and $\mu(\mathbb{R}) = \infty$. Then $\mu^*(C) = \infty$ for any $C \neq \emptyset$, hence, trivially, $\mu(C) = \mu(C \cap E) + \mu(C \cap E^c)$ for all $E \subseteq \mathbb{R}$, that is, $\mathcal{M}(\mu^*) = \mathcal{P}(\mathbb{R})$. On the other hand, since the only set of measure zero is the empty set, the completion of $\sigma(\mathcal{A}) = \mathcal{A}$ is just $\mathcal{A}$. ◊

Uniqueness of the Extension

Uniqueness is an immediate consequence of the following more general result:

1.6.8 Theorem. *Let $(Y, \mathcal{P})$ be a π-system and let μ_1 and μ_2 be measures on $\sigma(\mathcal{P})$ that are σ-finite on $\mathcal{P}$. If $\mu_1\big|_{\mathcal{P}} = \mu_2\big|_{\mathcal{P}}$, then $\mu_1 = \mu_2$.*

Proof. The proof uses Dynkin's π-λ theorem. Suppose first that $Y \in \mathcal{P}$ and $\mu_1(Y) < \infty$. Let $\mathcal{L} = \{E \in \sigma(\mathcal{P}) : \mu_2(E) = \mu_1(E)\}$. We claim that $\mathcal{L}$ is a λ-system. Indeed, property (a) of (1.5) holds by assumption and (c) holds by continuity from below. To verify (b), let A, $B \in \mathcal{L}$ with $A \subseteq B$. Then

$$\mu_2(B \setminus A) = \mu_2(B) - \mu_2(A) = \mu_1(B) - \mu_1(A) = \mu_1(B \setminus A),$$

verifying the claim. Since $\mathcal{P} \subseteq \mathcal{L}$, by Dynkin's theorem $\sigma(\mathcal{P}) \subseteq \mathcal{L}$. This proves the theorem for the case $\mu_1(Y) < \infty$.

Now let (Y_n) be a disjoint sequence in $\mathcal{P}$ with union Y and $\mu_1(Y_n) < \infty$ for all n. Applying the result of the first paragraph to the restriction of the measures to Y_n, we see that $\mu_1(A \cap Y_n) = \mu_2(A \cap Y_n)$ for all $A \in \sigma(\mathcal{P})$ and all n. Now use countable additivity to complete the proof. □

Applying Theorem 1.6.8 to the current setting we have

1.6.9 Theorem. *If $(\mathcal{A}, \mu)$ is σ-finite, then the extension of μ to $\sigma(\mathcal{A})$ is unique.*

1.6.10 Remarks. Without the σ-finite hypothesis the conclusion of 1.6.9 may fail. For example, let $\mathcal{A}$ be the semiring of all bounded intervals and take μ to be the measure on $\mathcal{B}(\mathbb{R}) = \sigma(\mathcal{A})$ that assigns the value ∞ to every nonempty set in $\mathcal{A}$ (hence $\mu^*(E) = \infty$ for every nonempty $E \subseteq \mathbb{R}$). If ν is counting measure on $\mathcal{B}(\mathbb{R})$, then $\mu \neq \nu$, yet the measures agree on $\mathcal{A}$. Note also that μ (vacuously) has the approximation property, but ν does not.

The conclusion of 1.6.9 may also fail if $\mathcal{A}$ is not a semiring. For example, let $\mathcal{A}$ be the collection of all intervals $(a, b]$ with $b - a = 1$. If $\mu(A)$ is the number of integers in $A \in \mathcal{B}(\mathbb{R})$ and λ is Lebesgue measure on $\mathcal{B}(\mathbb{R})$ (see §1.7), then $\mu = \lambda$ on $\mathcal{A}$ but not on $\mathcal{B}(\mathbb{R})$. ◊

The following consequence of 1.6.8 will be needed later.

1.6.11 Theorem. *Let ν be any measure on $\sigma(\mathcal{A})$ that is σ-finite on $\mathcal{A}$. Then*

$$\nu(E) = \inf\left\{\sum_{n=1}^{\infty} \nu(A_n) : \ A_n \in \mathcal{A} \text{ and } E \subseteq \bigcup_{n=1}^{\infty} A_n\right\}, \quad E \in \sigma(\mathcal{A}).$$

Proof. Let ν^* denote the outer measure generated by $(\nu|_{\mathcal{A}}, \mathcal{A})$. Then the measures $\nu^*|_{\sigma(\mathcal{A})}$ and ν agree on the π-system $\mathcal{A}$ and so are equal. □

Exercises

1.61 Let $\mathcal{A}_i$ be a semiring on X_i, $i = 1, 2$. Show that $\mathcal{A}_1 \times \mathcal{A}_2$ is a semiring.

1.62 Let μ be a measure on a semiring $\mathcal{A} \subseteq \mathcal{P}(X)$ and $E \in \mathcal{A}$

(a) Prove that $\mathcal{A} \cap E$ is a semiring consisting of the members of $\mathcal{A}$ that are subsets of E.

(b) Let ν be the restriction of μ to $\mathcal{A} \cap E$ and let μ^* and ν^* be the outer measures generated by $(X, \mathcal{A}, \mu)$ and $(E, \mathcal{A} \cap E, \nu)$. Show that ν^* is the restriction of μ^* to $\mathcal{P}(E)$.

(c) Prove that $\mathcal{M}(\nu^*) = \mathcal{M}(\mu^*) \cap E$.

1.63 Let μ be as in 1.6.4 and let ν be a measure on $\sigma(\mathcal{A})$ that equals μ on $\mathcal{A}$.

(a) Show that $\nu(E) \le \mu(E)$ for all $E \in \sigma(\mathcal{A})$. (1.6.10 shows equality may not hold.)

(b) Show that $\nu(E) = \mu(E)$ for all $E \in \sigma(\mathcal{A})$ with $\mu(E) < \infty$. ⟦ Assume that $\mathcal{A}$ is a ring (how?). Choose $A \in \mathcal{A}$ such that $E \subseteq A$ and $\mu(A) < \mu(E) + \varepsilon$. Then $\nu(E) + \nu(A \setminus E) < \nu(E) + \varepsilon$. ⟧

1.64 Let μ be a measure on a semiring $\mathcal{A} \subseteq \mathcal{P}(X)$ and let μ^* be the outer measure generated by $(\mathcal{A}, \mu)$. Prove that for any $E \subseteq X$ there exists $A \in \sigma(\mathcal{A})$ such that $E \subseteq A$ and $\mu^*(E) = \mu(A)$.

1.65 [↑1.64] Let μ be a measure on a semiring $\mathcal{A} \subseteq \mathcal{P}(X)$ and let μ^* be the outer measure generated by $(\mathcal{A}, \mu)$. Prove the *weak inclusion-exclusion principle*

$$\mu^*(E \cup F) + \mu^*(E \cap F) \le \mu^*(E) + \mu^*(F), \quad E, F \subseteq X.$$

1.66 [↑1.64] Let μ and ν be measures on a semiring $\mathcal{A} \subseteq \mathcal{P}(X)$ and let μ^* and ν^* be the outer measures generated by $(\mathcal{A}, \mu)$ and $(\mathcal{A}, \nu)$, respectively. Prove that $(\mu + \nu)^* = \mu^* + \nu^*$ and $\mathcal{M}(\mu^*) \cap \mathcal{M}(\nu^*) \subseteq \mathcal{M}(\mu^* + \nu^*)$. Show that the inclusion may be strict.

1.67 [↑1.64, 1.40] Let μ and μ_n be σ-finite measures on a semiring $\mathcal{A} \subseteq \mathcal{P}(X)$ with $\mu_n \uparrow \mu$ on $\sigma(\mathcal{A})$. Let μ^*, μ_n^* be the outer measures generated by $(\mathcal{A}, \mu)$ and $(\mathcal{A}, \mu_n)$ Prove that $\mu_n^* \uparrow \mu^*$ on $\mathcal{P}(X)$.

1.68 [↑1.66, 1.67] Let μ_n be measures on a semiring $\mathcal{A}$ on X and define $\mu(A) = \sum_{n=1}^{\infty} \mu_n(A)$ $(A \in \mathcal{A})$. Let μ^* and μ_n^* be the outer measures generated by $(\mathcal{A}, \mu)$ and $(\mathcal{A}, \mu_n)$, respectively. Prove that $\mu^* = \sum_{n=1}^{\infty} \mu_n^*$.

1.69 [↑1.64] Let μ be a measure on a semiring $\mathcal{A} \subseteq \mathcal{P}(X)$ and let μ^* be the outer measure generated by $(\mathcal{A}, \mu)$. Prove that μ^* is continuous from below. Why doesn't this imply that μ^* is a measure on $\mathcal{P}(X)$?

1.70 [↑1.64] Let μ be a measure on a semiring $\mathcal{A} \subseteq \mathcal{P}(X)$ and let μ^* be the outer measure generated by $(\mathcal{A}, \mu)$. Suppose that $\mu^*(X) < \infty$. Show that $E \in \mathcal{M}(\mu^*)$ iff $\mu(X) = \mu^*(E) + \mu^*(E^c)$.

1.7 Lebesgue Measure

The Volume Set Function

Recall that $\mathcal{H}_I$ denotes the semiring of bounded, left open d-dimensional intervals

$$(\boldsymbol{a}, \boldsymbol{b}] := (a_1, b_1] \times \cdots \times (a_d, b_d], \quad \boldsymbol{a} := (a_1, \dots, a_d), \quad \boldsymbol{b} := (b_1, \dots, b_d),$$

where $-\infty < a_j \le b_j < \infty$. Define the **$d$-dimensional volume** of $(\boldsymbol{a}, \boldsymbol{b}]$ by

$$\lambda(\boldsymbol{a}, \boldsymbol{b}] = \lambda^d(\boldsymbol{a}, \boldsymbol{b}] := \prod_{j=1}^{d} (b_j - a_j).$$

In this section we apply the results of §1.6 to the pair $(\mathcal{H}_I, \lambda)$ to construct **d-dimensional Lebesgue measure**. The following lemma is key to the construction.

1.7.1 Lemma. *Let $H, H_1, \ldots, H_m \in \mathcal{H}_I$.*

(a) *If $H_1, \ldots, H_m$ are disjoint and $H = \bigcup_{j=1}^m H_j$, then $\lambda(H) = \sum_{j=1}^m \lambda(H_j)$.*

(b) *If $H \subseteq \bigcup_{j=1}^m H_j$, then $\lambda(H) \leq \sum_{j=1}^m \lambda(H_j)$.*

(c) *If $H_1, \ldots, H_m$ are disjoint and $H \supseteq \bigcup_{j=1}^m H_j$, then $\lambda(H) \geq \sum_{j=1}^m \lambda(H_j)$.*

Proof. For ease of notation and exposition, we prove the lemma for $d = 2$, in which case the intervals are rectangles. Let $H = (a, b] \times (c, d]$. We may assume in (b) that $H = \bigcup_{j=1}^m H_j$,

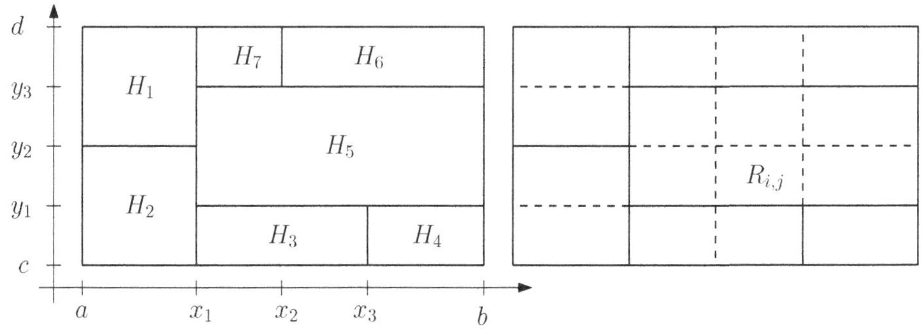

FIGURE 1.1: Pairwise disjoint interval grid of H.

otherwise replace H_j by $H_j \cap H$. Thus in each case the rectangles H_j are contained in H, hence the coordinates of their vertices form partitions

$$\{x_0 := a < x_1 < \ldots < x_p := b\} \text{ and } \{y_0 := c < y_1 < \ldots < y_q := d\}$$

of $[a, b]$ and $[c, d]$, respectively. These partitions generate a grid of disjoint subrectangles $R_{i,j} = (x_i, x_{i+1}] \times (y_j, y_{j+1}]$ with union H such that each H_k is a union of such subrectangles. The procedure for case (a) is illustrated in Figure 1.1. Since

$$b - a = \sum_{i=0}^{p-1}(x_{i+1} - x_i) \text{ and } d - c = \sum_{j=1}^{q-1}(y_{j+1} - y_j),$$

we have, by the definition of λ,

$$\lambda(H) = \left[\sum_{i=0}^{p-1}(x_{i+1} - x_i)\right]\left[\sum_{j=0}^{q-1}(y_{j+1} - y_j)\right] = \sum_{i=0}^{p-1}\sum_{j=0}^{q-1}\lambda(R_{i,j}). \tag{1.9}$$

Similarly,

$$\lambda(H_k) = \sum_{(i,j):R_{i,j} \subseteq H_k} \lambda(R_{i,j})$$

so that

$$\sum_k \lambda(H_k) = \sum_k \sum_{(i,j):R_{i,j} \subseteq H_k} \lambda(R_{i,j}). \tag{1.10}$$

Now compare (1.9) and (1.10). In (a), every $R_{i,j}$ is contained in exactly one H_k, hence the rectangles in (1.10) appear exactly once and so $\lambda(H) = \sum_{k=1}^m \lambda(H_k)$. In (b), a rectangle $R_{i,j}$ could be contained in more than one H_k, so $\lambda(H) \leq \sum_{k=1}^m \lambda(H_k)$. Finally, in (c) not every $R_{i,j}$ is necessarily contained in an H_k, hence $\lambda(H) \geq \sum_{k=1}^m \lambda(H_k)$. □

1.7.2 Lemma. *The volume set function λ is countably additivity on $\mathcal{H}_I$.*

Proof. Part (a) of 1.7.1 gives finite additivity. Let $\{H_j\}$ be a sequence of disjoint members of $\mathcal{H}_I$ such that $H := \bigcup_{j=1}^{\infty} H_j \in \mathcal{H}_I$. By 1.7.1(c), $\lambda(H) \geq \sum_{j=1}^{n} \lambda(H_j)$ for all n, hence $\lambda(H) \geq \sum_{j=1}^{\infty} \lambda(H_j)$.

For the reverse inequality, let $\varepsilon > 0$, and for each j let H_j^ε denote the member of $\mathcal{H}_I$ obtained by replacing each coordinate subinterval $(c, d]$ of H_j by $(c - \delta_j, d + \delta_j]$, where δ_j is chosen so that $\lambda(H_j^\varepsilon) < \lambda(H_j) + \varepsilon/2^j$. Then the collection of intervals $\operatorname{int} H_j^\varepsilon$ is an open covering of the compact set $\operatorname{cl} H$, so there exists an $m \in \mathbb{N}$ such that $H \subseteq \operatorname{int} H_1^\varepsilon \cup \cdots \cup \operatorname{int} H_m^\varepsilon \subseteq H_1^\varepsilon \cup \cdots \cup H_m^\varepsilon$. By 1.7.1(b),

$$\lambda(H) - \varepsilon < \lambda(H^\varepsilon) \leq \lambda(H_1^\varepsilon) + \cdots + \lambda(H_m^\varepsilon) \leq \sum_{j=1}^{\infty} \lambda(H_j) + \varepsilon.$$

Letting $\varepsilon \to 0$ yields $\lambda(H) \leq \sum_{j=1}^{\infty} \lambda(H_j)$, establishing countable additivity. □

Construction of the Measure

Since $\sigma(\mathcal{H}_I) = \mathcal{B}(\mathbb{R}^d)$, we may invoke 1.6.4 using the outer measure

$$\lambda^*(E) := \inf\left\{\sum_{n=1}^{\infty} \lambda(A_n) : \ A_n \in \mathcal{H}_I \text{ and } E \subseteq \bigcup_{n=1}^{\infty} A_n\right\}, \quad E \subseteq \mathbb{R}^d,$$

to obtain

1.7.3 Theorem. *The volume set function λ on $\mathcal{H}_I$ has a unique extension to $\mathcal{B}(\mathbb{R}^d)$. Moreover, $\mathcal{M}(\lambda^*)$ is the completion of $\mathcal{B}(\mathbb{R}^d)$.*

The members of $\mathcal{M}(\lambda^*)$ are called **Lebesgue measurable sets** and $\lambda := \lambda^*\big|_{\mathcal{M}(\lambda^*)}$ is called **Lebesgue measure** on $\mathbb{R}^d$.

Exercises

1.71 Let $I \in \mathcal{H}_I$. Show that $\lambda(I) = \lambda(\operatorname{int} I) = \lambda(\operatorname{cl} I)$. Also, in the definition

$$\lambda^*(E) := \inf\left\{\sum_{n=1}^{\infty} \lambda(A_n) : \ A_n \in \mathcal{A} \text{ and } E \subseteq \bigcup_{n=1}^{\infty} A_n\right\}, \quad E \subseteq \mathbb{R}^d,$$

where $\mathcal{A} = \mathcal{H}_I$, show that the infimum is unchanged if $\mathcal{A}$ is taken to be $\mathcal{O}_I$, $\mathcal{C}_I$, $\mathcal{O} :=$ the set of open sets of $\mathbb{R}^d$, or $\mathcal{K} :=$ the set of compact subsets of $\mathbb{R}^d$.

1.72 Let $N \subseteq \mathbb{R}^d$ with $\lambda(N) = 0$. Show that N^c is dense in $\mathbb{R}$.

1.73 (*Translation invariance of λ*). Let $E \subseteq \mathbb{R}^d$ and $x \in \mathbb{R}^d$. Show that

(a) $\lambda^*(x + E) = \lambda^*(E)$ (b) $E \in \mathcal{M}(\lambda) \Rightarrow x + E \in \mathcal{M}(\lambda)$.

1.74 (*Dilation property of λ*). Let $E \subseteq \mathbb{R}^d$ and $r \in \mathbb{R}$. Show that

(a) $\lambda^*(rE) = |r|^d \lambda^*(E)$ (b) $E \in \mathcal{M}(\lambda) \Rightarrow rE \in \mathcal{M}(\lambda)$.

1.75 Show that for any $\varepsilon > 0$ there exists an open set U dense in $\mathbb{R}^d$ such that $\lambda(U) < \varepsilon$.

1.76 Let A, $B \subseteq [0, 1]$, where $B \in \mathcal{M}(\lambda)$ and $\lambda(B) = 1$. Show that $\lambda^*(A) = \lambda^*(A \cap B)$.

1.77 Let $E \subseteq \mathbb{R}$ with $0 < \lambda(E) < \infty$ and let $0 < r < 1$. Show that there exists an interval $[a, b]$ such that $\lambda^*(E \cap [a, b]) > r(b - a)$. ⟦Let I_n be closed, bounded intervals that cover E with $\sum_n \lambda(I_n) < r^{-1}\lambda(E)$. ⟧

1.78 Show that the graph $G := \{(x, f(x)) : x \in \mathbb{R}\}$ of a continuous function f is a Borel set with two-dimensional Lebesgue measure zero.

1.8 Lebesgue-Stieltjes Measures

A measure on $\mathcal{B}(\mathbb{R}^d)$ that is finite on bounded, d-dimensional intervals is called a **Lebesgue-Stieltjes measure**. For example, Lebesgue measure λ^d is a Lebesgue-Stieltjes measure. Lebesgue-Stieltjes measures may be constructed from so-called *distribution functions*, discussed below. Before we describe the construction, we discuss some approximation properties possessed by these measures.

Regularity

The following theorem complements the approximation property 1.6.5.

1.8.1 Theorem. *Let μ be a Lebesgue-Stieltjes measure on $\mathbb{R}^d$ and let $E \in \mathcal{B}(\mathbb{R}^d)$. Then*

(a) $\mu(E) = \inf\{\mu(U) : U$ *open and* $U \supseteq E\}$.

(b) $\mu(E) = \sup\{\mu(K) : K$ *compact and* $K \subseteq E\}$.

Proof. Assume first that E is bounded. Let $\varepsilon > 0$. By 1.6.11 (taking $\mathcal{A} = \mathcal{O}_I$, say), there exists a sequence of bounded, open, d-dimensional intervals I_j with union $U \supseteq E$ such that $\mu(U) \le \sum_j \mu(I_j) < \mu(E) + \varepsilon$, verifying (a).

To verify (b) in the bounded case, let J be a bounded open interval with $\mathrm{cl}(E) \subseteq J$. Choose a sequence of open intervals V_k with union $V \supseteq J \setminus E$ such that $\sum_{k=1}^{\infty} \mu(V_k) < \mu(J \setminus E) + \varepsilon/2$. We may assume that $V_k \subseteq J$, otherwise replace V_k by $V_k \cap J$. By subadditivity

$$\mu(V) \le \sum_{k=1}^{\infty} \mu(V_k) \le \mu(J \setminus E) + \varepsilon/2 = \mu(J) - \mu(E) + \varepsilon/2.$$

Set $K = J \setminus V$. Since $K \subseteq E \subseteq \mathrm{cl}(E) \subseteq J$, $K = \mathrm{cl}(E) \setminus V$. Therefore, K is compact and $\mu(K) = \mu(J) - \mu(V) \ge \mu(J) - \big(\mu(J) - \mu(E) + \varepsilon/2\big) = \mu(E) - \varepsilon/2$, verifying (b).

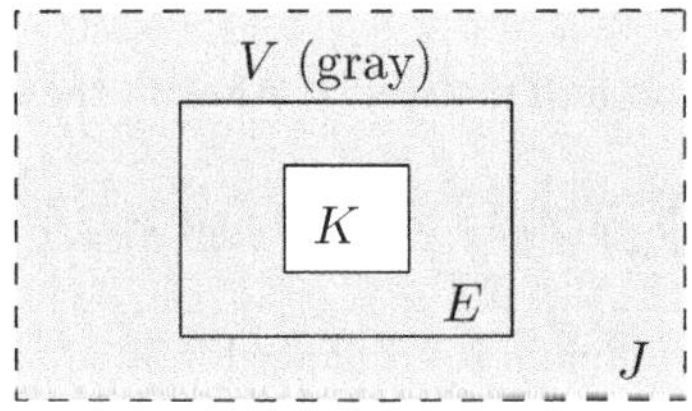

FIGURE 1.2: Construction of K.

Now suppose E is unbounded. Choose a sequence of bounded sets $E_n \in \mathcal{M}(\mu)$ such that $E_n \uparrow E$. Let $\varepsilon > 0$. For each n, use the first part of the proof to choose a compact set K_n and an open set U_n with finite measure such that

$$K_n \subseteq E_n \subseteq U_n, \quad \mu(U_n) - \mu(E_n) < \varepsilon/2^n \;\text{ and }\; \mu(E_n) - \mu(K_n) < \varepsilon.$$

Set $U := \bigcup_{n=1}^{\infty} U_n$. Then U is open, $E \subseteq U$, and $U \setminus E \subseteq \bigcup_n (U_n \setminus E_n)$. If $\mu(E) < \infty$, then

$$\mu(U) - \mu(E) = \mu(U \setminus E) \le \sum_n \mu(U_n \setminus E_n) < \varepsilon,$$

and for sufficiently large n,

$$\mu(E) - \mu(K_n) = \mu(E \setminus E_n) + \mu(E_n \setminus K_n) < \varepsilon,$$

verifying (a) and (b) in this case. On the other hand, if $\mu(E) = \infty$, then (a) clearly holds and (b) holds as well because then $\mu(E_n) \uparrow \infty$ and $\mu(K_n) > \mu(E_n) - \varepsilon$. $\square$

One-Dimensional Distribution Functions

A nondecreasing, right continuous function $F : \mathbb{R} \to \mathbb{R}$ is called a **distribution function.** Such functions arise naturally in probability theory (see Chapter 18). The connection between Lebesgue-Stieltjes measures and distribution functions is described in the following theorem, the proof of which is given below.

1.8.2 Theorem. *For every Lebesgue-Stieltjes measure μ on $\mathbb{R}$, there exists a distribution function F such that*

$$\mu(a, b] = F(b) - F(a) \quad \textit{for all } a < b. \tag{1.11}$$

Any two distributions that satisfy (1.11) *for the same μ differ by a constant. Conversely, every distribution function F gives rise to a unique Lebesgue-Stieltjes measure μ on $\mathcal{B}(\mathbb{R})$ satisfying* (1.11).

Here are three common examples:

1.8.3 Examples.

(a) The Dirac measure δ_0 on $\mathcal{B}(\mathbb{R})$ has distribution function $F = \mathbf{1}_{[0,\infty)}$.

(b) Let (c_n) and (p_n) be sequences in $\mathbb{R}$ with $p_n > 0$ and $\sum_n p_n < \infty$. Define

$$F(x) = \sum_{n:c_n \leq x} p_n,$$

where the sum is taken over all indices n for which $c_n \leq x$. (If there are no such indices, the sum is defined to be 0.) Note that because the order of summation is irrelevant, F is well-defined. The Lebesgue-Stieltjes measure corresponding to F is given by

$$\mu(B) = \sum_{n:c_n \in B} p_n \quad \text{for all Borel sets } B.$$

The distribution in (a) is a special case, obtained by taking $p_1 = 1$, $p_n = 0$ for $n \geq 2$, and $c_n = 0$ for all n.

(c) Let f be continuous and nonnegative on $\mathbb{R}$. Define

$$F(x) = F(0) + \int_0^x f(t)\, dt,$$

where $F(0)$ is arbitrary. The Lebesgue-Stieltjes measure corresponding to F is $d\mu = f\, dt$. (See Chapter 3.) ◊

Proof of Theorem 1.8.2. For the first part of the theorem, define $F : \mathbb{R} \to \mathbb{R}$ as follows: Let $F(0)$ be arbitrary and set

$$F(x) := \begin{cases} F(0) + \mu(0, x] & \text{if } x > 0, \\ F(0) - \mu(x, 0] & \text{if } x < 0. \end{cases}$$

By considering cases, we see that for $a < b$, $F(b) - F(a) = \mu(a, b]$. Therefore, F is nondecreasing and right continuous. If also $G(b) - G(a) = \mu(a, b]$ for all $a < b$, then $F(x) - F(0) = G(x) - G(0)$ for all x, hence $F = G + F(0) - G(0)$.

For the converse, let $F : \mathbb{R} \to \mathbb{R}$ be a distribution function. To construct the Lebesgue-Stieltjes measure defined by F, we apply the results of §1.6 to $(\mathcal{H}_I, \mu)$, where μ is the set function on $\mathcal{H}_I$ given by (1.11). Thus the proof of the theorem will be complete if we show that μ is countably additive on $\mathcal{H}_I$. The following lemmas, analogous to those of §1.7, establish this.

1.8.4 Lemma. *Let* $H, H_1, \ldots, H_m \in \mathcal{H}_I$.

(a) *If* $H_1, \ldots, H_m$ *are disjoint and* $H = \bigcup_{j=1}^m H_j$, *then* $\mu(H) = \sum_{j=1}^m \mu(H_j)$.

(b) *If* $H \subseteq \bigcup_{j=1}^m H_j$, *then* $\mu(H) \le \sum_{j=1}^m \mu(H_j)$.

(c) *If* $H_1, \ldots, H_m$ *are disjoint and* $H \supseteq \bigcup_{j=1}^m H_j$, *then* $\mu(H) \ge \sum_{j=1}^m \mu(H_j)$.

Proof. Let $H = (a, b]$ and $H_j = (a_j, b_j]$, where $a_1 < a_2 < \cdots < a_m$. In (a) there can be no "gaps" or "overlaps," that is, $a_1 = a$, $b_m = b$, and $b_j = a_{j+1}$. Therefore,

$$\sum_{j=1}^m \mu(H_j) = \sum_{j=1}^{m-1} [F(a_{j+1}) - F(a_j)] + F(b) - F(a_m) = F(b) - F(a) = \mu(H).$$

In (b), we may assume that $H = \bigcup_{j=1}^m H_j$, otherwise we could replace H_j by $H_j \cap H$. As in (a), $a_1 = a$, $b_m = b$, and $a_{j+1} \le b_j$. However, since the intervals are no longer disjoint it may happen that $a_{j+1} < b_j$ for some j, as illustrated in Figure 1.3. Form intersections of overlapping intervals, thus partitioning $(a, b]$ into a collection $\{I_i\}$ of disjoint half-open intervals, as shown in the figure. Each H_j is a union of some of these intervals so by (a)

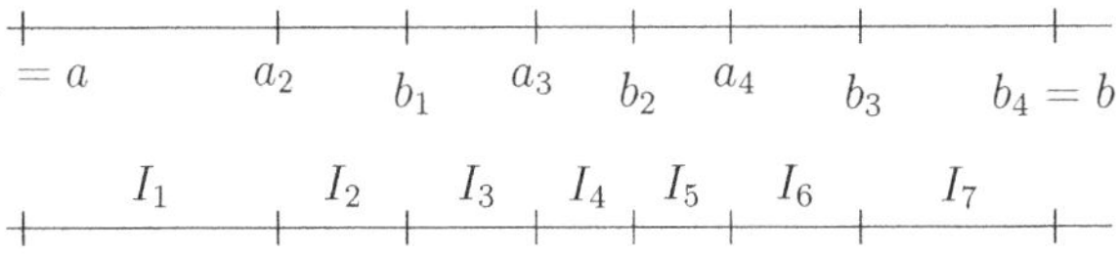

FIGURE 1.3: Construction of partition.

$$\mu(H) = \sum_i \mu(I_i) \quad \text{and} \quad \mu(H_j) = \sum_{i: I_i \subseteq H_j} \mu(I_i).$$

Since an I_i may be contained in more than one H_j

$$\sum_i \mu(I_i) \le \sum_j \sum_{i: I_i \subseteq H_j} \mu(I_i).$$

Therefore, $\mu(H) \le \sum_j \mu(H_j)$, proving (b). The proof of (c) is similar. □

1.8.5 Lemma. *The set function* μ *is countably additive on* $\mathcal{H}_I$.

Proof. By 1.8.4(a), μ is finitely additive. Let $H_j = (a_j, b_j]$ be disjoint members of $\mathcal{H}_I$ and let $H = (a, b] = \bigcup_{j=1}^\infty H_j$. By 1.8.4(c), $\mu(H) \ge \sum_{j=1}^m \mu(H_j)$ for all m, hence $\mu(H) \ge \sum_{j=1}^\infty \mu(H_j)$. For the reverse inequality, let $\varepsilon > 0$ and by right continuity at a choose $r \in (a, b)$ so that $F(r) \le F(a) + \varepsilon/2$. Then

$$\mu(r, b] = F(b) - F(r) \ge F(b) - F(a) - \varepsilon/2 = \mu(H) - \varepsilon/2. \tag{†}$$

Similarly, for each j choose $r_j > b_j$ such that $F(r_j) \le F(b_j) + \varepsilon/2^j$, so

$$\mu(a_j, r_j] = F(r_j) - F(a_j) \le F(b_j) - F(a_j) + \varepsilon/2^j = \mu(H_j) + \varepsilon/2^j. \tag{‡}$$

The open intervals (a_j, r_j) cover $[r, b]$, hence by compactness there exists an $m \in \mathbb{N}$ such that $(r, b] \subseteq \bigcup_{j=1}^m (a_j, r_j]$. By (†), (‡), and 1.8.4(b),

$$\mu(H) \le \varepsilon/2 + \mu(r, b] \le \varepsilon/2 + \sum_{j=1}^\infty \mu(a_j, r_j] \le \varepsilon + \sum_{j=1}^\infty \mu(H_j).$$

Letting $\varepsilon \to 0$ yields the desired inequality. □

*Higher Dimensional Distribution Functions

As in the one-dimensional case, there is a close connection between Lebesgue-Stieltjes measures on $\mathcal{B}(\mathbb{R}^d)$ and certain real-valued functions on $\mathbb{R}^d$. The technical details are more intricate, however, and depend on the following construct:

The ***i*th coordinate difference operator** on functions $F : \mathbb{R}^d \to \mathbb{R}$ is defined by

$$\triangle_{a_i}^{b_i} F(x_1, \ldots, x_d) = F(x_1, \ldots, x_{i-1}, b_i, x_{i+1}, \ldots, x_d) - F(x_1, \ldots, x_{i-1}, a_i, x_{i+1}, \ldots, x_d).$$

For example, consider the function $F(x_1, x_2, \ldots, x_d) = x_1 x_2 \ldots x_d$. For $1 \le i \le d$ and $a_i < b_i$, the difference operators may be applied successively to obtain the following:

$$\begin{aligned}\triangle_{a_1}^{b_1} \cdots \triangle_{a_d}^{b_d} F(x_1, x_2, \ldots, x_d) &= \triangle_{a_1}^{b_1} \cdots \triangle_{a_{d-1}}^{b_{d-1}} (x_1 \cdots x_{d-1})(b_d - a_d) \\ &= \triangle_{a_1}^{b_1} \cdots \triangle_{a_{d-2}}^{b_{d-2}} (x_1 \cdots x_{d-2})(b_{d-1} - a_{d-1})(b_d - a_d) \\ &\ \ \vdots \\ &= (b_1 - a_1) \cdots (b_d - a_d).\end{aligned}$$

Thus $\triangle_{a_1}^{b_1} \cdots \triangle_{a_d}^{b_d} F(x_1, x_2, \ldots, x_d)$ is the Lebesgue measure of the d-dimensional interval $(a_1, b_1] \times \cdots \times (a_d, b_d]$. This sort of connection holds more generally and is described in the theorem below. For the statement of the theorem we need the following definitions:

A function $F : \mathbb{R}^d \to \mathbb{R}$ is a **distribution function** if it is **nondecreasing** in the sense that

$$\triangle_{a_1}^{b_1} \cdots \triangle_{a_d}^{b_d} F(x_1, \ldots, x_d) \ge 0, \quad a_i < b_i, \ i = 1, \ldots, d,$$

and **right continuous** in the sense that

$$x_{i,n} \downarrow_n x_i, \quad i = 1, \ldots, d \ \Rightarrow F\big(x_{n,1}, \ldots, x_{n,d}\big) \to F(x_1, \ldots, x_d).$$

Here are some standard distribution functions:

1.8.6 Examples.

(a) Let F_i be a distribution function on $\mathbb{R}$, $i = 1, \ldots, d$. The function

$$F(x_1, x_2, \ldots, x_d) := F_1(x_1) F_2(x_2) \cdots F_d(x_d)$$

is a distribution function on $\mathbb{R}^d$ such that

$$\triangle_{a_1}^{b_1} \cdots \triangle_{a_d}^{b_d} F(x_1, \ldots, x_d) = \prod_{i=1}^{d} [F_i(b_i) - F_i(a_i)].$$

The function $F(x_1, x_2, \ldots, x_d) = x_1 x_2 \cdots x_d$ discussed above is a special case.

(b) Let f be a nonnegative, continuous function on $\mathbb{R}^d$. Then

$$F(x_1, \ldots, x_d) := \int_{-\infty}^{x_1} \cdots \int_{-\infty}^{x_d} f(t_1, \ldots, t_d)\, dt_d \cdots dt_1$$

is a distribution function on $\mathbb{R}^d$ (provided the improper integral is finite) such that

$$\triangle_{a_1}^{b_1} \cdots \triangle_{a_d}^{b_d} F(x_1, \ldots, x_d) = \int_{a_1}^{b_1} \cdots \int_{a_d}^{b_d} f(t_1, \ldots, t_d)\, dt_d \cdots dt_1.$$

(c) If μ is a finite measure on $\mathcal{B}(\mathbb{R}^d)$, then

$$F(x_1, \ldots, x_d) = \mu\big((-\infty, x_1] \times \cdots \times (-\infty, x_d]\big)$$

defines a distribution function on $\mathbb{R}^d$. ◊

The following theorem may be proved using a combination of ideas developed earlier in the construction of Lebesgue measure and Lebesgue-Stieltjes measures. For a proof, the reader is referred to [1] or [5]

1.8.7 Theorem. *Let μ be a Lebesgue-Stieltjes measure on $\mathcal{B}(\mathbb{R}^d)$. Then there exists a function $F : \mathbb{R}^d \to \mathbb{R}$ such that for all $a_i < b_i$*

$$\mu\big((a_1, b_1] \times \cdots \times (a_d, b_d]\big) = \Delta_{a_1}^{b_1} \cdots \Delta_{a_d}^{b_d} F(x_1, \ldots, x_d). \tag{1.12}$$

Conversely, given a distribution function $F : \mathbb{R}^d \to \mathbb{R}$, there exists a unique Lebesgue-Stieltjes measure on $\mathcal{B}(\mathbb{R}^d)$ such that (1.12) holds for all $a_i < b_i$ $(i = 1, \ldots, d)$.

Exercises

1.79 Describe the Lebesgue-Stieltjes measure for each of the following distribution functions.

(a) $F(x) = \lfloor x \rfloor$, the greatest integer function.

(b) $F(x) = x\mathbf{1}_{[0,1)} + \mathbf{1}_{[1,\infty]}$.

1.80 Show that the sum of finitely many distribution functions and the product of finitely many nonnegative distribution functions are distribution functions.

1.81 Verify that the function in 1.8.3(b) is a distribution function. Prove also that F is left continuous at a iff $a \neq c_n$ for every n.

1.82 For any monotone function $F : \mathbb{R} \to \mathbb{R}$ and $-\infty \le a < b \le \infty$, define

$$F(a+) := \lim_{x \to a^+} F(x) \quad \text{and} \quad F(b-) := \lim_{x \to b^-} F(x)$$

and set

$$F(-\infty) := F((-\infty)+) \quad \text{and} \quad F(\infty) := F(\infty-).$$

Let F be a distribution function and μ the associated Lebesgue-Stieltjes measure. Prove the following, when defined:

(a) $\mu(a, b) = F(b-) - F(a)$.

(b) $\mu[a, b) = F(b-) - F(a-)$.

(c) $\mu[a, b] = F(b) - F(a-)$.

Prove also that $\mu\{x\} = 0$ iff F is continuous at x.

1.83 Let μ be a finite Lebesgue-Stieltjes measure on $\mathcal{B}(\mathbb{R})$ such that $\mu(\{x\}) = 0$ for all x. Show that any distribution function F corresponding to μ is uniformly continuous on $\mathbb{R}$.

1.84 Show that a monotone function $f : \mathbb{R} \to \mathbb{R}$ has countably many discontinuities. Conclude that if μ is a Lebesgue-Stieltjes measure, then there exist at most countably many $x \in \mathbb{R}$ such that $\mu(\{x\}) > 0$. ⟦For each $t \in \mathbb{R}$, define $a_t = \lim_{x \to t^-} f(x)$ and $b_t = \lim_{x \to t^+} f(x)$. Then $a_t < b_t$ iff f is discontinuous at t.⟧

1.85 Let μ be a Lebesgue-Stieltjes measure on $\mathbb{R}$ with a continuous distribution function and let $A \in \mathcal{B}(\mathbb{R})$ with $\mu(A) > 0$. Prove that for each $b \in (0, \mu(A))$ there exists a Borel set $B \subseteq A$ such that $\mu(B) = b$. ⟦Use the intermediate value theorem on $G(x) = \mu\big(A \cap [-n, x]\big)$ for suitable n⟧.

*1.9 Some Special Sets

In this section we construct subsets of $\mathbb{R}$ that illustrate some of the finer points of Lebesgue and Borel measurability.

An Uncountable Set with Lebesgue Measure Zero

The **Cantor ternary set** C is constructed as follows: Remove from $I := [0,1] = I_{0,1}$ the "middle third" open interval $(1/3, 2/3)$, leaving closed intervals $I_{1,1}$ and $I_{1,2}$ with union C_1 and total length $2/3$. Next, remove from each of the intervals $I_{1,1}$ and $I_{1,2}$ the middle third open interval, leaving closed intervals $I_{2,1}$, $I_{2,2}$, $I_{2,3}$, and $I_{2,4}$ with union C_2 and total length $4/9 = (2/3)^2$. By induction, one obtains a decreasing sequence of closed sets $C_k = \bigcup_{j=1}^{2^k} I_{k,j}$ such that $\lambda(C_k) = (2/3)^k$. (See Figure 1.4.) Then $C := \bigcap_k C_k$ is closed and $\lambda(C) = 0$.

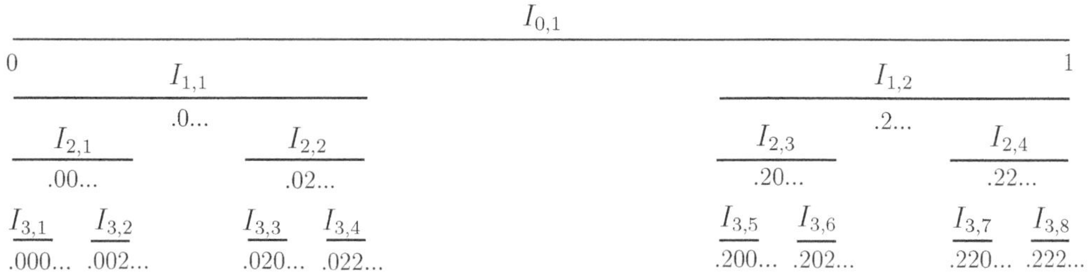

FIGURE 1.4: Middle thirds construction.

To show that C is uncountable, consider the ternary representation of a number $x \in [0,1]$:

$$x = .d_1 d_2 \ldots = \sum_{k=1}^{\infty} d_k 3^{-k}, \quad \text{where} \quad d_k \in \{0,1,2\}. \tag{1.13}$$

By induction, using the fact that $x \in I_{k-1,j} \Rightarrow I_{k,2j-1+d_k/2}$, one shows that $x \in C$ iff x has an expansion with even digits (see Figure 1.4). Define $\varphi : C \to [0,1]$ by

$$\varphi(.d_1 d_2 \ldots (\text{ternary})) = .e_1 e_2 \ldots (\text{binary}), \text{ where } d_k \in \{0,2\} \text{ and } e_k = d_k/2.$$

The function φ is not one-to-one, but by removing from C the countable set of all numbers with ternary representations ending in a sequence of zeros we obtain a set D on which φ is one-to-one. Since $\varphi(D) = (0,1)$, C is uncountable.

Non-Lebesgue-Measurable Sets

We show the following:

Every Lebesgue measurable set A with $\lambda(A) > 0$ contains a set that is not Lebesgue measurable.

Since $A = \bigcup_{n \in \mathbb{Z}} A \cap [n, n+1]$, we may suppose that A is bounded. Define an equivalence relation on A by $x \sim y$ iff $x - y \in \mathbb{Q}$. Let B be the subset of A obtained by choosing exactly one point from each distinct equivalence class. (The existence of B requires the axiom of choice.) Now observe that the sets $r + B$, $r \in \mathbb{Q}$, are disjoint. Indeed, if $(r+B) \cap (s+B) \neq \emptyset$, then $r + x = s + y$ for some x, $y \in B$, so $x = y$ and $r = s$. Moreover, since A is bounded so is $B + [0, 1]$. Let (r_n) be an enumeration of the rationals in $[0, 1]$ and assume that B is measurable. Then

$$\infty > \lambda\Big(\bigcup_n (B + r_n)\Big) = \sum_n \lambda(B + r_n) = \sum_n \lambda(B),$$

which implies that $\lambda(B) = 0$. But $A \subseteq B + \mathbb{Q}$, hence

$$\lambda(A) \le \lambda\Big(\bigcup_{r\in\mathbb{Q}} (B + r)\Big) = \sum_{r\in\mathbb{Q}} \lambda(B + r) = 0,$$

contradicting that $\lambda(A) > 0$. Therefore, B cannot be Lebesgue measurable.

A Lebesgue Measurable, Non-Borel Set

For this example, we first construct the **Cantor function** $f : I \to I$, where $I = [0, 1]$. The construction is based on the Cantor set C described earlier in the section. For each n, denote by $J_{n,k}$, $k = 1, \ldots, 2^{n-1}$, the open intervals in increasing order that were removed from I in the construction of C, that is, the intervals that form the complement of C_n in $[0, 1]$. For example, $J_{2,1} = (1/9, 2/9)$, $J_{2,2} = (1/3, 2/3)$, and $J_{2,3} = (7/9, 8/9)$, hence

$$[0, 1] = I_{2,1} \cup J_{2,1} \cup I_{2,2} \cup J_{2,2} \cup I_{2,3} \cup J_{2,3} \cup I_{2,4}.$$

Define a continuous function $f_n : I \to I$ so that

$$f_n(0) = 0, \quad f_n(1) = 1, \quad f_n = k/2^n \text{ on } J_{n,k},$$

and f_n is linear on the complementary intervals $I_{n,j}$. Since $|f_n(x) - f_{n+1}(x)| \le 1/2^{n+1}$, the

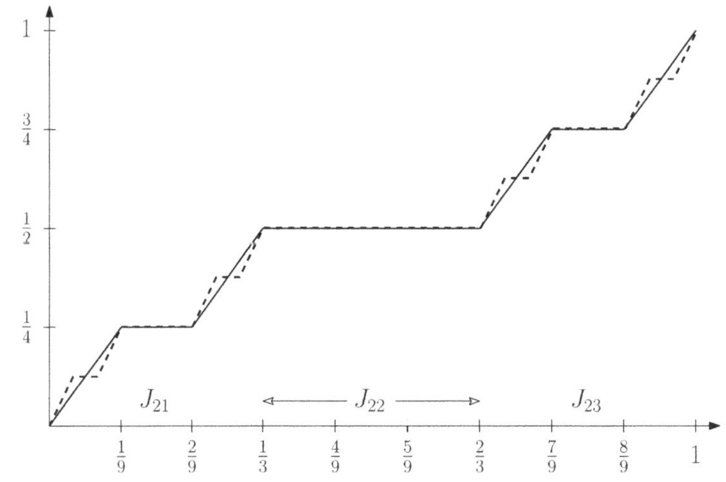

FIGURE 1.5: The functions f_2 and f_3.

sequence $\{f_n\}$ is uniformly Cauchy and so converges to a continuous function f, the Cantor function.

To construct the desired non-Borel set, note first that since $f_n(0) = 0$, $f_n(1) = 1$, and f_n is nondecreasing on $[0, 1]$, f also has these properties. Thus, by the intermediate value theorem, $f(I) = I$. Since the values of f on the intervals $J_{n,k}$ are already assumed at the endpoints and since these endpoints lie in C, $f(J_{n,k})$ contributes nothing additional to the range of f, hence $f(C) = I$. Now set $g(x) = (f(x)+x)/2$, $x \in I$. Then g is continuous, *strictly* increasing, $g(0) = 0$, and $g(1) = 1$, hence $g(I) = I$. It follows that $g : I \to I$ is a homeomorphism, hence $g(C)$ is closed. Thus $g(I \setminus C)$ is a proper nonempty open subset of I and so has positive Lebesgue measure. Moreover, g takes the interval $J_{n,k}$, on which f is constant, to an open interval half its length, so by countable additivity $\lambda\big(g(I \setminus C)\big) = \lambda(I \setminus C)/2 = 1/2$ and therefore $\lambda(g(C)) = 1/2$. Now let E be a subset of $g(C)$ that is not Lebesgue measurable and let $A := g^{-1}(E)$. Then $A \subseteq C$ and so is Lebesgue measurable with $\lambda(A) = 0$. However, A cannot be a Borel set since g maps Borel sets onto Borel sets. (This is proved in Chapter 2.)

1.9.1 Remark. While the intricate nature of the construction of A might lead one to believe that such sets are rare, there are in fact many more Lebesgue measurable sets than Borel sets. Indeed, since the Cantor set C is uncountable and every subset of C is Lebesgue measurable, the collection of Lebesgue measurable sets has cardinality $2^{\mathfrak{c}}$, where $\mathfrak{c}$ is the cardinality of the continuum. On the other hand, it may be shown that $\mathcal{B}(\mathbb{R})$ has only cardinality $\mathfrak{c}$. (See, for example, [38].) $\Diamond$

Exercises

1.86 Show that $(\mathbb{R}, \mathcal{B}(\mathbb{R}), \lambda)$ is not complete.

1.87 Carry out the steps below to prove following assertion: If $A \subseteq \mathbb{R}$ has positive Lebesgue measure then the set $A - A := \{x - y : x, y \in A\}$ contains an interval $(-r, r)$ for some $r > 0$.

(a) Show that it suffices to consider the case A compact.

(b) Choose an open set $U \supseteq A$ such that $\lambda(U) < 2\lambda(A)$ (how?). Define a distance function $d : U \to \mathbb{R}$ by $d(x) = \inf\{|x - y| : y \in U^c\}$. Show that d is continuous and positive. Conclude that d has a minimum $r > 0$ on A.

(c) Show that $|x| < r \Rightarrow x + A \subseteq U \Rightarrow (x + A) \cap A \neq \emptyset$. Conclude that $(-r, r) \subseteq A - A$.

1.88 [↑1.87] Show that the only subgroup of $(\mathbb{R}, +)$ that has positive Lebesgue measure is $\mathbb{R}$.

1.89 Let (a_n) be a sequence in $(0, 1)$ and set $b_n := 1 - a_n$. Mimic the construction of the Cantor ternary set by removing the middle part of $[0, 1]$ of length a_1, leaving two intervals with union E_1, each of length $b_1/2$, then removing the middle part of length $a_2 b_1/2$ from these leaving four intervals with union E_2, each of length $b_1 b_2/4$, and so forth. The intersection $E := \bigcap_n E_n$ is

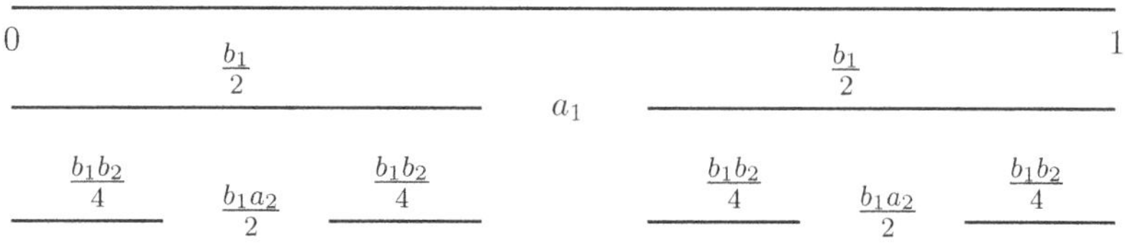

FIGURE 1.6: Generalized middle thirds construction.

called a **generalized Cantor set**. Verify the following:

(a) E is closed and $\lambda(E) = \prod_{n=1}^{\infty} b_n := \lim_n \prod_{j=1}^{n} b_j$.

(b) The interior of E is nonempty.

(c) If $r > 0$ and eventually $a_n \geq r$ (as in the Cantor ternary set), then $\lambda(E) = 0$.

(d) For each $a \in (0, 1)$, there exists a generalized Cantor set with Lebesgue measure a. ⟦Consider $\ln\left(\prod_{n=1}^{\infty} b_n\right) = \sum_{n=1}^{\infty} \ln b_n$. ⟧

1.90 Let A be the set of all $x \in [0, 1]$ having a decimal expansion $.d_1 d_2 \ldots$ with no digit equal to 3. Show that A is uncountable, $A \in \mathcal{B}(\mathbb{R})$, and $\lambda(A) = 0$.

Chapter 2

Measurable Functions

In this chapter we consider functions that are *measurable* with respect to a given σ-field $\mathcal{F}$, that is, functions f for which (in the real-valued case) the sets $\{x \in X : f(x) \in (a,b)\}$ are $\mathcal{F}$-measurable. As we shall see, such functions are natural candidates for integration with respect to Lebesgue measure. We begin with the more general notion of measurable transformation.

2.1 Measurable Transformations

Let $(X, \mathcal{F})$ and $(Y, \mathcal{G})$ be measurable spaces. By standard properties of preimages, $T^{-1}(\mathcal{G}) := \{T^{-1}(E) : E \in \mathcal{G}\}$ is a σ-field. If the inclusion $T^{-1}(\mathcal{G}) \subseteq \mathcal{F}$ holds, then T is said to be $\mathcal{F}/\mathcal{G}$**-measurable**, or simply **measurable** if the σ-fields $\mathcal{F}$ and $\mathcal{G}$ are understood. In this case, T is also called a **measurable transformation** or **measurable mapping**. To indicate such measurability we also say that $T : (X, \mathcal{F}) \to (Y, \mathcal{G})$ is **measurable**.

Constant functions are measurable transformations, since the only preimages are $\emptyset$ and X. Also, *every* function $T : X \to Y$ is both $\mathcal{P}(X)/\mathcal{G}$-measurable and $\mathcal{F}/\{\emptyset, X\}$-measurable.

General Properties

2.1.1 Proposition. *If $T : (X, \mathcal{F}) \to (Y, \mathcal{G})$ and $S : (Y, \mathcal{G}) \to (Z, \mathcal{H})$ are measurable, then $S \circ T : (X, \mathcal{F}) \to (Z, \mathcal{H})$ is measurable.*

Proof. This follows from $(S \circ T)^{-1}(A) = T^{-1}\big(S^{-1}(A)\big)$, $A \in \mathcal{H}$. □

The following result characterizes measurability in terms of the generators of a σ-field. It will play an important role in what follows.

2.1.2 Theorem. *Let $\mathcal{A} \subseteq \mathcal{P}(Y)$ and $T : X \to Y$. Then $\sigma\left(T^{-1}(\mathcal{A})\right) = T^{-1}\big(\sigma(\mathcal{A})\big)$. In particular, $T : (X, \mathcal{F}) \to (Y, \sigma(\mathcal{A}))$ is measurable iff $T^{-1}(A) \in \mathcal{F}$ for all $A \in \mathcal{A}$.*

Proof. Since $T^{-1}\big(\sigma(\mathcal{A})\big)$ is a σ-field and $T^{-1}(\mathcal{A}) \subseteq T^{-1}\big(\sigma(\mathcal{A})\big)$, it follows by minimality that $\sigma\left(T^{-1}(\mathcal{A})\right) \subseteq T^{-1}\big(\sigma(\mathcal{A})\big)$. For the reverse inclusion, observe that the set

$$\left\{A \subseteq \sigma(\mathcal{A}) : T^{-1}(A) \subseteq \sigma\left(T^{-1}(\mathcal{A})\right)\right\}$$

is a σ-field containing $\mathcal{A}$ and hence must equal $\sigma(\mathcal{A})$. □

2.1.3 Corollary. *Let X and Y be topological spaces. If $T : X \to Y$ is continuous, then T is $\mathcal{B}(X)/\mathcal{B}(Y)$-measurable.*

Proof. Let $\mathcal{O}_X$ and $\mathcal{O}_Y$ denote the collections of open subsets of X and Y, respectively. Then $T^{-1}(\mathcal{O}_Y) \subseteq \mathcal{O}_X \subseteq \mathcal{B}(X)$. □

For example, a linear transformation $T : \mathbb{R}^p \to \mathbb{R}^q$, being automatically continuous, is Borel measurable.

The inclusion $T^{-1}(\mathcal{B}(Y)) \subseteq \mathcal{B}(X)$ in the proof of 2.1.3 may be strict. For example, let X be any nontrivial set with the discrete topology, let $Y = X$ have the indiscrete topology, and take T to be the identity map.

2.1.4 Corollary. *Let X be a set and $\{(X_i, \mathcal{F}_i) : i \in \mathfrak{I}\}$ a family of measurable spaces. Given mappings $T_i : X \to X_i$, let*

$$\mathcal{E} := \bigcup_{i \in \mathfrak{I}} T_i^{-1}(\mathcal{F}_i) \quad \textit{and} \quad \mathcal{F} := \sigma(\mathcal{E}).$$

If $(X_0, \mathcal{F}_0)$ is a measurable space, then a mapping $T : X_0 \to X$ is $\mathcal{F}_0/\mathcal{F})$-measurable iff the mapping $T_i \circ T$ is $\mathcal{F}_0/\mathcal{F}_i$-measurable for every $i \in \mathfrak{I}$.

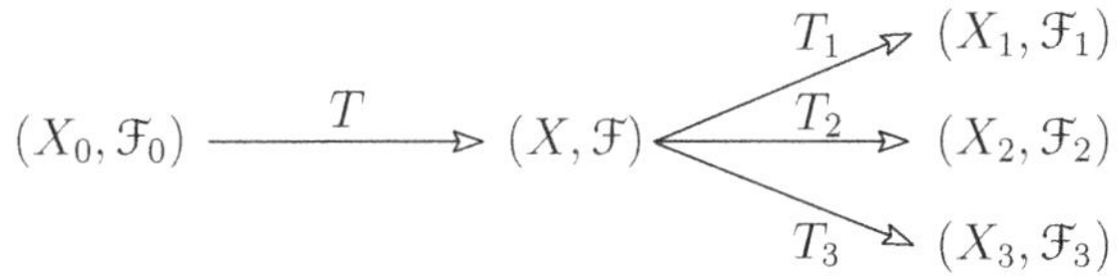

FIGURE 2.1: The mappings of 2.1.4.

Proof. Proposition 2.1.1 gives the necessity. For the sufficiency, if $T_i \circ T$ is $\mathcal{F}_0/\mathcal{F}_i$-measurable for every $i \in \mathfrak{I}$, then

$$T^{-1}(\mathcal{E}) = \bigcup_{i \in \mathfrak{I}} T^{-1}\left(T_i^{-1}(\mathcal{F}_i)\right) = \bigcup_{i \in \mathfrak{I}} (T_i \circ T)^{-1}(\mathcal{F}_i) \subseteq \mathcal{F}_0,$$

hence $\sigma\big(T^{-1}(\mathcal{E})\big) \subseteq \mathcal{F}_0$. But by the theorem, $\sigma\big(T^{-1}(\mathcal{E})\big) = T^{-1}(\mathcal{F})$. □

One of the most important applications of 2.1.4 is the following:

2.1.5 Corollary. *Let $(X_i, \mathcal{F}_i)$ be measurable spaces $(i = 1, \ldots, d)$ and let $(X, \mathcal{F})$ denote the product measurable space $(X_1 \times \cdots \times X_d, \mathcal{F}_1 \otimes \cdots \otimes \mathcal{F}_d)$. Then for each i the projection map*

$$\pi_i : X \to X_i, \quad \pi_i(x_1, \ldots, x_d) = x_i,$$

is $\mathcal{F}/\mathcal{F}_i$-measurable. Moreover, if $(X_0, \mathcal{F}_0)$ is a measurable space, then a mapping $T : X_0 \to X$ is $\mathcal{F}_0/\mathcal{F}$ measurable iff $\pi_i \circ T$ is $\mathcal{F}_0/\mathcal{F}_i$-measurable for every i.

Proof. If $A_i \in \mathcal{F}_i$, then

$$\pi_i^{-1}(A_i) = X_1 \times \cdots \times X_{i-1} \times A_i \times X_{i+1} \times \cdots \times X_d \in \mathcal{F},$$

hence π_i is $\mathcal{F}/\mathcal{F}_i$-measurable. The set $\mathcal{E}$ in 2.1.4 corresponding to the maps π_i is the collection of all such sets, and taking intersections produces $\mathcal{F}_1 \times \cdots \times \mathcal{F}_d$. Therefore, $\sigma(\mathcal{E}) = \mathcal{F}_1 \otimes \cdots \otimes \mathcal{F}_d$, and the conclusion of the theorem follows from 2.1.4. □

2.1.6 Corollary. *Let $(X_i, \mathcal{F}_i)$ be measurable spaces $(i = 0, 1, \ldots, d)$ and $T_i : X_0 \to X_i$ arbitrary mappings $(i = 1, \ldots, d)$. Define*

$$T = (T_1, \ldots, T_d) : X_0 \to X_1 \times \cdots \times X_d, \quad T(x) = \big(T_1(x), \ldots, T_d(x)\big).$$

Then T is $\mathcal{F}_0/(\mathcal{F}_1 \otimes \ldots \otimes \mathcal{F}_d)$-measurable iff each $T_i : (X_0, \mathcal{F}_0) \to (X_i, \mathcal{F}_i)$ is measurable.

Proof. The mappings $\pi_i \circ T$ of 2.1.5 are simply the given mappings T_i. □

Here is a complement to 2.1.4. The proof is left as an exercise.

2.1.7 Proposition. *Let X be any set and $\{(X_i, \mathcal{F}_i) : i \in \mathfrak{I}\}$ a family of measurable spaces. Given mappings $T_i : X_i \to X$, set*

$$\mathcal{F} := \bigcap_{i \in \mathfrak{I}} T_i^{-1}(\mathcal{F}_i).$$

Let $(X_0, \mathcal{F}_0)$ be a measurable space and $T : X \to X_0$. Then T is $\mathcal{F}/\mathcal{F}_0$-measurable iff $T \circ T_i$ is $\mathcal{F}_i/\mathcal{F}_0$-measurable for every $i \in \mathfrak{I}$.

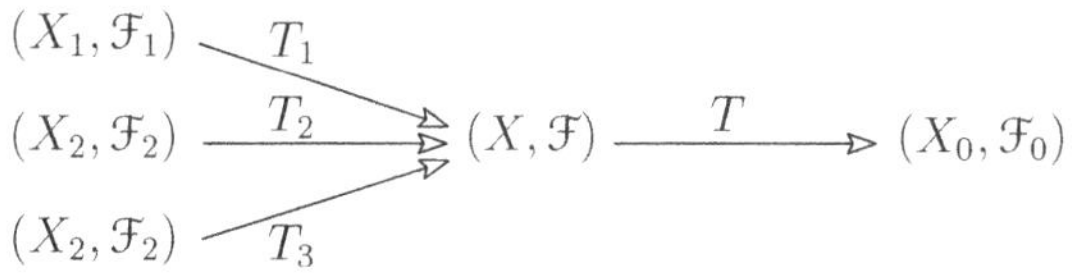

FIGURE 2.2: The mappings of 2.1.7.

Exercises

2.1 Show that for a measurable transformation $T : (X, \mathcal{F}) \to (Y, \mathcal{G})$ it is not necessarily the case that $T(\mathcal{F}) \subseteq \mathcal{G}$.

2.2 Let $(X, \mathcal{F})$ and $(Y, \mathcal{G})$ be measurable spaces and let $T : (X, \mathcal{F}) \to (Y, \mathcal{G})$ be measurable. If $E \subseteq X$, show that $T\big|_E$ is $(\mathcal{F} \cap E)/\mathcal{G}$-measurable.

2.3 Let $(X, \mathcal{F})$ and $(Y, \mathcal{G})$ be measurable spaces, $E_n \in \mathcal{F}$, and $X = \bigcup_{n=1}^{\infty} E_n$. Let $T : X \to Y$ have the property $T\big|_{E_n}$ is $(\mathcal{F} \cap E_n)/\mathcal{G}$-measurable for every n. Prove that T is $\mathcal{F}/\mathcal{G}$-measurable.

2.4 Let $(X, \mathcal{F})$, $(Y, \mathcal{G})$, and $(Z, \mathcal{H})$ be measurable spaces and let $T : X \to Y$ have countable range. Assume that $\mathcal{G}$ contains the singletons (e.g., a Borel σ-field). Show that

(a) T is $\mathcal{F}/\mathcal{G}$-measurable iff $T^{-1}(\{y\}) \in \mathcal{F}$ for every $y \in Y$.

(b) If T is $\mathcal{F}/\mathcal{G}$-measurable, then $S \circ T$ is $\mathcal{F}/\mathcal{H}$-measurable for *any* mapping $S : Y \to Z$.

2.5 Let $(X, \mathcal{F})$ and $(Y, \mathcal{G})$ be measurable spaces. Show that if $A \subseteq X$ and $B \subseteq Y$ are nonempty and $A \times B \in \mathcal{F} \otimes \mathcal{G}$, then $A \in \mathcal{F}$ and $B \in \mathcal{G}$.

2.6 Prove 2.1.7.

2.7 Let $\{(X_i, \mathcal{F}_i) : i \in \mathfrak{I}\}$ be a family of measurable spaces with union X and let $\mathcal{F}$ be the σ-field of all sets $E \subseteq X$ such that $E \cap X_i \in \mathcal{F}_i$ for all $i \in \mathfrak{I}$. Let $(X_0, \mathcal{F}_0)$ be a measurable space and $T : X \to X_0$. Show that T is $\mathcal{F}/\mathcal{F}_0$-measurable iff $T\big|_{X_i}$ is $\mathcal{F}_i/\mathcal{F}_0$-measurable for every $i \in \mathfrak{I}$.

2.8 Let $T, S : (X, \mathcal{F}) \to \big(\mathbb{R}^d, \mathcal{B}(\mathbb{R}^d)\big)$ be measurable. Prove that $\{x \in X : T(x) = S(x)\} \in \mathcal{F}$.

2.9 Let $(i_1, \ldots, i_d)$ be a permutation of $(1, \ldots, d)$ and define $T : \mathbb{R}^d \to \mathbb{R}^d$ by $T(x_1, \ldots, x_d) = (x_{i_1}, \ldots, x_{i_d})$. Prove that T is $\mathcal{B}(\mathbb{R}^d)/\mathcal{B}(\mathbb{R}^d)$-measurable.

2.10 Let $(X, \mathcal{F})$, $(Z, \mathcal{H})$ be measurable spaces, $T : X \to Y$ surjective, and $\mathcal{G} = \{A \subseteq Y : T^{-1}(A) \in \mathcal{F}\}$. Let $R : (X, \mathcal{F}) \to (Z, \mathcal{H})$ be measurable such that $T(x) = T(x') \Rightarrow R(x) = R(x')$. Show that there exists a measurable transformation $S : (Y, \mathcal{G}) \to (Z, \mathcal{H})$ such that $R = S \circ T$.

2.11 Let $(Y, \mathcal{F})$, $(Z, \mathcal{H})$ be measurable spaces, $T : X \to Y$, and set $\mathcal{F} := T^{-1}(\mathcal{G})$, so that the map $T : (X, \mathcal{F}) \to (Y, \mathcal{G})$ is measurable. Let $R : (X, \mathcal{F}) \to (Z, \mathcal{H})$ be measurable with countable range. Show that if $\mathcal{H}$ contains the singletons, then there exists a measurable $S : (Y, \mathcal{G}) \to (Z, \mathcal{H})$ such that $R = S \circ T$.

2.12 Prove that if $S, T : \mathbb{R}^p \to \mathbb{R}^q$ are continuous and $S = T$ λ-a.e., then $S = T$. What if only one of the mappings is continuous?

2.13 [$\downarrow$3.5.2] Let $(X, \mathcal{F})$, $(Y, \mathcal{G})$, and $(Z, \mathcal{H})$ be measurable spaces and $T : X \times Y \to Z$ an arbitrary mapping. We say that T is **separately measurable** if $T_y := T(\cdot, y)$ is $\mathcal{F}/\mathcal{H}$-measurable for each $y \in Y$ and $T_x := T(x, \cdot)$ is $\mathcal{G}/\mathcal{H}$-measurable for each $x \in X$. To distinguish from separate measurability, we sometimes refer to $\mathcal{F} \otimes \mathcal{G}/\mathcal{H}$-measurability of T as **joint measurability**. Show that if T is jointly measurable, then it is separately measurable.

2.2 Measurable Numerical Functions

In this section we consider functions $f : X \to \overline{\mathbb{K}}$, which we shall sometimes refer to as *numerical* or *numerically valued.* A numerical function f on a measurable space $(X, \mathcal{F})$ is said to be **$\mathcal{F}$-measurable**, or simply **measurable**, if f is $\mathcal{F}/\mathcal{B}(\overline{\mathbb{K}})$-measurable. Thus the σ-fields $\mathcal{B}(\overline{\mathbb{R}})$, $\mathcal{B}(\mathbb{R})$, or $\mathcal{B}(\mathbb{C})$ are always understood.

Since a function $f : X \to \mathbb{C}$ may be identified with the mapping $(\operatorname{Re} f, \operatorname{Im} f) : X \to \mathbb{R}^2$, it follows from 2.1.6 that f is measurable iff $\operatorname{Re} f$ and $\operatorname{Im} f$ are measurable. This fact frequently allows one to reduce arguments from the complex case to the real case.

The following notation for preimages will be convenient in further discussions involving measurability:

- $\{f \in A\} := \{x \in X : f(x) \in A\} = f^{-1}(A)$, where $f : X \to Y$, $A \subseteq Y$.
- $\{f > a\} := \{x \in X : f(x) > a\}$, where $f : X \to \overline{\mathbb{R}}$, $a \in \overline{\mathbb{R}}$.
- $\{f \le g\} := \{x \in X : f(x) \le g(x)\}$, where $f, g : X \to \overline{\mathbb{R}}$,

and so forth. Additionally, if μ is a set function we write $\mu(f \in A)$ rather than the more cumbersome $\mu(\{f \in A\})$, etc. These notational conventions are frequently used in probability theory, but they will be seen to have general utility.

Criteria for Measurability

If $\mathcal{A}$ is a generating class for $\mathcal{B}(\overline{\mathbb{K}})$, then measurability of $f : X \to \overline{\mathbb{K}}$ is equivalent to the assertion that $f^{-1}(\mathcal{A}) \subseteq \mathcal{F}$ (2.1.2). For $\overline{\mathbb{K}} = \overline{\mathbb{R}}$, it follows from 1.2.3 that f is $\mathcal{F}$-measurable iff

- $\{f = \infty\}$, $\{f = -\infty\} \in \mathcal{F}$

and any one of the following conditions holds:

- $\{f \in E\} \in \mathcal{F}$ for all open, (resp. closed, resp. Borel) sets $E \subseteq \mathbb{R}$.
- $\{f \le t\} \in \mathcal{F}$ for all $t \in \mathbb{R}$.
- $\{f < t\} \in \mathcal{F}$ for all $t \in \mathbb{R}$.
- $\{f > t\} \in \mathcal{F}$ for all $t \in \mathbb{R}$.
- $\{f \ge t\} \in \mathcal{F}$ for all $t \in \mathbb{R}$.

It follows easily that if the range of f is countable, say $\operatorname{ran} f = (a_n)$, then f is measurable iff $\{f = a_n\} \in \mathcal{F}$ for all n.

2.2.1 Example. Let $d_n(x)$ denote the nth digit of the decimal expansion of $x \in [0,1)$, where for definiteness we exclude expansions that end in a sequence of 9's, choosing for example $.500\cdots$ over $.499\cdots$. Let $e_n \in \{0,1,\ldots,9\}$. Then

$$\{x : d_1(x) = e_1\} = [e_1/10, (e_1+1)/10),$$

hence d_1 is Borel measurable. Similarly,

$$\{x : d_2(x) = e_2\} = \bigcup_{e_1=0}^{9} \{x : d_1(x) = e_1,\ d_2(x) = e_2\} = \bigcup_{e_1=0}^{9} \left[\frac{e_1}{10} + \frac{e_2}{10^2}, \frac{e_1}{10} + \frac{e_2+1}{10^2}\right),$$

hence d_2 is Borel measurable. By induction, d_n is Borel measurable for all n. ◊

Almost Everywhere Properties

2.2.2 Proposition. *Let $f, g : X \to \overline{\mathbb{K}}$. If g is $\mathcal{F}$-measurable and $f = g$ a.e., then f is measurable with respect to the completion $\mathcal{F}_\mu$.*

Proof. By considering real and imaginary parts, we may assume that f and g are $\overline{\mathbb{R}}$-valued. Let $N = \{x : f(x) \neq g(x)\}$ and $t \in \mathbb{R}$. Since N, $N^c \in \mathcal{F}_\mu$,

$$\{f < t\} = \Big[\{g < t\} \cap N^c\Big] \cup \Big[\{f < t\} \cap N\Big] \in \mathcal{F}_\mu.$$

Similarly, the sets $\{f = \infty\}$ and $\{f = -\infty\}$ are members of $\mathcal{F}_\mu$. □

2.2.3 Corollary. *Let $f, g : \mathbb{R}^d \to \mathbb{C}$. If g is continuous and $f = g$ a.e., then f is Lebesgue-measurable.*

The function f in 2.2.3 need not be Borel measurable. For example, let $A \in \mathcal{M}(\mathbb{R}) \setminus \mathcal{B}(\mathbb{R})$ with $\lambda(A) = 0$ (§ 1.7) and take $f = \mathbf{1}_A$, $g \equiv 0$.

2.2.4 Proposition. *If $f : \mathbb{R}^d \to \mathbb{K}$ is continuous except on a set E of Lebesgue measure zero, then f is Lebesgue measurable.*

Proof. Let $U \subseteq \mathbb{K}$ be open. Then $f^{-1}(U) = A \cup B$, where $A := f^{-1}(U) \cap E$ and $B := f^{-1}(U) \cap E^c$. Since $A \subseteq E$ and $\lambda(E) = 0$, $A \in \mathcal{M}(\lambda^d)$. Since f is continuous at each point of E^c, $B = V \cap E^c$ for some open subset V of $\mathbb{R}^d$. Therefore, $B \in \mathcal{M}(\lambda^d)$ and so $f^{-1}(U) \in \mathcal{M}(\lambda^d)$. □

By the preceding proposition, a function with at most countably many discontinuities, in particular a monotone function, is Lebesgue measurable. In fact, the proof of the proposition shows that such a function is Borel measurable.

Note that a function that is continuous except on a set of measure zero is not necessarily equal a.e. to a continuous function (Ex. 2.19). Conversely, a function equal a.e. to a continuous function need not be continuous anywhere (Ex. 2.14).

Combinatorial and Limit Properties of Measurable Functions

The following proposition shows that measurable $\overline{\mathbb{R}}$-valued functions may be combined in standard ways to produce new measurable functions.

2.2.5 Proposition. *If $f, g : X \to \mathbb{K}$ are measurable and $c \in \mathbb{C}$, then $f + g$, fg, cf, $\overline{f}$, and $|f|$ are measurable. Moreover, if $\mathbb{K} = \mathbb{R}$, then $f \vee g$ and $f \wedge g$ are measurable.*

Proof. Let $F : \mathbb{K} \times \mathbb{K}$ be defined by $F(x, y) = x + y$. Then F is Borel measurable, hence $f + g = F(f, g)$ is measurable by 2.1.1. The proofs of the remaining assertions are similar. $\square$

The limit properties of measurable functions are given in the next results.

2.2.6 Theorem. *Let $f_n : X \to \overline{\mathbb{R}}$ $(n \in \mathbb{N})$ be measurable. Then $\sup_n f_n$, $\inf_n f_n$, $\overline{\lim}_n f_n$, and $\underline{\lim}_n f_n$ are measurable.*

Proof. The assertions follow immediately from the relations

$$\{\sup_n f_n \le t\} = \bigcap_n \{f_n \le t\}, \quad \inf_n f_n = -\sup_n(-f_n)$$

and

$$\underline{\lim}_n f_n = \sup_n \inf_{k \ge n} f_k, \quad \overline{\lim}_n f_n = -\underline{\lim}_n(-f_n). \qquad \square$$

2.2.7 Corollary. *Let $f_n : X \to \overline{\mathbb{K}}$ be measurable and let $f : X \to \overline{\mathbb{K}}$.*

(a) *If $f_n \to f$, then f is $\mathcal{F}$-measurable.*

(b) *If $f_n \to f$ a.e., then f is $\mathcal{F}_\mu$-measurable.*

Proof. By considering real and imaginary parts, we may assume that f_n and f are $\overline{\mathbb{R}}$-valued. Part (a) follows from the fact that $f = \underline{\lim}_n f_n$. For (b), let $N = \{x : \lim_n f_n(x) \ne f(x)\}$ and set $g_n = f_n \mathbf{1}_{N^c}$ and $g = f \mathbf{1}_{N^c}$. Then g_n is $\mathcal{F}_\mu$-measurable and $g_n \to g$, hence g is $\mathcal{F}_\mu$-measurable by part (a). Since $g = f$ a.e., f is $\mathcal{F}_\mu$-measurable. $\square$

2.2.8 Example. Let $f : X \times \mathbb{R} \to \mathbb{C}$ have the property that $f(x, t)$ is left continuous in t for each x and $\mathcal{F}$-measurable in x for each t. We show that f is $\mathcal{F} \otimes \mathcal{B}(\mathbb{R})$-measurable. For this, it suffices to take f real-valued.

For each n, the collection of intervals of the form $I_{k,n} := \big(k/n, (k+1)/n\big]$, $k \in \mathbb{Z}$, partitions $\mathbb{R}$. Define

$$f_n(x, t) = f\big(x, k/n\big), \quad t \in I_{n,k},\ k \in \mathbb{Z}, \quad x \in X.$$

Then f_n is $\mathcal{F} \otimes \mathcal{B}(\mathbb{R})$-measurable, as may be seen by writing

$$f_n(x, t) = \sum_{n \in \mathbb{Z}} f\big(x, k/n\big) \mathbf{1}_{I_{n,k}}(t)$$

and using appropriate combinatorial properties of measurability. Now let $t \in \mathbb{R}$ and $x \in X$. For each $n \in \mathbb{N}$, there exists a unique $k = k(t, n)$ such that $t \in I_{k,n}$. Since $0 < t - k/n \le 1/n$, by left continuity $\lim_n f_n(x, t) = \lim_n f\big(x, k/n\big) = f(x, t)$. Therefore, f is a limit of $\mathcal{F} \otimes \mathcal{B}(\mathbb{R})$-measurable functions and hence is $\mathcal{F} \otimes \mathcal{B}(\mathbb{R})$-measurable.

By ignoring X in the preceding argument, we see that a left continuous function on $\mathbb{R}$ is Borel measurable. By taking $(X, \mathcal{F}) = (\mathbb{R}, \mathcal{B}(\mathbb{R}))$, we see that a function on $\mathbb{R}^2$ that is left continuous in each variable separately is a Borel function. It follows by induction that function on $\mathbb{R}^d$ that is left continuous in each variable separately is Borel measurable. Of course, a similar result holds for separately right continuous functions. (In this regard, see Ex. 2.25.) $\lozenge$

Exercises

2.14 Give an example of a nowhere continuous function equal a.e. to a continuous function.

2.15 Show that if $\mathcal{F} \neq \mathcal{P}(X)$, then there exists a nonmeasurable function f such that $|f|$ is measurable.

2.16 Let $f_n : X \to \mathbb{R}$ be $\mathcal{F}$-measurable for every n. Prove that the following sets are $\mathcal{F}$-measurable:

(a) $\{x : \lim_n f_n(x) \text{ exists in } \overline{\mathbb{R}}\}$. (b) $\{x : \lim_n f_n(x) \text{ exists in } \mathbb{R}\}$.

2.17 Let $f,\, g : X \to \mathbb{R}$ be $\mathcal{F}$-measurable. Prove:

(a) If f is never zero, then $1/f$ is $\mathcal{F}$-measurable. (b) If $f > 0$, then f^g is $\mathcal{F}$-measurable.

2.18 Let $f,\, g : X \to \overline{\mathbb{R}}$ be $\mathcal{F}$-measurable. Prove that $\{f > g\} \in \mathcal{F}$.

2.19 Prove that $f = \mathbf{1}_{[0,1]}$ is not equal a.e. to a continuous function on $\mathbb{R}$. Show, however, that f is a pointwise limit of continuous functions f_n such that for each $\varepsilon > 0$, $\lambda\{|f_n - f| \geq \varepsilon\} \to 0$.

2.20 Define $f : (0,1) \to \mathbb{R}$ by

$$f(x) = \begin{cases} 0 & \text{if } x \text{ is irrational} \\ 1/n & \text{if } x = m/n, \text{ reduced.} \end{cases}$$

Show that f is continuous λ-a.e. and is equal λ-a.e. to a continuous function.

2.21 Let $f : \mathbb{R} \to \mathbb{R}$ be differentiable. Prove that f' is Borel measurable.

2.22 Let $f : X \times [a,b] \to \mathbb{R}$ such that $f(x,t)$ is $\mathcal{F}$-measurable in x for each x and continuous in t for each t. Show that the Riemann integral $\int_a^b f(x,t)\,dt$ is $\mathcal{F}$-measurable in x.

2.23 [↑2.2.1] For $x \in (0,1)$ define $f(x)$ to be first digit in the decimal expansion of x that is greater than 5 and $f(x) = 0$ if there is no such digit. (For definiteness, use decimal expansions that do not end in a sequence of 9's.) Also, define $g(x)$ to be the first time a digit is greater than 5, and $g(x) = \infty$ if there is no such digit. Prove that f and g are Borel measurable.

2.24 Show that the supremum of an uncountable family of Borel functions on $\mathbb{R}$ need not be Lebesgue measurable.

2.25 [↑2.9] Let $f : \mathbb{R}^d \to \mathbb{R}$ have the property that for each i, $f(x_1, \ldots, x_i, \ldots x_d)$ is either left continuous or right continuous in x_i when the other variables are fixed. Show that f is Borel measurable.

2.26 Let $\mathcal{F}$ be a σ-field on $\mathbb{R}^d$ such that every continuous function $f : \mathbb{R}^d \to \mathbb{R}$ that vanishes outside a bounded interval is $\mathcal{F}$-measurable. Prove that $\mathcal{B}(\mathbb{R}^d) \subseteq \mathcal{F}$.

2.27 Let μ be a finite measure on $\mathcal{B}(\mathbb{R}^d)$ and $A \in \mathcal{B}(\mathbb{R}^d)$. Define $f(x) = \mu(A + x)$, $x \in \mathbb{R}^d$. Show that f is Borel measurable. ⟦Assume first that A is closed and show that $A_t := \{f \geq t\}$ is closed.⟧

2.28 A function $f : \mathbb{R} \to \overline{\mathbb{R}}$ is said to be *upper* (*lower*) *semicontinuous at* x_0 if

$$f(x_0) \geq \overline{\lim_{x \to x_0}}\, f(x) := \lim_{r \to 0^+} \sup_{0<|x-x_0|<r} f(x) \quad \Big(f(x_0) \leq \underline{\lim_{x \to x_0}}\, f(x) := \lim_{r \to 0^+} \inf_{0<|x-x_0|<r} f(x)\Big).$$

If f is upper (lower) semicontinuous at each point of $\mathbb{R}$, then f is said to be *upper* (*lower*) *semicontinuous on* $\mathbb{R}$. Prove the following:

(a) f is upper semicontinuous at x_0 iff $-f$ is lower semicontinuous at x_0.

(b) f is upper (lower) semicontinuous on $\mathbb{R}$ iff $\{f < t\}$ ($\{f > t\}$) is open for every t. In particular, upper semicontinuous and lower semicontinuous functions are Borel measurable.

(c) If $f(x_0)$ is finite, then f is continuous at x_0 iff it is upper and lower semicontinuous at x_0.

(d) For arbitrary f, the functions $g(x) := \overline{\lim}_{t\to x} f(t)$ and $h(x) := \underline{\lim}_{t\to x} f(t)$ are, respectively, upper and lower semicontinuous on $\mathbb{R}$.

(e) The set $\{x : \lim_{t\to x} f(t)$ exists in $\overline{\mathbb{R}}\}$ is Borel measurable.

(f) The set $\{x : \lim_{t\to x} f(t)$ exists in $\mathbb{R}\}$ is Borel measurable.

2.29 Let $\mathbf{0} \in A \subseteq \mathbb{R}^d$. Define the "radius function" $f_A : \mathbb{R}^d \to \overline{\mathbb{R}}$ by $f_A(x) := \sup\{t \geq 0 : tx \in A\}$.

(a) Let $\mathbf{0} \in A_k$ for all k and $A_k \uparrow A$. Show that $f_{A_k} \uparrow f_A$.

(b) Show that if A is open, then f_A is positive and Borel measurable.

(c) Use (b) to show that if A is compact, then f_A is Borel measurable.

(d) Conclude that f_A is Borel measurable for any Borel set $A \ni \mathbf{0}$.

2.30 [$\uparrow$ 1.20] Define a topology on $[-\infty, \infty]$ with open sets $\mathcal{O}$ such that $\mathcal{B}([-\infty, \infty]) = \sigma(\mathcal{O})$.

2.3 Simple Functions

We show in this section that measurable functions are generated by measurable indicator functions. This will enable us to construct the Lebesgue integral from measurable sets, a fundamental feature of the Lebesgue theory.

Let $(X, \mathcal{F})$ be a measurable space. A measurable function $f : (X, \mathcal{F}) \to \mathbb{K}$ with finite range is called an **$\mathcal{F}$-simple function**. If there is no chance of confusion, we drop the prefix $\mathcal{F}$. An indicator function $\mathbf{1}_E$ is simple iff $E \in \mathcal{F}$ (Ex. 2.31). If f is simple and $g : \mathbb{K} \to \mathbb{K}$ is Borel measurable, then $f \circ g$, being finite-valued and measurable, is simple. Similarly, if $F : \mathbb{K}^d \to \mathbb{K}$ is Borel-measurable and the functions $f_1, \ldots, f_d$ are simple, then $F(f_1, \ldots, f_d)$ is simple. In particular, a linear combination of measurable indicator functions is simple. Conversely, every simple function f may be written (not necessarily uniquely) as a linear combination of measurable indicator functions. For example, if $a_1, \ldots, a_n \in \mathbb{K}$ are the distinct values of f, then

$$f = \sum_{j=1}^{n} a_j \mathbf{1}_{A_j}, \quad A_j := \{f = a_j\} \in \mathcal{F}. \tag{2.1}$$

The sum in (2.1) is called the **standard representation of f**.

A Fundamental Convergence Theorem

As we shall see in Chapter 3, the following theorem is one of the key ingredients in the Lebesgue theory of integration, allowing the transition from measure to integral.

2.3.1 Theorem. *Let $(X, \mathcal{F})$ be a measurable space and $f : (X, \mathcal{F}) \to \overline{\mathbb{K}}$ measurable.*

(a) *If $f \geq 0$, then there exists a sequence of nonnegative, simple functions f_n such that $f_n \uparrow f$ on X.*

(b) *In the general case, there exists a sequence of simple functions such that $f_n \to f$ on X and $|f_n| \leq |f|$ for all n.*

(c) *The sequence (f_n) in (b) converges uniformly on sets E on which f is bounded.*

Proof. (a) Let $f_0 = 0$ and for each $n \in \mathbb{N}$ define

$$f_n = \sum_{j=1}^{n2^n} \frac{j-1}{2^n} \mathbf{1}_{A_{n,j}} + n\mathbf{1}_{A_n}, \text{ where}$$

$$A_{n,j} = \{(j-1)2^{-n} \le f < j2^{-n}\}, \quad j = 1, 2, \ldots, n2^n, \text{ and } \quad A_n = \{f \ge n\}.$$

We show that $f_n(x) \uparrow f(x)$ for each $x \in X$. This is clear if $f(x) = \infty$, since then $f_n(x) = n$ for all n. Suppose $f(x) \in \mathbb{R}$ and let $n \in \mathbb{N}$. If $f(x) \ge n+1$, then $f_{n+1}(x) = n+1 > n = f_n(x)$. If $n \le f(x) < n+1$, then $x \in A_{n+1,j}$ for some j with $2^{j-1}/2^{n+1} \ge n$, hence $f_{n+1}(x) \ge n = f_n(x)$. Finally, if $f(x) < n$, then $(j-1)2^{-n} \le f(x) < j2^{-n}$ for some $1 \le j \le n2^n$, hence

$$\frac{2j-2}{2^{n+1}} \le f(x) < \frac{2j-1}{2^{n+1}} \quad \text{or} \quad \frac{2j-1}{2^{n+1}} \le f(x) < \frac{2j}{2^{n+1}}.$$

(See Figure 2.3.) In either case,

$$f_{n+1}(x) \ge \frac{2j-2}{2^{n+1}} = \frac{j-1}{2^n} = f_n(x).$$

Thus $f_n \uparrow$ on X. Since $0 \le f(x) - f_n(x) < 2^{-n}$ for all sufficiently large n, $f_n(x) \to f(x)$.

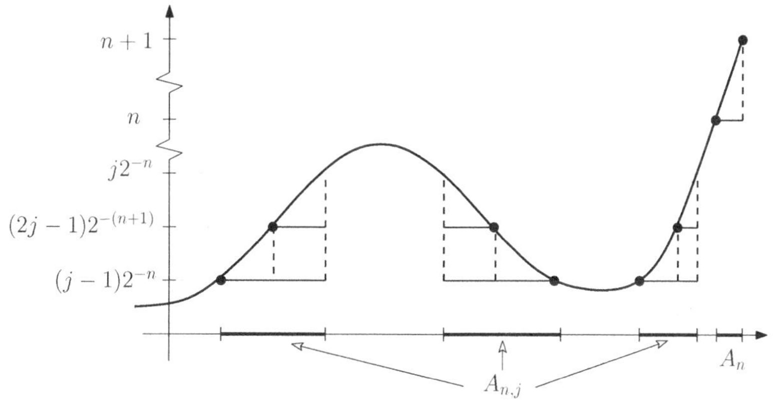

FIGURE 2.3: Components of f_n and f_{n+1}.

(b) If f is $\overline{\mathbb{R}}$-valued, apply (a) to f^+ and f^-. Suppose f is $\mathbb{C}$-valued. Let g_n and h_n be real-valued simple functions such that $g_n \to \operatorname{Re} f$, $h_n \to \operatorname{Im} f$, $|g_n| \le |\operatorname{Re} f|$, and $|h_n| \le |\operatorname{Im} f|$. Then $g_n + ih_n$ is a simple function, $g_n + ih_n \to f$, and

$$|g_n + ih_n|^2 = g_n^2 + h_n^2 \le (\operatorname{Re} f)^2 + (\operatorname{Im} f)^2 = |f|^2.$$

The proof of part (c) is left to the reader as an exercise (2.34). □

Applications of the Convergence Theorem

Theorem 2.3.1 is useful in establishing certain properties of measurable functions. The idea is to prove the property first for indicator functions, then for simple functions, then for nonnegative measurable functions via 2.3.1, then for $\overline{\mathbb{R}}$-valued functions using the identity $f = f^+ - f^-$, and finally for complex-valued functions by considering $\operatorname{Re} f$ and $\operatorname{Im} f$. The theorems in this subsection illustrate the technique.

2.3.2 Theorem. *Let X be a set, $(Y, \mathcal{G})$ a measurable space, $T : X \to Y$, and $\mathcal{F} = T^{-1}(\mathcal{G})$. If $f : X \to \mathbb{C}$ is $\mathcal{F}$-measurable, then there exists a $\mathcal{G}$-measurable function $g : Y \to \mathbb{C}$ such that $f = g \circ T$.*

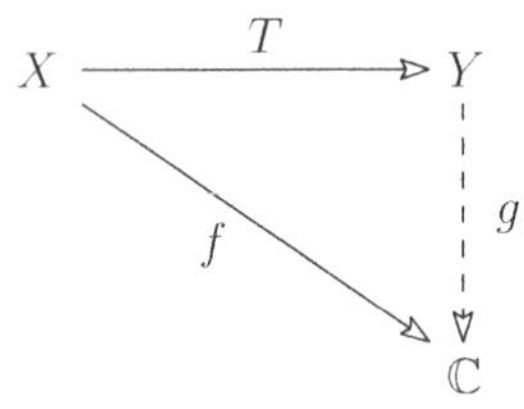

FIGURE 2.4: The mappings of 2.3.2.

Proof. Since $\mathbf{1}_{T^{-1}(A)} = \mathbf{1}_A \circ T$ $(A \in \mathcal{G})$, the assertion holds for $\mathcal{F}$-simple functions. If $f \geq 0$, let (f_n) be a sequence of nonnegative $\mathcal{F}$-simple functions such that $f_n \uparrow f$. For each n there exists a nonnegative $\mathcal{G}$-measurable function $h_n : Y \to \mathbb{R}$ such that $f_n = h_n \circ T$. Set $g_n = h_1 \vee \cdots \vee h_n$. Then g_n is $\mathcal{G}$-measurable and $g_n \uparrow$ on Y, hence $g := \lim_n g_n$ exists and is $\mathcal{G}$-measurable. Moreover, $h_n \uparrow$ on $T(X)$, hence $f_n = g_n \circ T$. Taking limits, we have $f = g \circ T$.

If f is real-valued, choose $\mathcal{G}$-measurable functions $g_1, g_2 : Y \to \mathbb{R}$ such that $f^+ = g_1 \circ T$ and $f^- = g_2 \circ T$. Then $f = (g_1 - g_2) \circ T$.

Finally, if f is complex-valued, choose $\mathcal{G}$-measurable functions $h_1, h_2 : Y \to \mathbb{R}$ such that $\operatorname{Re} f = h_1 \circ T$ and $\operatorname{Im} f = h_2 \circ T$. Then $f = (h_1 + ih_2) \circ T$. □

Hereafter, in arguments such as those in the preceding theorem we shall frequently omit the part of the proof that constitutes the transition from the nonnegative case to the complex case, this argument usually being straightforward.

2.3.3 Theorem. *Let $(X, \mathcal{F})$ be a measurable space and let $f : X \to \overline{\mathbb{K}}$ be $\mathcal{F}_\mu$-measurable. Then there exists an $\mathcal{F}$-measurable function g such that $f = g$ a.e.*

Proof. Let $f = \mathbf{1}_E$ $(E \in \mathcal{F}_\mu)$. Then $E = A \cup N$, where $A \in \mathcal{F}$, $N \subseteq M$, and $\mu(M) = 0$. Set $g = \mathbf{1}_A$. Then g is $\mathcal{F}$-measurable, and $\{f \neq g\} \subseteq N$, so $f = g$ a.e. Therefore, the assertion holds for indicator functions.

If f is a $\mathcal{F}_\mu$-simple function in standard form then, by the first paragraph, each of its terms is equal a.e. to an $\mathcal{F}$-measurable function. By considering a finite union of sets of measure zero we see that f has this property.

If f is nonnegative, there exists a sequence of nonnegative $\mathcal{F}_\mu$-simple functions f_n such that $f_n \to f$ on X. By the previous paragraph, for each n there exists an $\mathcal{F}$-measurable function g_n such that $f_n = g_n$ a.e. Let $N_n := \{x : f_n \neq g_n\}$ and $N := \bigcup_{n=1}^\infty N_n$. Then $N \in \mathcal{F}_\mu$, $\overline{\mu}(N) = 0$ and $f_n = g_n$ on N^c for all n. Let M denote the set of all x such that the sequence $(g_n(x))$ does not converge in $\overline{\mathbb{R}}$. Then $M \subseteq N$, and by Ex. 2.16, $M \in \mathcal{F}$. Let $g = \lim_n g_n \mathbf{1}_{M^c}$. Then g is measurable and $\{g \neq f\} \subseteq N$, so $g = f$ a.e. The general case $f : X \to \mathbb{K}$ follows by a standard argument. □

Exercises

2.31 Prove that $\mathbf{1}_E$ is measurable iff $E \in \mathcal{F}$.

2.32 Express the simple function $\lfloor x \rfloor \mathbf{1}_{[-n,n]}(x)$ in standard form.

2.33 [↓3.5] Let X be uncountable and let $\mathcal{F}$ the σ-field of countable or cocountable subsets of X. Show that a function $f : X \to \mathbb{C}$ is measurable iff f is constant on some cocountable set. ⟦Use 2.3.1.⟧

2.34 [↓4.2.1] Prove 2.3.1(c).

2.4 Convergence of Measurable Functions

Modes of Convergence

In this section we consider three important types of convergence of a sequence of functions on a measure space $(X, \mathcal{F}, \mu)$. A fourth type, L^p-convergence, is discussed in Chapter 4.

Let f, $f_n : (X, \mathcal{F}) \to \mathbb{C}$ be $\mathcal{F}$-measurable. The sequence (f_n) is said to **converge to** f

- **μ-almost everywhere**, written $f_n \overset{\text{a.e.}}{\to} f$, if $\mu\{\lim_n f_n \neq f\} = 0$.
- **in μ-measure**, written $f_n \overset{\mu}{\to} f$, if $\lim_n \mu\{|f_n - f)| \geq \varepsilon\} = 0 \ \ \forall\, \varepsilon > 0$.
- **μ-almost uniformly**, written $f_n \overset{\text{a.u.}}{\to} f$, if for each $\varepsilon > 0$ there exists a set A_ε in $\mathcal{F}$ such that $\mu(A_\varepsilon^c) < \varepsilon$ and $f_n \to f$ uniformly on A_ε.

For example, on $(\mathbb{R}, \mathcal{B}(\mathbb{R}), \lambda)$,

$$\begin{array}{lll} \mathbf{1}_{[0,1/n]} \overset{\text{a.e.}}{\to} 0, & \mathbf{1}_{[0,1/n]} \overset{\lambda}{\to} 0, & \mathbf{1}_{[0,1/n]} \overset{\text{a.u.}}{\to} 0, \\ \mathbf{1}_{[n,n+1]} \overset{\text{a.e.}}{\to} 0, & \mathbf{1}_{[n,n+1]} \overset{\lambda}{\not\to} 0, & \mathbf{1}_{[n,n+1]} \overset{\text{a.u.}}{\not\to} 0, \\ \mathbf{1}_{[n,n+1/n]} \overset{\text{a.e.}}{\to} 0, & \mathbf{1}_{[n,n+1/n]} \overset{\lambda}{\to} 0, & \mathbf{1}_{[n,n+1/n]} \overset{\text{a.u.}}{\not\to} 0\,. \end{array} \tag{2.2}$$

2.4.1 Proposition. *Let f, f_n, g, $g_n : (X, \mathcal{F}) \to \mathbb{K}$ be measurable, a, $b \in \mathbb{K}$, and let $\mathfrak{m}$ denote any of the three modes of convergence. Then*

(a) $f_n \overset{\mathfrak{m}}{\to} f$ *and* $g_n \overset{\mathfrak{m}}{\to} g \Rightarrow af_n + bg_n \overset{\mathfrak{m}}{\to} af + bg$.

(b) $f_n \overset{\mathfrak{m}}{\to} f \Rightarrow |f_n| \overset{\mathfrak{m}}{\to} |f|$ *and* $\overline{f}_n \overset{\mathfrak{m}}{\to} \overline{f}$.

(c) $f_n \overset{\mathfrak{m}}{\to} f$ *iff* $\operatorname{Re} f_n \overset{\mathfrak{m}}{\to} \operatorname{Re} f$ *and* $\operatorname{Im} f_n \overset{\mathfrak{m}}{\to} \operatorname{Im} f$.

(d) *If* $\mathbb{K} = \mathbb{R}$, *then* $f_n \overset{\mathfrak{m}}{\to} f$ *iff* $f_n^+ \overset{\mathfrak{m}}{\to} f^+$ *and* $f_n^- \overset{\mathfrak{m}}{\to} f^-$.

Proof. We prove the proposition for convergence in measure. Part (a) follows from

$$\big\{|(af + bg) - (af_n + bg_n)| \geq \varepsilon\big\} \subseteq \big\{|f - f_n| \geq \varepsilon/2(|a| + 1)\big\} \cup \big\{|g - g_n| \geq \varepsilon/2(|b| + 1)\big\},$$

which implies that

$$\mu\big\{|(af + bg) - (af_n + bg_n)| \geq \varepsilon\big\} \leq \mu\big\{|f - f_n| \geq \varepsilon/2(|a| + 1)\big\} + \mu\big\{|g - g_n| \geq \varepsilon/2(|b| + 1)\big\}.$$

For the first part of (b), use the inequality $\mu\{|\,|f_n| - |f|\,| \geq \varepsilon|\} \leq \mu\{|f_n - f| \geq \varepsilon|\}$. Part (c) follows from

$$\begin{gathered} \mu\{|\operatorname{Re} f_n - \operatorname{Re} f| \geq \varepsilon|\},\ \mu\{|\operatorname{Im} f_n - \operatorname{Im} f| \geq \varepsilon|\} \leq \mu\{|f_n - f| \geq \varepsilon\} \quad \text{and} \\ \mu\{|f_n - f| \geq \varepsilon|\} \leq \mu\{|\operatorname{Re} f_n - \operatorname{Re} f| \geq \varepsilon/2|\} + \mu\{|\operatorname{Im} f_n - \operatorname{Im} f| \geq \varepsilon/2|\}. \end{gathered}$$

The proof of (d) is similar, using the inequality $|x^+ - y^+| \leq |x - y|$. $\square$

Relationships Among the Modes of Convergence

The following theorems relate the three modes of convergence. The first shows that a.u. convergence is the strongest.

2.4.2 Theorem. *If* $f_n \overset{\text{a.u.}}{\to} f$*, then also* $f_n \overset{\mu}{\to} f$ *and* $f_n \overset{\text{a.e.}}{\to} f$*.*

Proof. For each $\delta > 0$, choose $A_\delta \in \mathcal{F}$ such that $\mu(A_\delta^c) < \delta$ and $f_n \to f$ uniformly on A_δ. Given $\varepsilon > 0$, choose m such that $|f - f_n| < \varepsilon$ on A_δ for all $n \geq m$. For such n, $\mu(|f_n - f| \geq \varepsilon) \leq \mu(A_\delta^c) < \delta$. Thus $f_n \overset{\mu}{\to} f$. Since $f_n \to f$ pointwise on $A := \bigcup_k A_{1/k}$ and $\mu(A^c) \leq 1/k$ for all k, $f_n \overset{\text{a.e.}}{\to} f$. □

Examples (2.2) show that $f_n \overset{\text{a.e.}}{\to} f$ does not necessarily imply $f_n \overset{\text{a.u.}}{\to} f$ or $f_n \overset{\mu}{\to} f$, and that $f_n \overset{\mu}{\to} f$ does not necessarily imply that $f_n \overset{\text{a.u.}}{\to} f$. The following example shows that $f_n \overset{\mu}{\to} f$ does not in general imply $f_n \overset{\text{a.e.}}{\to} f$.

2.4.3 Example. Let $f_1 = \mathbf{1}_{[0,1)}$ and for each $k \in \mathbb{N}$ set

$$f_{2^k+j} := \mathbf{1}_{[j/2^k,(j+1)/2^k)}, \quad 0 \leq j \leq 2^k - 1.$$

If $0 < \varepsilon < 1$ and $n = 2^k + j$, then

$$\lambda\{x \in [0,1) : f_n(x) \geq \varepsilon\} = \lambda[j/2^k, (j+1)/2^k) = 1/2^k,$$

hence $f_n \overset{\lambda}{\to} 0$ on $[0,1)$. On the other hand, for any $x \in [0,1)$, $f_n(x) = 1$ for infinitely many

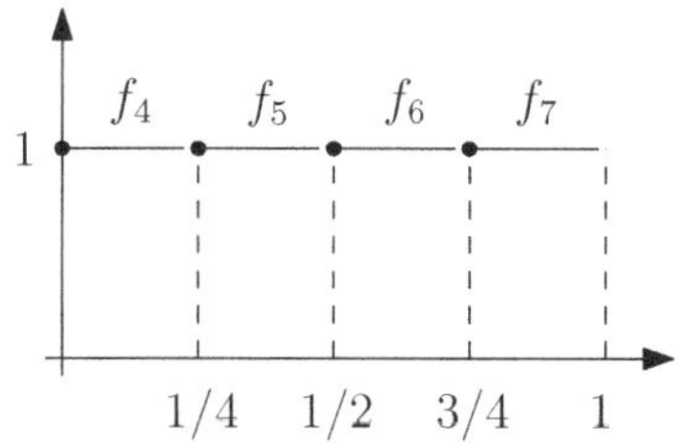

FIGURE 2.5: $f_n \overset{\mu}{\to} f \not\Rightarrow f_n \overset{\text{a.e}}{\to} f$.

n and $f_n(x) = 0$ for infinitely many n, so f_n cannot converge a.e. or a.u. ◊

While the sequence (f_n) in the last example does not converge a.e., there are *subsequences* that converge a.e. (for example, (f_{2^k})). This phenomenon holds generally:

2.4.4 Theorem. *If* $f_n \overset{\mu}{\to} f$*, then there exists a subsequence* (f_{n_k}) *such that* $f_{n_k} \overset{\text{a.u.}}{\to} f$*, hence also* $f_{n_k} \overset{\text{a.e.}}{\to} f$*.*

Proof. Since $\lim_n \mu\{|f_n - f| \geq \varepsilon\} = 0$, one may choose indices $n_1 < n_2 < \cdots$ such that for each k

$$\mu\{|f_n - f| \geq 1/2^k\} < 1/2^k \quad \text{for all } n \geq n_k. \tag{2.3}$$

Given ε, $\delta > 0$, choose $m \in \mathbb{N}$ such that $1/2^{m-1} < \min\{\varepsilon, \delta\}$ and set

$$B := \bigcup_{k \geq m} \{|f_{n_k} - f| \geq 1/2^k\}.$$

By (2.3), $\mu(B) \leq \sum_{k=m}^{\infty} 1/2^k = 1/2^{m-1} < \varepsilon$. Moreover, for $x \in B^c$ and $k \geq m$ we have $|f_{n_k}(x) - f(x)| < 1/2^k < \delta$. Therefore, $f_{n_k} \overset{\text{a.u}}{\to} f$. □

A converse of 2.4.2 holds, but with a restriction:

2.4.5 Egoroff's Theorem. *If $\mu(X) < \infty$ and $f_n \overset{a.e.}{\to} f$, then $f_n \overset{a.u.}{\to} f$, hence also $f_n \overset{\mu}{\to} f$.*

Proof. Let $E \in \mathcal{F}$ such that $\mu(E^c) = 0$ and $f_n \to f$ on E. For $x \in X$, define

$$g_n(x) = \sup_{j \geq n} |f(x) - f_j(x)|.$$

Then g_n is $\mathcal{F}$-measurable and $g_n \downarrow 0$ on E. For $n, k \in \mathbb{N}$, set

$$E_{n,k} = E \cap \{g_n < 1/k\}.$$

For each k, $E_{n,k} \uparrow_n E$, hence $\mu(E_{n,k}) \uparrow_n \mu(E) = \mu(X)$. Given $\varepsilon > 0$, we may therefore construct an increasing sequence $n_k \in \mathbb{N}$ such that

$$\mu(E^c_{n_k,k}) = \mu(X) - \mu(E_{n_k,k}) < \varepsilon/2^k.$$

Set $A_\varepsilon = \bigcap_k E_{n_k,k}$. Then $\mu(A^c_\varepsilon) < \varepsilon$ and $g_{n_k}(x) < 1/k$ for $x \in A_\varepsilon$. Thus $f_{n_k} \overset{\text{a.u.}}{\to} f$. □

From the above results we see that a.e. convergence and a.u. convergence are equivalent in a finite measure space. The sequence $f_n = \mathbf{1}_{[n,n+1]}$ on $\mathbb{R}$ shows that this is not true in general measure spaces.

Exercises

In these exercises, $(X, \mathcal{F}, \mu)$ denotes an arbitrary measure space.

2.35 Discuss the convergence behavior of $f_n(x) = x^n \mathbf{1}_{[0,1]}$ on $(\mathbb{R}, \mathcal{B}(\mathbb{R}), \lambda)$.

2.36 Let $E_n \in \mathcal{F}$ and let f be measurable. Suppose that $\mathbf{1}_{E_n} \overset{\mu}{\to} f$. Show that $f = \mathbf{1}_E$ a.e. for some $E \in \mathcal{F}$.

2.37 Let $E_n \in \mathcal{F}$, $A := \overline{\lim}_n E_n$, and $B := \underline{\lim}_n E_n$. Show that $\mathbf{1}_{E_n} \overset{\text{a.e.}}{\to} f$ for some f iff $\mu(A \setminus B) = 0$.

2.38 Let $f, f_n, g : X \to \mathbb{C}$ be measurable, $f_n \overset{\mu}{\to} f$ and $f_n \overset{\mu}{\to} g$. Show that $f = g$ a.e.

2.39 Let $f, f_n : X \to \mathbb{R}$ be measurable and $f_n \overset{\mu}{\to} f$. Show that if $f_n \uparrow$ then $f_n \overset{\text{a.e.}}{\to} f$.

2.40 Let $f, f_n : X \to \mathbb{C}$ be measurable. Show that $f_n \overset{\mu}{\to} f$ iff for each $\varepsilon > 0$ there exists m such that $\mu\{|f - f_n| \geq \varepsilon\} < \varepsilon$ for all $n \geq m$.

2.41 Let $f_n : X \to \mathbb{C}$ be measurable. Show that $f_n \overset{\text{a.e.}}{\to} f$ for some $\mathcal{F}_\mu$-measurable f iff $g_{m,n} := f_m - f_n \to 0$ a.e. as $m, n \to \infty$.

2.42 [↑ 2.41] Let $\mu(X) < \infty$, $a_n > 0$, and $\sum_n a_n < \infty$. Let $f_n : X \to \mathbb{C}$ be measurable and set $A_n := \{|f_n - f_{n+1}| \geq a_n\}$. Show that if $\sum_n \mu(A_n) < \infty$, then $f_n \overset{\text{a.e.}}{\to} f$ for some function $f : X \to \mathbb{C}$. ⟦By 1.37, $\mu(\overline{\lim}_n A_n) = 0$. Consider $\sum_{k=n}^{m-1}[f_k(x) - f_{k+1}(x)]$. ⟧

2.43 [↑ 2.42] Let $\mu(X) < \infty$ and $f_n : X \to \mathbb{C}$ measurable. Show that $f_n \overset{\mu}{\to} f$ for some $f : X \to \mathbb{C}$ iff $f_m - f_n \overset{\mu}{\to} 0$ as $m, n \to \infty$. ⟦For the sufficiency, modify the proof of 2.4.4 to obtain a strictly increasing sequence of positive integers n_k such that $\mu\{|f_n - f_m| \geq 1/2^k\} < 1/2^k$ for all $m, n \geq n_k$. ⟧

2.44 (Frechét). Let $\mu(X) < \infty$. Define

$$\rho(f, g) = \inf_{r>0} \left(r + \mu\{x : |f(x) - g(x)| \geq r\}\right)$$

Show that if functions that are equal μ-a.e are identified, then ρ becomes a metric on the space $L^0 = L^0(X, \mathcal{F}, \mu)$ of all measurable functions on X. Show also that $\rho(f, f_n) \to 0$ iff $f_n \overset{\mu}{\to} f$. Thus, by 2.43, the metric space is complete.

2.45 Let $\mu(X) < \infty$ and let $f, f_n : X \to \mathbb{C}$ be measurable. Set $A_n = \{|f_n - f| \geq a_n\}$, where $a_n > 0$ and $a_n \to 0$. Show that if $\sum_n \mu(A_n) < \infty$, then $f_n \stackrel{\text{a.e.}}{\to} f$

2.46 Let $\mu(X) < \infty$, let $f, g, f_n, g_n : X \to \mathbb{C}$ be measurable, $f_n \stackrel{\mu}{\to} f$, and $g_n \stackrel{\mu}{\to} g$. Show that $f_n g_n \stackrel{\mu}{\to} fg$.

2.47 Let $\mu(X) < \infty$, $f, f_n : X \to \mathbb{K}$ measurable, and $g : \mathbb{K} \to \mathbb{K}$ continuous. Show that $f_n \stackrel{\mu}{\to} f \Rightarrow g \circ f_n \stackrel{\mu}{\to} g \circ f$.

2.48 Let μ and ν be finite measures on $(X, \mathcal{F})$ with the same sets of measure zero and let $f, f_n : X \to \mathbb{C}$ be measurable. Show that $f_n \stackrel{\mu}{\to} f$ iff $f_n \stackrel{\nu}{\to} f$.

Chapter 3

Integration

In this chapter we construct the general Lebesgue integral. The construction proceeds in stages. The integral is first defined on the class of nonnegative simple functions and then extended to nonnegative measurable functions and ultimately to complex measurable functions. The basic properties of the integral are also developed in this chapter. Additional properties are discussed in subsequent chapters.

Throughout the chapter $(X, \mathcal{F}, \mu)$ denotes an arbitrary measure space.

3.1 Construction of the Integral

The **integral of a function** f **with respect to** μ is denoted variously by

$$\int f = \int f\, d\mu = \int_X f(x)\, d\mu(x) = \int_X f(x)\mu(dx) = \int f\, dF,$$

the last notation if μ is a Lebesgue-Stieltjes measure on $\mathbb{R}$ with distribution function F. The construction of the integral begins with nonnegative simple functions.

Integral of a Nonnegative Simple Function

Let f be a nonnegative simple function with standard representation

$$f = \sum_{j=1}^{m} a_j \mathbf{1}_{A_j}, \quad A_j := \{f = a_j\} \in \mathcal{F}.$$

The integral of f is then defined as

$$\int f\, d\mu := \sum_{j=1}^{m} a_j \mu(A_j).$$

Note that the above sum may contain terms of the form $a \cdot \infty$, where $a \in [0, \infty)$. Such terms have value either ∞ or 0, depending on whether $a > 0$ or $a = 0$ (see §0.1). In particular, the integral of the identically zero function is $0 \cdot \mu(X) = 0$, whether or not $\mu(X)$ is finite.

The following lemma summarizes the elementary properties of the integral of nonnegative simple functions. These will be used later to obtain analogous properties of the general integral.

3.1.1 Lemma. *Let f, g be nonnegative simple functions and $a \ge 0$. Then*

(a) $\displaystyle\int af\,d\mu = a\int f\,d\mu.$ **(b)** $\displaystyle\int (f+g)\,d\mu = \int f\,d\mu + \int g\,d\mu.$

(c) $\displaystyle\int f\,d\mu \le \int g\,d\mu$ *if* $f \le g$ *a.e.* **(d)** $\displaystyle\int f\,d\mu = \int g\,d\mu$ *if* $f = g$ *a.e.*

Proof. Part (a) is immediate from the definition of the integral, and (d) follows from (c). To prove (b), let f and g have standard representations

$$f = \sum_{i=1}^{m} a_i \mathbf{1}_{A_i} \quad \text{and} \quad g = \sum_{j=1}^{n} b_j \mathbf{1}_{B_j}.$$

Since $X = \bigcup_{i=1}^m A_i = \bigcup_{j=1}^n B_j$ (disjoint), we have $\mu(A_i) = \sum_{j=1}^n \mu(A_i \cap B_j)$ and $\mu(B_j) = \sum_{i=1}^m \mu(A_i \cap B_j)$, hence

$$\int f = \sum_{i=1}^{m} a_i\mu(A_i) = \sum_{i,j} a_i\mu(A_i \cap B_j) \quad \text{and} \quad \int g = \sum_{j=1}^{n} b_j\mu(B_j) = \sum_{i,j} b_j\mu(A_i \cap B_j). \qquad (\dagger)$$

Now let $c_1, \ldots, c_p$ be the distinct values of $f+g$ and set

$$C_k = \{f + g = c_k\}, \quad k = 1, \ldots, p.$$

Then

$$f + g = \sum_{k=1}^{p} c_k \mathbf{1}_{C_k} \quad \text{and} \quad C_k = \bigcup_{\{(i,j):a_i+b_j=c_k\}} A_i \cap B_j \quad \text{(disjoint)}$$

and so

$$\int (f+g) = \sum_{k=1}^{p} c_k\mu(C_k) = \sum_{k=1}^{p} c_k \sum_{a_i+b_j=c_k} \mu(A_i \cap B_j) = \sum_{i,j}(a_i + b_j)\mu(A_i \cap B_j) = \int f + \int g,$$

the last equality by (†).

For (c), let $E = \{f \le g\}$. Then $\mu(E^c) = 0$ and $a_i \le b_j$ for all i, j for which $A_i \cap B_j \cap E \ne \emptyset$. From (†) and the equalities

$$\mu(A_i \cap B_j) = \mu(A_i \cap B_j \cap E) + \mu(A_i \cap B_j \cap E^c) = \mu(A_i \cap B_j \cap E),$$

we have

$$\int f = \sum_{i=1}^{m}\sum_{j=1}^{n} a_i\mu(A_i \cap B_j \cap E) \le \sum_{j=1}^{n}\sum_{i=1}^{m} b_j\mu(A_i \cap B_j \cap E) = \int g. \qquad \square$$

Integral of a Real-Valued Function

For a measurable function $f : X \to [0, \infty]$, define

$$\int f\,d\mu := \sup\left\{\int f_s\,d\mu : 0 \le f_s \le f,\ f_s \text{ simple}\right\}. \tag{3.1}$$

Note that the integral is nonnegative and could be infinite. (For an extreme example, consider the measure μ on $\mathcal{P}(X)$ that assigns ∞ to every nonempty set. Then $\int f\,d\mu = \infty$ for all nonnegative functions except the identically zero function.)

The integral of a measurable function $f : X \to \overline{\mathbb{R}}$ is defined as

$$\int f\,d\mu := \int f^+\,d\mu - \int f^-\,d\mu,$$

provided at least one of the integrals on the right is finite. If both $\int f^+\,d\mu$ and $\int f^-\,d\mu$ are finite, then f is said to be **integrable**.

Integral of a Complex-Valued Function

To extend the integral to the complex case, recall that a complex-valued function f is measurable iff both $\operatorname{Re} f$ and $\operatorname{Im} f$ are measurable. Such a function is declared to be **integrable** if both $\operatorname{Re} f$ and $\operatorname{Im} f$ are integrable, in which case we define

$$\int f\,d\mu = \int \operatorname{Re} f\,d\mu + i\int \operatorname{Im} f\,d\mu,$$

that is,

$$\operatorname{Re}\int f\,d\mu = \int \operatorname{Re} f\,d\mu \quad\text{and}\quad \operatorname{Im}\int f\,d\mu = \int \operatorname{Im} f\,d\mu.$$

It follows that

$$\int \overline{f}\,d\mu = \overline{\int f\,d\mu}.$$

We have now constructed the integral with respect to μ on the class of all (suitably restricted) measurable functions $f : X \to \overline{\mathbb{K}}$. The special cases of the integral with respect to Lebesgue measure on $\mathbb{R}^d$ and Lebesgue-Stieltjes measures on $\mathbb{R}^d$ are important examples. Here is another example:

3.1.2 Example. Let $x \in X$ and let δ_x be the Dirac measure defined in 1.3.3(d). Then

$$\int f\,d\delta_x = f(x) \tag{3.2}$$

for every $\mathcal{F}$-measurable function $f : X \to \overline{\mathbb{K}}$. Indeed, this clearly holds for indicator functions f, and, by 3.1.1(a) and (b), it holds for nonnegative $\mathcal{F}$-simple functions. If $f \geq 0$, then

$$\int f\,d\delta_x = \sup\Big\{f_s(x) : 0 \leq f_s \leq f,\ f_s \text{ simple}\Big\} = f(x),$$

the last equality by 2.3.1. For the general real-valued case, use the positive and negative parts of f. For the complex case, consider the real and imaginary parts of f. ◊

Integral over a Measurable Set

Let $f : X \to \mathbb{C}$ be measurable and $E \in \mathcal{F}$. The **integral of f on E** is defined by

$$\int_E f\,d\mu := \int f\cdot \mathbf{1}_E\,d\mu$$

if the right side is defined. If $f \cdot \mathbf{1}_E$ is integrable, then f is said to be **integrable on E**.

3.1.3 Remark. It is worth noting that the integral of f on E is simply the integral of $f\big|_E$ with respect to the measure $\nu := \mu\big|_{\mathcal{F}\cap E}$ on E, that is,

$$\int_E f\,d\mu = \int f\big|_E\,d\nu. \tag{3.3}$$

To see this, first take $f = \mathbf{1}_A$, $A \in \mathcal{F}$. Then the left side of (3.3) is simply $\mu(A\cap E)$, and since $\mathbf{1}_A\big|_E$ is the indicator function of $E\cap A$ on the domain E, the right side is $\nu(A\cap E) = \mu(A\cap E)$. Thus (3.3) holds for indicator functions, hence for nonnegative $\mathcal{F}$-simple functions. Taking suprema over integrals of simple functions shows that the equation holds for nonnegative measurable functions, hence for arbitrary measurable $\overline{\mathbb{R}}$-valued functions via $f = f^+ - f^-$, and finally for measurable $\mathbb{C}$-valued functions using $f = \operatorname{Re} f + i\operatorname{Im} f$. ◊

The preceding remark implies that general properties of integrals $\int f\,d\mu$ are immediately valid for $\int_E f\,d\mu$—no special argument is necessary.

3.2 Basic Properties of the Integral

Almost Everywhere Properties

3.2.1 Proposition. *If $f, g : X \to \overline{\mathbb{R}}$ are measurable, $f \le g$ a.e., and $\int f\,d\mu$, $\int g\,d\mu$ are defined, then $\int f\,d\mu \le \int g\,d\mu$. In particular, if $f \ge 0$ and g is integrable, then f is integrable*

Proof. Assume first that $f, g \ge 0$. Let f_s be simple with $0 \le f_s \le f$ and set $g_s := \mathbf{1}_E f_s$, where $E := \{f \le g\}$. Then g_s is simple, $f_s = g_s$ a.e., and $0 \le g_s \le \mathbf{1}_E f \le \mathbf{1}_E g \le g$. By 3.1.1, $\int f_s\,d\mu = \int g_s\,d\mu \le \int g\,d\mu$. Since f_s was arbitrary, $\int f\,d\mu \le \int g\,d\mu$.

In the general case, $f^+ \le g^+$ and $f^- \ge g^-$ a.e., hence, by the first paragraph,

$$\int f\,d\mu = \int f^+\,d\mu - \int f^-\,d\mu \le \int g^+\,d\mu - \int g^-\,d\mu = \int g\,d\mu. \qquad \square$$

Applying the proposition to the real and imaginary parts of f and g, we have

3.2.2 Corollary. *Let $f, g : X \to \overline{\mathbb{K}}$ be measurable, f integrable, and $f = g$ a.e. Then g is integrable and $\int f\,d\mu = \int g\,d\mu$.*

3.2.3 Remark. In view of the last corollary, it makes sense to integrate functions that are defined only a.e., that is, defined on a set E with $\mu(E^c) = 0$. More precisely, such a function h is said to be **integrable** if it has an integrable extension f to X, in which case we define $\int h\,du := \int f\,du$. By the corollary, the integral is well-defined, that is, does not depend on the particular extension of h. ◊

3.2.4 Proposition. *If $f : X \to \overline{\mathbb{R}}$ is integrable, then f is finite a.e.*

Proof. Suppose first that $f \ge 0$. Let $A := \{f = \infty\}$. Since $n\mathbf{1}_A \le f$, $\mu(A) \le n^{-1} \int f\,d\mu$. Letting $n \to \infty$ shows that $\mu(A) = 0$. In the general case, apply the preceding to f^+ and f^- to conclude that $\mu(|f| = \infty) = 0$. $\square$

3.2.5 Proposition. *Let $f \ge 0$ be measurable. Then $\int f\,d\mu = 0$ iff $f = 0$ a.e.*

Proof. The sufficiency follows from 3.2.2. For the necessity, suppose that $\int f\,d\mu = 0$ and let $B := \{f > 0\}$ and $B_n := \{f \ge 1/n\}$. Then $B = \bigcup_{n=1}^{\infty} B_n$ and $n^{-1}\mathbf{1}_{B_n} \le f\mathbf{1}_{B_n} \le f$, so $\mu(B_n) \le n \int f\,d\mu = 0$ for all n. By countable subadditivity, $\mu(B) = 0$, that is, $f \le 0$ a.e. $\square$

By 3.2.1, if $f \ge 0$ then $\int_A f\,d\mu \ge 0$ for all $A \in \mathcal{F}$. Here is a converse:

3.2.6 Proposition. *Let $f : X \to \overline{\mathbb{R}}$ be measurable and let $\int_A f\,d\mu$ be defined for all $A \in \mathcal{F}$.*

(a) *If $\int_A f\,d\mu \ge 0$ for all $A \in \mathcal{F}$, then $f \ge 0$ a.e.*

(b) *If $\int_A f\,d\mu = 0$ for all $A \in \mathcal{F}$, then $f = 0$ a.e.*

Proof. Part (b) follows from part (a). To prove (a), let $A_n = \{f \le -n^{-1}\}$ and $A = \{f < 0\}$. Then $A = \bigcup_{n=1}^{\infty} A_n$ and $\mathbf{1}_{A_n} \le -nf\mathbf{1}_{A_n}$, hence $\mu(A_n) \le -n \int_{A_n} f\,d\mu$. Since $\int_{A_n} f \ge 0$, $\mu(A_n) = 0$ and so $\mu(A) = 0$. $\square$

Monotone Convergence Theorem

The following result is one of the key theorems in integration theory, underlying many of the deeper properties of the integral. A generalization is given in §3.4.

3.2.7 Theorem. *If (f_n) is a sequence of nonnegative measurable functions such that $f_n \uparrow f$ on X, then*

$$\int f\,d\mu = \lim_n \int f_n\,d\mu.$$

Proof. By 2.2.6, f is measurable. Moreover, $\int f_n\,d\mu \le \int f_{n+1}\,d\mu \le \int f\,d\mu$ for all n, hence $L := \lim \int f_n\,d\mu$ exists in $\overline{\mathbb{R}}$ and $L \le \int f\,d\mu$. For the reverse inequality, it suffices to show that $\int g\,d\mu \le L$ for any simple function g with $0 \le g \le f$. Let g have the standard representation $\sum_{j=1}^m a_j \mathbf{1}_{A_j}$ and set $E_n := \{f_n \ge rg\}$, where $0 < r < 1$. Then

$$f_n \ge rg\mathbf{1}_{E_n} = r\sum_{j=1}^m a_j \mathbf{1}_{E_n \cap A_j}$$

and so

$$\int f_n\,d\mu \ge r\sum_{j=1}^m a_j \mu(E_n \cap A_j).$$

Letting $n \to \infty$ in the last inequality and noting that $E_n \uparrow X$, we obtain

$$L \ge r\sum_{j=1}^m a_j \mu(A_j) = r\int g\,d\mu.$$

Finally, letting $r \uparrow 1$ we see that $L \ge \int g\,d\mu$. □

Linearity of the Integral

To simplify the development in this subsection, we divide the verification of linearity into three parts. The first part treats the nonnegative case; the second and third parts treat the real and complex cases, respectively.

3.2.8 Theorem. *Let $f, g : X \to [0, \infty]$ be measurable and $a, b \in \mathbb{R}^+$. Then*

$$\int (af + bg)\,d\mu = a\int f\,d\mu + b\int g\,d\mu.$$

In particular, if f and g are integrable then so is $af + bg$.

Proof. Choose sequences (f_n) and (g_n) of nonnegative simple functions such that $f_n \uparrow f$ and $g_n \uparrow g$. Then $af_n + bg_n \uparrow af + bg$, hence, by 3.2.7 and 3.1.1,

$$\int (af + bg) = \lim_n \int (af_n + bg_n) = a\lim_n \int f_n + b\lim_n \int g_n = a\int f + b\int g. \qquad \square$$

3.2.9 Corollary. *If g_n is measurable and nonnegative for every n, then*

$$\int \left(\sum_{n=1}^\infty g_n\right) d\mu = \sum_{n=1}^\infty \int g_n\,d\mu.$$

Proof. Let $f_n = \sum_{j=1}^n g_j$ and $f = \sum_{n=1}^\infty g_n$. Then $0 \le f_n \uparrow f$, so by the monotone convergence theorem and linearity,

$$\int f = \lim_n \int f_n = \lim_n \sum_{j=1}^n \int g_j = \sum_{j=1}^\infty \int g_j. \qquad \square$$

3.2.10 Corollary. *Let $h \geq 0$ be measurable. Define a set function ν on $\mathcal{F}$ by*

$$\nu(E) := \int_E h\,d\mu, \quad E \in \mathcal{F}.$$

Then ν is a measure on $\mathcal{F}$.

Proof. For countable additivity, apply 3.2.9 to $g_n = \mathbf{1}_{E_n} \cdot h$. □

3.2.11 Corollary. *Let $f, g : X \to \overline{\mathbb{K}}$ be measurable.*

(a) *f is integrable iff $|f|$ is integrable.*

(b) *If f is integrable and $|g| \leq |f|$, then g is integrable.*

(c) *If f is integrable and $E \in \mathcal{F}$, then f is integrable on E.*

Proof. (a) Suppose first that f is $\overline{\mathbb{R}}$-valued. If f is integrable, then, by definition, f^+ and f^- are integrable, hence by the theorem $|f| = f^+ + f^-$ is integrable. Conversely, if $|f|$ is integrable, then $0 \leq \int f^\pm\,d\mu \leq \int |f|\,d\mu$, hence f^+ and f^- are integrable.

Now let f be $\mathbb{C}$-valued. If f is integrable then by definition $\operatorname{Re} f$ and $\operatorname{Im} f$ are integrable. By the first paragraph, $|\operatorname{Re} f|$ and $|\operatorname{Im} f|$ are integrable, hence, by the theorem $|\operatorname{Re} f| + |\operatorname{Im} f|$ is integrable. Since $|f| \leq |\operatorname{Re} f| + |\operatorname{Im} f|$, $|f|$ is integrable. This proves the necessity of (a). A similar argument shows that if $|f|$ is integrable, then $\operatorname{Re} f$ and $\operatorname{Im} f$ are integrable, verifying the sufficiency.

(b) By part (a), $|f|$ is integrable. The inequality $|g| \leq |f|$ then implies that $|g|$ is integrable. By (a) again, g is integrable.

(c) This follows from (b), since $|f\mathbf{1}_E| \leq |f|$. □

We may now prove linearity for the real-valued case:

3.2.12 Theorem. *Let $f, g : X \to \mathbb{R}$ be measurable, g integrable, and $a, b \in \mathbb{R}$. If $\int f\,du$ exists, then $\int (af + bg)\,d\mu$ exists and*

$$\int (af + bg)\,d\mu = a \int f\,d\mu + b \int g\,d\mu.$$

Proof. Suppose first that f is integrable. The identity

$$(f+g)^+ + f^- + g^- = (f+g)^- + f^+ + g^+$$

and 3.2.8 imply that

$$\int (f+g)^+ + \int f^- + \int g^- = \int (f+g)^- + \int f^+ + \int g^+.$$

Since these terms are finite we may rearrange them to obtain

$$\int (f+g) = \int (f+g)^+ - \int (f+g)^- = \int f^+ - \int f^- + \int g^+ - \int g^- = \int f + \int g,$$

proving additivity.

If $a \geq 0$, then $(af)^+ = af^+$ and $(af)^- = af^-$, hence, by 3.2.8,

$$\int af = \int (af)^+ - \int (af)^- = a \int f^+ - a \int f^- = a \int f.$$

Also, since $(-f)^+ = f^-$ and $(-f)^- = f^+$,

$$\int(-f) = \int(-f)^+ - \int(-f)^- = \int f^- - \int f^+ = -\int f.$$

Therefore, if $a < 0$,

$$\int af = \int(-a)(-f) = -a\int(-f) = a\int f.$$

This proves linearity if both f and g are integrable.

Now suppose that f is not integrable but that the integral of f exists. There are two possibilities:

(i) $\int f^- < \infty$ and $\int f^+ = \infty$.

(ii) $\int f^+ < \infty$ and $\int f^- = \infty$.

Suppose (i) holds. Since

$$\int(f+g)^- \le \int(f^- + g^-) < \infty,$$

$\int(f+g)$ is defined. If $\int(f+g)^+ < \infty$, then $(f+g)$ would be integrable, hence, by the first part of the proof, so would $f^+ = (f+g) + f^- - g$, contrary to our assumption. Therefore,

$$\int(f+g) = \infty = \int f + \int g.$$

Also, if $a > 0$ $(a < 0)$, then $\int af$ and $a\int f$ both equal ∞ $(-\infty)$, hence $\int af = a\int f$. This proves linearity in case (i). Case (ii) is similar (or apply case (i) to $-f$). □

Linearity in the complex case follows from the preceding theorem by considering real and imaginary parts of f:

3.2.13 Theorem. *Let $f, g : X \to \mathbb{C}$ be integrable and let $\alpha, \beta \in \mathbb{C}$. Then $\alpha f + \beta g$ is integrable and*

$$\int(\alpha f + \beta g)\, d\mu = \alpha\int f\, d\mu + \beta\int g\, d\mu.$$

Proof. By 3.2.8, $|\alpha|\,|f| + |\beta|\,|g|$ is integrable. Since $|\alpha f + \beta g| \le |\alpha|\,|f| + |\beta|\,|g|$, by 3.2.11 $\alpha f + \beta g$ is integrable. Now let $\alpha = a + i\,b$ and set $f_r = \operatorname{Re} f$ and $f_i = \operatorname{Im} f$. Then

$$\alpha f = a\, f_r - b\, f_i + i\,[b\, f_r + a\, f_i],$$

hence, by 3.2.12 and the definition of the complex integral,

$$\int \alpha f = a\int f_r - b\int f_i + i\left(b\int f_r + a\int f_i\right) = (a + i\,b)\left(\int f_r + i\int f_i\right) = \alpha\int f.$$

A similar argument, using 3.2.12 again, shows that $\int(f+g) = \int f + \int g$. □

3.2.14 Corollary. *If $f : X \to \mathbb{C}$ is integrable, then $\left|\int f\, d\mu\right| \le \int |f|\, d\mu$.*

Proof. Write $\int f\, d\mu$ in polar form $e^{i\,\theta}\left|\int f\, d\mu\right|$, so that

$$\left|\int f\, d\mu\right| = e^{-i\theta}\int f\, d\mu = \int e^{-i\theta} f\, d\mu = \int \operatorname{Re}(e^{-i\theta} f)\, d\mu \le \int |f|\, d\mu,$$

the third equality because $\int e^{-i\theta} f\, d\mu$ is real. □

Integration Against an Image Measure

Let $(Y, \mathcal{G})$ be a measurable space and $T : (X, \mathcal{F}) \to (Y, \mathcal{G})$ a measurable transformation. The **image of** μ **under** T is the measure $T(\mu)$ on $(Y, \mathcal{G})$ defined by

$$T(\mu)(E) = \mu\big(T^{-1}(E)\big), \quad E \in \mathcal{G}.$$

Image measures occur frequently in probability theory as distributions of random variables (see §18.1).

3.2.15 Theorem. *Let $g : Y \to \overline{\mathbb{K}}$ be $\mathcal{G}$-measurable. Then*

$$\int_Y g\, dT(\mu) = \int_X g \circ T\, d\mu \tag{3.4}$$

in the sense that if one side is defined, then so is the other and equality holds.

Proof. Since $\mathbf{1}_{T^{-1}(A)} = \mathbf{1}_A \circ T$, (3.4) holds for indicator functions g, hence by linearity for simple functions. Taking a sequence of nonnegative simple functions increasing to g and applying the monotone convergence theorem yields (3.4) for nonnegative measurable functions g. The general case follows by standard arguments. □

Applying the theorem to the transformations $x \to x + z$ and $x \to rx$ on $\mathbb{R}^d$, we have

3.2.16 Corollary. *The following are valid in the sense that if one side of an equation is defined, then so is the other and equality holds.*

$$\int f(x+y)\, d\lambda^d(x) = \int f(x)\, d\lambda^d(x) \quad \text{and} \quad \int f(rx)\, d\lambda^d(x) = |r|^{-d} \int f(x)\, d\lambda^d(x). \tag{3.5}$$

Properties (3.5) express, respectively, the **translation invariance** and **dilation** properties of the Lebesgue integral. The special case $r = -1$ gives the **reflection invariance** of the integral.

Integration Against a Measure with Density

Let h be a nonnegative $\mathcal{F}$-measurable function. The **measure with density** h **with respect to** μ is defined by

$$(h\mu)(E) := \int_E h\, d\mu, \quad E \in \mathcal{F}.$$

(See 3.2.10.) We also express this by writing $d(h\mu) = h\, d\mu$. Densities arise as Radon-Nikodym derivatives and in particular as conditional expectations in probability theory. The proof of the following theorem is similar to that of 3.2.15. The details are left to the reader (Ex. 3.15).

3.2.17 Theorem. *Let f be $\mathcal{F}$-measurable. Then*

$$\int f\, d(h\mu) = \int (f \cdot h)\, d\mu$$

in the sense that if one side is defined, then so is the other and equality holds.

Note that the Dirac measure δ_x on $\mathbb{R}$ (1.3.3(d)) has no density with respect to λ. Nevertheless, it is customary in physics and elsewhere to write

$$f(x) = \int_{-\infty}^{\infty} f(y)\delta(y-x)\, dy = \int_{-\infty}^{\infty} f(x+y)\delta(y)\, dy$$

for a symbolic density function $\delta(\cdot)$, the so-called *Dirac delta function*. This interpretation can be made rigorous using distribution theory. (See §15.1.)

Change of Variables Theorem

For the next theorem, recall that a vector function $\varphi = (\varphi_1, \ldots, \varphi_d)$ on an open subset U of $\mathbb{R}^d$ is said to be C^1 if the components φ_i have continuous first partial derivatives on U. The **derivative** $\varphi'(x)$ **of** φ **at** x is the $d \times d$ matrix with (i,j)-entry $\partial_j \varphi_i$. The determinant of this matrix is called the **Jacobian of** φ **at** x.

3.2.18 Theorem. *Let U and V be open subsets of $\mathbb{R}^d$ and let $\varphi : U \to V$ be C^1 on U with C^1 inverse $\varphi^{-1} : V \to U$. If f is Lebesgue measurable on V, then*

$$\int_V f(y)\, d\lambda^d(y) = \int_U (f \circ \varphi)(x) |\det \varphi'(x)|\, d\lambda^d(x) \tag{3.6}$$

in the sense that if one side is defined, then so is the other and then equality holds.

A proof of the theorem is given in Appendix A. Note that for all Lebesgue measurable functions $f \geq 0$ on V,

$$\begin{aligned} \int_V f\, d\lambda^d &= \int_U (f \circ \varphi) \cdot |\det \varphi'|\, d\lambda^d = \int_U (f \circ \varphi) \cdot |\det \varphi'| \circ \varphi^{-1} \circ \varphi\, d\lambda^d \\ &= \int_V f \cdot \left(|\det \varphi'| \circ \varphi^{-1}\right) d\varphi(\lambda^d). \end{aligned}$$

Replacing f by the $f \cdot \left(|\det \varphi'| \circ \varphi^{-1}\right)^{-1}$ we have

$$\int_V f \cdot \left(|\det \varphi'| \circ \varphi^{-1}\right)^{-1} d\lambda^d = \int_V f\, d\varphi(\lambda^d)$$

hence

$$d\varphi(\lambda^d) = \frac{d\lambda^d}{|\det \varphi'| \circ \varphi^{-1}}, \tag{3.7}$$

which combines the notions of density and image measure.

Exercises

3.1 Let $f \in L^1(\mathbb{R})$ be positive and $a \neq 0 \in \mathbb{R}$. Prove that

$$\left| \int e^{iat} f(t)\, dt \right| < \int f(t)\, dt.$$

3.2 Let μ be the infinite series measure of 1.3.3(h). Prove that the equation

$$\int f\, d\mu = \sum_{k=1}^{\infty} f(k) p_k$$

holds for any function $f : \mathbb{N} \to \mathbb{C}$ in the sense that if one side is defined, then so is the other and equality holds. What is the significance of the case $p_k = 1$?

3.3 [↑1.40] Let $(X, \mathcal{F})$ be a measurable space and $\{\mu_n\}$ a sequence of measures such that $(\mu_n(A))$ is a nondecreasing sequence for each $A \in \mathcal{F}$. Then $\mu(A) = \lim_n \mu_n(A)$ defines a measure on $\mathcal{F}$. Prove that f is a nonnegative μ-integrable Borel measurable function on X then f is μ_n integrable for every n and $\int f\, d\mu = \lim_n \int f\, d\mu_n$.

3.4 Let $f, g : X \to \overline{\mathbb{R}}$ be integrable. Prove:

(a) $\int_A f\, d\mu \geq \int_A g\, d\mu\ \forall\, A \in \mathcal{F} \Rightarrow f \geq g$ a.e. (b) $\int_A f\, d\mu = \int_A g\, d\mu\ \forall\, A \in \mathcal{F} \Rightarrow f = g$ a.e.

3.5 Let X be uncountable and let $\mathcal{F}$ be the σ-field consisting of the countable and cocountable subsets of X. Let μ be the probability measure on $\mathcal{F}$ assigns 0 to countable sets and 1 to cocountable sets. By 2.33, an $\mathcal{F}$-measurable function f is constant on some cocountable set. Show that the constant is unique and equals $\int f\,d\mu$.

3.6 Let $f : X \to \mathbb{C}$ be measurable with countable range $\{a_1, a_2, \ldots\}$. Set $A_n := \{f = a_n\}$. Prove that f is integrable iff the series $\sum_{n=1}^{\infty} a_n \mu(A_n)$ converges absolutely, in which case the value of the series equals $\int f\,d\mu$.

3.7 Let $f(x) := \lfloor x^{-1} \rfloor$, $0 < x \le 1$. Find

(a) $\int_{(0,1)} [1 + (-1)^{f(x)}]\,d\lambda(x)$.

(b) $\int_{(0,1)} f^p(x)\,d\lambda(x)$, $0 < p < 1$.

3.8 [↑2.2.1] Let $d_n(x)$ denote the nth digit in the decimal expansion of a number $x \in [0,1)$. Find $\int_{[0,1]} d_n(x)\,d\lambda(x)$.

3.9 Let $a_n \in \mathbb{C}$, $n = 1, \ldots, 9$. Define $f : [0,1] \to \mathbb{C}$ by $f(x) = 0$ if x is rational and $f(x) = a_n$ if x is irrational, where n is the first nonzero digit in the decimal expansion of x. (Assume for definiteness that repeated 9's are not allowed.) Prove that f is Borel measurable and calculate $\int_{[0,1]} f\,d\lambda$.

3.10 Let $a_n \in \mathbb{C}$, $n = 1, 2, \ldots$ and define $f : [0,1) \to \mathbb{C}$ by $f(x) = 0$ if x is rational and $f(x) = a_n$ if x is irrational and the first nonzero digit in the decimal expansion of x occurs at the nth place. (Assume for definiteness that repeated 9's are not allowed.) Prove that f is Borel measurable and calculate $\int_{[0,1)} f\,d\lambda$ whenever it is defined. In particular, show that if $a_n = 2^n$ then the integral is $9/4$.

3.11 Let f be Lebesgue integrable on $\mathbb{R}^d$ and $\int |f| > 0$. Prove that the series $\sum_{n=1}^{\infty} n^{-p} f(nx)$ converges absolutely a.e. on $\mathbb{R}^d$ iff $p > 1 - d$.

3.12 Let μ be a Lebesgue-Stieltjes measure on R such that for all integrable functions f,

$$\int f(x+y)\,d\mu(x) = 2^y \int f(x)\,d\mu(x) \quad y \in \mathbb{R}.$$

Find the distribution function for μ.

3.13 [↓4.3.4] Let $f \ge 0$ be μ-integrable. Prove that for each $\varepsilon > 0$ there exists a $\delta > 0$ such that

$$E \in \mathcal{F} \text{ and } \mu(E) < \delta \Rightarrow \int_E f\,d\mu < \varepsilon.$$

Conclude that if $\{E_n\}$ is a sequence in $\mathcal{F}$ with $\mu(E_n) \to 0$, then $\int_{E_n} f\,d\mu \to 0$. ⟦Begin with simple functions.⟧

3.14 Let $f, g : X \to (0, \infty)$ be measurable. Prove:

$$\int (1 + k^{-1} f)^k g \exp(-f)\,d\mu \to \int g\,d\mu.$$

3.15 Prove 3.2.17.

3.16 Let $d\mu = h\,d\nu$, where h is positive, finite and measurable. Show that $d\nu = h^{-1}\,d\mu$

3.17 [↓4.3.4] Let $f : X \to \overline{\mathbb{K}}$ be μ integrable and $\varepsilon > 0$. Prove:

(a) The set $A = \{|f| \ge \varepsilon\}$ has finite measure.

(b) There exists $B \in \mathcal{F}$ with $\mu(B) < \infty$ such that $\left| \int f\,d\mu - \int_B f\,d\mu \right| < \varepsilon$.

3.18 Let μ be a probability measure, f positive and integrable, and g_n measurable such that $0 \le g_n \le C$ for all n. Prove that $\int f g_n \, d\mu \to 0$ iff $\int g_n \, d\mu \to 0$. [[Choose a and b so that $\mu\{f < a\}$ and $\mu\{f > b\}$ are small.]]

3.19 [↑3.13] Let $f_n, f : X \to \mathbb{C}$ be integrable and $E_n, E \in \mathfrak{F}$ such that $\lim_n \int |f_n - f| \, d\mu = 0$ and $\lim_n \mu(E_n \triangle E) = 0$. Prove that

$$\lim_n \int_{E_n} f_n \, d\mu = \int_E f \, d\mu.$$

3.20 Let $f_1, f_2 \ldots, f_n$ be positive and measurable and set $f := f_1 \vee \cdots \vee f_n$. Prove:

$$\int_{\{f>t\}} f \, d\mu \le \sum_{j=1}^{n} \int_{\{f_j>t\}} f_j \, d\mu, \quad t > 0.$$

[[Consider $B_j := \{f = f_j\} \cap \{f_j > t\}$.]]

3.21 Let f be measurable, $\varepsilon > 0$, and $p > 0$. Prove that

$$\mu\{|f| \ge \varepsilon\}) \le \frac{1}{\varepsilon^p} \int |f|^p \, d\mu.$$

3.22 Let $(X, \mathfrak{F}, \mu)$ be a finite measure space and let $L^0 = L^0(X, \mathfrak{F}, \mu)$ denote the linear space of all measurable functions $f : X \to \mathbb{K}$. Show that

$$d(f, g) = \int \frac{|f - g|}{1 + |f - g|} \, d\mu, \quad f, g \in L^0,$$

defines a metric on L^0, where we identify functions equal a.e. Show also that convergence in this metric is convergence in measure.

3.23 Let $f : X \to \mathbb{R}$ be μ-integrable. Prove that $a \le f \le b$ a.e iff $\dfrac{1}{\mu(A)} \displaystyle\int_A f \, d\mu \in [a, b]$ for all $A \in \mathfrak{F}$ with $0 < \mu(A) < \infty$.

3.24 [↑3.21] Let f be μ-integrable. Show that the set $\{f \ne 0\}$ is σ-finite.

3.25 Show that $(X, \mathfrak{F}, \mu)$ is σ-finite iff there exists a positive integrable function f on X.

3.26 Let $\mathfrak{I}$ be an arbitrary index set. For $i \in \mathfrak{I}$ and $a_i \in [0, \infty]$, define the extended real number

$$\sum_{i \in \mathfrak{I}} a_i := \sup \left\{ \sum_{j \in F} a_j : F \subseteq \mathfrak{I}, \ F \text{ finite} \right\}.$$

(a) Show that there exists a sequence $\{i_n\}$ in $\mathfrak{I}$ such that $\sum_{i \in \mathfrak{I}} a_i = \sum_{n=1}^{\infty} a_{i_n}$.

(b) Let $(X_i, \mathfrak{F}_i, \mu_i)$, $i \in \mathfrak{I}$, be a family of measure spaces, where the sets X_i are disjoint. The **direct sum** of these measure spaces is the triple $(X, \mathfrak{F}, \mu)$, where

$$X := \bigcup_{i \in \mathfrak{I}} X_i, \quad \mathfrak{F} := \{E \subseteq X : E \cap X_i \in \mathfrak{F}_i \ \forall\, i \in \mathfrak{I}\}, \quad \mu(E) := \sum_{i \in \mathfrak{I}} \mu_i(E \cap X_i)$$

Verify that $(X, \mathfrak{F}, \mu)$ is a measure space.

(c) Show that for a nonnegative, $\mathfrak{F}$-measurable function f,

$$\int f \, d\mu = \sum_{i \in \mathfrak{I}} \int f\big|_{X_i} \, d\mu_i.$$

(d) Show that a σ-finite measure space is a direct sum.

3.27 Let μ be a Lebesgue-Stieltjes measure on $\mathcal{B}(\mathbb{R}^d)$ and let $f : \mathbb{R}^d \to \mathbb{C}$ be Borel measurable and μ-integrable such that $\int_I f\,d\mu = 0$ for all closed bounded intervals I. Prove that $f = 0$ a.e. ⟦Reduce to the real case and consider $f^{\pm}$. Use 3.2.10 and 1.8.1.⟧

3.28 (Weighted mean value theorem for integrals). Let μ be a Lebesgue-Stieltjes measure on $\mathcal{B}(\mathbb{R}^d)$ and $E \subseteq \mathbb{R}^d$ compact and connected. Let $f, g : E \to \mathbb{R}$ with g μ-integrable and f continuous. If g does not change sign on E, show that for some $c \in E$.

$$\int_E fg\,d\mu - f(c)\int_E g\,d\mu.$$

3.29 Let f be measurable and $\mu(X) < \infty$. Set $A_n = \{|f| \geq n\}$. Prove that f is integrable iff $\sum_{n=1}^{\infty} \mu(A_n)$ converges, in which case $\lim_n n\mu(A_n) = 0$. ⟦Consider $B_n := \{n \leq |f| < n+1\}$.⟧

3.30 Let $f \geq 0$ be Lebesgue integrable on $[1, \infty)$. Prove that $\sum_{n=1}^{\infty} f(x+n)$ is integrable on $[0, 1]$. Conclude that the series converges a.e. on $[1, \infty)$.

3.31 Let μ be a Lebesgue-Stieltjes measure on R with distribution function F and let $T : \mathbb{R} \to \mathbb{R}$ be continuous and strictly increasing with $T(\mathbb{R}) = \mathbb{R}$. Find the distribution function of $T(\mu)$.

3.32 Let $(X_i, \mathcal{F}_i)$, $i = 1, 2$, be measurable spaces and $T : (X_1, \mathcal{F}_1) \to (X_2, \mathcal{F}_2)$ measurable with measurable inverse. Let μ be a measure on $(X_1, \mathcal{F}_1)$ and let $h \geq 0$ be $\mathcal{F}_1$-measurable. Show that $T(h\mu) = (h \circ T^{-1})T(\mu)$.

3.33 Let $\mathscr{V}$ is a linear subspace of $\mathbb{R}^d$ of dimension $m < d$. Use the change of variables theorem to show that $\lambda^d(\mathscr{V}) = 0$. ⟦Construct a suitable linear transformation.⟧

3.34 Let X be a metric space and μ, μ_1, $\mu_2, \ldots$ finite measures on $\mathcal{B}(X)$ such that $\lim_n \int f\,d\mu_n = \int f\,d\mu$ for all bounded continuous $f : X \to \mathbb{R}$. Carry out the following steps to show that $\lim_n \mu_n(E) = \mu(E)$ for all $E \in \mathcal{B}(X)$ with $\mu(\text{bd}(E)) = 0$.

(a) Show that for each open $U \subseteq X$, there exists a sequence of closed sets $C_n \uparrow U$.

(b) Referring to (a), show that there exist bounded continuous functions $f_k \uparrow \mathbf{1}_U$.

(c) Show that $\int f_k\,du \leq \underline{\lim}_n\, \mu_n(U)$ and hence $\mu(U) \leq \underline{\lim}_n\, \mu_n(U)$.

(d) Apply (c) to $U = \text{int}(E)$ and $U = X \setminus \text{cl}(E)$ to obtain the desired conclusion.

3.3 Connections with the Riemann Integral on $\mathbb{R}^d$

As noted in the introduction to Chapter 1, the Lebesgue integral has several distinct advantages over the Riemann integral. First (proper) Riemann integration takes place on compact subintervals of $\mathbb{R}^d$ while no such restriction is placed on the Lebesgue integral. Second, the class of functions that are Lebesgue integrable on compact intervals is much larger than the class of Riemann integrable functions. Third, and perhaps most importantly, the Lebesgue theory makes available powerful tools in the form of limit theorems such as the monotone convergence theorem and the dominated convergence theorem, leading to many important results in analysis and its applications. Nevertheless, the Riemann integral still plays an important role in mathematics and the sciences and as such is worthy of discussion here. In this section we give a brief description of the d-dimensional Riemann integral and compare it to the Lebesgue integral.

The Darboux Integral

Let f be a bounded, real-valued function on a d-dimensional interval $[a, b]$. For each j, let $\mathcal{P}_j$ be a partition of the jth coordinate interval $[a_j, b_j]$ of $[a, b]$. The collection of subintervals $[x, y]$ of $[a_j, b_j]$ produced by the partition will also be denoted by $\mathcal{P}_j$. Points of $\mathcal{P}_j$ in (a_j, b_j) are called **interior points** of $\mathcal{P}_j$. Taking Cartesian products of subintervals of the partitions $\mathcal{P}_j$ produces what we shall loosely call a **partition of** $[a, b]$ and denote by $\mathcal{P} = \mathcal{P}_1 \times \cdots \times \mathcal{P}_d$. For ease of notation, we set $|I| := \lambda^d(I)$ for intervals I. The **lower and upper (Darboux) sums of** f **over** $\mathcal{P}$ are defined, respectively, by

$$\underline{S}(f, \mathcal{P}) = \sum_{I \in \mathcal{P}} m_I |I|, \quad m_I := \inf_{x \in I} f(x) \quad \text{and} \quad \overline{S}(f, \mathcal{P}) = \sum_{I \in \mathcal{P}} M_I |I|, \quad M_I := \sup_{x \in I} f(x).$$

The **lower and upper (Darboux) integrals of** f are defined, respectively, by

$$\underline{\int_a^b} f := \sup_{\mathcal{P}} \underline{S}(f, \mathcal{P}) \quad \text{and} \quad \overline{\int_a^b} f := \inf_{\mathcal{P}} \overline{S}(f, \mathcal{P}),$$

where the supremum and infimum are taken over all partitions $\mathcal{P}$ of $[a, b]$. If the upper and lower integrals are equal, then f is said to be **Darboux-integrable on** $[a, b]$, the common value of these integrals then being denoted by $\int_a^b f$.

For a limit description of $\int_a^b f$, we need the following notions: A **refinement** of $\mathcal{P} = \mathcal{P}_1 \times \cdots \times \mathcal{P}_d$ is a partition $\mathcal{Q} = \mathcal{Q}_1 \times \cdots \times \mathcal{Q}_d$ of $[a, b]$ such that, as a sets of points, $\mathcal{Q}_j \supseteq \mathcal{P}_j$ for each j. Every member I of $\mathcal{P}$ is then a union of members J of $\mathcal{Q}$, and because boundaries of intervals have Lebesgue measure zero,

$$|I| = \sum_{J \in \mathcal{Q},\ J \subseteq I} |J|.$$

The **common refinement** of partitions $\mathcal{P}$ and $\mathcal{Q}$ is the partition of $[a, b]$ whose jth coordinate partition consists of the points in $\mathcal{P}_j \cup \mathcal{Q}_j$. The following lemma shows that taking refinements decreases the difference of upper and lower sums.

3.3.1 Lemma. *If $\mathcal{Q}$ is a refinement of $\mathcal{P}$, then*

$$\underline{S}(f, \mathcal{P}) \le \underline{S}(f, \mathcal{Q}) \le \overline{S}(f, \mathcal{Q}) \le \overline{S}(f, \mathcal{P}).$$

Proof. The second inequality is clear, and the first inequality follows from the third by considering $-f$. For the third inequality, we have

$$\overline{S}(f, \mathcal{P}) = \sum_{I \in \mathcal{P}} M_I |I| = \sum_{I \in \mathcal{P}} M_I \sum_{J \in \mathcal{Q}, J \subseteq I} |J| \ge \sum_{I \in \mathcal{P}} \sum_{J \in \mathcal{Q}, J \subseteq I} M_J |J| = \overline{S}(f, \mathcal{Q}). \qquad \square$$

3.3.2 Lemma. *For any partition $\mathcal{P}$ of $[a, b]$, $\underline{S}(f, \mathcal{P}) \le \underline{\int_a^b} f \le \overline{\int_a^b} f \le \overline{S}(f, \mathcal{P})$.*

Proof. The first and last inequalities are immediate from the definition of lower and upper integrals. For the middle inequality, let $\mathcal{P}$ and $\mathcal{Q}$ be partitions of $[a, b]$, and let $\mathcal{R}$ be a refinement of both $\mathcal{P}$ and $\mathcal{Q}$. By 3.3.1,

$$\underline{S}(f, \mathcal{P}) \le \underline{S}(f, \mathcal{R}) \le \overline{S}(f, \mathcal{R}) \le \overline{S}(f, \mathcal{Q}).$$

Taking the supremum over $\mathcal{P}$ and the infimum over $\mathcal{Q}$ yields the desired inequality. $\square$

3.3.3 Corollary. *A bounded function $f : [a, b] \to \mathbb{R}^d$ is Darboux integrable iff for each $\varepsilon > 0$ there exists a partition $\mathcal{P}$ of $[a, b]$ such that $\overline{S}(f, \mathcal{P}) - \underline{S}(f, \mathcal{P}) < \varepsilon$.*

We may now describe the integral as a limit of Darboux sums. Given $L \in \mathbb{R}$ and a real-valued function $F(\mathcal{P})$ of partitions $\mathcal{P}$ of $[a, b]$, we write

$$L = \lim_{\mathcal{P}} F(\mathcal{P})$$

if, given $\varepsilon > 0$, there exists a partition $\mathcal{P}_\varepsilon$ such that $|F(\mathcal{P}) - L| < \varepsilon$ for all partitions $\mathcal{P}$ that refine $\mathcal{P}_\varepsilon$. By applying standard techniques, one easily shows that such limits are unique and have the usual combinatorial properties.[1] Using this notion we can give the following characterization of the Darboux integral. The proof is left as an exercise.

3.3.4 Theorem. *A bounded function $f : [a, b] \to \mathbb{R}^d$ is Darboux integrable iff the limits $\lim_{\mathcal{P}} \overline{S}(f, \mathcal{P})$ and $\lim_{\mathcal{P}} \underline{S}(f, \mathcal{P})$ exist and are equal. In this case, their common value is $\int_a^b f$.*

A more useful limit characterization of the Darboux integral may be given in terms of the following. The **mesh** of a partition $\mathcal{P}$ is the value

$$\|\mathcal{P}\| := \max\{y - x : [x, y] \in \mathcal{P}_j,\ 1 \le j \le d\}.$$

The Darboux integral may be expressed as a limit of Darboux sums as $\|\mathcal{P}\| \to 0$. For this we need the following technical lemma:

3.3.5 Lemma. *Let $\mathcal{P}' = \mathcal{P}_1' \times \cdots \times \mathcal{P}_d'$ be a partition of $[a, b]$. Then there exist a positive constant C such that for all partitions $\mathcal{P}$ with $\|\mathcal{P}\|$ sufficiently small,*

$$\overline{S}(f, \mathcal{P}) \le \overline{S}(f, \mathcal{P}') + C\|\mathcal{P}\|.$$

Proof. Let $\mathcal{P}' = \mathcal{P}_1' \times \cdots \times \mathcal{P}_d'$ and let $\mathcal{P} = \mathcal{P}_1 \times \cdots \times \mathcal{P}_d$ with $\|\mathcal{P}\|$ sufficiently small so that each interval $I = I_1 \times \cdots \times I_d$ of $\mathcal{P}$ has the property that either some I_j contains exactly one interior point of $\mathcal{P}_j'$, or no I_j contains such a point. Let J_α denote the d-dimensional intervals of $\mathcal{P}$ of the former type and J_β the intervals of the latter type. The construction is illustrated in the figure, where $[x, y]$ is a coordinate interval of several J_α's and z is an interior point of $\mathscr{P}_2'$. Let N be the number of intervals of type J_α and note that N depends only on $\mathcal{P}'$. Let $\mathcal{P}''$ denote the common refinement of $\mathcal{P}$ and $\mathcal{P}'$. An interval in $\mathcal{P}''$ is either a

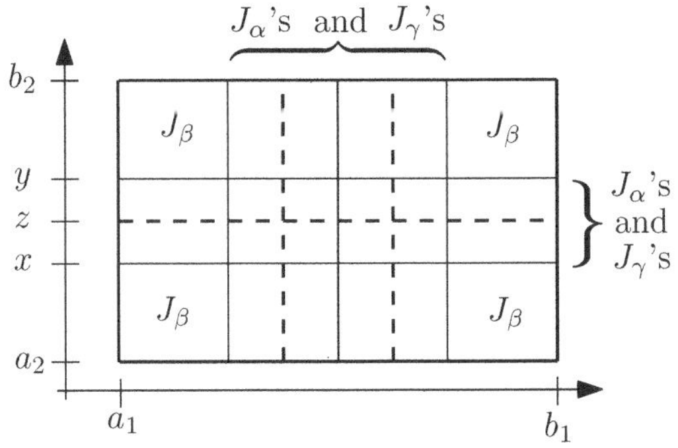

FIGURE 3.1: The intervals of $\mathcal{P}$ (solid), $\mathcal{P}'$ (dotted), and $\mathcal{P}''$.

J_β or was formed from a J_α. Denote intervals of the latter type by J_γ. Since the introduction of a point into a jth coordinate interval of a J_α results in two jth coordinate intervals, each

[1]If the set of partitions of $[a, b]$ is partially ordered by refinement, then the described convergence is simply convergence of the net $(F(\mathcal{P}))_{\mathcal{P}}$.

J_α can produce at most 2^d J_γ's. Thus the number of J_γ's is at most 2^dN. Since the terms of $\overline{S}(f,\mathcal{P})$ and $\overline{S}(f,\mathcal{P}'')$ corresponding to intervals J_β are identical, we have

$$\overline{S}(f,\mathcal{P}) - \overline{S}(f,\mathcal{P}'') = \overline{S}(f,\mathcal{P}) + \underline{S}(-f,\mathcal{P}'') = \sum_\alpha M_{J_\alpha}(f)\,|J_\alpha| + \sum_\gamma m_{J_\gamma}(-f)\,|J_\gamma|. \quad (\dagger)$$

Since the number of terms in the first sum in (†) is no more than N, we see that this sum is majorized by $MN\,\|\mathcal{P}\|^d \le MN\,\|\mathcal{P}\|$, where $M = \|f\|_\infty$ and $\|\mathcal{P}\|$ is taken < 1. Similarly, the second sum in (†) is majorized by $MN2^d\,\|\mathcal{P}''\| \le MN2^d\,\|\mathcal{P}\|$, the inequality following from the fact that $\mathcal{P}''$ is a refinement of $\mathcal{P}$. Thus there exists a constant C depending only on $\mathcal{P}'$ and f such that $\overline{S}(f,\mathcal{P}) - \overline{S}(f,\mathcal{P}'') \le C\,\|\mathcal{P}\|$. Since $\mathcal{P}''$ is a refinement of $\mathcal{P}'$, $\overline{S}(f,\mathcal{P}) - \overline{S}(f,\mathcal{P}') \le C\|\mathcal{P}\|$. □

We may now prove the following complement to 3.3.4.

3.3.6 Theorem. *A bounded function $f : [a,b] \to \mathbb{R}$ is Darboux integrable iff the limits $\lim_{\|\mathcal{P}\|\to 0} \overline{S}(f,\mathcal{P})$ and $\lim_{\|\mathcal{P}\|\to 0} \underline{S}(f,\mathcal{P})$ exist and are equal. In this case, their common value is $\int_a^b f$.*

Proof. It suffices to prove that

$$\overline{\int_a^b} f = \lim_{\|\mathcal{P}\|\to 0} \overline{S}(f,\mathcal{P}) \quad \text{and} \quad \underline{\int_a^b} f = \lim_{\|\mathcal{P}\|\to 0} \underline{S}(f,\mathcal{P}).$$

Given $\varepsilon > 0$, choose a partition $\mathcal{P}'$ such that $\overline{S}(f,\mathcal{P}') < \overline{\int_a^b} f + \varepsilon$. In the notation of 3.3.5, for all partitions $\mathcal{P}$ with sufficiently small mesh,

$$\overline{\int_a^b} f \le \overline{S}(f,\mathcal{P}) \le \overline{S}(f,\mathcal{P}') + C\,\|\mathcal{P}\| < \overline{\int_a^b} f + \varepsilon + C\,\|\mathcal{P}\|.$$

Therefore $\overline{S}(f,\mathcal{P}) - \overline{\int_a^b} f < 2\varepsilon$ for all $\mathcal{P}$ with sufficiently small mesh. This establishes the first limit. The second follows from the first by considering $-f$. □

The Riemann Integral

Let f be a bounded, real-valued function on a d-dimensional interval $[a,b]$. Let $\mathcal{P}$ be a partition of $[a,b]$, and for each interval I in $\mathcal{P}$ choose an arbitrary point $\xi_I \in I$. Set $\xi := \{\xi_I : I \in \mathcal{P}\}$. The pair $(\mathcal{P},\xi)$ is called a **tagged partition of** $[a,b]$ and the quantity

$$S(f,\mathcal{P},\xi) := \sum_{I\in\mathcal{P}} f(\xi_I)|I|$$

is called a **Riemann sum of** f. A bounded function $f : [a,b] \to \mathbb{R}$ is said to be **Riemann integrable** iff the limit

$$R(f) := \lim_{\|\mathcal{P}\|\to 0} S(f,\mathcal{P},\xi)$$

exists in the sense that, given $\varepsilon > 0$ there exists a $\delta > 0$ such that

$$|S(f,\mathcal{P},\xi) - R(f)| < \varepsilon \text{ for all partitions } \mathcal{P} \text{ with } \|\mathcal{P}| < \varepsilon \text{ and all choices of } \xi.$$

The connection between the Darboux and Riemann integrals is given in the following result.

3.3.7 Theorem. *A bounded function $f : [a,b] \to \mathbb{R}$ is Darboux integrable iff it is Riemann integrable. In this case, $R(f) = \int_a^b f$.*

Proof. Since $\underline{S}(f,\mathcal{P}) \le S(f,\mathcal{P},\xi) \le \overline{S}(f,\mathcal{P})$ for all ξ, the necessity follows from 3.3.6. For the sufficiency, given $\varepsilon > 0$ choose a partition $\mathcal{P}_\varepsilon$ such that

$$R(f) - \varepsilon < S(f,\mathcal{P}_\varepsilon,\xi) < R(f) + \varepsilon \text{ for all choices of } \xi.$$

Since ξ is arbitrary, the approximation properties of suprema and infima imply that

$$R(f) - \varepsilon \le \underline{S}(f,\mathcal{P}_\varepsilon) \le \overline{S}(f,\mathcal{P}_\varepsilon) \le R(f) + \varepsilon.$$

Therefore, $\overline{S}(f,\mathcal{P}_\varepsilon) - \underline{S}(f,\mathcal{P}_\varepsilon) < 2\varepsilon$, hence f is Darboux integrable by 3.3.3. □

Measure Zero Criterion for Riemann Integrability

Here is the connection between the Riemann integral and the Lebesgue integral on $[a,b]$.

3.3.8 Theorem. *Let $f : [a,b] \to \mathbb{R}$ be bounded and let D be the set of discontinuities of f. Then f is Riemann integrable iff $\lambda(D) = 0$. In this case, f is Lebesgue measurable and the Riemann and Lebesgue integrals of f on $[a,b]$ are equal.*

Proof. We carry out the proof in the following steps:

(1) *There exists a sequence of partitions $\mathcal{P}_n$ of $[a,b]$ such that $\mathcal{P}_{n+1}$ is a refinement of $\mathcal{P}_n$, $\|\mathcal{P}_n\| \to 0$, and the following hold:*

$$\lim_{n\to\infty} \underline{S}(f,\mathcal{P}_n) = \underline{\int_a^b} f \quad \text{and} \quad \lim_{n\to\infty} \overline{S}(f,\mathcal{P}_n) = \overline{\int_a^b} f.$$

⟦By the approximation property of infima and suprema, for each n there exist partitions $\mathcal{P}'_n$ and $\mathcal{P}''_n$ of $[a,b]$ such that

$$\underline{\int_a^b} f - \frac{1}{n} < \underline{S}(f,\mathcal{P}'_n) \le \underline{\int_a^b} f \le \overline{\int_a^b} f \le \overline{S}(f,\mathcal{P}''_n) < \overline{\int_a^b} f + \frac{1}{n}.$$

Since refinements decrease upper sums and increase lower sums, the inequalities still hold if $\mathcal{P}'_n$ and $\mathcal{P}''_n$ are replaced by refinements. Now let $\mathcal{P}_1$ be a refinement of $\mathcal{P}'_1$ and $\mathcal{P}''_1$ with $\|\mathcal{P}_1\| < 1$, then let $\mathcal{P}_2$ be a refinement of $\mathcal{P}_1$, $\mathcal{P}'_1$, and $\mathcal{P}''_1$ with $\|\mathcal{P}_2\| < 1/2$, etc.⟧

(2) *Define Borel simple functions*

$$h_n = \sum_{I\in\mathcal{P}_n} m_I \mathbf{1}_I \quad \text{and} \quad g_n = \sum_{I\in\mathcal{P}_n} M_I \mathbf{1}_I.$$

Then $h_n \uparrow h$, $g_n \downarrow g$, h and g are bounded Borel functions, $h \le f \le g$, and

$$\underline{\int_a^b} f = \int_{[a,b]} h\, d\lambda \quad \text{and} \quad \overline{\int_a^b} f = \int_{[a,b]} g\, d\lambda. \tag{†}$$

⟦Clearly, $h_1 \le h_2 \le \cdots \le f \le \ldots \le g_2 \le g_1$, hence h and g exist, are bounded, and $h \le f \le g$. Moreover, as limits of Borel functions, h and g are themselves Borel functions. Since $\underline{S}(f,\mathcal{P}_n) = \int_{[a,b]} h_n$ and $\overline{S}(f,\mathcal{P}_n) = \int_{[a,b]} g_n$, (†) follows from step (1) and the monotone convergence theorem applied to the (nonnegative) differences $h - h_n$ and $g_n - g$.⟧

(3) *f is Riemann integrable on $[a, b]$ iff $g = h$ a.e. In this case, f is Lebesgue measurable and $\int_a^b f = \int_{[a,b]} f$.*

⟦From (†), f is Riemann integrable iff $\int_{[a,b]}(g - h) = 0$, which is equivalent to $g = h$ a.e. If the latter holds, then $\{f \neq h\}$ and $\{f \neq g\}$ are null sets, hence f is Lebesgue measurable and the integrals are equal.⟧

(4) *If f is continuous at $x \in [a, b]$, then $h(x) = g(x)$.*

⟦Given $\varepsilon > 0$, choose $\delta > 0$ such that $d(x, y) < \delta$ implies $|f(x) - f(y)| < \varepsilon$, where d is the metric on $\mathbb{R}^d$ defined by $d(x, y) = \max_j |x_j - y_j|$. Choose m so that $\|\mathcal{P}_n\| < \delta$ for all $n \geq m$. For such n and for $x \in I \in \mathcal{P}_n$,

$$f(x) - \varepsilon < f(y) < f(x) + \varepsilon \text{ for all } y \in I.$$

Taking the infimum and supremum of $f(y)$, we have

$$f(x) - \varepsilon \leq h_n(x) = m_I \leq M_I = g_n(x) \leq f(x) + \varepsilon.$$

Letting $n \to \infty$ yields

$$f(x) - \varepsilon \leq h(x) \leq g(x) \leq f(x) + \varepsilon.$$

Therefore, $g(x) = h(x)$.⟧

(5) *Let $x \in [a, b]$ such that x is not on the boundary of any subinterval of any $\mathcal{P}_n$. If $h(x) = g(x)$, then f is continuous at x.*

⟦Given $\varepsilon > 0$, choose n such that

$$|g_n(x) - g(x)| < \varepsilon/2 \text{ and } |h_n(x) - h(x)| < \varepsilon/2.$$

Suppose that x is in the interior of $I \in \mathcal{P}_n$. For all y in the interior of I, $h_n(y) = m_I$ and $g_n(y) = M_I$, hence

$$h(x) - \varepsilon/2 < h_n(x) \leq f(y) \leq g_n(x) < g(x) + \varepsilon/2 = h(x) + \varepsilon/2.$$

Therefore, $|f(x) - f(y)| < \varepsilon$, hence f is continuous at x.⟧

(6) *Set $A = \{x : g(x) \neq h(x)\}$. Then $\lambda^d(A) = \lambda^d(D)$.*

⟦Let B be the union of all the boundaries in the partitions $\mathcal{P}_n$. Since there are countably many boundaries, $\lambda(B) = 0$. By steps (4) and (5), $A \subseteq D \subseteq A \cup B$. Therefore, $\lambda^d(A) = \lambda^d(D)$.⟧

To complete the proof of the theorem, observe that, by step (3), f is Riemann integrable iff $\lambda^d(A) = 0$. By step (6), this occurs iff $\lambda^d(D) = 0$. □

3.3.9 Corollary. *If $F : \mathbb{R}^n \to \mathbb{R}$ is continuous and the functions $f_1 \ldots, f_n$ are Riemann integrable on $[a, b]$, then $F(f_1, \ldots, f_n)$ is Riemann integrable on $[a, b]$. In particular, if $c \in \mathbb{R}$ and $p > 0$, then the functions cf_1, $f_1 + f_2$, $f_1 f_2$, $f_1 \vee f_2$, $f_2 \wedge f_2$, $f_1^{\pm}$, and $|f_1|^p$ are Riemann integrable on $[a, b]$.*

Improper Riemann Integrals

A real-valued function g is said to be **locally Riemann integrable** on an interval I if g is Riemann integrable on every compact subinterval of I. For example, a continuous function is locally integrable.

3.3.10 Theorem. *Let g be locally Riemann integrable on $[a, b)$ (where b could be infinite). Then g is Lebesgue measurable on $[a, b)$. Moreover:*

(a) *If $g \geq 0$ and g is improperly integrable on $[a, b)$, then g is Lebesgue integrable on $[a, b)$ and*

$$\int_a^b g(x)\,dx = \int_{[a,b)} g\,d\lambda. \tag{3.8}$$

(b) *If g is Lebesgue integrable on $[a, b)$, then g is improperly integrable on $[a, b)$ and (3.8) holds.*

Proof. That g is Lebesgue measurable on $[a, b)$ follows from 3.3.8. To prove (a), let $b_n \uparrow b$ and let D denote the set of discontinuities of g on $[a, b)$. Since g is Riemann integrable on $[a, b_n]$, $\lambda([a, b_n] \cap D) = 0$. Then $\mathbf{1}_{[a,b_n]}g$ is Lebesgue measurable for every n and

$$\int_a^{b_n} g(x)\,dx = \int \mathbf{1}_{[a,b_n]}g\,d\lambda.$$

Taking limits, using 3.2.7, we see that g is Lebesgue measurable on $[a, b)$ and (3.8) holds.

For (b) note that by 3.3.9 the functions $g^{\pm}$ are locally Riemann integrable on $[a, b)$. Therefore, by (a), $\int_a^n g^{\pm}(x)\,dx = \int_{[a,n]} g^{\pm}\,d\lambda$ for all n, and an application of the monotone convergence theorem completes the proof. □

Notation. In view of 3.3.8 and 3.3.10 one frequently writes $\int_a^b g(x)\,dx$ for the integral $\int_{[a,b]} g\,d\lambda$ and $\int_a^\infty g(x)\,dx$ for the integral $\int_{[a,\infty)} g\,d\lambda$.

The following example show that part (a) of the above theorem does not necessarily hold for functions that are not nonnegative.

3.3.11 Example. Consider the function $g(x) = x^{-1}\sin x$ on the interval $[1, \infty)$. Integrating by parts, we have

$$\int_1^t g(x)\,dx = -\frac{\cos x}{x}\Big|_1^t + \int_1^t \frac{\sin x}{x^2}\,dx.$$

Since the integral $\int_1^\infty x^{-2}\,dx$ converges, the right side converges as $t \to \infty$, hence g is improperly integrable on $[1, \infty)$. On the other hand,

$$\int_{[\pi,n\pi]} |g|\,d\lambda = \sum_{k=2}^n \int_{(k-1)\pi}^{k\pi} |g(x)|\,dx > \sum_{k=2}^n \frac{1}{k\pi}\int_{(k-1)\pi}^{k\pi} |\sin x|\,dx = \frac{2}{\pi}\sum_{k=2}^n \frac{1}{k},$$

hence g is not Lebesgue integrable. It may be shown (Ex. 3.64) that

$$\int_0^\infty \frac{\sin x}{x}\,dx = \frac{\pi}{2}.$$ ◊

Exercises

3.35 Let μ be a Lebesgue-Stieltjes measure whose distribution function F has a positive continuous derivative on $\mathbb{R}$. Show that $d\mu = F'\,d\lambda$.

3.36 For a bounded function $f : [a,b] \to \mathbb{R}$, set

$$i(f) = \inf\left\{\int_a^b g\,d\lambda : g \in C[a,b],\ g \ge f\right\},\quad s(f) = \sup\left\{\int_a^b h\,d\lambda : h \in C[a,b],\ h \le f\right\}.$$

Show that $i(f) = \underline{\int_a^b} f\,d\lambda$ and $s(f) = \overline{\int_a^b} f\,d\lambda$.

3.37 Show that a bounded function f on $[a,b]$ is Riemann integrable iff there exists a real number L such that $S(f, \mathcal{P}_n, \xi_n) \to L$ for each sequence of tagged partitions $(\mathcal{P}_n, \xi_n)$ with $\|\mathcal{P}_n\| \to 0$.

3.38 The **gamma function** is defined by $\Gamma(x) = \int_0^\infty t^{x-1}e^{-t}\,dt,\quad x > 0$.

(a) Show that the integral converges.

(b) Integrate by parts to show that $\Gamma(x+1) = x\Gamma(x)$ for all $x > 0$.

(c) Show that $\Gamma(n+1) = n!$ for all $n \in \mathbb{N}$.

(d) Given that $\int_0^\infty e^{-t^2}\,dt = \sqrt{\pi}/2$, evaluate $\Gamma\left(\frac{1}{2}\right)$, $\Gamma\left(\frac{3}{2}\right)$, and $\Gamma\left(\frac{5}{2}\right)$.

(e) The formula $\Gamma(x) = x^{-1}\Gamma(x+1)$ may be used to extend the gamma function to noninteger values $x < 0$. Use this to show that $\Gamma\left(-\frac{1}{2}\right) = -2\sqrt{\pi}$ and $\Gamma\left(-\frac{3}{2}\right) = \frac{4\sqrt{\pi}}{3}$.

3.39 Show that for $n \ge 2$,

$$\int_0^{\pi/2} \sin^n x\,dx = \int_0^{\pi/2} \cos^n x\,dx = \int_0^1 \frac{x^n}{\sqrt{1-x^2}}\,dx = \begin{cases} \dfrac{(n-1)(n-3)\cdots 4\cdot 2}{n(n-2)\cdots 5\cdot 3}, & n \text{ odd}, \\[2ex] \dfrac{\pi}{2}\,\dfrac{(n-1)(n-3)\cdots 5\cdot 3}{n(n-2)\cdots 4\cdot 2}, & n \text{ even}. \end{cases}$$

3.40 Let $n \in \mathbb{Z}^+$. Verify the formulas

(a) $\displaystyle\int_0^\infty x^n e^{-x}\,dx = n!$

(b) $\displaystyle\int_0^\infty x^n e^{-x^2/2}\,dx = (n-1)(n-3)\cdots 1\cdot\sqrt{\pi/2}$ if $n \ge 2$ is even,

$= (n-1)(n-3)\cdots 2$ if $n \ge 3$ is odd.

(c) $\displaystyle\int_0^\infty x^n e^{-x^2}\,dx = 2^{-(n+1)/2}\int_0^\infty x^n e^{-x^2/2}\,dx.$

⟦For (b), use $\int_0^\infty e^{-x^2/2}\,dt = \sqrt{\pi/2}$.⟧

3.41 Show that if f_n is Riemann integrable on $[a,b]$ and $f_n \to f$ uniformly on $[a,b]$, then f is Riemann integrable and $\int_a^b f_n \to \int_a^b f$. Show also that the assertion is false if the convergence is merely pointwise.

3.42 Show that the function

$$f = \sum_{n=1}^\infty \left[(2n)^{-p}\mathbf{1}_{[n,n+1)} - (2n+1)^{-p}\mathbf{1}_{[n+1,n+2)}\right]$$

is improperly Riemann integrable on $[1,\infty)$ for any $p > 0$, but is Lebesgue integrable iff $p > 1$.

3.43 Show that $(x^{-1}\sin x)^2$ extended continuously to $[0,\infty)$ is Lebesgue integrable and improperly Riemann integrable on $[0,\infty)$ and

$$\int_0^\infty \left(\frac{\sin x}{x}\right)^2 dx = \int_0^\infty \frac{\sin x}{x}\,dx.$$

3.44 Let $p > 0$. Show that

$$\int_0^1 \frac{dx}{1+x^p} = \sum_{n=0}^\infty \left(\frac{1}{2np+1} - \frac{1}{(2n+1)p+1}\right).$$

Show that for suitable p the formula yields

$$\ln 2 = 1 - \frac{1}{2} + \frac{1}{3} - \frac{1}{4} + \cdots \quad \text{and} \quad \frac{\pi}{4} = 1 - \frac{1}{3} + \frac{1}{5} - \frac{1}{7} + \cdots$$

⟦Use the identity $(1+y)^{-1} = \sum_{n=0}^\infty \left[y^{2n} - y^{(2n+1)}\right]$, $0 \le y < 1$.⟧

3.45 [↑3.39] Show that

$$\int_0^{\pi/2} \frac{dx}{1+\sin x} = \pi \sum_{n=0}^\infty \frac{(2n)!}{4^{n+1}(n+1)!\,n!}.$$

3.46 [↑3.40] Let $p \in \mathbb{N}$.

(a) Show that $x^p/(e^x - 1)$, extended continuously to $[0,\infty)$, is Lebesgue integrable and

$$\int_0^\infty \frac{x^p}{e^x - 1}\,dx = p! \sum_{n=1}^\infty \frac{1}{n^{p+1}}.$$

(b) Show that $x^{2p}/(e^{x^2} - 1)$, extended continuously to $[0,\infty)$, is Lebesgue integrable and

$$\int_0^\infty \frac{x^{2p}}{e^{x^2} - 1}\,dx = \frac{\sqrt{\pi}(2p-1)!}{4^p(p-1)!} \sum_{n=1}^\infty \frac{1}{n^{p+1/2}}.$$

⟦Use $(z-1)^{-1} = \sum_{n=1}^\infty z^{-n}$, $z > 1$.⟧

3.4 Convergence Theorems

The General Monotone Convergence Theorem

The monotone convergence theorem established in § 3.2 is one of several theorems that give sufficient conditions for the convergence of a sequence of integrals. In this section we consider three such theorems. Other convergence theorems are treated in the next chapter. We begin with the following extension of 3.2.7.

3.4.1 Theorem. *Let f_n and f be measurable, $f_n \uparrow f$ a.e., and $\int f_1^- \,d\mu < \infty$. Then*

$$\int f\,d\mu = \lim_n \int f_n\,d\mu. \tag{3.9}$$

Proof. By hypothesis, there exists a set $E \in \mathcal{F}$ such that $f_n \uparrow f$ on E and $\mu(E^c) = 0$. Replacing f_n by $f_n\mathbf{1}_E$, we may assume that $f_n \uparrow f$ on X.

Since $0 \le f^- \le f_n^- \le f_1^-$ and $\int f_1^- \,d\mu < \infty$, the integrals in (3.9) are defined. If $\int f_1^+ \,d\mu = \infty$, then from $f_1^+ \le f_n^+ \le f^+$ we see that each side of (3.9) is ∞. If $\int f_1^+ \,d\mu < \infty$, then f_1 is integrable and we may apply 3.2.7 to the sequence of nonnegative functions $f_n - f_1$ to obtain

$$\int f_n = \int (f_n - f_1) + \int f_1 \to \int (f - f_1) + \int f_1 = \int f. \qquad \square$$

Fatou's Lemma

The following result is useful in cases where $\lim_n f_n$ does not exist.

3.4.2 Theorem. *If f_n and g are measurable, $f_n \geq g$ a.e. for all n, and $\int g^- \, du < \infty$, then*

$$\int \varliminf_n f_n \, d\mu \leq \varliminf_n \int f_n \, d\mu. \tag{3.10}$$

In particular, the inequality holds if $f_n \geq 0$ a.e. for all n.

Proof. Let $h_n = \inf_{j \geq n} f_j$ and $h = \varliminf_n f_n$. Then $\int h_1^- \leq \int f_1^- \leq \int g^- < \infty$, $h_n \leq f_n$, and $h_n \uparrow h$ a.e. By 3.4.1 applied to the sequence (h_n),

$$\int \varliminf_n f_n = \int h = \lim_n \int h_n = \varliminf_n \int h_n \leq \varliminf_n \int f_n. \qquad \square$$

The inequality in (3.10) may be strict. For example, if $\mu = \lambda$ and $f_n = n^2 \mathbf{1}_{[0,1/n]}$, then the left side of (3.10) is zero while the right side is ∞.

The Dominated Convergence Theorem

The third convergence theorem in this section is typically used in cases where a sequence of functions may not be monotone.

3.4.3 Theorem. *Let $g \geq 0$ be integrable and let $(f_n : X \to \overline{\mathbb{K}})$ be a sequence of measurable functions such that $|f_n| \leq g$ a.e. for all n. If $f_n \to f$ a.e., then f is integrable and $\int f_n \, d\mu \to \int f \, d\mu$.*

Proof. By considering real and imaginary parts, we may assume that the functions f_n and f are $\overline{\mathbb{R}}$-valued. Since $|f_n| \leq g$ a.e. and $|f| \leq g$ a.e., f_n and f are integrable by 3.2.11. From Fatou's lemma applied to $g \pm f_n$ (≥ 0), we have

$$\int (g + f) \, d\mu \leq \varliminf_n \int (g + f_n) \, d\mu = \int g \, d\mu + \varliminf_n \int f_n \, d\mu$$

and

$$\int (g - f) \, d\mu \leq \varliminf_n \int (g - f_n) \, d\mu = \int g \, d\mu - \varlimsup_n \int f_n \, d\mu.$$

Subtracting $\int g \, d\mu$ in each inequality yields

$$\int f \, d\mu \leq \varliminf_n \int f_n \, d\mu \leq \varlimsup_n \int f_n \, d\mu \leq \int f \, d\mu. \qquad \square$$

We note that the hypothesis that the functions $|f_n|$ be dominated by an integrable function cannot be omitted. For example, $\mathbf{1}_{[n,2n]} \to 0$ on $\mathbb{R}$, but $\int \mathbf{1}_{[n,2n]} \, d\lambda \to \infty$.

We conclude this section with two applications of the dominated convergence theorem. The first, whose proof is an exercise for the reader, describes a continuity property of integrals. The second gives sufficient conditions for differentiating "under the integral sign."

3.4.4 Corollary. *Let I be an open d-dimensional interval and let f be $\mathcal{B}(I) \otimes \mathcal{F}$-measurable on $I \times X$ such that $f(t,x)$ is continuous in t for each $x \in X$. If there exists an integrable function g on X such that $|f(t,x)| \leq g(x)$ for all t and x, then $\int f(t,x) \, d\mu(x)$ is continuous in t.*

3.4.5 Corollary. *Let I be an open d-dimensional interval and let f be $\mathcal{B}(I)\otimes\mathcal{F}$-measurable on $I\times X$ such that for each t in I the function $f(t,\cdot)$ is μ-integrable. Let α be a fixed multi-index and assume that for all multi-indices β with $|\beta|\le|\alpha|$ the derivative $\partial_t^\beta f(t,x)$ exists for each t and x and is measurable in x for each fixed t. If there exists an integrable function g on X such that $|\partial_t^\beta f(t,x)|\le g(x)$ for all such β, t and x, then*

$$\partial_t^\alpha \int f(t,x)\,d\mu(x) = \int \partial_t^\alpha f(t,x)\,d\mu(x).$$

Proof. We prove the right-hand derivative version for the case $d=1$. The general formula follows by induction. Fix $t\in I$ and let $t_n\downarrow t$. Set

$$H(t)=\int f(t,x)\,d\mu(x) \quad\text{and}\quad h_n(x)=\frac{f(t_n,x)-f(t,x)}{t_n-t}.$$

By the mean value theorem, $h_n(x)=f_t(s,x)$ for some $s=s(n,x)\in(t,t_n)$. Then $|h_n|\le g$ and $h_n(x)\to f_t(t,x)$, hence

$$\frac{H(t_n)-H(t)}{t_n-t}=\int h_n(x)\,d\mu(x)\to\int f_t(t,x)\,d\mu(x).$$

This shows that the right-hand derivative of H exists at x and equals $\int f_t(t,x)\,d\mu(x)$. □

Exercises

3.47 Find all $p>0$ for which there is an λ integrable function g on $\mathbb{R}^+$ such that $n^{-p}I_{[0,n]}\le g$ for all n.

3.48 Let μ be a Lebesgue-Stieltjes measure on $\mathcal{B}(\mathbb{R})$ and f μ-integrable. Show that $\lim_n\int_{a_n}^{b_n} f\,d\mu=0$ for any pair of sequences (a_n) and (b_n) with $a_n<b_n$ and $a_n\to\infty$. Show that this may not hold if $f\ge 0$ is not integrable.

3.49 Let μ be a Lebesgue-Stieltjes measure on $\mathcal{B}(\mathbb{R})$ and f integrable. Let g be measurable and bounded on $\mathbb{R}$ such that $r:=\lim_{t\to\infty} g(t)$ exists and is finite. Show that

$$\lim_{x\to\infty}\int g(x+t)f(t)\,d\mu(t)=r\int f\,d\mu.$$

3.50 Let μ be a Lebesgue-Stieltjes measure on $\mathcal{B}(\mathbb{R}^d)$ and $f>0$ μ-integrable. Prove that

(a) $\displaystyle\int_{1/n}^{n} n\ln(1+n^{-1}f)\,d\mu\to\int f\,d\mu.$ (b) $\displaystyle\int n\ln(1+n^{-2}f)\,d\mu\to 0.$

(c) $\displaystyle\int n^p\sin^p\left(n^{-1}f^{1/p}\right)d\mu\to\int f\,d\mu.$ (d) $\displaystyle\int_E f^{1/n}\,d\mu\to\mu(E).$

3.51 [↑3.40] Show that

(a) $\displaystyle\int_0^\infty e^{-t^2}\cos(xt)\,dt=\frac{\sqrt{\pi}}{2}e^{-x^2/4}.$ (b) $\displaystyle\int_0^\infty e^{-t^2}\sin^2(xt)\,dt=\frac{\sqrt{\pi}}{4}(1-e^{-x^2}).$

⟦For (a) use the power series for $\cos(xt)$.⟧

3.52 [↑3.46] Show that

(a) $\displaystyle\int_0^\infty\frac{\sin(xt)}{e^t-1}\,dt=\sum_{n=1}^\infty\frac{x}{n^2+x^2}.$ (b) $\displaystyle\int_0^\infty\frac{t\sin(xt)}{e^{t^2}-1}\,dt=\frac{\sqrt{\pi}}{2}\sum_{n=1}^\infty\frac{x}{2n^{3/2}}e^{-x^2/4n}.$

3.53 [↑3.38] Prove that the kth derivative of the gamma function is

$$\Gamma^{(k)}(x) = \int_0^\infty t^{x-1}e^{-t}\ln^k t\,dt, \quad x > 0.$$

3.54 Let $g_n : X \to \mathbb{C}$ such that $\sum_n \int |g_n|\,d\mu < \infty$. Show that the series $g := \sum_n g_n$ converges a.e., is integrable and $\int \sum_n g_n\,d\mu = \sum_n \int g_n\,d\mu$.

3.55 Let $f : \mathbb{R} \to \overline{\mathbb{K}}$ be λ-integrable. Show that the series $\sum_{k=-\infty}^{\infty} f(k+x) := \lim_n \sum_{k=-n}^{n} f(k+x)$ converges absolutely a.e. on $\mathbb{R}$.

3.56 Let $f : \mathbb{R} \to \mathbb{R}$ be Lebesgue integrable on every interval and satisfy $f(x+y) = f(x) + f(y)$ for all x, y. Show that $f(x) = f(1)x$ for all x. ⟦Show first that f is continuous.⟧

3.57 Let $f_n : X \to [0,\infty)$ be integrable and $f_{n+1} \le f_n$ a.e. for all n. Show that $\sum_n (-1)^{n+1} f_n$ is integrable and that $\int \left(\sum_n f_n\right) d\mu = \sum_n \int f_n\,d\mu$.

3.58 Let g be integrable on X and let (f_n) be a sequence of real-valued measurable functions on X such that $|f_n| \le g$. Prove that

$$\int \underline{\lim}_n f_n\,d\mu \le \underline{\lim}_n \int f_n\,d\mu \le \overline{\lim}_n \int f_n\,d\mu \le \int \overline{\lim}_n f_n\,d\mu.$$

3.59 Let f, g, f_n, g_n be real valued and integrable such that $f_n \to f$, $g_n \to g$ a.e., $|f_n| \le g_n$ a.e., and $\int g_n\,d\mu \to \int g\,d\mu$. Prove that $\int f_n\,d\mu \to \int f\,d\mu$.

3.60 [↑3.59] Let f_n, g_n, h_n, f, g, h be integrable, $f_n \le g_n \le h_n$ a.e. for all n, $f_n \stackrel{\text{a.e.}}{\to} f$, $g_n \stackrel{\text{a.e.}}{\to} g$, and $h_n \stackrel{\text{a.e.}}{\to} h$. Show that if $\int f_n\,d\mu \to \int f\,d\mu$ and $\int h_n\,d\mu \to \int h\,d\mu$, then $\int g_n\,d\mu \to \int g\,d\mu$.

3.61 Show that the dominated convergence theorem holds if the hypothesis $f_n \stackrel{\text{a.e.}}{\to} f$ is replaced by $f_n \stackrel{\mu}{\to} f$.

3.5 Integration against a Product Measure

In this section we construct the product of finitely many measures and prove a theorem that give conditions under which the corresponding integral may be expressed as an iterated integral. This result, known as *Fubini's theorem*, is fundamental in integration theory, having both concrete and theoretical applications. Some of these are given in the next section.

Construction of the Product of Two Measures

Let $(X, \mathcal{F}, \mu)$ and $(Y, \mathcal{G}, \nu)$ be arbitrary measure spaces. Recall that $\mathcal{F} \otimes \mathcal{G}$ denotes the σ-field generated by the semiring $\mathcal{R}$ of measurable rectangles $A \times B$, $A \in \mathcal{F}$ and $B \in \mathcal{G}$. The following theorem shows that a measure on $\mathcal{R}$ may be constructed from the measures μ and ν in a natural way and then extended to a measure $\mu \otimes \nu$ on $\mathcal{F} \otimes \mathcal{G}$, called the **product of the measures μ and ν**. For the statement of the theorem, we remind the reader of the convention $0 \cdot \infty = \infty \cdot 0 = 0$.

3.5.1 Theorem. *There exists a measure $\mu \otimes \nu$ on $(X \times Y, \mathcal{F} \otimes \mathcal{G})$ such that*

$$(\mu \otimes \nu)(A \times B) = \mu(A) \cdot \nu(B) \quad \text{for all } A \in \mathcal{F} \text{ and } B \in \mathcal{G}. \tag{3.11}$$

Moreover, if the measure spaces $(X, \mathcal{F}, \mu)$ and $(Y, \mathcal{G}, \nu)$ are σ-finite, then the measure $\mu \otimes \nu$ is unique with respect to property (3.11).

Proof. Define $\mu \otimes \nu$ on the semiring $\mathcal{R}$ by Equation (3.11). We claim that $\mu \otimes \nu$ is a measure on $\mathcal{R}$. Clearly, $(\mu \otimes \nu)(\emptyset) = 0$. For countable additivity, let $(A_n \times B_n)$ be a disjoint sequence in $\mathcal{R}$ such that $\bigcup_n A_n \times B_n = A \times B \in \mathcal{R}$. Then for $(x, y) \in X \times Y$, $\mathbf{1}_A(x)\mathbf{1}_B(y) = \sum_n \mathbf{1}_{A_n}(x)\mathbf{1}_{B_n}(y)$. For fixed x we can integrate with respect to y and use 3.2.9 to obtain $\mathbf{1}_A(x)\nu(B) = \sum_n \mathbf{1}_{A_n}(x)\nu(B_n)$. Integrating with respect to x yields $(\mu \otimes \nu)(A \times B) = \sum_n (\mu \otimes \nu)(A_n \times B_n)$, verifying the claim. By 1.6.4, $\mu \otimes \nu$ may be extended to a measure on $\mathcal{F} \otimes \mathcal{G}$. If $(X, \mathcal{F}, \mu)$ and $(Y, \mathcal{G}, \nu)$ are σ-finite, then $(\mathcal{R}, \mu \otimes \nu)$ is σ-finite, hence uniqueness follows from 1.6.9. □

The measure space $(X \times Y, \mathcal{F} \otimes \mathcal{G}, \mu \otimes \nu)$ is called the **product of the measure spaces** $(X, \mathcal{F}, \mu)$ and $(Y, \mathcal{G}, \nu)$.

Fubini's Theorem

3.5.2 Theorem (Fubini-Tonelli). *Let $(X, \mathcal{F}, \mu)$ and $(Y, \mathcal{G}, \nu)$ be σ-finite measure spaces and let $f : X \times Y \to \overline{\mathbb{K}}$ be $\mathcal{F} \otimes \mathcal{G}$-measurable.*

(a) *If $f \geq 0$, then the functions $\int_X f(x, y)\, d\mu(x)$ and $\int_Y f(x, y)\, d\nu(y)$ are measurable in y and x, respectively, and*

$$\int_{X \times Y} f(x, y)\, d(\mu \otimes \nu)(x, y) = \int_Y \int_X f(x, y)\, d\mu(x)\, d\nu(y) = \int_X \int_Y f(x, y)\, d\nu(y)\, d\mu(x). \tag{3.12}$$

(b) *If one of the quantities*

$$\int_{X \times Y} |f(x, y)|\, d(\mu \otimes \nu)(x, y), \quad \int_Y \int_X |f(x, y)|\, d\mu(x)\, d\nu(y), \quad \int_X \int_Y |f(x, y)|\, d\nu(y)\, d\mu(x)$$

is finite, then so are the other two and (3.12) *holds in the sense that $\int f(x, y)\, d\mu(x)$ is defined and finite for a.a y and is integrable with respect to y, and $\int f(x, y)\, d\nu(y)$ is defined and finite for a.a x and is integrable with respect to x.*

Proof. Recall that a measurable function $f(x, y)$ is separably measurable, that is, measurable in x for each fixed y and measurable in y for each fixed x (Ex. 2.13). Thus the inner integrals in (3.12) are legitimate.

We now make the following reductions. First, part (b) of the theorem is a consequence of part (a). Indeed, if one of the inequalities in (b) holds, then f is integrable by part (a) applied to $|f|$. By considering real, imaginary, positive, and negative parts, we see that (3.12) holds. Second, to prove (a) we may assume by the usual arguments that f is an indicator function. Thus to prove the theorem it suffices to show that for any $C \in \mathcal{F} \otimes \mathcal{G}$,

$$\eta(C) = \int_Y \int_X \mathbf{1}_C(x, y)\, d\mu(x)\, d\nu(y) = \int_X \int_Y \mathbf{1}_C(x, y)\, d\nu(y)\, d\mu(x), \text{ where } \eta := \mu \otimes \nu. \tag{†}$$

For this we may assume that the measure spaces $(X, \mathcal{F}, \mu)$ and $(Y, \mathcal{G}, \nu)$ are finite. Indeed, if (†) holds in the finite case and if $X_n \uparrow X$, where $\mu(X_n) < \infty$ and $\nu(Y) < \infty$, then by considering the measures restricted to $\mathcal{F} \cap X_n$, and $(\mathcal{F} \otimes \mathcal{G}) \cap (X_n \times Y)$ and applying the monotone convergence theorem, we see that (†) holds for σ-finite and Y finite. Repeating the argument with $Y_n \uparrow Y$ shows that (†) holds for σ-finite X and Y.

To prove (†) for finite measures μ and ν, let $\mathcal{H}$ denote the collection of all $C \in \mathcal{F} \otimes \mathcal{G}$ for which first equality in (†) holds and the inner integral in that equality is measurable in y. We show that $\mathcal{H}$ is a λ-system (1.5) containing all measurable rectangles $A \times B$. It will follow from the π-λ theorem (1.2.6) that $\mathcal{H} = \mathcal{F} \otimes \mathcal{G}$, verifying the first equality in (†). The second is proved in a similar manner.

Let $A \in \mathcal{F}$, $B \in \mathcal{G}$ and set $C := A \times B$. Then

$$\int \mathbf{1}_C(x, y)\, d\mu(x) = \int \mathbf{1}_A(x)\mathbf{1}_B(y)\, d\mu(x) = \mu(A)\mathbf{1}_B(y),$$

which is measurable in y, hence

$$\iint \mathbf{1}_C(x, y)\, d\mu(x)\, d\nu(y) = \mu(A)\nu(B) = \eta(C).$$

Thus $\mathcal{H}$ contains all measurable rectangles.

Now let C, $D \in \mathcal{H}$ with $C \subseteq D$ and let $E = D \setminus C$. Then

$$\int \mathbf{1}_E(x, y)\, d\mu(x) = \int \mathbf{1}_D(x, y)\, d\mu(x) - \int \mathbf{1}_C(x, y)\, d\mu(x),$$

which is measurable in y and implies that

$$\begin{aligned}\iint \mathbf{1}_E(x, y)\, d\mu(x)\, d\nu(y) &= \iint \mathbf{1}_D(x, y)\, d\mu(x)\, d\nu(y) - \iint \mathbf{1}_C(x, y)\, d\mu(x)\, d\nu(y)\\ &= \eta(D) - \eta(C) = \eta(E).\end{aligned}$$

Therefore, $\mathcal{H}$ is closed under relative differences.

Finally, let $C_n \in \mathcal{H}$ and $C_n \uparrow C$. Then $\mathbf{1}_{C_n} \uparrow \mathbf{1}_C$, hence, by the monotone convergence theorem,

$$\int \mathbf{1}_{C_n}(x, y)\, d\mu(x) \uparrow \int \mathbf{1}_C(x, y)\, d\mu(x).$$

Thus $\int \mathbf{1}_C(x, y)\, d\mu(x)$ is measurable in y. Applying the monotone convergence theorem again yields

$$\eta(C) = \lim_n \eta(C_n) = \lim_n \iint \mathbf{1}_{C_n}(x, y)\, d\mu(x)\, d\nu(y) = \iint \mathbf{1}_C(x, y)\, d\mu(x)\, d\nu(y).$$

Therefore, $\mathcal{H}$ is closed under increasing unions, completing the proof that $\mathcal{H}$ is a λ-system and establishing the theorem. □

Note that a special case of part (a) is the interchange of summation in $\sum_n \sum_m a_{mn}$, where $a_{mn} \geq 0$, even when the double sum $\sum_{m,n} a_{mn}$ is infinite.

3.5.3 Remarks. **(a)** The σ-finiteness hypothesis in Fubini's theorem is essential: Consider Lebesgue measure λ and counting measure ν on $([0, 1], \mathcal{B}[0, 1])$. The diagonal $E = \{(t, t) : t \in [0, 1]\}$ is closed and so is a member of $\mathcal{B}[0, 1] \otimes \mathcal{B}[0, 1]$. But for all x and y

$$\int \mathbf{1}_E(t, y)\, d\lambda(t) = \lambda\{y\} = 0 \quad\text{and}\quad \int \mathbf{1}_E(x, t)\, d\nu(t) = \nu(\{x\}) = 1,$$

hence the iterated integrals are unequal.

(b) Part (b) of the theorem fails if the absolute values on the integrands are removed. Indeed for Lebesgue measure on $[0,1]$ we have

$$\int_0^1 \int_0^1 \frac{x^2 - y^2}{(x^2+y^2)^2}\, dy\, dx = -\int_0^1 \int_0^1 \frac{x^2 - y^2}{(x^2+y^2)^2}\, dx\, dy = \frac{\pi}{4}.$$

Thus $(x^2 - y^2)(x^2 + y^2)^{-2}$ is not integrable on $[0,1] \times [0,1]$. (See Ex. 3.71.) ◊

The d-Dimensional Case

The above concepts generalize by induction to finitely many measure spaces $(X_i, \mathcal{F}_i, \mu_i)$ $(i = 1, \ldots, d)$. We state the generalizations and leave the verifications to the reader.

3.5.4 Theorem. *There exists a measure* $\mu = \mu_1 \otimes \cdots \otimes \mu_d$ *on the product measurable space* $(X_1 \times \cdots \times X_d, \mathcal{F}_1 \otimes \cdots \otimes \mathcal{F}_d)$ *such that*

$$\mu(A_1 \times \cdots \times A_d) = \mu_1(A_1) \cdots \mu_d(A_d) \quad \textit{for all } A_i \in \mathcal{F}_i. \tag{3.13}$$

Moreover, if the measure spaces $(X_i, \mathcal{F}_i, \mu_i)$ *are* σ*-finite, then* μ *is unique with respect to property* (3.13).

3.5.5 Example. Consider the measure spaces $(\mathbb{R}^{p_i}, \mathcal{B}(\mathbb{R}^{p_i}), \lambda^{p_i})$ $(i = 1, \ldots, d)$ and $(\mathbb{R}^p, \mathcal{B}(\mathbb{R}^p), \lambda^p)$, where $p = p_1 + \cdots + p_d$. Since $\lambda^{p_1} \otimes \cdots \otimes \lambda^{p_d} = \lambda^p$ on the semiring of half-open intervals, the measures must be equal on $\mathcal{B}(\mathbb{R}^{p_1}) \otimes \cdots \otimes \mathcal{B}(\mathbb{R}^{p_d}) = \mathcal{B}(\mathbb{R}^p)$. ◊

3.5.6 Theorem. *Let the measure spaces* $(X_i, \mathcal{F}_i, \mu_i)$ *be* σ*-finite and let* $f : X \to \overline{\mathbb{R}}$ *be* $\mathcal{F}$*-measurable.*

(a) *If* $f \geq 0$, *then*

$$\int f\, d\mu = \int \cdots \int f(x_1, \ldots, x_d)\, d\mu_1(x_1) \ldots d\mu_d(x_d), \tag{3.14}$$

where $\int f(x_1, \ldots, x_i, \ldots, x_d)\, d\mu_i(x_i)$ *is measurable in* $(x_1, \ldots, x_{i-1}, x_{i+1}, \ldots, x_d)$, *and the iterated integration may be carried out in any of the* $d!$ *orders.*

(b) *If for some permutation* $(i_1, i_2, \ldots, i_d)$ *of the indices* $1, 2, \ldots, d$

$$\int \cdots \int |f(x_{i_1}, \ldots, x_{i_d})|\, d\mu_{i_1}(x_{i_1}) \ldots d\mu_{i_d}(x_{i_d}) < \infty, \tag{3.15}$$

then f *is* μ*-integrable and* (3.14) *holds, where* $\int \cdots \int f(x_1, \ldots, x_d)\, d\mu_1(x_1) \ldots d\mu_i(x_i)$, *is defined and finite for a.a values of* $x_{i+1}, \ldots, x_d$ *and is integrable in these variables.*

3.5.7 Example. In elementary calculus, integration is sometimes carried out on regions in $\mathbb{R}^3$ bounded by surfaces. This idea generalizes to higher dimensions as follows: Given continuous functions $u_2(x_1) \leq v_2(x_1)$ on $E_1 := [a, b]$, and in general continuous functions $u_{k+1}(x_1, \ldots, x_k) \leq v_{k+1}(x_1, \ldots, x_k)$ defined on the set

$$E_k := \{(x_1, \ldots, x_k) : a \leq x_1 \leq b,\ u_2(x_1) \leq x_2 \leq v_2(x_1), \ldots, \\ u_k(x_1, \ldots, x_{k-1}) \leq x_k \leq v_k(x_1, \ldots, x_{k-1})\},$$

then for any integrable f on E_d,

$$\int_{E_d} f\, d\lambda^d = \int_a^b \int_{u_2(x_1)}^{v_2(x_1)} \cdots \int_{u_d(x_1, \ldots, x_d)}^{v_d(x_1, \ldots, x_{d-1})} f(x_1, \ldots, x_d)\, dx_d \ldots dx_2\, dx_1.$$ ◊

Exercises

3.62 Show that the product of complete measure spaces need not be complete.

3.63 Let μ be a probability measure on $\mathcal{B}(\mathbb{R}^d)$. Find $\int \mu(I_x)\,dx$, where

$$I_x := [x_1, x_1 + a_1] \times \cdots \times [x_d, x_d + a_d], \quad a_j > 0.$$

3.64 Let a, $b > 0$. Use Fubini's theorem, the dominated convergence theorem, and the identity $1/x = \int_0^\infty e^{-xt}\,dt$, $x > 0$, to prove that

(a) $\displaystyle\int_0^\infty \frac{\sin x}{x}\,dx = \frac{\pi}{2}$. (b) $\displaystyle\int_0^\infty \frac{e^{-ax} - e^{-bx}}{x}\,dx = \ln(b) - \ln(a)$.

3.65 Let μ be a Lebesgue-Stieltjes measure on $\mathbb{R}$. Show that if $0 < \mu(E) < \infty$ and $a > 0$, then

$$\frac{1}{\mu(E)} \int_{-\infty}^{\infty} \mu\big((x, x+a] \cap E\big)\,dx = a.$$

3.66 Let $(X_i, \mathcal{F}_i, \mu_i)$ $(i = 1, 2)$ be σ-finite measure spaces and let $f_i \geq 0$ be $\mathcal{F}_i$-measurable. Find a density function for the product measure $(f_1\mu_1) \otimes (f_2\mu_2)$.

3.67 Let $(X, \mathcal{F}, \mu)$ be σ-finite and $f : X \to [0, \infty)$ measurable. Prove that the integral of f is the "area under the graph," that is,

$$\int f\,d\mu = (\mu \otimes \lambda)\{(x,t) : 0 < t < f(x)\} = (\mu \otimes \lambda)\{(x,t) : 0 < t \leq f(x)\}.$$

Conclude that if f is integrable, then the graph $\{(x,t) : t = f(x)\}$ has measure zero.

3.68 (Cavalieri's principle). For $E \in \mathcal{B}(\mathbb{R}^d)$ and $t \in \mathbb{R}$, define

$$E_t := \{x = (x_1, \ldots, x_{d-1}) \in \mathbb{R}^{d-1} : (x,t) \in E\}.$$

Show that $E_t \in \mathcal{B}(\mathbb{R}^d)$ for all $t \in [a,b]$ and prove that

$$\lambda^d\Big[E \cap \big(\mathbb{R}^{d-1} \times [a,b]\big)\Big] = \int_a^b \lambda^{d-1}(E_t)\,dt.$$

Thus the "volume" of the portion of E between the hyperplanes $x_d = a$ and $x_d = b$ is the integral from a to b of the "cross-sectional areas" $\lambda^{d-1}(E_t)$.

3.69 Let $(X, \mathcal{F}, \mu)$ be σ-finite and $f : X \to [0, \infty)$ measurable. Suppose that $\varphi : [0, \infty) \to [0, \infty)$ has a positive continuous derivative and $\varphi(0) = 0$. Prove that

$$\int_X \varphi \circ f\,d\mu = \int_0^\infty \varphi'(x)\mu\{f \geq x\}\,dx.$$

Deduce, in particular, that $\int_X f^p\,d\mu = \int_0^\infty p x^{p-1} \mu(f \geq x)\,dx$, $(p \geq 1)$.

3.70 Let $a > 0$. Define the **d-dimensional simplex in $\mathbb{R}^d$** by

$$S(a,n) = \Big\{x : \sum_{j=1}^{n} x_j \leq a \ \text{ and } x_j \geq 0\Big\}.$$

Use Fubini's theorem and induction to show that $\lambda^n\big(S(a,n)\big) = \dfrac{a^n}{n!}$.

3.71 Verify the assertions in 3.5.3. Also, show directly that $\int_{[0,1]^2} |x^2 - y^2|(x^2 + y^2)^{-2}\,d\lambda^2(x,y) = \infty$.

3.72 Let μ be a translation invariant Lebesgue-Stieltjes measure on $\mathcal{B}(\mathbb{R}^d)$ and set $E = [0,1]^d$. Use Fubini's theorem to show that for all $B \in \mathcal{B}(\mathbb{R}^d)$,

$$\int \mathbf{1}_E(x)\mathbf{1}_B(y)\,d\lambda^d(x)\,d\mu(y) = \int \mathbf{1}_E(y)\mathbf{1}_B(x)\,d\lambda^d(x)\,d\mu(y),$$

hence $\mu(B) = \mu(E)\lambda(B)$. Conclude that Lebesgue measure λ^d is the only σ-finite translation invariant measure μ on $\mathcal{B}(\mathbb{R}^d)$ with $\mu[0,1]^d = 1$. [[Consider $\int \mathbf{1}_E(x+y)\mathbf{1}_B(y)\,d\lambda^d(x)\,d\mu(y)$.]]

3.6 Applications of Fubini's Theorem

Gaussian Density

We show that

$$\int_{-\infty}^{\infty} \frac{1}{\sqrt{2\pi}} e^{-t^2/2}\, dt = 1. \tag{3.16}$$

The integrand in (3.16) is called the **Gaussian density** or the **standard normal density**; it is the familiar "bell-curve" of statistics.

By a change of variable, it suffices to prove that $\int_0^\infty e^{-t^2}\, dt = \frac{\sqrt{\pi}}{2}$. Denoting the latter integral by I we have

$$\begin{aligned} I^2 &= \int_0^\infty e^{-y^2} \int_0^\infty e^{-t^2}\, dt\, dy = \int_0^\infty e^{-y^2} \int_0^\infty y e^{-x^2y^2}\, dx\, dy && (t = xy) \\ &= \int_0^\infty \int_0^\infty y e^{-y^2(1+x^2)}\, dy\, dx, && \text{(Fubini's theorem)} \\ &= \frac{1}{2} \int_0^\infty (1+x^2)^{-1} \int_0^\infty e^{-u}\, du\, dx && (u = y^2(1+x^2).) \end{aligned}$$

The last expression evaluates to $\pi/4$, establishing the formula.

Note that, by a suitable substitution, for $m \in \mathbb{R}$ and $\sigma > 0$ we have

$$\int_{-\infty}^{\infty} \frac{1}{\sigma\sqrt{2\pi}} \exp\left[-\frac{1}{2}\left(\frac{x-m}{\sigma}\right)^2\right] = 1.$$

The integrand here is the density of a normal random variable with mean m and standard deviation σ.

Integration by Parts

Let F and G be distribution functions on $\mathbb{R}$ with $\lim_{x\to-\infty} F(x) = \lim_{x\to-\infty} G(x) = 0$, and let μ and ν be the corresponding Lebesgue-Stieltjes measures:

$$\mu(x,y] = F(y) - F(x) \quad \text{and} \quad \nu(x,y] = G(y) - G(x), \quad x < y.$$

We establish the formula

$$\int_{(a,b]} \frac{G(x)+G(x-)}{2}\, dF(x) + \int_{(a,b]} \frac{F(x)+F(x-)}{2}\, dG(x) = F(b)G(b) - F(a)G(a). \tag{3.17}$$

For the verification, let $R := (a,b] \times (a,b]$ and define

$$R_1 := \{(x,y) : a < x \le b,\ a < y \le x\}, \quad R_2 := \{(x,y) : a < y \le b,\ a < x < y\}.$$

Since R is the disjoint union of R_1 and R_2,

$$(\mu\otimes\nu)(R_1) + (\mu\otimes\nu)(R_2) = (\mu\otimes\nu)(R) = \big[F(b) - F(a)\big]\big[G(b) - G(a)\big]. \tag{†}$$

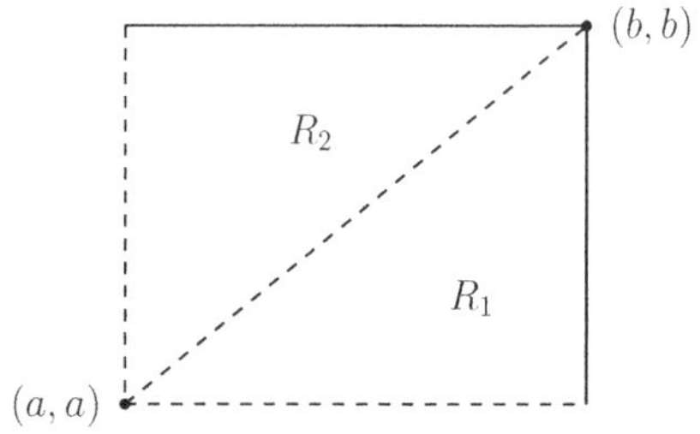

FIGURE 3.2: Rectangles for integration by parts formula.

By Fubini's theorem,

$$
\begin{aligned}
(\mu \otimes \nu)(R_1) &= \iint \mathbf{1}_{(a,b]}(x)\mathbf{1}_{(a,x]}(y)\, d\nu(y)\, d\mu(x) = \int_{(a,b]} [G(x) - G(a)]\, d\mu(x) \\
&= \int_{(a,b]} G(x)\, d\mu(x) - G(a)\big[F(b) - F(a)\big], \quad \text{and} \\
(\mu \otimes \nu)(R_2) &= \iint \mathbf{1}_{(a,b]}(y)\mathbf{1}_{(a,y)}(x)\, d\mu(x)\, d\nu(y) = \int_{(a,b]} [F(y-) - F(a)]\, d\nu(y) \\
&= \int_{(a,b]} F(y-)\, d\nu(y) - F(a)\big[G(b) - G(a)\big].
\end{aligned}
$$

Adding these equations and using (†), we find after cancellations that

$$F(b)G(b) - F(a)G(a) = \int_{(a,b]} G(x)\, d\mu(x) + \int_{(a,b]} F(y-)\, d\nu(y).$$

Since F and G play symmetrical roles,

$$F(b)G(b) - F(a)G(a) = \int_{(a,b]} G(x-)\, d\mu(x) + \int_{(a,b]} F(y)\, d\nu(y).$$

Averaging yields (3.17).

A simpler formula is available if F and G have no common points of discontinuity. In this case, $G(x-) \neq G(x)$ iff $F(x-) = F(x)$ iff $\mu\{x\} = 0$. Since the set S of such points is countable, $\mu(S) = 0$, hence the first integral on the left in (3.17) is $\int_{(a,b]} G\, dF$. Similarly, the second integral is $\int_{(a,b]} F\, dG$. Therefore,

$$\int_{(a,b]} G(x)\, dF(x) + \int_{(a,b]} F(x)\, dG(x) = F(b)G(b) - F(a)G(a).$$

For a special case, suppose that g is continuously differentiable and zero outside a bounded interval. Taking first $G(x) = \int_{-\infty}^{x} (g')^+(t)\, dt$ then $G(x) = \int_{-\infty}^{x} (g')^-(t)\, dt$ in the preceding formula and then subtracting we obtain

$$\int_{\mathbb{R}} g(x)\, dF(x) + \int_{\mathbb{R}} g'(x)F(x)dx = 0.$$

Taking $F(x) = \int_{-\infty}^{x} f(t)\, dt$ for a continuously differentiable f we obtain the formula

$$\int_{\mathbb{R}} g(x)f'(x)\, dx = -\int_{\mathbb{R}} g'(x)f(x)\, dx.$$

Spherical Coordinates

Define open sets U, $V \subseteq \mathbb{R}^d$ by

$$U := (0,\infty)\times(0,\pi)^{d-2}\times(0,2\pi) \text{ and } V := \mathbb{R}^d \setminus \big(\mathbb{R}^{d-2}\times[0,\infty)\times\{0\}\big),$$

and define a transformation $\varphi : U \to V$ from spherical to rectangular coordinates by

$$x = \varphi(s), \quad x := (x_1,\ldots,x_d), \quad s = (r,\theta_1,\ldots,\theta_{d-1}), \text{ where}$$

$$\begin{aligned} x_1 = r\cos\theta_1, \quad x_2 = r\sin\theta_1\cos\theta_2, \quad \ldots, \quad x_3 = r\sin\theta_1\sin\theta_2\cos\theta_3 \;\ldots, \\ x_{d-1} = r\sin\theta_1\cdots\sin\theta_{d-2}\cos\theta_{d-1}, \text{ and } x_d = r\sin\theta_1\cdots\sin\theta_{d-2}\sin\theta_{d-1}. \end{aligned} \tag{3.18}$$

Clearly φ is C^∞ on U. Moreover, φ is a bijection, that is, for each $(x_1,\ldots,x_n)\in V$ the system (3.18) has a unique solution. To see this, note that $r = \sqrt{\sum_{j=1}^d x_j^2}$ and $\theta_1 = \arccos(x_1/r)$, hence (3.18) has a unique solution for $d = 2$. Setting

$$y_j = x_j/(r\sin\theta_1), \quad 2 \le j \le d,$$

we may assume by induction that the reduced system

$$\begin{aligned} &y_2 = \cos\theta_2, \; y_3 = \sin\theta_2\cos\theta_3, \ldots, \; y_{d-1} = \sin\theta_2\cdots\sin\theta_{d-2}\cos\theta_{d-1}, \\ &y_d = \sin\theta_2\cdots\sin\theta_{d-2}\sin\theta_{d-1} \end{aligned}$$

has a unique solution $(\theta_2,\ldots,\theta_{d-1})$. Then (3.18) has the unique solution $(r,\theta_1,\ldots,\theta_{d-1})$.

By standard properties of determinants and a reduction argument,

$$\det\varphi'(s) = r^{d-1}\sin^{d-2}\theta_1\sin^{d-3}\theta_2\cdots\sin^2\theta_{d-3}\sin\theta_{d-2}.$$

Since the determinant is positive on U, φ has a C^∞ inverse on U by the inverse function theorem. By the change of variables theorem and Fubini's theorem, if f is Lebesgue measurable on $\mathbb{R}^d$ and either $f \ge 0$ or f is integrable, then since V differs from $\mathbb{R}^d$ by a set of measure zero we have for $s = (r,\theta_1,\ldots,\theta_{d-1})$,

$$\int_{\mathbb{R}^d} f\,d\lambda^d = \int_U (f\circ\varphi)(s)|r^{d-1}\sin^{d-2}\theta_1\cdots\sin^2\theta_{d-3}\sin\theta_{d-2}|\,dr\,d\theta_1\cdots d\theta_{d-1}. \tag{3.19}$$

Volume of a d-Dimensional Ball

For $d \ge 1$, let $C_r^d(x)$ denote the closed ball in $\mathbb{R}^d$ with center x and radius r. We show that $\lambda^d\big(C_r^d(x)\big) = r^d\alpha_d$, where

$$\alpha_d = \left\{\begin{array}{ll} \dfrac{(2\pi)^{d/2}}{d(d-2)\cdots 4\cdot 2} & \text{if } d \text{ is even,} \\ \dfrac{2(2\pi)^{(d-1)/2}}{d(d-2)\cdots 3\cdot 1} & \text{if } d \text{ is odd} \end{array}\right\} = \text{volume of } C_1^d(\mathbf{0}) \text{ in } \mathbb{R}^d. \tag{3.20}$$

By translation invariance and the dilation property of Lebesgue measure, $\lambda^d\big(C_r^d(x)\big) = r^d\lambda^d\big(C_1^d(\mathbf{0})\big)$, hence it suffices to establish the formula for the case $r = 1$ and $x = \mathbf{0}$, which is the version expressed in (3.20).

To simplify notation, for $1 \le k \le d$ let $C^k(r) := C_r^k(\mathbf{0})$ and let $\mathbf{1}_k(r; x_1,\ldots,x_k)$ denote the indicator function of $C^k(r)$. Formula (3.20) is easily verified for $d = 1$ and 2, so we assume that $d > 2$. Since

$$C^d(1) = \{(x_1,\ldots,x_d) : x_3^2+\cdots+x_d^2 \le 1 - x_1^2 - x_2^2, \;\; (x_1,x_2)\in C^2(1)\},$$

by Fubini's theorem we have

$$\lambda(C^d(1)) = \int_{\mathbb{R}^2} \mathbf{1}_2(1;x_1,x_2) \int_{\mathbb{R}^{d-2}} \mathbf{1}_{d-2}\left(\sqrt{1-x_1^2-x_2^2};x_3,\ldots,x_d\right) d\lambda^{d-2}(x_3,\ldots,x_d)\,dx_1\,dx_2.$$

The inner integral is

$$\lambda\left(C^{d-2}\left(\sqrt{1-x_1^2-x_2^2}\right)\right) = (1-x_1^2-x_2^2)^{(d-2)/2}\lambda^{d-2}\big(C^{d-2}(1)\big),$$

hence, changing to polar coordinates,

$$\begin{aligned}
\lambda^d\big(C^d(1)\big) &= \lambda^{d-2}\big(C^{d-2}(1)\big)\int_{x_1^2+x_2^2\le 1}(1-x_1^2-x_2^2)^{(d-2)/2}\,dx_1\,dx_2 \\
&= \lambda\big(C^{d-2}(1)\big)\int_0^{2\pi}\int_0^1 (1-r^2)^{(d-2)/2} r\,dr\,d\theta \\
&= \frac{2\pi}{d}\lambda^{d-2}\big(C^{d-2}(1)\big).
\end{aligned}$$

Iterating, we obtain (3.20).

Integration of Radial Functions

Let $S^{d-1} := \{x \in \mathbb{R}^d : |x| = 1\}$, where $|x|$ is the Euclidean norm of x. The theorem in this subsection asserts that the Lebesgue integral of a function on $\mathbb{R}^d$ may be calculated by a two-stage process, integrating first over S^{d-1} with respect to a surface measure μ and then radially outward. The surface measure is constructed as follows: Set $\mathbb{R}^d_* := \mathbb{R}^d \setminus \{\mathbf{0}\}$ and define a mapping

$$T : (0,\infty)\times S^{d-1} \to \mathbb{R}^d_* \ \text{ by } \ T(r,x) = rx.$$

Then T is continuous with continuous inverse

$$T^{-1} : \mathbb{R}^d_* \to (0,\infty)\times S^{d-1}, \quad T^{-1}(x) = (|x|, x/|x|).$$

Now define a measure μ on $\mathcal{B}(S^{d-1})$ by

$$\mu(B) := d\cdot\lambda^d\ \big(T((0,1]\times B)\big), \quad B \in \mathcal{B}(S^{d-1}).$$

We then have

3.6.1 Theorem. *If $f : \mathbb{R}^d_* \to \overline{\mathbb{K}}$ is Borel measurable, then*

$$\int_{\mathbb{R}^d_*} f(x)\,d\lambda^d(x) = \int_0^\infty \int_{S^{d-1}} r^{d-1} f(rx)\,d\mu(x)\,dr$$

in the sense that if one side of the equation is defined, then so is the other and equality holds.

Proof. Define a measure ρ on $\mathcal{B}(0,\infty)$ by $d\rho := r^{d-1}\,d\lambda^d$. By Fubini's theorem, the desired equation may be written

$$\int_{\mathbb{R}^d_*} f(x)\,d\lambda^d(x) = \int_{(0,\infty)\times S^{d-1}} (f\circ T)(r,x)\,d(\rho\times\mu)(r,x).$$

By 3.2.15, this is equivalent to $\lambda^d = T(\rho \otimes \mu)$ or $T^{-1}(\lambda^d) = \rho \otimes \mu$. By the uniqueness theorem for product measures it therefore suffices to show that

$$\lambda^d\big[T(A \times B)\big] = \rho(A)\mu(B), \quad A \in \mathcal{B}(0,\infty)), \quad B \in \mathcal{B}(S^{d-1}).$$

Since the collection of intervals is a π-system, by the measure uniqueness theorem it suffices to take $A = (a, b]$. The above equation then reduces to

$$\lambda^d\big(T((a,b] \times B))\big) = d^{-1}(b^d - a^d)\,\mu(B).$$

But this follows from the dilation property of λ^d, using the relations

$$T\big((a,b] \times B\big) = T\big((0,b] \times B\big) - T\big((0,a] \times B\big), \quad T\big((0,c] \times B\big) = cT\big((0,1] \times B\big). \qquad \square$$

Theorem 3.6.1 is useful for calculating integrals of **radial functions**, that is, functions f on $\mathbb{R}^d$ of the form $f(x) = g(|x|)$.

3.6.2 Corollary. *Let g be a Borel function on $(0, \infty)$. Then*

$$\int_{\mathbb{R}^d_*} g(|x|)\,d\lambda(x) = d\alpha_d \int_0^\infty r^{d-1} g(r)\,dr$$

whenever the side is defined, where α_d is given in (3.20).

Proof. Let $f(x) = g(|x|)$ and note that $f(rx) = g(r)$ on S^{d-1}. By the theorem,

$$\int_{\mathbb{R}^d_*} g(|x|)\,d\lambda(x) = \mu\big(S^{d-1}\big) \int_0^\infty r^{d-1} g(r)\,dr.$$

Taking $g(r) = e^{-r^2}$ we have

$$\int_{\mathbb{R}^d_*} e^{-|x|^2}\,dx = \mu(S^{d-1}) \int_0^\infty r^{d-1} e^{-r^2}\,dr = \mu(S^{d-1}) \frac{\pi^{d/2}}{d\alpha_d},$$

where the last equality is from Ex 3.74. By Fubini's theorem and (3.16), the integral on the left is $\pi^{d/2}$. Therefore, $\mu(S^{d-1}) = d\alpha_d$, completing the proof. $\square$

Corollary 3.6.2 may be used to establish the integrability of certain functions on $\mathbb{R}^d$:

3.6.3 Example. Let $f(x) = (1 + c|x|^s)^{-t}$, where c, s, $t > 0$. Then

$$(d\alpha_d)^{-1} \int_{-\infty}^\infty f\,d\lambda = \int_0^\infty \frac{r^{d-1}}{(1 + cr^s)^t}\,dr \le \int_0^1 r^{d-1}\,dr + c^{-t} \int_1^\infty \frac{1}{r^{st-d+1}}\,dr.$$

Hence if $st > d$, then f is integrable on $\mathbb{R}^d$. $\Diamond$

Surface Area of a d-Dimensional Ball

We use Corollary 3.6.2 to derive the following formula for the surface area of the sphere $S = S_r^{d-1} = \{x \in \mathbb{R}^d : |x| = r\}$:

$$\text{area}(S) = dr^{d-1}\alpha_d = dr^{-1}\lambda^d\big(C^d(0,r)\big), \tag{3.21}$$

where α_d is given in (3.20).

As a starting point, we take as the definition of the surface area of a graph

$$\{x \in \mathbb{R}^d : x_d = h(x_1, \ldots, x_{d-1}),\ \ (x_1, \ldots, x_{d-1}) \in U\},\ \ U \subseteq \mathbb{R}^{d-1} \text{ open},$$

the integral

$$\int_U \sqrt{1 + |\nabla h(\boldsymbol{u})|^2}\, d\lambda^{d-1}(\boldsymbol{u}), \tag{†}$$

where h is C^1 on U. Let $H := \{x \in S : x_d > 0\}$, the upper hemisphere of S, and for $0 < t < r$ let H_t denote the part of H for which $(x_1^2 + \cdots x_{d-1}^2)^{1/2} < t$. Then H_t is the graph of

$$h_t(u_1, \ldots, u_{d-1}) = \sqrt{r^2 - |\boldsymbol{u}|^2}, \quad |\boldsymbol{u}| = (u_1^2 + \cdots u_{d-1}^2)^{1/2} < t,$$

hence from (†)

$$\text{area}(H_t) = r \int_{|\boldsymbol{u}|<t} \left(r^2 - |\boldsymbol{u}|^2\right)^{-1/2} d\boldsymbol{u} = r(d-1)\alpha_{d-1} \int_0^t \frac{s^{d-2}}{\sqrt{r^2 - s^2}}\, ds,$$

the last equality by 3.6.2. By the substitution $s = xr$, we have

$$\text{area}(H_t) = (d-1)r^{d-1}\alpha_{d-1} \int_0^{t/r} \frac{x^{d-2}}{\sqrt{1 - x^2}}\, dx.$$

Therefore, defining $\text{area}(H) := \lim_{t \to r} \text{area}(H_t)$, [2] we have

$$\text{area}(H) = (d-1)r^{d-1}\alpha_{d-1} \int_0^1 \frac{x^{d-2}}{\sqrt{1 - x^2}}\, dx.$$

By Exercises 3.75 and 3.76, the integral on the right is

$$\frac{\alpha_{d-2}}{2\alpha_{d-3}} = \frac{d}{d-1} \cdot \frac{\alpha_d}{2 \cdot \alpha_{d-1}}.$$

(Here we assume $d \geq 3$, the desired formula for the cases $d = 1$, 2 being clear.) Therefore, $\text{area}(S) = 2 \cdot \text{area}(H) = dr^{d-1}\alpha_d$. Recalling that $r^d\alpha_d$ is the volume of the ball of radius r in $\mathbb{R}^d$, (3.21) follows.

Exercises

3.73 Let F be a continuous distribution function on $\mathbb{R}$ with finite limits

$$\alpha := \lim_{t \to -\infty} F(t) \quad \text{and} \quad \beta = \lim_{t \to \infty} F(t).$$

Calculate $\int F(t)\, dF(t)$ in terms of α and β.

3.74 [↑3.40] Establish the formula $\displaystyle\int_0^\infty x^{n-1} e^{-x^2}\, dx = \frac{\pi^{n/2}}{n\alpha_n}$.

3.75 Show that $\alpha_1 = 2$ and $\alpha_2 = \pi$. Define $\alpha_0 = 1$. Prove that for $n \geq 3$,

$$\frac{(n-1)(n-3)\cdots 2}{n(n-2)\cdots 3} = \frac{\alpha_n}{2\alpha_{n-1}} = \frac{n-1}{n} \cdot \frac{\alpha_{n-2}}{2\alpha_{n-3}} \quad \text{if } n \text{ is odd}$$

$$\frac{(n-1)(n-3)\cdots 3}{n(n-2)\cdots 2} = \frac{\alpha_n}{\pi\alpha_{n-1}} = \frac{n-1}{n} \cdot \frac{\alpha_{n-2}}{\pi\alpha_{n-3}} \quad \text{if } n \text{ is even.}$$

Thus for all $n \geq 3$,

$$\frac{\alpha_n}{\alpha_{n-1}} = \frac{n-1}{n} \cdot \frac{\alpha_{n-2}}{\alpha_{n-3}}$$

[2] This may be justified from general measure theory on surfaces.

3.76 [↑3.39] Show that for $n \geq 1$, $\displaystyle\int_0^1 \frac{x^n}{\sqrt{1-x^2}}\,dx = \frac{\alpha_n}{2\alpha_{n-1}}$, where $\alpha_0 := 1$.

3.77 Let $1 \leq d < m$, $U \subseteq \mathbb{R}^d$ open and $f : U \to \mathbb{R}^m$ such that $|f(x) - f(y)| \leq C|x - y|^p$, where $C > 0$ and $p > d/m$. Verify the following to prove that $\lambda^m\big(f(U)\big) = 0$. What if $p \leq d/m$?

(a) It suffices to prove that $\lambda^m\big(f(I)\big) = 0$ for a d-dimensional interval $I = [a, b] \subseteq U$.

(b) For fixed $n \in \mathbb{N}$ and each k form the partition $\mathcal{P}_{k,n} = \{a_k + j(b_k - a_k)/n : j = 0, \ldots, n\}$ of the kth coordinate interval $[a_k, b_k]$ of I, $k = 1, \ldots, d$. Let $J_k \in \mathcal{P}_{k,n}$, $J := J_1 \times \cdots \times J_d$ and y the midpoint of J. Then for all $x \in J$

$$|f(x) - f(y)| \leq Cn^{-p}\big[(a_1 - b_1)^2 + \cdots + (a_d - b_d)^2\big]^{p/2} := Mn^{-p}.$$

(c) $\lambda^m(f(J)) \leq \alpha_d(n^{-p}M)^m$.

(d) $\lambda^m\big(f(I)\big) \leq n^{d-mp}M'$.

3.78 Show that the measure μ in 3.6.1 satisfies $L(\mu) = |\det(L^{-1})|\,\mu$ for any 1-1 linear transformation L on $\mathbb{R}^d$ for which $|L(x)| = |x|$. In particular, μ is invariant under rotations.

3.79 Let M, a, and ε be positive constants. Suppose f is Borel measurable on $\mathbb{R}^d$ and satisfies

$$|f(x)| \leq \begin{cases} M|x|^{\varepsilon - d} & \text{if } |x| \leq a, \\ M|x|^{-\varepsilon - d} & \text{if } |x| > a. \end{cases}$$

Prove that f is integrable on $\mathbb{R}^d$.

3.80 Let $0 \leq a < b \leq \infty$ and set $A(a, b) = \{x \in \mathbb{R}^d : a < |x| < b\}$. Prove that

$$\int_{A(a,b)} f(x)\,d\lambda^d(x) = \int_a^b \int_{S^{d-1}} r^{d-1} f(rx)\,d\mu(x)\,dr,$$

where μ is the measure in 3.6.1.

3.81 [↑(3.19)] Show that $\displaystyle\int_0^\pi \cdots \int_0^\pi (\sin^{d-2}\theta_1)(\sin^{d-3}\theta_2)\cdots(\sin\theta_{d-2})\,d\theta_1 \cdots d\theta_{d-2} = \frac{d\alpha_d}{2\pi}$.

Chapter 4

L^p Spaces

In this chapter we examine the properties of spaces of measurable functions f for which $|f|^p$ $(p > 0)$ is integrable, the so-called L^p spaces. These are among the most important examples of Banach spaces. In particular, the case $p = 2$ is of critical importance in Fourier analysis.

Throughout the chapter, unless otherwise stated,
$(X, \mathcal{F}, \mu)$ denotes an arbitrary measure space.

4.1 Definition and General Properties

The L^p spaces fall naturally into three categories depending on the range of values of p.

The Case $1 \le p < \infty$

The space of L^p **functions on** X is defined by

$$L^p(X, \mathcal{F}, \mu) := \{f : X \to \mathbb{K} : f \text{ is } \mathcal{F}\text{-measurable and } \|f\|_p < \infty\}, \text{ where}$$
$$\|f\|_p := \left(\int |f|^p \, d\mu \right)^{1/p}.$$

If there is no ambiguity, we write $L^p(X)$, $L^p(\mu)$, or L^p instead of $L^p(X, \mathcal{F}, \mu)$. Note that $L^1(\mu)$ is just the space of μ-integrable functions.

The quantity $\|f\|_p$ is called the L^p **norm of** f. This terminology is a slight abuse of language, since the property of positivity of a norm does not always hold. Indeed, $\|f\|_p = 0$ implies only that $f = 0$ a.e. We resolve this discrepancy informally by identifying functions that are equal a.e. This will cause no problems as long as the reader keeps in mind that the symbol f has the dual interpretation of a function as well as the equivalence class of all measurable functions equal a.e. to f. A precise resolution may be given in terms of quotient spaces. (See Ex. 8.56.)

The following inequality will be needed to establish that $\|\cdot\|_p$ is indeed a norm (subject to the aforementioned convention of identifying functions that are equal a.e.).

4.1.1 Lemma. *Let $a, b > 0$ and $0 < t < 1$. Then $a^t b^{1-t} \le ta + (1-t)b$, equality holding iff $a = b$.*

Proof. Equality clearly holds if $a = b$. Assume $a < b$ and set $x = ta + (1-t)b$. To prove that $a^t b^{1-t} < x$ we use the strict concavity of $\ln x$ established as follows: By the mean value theorem there exist $y \in (a, x)$ and $z \in (x, b)$ such that

$$\frac{\ln b - \ln x}{b - x} = \frac{1}{z} < \frac{1}{y} = \frac{\ln x - \ln a}{x - a}.$$

Solving for $\ln x$ we have

$$\frac{b-x}{b-a}\ln a + \frac{x-a}{b-a}\ln b < \ln x.$$

Since $x-a = (1-t)(b-a)$ and $b-x = t(b-a)$, the last inequality becomes $\ln a^t + \ln b^{1-t} < \ln x$, proving the lemma. $\square$

We may now prove the following fundamental inequality:

4.1.2 Hölder's Inequality. *Let $1 < p, q < \infty$ with $p^{-1} + q^{-1} = 1$. If $f \in L^p$ and $g \in L^q$, then $fg \in L^1$ and $\|fg\|_1 \le \|f\|_p \|g\|_q$. Moreover, equality holds iff there exist nonnegative constants a and b, not both zero, such that $a|f|^p = b|g|^q$ a.e.*

Proof. If the right side of the desired inequality holds then either f or g is zero a.e. and the inequality hold trivially. So assume $\|f\|_p\|g\|_q > 0$. Fix x and set

$$F(x) = \frac{|f(x)|^p}{\|f\|_p^p} \quad\text{and}\quad G(x) = \frac{|g(x)|^q}{\|g\|_q^q}.$$

By the lemma with $t = 1/p$,

$$\frac{|f(x)|}{\|f\|_p}\cdot\frac{|g(x)|}{\|g\|_q} = F(x)^{1/p}G(x)^{1/q} \le \frac{F(x)}{p} + \frac{G(x)}{q} = \frac{|f(x)|^p}{p\|f\|_p^p} + \frac{|g(x)|^q}{q\|g\|_q^q}. \tag{$\dagger$}$$

Integrating we obtain

$$\frac{1}{\|f\|_p\|g\|_q}\int |f(x)g(x)|\,d\mu(x) \le \frac{\|f\|_p^p}{p\|f\|_p^p} + \frac{\|g\|_q^q}{q\|g\|_q^q} = 1,$$

which gives the desired inequality.

For the second part of the theorem, we may again assume that $\|f\|_p\|g\|_q > 0$, since if $\|f\|_p = 0$, say, then the desired equality holds with $a = 1$ and $b = 0$. With this assumption we see that $\|fg\|_1 = \|f\|_p\|g\|_q$ iff the integral of the right side of ($\dagger$) equals the integral of the left side iff the two sides are equal a.e. iff $F(x) = G(x)$ a.e. (by the lemma). $\square$

The numbers p and q in the theorem are called **conjugate exponents**. In the special case $p = q = 2$, the inequality in 4.1.2 is known as the *Cauchy-Schwarz inequality*.

4.1.3 Minkowski's Inequality. *Let $f, g \in L^p(\mu)$ $(1 < p < \infty)$. Then $f + g \in L^p$ and $\|f+g\|_p \le \|f\|_p + \|g\|_p$. Moreover, equality holds iff there exist nonnegative constants a and b not both zero such that $af = bg$ a.e.*

Proof. Since $|f+g|^p \le 2^p(|f|^p + |g|^p)$, $f + g \in L^p$. For the inequality, apply Hölder's inequality to the conjugate exponents p and $q := p/(p-1)$ to obtain

$$\begin{aligned}\|f+g\|_p^p = \int |f+g|^p &\le \int |f|\,|f+g|^{p-1} + \int |g|\,|f+g|^{p-1}\\ &\le \left(\int |f|^p\right)^{1/p}\left(\int |f+g|^p\right)^{1/q} + \left(\int |g|^p\right)^{1/p}\left(\int |f+g|^p\right)^{1/q}\\ &= (\|f\|_p + \|g\|_p)\,\|f+g\|_p^{p-1},\end{aligned} \tag{α}$$

which is equivalent to $\|f+g\|_p \le \|f\|_p + \|g\|_p$.

Now suppose that $\|f+g\|_p = \|f\|_p + \|g\|_p$. Then the inequalities in (α) are equalities. From the second of these we have

$$\int |f|\,|f+g|^{p-1} = \left(\int |f|^p\right)^{1/p}\left(\int |f+g|^p\right)^{1/q}$$

and

$$\int |g|\,|f+g|^{p-1} = \left(\int |g|^p\right)^{1/p}\left(\int |f+g|^p\right)^{1/q},$$

hence, by the second part of 4.1.2, there exist nonnegative constants a_1, b_1 not both zero and nonnegative constants a_2, b_2 not both zero such that

$$a_1|f|^p = b_1|f+g|^p \text{ and } a_2|g|^p = b_2|f+g|^p \text{ a.e.} \tag{β}$$

Now, if $f+g=0$ a.e., then $0 = \|f\|_p + \|g\|_p$, hence $1\cdot f = 0 = 1\cdot g$ a.e. Also, if $f=0$ a.e., then $1\cdot f = 0\cdot g$, and similarly if $g=0$ a.e. Thus we may suppose that $\mu\{f+g\neq 0\}>0$, $\mu\{f\neq 0\}>0$, and $\mu\{g\neq 0\}>0$. It follows that none of the constants a_j and b_j is zero and so by (β)

$$|f| = a|f+g| = b|g| \text{ a.e.}, \tag{γ}$$

for suitable positive constants a and b. Since the first inequality in (α) is an equality,

$$|f|\,|f+g|^{p-1} + |g|\,|f+g|^{p-1} = |f+g|^p \text{ a.e.},$$

hence

$$|f+g| = |f| + |g| = (1+b)|g| \text{ a.e. on the set } E := \{f+g\neq 0\}.$$

Therefore,

$$\left|1+\frac{f}{g}\right| = 1+b = 1+\left|\frac{f}{g}\right| \text{ a.e. on } E.$$

It follows that f/g is real and nonnegative and so $f = bg$ a.e. on E. But by (γ), $f=g=0$ a.e. on E^c. Therefore, $f = bg$ a.e. on X. □

4.1.4 Theorem. *L^p is a linear space over $\mathbb{K}$ with respect to pointwise addition and scalar multiplication of functions. Moreover, $\|\cdot\|_p$ is a norm relative to which L^p is a Banach space.*

Proof. That L^p is closed under scalar multiplication is clear. The triangle inequality for the case $1<p<\infty$ is Minkowski's inequality. In particular, L^p is closed under addition and so is a linear space. The remaining properties of a norm are clear. It remains to prove that L^p is complete. For this we use the series characterization of completeness given in 0.4.3. Let (f_n) be a sequence in L^p such that $\sum_{n=1}^\infty \|f_n\|_p < \infty$ and set

$$g_n = \sum_{k=1}^n |f_k| \text{ and } g := \sum_{k=1}^\infty |f_k|,$$

so that $g_n \uparrow g$ and $\|g_n\|_p \le \sum_{k=1}^n \|f_k\|_p$. By the monotone convergence theorem, $\|g_n\|_p \uparrow \|g\|_p$, hence $\|g\|_p \le \sum_{k=1}^\infty \|f_k\|_p < \infty$, Therefore, g^p is integrable, hence the series $f := \sum_{k=1}^\infty f_k$ is finite a.e. Defining f to be zero where the series fails to converge, we see that f is measurable and satisfies

$$\left|f - \sum_{k=1}^n f_k\right|^p \le g^p \text{ and } \lim_n \left|f - \sum_{k=1}^n f_k\right|^p = 0 \text{ a.e.}$$

By the dominated convergence theorem, $\lim_n \|f - \sum_{k=1}^n f_k\|_p = 0$. □

The following generalization of Minkowski's inequality will be needed in Chapter 6.

4.1.5 Minkowski's Integral Inequality. *Let $(X, \mathcal{F}, \mu)$ and $(Y, \mathcal{G}, \nu)$ be σ-finite measure spaces, f a nonnegative $\mathcal{F} \otimes \mathcal{G}$-measurable function, and $1 \le p < \infty$. Then*

$$\left[\int \left(\int f(x,y)\,d\nu(y)\right)^p d\mu(x)\right]^{1/p} \le \int \left[\int f(x,y)^p\,d\mu(x)\right]^{1/p} d\nu(y), \tag{4.1}$$

provided the integrals in these expressions are finite.

Proof. For $p = 1$ the inequality is actually equality and is a consequence of Fubini's theorem. Now let $1 < p < \infty$ and let q be conjugate to p. Set $h(x) := \int f(x,y)\,d\nu(y)$, so that the left side of (4.1) is $\|h\|_p$. We may assume that $\|h\|_p > 0$. Set

$$g(x) := \begin{cases} \|h\|_p^{1-p}\,h(x)^{p-1} & \text{if } h(x) \ne 0, \\ 0 & \text{otherwise.} \end{cases}$$

Then

$$g(x)^q = \|h\|_p^{q-qp}\,h(x)^{qp-q} = \|h\|_p^{-p}\,h(x)^p, \quad \|g\|_q = 1, \text{ and } \int hg\,d\mu = \|h\|_p. \tag{†}$$

By Fubini's theorem and Hölder's inequality,

$$\int hg\,d\mu = \iint f(x,y)g(x)\,d\mu(x)\,d\nu(y) \le \int \left[\int f(x,y)^p\,d\mu(x)\right]^{1/p} \|g\|_q\,d\nu(y),$$

which, by virtue of (†), reduces to the desired inequality. □

The Case $p = \infty$

The space of L^∞ **functions** on X is defined by

$$L^\infty(X, \mathcal{F}, \mu) = \{f : X \to \mathbb{K} : f \text{ is } \mathcal{F}\text{-measurable and } \|f\|_\infty < \infty\}, \text{ where}$$
$$\|f\|_\infty := \sup\{t : \mu\{|f| > t\} > 0\}.$$

The quantity $\|f\|_\infty$ is called the L^∞ **norm of** f. As in the case $p < \infty$, for $\|f\|_\infty$ to be an actual norm we must (and do) identify functions that agree a.e. (see 4.1.7 below). The following properties of $\|\cdot\|_\infty$ will be needed.

4.1.6 Proposition. *Let f be measurable. Then*

(a) $|f| \le \|f\|_\infty$ *a.e.*

(b) $\|f\|_\infty = \inf\{t > 0 : |f| \le t$ *a.e.*$\}$.

(c) $f \in L^\infty$ *iff there exists* $0 < t < \infty$ *such that* $|f| \le t$ *a.e.*

Proof. (a) We may assume that $\|f\|_\infty < \infty$. Set $A_n := \{|f| > \|f\|_\infty + 1/n\}$ and $A = \bigcup_n A_n$. By definition of $\|f\|_\infty$, $\mu(A_n) = 0$ for all n, hence $\mu(A) = 0$. Since $|f| \le \|f\|_\infty + 1/n$ on A^c, $|f| \le \|f\|_\infty$ a.e.

(b) Let α denote the infimum. By (a), $\alpha \le \|f\|_\infty$. For the reverse inequality, let $|f| \le t$ a.e. If $\|f\|_\infty > t$, there would exist x with $t < x \le \|f\|_\infty$ and $\mu(|f| > t) \ge \mu(|f| > x) > 0$, impossible. Thus $\|f\|_\infty \le t$, and taking infima over all such t yields $\|f\|_\infty \le \alpha$.

Part (c) follows from (a) and (b). □

4.1.7 Theorem. *L^∞ is a linear space over $\mathbb{K}$ with respect to pointwise addition and scalar multiplication of functions. Moreover, $\|\cdot\|_\infty$ is a norm relative to which L^∞ is a Banach space.*

Proof. Let $f,\, g \in L^\infty$ and $c \in \mathbb{K}$. The inequalities

$$|cf| = |c|\,|f| \le |c|\,\|f\|_\infty \quad \text{and} \quad |f+g| \le |f| + |g| \le \|f\|_\infty + \|g\|_\infty \quad \text{(a.e.)}$$

show that L^∞ is a linear space and that $\|f\|_\infty$ satisfies the triangle inequality. Moreover, from 4.1.6, $\|f\|_\infty \ge 0$, equality holding iff $f = 0$ a.e.

To see that L^∞ is complete, we use 0.4.3 again. Let (f_n) be a sequence in L^∞ such that $\sum_{n=1}^\infty \|f_n\|_\infty < \infty$. By 4.1.6, the sets $N_k := \{|f_k| > \|f_k\|_\infty\}$ have measure zero, hence so does $N := \bigcup_k N_k$. Moreover, the series $\sum_{n=1}^\infty |f_n|$ converges on N^c, hence the function $f := \mathbf{1}_{N^c} \sum_{n=1}^\infty f_n$ is finite a.e., measurable, and is a version of $\sum_{n=1}^\infty f_n$. Since

$$\left| f - \sum_{k=1}^n f_k \right| = \left| \sum_{k>n} f_k \right| \le \sum_{k>n} \|f_k\|_\infty \quad \text{a.e.},$$

by 4.1.6(b) we have

$$\left\| f - \sum_{k=1}^n f_k \right\|_\infty \le \sum_{k>n} \|f_k\|_\infty .$$

This shows that $f \in L^\infty$ and that $\sum_{n=1}^\infty f_n$ converges to f in the L^∞ form. □

Hölder's inequality may now be extended to the case $1 \le p \le \infty$, where $\frac{1}{\infty} := 0$:

$$\|fg\|_1 \le \|f\|_1\, \|g\|_\infty, \quad f \in L^1,\ g \in L^\infty.$$

The verification is left to the reader.

The Case $0 < p < 1$

The definitions of $\|f\|_p$ and $L^p(\mu)$ for the case $1 \le p < \infty$ clearly make sense for $0 < p < 1$. However, for such p the notion of conjugate exponents is not possible, hence Hölder's inequality is not available. Furthermore, it easy to see that Minkowski's inequality does not hold. Indeed, if A and B are disjoint sets of positive measure a and b, respectively, then the triangle inequality for $f = \mathbf{1}_A$ and $g = \mathbf{1}_B$ reduces to $(a+b)^{1/p} \le a^{1/p} + b^{1/p}$, which is clearly false. On the other hand, it *is* the case that

$$\|f+g\|_p^p \le \|f\|_p^p + \|g\|_p^p,$$

(Ex. 4.2), which implies that $L^p(\mu)$ is a linear space and $d(f,g) = \|f-g\|$ is a metric. One may prove, as in the case $p \ge 1$, that $L^p(\mu)$ is complete in this metric.

ℓ^p-Spaces

An important special case of an L^p space is obtained by taking $X = \mathbb{N}$ and $\mu =$ counting measure on $\mathbb{N}$. In this case we write $\ell^p(\mathbb{N})$ instead of $L^p(\mathbb{N})$. Thus for $1 \le p < \infty$,

$$\ell^p(\mathbb{N}) := \left\{ \boldsymbol{x} := (x_n) : x_n \in \mathbb{K},\ \|\boldsymbol{x}\|_p^p = \sum_{n=1}^\infty |x_n|^p < \infty \right\},$$

and for $p = \infty$

$$\ell^\infty(\mathbb{N}) := \left\{ \boldsymbol{x} := (x_n) : x_n \in \mathbb{K},\ \|\boldsymbol{x}\|_\infty = \sup_n |x_n| < \infty \right\}.$$

Note that $\mathbb{K}^d$ may be identified with a linear subspace of $\ell^p(\mathbb{N})$ and, as such, inherits the ℓ^p norm. The case $p = 2$ is simply the Euclidean norm.

Exercises

4.1 Let $a, b > 0$ and $p \geq 1$. Prove that $(a+b)^p \leq 2^{p-1}(a^p + b^p)$. ⟦Consider $\varphi(x) = x^p$.⟧

4.2 Let $a, b > 0$ and $0 < p < 1$. Prove that $(a+b)^p \leq a^p + b^p$. ⟦Consider the function $\varphi(x) = a^p + x^p - (a+x)^p$.⟧

4.3 Show that the mapping $(f, g) \to \int fg : L^2(\mu) \times L^2(\mu) \to \mathbb{C}$ is continuous in the L^2 norm.

4.4 Show that $f, g \in L^p(\mu) \Rightarrow f \vee g, f \wedge g \in L^p(\mu)$.

4.5 Prove *Hölder's equality*: If $1 < p < \infty$, $p^{-1} + q^{-1} = 1$, and $f \in L^q(\mu)$, then

$$\int |f| \cdot |f|^{q/p}\, d\mu = \left(\int |f|^q\, d\mu\right)^{1/p} \left(\int |f|^q\, d\mu\right)^{1/q}.$$

4.6 Let f_i be measurable and $p_i > 1$ such that $\sum_{i=1}^n 1/p_i = 1$ (*generalized conjugate exponents*). Prove the *generalized Hölder's inequality*

$$\int \prod_{i=1}^n |f_i|\, d\mu \leq \prod_{i=1}^n \|f_i\|_{p_i}.$$

4.7 Let f be continuous and bounded on $\mathbb{R}^d$. Show that $\|f\|_\infty = \sup\{|f(x)| : x \in \mathbb{R}^d\}$ (relative to Lebesgue measure).

4.8 Let $f : X \to \mathbb{C}$ be measurable. The **essential range of** f is defined as

$$\operatorname{ran_e}(f) = \{z \in \mathbb{C} : \mu\{|f - z| < \varepsilon\} > 0 \text{ for all } \varepsilon > 0\}.$$

Prove:

(a) $\operatorname{ran_e}(f)$ is closed and contained in $\operatorname{cl} f(X)$.

(b) $f = g$ a.e. $\Rightarrow \operatorname{ran_e}(f) = \operatorname{ran_e}(g)$.

(c) $\operatorname{ran_e}(f) = \bigcap_{f=g \text{ a.e.}} \operatorname{cl}\big(g(X)\big)$.

(d) If $f \in L^\infty$, then $\operatorname{ran_e}(f)$ is compact and $\|f\|_\infty = \sup\{x : x \in \operatorname{ran_e}(|f|)\}$.

4.9 Let $1 < p < \infty$, $0 < r < 1$, and $f \in L^p\big((0, \infty), \lambda\big)$. Define $g(x, y) := f(x)x^{-1}\sin(xy)$.

(a) Show that $g(\cdot, y)$ is integrable for each $y > 0$.

(b) Define $h(y) = \int_0^\infty g(x, y)\, dx$. Prove that

$$\lim_{t \downarrow 0} \frac{|h(y + t^p) - h(y)|}{t^r} \quad \text{uniformly in } y > 0.$$

4.10 Let $1 \leq p < q < \infty$. Prove:

(a) If $\mu(X) < \infty$, then $L^\infty \subseteq L^q \subseteq L^p \subseteq L^1$, where the inclusions may be strict.

(b) $\ell^\infty \supseteq \ell^q \supseteq \ell^p \supseteq \ell^1$, where the inclusions are strict.

(c) For $1 \leq p < q \leq \infty$, neither of the spaces $L^p(\lambda^1)$ or $L^q(\lambda^1)$ is contained in the other.

4.11 Let μ be a probability measure and let f and g be positive and measurable such that $fg \geq 1$. Prove that $\int f\, d\mu \int g\, d\mu \geq 1$.

4.12 Let $\mu(X) < \infty$ and f bounded and measurable. Show that $\lim_{p\to\infty} \|f\|_p = \|f\|_\infty$ via the following steps.

(a) For $1 \leq q < p < \infty$, $\|f\|_p^p \leq \|f\|_q^q \|f\|_\infty^{p-q}$.

(b) $\overline{\lim}_{p\to\infty} \|f\|_p \leq \|f\|_\infty$.

(c) Assume $\|f\|_\infty > 0$. Let $0 < r < \|f\|_\infty$ and $r < t \leq \|f\|_\infty$ such that $\mu(E_t) > 0$, where $E_t = \{|f| > t\} > 0$. Then $\underline{\lim}_{p\to\infty} \|f\|_p \geq r$.

(d) Conclude that $\underline{\lim}_{p\to\infty} \|f\|_p \geq \|f\|_\infty$.

4.13 Let $1 \le p,\ q,\ r < \infty$, $r^{-1} = p^{-1} + q^{-1}$. Prove that if $f \in L^p$ and $g \in L^q$, then $fg \in L^r$ and $\|fg\|_r \le \|f\|_p \|g\|_q$.

4.14 Let f and g be nonnegative and measurable and $0 < p < q < r < \infty$. Prove:

(a) $\left(\int fg^q\, d\mu\right)^{r-p} \le \left(\int fg^p\, d\mu\right)^{r-q} \left(\int fg^r\, d\mu\right)^{q-p}$.

(b) $\left(\int fg\, d\mu\right)^{r} \le \left(\int f\, d\mu\right)^{r-1} \left(\int fg^r\, d\mu\right)$ for $r > 1$.

4.15 Let $0 \le p < r < q \le \infty$. Prove the following:

(a) $L^r \subseteq L^p + L^q$. ⟦For $f \in L^r$, let $A = \{|f| > 1\}$ and consider $f \cdot \mathbf{1}_A$ and $f \cdot \mathbf{1}_{A^c}$.⟧

(b) $L^p \cap L^q \subseteq L^r$ and for $f \in L^p \cap L^q$,

$$\|f\|_r \le \|f\|_p^s \|f\|_q^t, \quad \text{where } s := \frac{r^{-1} - q^{-1}}{p^{-1} - q^{-1}} \quad \text{and } t := \frac{p^{-1} - r^{-1}}{p^{-1} - q^{-1}}.$$

⟦$s + t = 1$. If $q = \infty$, then $rs/p = 1$; if $q < \infty$, then p/sr and q/tr are conjugate exponents.⟧

(c) $\|f\|_r \le \max\{\|f\|_p, \|f\|_q\}$.

(d) If $f \in L^p \cap L^\infty$, then $\lim_{r\to\infty} \|f\|_r = \|f\|_\infty$. ⟦Use (b) for one inequality. For the reverse inequality, note that $\|f\|_r^r \ge M^r \mu\{|f| \ge M\}$.⟧

4.16 Let $T : L^1(\mu) \to L^1(\mu)$ be a continuous linear transformation, and let $g(t, x)$ be continuous in $t \in [a, b]$ for each $x \in X$ and measurable in $x \in X$ for each t. and set $g_t = g(t, \cdot)$. Suppose that there exists an integrable function $h \ge 0$ such that $|g(t, x)| \le h(x)$ for all t and x. Let $\int_a^b g_t\, dt$ denote the function $x \mapsto \int_a^b g_t(x)\, dt$. Assume that $[Tg_t](x)$ is continuous in t for each $x \in X$. Carry out the following to show that $\int_a^b g_t\, dt$ is in L^1 and

$$T \int_a^b g_t\, dt = \int_a^b Tg_t\, dt. \qquad (\dagger)$$

(a) Let $(\mathcal{P}_n, t_n)$ be any sequence of tagged partitions of $[a, b]$ with $\|\mathcal{P}_n\| \to 0$ and let $S(g, \mathcal{P}_n, t_n)$ denote the function

$$x \mapsto S(g(\cdot, x), \mathcal{P}_n, t_n) = \sum_{I \in \mathcal{P}_n} g(t_{j,n}, x)\, |I|$$

Then $|S(g(\cdot, x), \mathcal{P}_n, t_n)| \le (b - a)h(x)$.

(b) $\int_a^b g_t\, dt \in L^1$ and $\lim_n S(g, \mathcal{P}_n, t_n) = \int_a^b g_t\, dt$ in the L^1 norm.

(c) $TS(g, \mathcal{P}_n, t_n) = \sum_{I \in \mathcal{P}_n} [Tg_{t_{j,n}}]\, |I| \to T \int_a^b g_t\, dt$ in the L^1 norm.

4.2 L^p Approximation

In this section we prove three approximation theorems that are useful in establishing certain properties of L^p functions, as illustrated by Corollary 4.2.3 below.

Approximation by Simple Functions

4.2.1 Theorem. *Let $(X, \mathcal{F}, \mu)$ be a measure space and $1 \le p \le \infty$. For each $f \in L^p(\mu)$ and $\varepsilon > 0$ there exists a simple function f_s such that $|f_s| \le |f|$ and $\|f - f_s\|_p < \varepsilon$. Moreover, if $p < \infty$, then f_s may be chosen to vanish outside a set of finite measure.*

Proof. Let $\{f_n\}$ be a sequence of simple functions such that $f_n \to f$ and $|f_n| \le |f|$ (2.3.1). The case $p = \infty$ follows from part (c) of that theorem. Assume that $p < \infty$. Then $|f_n - f|^p \le 2^{p+1}|f|^p$, hence $\|f_n - f\|_p \to 0$ by the dominated convergence theorem. The first assertion of the theorem follows by taking $f_s = f_n$ for sufficiently large n. For the second assertion, let $f_s = \sum_{k=1}^m a_k \mathbf{1}_{A_k}$, where $a_k \ne 0$ and the sets A_k are disjoint. Then

$$\int |f_s|^p \, d\mu = \sum_{k=1}^m |a_k|^p \mu(A_k).$$

Since the integral is finite and $a_k \ne 0$, $\mu(A_k) < \infty$. Therefore, $f_s = 0$ outside $\bigcup_{k=1}^m A_k$, a set of finite measure. □

Approximation by Continuous Functions

4.2.2 Theorem. *Let $1 \le p < \infty$, $f \in L^p(\lambda^d)$, and $\varepsilon > 0$. Then there exists a continuous function g vanishing outside a bounded interval such that $\|f - g\|_p < \varepsilon$.*

Proof. By 4.2.1, we may assume that f is simple with standard representation $\sum_{k=1}^m a_k \mathbf{1}_{A_k}$, where $a_k \ne 0$ and $\lambda^d(A_k) < \infty$. We may further assume that A_k is bounded, otherwise replace A_k by $A_k \cap I$, where I is a bounded interval with $\lambda^d(A_k) - \lambda^d(A_k \cap I)$ sufficiently small so that f may be approximated by $\sum_{k=1}^m a_k \mathbf{1}_{A_k \cap I}$.

Now let $\alpha > 0$. By 1.8.1 we may choose for each k a compact set C_k and a bounded open set U_k such that $C_k \subseteq A_k \subseteq U_k$ and $\lambda^d(U_k \setminus C_k) < \alpha$. By 0.3.10, there exists a continuous function $g_k : \mathbb{R}^d \to [0, 1]$ such that $g_k = 1$ on C_k and $g_k = 0$ on U_k^c. Since $g_k = \mathbf{1}_{A_k}$ on $U_k^c \cup C_k = (U_k \setminus C_k)^c$,

$$\|a_k \mathbf{1}_{A_k} - a_k g_k\|_p^p = |a_k|^p \int_{U_k \setminus C_k} |\mathbf{1}_{A_k} - g_k|^p \, d\lambda^d \le 2^p |a_k|^p \lambda(U_k \setminus C_k) < (2M)^p \alpha,$$

where $M := \sup_k |a_k|$. The function $g := \sum_{k=1}^m a_k g_k$ is continuous, and by the triangle inequality $\|f - g\|_p < 2mM\alpha^{1/p}$. We then have $\|f - g\|_p < \varepsilon$ for sufficiently small α. Furthermore, $g = 0$ on $\bigcup_k U_k$, which is contained in a bounded interval. □

Here is an important application of 4.2.2.

4.2.3 Corollary. *Let $1 \le p < \infty$, and for $y \in \mathbb{R}^d$ let T_y be the translation operator $T_y f(x) = f(x + y)$. Then for each $f \in L^p(\mathbb{R}^d, \lambda)$, $\lim_{y \to y_0} \|T_y f - T_{y_0} f\|_p = 0$.*

Proof. By translation invariance of the integral, we may take $\mathbf{y}_0 = \mathbf{0}$. By the theorem, given $\varepsilon > 0$ there exists continuous function g such that $\|f - g\|_p < \varepsilon$ and $g = 0$ on the complement of some interval $[\mathbf{a}, \mathbf{b}]$. By translation invariance, $\|T_y f - T_y g\|_p = \|f - g\|_p$, hence

$$\|T_y f - f\|_p \le \|T_y f - T_y g\|_p + \|T_y g - g\|_p + \|g - f\|_p < 2\varepsilon + \|T_y g - g\|_p.$$

It now suffices to prove that $\lim_{y \to 0} \|T_y g - g\|_p = 0$. Let $\mathbf{c} = (1, \ldots, 1)$ and let $\mathbf{y}_n \to \mathbf{0}$ such that $|y_{n,j}| < 1$ $(1 \le j \le d)$. For $\mathbf{x} \in [\mathbf{a} - \mathbf{c}, \mathbf{b} + \mathbf{c}]^c$, $\mathbf{x} + \mathbf{y}_n \in [\mathbf{a}, \mathbf{b}]^c$, hence $g(\mathbf{x} + \mathbf{y}_n) = 0$. Thus if M is a bound for $|g|$, then

$$|g(\mathbf{x} + \mathbf{y}_n) - g(\mathbf{x})|^p \le 2M^p \mathbf{1}_{[\mathbf{a}-\mathbf{c}, \mathbf{b}+\mathbf{c}]}, \quad \mathbf{x} \in \mathbb{R}^d.$$

By continuity of g, the left side of the inequality tends to zero so, by the dominated convergence theorem, $\int |g(\mathbf{x} + \mathbf{y}_n) - g(\mathbf{x})|^p \, d\lambda^d \to 0$. □

Approximation by Step Functions

A **step function** on $\mathbb{R}^d$ is a simple function of the form $\sum_{i=1}^{n} a_i \mathbf{1}_{I_j}$, where I_j is a bounded open d-dimensional interval. The following result complements 4.2.2.

4.2.4 Theorem. *Let $1 \le p < \infty$, $f \in L^p(\lambda^d)$, and $\varepsilon > 0$. Then there exists a step function h vanishing outside a bounded interval such that $\|f - h\|_p < \varepsilon$.*

Proof. Refer to the proof of 4.2.2. Let $\beta > 0$. Since U_k is a countable disjoint union of open intervals, we may choose disjoint open intervals with union $V_k \subseteq U_k$ such that $\lambda(U_k \setminus V_k) < \beta$. Now,

$$A_k \triangle V_k = (A_k \setminus V_k) \cup (V_k \setminus A_k) \subseteq (U_k \setminus V_k) \cup (U_k \setminus C_k),$$

hence

$$\|\mathbf{1}_{A_k} - \mathbf{1}_{V_k}\|_p^p = \int |\mathbf{1}_{A_k} - \mathbf{1}_{V_k}|^p \, d\lambda^d = \lambda^d(A_k \triangle V_k) < \alpha + \beta.$$

The function $h := \sum_{k=1}^{m} a_k \mathbf{1}_{V_k}$ is a step function, and by the triangle inequality we have $\|f - h\|_p < 2mM(\alpha + \beta)^{1/p}$. Thus for sufficiently small α and β, $\|f - h\|_p < \varepsilon$. □

Exercises

4.17 [↑3.2.16] Let D_r be the dilation operator $D_r f(x) = f(rx)$ on $L^p(\mathbb{R}^d)$, $1 \le p < \infty$. Show that $\lim_{r \to s} \|D_r f - D_s f\|_p = 0$, $r, s > 0$.

4.18 Let $f \in L^1(\mathbb{R})$ and let g be bounded with bounded continuous derivative. Prove that

$$\lim_n \int f(x) g'(nx) \, d\lambda(x) = 0.$$

4.19 Show that the last assertion of 4.2.1 fails for the case $p = \infty$. Show also that 4.2.2 does not hold for $p = \infty$.

4.3 L^p Convergence

Let $f_n, f \in L^p(X, \mathcal{F}, \mu)$ $(p \ge 1)$. Convergence of f_n to f in the L^p norm is called L^p **convergence** and is written $f_n \xrightarrow{L^p} f$. For example, the approximation theorems in the preceding section may be phrased in terms of L^p convergence. The results in the present section relate L^p convergence to various modes of convergence considered in §2.4. The case $p = \infty$ is easy to treat:

4.3.1 Theorem. *Let $f_n, f \in L^\infty$. Then $f_n \xrightarrow{L^\infty} f$ iff there exists a set A of measure zero such that $f_n \to f$ uniformly on A^c. In particular, $f_n \xrightarrow{a.u.} f$.*

Proof. Let $f_n \xrightarrow{L^\infty} f$ and let A_n be a set of measure zero such that $|f_n - f| \le \|f_n - f\|_\infty$ on A_n^c (4.1.6). Set $A = \bigcup_n A_n$. Then on A^c, $|f_n - f| \le \|f_n - f\|_\infty$ for all n, hence $f_n \to f$ uniformly on A^c.

Conversely, let $\mu(A) = 0$ and $f_n \to f$ uniformly on A^c. Given $\varepsilon > 0$, choose N so that $|f_n - f| \le \varepsilon$ on A^c for all $n \ge N$. By 4.1.6(b), for such n we have $\|f_n - f\|_\infty \le \varepsilon$. Therefore, $f_n \xrightarrow{L^\infty} f$. □

The case $1 \le p < \infty$ is more delicate. We shall need the following lemma.

4.3.2 Lemma. *Let $1 \le p < \infty$ and f, $f_n \in L^p$. If $\|f_n\|_p \to \|f\|_p$ and $f_n \overset{a.e.}{\to} f$ then $f_n \overset{L^p}{\to} f$.*

Proof. From the inequality $|f_n - f|^p \le 2^p(|f_n|^p + |f|^p)$ we have $2^p(|f_n|^p + |f|^p) - |f_n - f|^p \ge 0$. Moreover,

$$\lim_n \left[2^p(|f_n|^p + |f|^p) - |f_n - f|^p\right] = 2^{p+1}|f|^p \quad \text{a.e.}$$

Thus by Fatou's lemma

$$\begin{aligned} 2^{p+1}\int |f|^p\, d\mu &\le \underline{\lim}_n \int \left[2^p(|f_n|^p + |f|^p) - |f_n - f|^p\right] d\mu \\ &= 2^{p+1}\int |f|^p\, d\mu - \overline{\lim}_n \int |f_n - f|^p\, d\mu. \end{aligned}$$

Therefore, $\overline{\lim}_n \int |f_n - f|^p\, d\mu = 0$, hence $f_n \overset{L^p}{\to} f$. □

The following result characterizes L^p convergence in terms of convergence in measure.

4.3.3 Theorem. *Let $1 \le p < \infty$ and f, $f_n \in L^p$. Then $f_n \overset{L^p}{\to} f$ iff both $f_n \overset{\mu}{\to} f$ and $\|f_n\|_p \to \|f\|_p$. In this case, there exists a subsequence $f_{n_k} \overset{a.e.}{\to} f$.*

Proof. The necessity follows from the inequalities $\big|\|f\|_p - \|f_n\|_p\big| \le \|f - f_n\|_p$ and

$$\mu\big(|f_n - f| \ge \varepsilon\big) = \int \mathbf{1}_{\{|f_n - f| \ge \varepsilon\}}\, d\mu = \int \mathbf{1}_{\{|f_n - f|^p \ge \varepsilon^p\}}\, d\mu \le \varepsilon^{-p}\int |f_n - f|^p\, d\mu.$$

For the sufficiency, suppose for a contradiction that $\|f_n - f\|_p \not\to 0$. Then there exists an $\varepsilon > 0$ and an infinite subset S of $\mathbb{N}$ such that $\|f_n - f\|_p \ge \varepsilon$ for all $n \in S$. Since $f_n \overset{\mu}{\to} f$ holds for subsequences and since convergence in measure implies a.e. convergence for some subsequence (2.4.4), we may choose a subsequence (f_{n_k}) of (f_n) with indices in S such that $f_{n_k} \overset{a.e}{\to} f$. But then by 4.3.2, $f_{n_k} \overset{L^p}{\to} f$, which is impossible by definition of S. □

A deeper result is the following, whose proof brings together some earlier results on convergence.

4.3.4 Vitali Convergence Theorem I. *Let $1 \le p < \infty$ and f_n, $f \in L^p$ such that $f_n \overset{a.e.}{\to} f$. Then $f_n \overset{L^p}{\to} f$ iff for each $\varepsilon > 0$ the following conditions hold:*

(a) *There exists $A \in \mathcal{F}$ with finite measure such that $\sup_n \|f_n \mathbf{1}_{A^c}\|_p \le \varepsilon$.*

(b) *There exists $\delta > 0$ such that $E \in \mathcal{F}$ and $\mu(E) < \delta \Rightarrow \overline{\lim}_n \|f_n \mathbf{1}_E\|_p \le \varepsilon$.*

Proof. Suppose $f_n \overset{L^p}{\to} f$. To establish (a), choose m so that $\|f_n - f\|_p < \varepsilon/2$ for all $n > m$. For such n and any $E \in \mathcal{F}$,

$$\|f_n \mathbf{1}_E\|_p \le \|f_n - f\|_p + \|f \mathbf{1}_E\|_p \le \varepsilon/2 + \|f \mathbf{1}_E\|_p. \tag{α}$$

By Ex. 3.17, we may choose $E, E_1, \ldots, E_m \in \mathcal{F}$ with finite measure such that

$$\|\mathbf{1}_{E^c} f\|_p, \quad \|\mathbf{1}_{E_n^c} f_n\|_p < \varepsilon/2, \quad n = 1, \ldots, m. \tag{β}$$

Set $A = E \cup E_1 \cup \cdots \cup E_m$. Then, by ($\alpha$) and ($\beta$), $\|f_n \mathbf{1}_{A^c}\|_p \le \varepsilon$ for all n, verifying (a).

To establish (b), choose δ so that $\|\mathbf{1}_E f\|_p < \varepsilon$ for all E with $\mu(E) < \delta$ (Ex. 3.13). For such E and all n,

$$\|f_n \mathbf{1}_E\|_p \le \|(f_n - f)\mathbf{1}_E\|_p + \|f \mathbf{1}_E\|_p \le \|f_n - f\|_p + \varepsilon,$$

hence $\overline{\lim}_n \|f_n\mathbf{1}_E\| \le \varepsilon$.

Conversely, assume that (a) and (b) hold and set $g_n = f - f_n$. For measurable sets $E \subseteq A$, $\mathbf{1}_E + \mathbf{1}_{A\setminus E} + \mathbf{1}_{A^c} = 1$, hence multiplying by g_n and integrating we have

$$\|g_n\|_p \le \|f\mathbf{1}_E\|_p + \|f_n\mathbf{1}_E\|_p + \|g_n\mathbf{1}_{A\setminus E}\|_p + \|f\mathbf{1}_{A^c}\|_p + \|f_n\mathbf{1}_{A^c}\|_p. \qquad (\gamma)$$

We show that the right side of (γ) may be made arbitrarily small. Enlarging A if necessary, we may assume by Ex. 3.17 that $\|f\mathbf{1}_{A^c}\|_p < \varepsilon$. By (a) we then have

$$\|f\mathbf{1}_{A^c}\|_p + \|f_n\mathbf{1}_{A^c}\|_p < 2\varepsilon \text{ for all } n.$$

Now let δ be as in (b). Since $f_n \overset{\text{a.e.}}{\to} f$, by Egoroff's theorem there exists a measurable subset E of A with $\mu(E) < \delta$ such that $g_n \to 0$ uniformly on $A \setminus E$. Therefore,

$$\lim_n \|g_n\mathbf{1}_{A\setminus E}\|_p = 0.$$

Finally, applying (b) and Fatou's lemma, we have

$$\|f\mathbf{1}_E\|_p^p \le \underline{\lim}_n \int_E |f_n|^p \, d\mu \le \overline{\lim}_n \int_E |f_n|^p \, d\mu \le \varepsilon^p.$$

Therefore, by (γ), $\overline{\lim}_n \|g_n\|_p \le 3\varepsilon$, proving that $\|g_n\|_p \to 0$. □

We shall call properties (a) and (b) in the theorem the *Vitali convergence conditions.*

Exercises

4.20 Let μ be a probability measure and $f_n \in L^p(\mu)$. Prove:

(a) If $f_n \to f$ uniformly on X, then $f \in L^p(\mu)$ and $f_n \to f$ in L^p.

(b) If $f_n \overset{\text{a.u.}}{\to} f$ and the functions f_n and f are uniformly bounded, then $f_n \to f$ in L^p.

4.21 Let $\|f_n\|_\infty \le C < \infty$ for all n and $f_n \overset{\text{a.u.}}{\to} f$. Show that $f \in L^\infty$.

4.22 Let $1 \le p < \infty$ and $f, f_n \in L^p$. Show that

(a) $f_n \overset{\text{a.e.}}{\to} f$ does not necessarily imply that $f_n \overset{L^p}{\to} f$.

(b) $f_n \overset{L^p}{\to} f$ does not necessarily imply that $f_n \overset{\text{a.e.}}{\to} f$.

(c) $f_n \overset{\mu}{\to} f$ does not necessarily imply that $f_n \overset{L^p}{\to} f$.

(d) $f_n \overset{L^q}{\to} 0$ for all $1 \le q < p$ does not necessarily imply that $f_n \overset{L^p}{\to} 0$.

(e) $f_n \overset{L^q}{\to} 0$ for all $q > p$ does not necessarily imply that $f_n \overset{L^p}{\to} 0$.

4.23 Show that the hypothesis $f_n \overset{\text{a.e.}}{\to} f$ in the Vitali convergence theorem may be replaced by $f_n \overset{\mu}{\to} f$.

*4.4 Uniform Integrability

Throughout this section, $(X, \mathcal{F}, \mu)$ is a finite measure space.

For a finite measure, additional convergence results may be obtained via the notion of *uniform integrability.* The following proposition motivates the definition.

4.4.1 Proposition. *A measurable function $f : X \to \mathbb{C}$ is integrable iff*

$$\lim_{t\to\infty} \int_{\{|f|\geq t\}} |f|\, d\mu = 0. \tag{4.2}$$

Proof. Suppose that f is integrable. Then $A := \{|f| = \infty\}$ has measure zero. Set $A_n := \{|f| > n\}$. Then $|f| \geq \mathbf{1}_{A_n}|f| \downarrow \mathbf{1}_A|f|$, so by the dominated convergence theorem, $\int_{A_n} |f| \to \int \mathbf{1}_A|f|\, d\mu = 0$, which implies (4.2).

Conversely, suppose that (4.2) holds. Choose t so that $\int_{\{|f|\geq t\}} |f|\, d\mu < 1$. Then

$$\int |f|\, d\mu = \int_{\{|f|\geq t\}} |f|\, d\mu + \int_{\{|f|<t\}} |f|\, d\mu \leq 1 + t\cdot\mu(X) < \infty,$$

hence f is integrable. □

Note that on an infinite measure space a nonzero constant function trivially satisfies (4.2) yet is not integrable. Thus the sufficiency of the proposition fails on infinite measure spaces.

With the preceding proposition in mind, we say that a family $\mathscr{F}$ of measurable functions $f : X \to \overline{\mathbb{K}}$ is **uniformly integrable (u.i.)**, if

$$\lim_{t\to\infty} \sup\left\{\int_{\{|f|\geq t\}} |f|\, d\mu \ :\ f \in \mathscr{F}\right\} = 0. \tag{4.3}$$

By 4.4.1, each member of such a family is integrable. Conversely, by the same proposition, any finite family of integrable functions is u.i. Moreover, it is trivially the case that any uniformly bounded family of measurable functions is uniformly integrable.

The following result is sometimes useful for establishing uniform integrability of a family of functions. The reader should compare the conditions in the theorem with the Vitali convergence conditions (4.3.4).

4.4.2 Theorem. *A family $\mathscr{F}$ of measurable functions is u.i. iff the following conditions hold:*

(a) $\sup\{\|f\|_1 : f \in \mathscr{F}\} < \infty$.

(b) *For each $\varepsilon > 0$ there exists $\delta > 0$ such that* $\sup\{\|f\mathbf{1}_E\|_1 : f \in \mathscr{F}\} < \varepsilon$ *for all $E \in \mathfrak{F}$ with $\mu(E) < \delta$.*

Proof. Suppose that $\mathscr{F}$ is u.i. Given $\varepsilon > 0$, choose t so that

$$\int_{\{|f|\geq t\}} |f|\, d\mu < \frac{\varepsilon}{2} \quad \text{for all } f \in \mathscr{F}.$$

Then for $E \in \mathfrak{F}$ and all $f \in \mathscr{F}$,

$$\int_E |f|\, d\mu = \int_{E\cap\{|f|\geq t\}} |f|\, d\mu + \int_{E\cap\{|f|<t\}} |f|\, d\mu \leq \frac{\varepsilon}{2} + t\cdot\mu(E).$$

Taking $E = X$ establishes (a). To establish (b), choose $\delta < \varepsilon/2t$.

Conversely, assume that (a) and (b) hold. Let $C := \sup\{\|f\|_1 : f \in \mathscr{F}\}$. Given $\varepsilon > 0$, choose t_0 so that $C/t_0 < \delta$, where δ is as in (b). Then for $t \geq t_0$ and all $f \in \mathscr{F}$,

$$\mu\{|f| \geq t\} \leq \frac{1}{t}\int |f|\, d\mu \leq \frac{C}{t} < \delta.$$

Applying (b) with $E := \{|f| \geq t\}$, we have $\int_{\{|f|\geq t\}} |f|\, d\mu < \varepsilon$ for all $f \in \mathscr{F}$. Therefore, $\mathscr{F}$ is u.i. □

The proof of the following corollary is left as an exercise.

4.4.3 Corollary. *Let $\mathscr{F}$ and $\mathscr{G}$ be families of measurable functions such that $\mathscr{G}$ is u.i. Then the following hold:*

(a) $\{\alpha f + \beta g : f, g \in \mathscr{G}\}$ *is u.i.*, $\alpha, \beta \in \mathbb{C}$.

(b) *If for each $f \in \mathscr{F}$ there exists $g \in \mathscr{G}$ such that $|f| \le |g|$ a.e., then $\mathscr{F}$ is u.i.*

Here is a general method for constructing u.i. functions.

4.4.4 Proposition. *Let $\phi : (0, \infty) \to \mathbb{R}$ be Borel measurable such that $\lim_{x\to\infty} \phi(x)/x = \infty$. If $\mathscr{F}$ is a family of measurable functions such that $\sup\{\|\phi(|f|)\|_1 : f \in \mathscr{F}\} < \infty$, then $\mathscr{F}$ is u.i.*

Proof. Let s denote the supremum in the statement of the theorem. Given $\varepsilon > 0$, choose t_0 such that

$$\frac{\phi(t)}{t} > \frac{s\mu(X)}{\varepsilon} \quad \text{for all } t \ge t_0.$$

For such t, $|f| \le \varepsilon(s\mu(X))^{-1}\phi(|f|) \le \varepsilon\mu(X)^{-1}$ on the set $\{|f| \ge t\}$, hence $\int_{\{|f|\ge t\}} |f|\,d\mu \le \varepsilon$ for all $f \in \mathscr{F}$ and $t \ge t_0$. □

For example, taking $\phi(x) = x^p$ $(1 < p < \infty)$ we see that any bounded subset of L^p is uniformly integral.

Here is the main connection between uniform integrability and L^p convergence.

4.4.5 Vitali Convergence Theorem II. *Let f and f_n be measurable and $1 \le p < \infty$. Then $f \in L^p$ and $f_n \xrightarrow{L^p} f$ iff the sequence $(|f_n|^p)$ is u.i. and $f_n \xrightarrow{\mu} f$.*

Proof. Suppose that $(|f_n|^p)$ is u.i. and $f_n \xrightarrow{\mu} f$. By 4.4.2 applied to the functions $|f_n|^p$, we see that the Vitali convergence conditions hold. Therefore, if we show that $f \in L^p$, then, by Ex. 4.23, $f_n \xrightarrow{L^p} f$. By 2.4.4, there exists a subsequence (g_n) of (f_n) such that $g_n \xrightarrow{\text{a.e}} f$. By Fatou's lemma,

$$\int |f|^p\,d\mu \le \varliminf_n \int |g_n|^p\,d\mu.$$

Since the right side is finite (4.4.2(a)), $f \in L^p$.

Conversely, suppose $f \in L^p$ and $f_n \xrightarrow{L^p} f$. Given $\varepsilon > 0$, choose m such that $\|f - f_n\|_p^p < \varepsilon$ for all $n > m$, By 4.4.1, there exists t_0 such that

$$\int_{\{|f_n - f|^p \ge t\}} |f_n - f|^p\,d\mu < \varepsilon \quad \text{for } n = 1, \ldots, m \text{ and } t \ge t_0.$$

For such t and all n we then have $\int_{\{|f_n-f|^p\ge t\}}|f_n - f|^p\,d\mu < \varepsilon$, which shows that the sequence $(|f_n - f|^p)_n$ is u.i. Since $|f_n|^p \le 2^p(|f_n - f|^p + |f|^p)$, it follows from 4.4.3 that $(|f_n|^p)_n$ is u.i. That $f_n \xrightarrow{\mu} f$ follows from 4.3.3. □

Exercises

4.24 Consider Lebesgue measure on $\mathcal{B}(0, 1]$. Show that the sequence of functions $f_n = n\mathbf{1}_{(0,1/n]}$ is not u.i. even though the sequence $(\|f_n\|_1)$ converges.

4.25 Show that $\{f_n\}$ is u.i. iff $\{f_n^+\}$ and $\{f_n^-\}$ are u.i.

4.26 Let (f_n) be a sequence of $\mathcal{F}$-measurable functions such that $\sup_n \int |f_n|^r\,d\mu < \infty$ for some $r > 1$. Show that (f_n) is u.i.

4.27 Let $\{f_n\}$ be u.i. and set $g_n := |f_1| \vee \cdots \vee |f_n|$. Use Ex. 3.20 to show that $\frac{1}{n}\int g_n\,d\mu \to 0$.

*4.5 Convex Functions and Jensen's Inequality

In this section we develop the basic properties of convex functions on intervals (a,b), where $-\infty \le a < b \le \infty$. These properties are then used to establish an important integral inequality.

A real-valued function φ on (a,b) is said to be **convex** if

$$\varphi\big((1-t)u+tv\big) \le (1-t)\varphi(u)+t\varphi(v) \quad \text{for } a<u<v<b \text{ and } t\in(0,1). \tag{4.4}$$

Strict convexity is defined by replacing weak inequality by strict inequality. Thus a function

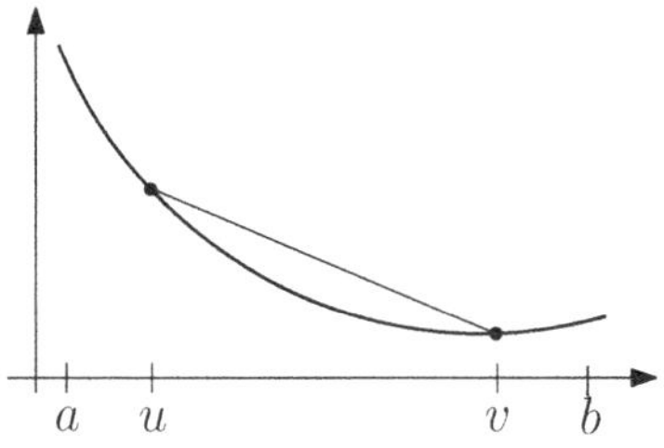

FIGURE 4.1: A strictly convex function.

is convex iff the line segment connecting any two points on its graph lies above the part of the graph between the two points. A function f is **(strictly) concave** if $-f$ is (strictly) convex.

A function φ with an increasing derivative (in particular, a function with a nonnegative second derivative) is convex. Indeed, if $x=(1-t)u+tv$ $(0<t<1)$ then, by the mean value theorem, there exist points $y\in\big(u,x\big)$ and $z\in\big(x,v\big)$ such that

$$\frac{\varphi(x)-\varphi(u)}{x-u}=\varphi'(y)\le\varphi'(z)=\frac{\varphi(v)-\varphi(x)}{v-x}.$$

Solving the inequality for $\varphi(x)$ yields

$$\varphi(x)\le\frac{v-x}{v-u}\varphi(u)+\frac{x-u}{v-u}\varphi(v)=(1-t)\varphi(u)+t\varphi(v).$$

Analogous results hold for the strict case and the concave case. For example, taking second derivatives we see that e^x, and e^{-x} are strictly convex on $\mathbb{R}$, $x\ln x$ and $1/x^p$ $(p>0)$ are strictly convex on $(0,\infty)$, and $\ln x$ and x^p $(0<p<1)$ are strictly concave on $(0,\infty)$.

The basic properties of convex functions are summarized in the following theorem.

4.5.1 Theorem. *Let φ be convex on (a,b).*

(a) *For fixed $z\in(a,b)$, the difference quotients $\dfrac{\varphi(t)-\varphi(z)}{t-z}$ increase in t on the intervals (a,z) and (z,b).*

(b) *The left- and right-hand derivatives*

$$\varphi'_\ell(u)=\lim_{x\to u^-}\frac{\varphi(x)-\varphi(u)}{x-u},\quad \varphi'_r(u):=\lim_{y\to u^+}\frac{\varphi(y)-\varphi(u)}{y-u}$$

exist, are nondecreasing, and satisfy $\varphi'_\ell(x)\le\varphi'_r(x)$.

(c) $\varphi'_\ell(z) \ge \dfrac{\varphi(z)-\varphi(t)}{z-t}$ $(z>t)$ *and* $\varphi'_r(z) \le \dfrac{\varphi(z)-\varphi(t)}{z-t}$ $(z<t)$.

Proof. Let $a<u<x<y<v<b$. The assertions are a consequence of the following numbered inequalities, which are verified below.

$$\frac{\varphi(x)-\varphi(u)}{x-u} \overset{(1)}{\le} \frac{\varphi(y)-\varphi(u)}{y-u}, \quad \frac{\varphi(y)-\varphi(u)}{y-u} \overset{(2)}{\le} \frac{\varphi(v)-\varphi(y)}{v-y}, \quad \frac{\varphi(v)-\varphi(x)}{v-x} \overset{(3)}{\le} \frac{\varphi(v)-\varphi(y)}{v-y}.$$

To prove (a), take $u=z$ in (1) and $v=z$ in (3). For (b), observe that because the difference quotients $[\varphi(x)-\varphi(u)]/(x-u)$ decrease as $x\downarrow u$, $\varphi'_r(u)$ exists in $\overline{\mathbb{R}}$, and by (1) and (2)

$$\varphi'_r(u) \le \frac{\varphi(v)-\varphi(y)}{v-y} < \infty.$$

Letting $v\downarrow y$ shows that $\varphi'_r(u) \le \varphi'_r(y)$. Therefore, φ'_r is increasing. Similarly, since the difference quotients $[\varphi(v)-\varphi(y)]/(v-y)$ increase as $y\uparrow v$, $\varphi'_\ell(v)$ exists in $\overline{\mathbb{R}}$ and by (1) and (2),

$$\varphi'_\ell(v) \ge \frac{\varphi(x)-\varphi(u)}{x-u} > -\infty.$$

Letting $u\uparrow x$ shows that $\varphi'_\ell(v) \ge \varphi'_\ell(x)$. Therefore, φ'_ℓ is increasing. Taking $x=y$ in (2), we have

$$\frac{\varphi(x)-\varphi(u)}{x-u} \le \frac{\varphi(v)-\varphi(x)}{v-x}.$$

Letting $u\uparrow x$ and $v\downarrow x$, we obtain $\varphi'_\ell(x) \le \varphi'_r(x)$. In particular, $\varphi'_\ell(x)$ and $\varphi'_r(x)$ are finite. This proves (a) and (b). Part (c) follows from these.

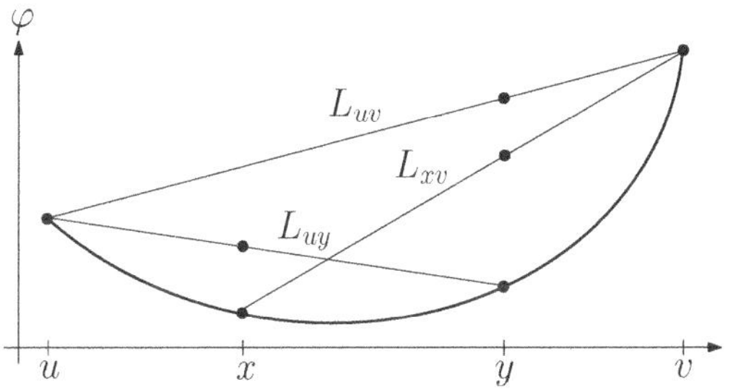

It remains to verify the inequalities (1)–(3) above. For $a<c<d<b$, let L_{cd} denote the function whose graph is the line segment from $(c,\varphi(c))$ to $(d,\varphi(d))$. Since $u<x<y<v$, convexity implies that $\varphi(x) \le L_{uy}(x)$ and $\varphi(y) \le L_{uv}(y)$, hence

$$\frac{\varphi(x)-\varphi(u)}{x-u} \le \frac{L_{uy}(x)-\varphi(u)}{x-u} = \text{slope of } L_{uy} = \frac{\varphi(y)-\varphi(u)}{y-u}$$
$$\frac{\varphi(y)-\varphi(u)}{y-u} \le \frac{L_{uv}(y)-\varphi(u)}{y-u} = \text{slope of } L_{uv} = \frac{L_{uv}(v)-L_{uv}(y)}{v-y}, \text{ and}$$
$$\frac{L_{uv}(v)-L_{uv}(y)}{v-y} < \frac{L_{uv}(v)-\varphi(y)}{v-y} = \frac{\varphi(v)-\varphi(y)}{v-y},$$

verifying (1) and (2). A similar argument establishes (3). □

4.5.2 Corollary. *A convex function is continuous.*

4.5.3 Corollary. *If a convex function φ is differentiable at $x\in(u,v)$, then*

$$\varphi'(x)(t-x)+\varphi(x) \le \varphi(t) \quad \textit{for all } t\in(u,v).$$

That is, the tangent line at $(x,\varphi(x))$ lies below the graph of φ on (u,v).

We may now prove

4.5.4 Jensen's Inequality. *Let $(X, \mathcal{F}, \mu)$ be a probability space and let $\varphi : (a,b) \to \mathbb{R}$ be convex. If $f : X \to (a,b)$ and f, $\varphi \circ f \in L^1$, then*

$$\varphi\left(\int f\,d\mu\right) \le \int \varphi \circ f\,d\mu.$$

Proof. By 4.5.1(c), for fixed $z \in (a,b)$ there exists a constant c such that

$$c(t-z) \le \varphi(t) - \varphi(z) \quad \text{for all } t \in (a,b).$$

Taking $t = f(x)$ and integrating with respect to x yields

$$c\left(\int f\,d\mu - z\right) \le \int \varphi \circ f\,d\mu - \varphi(z).$$

Taking $z = \int f\,d\mu$ produces the desired inequality. □

Note that the inequality in 4.5.4 reverses for concave functions, as may be seen by considering $-\varphi$.

Exercises

4.28 Prove that for $x_j,\ t_j > 0$ and $\sum_{j=1}^n t_j = 1$,

$$\prod_{j=1}^n x_j^{t_j} \le \sum_{j=1}^n t_j x_j.$$

⟦Use the concavity of ln.⟧

4.29 Use Jensen's inequality to verify the following for a probability measure μ:

(a) $\|f\|_p$ is increasing on $(0, \infty]$.

(b) $\|f\|_1 \|1/f\|_p \ge 1 \quad (p > 0)$.

(c) $\|\ln f\|_1 \le \ln \|f\|_1, \quad (f > 0)$.

(d) $\|f\|_1 \ln \|f\|_1 \le \ln \|f \ln f\|_1, \quad (f > 0)$.

4.30 Let μ be a probability measure, $0 < q < \infty$, and $f \in L^q$ with $\|f\|_q > 0$. Verify (a)–(e) and conclude that

$$\lim_{p \to 0} \ln\left(\|f\|_p\right) = \int \ln|f|\,d\mu.$$

(a) The function $p \mapsto \|f\|_p$ is finite and decreasing on $(0, q]$.

(b) $\int \ln|f|\,d\mu \le \ln \|f\|_p$.

(c) $\ln x \le x - 1 \ (x > 0)$, hence $\ln \|f\|_p \le p^{-1} \int (|f|^p - 1)\,d\mu$.

(d) $(a^p - 1)/p \to \ln a$ monotonically as $p \downarrow 0$.

(e) $\lim_{p \to 0} p^{-1} \int (|f|^p - 1)\,d\mu = \int \ln|f|\,d\mu$.

Chapter 5

Differentiation

In this chapter we consider countably additive set functions that take values in $\overline{\mathbb{K}}$, so-called *signed* and *complex measures*. These set functions play an important role in the description of linear functionals on spaces of continuous functions a topic considered in Chapter 7, as well as in harmonic analysis, developed in Chapter 16. The main result of the chapter is the Radon-Nikodym theorem, which establishes in terms of integrals the existence of a derivative of one measure with respect to another. This notion of measure differentiation is made concrete for Lebesgue-Stieltjes measures on $\mathbb{R}^d$.

Throughout the chapter, $(X, \mathcal{F})$ denotes an arbitrary measurable space.

5.1 Signed Measures

In this section we discuss countably additive, $\overline{\mathbb{R}}$-valued set functions μ on $(X, \mathcal{F})$. An example is the difference of two measures, one of which is finite. It turns out that every such set function μ is of this form, a fact of considerable importance and whose proof is the main goal of the current section.

Definition and a Fundamental Example

A **signed measure** on $(X, \mathcal{F})$ is a $\overline{\mathbb{R}}$-valued set function μ with the following properties:

(a) $\mu(\emptyset) = 0$.

(b) The range of μ contains at most one of the values $-\infty$, ∞.

(c) If (A_n) is a sequence of disjoint sets in $\mathcal{F}$, then $\mu\left(\bigcup_n A_n\right) = \sum_n \mu(A_n)$.

Property (b) is needed to avoid expressions such as $\infty - \infty$. Property (c) asserts that μ is **countably additive**. This property, together with (a), implies that μ is also **finitely additive**, which may be verified by considering sequences with a "tail end" of empty sets. Note that because the left side of (c) is invariant under permutations of the sequence (A_n), the right side must also have this property. We shall therefore make that assumption.

To emphasize the distinction between signed measures and the set functions considered in Chapters 1–4, we sometimes refer to the latter as *nonnegative measures*. Nevertheless, the unadorned term *measure* will continue to refer to the nonnegative set functions studied in previous chapters.

The **sum** $\mu_1 + \mu_2$ of signed measures μ_1 and μ_2 is defined by

$$(\mu_1 + \mu_2)(E) = \mu_1(E) + \mu_2(E), \quad E \in \mathcal{F}.$$

For this to be well-defined, the right side must not be of the form $\infty - \infty$ or $-\infty + \infty$. When dealing with such sums we shall therefore tacitly assume that this restriction holds.

A signed measure μ is said to be **finite** if $\mu(X)$ is finite. Note that in this case $\mu(A)$ is finite for all $A \in \mathcal{F}$ (use additivity on the sequence $\{A, A^c\}$).

5.1.1 Example. Let ν and η be measures on $\mathcal{F}$ at least one which, say η, is finite. We show that $\mu := \nu - \eta$ is a signed measure. Properties (a) and (b) are clear. For (c) we consider two cases: If $\mu(A) = \infty$, then $\nu(A) = \infty$ and $\eta(A) < \infty$, hence $\sum_n \nu(A_n) = \infty$ and $\sum_n \eta(A_n) < \infty$, which implies that $\sum_n \mu(A_n) = \infty$. If $\mu(A)$ is finite, then both $\nu(A)$ and $\eta(A)$ are finite and are equal to $\sum_n \nu(A_n)$ and $\sum_n \eta(A_n)$, respectively. In each case, $\mu(A) = \sum_n \mu(A_n)$. ◊

The Hahn-Jordan Decomposition

Two measures ν and η on $\mathcal{F}$ are said to be **mutually singular**, written $\nu \perp \eta$, if $\nu(A) = \eta(A^c) = 0$ if for some $A \in \mathcal{F}$. The following theorem shows that a signed measure μ may always be decomposed in the manner described in 5.1.1, with the additional property that the measures ν and η are mutually singular.

5.1.2 Hahn-Jordan Theorem. *Let μ be a signed measure on $\mathcal{F}$. Then there exists a set $P \in \mathcal{F}$ such that the following hold:*

(a) $\mu^+(E) := \mu(E \cap P) \geq 0$ *and* $\mu^-(E) := -\mu(E \cap P^c) \geq 0$ *for all* $E \in \mathcal{F}$.

(b) $\mu^+(E) = \sup\{\mu(A) : A \in \mathcal{F} \cap E\}$ *and* $\mu^-(E) = -\inf\{\mu(A) : A \in \mathcal{F} \cap E\}$.

(c) $\mu^\pm$ *are the unique mutually singular measures that satisfy* $\mu = \mu^+ - \mu^-$.

(d) *If also* $\mu^+(E) = \mu(E \cap P_1)$ *and* $\mu^-(E) = -\mu(E \cap P_1^c)$ *for some* $P_1 \in \mathcal{F}$, *then* $\mu^+(P \triangle P_1) = \mu^-(P \triangle P_1) = 0$.

We give the proof below. The measures μ^+ and μ^- in the statement of the theorem are called the **positive and negative variations of** μ, and the measure

$$|\mu| := \mu^+ + \mu^-$$

is called the **total variation measure of** μ. The quantity $|\mu|(X)$, which may be infinite, is called the **total variation** of μ. The equation $\mu = \mu^+ - \mu^-$ in (b) is called the **Jordan decomposition of** μ. The decomposition of X into a disjoint union of measurable sets P and P^c such that $\mu \geq 0$ on $\mathcal{F} \cap P$ and $\mu \leq 0$ on $\mathcal{F} \cap P^c$ is called a **Hahn decomposition for** μ. Thus part (a) guarantees the existence of a Hahn decomposition (P, P^c) and part (d) asserts that the decomposition is unique up to a set of total variation measure zero.

5.1.3 Example. Let ν be a measure on $\mathcal{F}$ and let f be measurable such that $\int f\,d\nu$ is defined (hence $\int_E f\,d\nu$ is defined for all $E \in \mathcal{F}$). Set

$$\mu(E) = \int_E f\,d\nu, \quad \mu_1(E) = \int_E f^+\,d\nu, \text{ and } \mu_2(E) = \int_E f^-\,d\nu.$$

Then μ_1 and μ_2 are measures and $\mu = \mu_1 - \mu_2$. Moreover, $f^- = 0$ on $P := \{f \geq 0\}$ and $f^+ = 0$ on $P^c = \{f < 0\}$, hence $\mu_1(P^c) = \mu_2(P) = 0$. Therefore, $\mu_1 \perp \mu_2$. By uniqueness, $\mu_1 = \mu^+$ and $\mu_2 = \mu^-$, hence the total variation measure of μ is $|\mu| = |f|\nu$. ◊

5.1.4 Corollary. *If μ is a signed measure, then for all* $E \in \mathcal{F}$,

$$|\mu|(E) = \sup\left\{\sum_{j=1}^{n} |\mu(E_j)| : E_1, \ldots, E_n \text{ is a measurable partition of } E\right\}.$$

Proof. Let $\nu(E)$ denote expression on the right. We show that $\nu(E) = \mu^+(E) + \mu^-(E)$. This is clear if $|\mu(E)| = \infty$, since then $\nu(E) = \infty$ and either $\mu^+(E) = \infty$ or $\mu^-(E) = \infty$. Now let $|\mu(E)| < \infty$, so $\mu^+(E) < \infty$ and $\mu^-(E) < \infty$. Let A and B be measurable subsets of E and set $C := A \cap B$. Then

$$\mu(A) = \mu(A \setminus C) + \mu(C) \quad \text{and} \quad \mu(B) = \mu(B \setminus C) + \mu(C).$$

and since $\mu(A)$ and $\mu(B)$ are finite,

$$\mu(A) - \mu(B) = \mu(A \setminus C) - \mu(B \setminus C) \leq |\mu(A \setminus C)| + |\mu(B \setminus C)| \leq \nu(E),$$

the last inequality because $A \setminus C$ and $B \setminus C$ are disjoint subsets of E and so are members of a measurable partition of E. Therefore, by (b) of the theorem,

$$\mu^+(E) + \mu^-(E) = \sup\{\mu(A) - \mu(B) : A,\, B \in \mathcal{F} \cap E\} \leq \nu(E).$$

For the reverse inequality, let $E_1, \ldots, E_n$ be a measurable partition of E and let A be the union of those E_j for which $\mu(E_j) \geq 0$ and B the union of the remaining E_j. Then

$$\sum_{j=1}^{n} |\mu(E_j)| = \sum_{j:\mu(E_j)\geq 0} \mu(E_j) - \sum_{j:\mu(E_j)<0} \mu(E_j) = \mu(A) - \mu(B) \leq \mu^+(E) + \mu^-(E).$$

Since the partition $\{E_1, \ldots, E_n\}$ was arbitrary, $\nu(E) \leq \mu^+(E) + \mu^-(E)$. □

Proof of the Hahn-Jordan theorem:

We may assume that $-\infty \leq \mu(E) < +\infty$ for all $E \in \mathcal{F}$ (otherwise replace μ by $-\mu$).
Proof of (a): Define

$$\mathcal{A}^+ := \{A \in \mathcal{F} : \mu \geq 0 \text{ on } \mathcal{F} \cap A\} \quad \text{and} \quad \mathcal{A}^- := \{A \in \mathcal{F} : \mu \leq 0 \text{ on } \mathcal{F} \cap A\}.$$

Note that the sets contain $\emptyset$ and hence are nonempty. Set $a := \sup\{\mu(A) : A \in \mathcal{A}^+\}$ and choose $A_n \in \mathcal{A}^+$ such that $\mu(A_n) \to a$. Define $P := \bigcup_n A_n$. We claim that $P \in \mathcal{A}^+$, that is, $\mu(E) \geq 0$ for all $E \in \mathcal{F} \cap P$. To see this, set $B_1 := A_1 \cap E$, $B_n := A_n \cap A_{n-1}^c \cap \cdots \cap A_1^c \cap E$, $n \geq 2$. The sets B_n are disjoint members of $\mathcal{A}^+$ with union E, hence $\mu(E) = \sum_n \mu(B_n) \geq 0$.

We show next that $P^c \in \mathcal{A}^-$, which will complete the proof of (a). Since $P \in \mathcal{A}^+$, $a \geq \mu(P) = \mu(A_n) + \mu(P \cap A_n^c) \geq \mu(A_n)$ for all n. Taking limits we see that $\mu(P) = a$. Since $a \geq 0$ and μ never takes on the value ∞, a is finite. These facts imply that $P^c \in \mathcal{A}^-$. Indeed, let $E \subseteq P^c$ be measurable and suppose for a contradiction that $\mu(E) > 0$. We claim that there exists and $F \in \mathcal{A}^+$ such that $F \subseteq E$ and $\mu(F) > 0$. Assuming this for the moment, we then have $\mu(P \cup F) = \mu(P) + \mu(F) > \mu(P) = a$. On the other hand, since P and F are disjoint, $P \cup F \in \mathcal{A}^+$ and so $\mu(P \cup F) \leq a$. With this contradiction we see that $\mu(E) \leq 0$, hence $P^c \in \mathcal{A}^-$.

It remains to verify the claim, namely:

If $E \in \mathcal{F}$ with $\mu(E) > 0$, then there exists a set $F \in \mathcal{A}^+$ such that $F \subseteq E$ and $\mu(F) > 0$.

If $E \in \mathcal{A}^+$, take $F = E$. Otherwise, E contains a set with negative measure, and in particular E contains sets E' with $\mu(E') < -1/n$ for some $n \in \mathbb{N}$. Let n_1 be the smallest positive integer for which $\mu(E_1) < -1/n_1$ for some $E_1 \in \mathcal{F} \cap E$. We then have

$$\mu(E \setminus E_1) > \mu(E \setminus E_1) + \mu(E_1) = \mu(E) > 0.$$

If $E \setminus E_1 \in \mathcal{A}^+$, take $F = E \setminus E_1$. Otherwise, apply the same argument to $E \setminus E_1$, obtaining a set

$E_2 \in \mathcal{F} \cap (E \setminus E_1)$ and $n_2 \geq n_1$ such that $\mu(E_2) < -1/n_2$. Continue inductively. If the process stops at some point, we are done. Otherwise, we generate a sequence $1 \leq n_1 \leq n_2 \leq \ldots$ in $\mathbb{N}$ and disjoint $E_1, E_2, \ldots \in \mathcal{F}$ such that

$$E_k \subseteq E \setminus \bigcup_{j=1}^{k-1} E_j, \quad \mu(E_k) < -1/n_k, \quad \text{and}$$

$$\mu(A) \geq -1/(n_k - 1) \text{ for all } k \text{ with } n_k > 1 \text{ and all } A \subseteq E \setminus \bigcup_{j=1}^{k-1} E_j. \tag{†}$$

Set $F := E \setminus \bigcup_{k=1}^{\infty} E_k$. Then $E \setminus F = \bigcup_{k=1}^{\infty} E_k$, hence $\mu(E \setminus F) = \sum_{k=1}^{\infty} \mu(E_k) < 0$. Because $\mu(E)$ is finite, so is $\mu(E \setminus F)$, hence the series converges and so $\mu(E_k) \to 0$. Since $-\mu(E_k) > 1/n_k$, $n_k \to \infty$. Also $\mu(F) > \mu(F) + \mu(E \setminus F) = \mu(E) > 0$. Finally, if $A \in \mathcal{F} \cap F$, letting $k \to \infty$ in (†) yields $\mu(A) \geq 0$. Therefore, $F \in \mathcal{A}^+$.

Proof of (b): Let s denote the supremum in (b). Since $\mu^+(E) = \mu(E \cap P)$, $\mu^+(E) \leq s$. For the reverse inequality, let $A \in \mathcal{F}$ with $A \subseteq E$. By definition of $\mu^{\pm}$,

$$\mu(A) = \mu^+(A) - \mu(A^-) \leq \mu^+(A) \leq \mu^+(E).$$

Taking the sup over all such A yields $s \leq \mu^+(E)$. This proves the first part of (b). The proof of the second part is similar.

Proof of (c): The set functions $\mu^{\pm}$ are clearly mutually singular measures and $\mu = \mu^+ - \mu^-$. Suppose also that $\mu = \mu_1 - \mu_2$, where μ_1 and μ_2 are nonnegative singular measures. Let $E \in \mathcal{F}$. For any measurable $A \subseteq E$, $\mu_1(E) \geq \mu_1(A) \geq \mu(A)$, hence, taking the sup over all such A, we have $\mu_1(E) \geq \mu^+(E)$ by (b). For the reverse inequality, use the mutual singularity to obtain $B \in \mathcal{F}$ such that $\mu_1(B^c) = \mu_2(B) = 0$. Then

$$\mu_1(E) = \mu_1(B \cap E) - \mu_2(B \cap E) = \mu(B \cap E) \leq \mu^+(B \cap E) \leq \mu^+(E)$$

and so $\mu_1(E) = \mu^+(E)$. Similarly, $\mu_2(E) = \mu^-(E)$.

Proof of (d): For any $A \in \mathcal{F}$, $\mu(A \cap P) = \mu^+(A) = \mu(A \cap P_1)$, hence $\mu(A) = 0$ for $A \subseteq P \cap P_1^c$. Therefore, $\mu^+(P \cap P_1^c) = 0$. Switching P and P_1, $\mu^+(P_1 \cap P^c) = 0$. Therefore, $\mu^+(P \triangle P_1) = 0$. Similarly, $\mu^-(P \triangle P_1) = 0$. □

Exercises

5.1 Show that a signed measure μ is finite iff $|\mu|$ is finite.

5.2 Let μ_1 and μ_2 be signed measures such that $\mu_1 + \mu_2$ is defined. Prove that

$$(\mu_1 + \mu_2)^+ + \mu_1^- + \mu_2^- = (\mu_1 + \mu_2)^- + \mu_1^+ + \mu_2^+.$$

5.3 Let $Q \in \mathcal{F}$ have the property that $\mu(E \cap Q) \geq 0$ and $\mu(E \cap Q^c) \leq 0$ for all $E \in \mathcal{F}$. Show that $\mu^+(E) = \mu(E \cap Q)$ and $\mu^-(E) = \mu(E \cap Q^c)$.

5.4 Let μ be a finite measure and $x \in X$. Find the Hahn decomposition of $\mu - a\delta_x$, where $a = \mu(X)$.

5.5 Let μ be a signed measure with Hahn decomposition (P, P^c). Show that $(-\mu)^+ = \mu^-$ and $(-\mu)^- = \mu^+$ and that (P^c, P) is a Hahn decomposition of $-\mu$.

5.6 Let μ and ν be probability measures and $\eta := \mu - \nu$. Show that

$$|\eta|(X) = 2 \sup \{|\eta(E)| : E \in \mathcal{F}\}.$$

5.7 Let $\mu = \mu_1 - \mu_2$, where μ_1 and μ_2 are measures. Show that $\mu_1 \geq \mu^+$ and $\mu_2 \geq \mu^-$.

5.8 [↑5.7] Show that if μ_1 and μ_2 are measures and $\mu_1(Q) = \mu_2(Q^c) = 0$, then (Q, Q^c) is a Hahn decomposition for $\mu_2 - \mu_1$.

5.9 Let μ be a finite signed measure such that $|\mu|(X) = \mu(X)$. Show that $|\mu| = \mu$.

5.10 Show that for a signed measure μ and real-valued $h \in L^1(\mu)$,

$$|h\mu| = |h|\,|\mu|, \quad (h\mu)^+ = h^+\mu^+ + h^-\mu^-, \text{ and } (h\mu)^- = h^+\mu^- + h^-\mu^+.$$

5.11 Let $\mu, \mu_1, \mu_2, \ldots$ be signed measures with $\mu_n \to \mu$. Show that $\mu^\pm(E) \leq \underline{\lim}_n\, \mu_n^\pm(E)$, and $|\mu|(E) \leq \underline{\lim}_n\, |\mu_n|(E)$.

5.12 Let μ be a signed measure and ν a measure. Show that $|\mu| \leq \nu$ iff $-\nu \leq \mu \leq \nu$.

5.13 [↓5.29] For finite signed measures μ_1 and μ_2, define

$$\mu_1 \vee \mu_2 = \mu_1 + (\mu_2 - \mu_1)^+ \quad \text{and} \quad \mu_1 \wedge \mu_2 = \mu_2 - (\mu_1 - \mu_2)^-.$$

Prove:

(a) $\mu_1 \vee \mu_2 = \mu_2 \vee \mu_1$ and $\mu_1 \wedge \mu_2 = \mu_2 \wedge \mu_1$.

(b) $\mu_1,\ \mu_2 \leq \mu_1 \vee \mu_2$, and if ν is a signed measure such that $\mu_1,\ \mu_2 \leq \nu$, then $\mu_1 \vee \mu_2 \leq \nu$. Thus $\mu_1 \vee \mu_2$ is the smallest signed measure larger than μ_1 and μ_2.

(c) $\mu_1 \wedge \mu_2 = -\big[(-\mu_1) \vee (-\mu_2)\big]$. Use this to formulate and prove assertions for $\mu_1 \wedge \mu_2$ analogous to those in (b).

(d) $\mu_1 \vee (-\mu_1) = |\mu_1|$ and $\mu_1 \wedge (-\mu_1) = -|\mu_1|$.

(e) $\mu_1 \wedge \mu_2 + \mu_1 \vee \mu_2 = \mu_1 + \mu_2$.

(f) If μ is a measure and $f,\ g \in L^1(\mu_1)$, then $(f \vee g)\mu_1 = (f\mu_1) \vee (g\mu_1)$ and $(f \wedge g)\mu_1 = (f\mu_1) \wedge (g\mu_1)$.

5.2 Complex Measures

A **complex measure** is a $\mathbb{C}$-valued set function μ on a measurable space $(X, \mathcal{F})$ with the following properties:

(a) $\mu(\emptyset) = 0$.

(b) If (A_n) is a sequence of pairwise disjoint sets in $\mathcal{F}$, then $\mu\Big(\bigcup_n A_n\Big) = \sum_n \mu(A_n)$, where the sequence is assumed to converge absolutely.

For a complex measure μ, define

$$\mu_r(E) := \mathrm{Re}\big(\mu(E)\big) \quad \text{and} \quad \mu_i(E) := \mathrm{Im}\big(\mu(E)\big), \quad E \in \mathcal{F}.$$

Then μ_r and μ_i are finite signed measures, called the **real and imaginary parts of** μ, and

$$\mu = \mu_r + i\mu_i = \mu_r^+ - \mu_r^- + i(\mu_i^+ - \mu_i^-). \tag{5.1}$$

Thus complex-valued measures are linear combinations over $\mathbb{C}$ of finite measures. This observation will be useful in reducing some arguments involving complex measures to the nonnegative case. We also define the **complex conjugate of** μ in the expected way:

$$\overline{\mu} = \mu_r - i\mu_i. \tag{5.2}$$

The notation conflicts with that for the completion of a measure, but this should not be a problem, as context will make clear which meaning is intended.

The Total Variation Measure

The **total variation measure** $|\mu|$ of a complex measure μ is defined by

$$|\mu|(E) = \sup\left\{\sum_{j=1}^{n} : |\mu(E_j)| : E_1, \ldots, E_n \text{ is a measurable partition of } E\right\}, \quad E \in \mathcal{F}. \quad (5.3)$$

This definition is compatible with the corresponding notion for signed measures (5.1.4). In the latter case, however, the total variation was immediately seen to be a measure. In the complex case, some work is required to verify this.

5.2.1 Theorem. *If μ is a complex measure, then $|\mu|$ is a finite measure and $|\mu(E)| \le |\mu|(E)$ for all $E \in \mathcal{F}$. Moreover, if ν is a complex measure, then $|\mu + \nu| \le |\mu| + |\nu|$.*

Proof. To show that $|\mu|$ is finite, let $E_1, \ldots, E_n$ be an arbitrary measurable partition of $E \in \mathcal{F}$. From (5.1),

$$|\mu(E_j)| \le \mu_r^+(E_j) + \mu_r^-(E_j) + \mu_i^+(E_j) + \mu_i^-(E_j),$$

hence

$$\sum_{j=1}^{n} |\mu(E_j)| \le \mu_r^+(E) + \mu_r^-(E) + \mu_i^+(E) + \mu_i^-(E) < \infty.$$

To show countable additivity, let (A_n) be a sequence of disjoint measurable sets with union A and let $\{E_1, \ldots, E_n\}$ be a measurable partition of A. Then

$$\begin{aligned}\sum_{j=1}^{n} |\mu(E_j)| = \sum_{j=1}^{n} |\mu(A \cap E_j)| = \sum_{j=1}^{n} \left|\sum_{k=1}^{\infty} \mu(A_k \cap E_j)\right| &\le \sum_{k=1}^{\infty}\sum_{j=1}^{n} |\mu(A_k \cap E_j)| \\ &\le \sum_{k=1}^{\infty} |\mu|(A_k),\end{aligned}$$

the last inequality because $\{A_k \cap E_1, \ldots, A_k \cap E_n\}$ is a measurable partition of A_k. Therefore, $|\mu|(A) \le \sum_{k=1}^{\infty} |\mu|(A_k)$.

For the reverse inequality, for each k let $\mathcal{P}_k := \{E_{k,1}, \ldots, E_{k,n_k}\}$ be a measurable partition of A_k. For each m, the sets $E_{k,j}$ $(1 \le k \le m,\ 1 \le j \le n_k)$ are part of a finite measurable partition of A, hence

$$|\mu|(A) \ge \sum_{k=1}^{m}\sum_{j=1}^{n_k} |\mu(E_{k,j})| = \sum_{k=1}^{m}\sum_{E \in \mathcal{P}_k} |\mu(E)|.$$

Taking the suprema over each of the partitions $\mathcal{P}_k$ yields $|\mu|(A) \ge \sum_{k=1}^{m} |\mu|(A_k)$. Since m was arbitrary, $|\mu|(A) \ge \sum_{k=1}^{\infty} |\mu|(A_k)$. This establishes countable additivity of $|\mu|$.

The inequality $|\mu(E)| \le |\mu|(E)$ follows directly from the definition of $|\mu|(E)$. The proof of the triangle inequality is an exercise (5.15). □

5.2.2 Example. Let ν be a measure on $\mathcal{F}$ and let $f : X \to \mathbb{C}$ be ν-integrable. Define the complex measure μ by $\mu(E) = \int_E f\, d\nu$, that is, $d\mu = f\, d\nu$. We show that $d|\mu| = |f|\, d\nu$.

Let $\{E_1, \ldots, E_n\}$ be an arbitrary measurable partition of E. Then

$$\sum_{j=1}^{n} |\mu(E_j)| = \sum_{j=1}^{n} \left|\int_{E_j} f\, d\nu\right| \le \sum_{j=1}^{n} \int_{E_j} |f|\, d\nu = \int_E |f|\, d\nu,$$

hence $|\mu|(E) \le \int_E |f|\,d\nu$. In particular, $|\mu|\{f = 0\} = 0$, so for the reverse inequality $\int_E |f|\,d\nu \le |\mu|(E)$ we may assume that f is never zero, otherwise remove the part of E on which $f = 0$. Consider the polar form of $z \neq 0$, written as $|z| = ze^{i\theta(z)}$, where $-\pi \le \theta(z) < \pi$. For each n define

$$g_n(z) = \sum_{k=1}^{n} e^{i\theta_k} \mathbf{1}_{[\theta_k, \theta_{k+1})}\big(\theta(z)\big), \quad \theta_k = -\pi + 2\pi k/n, \quad k = 0, 1, \ldots, n-1.$$

Then $zg_n(z) \to |z|$ and $|g_n(z)| = 1$. Therefore, $f_n := g_n \circ f$ is an $\mathcal{F}$-simple function satisfying $|f_n| = 1$ and $f \cdot f_n \to |f|$ on X. Let $E \in \mathcal{F}$ and let f_n have standard form

$$f_n = \sum_{j=1}^{m_n} c_j \mathbf{1}_{A_j}, \quad |c_j| = 1.$$

Then $E \cap A_1, \ldots, E \cap A_{m_n}$ is a measurable partition of E and

$$\int_E f \cdot f_n\,d\nu = \sum_{j=1}^{m_n} c_j \int_{E\cap A_j} f\,d\nu = \sum_{j=1}^{m_n} c_j \mu(E \cap A_j),$$

hence

$$\left| \int_E f \cdot f_n\,d\nu \right| \le \sum_{j=1}^{m_n} |c_j|\,|\mu(E \cap A_j)| \le |\mu|(E).$$

Letting $n \to \infty$ and applying the dominated convergence theorem yields $\int_E |f|\,d\nu \le |\mu|(E)$, as required. ♢

The Vitali-Hahn-Saks Theorem

We give an application of the total variation measure which asserts that the limit of a sequence of complex measures on a measurable space $(X, \mathcal{F})$ is a complex measure. For this we need the following lemma.

5.2.3 Lemma. *Let η be a finite measure on $\mathcal{F}$ and define*

$$d(A, B) := \eta\big(A \triangle B\big) = \|\mathbf{1}_A - \mathbf{1}_B\|_1, \quad A, B \in \mathcal{F}.$$

If we identify A and A' whenever $\eta\big(A \triangle A'\big) = 0$, then d is a metric on $\mathcal{F}$ and $(\mathcal{F}, d)$ is complete.

Proof. To see that d is well-defined, let $\eta\big(A \triangle A'\big) = \eta\big(B \triangle B'\big) = 0$. Then

$$d(A, B) = \|\mathbf{1}_A - \mathbf{1}_B\|_1 \le \|\mathbf{1}_A - \mathbf{1}_{A'}\|_1 + \|\mathbf{1}_{A'} - \mathbf{1}_{B'}\|_1 + \|\mathbf{1}_{B'} - \mathbf{1}_B\|_1 = d(A', B'),$$

and similarly $d(A', B') \le d(A, B)$. That d is a metric follows easily from the properties of the L^1 norm. To show completeness, let (A_n) be a Cauchy sequence in $\mathcal{F}$. Then $(\mathbf{1}_{A_n})$ is a Cauchy sequence in L^1, hence there exists $f \in L^1$ such that $\|f - \mathbf{1}_{A_n}\|_1 \to 0$. Choose a subsequence $(\mathbf{1}_{A_{n_k}})$ that converges a.e. to f. Then f takes on the values 0 and 1 a.e., hence $f = \mathbf{1}_A$ a.e., where $A = \{f = 1\}$. Therefore, $d(A_n, A) = \|\mathbf{1}_{A_n} - \mathbf{1}_A\|_1 \to 0$. □

5.2.4 Vitali-Hahn-Saks Theorem. *Let $(X, \mathcal{F})$ be a measurable space and (μ_n) a sequence of complex measures on $\mathcal{F}$ such that the limit*

$$\mu(A) := \lim_n \mu_n(A)$$

exists for every $A \in \mathcal{F}$. Then μ is countably additive and hence is a complex measure.

Proof. The set function μ is clearly finitely additive, and $\mu(\emptyset) = 0$. It remains to show that μ is continuous from below. To this end, apply the lemma to the finite measure

$$\eta(A) := \sum_{n=1}^{\infty} \frac{1}{2^n} \frac{|\mu_n|(A)}{1 + |\mu_n|(X)}, \quad A \in \mathcal{F}.$$

For each n, the function $A \mapsto \mu_n(A)$ on $(\mathcal{F}, d)$ (viewed as a collection of equivalence classes) is well-defined, since $d(A, A') = 0 \Rightarrow |\mu_n|(A \triangle A') = 0 \Rightarrow \mu_n(A) = \mu_n(A')$. Moreover, from

$$|\mu_n(A) - \mu_n(B)| \le |\mu_n|(A \triangle B) \le 2^n\big(1 + |\mu_n|(X)\big)\eta(A \triangle B) = 2^n\big(1 + |\mu_n|(X)\big)d(A, B)$$

we see that the mapping $A \mapsto \mu_n(A)$ is continuous in the metric topology.

Now let $\varepsilon > 0$ and set

$$\mathcal{C}_n = \bigcap_{k=1}^{\infty} \{A \in \mathcal{F} : |\mu_k(A) - \mu_{n+k}(A)| \le \varepsilon\}.$$

Then $\mathcal{C}_n$ is closed in the metric topology, and since $\mu_n(A) \to \mu(A)$, $\mathcal{F} = \bigcup_n \mathcal{C}_n$. Since $(\mathcal{F}, d)$ is complete, the Baire category theorem implies that some $\mathcal{C}_m$ contains an open ball. Thus for some $A_0 \in \mathcal{F}$ and $\delta > 0$,

$$|\mu_k(A) - \mu_{m+k}(A)| \le \varepsilon \ \text{ for all } k \ge 1 \text{ and all } A \text{ with } d(A, A_0) < \delta. \tag{†}$$

Let $\eta(A) < \delta$ and set $B = A \cup A_0$ and $C = A_0 \setminus A$. Then

$$A = B \setminus C, \quad C \subseteq B, \quad B \triangle A_0 \subseteq A, \text{ and } \ C \triangle A_0 \subseteq A,$$

hence $d(B, A_0)$, $d(C, A_0) < \delta$ and $\mu_n(A) = \mu_n(B) - \mu_n(C)$. Therefore, for all $n \ge m$,

$$\begin{aligned} |\mu_n(A)| &\le |\mu_m(A)| + |\mu_n(A) - \mu_m(A)| \\ &\le |\mu_m(A)| + |\mu_n(B) - \mu_m(B)| + |\mu_n(C) - \mu_m(C)| \\ &\le |\mu_m(A)| + 2\varepsilon, \end{aligned}$$

the last inequality by (†). Letting $n \to \infty$ we have

$$|\mu(A)| \le |\mu_m(A)| + 2\varepsilon \ \text{ for all } A \text{ with } \eta(A) \le \delta. \tag{‡}$$

Now let $E_n \uparrow E$. Then $\eta(E \setminus E_n) \to 0$, hence also $\mu_m(E \setminus E_n) \to 0$. Thus from (‡),

$$\overline{\lim_n}\, |\mu(E \setminus E_n)| \le \overline{\lim_n}\, |\mu_m(E \setminus E_n)| + 2\varepsilon \le 2\varepsilon.$$

Since ε was arbitrary, $\lim_n \mu(E \setminus E_n) = 0$, which shows that μ is continuous from below. □

The Banach Space of Complex Measures

Let $M = M(X, \mathcal{F})$ denote the linear space of complex measures on $\mathcal{F}$ under pointwise addition and scalar multiplication. It is easy to check that the **total variation norm** $\|\mu\| := |\mu|(X)$ is a norm on M. For example, the triangle inequality follows from the second part of 5.2.1. We claim that M is complete in this norm. For this we use 0.4.3. Let (μ_n) be a sequence M such that $\sum_n \|\mu_n\| < \infty$. The sequence of complex measures $\nu_n := \sum_{k=1}^{n} \mu_k$ then converges to a complex measure in the total variation norm. To see this, let $E \in \mathcal{F}$ and note that for $m > n$,

$$|\nu_m(E) - \nu_n(E)| \le \sum_{k=n+1}^{m} |\mu_k(E)| \le \sum_{k>n} \|\mu_k\|,$$

so $(\nu_n(E))$ is a Cauchy sequence in $\mathbb{C}$. Let $\nu_n(E) \to \nu(E)$. By the Vitali-Hahn-Saks theorem, ν is a complex measure. For any measurable partition $E_1, \ldots, E_p$ of X,

$$\sum_{j=1}^{p} |\nu_m(E_j) - \nu_n(E_j)| \le \sum_{k=n+1}^{m} \sum_{j=1}^{p} |\mu_k(E_j)| \le \sum_{k>n} \|\mu_k\|,$$

and letting $m \to \infty$ we have $\sum_{j=1}^{p} |\nu(E_j) - \nu_n(E_j)| \le \sum_{k>n} \|\mu_k\|$. Since the partition was arbitrary, $\|\nu - \nu_n\| \le \sum_{k>n} \|\mu_k\|$. Therefore, $\|\nu_n - \nu\| \to 0$, proving that M is complete. We summarize this discussion in

5.2.5 Proposition. *The linear space $M(X, \mathcal{F})$ of complex measures on a measurable space $(X, \mathcal{F})$ is a Banach space under the total variation norm.*

Integration against a Signed or Complex Measure

If μ is a signed measure, define

$$\int f\,d\mu := \int f\,d\mu^+ - \int f\,d\mu^-, \quad f \in L^1(|\mu|).$$

For a complex measure μ, define

$$\int f\,d\mu := \int f\,d\mu_r + i \int f\,d\mu_i, \quad f \in L^1(|\mu|).$$

It is straightforward to check that in each case the integrals are well-defined, linear, and satisfy

$$\left| \int f\,d\mu \right| \le \int |f|\,d|\mu| \quad \text{and} \quad \overline{\int f\,d\mu} = \int \overline{f}\,d\overline{\mu} \tag{5.4}$$

(Ex. 5.16). Moreover, the dominated convergence theorem holds for a signed or complex measure μ, as may be seen by decomposing μ into a linear combination of the measures $\mu_{r,i}^{\pm}$.

Exercises

5.14 Show that if μ is a complex measure and $\|\mu\| = \mu(X) < \infty$, then μ is a nonnegative measure.

5.15 Verify the inequality $|\mu + \nu| \le |\mu| + |\nu|$ for complex measures.

5.16 Let μ be a signed or complex measure. Verify that the integral with respect to μ is well-defined and linear. Also, verify the assertions in (5.4).

5.17 Show that in the definition of $|\mu|$, the finite measurable partition $E_1, \ldots, E_n$ may be replaced a countable measurable partition.

5.18 Let μ be a complex measure and $E \in \mathcal{F}$. Prove that

$$|\mu|(E) = \sup\left\{ \left| \int_E f\,d\mu \right| : f \text{ is measurable and } |f| \le 1 \right\}.$$

5.19 Let μ and ν be complex measures on measurable spaces $(X, \mathcal{F})$ and $(Y, \mathcal{G})$, respectively.

(a) Show that there exists a unique complex measure $\mu \times \nu$ on $\mathcal{F} \otimes \mathcal{G}$ such that $(\mu \otimes \nu)(A \times B) = \mu(A)\nu(B)$ for all $A \in \mathcal{F}$ and $B \in \mathcal{G}$.

(b) Prove the complex version of Fubini's theorem for a $(\mathcal{F} \times \mathcal{G})$-measurable function f: If $\iint |f(x,y)|\,d|\mu|(x)\,d|\nu|(y) < \infty$, then

$$\int_{X \times Y} f\,d(\mu \otimes \nu) = \int_Y \int_X f(x,y)\,d\mu(x)\,d\nu(y) = \int_X \int_Y f(x,y)\,d\nu(y)\,d\mu(x).$$

(c) Show that $|\mu \otimes \nu| = |\mu| \otimes |\nu|$.

5.20 Let $(X, \mathcal{F})$ and $(Y, \mathcal{G})$ be measurable spaces and $T : (X, \mathcal{F}) \to (Y, \mathcal{G})$ measurable. If μ is a signed or complex measure on $(X, \mathcal{F})$, then the image of μ under T is the signed of complex measure $T(\mu)$ on $(Y, \mathcal{G})$ defined as before by $T(\mu)(E) = \mu(T^{-1}(E))$, $E \in \mathcal{G}$. Show that in the signed case $|T(\mu)| \leq |T|(\mu)$, $(T\mu)^+ \leq T(\mu^+)$, and $(T\mu)^- \leq T\mu^-$, and in the complex case $|T(\mu)| \leq |T|(\mu)$.

5.3 Absolute Continuity of Measures

Let $(X, \mathcal{F}, \mu)$ be a measure space. A signed or complex measure ν on $(X, \mathcal{F})$ is said to be **absolutely continuous with respect to μ**, written $\nu \ll \mu$, if

$$E \in \mathcal{F} \text{ and } \mu(E) = 0 \Rightarrow \nu(E) = 0.$$

For example, if $g \in L^1(\mu)$ then $\nu(E) := \int_E g\, d\mu$ defines ν a signed or complex measure on $(X, \mathcal{F})$ with $\nu \ll \mu$. The main goal of this section is to prove the converse: if μ is σ-finite, then every signed or complex measure ν that is absolutely continuous with respect to μ is of the form $g\, d\mu$ for some g. This result, known as the Radon-Nikodym theorem, has many important applications in analysis and probability theory, some of which are given below. Before we state the theorem we develop a few preliminary results.

General Properties of Absolute Continuity

The following proposition is useful in reducing arguments involving absolute continuity of signed and complex measures to the nonnegative case.

5.3.1 Proposition. *Let μ be a measure and ν, η signed or complex measures.*

(a) *If ν is signed, then $\nu \ll \mu$ iff $|\nu| \ll \mu$ iff both $\nu^+ \ll \mu$ and $\nu^- \ll \mu$.*

(b) *If ν is complex, then $\nu \ll \mu$ iff $|\nu| \ll \mu$ iff both $\nu_r \ll \mu$ and $\nu_i \ll \mu$.*

(c) *$\nu \ll \mu$ and $|\eta| \perp \mu \Rightarrow |\nu| \perp |\eta|$.*

(d) *$\nu \ll \mu$ and $|\nu| \perp \mu \Rightarrow |\nu| = 0$.*

Proof. (a) Let $\nu \ll \mu$, $\mu(E) = 0$, and $A \subseteq E$ measurable. Then, $\mu(A) = 0$, hence $\nu(A) = 0$. Therefore, by 5.1.2, $\nu^+(E) = \nu^-(E) = 0$, hence also $|\nu|(E) = 0$. The converses are clear.

(b) This follows from $|\nu_{r,i}| \leq |\nu| \leq |\nu_r| + |\nu_i|$ and (a).

(c) For the signed case, let $|\eta|(E^c) = \mu(E) = 0$. Since $\nu \ll \mu$, $\nu(A) = 0$ for all measurable $A \subseteq E$, hence $|\nu|(E) = 0$. Therefore, $|\nu| \perp |\eta|$. The complex case is obtained by using $\nu_{r,i}$.

(d) By (c), $|\nu| \perp |\nu|$, that is, $|\nu|(E) = |\nu|(E^c) = 0$ for some E. Therefore, $|\nu| = 0$. □

5.3.2 Proposition. *Let μ be a measure and ν a complex measure. Then $\nu \ll \mu$ iff $\lim_{\mu(E)\to 0} \nu(E) = 0$ for all $E \in \mathcal{F}$.*

Proof. The limit assertion means that for every $\varepsilon > 0$ there exists $\delta > 0$ such that $|\nu(E)| < \varepsilon$ for all $E \in \mathcal{F}$ with $\mu(E) < \delta$. Suppose this holds. If $\mu(E) = 0$, then the δ-inequality holds trivially, hence $|\nu(E)| < \varepsilon$ for all ε and so $\nu(E) = 0$. Therefore, $\nu \ll \mu$.

Conversely, suppose $\nu \ll \mu$. Then by 5.3.1(b), $|\nu| \ll \mu$. If we show that $\lim_{\mu(E)\to 0} |\nu|(E) = 0$, then the inequality, $|\nu(E)| \leq |\nu|(E)$ will imply that $\lim_{\mu(E)\to 0} \nu(E) = 0$. Thus we may assume without loss of generality that ν is nonnegative. Suppose that the ε-δ condition does not hold. Then there exists $\varepsilon > 0$ and for each $n \in \mathbb{N}$ a measurable set E_n such that

$\mu(E_n) < 1/2^n$ and $\nu(E_n) \geq \varepsilon$. Let $E = \overline{\lim}_n E_n$. Since $\sum_n \mu(E_n) < +\infty$, $\mu(E) = 0$ (Ex. 1.37). But by continuity from above (since ν is finite), $\nu(E) \geq \overline{\lim}_n \nu(E_n) \geq \varepsilon$, contradicting the assumption that $\nu \ll \mu$. □

5.3.3 Remark. The necessity of 5.3.2 does not necessarily hold if ν is not finite. For example, let ν be counting measure on $\mathbb{N}$ and let $\mu(E) := \sum_{n\in E} 1/2^n$. Clearly $\nu \ll \mu$. On the other hand, if $A_n := \{n, n+1, \ldots\}$ then $\mu(A_n) \to 0$ but $\nu(A_n) = \infty$ for all n. ◊

The Radon-Nikodym Theorem

5.3.4 Theorem. *Let $(X, \mathcal{F}, \mu)$ be a σ-finite measure space and ν a complex or signed measure on $\mathcal{F}$ such that $\nu \ll \mu$. Then there exists a measurable function $h : X :\to \overline{\mathbb{K}}$, unique up to a set of μ-measure zero, such that*

$$\nu(E) = \int_E h\, d\mu \quad \textit{for all } E \in \mathcal{F}.$$

Moreover, we may take h to be $\overline{\mathbb{R}}$-valued if ν is a signed measure and nonnegative if ν is a measure.

We prove the theorem below. The function h in the theorem is called the **Radon-Nikodym derivative of ν with respect to μ** and is denoted by $d\nu/\,d\mu$. By the usual arguments,

$$\int f\, d\nu = \int f \frac{d\nu}{d\mu}\, d\mu \tag{5.5}$$

in the sense that if one side is defined, then so is the other and then equality holds.

5.3.5 Remark. The theorem is false if μ is not σ-finite. For example, if μ is counting measure on $\mathcal{B}[0,1]$, then, trivially, $\lambda \ll \mu$, but there is no function g such that $\lambda(E) = \int_E g\, d\mu$ for all $E \in \mathcal{F}$. Otherwise, taking $E = \{x\}$, we would have $0 = g(x)$ for all x implying that $\lambda[0,1] = 0$. ◊

5.3.6 Examples. In the following, $(X, \mathcal{F}, \mu)$ is a σ-finite measure space.

(a) Let ν be a measure such that $\nu(E) \leq \mu(E)$ for all $E \in \mathcal{F}$. Then evidently $\nu \ll \mu$, hence (5.5) holds for all suitable f and so $0 \leq d\nu/\,d\mu \leq 1$ a.e. In particular, if μ_1 and μ_2 are σ-finite measures, then

$$\int f\, d\mu_1 = \int f \frac{d\mu_1}{d(\mu_1+\mu_2)}\, d(\mu_1+\mu_2), \quad \text{where } 0 \leq \frac{d\mu_1}{d(\mu_1+\mu_2)} \leq 1 \textit{ a.e.}$$

(b) Let ν be a complex measure. Since $\nu \ll |\nu|$, we have

$$\nu(E) = \int_E \frac{d\nu}{d|\nu|}\, d|\nu|, \quad E \in \mathcal{F}.$$

Recalling 5.2.2, we see that for all $E \in \mathcal{F}$

$$\int_E 1\, d|\nu| = |\nu|(E) = \int_E \left|\frac{d\nu}{d|\nu|}\right| d|\nu|.$$

Therefore,

$$\left|\frac{d\nu}{d|\nu|}\right| = 1, \ |\nu| \text{ a.e.}$$

As the function $\operatorname{Arg}(z)$ is continuous on $\mathbb{C} \setminus (-\infty, 0]$, the function

$$\theta := \operatorname{Arg}\Big(\frac{d\nu}{d|\nu|}\Big)$$

is measurable, and we have

$$d\nu = \frac{d\nu}{d|\nu|} d|\nu| = e^{i\theta} d|\nu|.$$

This equation is called the **polar decomposition of** ν.

(c) Let U and V be open subsets of $\mathbb{R}^d$ and let $\varphi : U \to V$ be C^1 on U with C^1 inverse $\varphi^{-1} : V \to U$. By (3.7)

$$\frac{d\varphi(\lambda)}{d\lambda} = \frac{1}{|\det \varphi'| \circ \varphi^{-1}}.$$

(d) Let $\mathcal{G}$ be a sub-σ-field of $\mathcal{F}$ and let $f \in L^1(\mu)$. Define

$$\mu_f(E) = \int_E f\, d\mu, \quad E \in \mathcal{G}.$$

Then $\mu_f \ll \mu$ on $\mathcal{G}$ and

$$\int_E \frac{d\mu_f}{d\mu}\, d\mu = \int_E f\, d\mu \quad \text{for all} \quad E \in \mathcal{G}. \tag{5.6}$$

The salient point here is that $d\mu_f/\, d\mu$ has the same integral property as f but is $\mathcal{G}$-measurable, while, of course, f need not be. If μ is a probability measure, then $d\mu_f/\, d\mu$ is called the **conditional expectation of** f **given** $\mathcal{G}$, studied in detail in Chapter 18. For a concrete example, let $(X, \mathcal{F}, \mu)$ be the product of the probability spaces $(X_1, \mathcal{F}_1, \mu_1)$ and $(X_2, \mathcal{F}_2, \mu_2)$ and take $\mathcal{G} = \mathcal{F}_1 \times X_2$ so that

$$\int_{E_1 \times X_2} f\, d\mu = \int_{E_1} \int_{X_2} f(x_1, x_2)\, d\mu_2(x_1)\, d\mu_1(x_1), \quad E_1 \in \mathcal{F}_1.$$

Identifying $\mathcal{G}$ with $\mathcal{F}_1$ and comparing with (5.6), we see that

$$\frac{d\mu_f}{d\mu}(x_1) = \int_{X_2} f(x_1, x_2)\, d\mu_2(x_2),$$

where we have omitted the redundant argument x_2 in $d\mu_f/\, d\mu$. Viewing $X_1 \times X_2$ as the set of outcomes of a two-stage experiment and taking x_1 as the outcome of stage one, we see that $d\mu_f/\, d\mu(x_1)$ is the average of f over the possible outcomes of stage two. Thus if the σ-field $\mathcal{F}_1$ is interpreted as "given information," namely as the information revealed after the first stage, then $d\mu_f/\, d\mu$ incorporates both the "known" and an average over the "unknown." Therefore, $d\mu_f/\, d\mu$ may be interpreted as the best information regarding f that is available after stage one but before stage two. ◊

Proof of the Radon-Nikodym theorem:

The uniqueness of h follows from 3.4. For proof of existence, we consider several cases.

Case I. *μ is finite and ν is a finite measure.*

Let $\mathscr{F}$ denote the collection of $\mathcal{F}$-measurable functions $f \geq 0$ such that $\int_E f\, d\mu \leq \nu(E)$ for all $E \in \mathcal{F}$. Clearly, $0 \in \mathscr{F}$. Furthermore, $f,\, g \in \mathfrak{F} \Rightarrow f \vee g \in \mathscr{F}$. Indeed, if $A = \{f > g\}$ and $E \in \mathcal{F}$, then

$$\int_E f \vee g\, d\mu = \int_{E \cap A} f\, d\mu + \int_{E \cap A^c} g\, d\mu \leq \nu(E \cap A) + \nu(E \cap A^c) = \nu(E).$$

We claim there exists an $h \in \mathscr{F}$ such that

$$\int h\,d\mu = \sup\left\{\int f\,d\mu : f \in \mathscr{F}\right\}. \tag{†}$$

To see this, let s denote the supremum on the right and let $f_n \in \mathscr{F}$ such that $\int f_n\,d\mu \to s$. Replacing f_n by $f_1 \vee \cdots \vee f_n$ if necessary, we may assume that $f_n \uparrow h$ for some measurable $h \geq 0$. By the monotone convergence theorem,

$$\int_E h\,d\mu = \lim_n \int_E f_n\,d\mu \leq \nu(E), \quad E \in \mathfrak{F}.$$

Therefore, $h \in \mathscr{F}$ and $s = \int_E h\,d\mu$, verifying the claim.

Now define

$$\eta(E) := \nu(E) - \int_E h\,d\mu, \quad E \in \mathfrak{F}.$$

Since $h \in \mathscr{F}$, $\eta(E) \geq 0$ for every $E \in \mathfrak{F}$. Therefore, η is a finite measure. The proof of the theorem for Case I will be complete once we show that $\eta(X) = 0$. Let $r > 0$ and let (P, P^c) be a Hahn decomposition for the signed measure $\eta - r\mu$. Since $(\eta - r\mu)(E \cap P) \geq 0$,

$$\nu(E) = \int_E h\,d\mu + \eta(E) \geq \int_E h\,d\mu + r\mu(E \cap P) = \int_E (h + r\mathbf{1}_P)\,d\mu, \quad E \in \mathfrak{F},$$

hence $h + r\mathbf{1}_P \in \mathscr{F}$. By (†), $\mu(P) = 0$, hence, by absolute continuity, $\nu(P) = 0$. Therefore, $(\eta - r\mu)(P) = \eta(P) = -\int_P h\,d\mu = 0$ and so

$$0 \leq \eta(X) = r\mu(X) + (\eta - r\mu)(X) = r\mu(X) + (\eta - r\mu)(P^c) \leq r\mu(X).$$

Letting $r \to 0$ we conclude that $\eta(X) = 0$.

Case II. *μ is finite and ν is a σ-finite measure.*

Let $\{X_n\}$ be a sequence of pairwise disjoint sets in $\mathscr{F}$ such that $X = \bigcup_n X_n$ and $\nu(X_n) < \infty$. For each n, $\nu \ll \mu$ on $\mathfrak{F} \cap X_n$, hence by Case I there exists a finite $\mathfrak{F} \cap X_n$-measurable function $h_n \geq 0$ on X_n such that $\nu(E) = \int_E h_n\,d\mu$ for all $E \in \mathfrak{F} \cap X_n$. Extend h_n to an $\mathfrak{F}$-measurable function on X by setting $h_n = 0$ on X_n^c. Then $h := \sum_n h_n$ is $\mathfrak{F}$-measurable, $h \geq 0$, and

$$\nu(E) = \bigcup_n \nu(E \cap X_n) = \sum_n \int_E h_n\,d\mu = \int_E h\,d\mu, \quad E \in \mathfrak{F}.$$

Case III. *μ is finite and ν is an arbitrary measure.*

Let $\mathcal{A}$ be the collection of all sets $A \in \mathfrak{F}$ such that ν is σ-finite on $\mathfrak{F} \cap A$. Since $\emptyset \in \mathcal{A}$, $\mathcal{A} \neq \emptyset$. Choose a sequence (A_n) in $\mathcal{A}$ such that $\mu(A_n) \to s := \sup\{\mu(A) : A \in \mathcal{A}\}$. Then $B := \bigcup_n A_n \in \mathcal{A}$, and letting $n \to \infty$ in $\mu(A_n) \leq \mu(B) \leq s$ we see that $\mu(B) = s$. By Case II, applied to μ and ν on $\mathfrak{F} \cap B$, there exists a nonnegative measurable function g on B such that

$$\nu(E) = \nu(B^c \cap E) + \nu(B \cap E) = \nu(B^c \cap E) + \int_{B \cap E} g\,d\mu \quad \text{for all } E \in \mathfrak{F}.$$

Extend g to a measurable function on X by setting $g = \infty$ on B^c. It remains to show that

$$\nu(B^c \cap E) = \int_{B^c \cap E} g\,d\mu, \quad E \in \mathfrak{F}. \tag{‡}$$

But if $\mu(B^c \cap E) = 0$, then by absolute continuity $\nu(B^c \cap E) = 0$, hence both sides of (‡) are zero. On the other hand, if $\mu(B^c \cap E) > 0$, then the right side is ∞. In this case the left side must also be ∞, since otherwise $B \cup (B^c \cap E)$ would be in $\mathcal{A}$, impossible because $\mu\big[B \cup (B^c \cap E)\big] = s + \mu(B^c \cap E) > s$.

Case IV. *μ is σ-finite and ν is an arbitrary measure.*

The proof is similar to that of Case II. The details are left to the reader.

Case V. *μ is σ-finite and ν is an arbitrary signed measure.*

Apply Case IV to ν^+ and ν^- to obtain nonnegative measurable functions h_1 and h_2 such that

$$\nu^+(E) = \int_E h_1\,d\mu \quad \text{and} \quad \nu^-(E) = \int_E h_2\,d\mu, \qquad E \in \mathcal{F}.$$

Since $\nu^+(X)$ and $\nu^-(X)$ are not both infinite, one of the h_j is μ-integrable. Taking $h := h_1 - h_2$ produces the desired result.

Case VI. *μ is σ-finite and ν is an arbitrary complex measure.*

Apply Case V to ν_r and ν_i. The details are left to the reader. □

Lebesgue-Decomposition of a Measure

The following result, a consequence of the Radon-Nikodym theorem, asserts that for a suitable pair of measures μ and ρ, the former may be decomposed into parts that are, respectively, absolutely continuous and singular with respect to the latter. This decomposition will lead to an important result in the next section regarding the derivative of a Lebesgue-Stieltjes measure on $\mathbb{R}^d$.

5.3.7 Lebesgue Decomposition Theorem. *Let ρ be a σ-finite measure and μ a signed (resp., complex) measure on $(X, \mathcal{F})$ such that $|\mu|$ is σ-finite. Then there exist unique signed (resp., complex) measures μ_a and μ_s such that $\mu = \mu_a + \mu_s$, $\mu_a \ll \rho$, and $|\mu_s| \perp \rho$. Furthermore, if μ is a measure, then so are μ_a and μ_s.*

Proof. Suppose first that μ is a measure. Consider the σ-finite measure $m = \rho + \mu$. By 5.3.6(a) there exists a measurable function h $(0 \le h \le 1)$ such that for all $E \in \mathcal{F}$

$$\mu(E) = \int_E h\,dm \quad \text{and} \quad \rho(E) = \int_E (1-h)\,dm.$$

Define

$$\mu_a(E) = \mu\big(E \cap \{h < 1\}\big) \quad \text{and} \quad \mu_s(E) = \mu\big(E \cap \{h = 1\}\big).$$

Clearly, $\mu_a + \mu_s = \mu$. If $\rho(E) = 0$, then $h = 1$ m-a.e. and hence also μ-a.e. on E and so $\mu_a(E) = 0$. Therefore, $\mu_a \ll \rho$. Since $\mu_s(h < 1) = 0 = \rho(h = 1)$, $\mu_s \perp \rho$. This proves the theorem for the case μ a measure.

If μ is a signed measure, then, by the previous paragraph, there exist measures μ_{a1}, μ_{a2} and μ_{s1}, μ_{s2} such that

$$\mu^+ = \mu_{a1} + \mu_{s1}, \quad \mu_{a1} \ll \rho, \quad |\mu_{s1}| \perp \rho, \quad \text{and} \quad \mu^- = \mu_{a2} + \mu_{s2}, \quad \mu_{a2} \ll \rho, \quad |\mu_{s2}| \perp \rho.$$

Set $\mu_a = \mu_{a1} - \mu_{a2}$ and $\mu_s = \mu_{s1} - \mu_{s2}$. Clearly, $\mu_a \ll \rho$. Also, if $|\mu_{sj}|(E_j) = \rho(E_j^c) = 0$, then $|\mu_s|(E_1 \cap E_2) \le |\mu_{s1}|(E_1 \cap E_2) + |\mu_{s2}|(E_1 \cap E_2) = 0$ and $\rho((E_1 \cap E_2)^c) = 0$, so $|\mu_s| \perp \rho$. A similar argument proves the complex case.

For uniqueness, assume that

$$\mu = \mu_a' + \mu_s', \quad \text{where } \mu_a' \ll \rho \text{ and } |\mu_s'| \perp \rho.$$

Then, $\mu_a - \mu_a' = \mu_s - \mu_s'$, hence the common value is both absolutely continuous and singular with respect to ρ and so must be zero (5.3.1(d)). □

5.3.8 Remark. The conclusion of the theorem is false if $|\mu|$ is not σ-finite. For example, take $\rho = \lambda$ and let $\mu =$ counting measure on $\mathcal{B}[0,1]$. Suppose that $\mu = \mu_a + \mu_s$, where $\mu_a \ll \lambda$ and $\mu_s \perp \lambda$. Then $\mu_s(A^c) = \lambda(A) = 0$ for some $A \in \mathcal{B}[0,1]$ and $\mu_a\{x\} = 0$ for all x. Since $\mu_s\{x\} = \mu_s\{x\} + \mu_a\{x\} = \mu\{x\} = 1$, $A^c = \emptyset$. But then $A = [0,1]$, impossible. ◊

Exercises

5.21 Let μ and ν be finite measures with $\nu \ll \mu$ and let $a > 0$. Find a Hahn decomposition of $\nu - a\mu$ in terms of $h = d\nu/d\mu$.

5.22 Let $p > 0$ and define $\nu(E) = \int_E x^p \, d\lambda(x)$, $E \in \mathcal{B}[1,\infty)$. Show that $\nu \ll \lambda$, but the limit $\lim_{\lambda(E)\to 0} |\nu(E)|$ is not zero.

5.23 Let f be the (increasing, continuous) Cantor function and let ν be the probability measure on $\mathcal{B}[0,1]$ with distribution function f. Show that $\nu \perp \lambda$.

5.24 Let ν_1 and ν_2 be complex measures, μ a σ-finite measure, and $c_1, c_2 \in \mathbb{C}$. Show that if $\nu_1 \ll \mu$ and $\nu_2 \ll \mu$, then $c_1\nu_1 + c_2\nu_2 \ll \mu$ and

$$\frac{d(c_1\mu_1 + c_2\mu_2)}{d\mu} = c_1\frac{d\mu_1}{d\mu} + c_2\frac{d\mu_2}{d\mu}.$$

5.25 Let μ_1 and μ_2 be signed measures. Find a Hahn decomposition for $\mu_1 + \mu_2$. ⟦Consider Radon-Nikodym derivatives.⟧

5.26 Let μ be a σ-finite measure and ν a signed or complex measure with $\nu \ll \mu$. Show that $\dfrac{d|\nu|}{d\mu} = \left|\dfrac{d\nu}{d\mu}\right|$.

5.27 Let μ_j be σ-finite measures with $\mu_1 \ll \mu_2$ and $\mu_2 \ll \mu_3$. Prove:

$$\frac{d\mu_1}{d\mu_3} = \frac{d\mu_1}{d\mu_2}\frac{d\mu_2}{d\mu_3}, \quad \mu_3\text{-a.e.}$$

5.28 Let σ-finite measures with $\nu \ll \mu$. Show that $\dfrac{d\nu}{d(\mu+\nu)} = \dfrac{d\nu}{d\mu}\left(1 + \dfrac{d\nu}{d\mu}\right)^{-1}$.

5.29 [↑5.13] Let μ be a measure and let μ_1 and μ_2 be a finite signed measure with $\mu_j \ll \mu$. Show that $(\mu_1 \vee \mu_2) \ll \mu$ and

$$\frac{d(\mu_1 \vee \mu_2)}{d\mu} = \frac{d\mu_1}{d\mu} \vee \frac{d\mu_2}{d\mu}.$$

Show conversely that if $(\mu_1 \vee \mu_2) \ll \mu$, then $\mu_1 \ll \mu$ and $\mu_2 \ll \mu$. Formulate and prove the analogous assertions for $\mu_1 \wedge \mu_2$.

5.30 Let μ_1 and μ_2 be finite measures. Show that $\mu_1 \perp \mu_2$ iff $\mu_1 \wedge \mu_2 = 0$ iff $\mu_1 \vee \mu_2 = \mu_1 + \mu_2$. ⟦In one direction use Ex. 5.8.⟧

5.31 Two σ-finite measures μ and ν are said to be **equivalent** if $\mu \ll \nu$ and $\nu \ll \mu$, that is, μ and ν have the same sets of measure zero.

(a) Show that μ and ν are equivalent iff there exists a finite, positive, measurable function h such that $\nu = h\mu$.

(b) Show that every σ-finite measure μ is equivalent to some probability measure ν. ⟦Consider an infinite series of measures.⟧

5.32 For $j = 1, 2$, let μ_j and ν_j be nontrivial σ-finite measures on $(X_j, \mathcal{F}_j)$. Show that $\nu_1 \otimes \nu_2 \ll \mu_1 \otimes \mu_2$ iff $\nu_1 \ll \mu_1$ and $\nu_2 \ll \mu_2$, in which case

$$\frac{d(\nu_1 \otimes \nu_2)}{d(\mu_1 \otimes \mu_2)}(x_1, x_2) = \frac{d\nu_1}{d\mu_1}(x_1)\frac{d\nu_2}{d\mu_2}(x_2).$$

5.33 Let $T : (X, \mathcal{F}) \to (Y, \mathcal{G})$ be measurable, μ a σ-finite measure and ν a complex or signed measure on X such that $\nu \ll \mu$. (a) Show that $T(\nu) \ll T(\mu)$. (b) Suppose also that $T^{-1} : (Y, \mathcal{G}) \to (X, \mathcal{F})$ exists and is measurable. Prove that

$$\frac{dT(\nu)}{dT(\mu)} = \frac{d\nu}{d\mu} \circ T^{-1}.$$

5.34 Let ν be a σ-finite measure.

(a) Let μ_n, μ be measures such that $\mu_n \uparrow \mu$ and $\mu \ll \nu$. Show that

$$\frac{d\mu_n}{d\nu} \uparrow \frac{d\mu}{d\nu}, \quad \nu\text{-a.e.}$$

In particular, $\dfrac{d\mu_n}{d\mu} \uparrow 1$.

(b) Let η_n be measures, $\mu := \sum_{n=1}^{\infty} \eta_n$ and $\mu \ll \nu$. Show that

$$\frac{d\mu}{d\nu} = \sum_{n=1}^{\infty} \frac{d\eta_n}{d\nu} \quad \nu\text{-a.e.}$$

5.35 Let μ, μ_n be probability measures.

(a) Show that there exists a probability measure ν such that $\mu \ll \nu$ and $\mu_n \ll \nu$ for all n. (Consider an infinite series of measures.)

(b) Suppose that $|\mu - \mu_n|(X) \to 0$. Let ν be any finite measure such that $\mu \ll \nu$ and $\mu_n \ll \nu$ for all n. Show that

$$\frac{d\mu_n}{d\nu} \xrightarrow{\nu} \frac{d\mu}{d\nu}.$$

⟦Let $h_n = d\mu_n/d\nu - d\mu/d\nu$. Consider $A_n := \{|h_n| \geq \varepsilon\}$, $B_n := \{h_n \geq \varepsilon\}$, $C_n := \{h_n \leq -\varepsilon\}$.⟧

5.36 Let μ, ν, ν_n be measures with μ σ-finite and $\nu_n(E) \uparrow \nu(E)$ for every $E \in \mathcal{F}$. Show that $\nu \ll \mu$ iff $\nu_n \ll \mu$ for all n, in which case

$$\frac{d\nu}{d\mu} = \lim_n \frac{d\nu_n}{d\mu} \quad \mu \text{ a.e.}$$

5.37 [↓5.5.9] Let μ and μ_n be finite measures with $\mu = \sum_n \mu_n$ and let $\mu_n = \mu_{na} + \mu_{ns}$ and $\mu = \mu_a + \mu_s$ be the Lebesgue decompositions with respect to a σ-finite measure ρ. Show that

$$\mu_a = \sum_n \mu_{na} \quad \text{and} \quad \mu_s = \sum_n \mu_{ns}.$$

5.4 Differentiation of Measures

In this section we show that a Radon-Nikodym derivative on $\mathbb{R}^d$ can be expressed as a limit of difference quotients, thus establishing a connection with the classical derivative.

Definition and Properties of the Derivative

Let μ be a signed measure on $\mathcal{B}(\mathbb{R}^d)$ which is finite on bounded sets. We shall call such a measure a **Lebesgue-Stieltjes signed measure**. For each $x \in \mathbb{R}^d$ and $r > 0$, let $\mathcal{B}(x, r)$ denote the collection of all open balls containing x and with radius less than r. Define

$$\overline{D}(\mu; x, r) := \sup\left\{\frac{\mu(B)}{\lambda(B)} : B \in \mathcal{B}(x, r)\right\} \text{ and } \underline{D}(\mu; x, r) := \inf\left\{\frac{\mu(B)}{\lambda(B)} : B \in \mathcal{B}(x, r)\right\},$$

where for simplicity of notation we set $\lambda = \lambda^d$. Note that for fixed x, the functions $\overline{D}(\mu; x, r)$ and $\underline{D}(\mu; x, r)$ decrease and increase, respectively, as $r \downarrow 0$. Moreover, for each c and r the sets $\{x : \overline{D}(\mu; x, r) > c\}$ and $\{x : \underline{D}(\mu; x, r) < c\}$ are open (Ex. 5.38). Thus $\overline{D}(\mu; x, r)$ and $\underline{D}(\mu; x, r)$ are Borel measurable in x for fixed r.

Now define the **upper and lower derivates** $\overline{D}\,\mu$ and $\underline{D}\,\mu$ of μ by

$$\overline{D}\,\mu(x) := \lim_{r\to 0^+} \overline{D}(\mu; x, r) = \inf_{r>0} \overline{D}(\mu; x, r), \quad \underline{D}\,\mu(x) := \lim_{r\to 0^+} \underline{D}(\mu; x, r) = \sup_{r>0} \underline{D}(\mu; x, r).$$

Then

$$\underline{D}(\mu; x, r) \uparrow \underline{D}\,\mu(x) \text{ and } \overline{D}(\mu; x, r) \downarrow \overline{D}\,\mu(x) \text{ as } r \downarrow 0,$$

so by the preceding observations the functions $\overline{D}\,\mu$ and $\underline{D}\,\mu$ are Borel measurable. If $\overline{D}\,\mu(x)$ and $\underline{D}\,\mu(x)$ are finite and equal, then μ is said to be **differentiable at** x. In this case, the common value is denoted by $D\mu(x)$ and is called the **derivative of** μ **at** x. Note that the inequalities

$$\underline{D}(\mu; x, r) \le \frac{\mu(B(x,r))}{\lambda(B(x,r))} \le \overline{D}(\mu; x, r)$$

imply that

$$\underline{D}\,\mu(x) \le \varliminf_{r\to 0} \frac{\mu(B(x,r))}{\lambda(B(x,r))} \le \varlimsup_{r\to 0} \frac{\mu(B(x,r))}{\lambda(B(x,r))} \le \overline{D}\,\mu(x).$$

It follows that if $D\mu$ is differentiable at x, then

$$D\,\mu(x) = \lim_{r\to 0} \frac{\mu(B(x,r))}{\lambda(B(x,r))}. \tag{5.7}$$

Here is a related sequential characterization of differentiability:

5.4.1 Proposition. *A Lebesgue-Stieltjes signed measure μ is differentiable at $x \in \mathbb{R}^d$ iff there exists a real number a such that for any sequence of open balls $B_n \in \mathcal{B}(x, r_n)$ with $r_n \to 0$,*

$$\lim_n \frac{\mu(B_n)}{\lambda(B_n)} = a.$$

In this case, $a = D\mu(x)$.

Proof. The inequality $\underline{D}(\mu; x, r) \le \mu(B_n)/\lambda(B_n) \le \overline{D}(\mu; x, r)$ $(r_n \le r)$ implies that

$$\underline{D}(\mu; x, r) \le \varliminf_n \frac{\mu(B_n)}{\lambda(B_n)} \le \varlimsup_n \frac{\mu(B_n)}{\lambda(B_n)} \le \overline{D}(\mu; x, r).$$

Letting $r \to 0$, we obtain

$$\underline{D}\mu(x) \le \varliminf_n \frac{\mu(B_n)}{\lambda(B_n)} \le \varlimsup_n \frac{\mu(B_n)}{\lambda(B_n)} \le \overline{D}\mu(x).$$

Therefore, if $D\mu$ is differentiable at x, then $\lim_n \mu(B_n)/\lambda(B_n) = D\mu(x)$.

Conversely, suppose that $D\mu$ is not differentiable at x, so $\underline{D}\,\mu(x) < \overline{D}\,\mu(x)$. Let $t_n \uparrow \underline{D}\,\mu(x)$. For each n, $\underline{D}(\mu; x, r) > t_n$ for all sufficiently small r, hence for each $n \in \mathbb{N}$ we may choose $a_n < 1/n$ and $A_n \in \mathcal{B}(x, a_n)$ such that $t_n < \mu(A_n)/\lambda(A_n) \le \underline{D}\,\mu(x)$. Thus $\lim_n \mu(A_n)/\lambda(A_n) = \underline{D}\,\mu(x)$. Similarly, there exists $b_n < 1/n$ and $B_n \in \mathcal{B}(x, b_n)$ such that $\lim_n \mu(B_n)/\lambda(B_n) = \overline{D}\,\mu(x)$. Since these limits are unequal, the sequential criterion fails. □

5.4.2 Corollary. *The differential operator D is linear: If $D\mu(x)$ and $D\nu(x)$ exist and $a, b \in \mathbb{R}$, then $D(a\mu + b\nu)(x)$ exists and $D(a\mu + b\nu)(x) = aD(\mu)(x) + bD(\nu)(x)$.*

Connections with the Classical Derivative

Equation (5.7) expresses the derivative of a measure as a limit of quotients, analogous to the definition of the classical derivative of a function of a real variable. This connection is crystallized in Proposition 5.4.4 below. For the proof we need the following lemma.

5.4.3 Lemma. *Let μ be a Lebesgue-Stieltjes signed measure on $\mathbb{R}$ and let F be a function on $\mathbb{R}$ such that $\mu(a, b] = F(b) - F(a)$ for all $a < b$. Then at each continuity point x of F,*

$$\underline{D}\,\mu(x) = \lim_{r \to 0+} \inf_{0<|h|\leq r} \frac{F(x+h) - F(x)}{h} \quad \text{and} \quad \overline{D}\,\mu(x) = \lim_{r \to 0+} \sup_{0<|h|\leq r} \frac{F(x+h) - F(x)}{h}.$$

Proof. Note first that F has at most countably many discontinuities. This follows from $\mu = \mu^+ - \mu^-$, allowing us to write $F(b) - F(a) = [F_+(b) - F_+(a)] - [F_-(b) - F_-(a)]$, where $F_\pm$ are distribution functions, hence nondecreasing, and so have at most countably many discontinuities (Ex. 1.84).

We prove only the equality for $\underline{D}\,\mu(x)$. (The other equality follows by considering $-\mu$.) Define

$$f(r) := \inf_{0<|h|\leq r} \frac{F(x+h) - F(x)}{h} \quad \text{and} \quad f(0+) := \lim_{r \to 0^+} f(r).$$

We show first that $f(0+) \geq \underline{D}\,\mu(x)$. Suppose for a contradiction that $f(0+) < \underline{D}\,\mu(x)$. Let $f(0+) < a < \underline{D}\,\mu(x))$. Since $f(1/n) \uparrow f(0+) < a$, we may choose a sequence $h_n \to 0$ such that

$$\frac{F(x+h_n) - F(x)}{h_n} < a.$$

By considering subsequences, it suffices to consider two cases: (a) $h_n \downarrow 0$ and (b) $h_n \uparrow 0$.

Suppose that (a) holds. Since F is continuous at x and right continuous at $x + h_n$, for each n we may choose $0 < t_n < 1/n$ such that

$$\frac{F(x + (1+t_n)h_n) - F(x - t_n h_n)}{h_n} < a.$$

Since F has at most countably many discontinuities, we may take t_n so that F is continuous at $x + (1+t_n)h_n$. Setting $B_n = (x - t_n h_n, x + (1+t_n)h_n)$ and $r_n = h_n(1+2t_n)/2$ (the radius of the interval B_n), we then have

$$\underline{D}(\mu, x, r_n) \leq \frac{\mu(B_n)}{\lambda(B_n)} = \frac{F(x + (1+t_n)h_n) - F(x - t_n h_n)}{h_n(1+2t_n)} < \frac{a}{1+2t_n}.$$

Letting $n \to \infty$ produces the contradiction $a < \underline{D}\,\mu(x) \leq a$.

Now assume that (b) holds. Let $k_n = -h_n$, so $k_n > 0$ and

$$\frac{F(x) - F(x - k_n)}{k_n} < a.$$

By right continuity at x, there exists $0 < t_n < 1/n$ such that

$$\frac{F(x + t_n k_n) - F(x - k_n)}{k_n} < a.$$

Setting $B_n = (x - k_n, x + t_n k_n)$ and $r_n = k_n(1+t_n)/2$ we then have

$$\underline{D}(\mu, x, r_n) \leq \frac{\mu(B_n)}{\lambda(B_n)} = \frac{F(x + t_n k_n) - F(x - k_n)}{k_n(1+t_n)} < \frac{a}{1+t_n}.$$

Letting $n \to \infty$ gives the contradiction $a < \underline{D}\,\mu(x) \le a$.

We have shown that $f(0+) \ge \underline{D}\,\mu(x)$. To show equality, let $b > \underline{D}\,\mu(x)$ and choose r_n, $s_n \downarrow 0$ such that for $B_n := (x - s_n, x + r_n)$

$$\frac{F(x+r_n) - F(x-s_n)}{r_n + s_n} = \frac{\mu(B_n)}{\lambda(B_n)} < b. \tag{†}$$

Setting $q_n := r_n + s_n$, we may write the left side as

$$\frac{F(x+r_n) - F(x)}{r_n}\frac{r_n}{q_n} + \frac{F(x-s_n) - F(x)}{-s_n}\frac{s_n}{q_n} \ge \min\{f(r_n), f(s_n)\}. \tag{‡}$$

Letting $n \to \infty$ we see from (†) and (‡) that $b \ge f(0+)$. Since $b > \underline{D}\,\mu(x)$ was arbitrary, $\underline{D}\,\mu(x) \ge f(0+)$. □

Here is the promised connection between differentiability of measures and differentiability of functions.

5.4.4 Theorem. *Let μ be a Lebesgue-Stieltjes signed measure on $\mathbb{R}$ and F a function on $\mathbb{R}$ such that $\mu(a,b] = F(b) - F(a)$ for all $a < b$. Then μ is differentiable at $x \in \mathbb{R}$ iff F is differentiable at x. In this case, $D\,\mu(x) = F'(x)$.*

Proof. If μ is differentiable at x, then F is continuous at x (Ex. 5.39), hence the expressions in the preceding lemma are all equal. This implies that F is differentiable at x with derivative $D\mu(x)$. The converse is similar. □

Existence of the Measure Derivative

The following theorem shows that $D\mu$ exists λ-a.e. and is in fact a Radon-Nikodym derivative, where as before we set $\lambda = \lambda^d$.

5.4.5 Theorem. *Let μ be a Lebesgue-Stieltjes signed measure on $\mathcal{B}(\mathbb{R}^d)$ with Lebesgue decomposition $\mu = \mu_a + \mu_s$, where $\mu_a \ll \lambda$ and $|\mu_s| \perp \lambda$. Then*

$$D\mu(x) = \frac{d\mu_a}{d\lambda}(x) = D\mu_a(x) \text{ and } D\mu_s(x) = 0 \ \lambda\text{-a.e.}$$

In particular, $D\mu(x)$ exists and is finite λ-a.e.

Proof. It is enough to prove the first equality. The proof consists of several steps, the first of which is called the *Vitali covering lemma*.

(1) *Let $\{B_1, \ldots, B_n\}$ be a collection of open balls in $\mathbb{R}^d$. Then there is a subcollection of disjoint balls $\{B_{k_1}, \ldots, B_{k_m}\}$ such that*

$$\lambda\Big(\bigcup_{i=1}^{n} B_i\Big) \le 3^d \sum_{i=1}^{m} \lambda(B_{k_i}).$$

[Choose the notation so that the radius of B_i decreases as i increases. Let $k_1 = 1$, and successively choose $k_i \in \mathbb{N}$ such that k_{i+1} is the smallest index $j > k_i$ for which B_j is disjoint from $B_{k_1} \cup \cdots \cup B_{k_i}$. Let k_m be the index for which the process stops, so that the collection $\{B_{k_1}, \ldots, B_{k_m}\}$ is disjoint. By choice of k_m, if $j > k_m$ or $k_i < j < k_{i+1}$, then $B_j \cap B_{k_q} \ne \emptyset$ for some q with $k_q < j$. Thus for each $j = 1, \ldots, n$ there exists $k_q \le j$ such that $B_j \cap B_{k_q} \ne \emptyset$. Now let A_{k_q} be the ball with the same center as B_{k_q} and with triple the radius.

Since $j \ge k_q$, the radius of B_j is no larger than that of B_{k_q}, hence $B_j \subseteq A_{k_q}$. Therefore $\bigcup_{j=1}^{n} B_j \subseteq \bigcup_{i=1}^{m} A_{k_i}$, and the desired inequality follows from the dilation property of λ.]

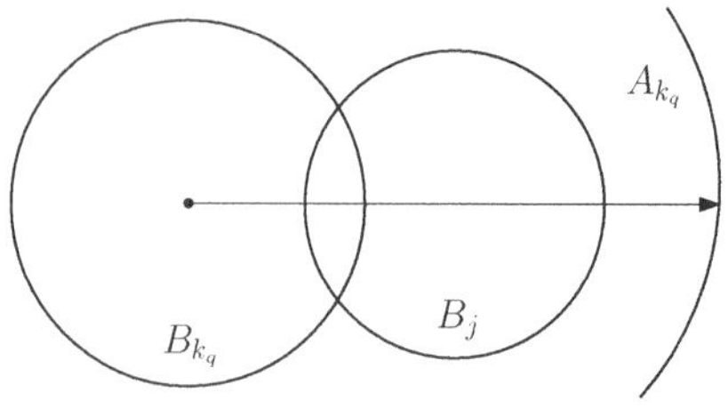

FIGURE 5.1: Construction of A_{k_q}.

(2) *Let μ be nonnegative, $c > 0$, and $K \subseteq \{\overline{D}\mu > c\}$ compact. Then $c\lambda(K) \le 3^d\mu(K)$.*

⟦Let $r > 0$. For each $x \in K$, choose $B \in \mathcal{B}(x, r)$ such that $\mu(B)/\lambda(B) > c$. By compactness, there exists a finite subcover $B_1, \ldots, B_n$ of K of such balls. Choose $\{B_{k_1}, \ldots, B_{k_m}\}$ as in step (1). Then

$$\lambda(K) \le 3^d \sum_{i=1}^m \lambda(B_{k_i}) \le \frac{3^d}{c} \sum_{i=1}^m \mu(B_{k_i}) = \frac{3^d}{c}\mu\Big(\bigcup_{i=1}^m B_{k_i}\Big).$$

Since B_{k_i} has radius $< r$ and meets K it must be contained in $U_r := \{x : d(x, K) < 2r\}$. Therefore, $\lambda(K) \le (3^d/c)\mu(U_r)$. Letting $r \downarrow 0$ yields $\lambda(K) \le (3^d/c)\mu(K)$.⟧

(3) *If μ is nonnegative and $\mu(E) = 0$, then $D\mu(x) = 0$ for λ-a.a. $x \in E$.*

⟦Let $c > 0$ and $B := E \cap \{\overline{D}\,\mu > c\}$. We show that $\lambda(B) = 0$. By regularity, it suffices to show that $\lambda(K) = 0$ for any compact $K \subseteq B$. But this follows from step (2), since $\lambda(K) \le (3^d/c)\mu(K) \le (3^d/c)\mu(E) = 0$.⟧

(4) *If $|\mu| \perp \lambda$, then $D\mu(x) = 0$ for λ-a.a. $x \in \mathbb{R}^d$.*

⟦Let $E \in \mathcal{B}(\mathbb{R}^d)$ such that $\lambda(E^c) = |\mu(E)| = 0$. Then also $\mu^+(E) = 0$, so by step (3), $D\mu^+(x) = 0$ for λ-a.a. $x \in E$, hence for λ-a.a. $x \in \mathbb{R}^d$. Similarly $D\mu^-(x) = 0$ for λ-a.a. $x \in \mathbb{R}^d$. Therefore, $D\mu = D\mu^+ - D\mu^- = 0$ λ-a.e.⟧

(5) *If $\mu \ll \lambda$, then $D\mu = d\mu/d\lambda$ and is finite λ-a.e.*

⟦Since μ is finite on bounded sets, $h := d\mu/d\lambda$ is finite λ-a.e. For the desired equality it suffices to show that $\overline{D}\,\mu \le h$ λ-a.e.; the reverse inequality will follow by considering $-\mu$. It therefore must be shown that $\lambda\{h < t < \overline{D}\,\mu\} = 0$ for all $t \in \mathbb{R}$. To this end define a measure ρ on $\mathcal{B}(\mathbb{R}^d)$ by

$$\rho(A) = \int_{A \cap \{h \ge t\}} (h - t)\, d\lambda.$$

For any open ball B,

$$\mu(B) - t\lambda(B) = \int_B (h - t)\, d\lambda \le \int_{B \cap \{h \ge t\}} (h - t)\, d\lambda = \rho(B),$$

hence

$$\frac{\mu(B)}{\lambda(B)} \le \frac{\rho(B)}{\lambda(B)} + t.$$

Since B was arbitrary, $\overline{D}\,\mu \le \overline{D}\,\rho + t$. Set $E := \{h < t\}$. Then $\rho(E) = 0$, so by step (3) applied to ρ, $\overline{D}\rho(x) = 0$ for λ-a.a $x \in E$. Therefore, $\overline{D}\,\mu(x) \le t$ for λ-a.a. $x \in E$, as required.⟧

The desired equality now follows by applying step (4) to μ_s and step (5) to μ_a. □

5.4.6 Corollary. *Let μ and F be as in 5.4.4. Then $F' = \dfrac{d\mu_a}{d\lambda}$ λ-a.e.*

Exercises

5.38 Let μ be a Lebesgue-Stieltjes signed measure on $\mathcal{B}(\mathbb{R}^d)$. Show that the sets $\{x : \overline{D}(\mu; x, r) > c\}$ and $\{x : \underline{D}(\mu; x, r) < c\}$ are open.

5.39 Let μ be a Lebesgue-Stieltjes signed measure on $\mathcal{B}(\mathbb{R}^d)$. Prove that if μ is differentiable at x, then $\mu\{x\} = 0$. In particular, if $d = 1$ and μ has distribution function F, then F is continuous at x.

5.40 Prove that if μ is a nonnegative measure, then $\int_E (D\mu)\, d\lambda \le \mu(E)$ for all $E \in \mathcal{B}(\mathbb{R}^d)$.

5.41 Define the *metric density of* $E \in \mathcal{B}(\mathbb{R}^d)$ by

$$d_E(x) = \lim_{r \to 0} \frac{\lambda\big(E \cap B(x, r)\big)}{\lambda\big(B(x, r)\big)}.$$

Show that d_E exists and $= \mathbf{1}_E$ λ-a.e.

5.42 Let f be locally integrable on $\mathbb{R}^d$. Verify (a)–(c) to prove the *Lebesgue differentiation theorem*: For λ-a.e. x,

$$\lim_{r \to 0} \frac{1}{\lambda\big(B(x, r)\big)} \int_{B(x,r)} |f(y) - f(x)|\, dy = 0$$

Points x for which this holds are called **Lebesgue points**.

(a) $\displaystyle\lim_{r \to 0} \frac{1}{\lambda\big(B(x, r)\big)} \int_{B(x,r)} f(y)\, dy = f(x)$.

(b) Let $a \in \mathbb{R}$. Then there exists N_a with $\lambda(N_a) = 0$ such that

$$\lim_{r \to 0} \frac{1}{\lambda\big(B(x, r)\big)} \int_{B(x,r)} |f(y) - a|\, dy = |f(x) - a|, \quad x \in N_a^c.$$

(c) Set $N = \bigcup_{a \in \mathbb{Q}} N_a$. Let $\varepsilon > 0$, $x \in N^c$ and choose $a \in \mathbb{Q}$ such that $|f(x) - a| < \varepsilon$. There exists $\delta > 0$ such that for all $r < \delta$,

$$\frac{1}{\lambda\big(B(x, r)\big)} \int_{B(x,r)} |f(y) - f(x)|\, dy < 2\varepsilon.$$

5.43 [$\uparrow$ 5.42] For each $x \in \mathbb{R}^d$, let $\{E(x, r) : r > 0\}$ be a collection in $\mathcal{B}(\mathbb{R}^d)$ such that $E(x, r) \subseteq B(x, r)$ and $\lambda\big(E(x, r)\big) \ge c\lambda\big(B(x, r)\big)$ for all r, where $c > 0$ is a constant that does not depend on r. (Such a collection is said to **shrink to x nicely**. For example, the collection $\{(x, x + r) : r > 0\}$ shrinks to x nicely.) Let μ be a measure with $\mu \ll \lambda$. Prove that for λ-a.a. x,

$$D\mu(x) = \lim_{r \to 0} \frac{\mu\big(E(x, r)\big)}{\lambda\big(E(x, r)\big)}.$$

5.5 Functions of Bounded Variation

Definition and Basic Properties

Let $I \subseteq \mathbb{R}$ be an interval and $f : I \to \mathbb{C}$. For an ordered subset $\mathcal{P} = \{x_0 < x_1 < \cdots < x_n\}$ of I, define the **$\mathcal{P}$-variation of f on I** by

$$V_{I,\mathcal{P}}(f) = V_{\mathcal{P}}(f) = \sum_{j=1}^{n} |f(x_j) - f(x_{j-1})|. \tag{5.8}$$

Note that if $\mathcal{Q} \supseteq \mathcal{P}$, then, by the triangle inequality, $V_{I,\mathcal{Q}}(f) \geq V_{I,\mathcal{P}}(f)$. The **total variation of f on I** is the extended real number

$$V_I(f) := \sup_{\mathcal{P}} V_{I,\mathcal{P}}(f), \tag{5.9}$$

where the supremum is taken over all ordered subsets $\mathcal{P}$ of I. We say that f has **bounded variation on I** if $V_I(f) < \infty$. We shall mainly be concerned with the cases $I = \mathbb{R}$ and $I = [a, b]$. For the latter, we may assume in (5.9) that $\mathcal{P}$ is a partition of $[a, b]$, as the supremum does not change by adjoining the points a and b. We denote set of all functions with bounded variation on I by $BV(I)$.

By the mean value theorem, a real-valued function with a bounded derivative has bounded variation on bounded intervals. In particular, $\sin x$ has bounded variation on any bounded interval (but not on $\mathbb{R}$: consider partition points $(2k + 1)\pi/2$).

The following proposition summarizes the elementary properties of $BV(I)$. The proof is left as an exercise for the reader (5.44).

5.5.1 Proposition. *Let I be any interval.*

(a) *A bounded, monotone function f on $\mathbb{R}$ has bounded variation.*

(b) *If $f \in BV(I)$, then f is bounded.*

(c) $f \in BV(I)$ *iff* $\text{Re}(f), \text{Im}(f) \in BV(I)$.

(d) *If f is real-valued, then $f \in BV(I)$ iff $f^{\pm} \in BV(I)$.*

(e) *If $f, g \in BV(I)$ and $c \in \mathbb{C}$ then $f + g$, cf, ,$\overline{f}$, fg, $|f| \in BV(I)$.*

By the proposition, the difference of two bounded monotone functions on I has bounded variation on I. The converse also holds:

5.5.2 Proposition. *If $f \in BV(I)$ is real-valued, then there exist nondecreasing functions g and h on I such that $f = g - h$. In particular, f is a Borel function.*

Proof. For definiteness, we take $I = \mathbb{R}$. For $x \in \mathbb{R}$, define

$$g(x) := V_{(-\infty,x]}(f) \quad \text{and} \quad h(x) := g(x) - f(x),$$

so that $f = g - h$. Clearly, g is increasing. To see that h is increasing, let $a < x < y$, let $\mathcal{P}_x$ be an arbitrary partition of $[a, x]$, and set $\mathcal{P}_y := \mathcal{P}_x \cup \{y\}$. Then

$$V_{\mathcal{P}_x}(f) + f(y) - f(x) \leq V_{\mathcal{P}_x}(f) + |f(y) - f(x)| = V_{\mathcal{P}_y}(f) \leq g(y).$$

Taking the supremum over all partitions $\mathcal{P}_x$ yields $V_{[a,x]}(f) + f(y) - f(x) \leq g(y)$. Since a was arbitrary, $g(x) + f(y) - f(x) \leq g(y)$, that is, $h(x) \leq h(y)$. □

Since monotone functions have at most countably many discontinuities (Ex. 1.84), we have

5.5.3 Corollary. *If $f \in BV(\mathbb{R})$, then f has at most countably many discontinuities.*

The Total Variation Function

Let $f \in BV(\mathbb{R})$. The function $x \mapsto V_{(-\infty,x]}(f)$ in the proof of 5.5.2 is called the **total variation function of** f and is denoted by T_f:

$$T_f(x); = V_{(-\infty,x]}(f), \quad x \in \mathbb{R}.$$

Clearly, T_f is increasing, hence has bounded variation on any bounded interval. The theorem below makes a connection between the total variation function and the total variation measure of a complex measure. For the proof we need the following lemmas.

5.5.4 Lemma. *Let $f \in BV(\mathbb{R})$. Then for $x < y$, $T_f(y) - T_f(x) = V_{(x,y]}(f)$.*

Proof. Set $T := T_f$. Note first that for the sets $\mathcal{P} = \{x_0 < x_1 < \cdots < x_n = y\} \subseteq (-\infty, y]$ implicit in the definition of $T(y)$, we may assume that $x_0 \le x$, otherwise simply adjoin a suitable point to $\mathcal{P}$, increasing the $\mathcal{P}$-variation of f but not altering $T(y)$. Choose k so that $x_k \le x < x_{k+1}$ and set

$$\mathcal{Q} = \{x_0 < x_1 < \cdots < x_k\} \text{ and } \mathcal{R} = \{x_{k+1} < \cdots < x_n = y\}.$$

Then $V_{\mathcal{P}}(f) \le V_{\mathcal{Q}}(f) + V_{\mathcal{R}}(f) \le T(x) + V_{(x,y]}(f)$. Taking the supremum over all $\mathcal{P}$ yields $T(y) \le T(x) + V_{(x,y]}(f)$. For the reverse inequality, for arbitrary $\mathcal{Q}$ and $\mathcal{R}$ as above we have $V_{\mathcal{Q}} + V_{\mathcal{R}} \le T(y)$. Taking the supremum over $\mathcal{Q}$ and $\mathcal{R}$ yields $T(x) + V_{(x,y]}(f) \le T(y)$. □

5.5.5 Lemma. *Let $f \in BV(\mathbb{R})$. If f is right continuous at x, then so is T_f.*

Proof. Set $T := T_f$ and let $y > x$. By right continuity at x, given $\varepsilon > 0$ we may choose $\delta > 0$ so that $0 < y - x < \delta \Rightarrow |f(x) - f(y)| < \varepsilon$. For such a fixed y, choose an ordered set $\mathcal{P} = \{x_0 < x_1 < \cdots < x_n = y\} \subseteq (x, y]$ such that

$$\sum_{j=1}^{n} |f(x_j) - f(x_{j-1})| \ge V_{(x,y]} - \varepsilon = T(y) - T(x) - \varepsilon.$$

Since $|f(x_1) - f(x_0)| \le |f(x_1) - f(x)| + |f(x_0) - f(x)| < 2\varepsilon$, we have

$$\sum_{j=2}^{n} |f(x_j) - f(x_{j-1})| = \sum_{j=1}^{n} |f(x_j) - f(x_{j-1})| - |f(x_1) - f(x_0)| \ge T(y) - T(x) - 3\varepsilon. \quad (\dagger)$$

Next, choose $\mathcal{Q} = \{t_0 < t_1 < \cdots < t_m = x_1\} \subseteq (x, x_1]$ such that

$$\sum_{j=1}^{m} |f(t_j) - f(t_{j-1})| \ge V_{(x,x_1]} - \varepsilon = T(x_1) - T(x) - \varepsilon. \quad (\ddagger)$$

Adding (†) and (‡), we have

$$\sum_{j=1}^{m} |f(t_j) - f(t_{j-1})| + \sum_{j=2}^{n} |f(x_j) - f(x_{j-1})| \ge T(y) + T(x_1) - 2T(x) - 4\varepsilon.$$

Since the left side is $\le V_{(x,y]} = T(y) - T(x)$, we see that $T(x_1) - T(x) \le 4\varepsilon$. Letting $x_1 \downarrow x$ yields $T(x+) - T(x) \le 4\varepsilon$. Therefore, $T(x+) = T(x)$, as required. □

5.5.6 Corollary. *Let $f \in BV(\mathbb{R})$ be right continuous. Then $f = g - h$, where g and h are distribution functions.*

Proof. In the proof of 5.5.2, g is the function T_f. $\square$

5.5.7 Theorem. *Let μ be a complex measure on $\mathcal{B}(\mathbb{R})$ and set $f(x) := \mu(-\infty, x]$. Then $f \in BV(\mathbb{R})$ and T_f is a distribution function for $|\mu|$.*

Proof. That $f \in BV(\mathbb{R})$ follows from

$$\sum_{j=1}^{n} \big|\mu(x_j, x_{j+1}]\big| \le \sum_{j=1}^{n} |\mu|(x_j, x_{j+1}] \le |\mu|(\mathbb{R}) < \infty.$$

Since f is right continuous, $T := T_f$ is a distribution function by 5.5.5. By the measure uniqueness theorem (1.6.8), it remains to show $\mu_T(a,b] = |\mu|(a,b]$ for all $a < b$. For the inequality $\mu_T(a,b] \le |\mu|(a,b]$ we have

$$\begin{aligned} \mu_T(a,b] &= V_{(a,b]}(f) = \sup\Big\{\sum\nolimits_j |\mu(x_j, x_{j+1}]| : a < x_1 < \cdots < x_n = b\Big\} \\ &\le \sup\Big\{\sum\nolimits_j |\mu(E_j)| : E_1, \ldots, E_k \text{ a measurable partition of } (a,b]\Big\} \\ &= |\mu|(a,b]. \end{aligned}$$

For the reverse inequality, let $E_1, \ldots, E_k$ be a measurable partition of $(a,b]$. Given $\varepsilon > 0$, by regularity of $|\mu|$ (1.8.1) there exist compact sets $K_j \subseteq E_j$ such that $|\mu|(E_j \setminus K_j) < \varepsilon/k$, hence

$$\sum_j |\mu(E_j)| \le \varepsilon + \sum_j |\mu(K_j)|. \qquad (\dagger)$$

Since the sets K_j are disjoint, there exist disjoint open sets $U_j \supseteq K_j$. Each U_j is a countable union of disjoint open intervals (a_{jn}, b_{jn}), hence

$$\sum_j |\mu(K_j)| \le \sum_j \sum_n \big|\mu\big((a_{jn}, b_{jn}] \cap (a,b]\big)\big|.$$

The partial sums of the double sum on the right are terms of $V_{(a,b],\mathcal{P}}(f)$ for suitable partitions $\mathcal{P}$ and hence are $\le \mu_T(a,b]$. Therefore, by (†), $\sum_j |\mu(E_j)| \le \varepsilon + \mu_T(a,b]$. Since $E_1, \ldots, E_k$ and ε were arbitrary, $|\mu|(a,b] \le \mu_T(a,b]$. $\square$

Differentiation of Functions of Bounded Variation

We may now prove the following fundamental property of functions of bounded variation.

5.5.8 Theorem. *If $f \in BV(\mathbb{R})$, then $f'(x)$ exists for λ-a.a. $x \in \mathbb{R}$. Moreover, if $g(x) = f(x+)$ $(= \lim_{t \to x^+} f(t))$, then $g' = f'$ a.e.*

Proof. By decomposing f as in 5.5.2, we may assume that f is real-valued and nondecreasing. Then g has these properties, and since g is right continuous it is a distribution function and so is differentiable λ-a.e. (5.4.6). Moreover, $g(x) = f(x)$ except possibly at a sequence of points x_n, the points at which f is not continuous. Set $h(x) := g(x) - f(x)$. The theorem will be proved if we show that $h' = 0$ a.e.

Since the intervals $\big(f(x_n), f(x_n+)\big)$ $(|x_n| < m)$ are pairwise disjoint and contained in the interval $\big(f(-m), f(m)\big)$,

$$\sum_{|x_n|<m} h(x_n) = \sum_{|x_n|<m} [f(x_n+) - f(x_n)] < \infty. \qquad (\dagger)$$

Define

$$\nu := \lim_{m\to\infty} \sum_{|x_n|<m} h(x_n)\delta_{x_n}.$$

By (†), ν is a Lebesgue-Stieltjes measure on $\mathbb{R}$. Moreover, for any $a < x < y < b$, the quantity $h(x) + h(y)$ is zero unless x or y is one of the discontinuity points of f, in which case its value is at most $h(x_m) + h(x_n)$ for some m and n. Therefore, $h(x) + h(y) \le \nu(a, b)$ for all $a < x < y < b$. For $r > 0$, we then have

$$\left|\frac{h(x+r) - h(x)}{r}\right| \le \frac{h(x+r) + h(x)}{r} \le \frac{\nu(x+2r, x-2r)}{r} = 4\,\frac{\nu\big(B(x,2r)\big)}{\lambda\big(B(x,2r)\big)},$$

hence

$$\overline{\lim_{r\to 0}} \left|\frac{h(x+r) - h(x)}{r}\right| \le 4\,\overline{\lim_{r\to 0}}\, \frac{\nu\big(B(x,r)\big)}{\lambda\big(B(x,r)\big)}.$$

Since $\lambda\{x_1, x_2, \ldots\} = 0 = \nu\{x_1, x_2, \ldots\}^c$, $\nu \perp \lambda$, hence the right side of the preceding inequality is zero for λ-a.a x (5.4.5). Therefore, $\lim_{r\to 0} r^{-1}|h(x+r) - h(x)| = 0$ for λ-a.a. x, completing the proof. □

5.5.9 Corollary. *Let (f_n) be a sequence of nonnegative, nondecreasing functions on $\mathbb{R}$ such that the series $f(x) := \sum_n f_n(x)$ converges for all x. Then $f'(x) = \sum_n f_n'(x)$ λ-a.e.*

Proof. Let g and g_n correspond to f and f_n as in the theorem. Then $g(x) = \sum_n g_n(x)$, as is easily verified, hence we may assume that f and f_n are distribution functions on $\mathbb{R}$. Let μ_n and μ be the corresponding Lebesgue-Stieltjes measures. By the hypothesis, $\mu = \sum \mu_n$ on intervals $(a, b]$ hence, by the uniqueness theorem for measures, $\mu = \sum \mu_n$ on $\mathcal{B}(\mathbb{R})$. Let $\mu_n = \mu_{na} + \mu_{ns}$ and $\mu = \mu_a + \mu_s$ be the Lebesgue decompositions with respect to λ. By Ex. 5.37, $\mu_a = \sum_n \mu_{na}$ so

$$\int_E \frac{d\mu_a}{d\lambda}\, d\lambda = \mu_a(E) = \sum_n \mu_{na}(E) = \int_E \sum_n \frac{d\mu_{na}}{d\lambda}\, d\lambda \quad \text{for all } E \in \mathcal{B}(\mathbb{R}),$$

hence

$$\frac{d\mu_a}{d\lambda} = \sum_n \frac{d\mu_{na}}{d\lambda} \quad \lambda\text{-a.e.}$$

The assertion now follows from 5.4.6. □

Exercises

5.44 Prove 5.5.1.

5.45 Let $f,\, g \in BV(I)$ be real-valued. Prove that $f \vee g,\ f \wedge g \in BV(I)$. Show also that if $|f(x)| \ge c > 0$ for all x, then $1/f \in BV(I)$.

5.46 Show that if $E \subset \mathbb{R}$ and E^c are dense in $\mathbb{R}$, then $\mathbf{1}_E \notin BV(I)$ for all intervals I.

5.47 Let f be real-valued. Prove:

(a) If f' exists and is Riemann integrable on $[a, b]$, then $V_{[a,b]}(f) = \int_a^b |f'(x)|\, dx$.

(b) If f is right continuous at a and f' is locally Riemann integrable on $(a, b]$, then the equation in (a) holds, where the integral is improper. Thus $f \in BV[a, b]$ iff $\int_a^b |f'|$ converges.

5.48 Let $F \in BV(\mathbb{R})$ be right continuous and suppose that $F(-\infty)$ exists and is finite. Show that $F' \in L^1(\lambda)$. ⟦May assume F is a distribution function. Define a finite measure μ on $\mathcal{B}(\mathbb{R})$ so that $\mu(-\infty, x] = F(x) - F(-\infty)$. ⟧

5.6 Absolutely Continuous Functions

Definition and Basic Properties

A complex-valued function f on an interval I is said to be **absolutely continuous** if for each $\varepsilon > 0$ there exists a $\delta > 0$ such that

$$\sum_{j=1}^{n} |f(b_j) - f(a_j)| < \varepsilon$$

for any collection of disjoint subintervals $(a_1, b_1), \ldots, (a_n, b_n)$ of I with $\sum_{j=1}^{n}(b_j - a_j) < \delta$. We denote the collection of all absolutely continuous functions on I by $AC(I)$.

By taking a single open interval in the definition, we see that an absolutely continuous function is necessarily uniformly continuous. The converse is false: The function f of Example 5.6.8 below is uniformly continuous on $[0,1]$ but is not absolutely continuous if $\alpha < 1$.

The straightforward proof of the following proposition is left to the reader.

5.6.1 Proposition. *Let I be any interval.*

(a) $f, g \in AC(I)$ *and* $c \in \mathbb{C} \Rightarrow f + g, cf, \overline{f}, |f| \in AC(I)$.

(b) $f \in AC(I)$ *iff* $\operatorname{Re}(f), \operatorname{Im}(f) \in AC(I)$.

(c) *If f is real-valued, then $f \in AC(I)$ iff $f^{\pm} \in AC(I)$.*

5.6.2 Proposition. *If I is a bounded interval, then $AC(I) \subseteq BV(I)$.*

Proof. Let $f \in AC(I)$, $[a,b] \subseteq I$, and let $\delta > 0$ correspond to $\varepsilon = 1$ in the definition of absolute continuity. Give a partition $\mathcal{P}$ of $[a,b]$, let $\mathcal{Q} = \{a = x_0 < x_1 < \cdots < x_n = b\}$ be a refinement such that $x_k - x_{k-1} < \delta/2$ for all k. Set $k_0 = 0$ and let k_1 be the largest index $> k_0$ such that $x_{k_1} - x_{k_0} < \delta$. In general, let k_i be the largest index $> k_{i-1}$ such that $x_{k_i} - x_{k_{i-1}} < \delta$. The process will terminate with an index $k_m = n$. Figure 5.2 illustrates the construction. Note that for $i = 1\ldots, m-1$, we have $x_{k_i} - x_{k_{i-1}} \geq \delta/2$, otherwise we could

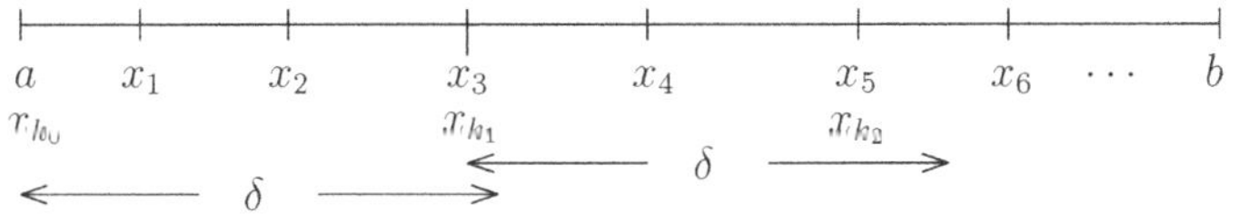

FIGURE 5.2: Construction of the sequence x_{k_i}.

choose an index $k > k_i$ for which $x_k - x_{k_{i-1}} < \delta$. Thus if ℓ denotes the length of I, then

$$\ell \geq b - a \geq x_{k_{m-1}} - a = \sum_{i=1}^{m-1}(x_{k_i} - x_{k_{i-1}}) \geq (m-1)\delta/2. \tag{†}$$

Since $\sum_{j=k_{i-1}}^{k_i - 1}(x_{j+1} - x_j) = x_{k_i} - x_{k_{i-1}} < \delta$, we have by absolute continuity

$$\sum_{j=k_{i-1}}^{k_i-1} |f(x_{j+1}) - f(x_j)| < 1.$$

Since $m-1$ such sums comprise $V_{\mathcal{Q}}(f)$, $V_{\mathcal{P}}(f) \leq V_{\mathcal{Q}}(f) \leq (m-1) \leq 2\ell/\delta$, the last inequality by (†). Since $\mathcal{P}$ and $[a,b]$ were arbitrary, $V_I(f) \leq 2\ell/\delta$. □

Note that the inclusion in the proposition is always strict (see 5.6.7) and is clearly false for unbounded intervals.

The next result complements 5.5.2.

5.6.3 Proposition. *Let I be an arbitrary interval and let $f \in AC(I) \cap BV(I)$ be real-valued. Then there exist monotone increasing functions g, $h \in AC(I) \cap BV(I)$ such that $f = g - h$.*

Proof. As in the proof of 5.5.2, for definiteness we take $I = \mathbb{R}$. It suffices to show that the function $g(x) := V_{(-\infty,x]}(f) = T_f(x)$ in that proof is absolutely continuous, since then $h = g - f$ will also be absolutely continuous. Let $\varepsilon > 0$ and let δ correspond to ε in the definition of absolute continuity of f. Let $(a_1, b_1), \ldots, (a_n, b_n)$ be disjoint with $\sum_{j=1}^n (b_j - a_j) < \delta$. For each j, let $\mathcal{P}_j$ be a partition of $[a_j, b_j]$. The open intervals formed by the totality of these partitions are disjoint and have total length $< \delta$, hence, by absolute continuity, $\sum_{j=1}^n V_{\mathcal{P}_j}(f) < \varepsilon$. Taking the supremum over $\mathcal{P}_1, \ldots, \mathcal{P}_n$ and using 5.5.4 yields

$$\sum_{j=1}^n [g(b_j) - g(a_j)] = \sum_{j=1}^n V_{(a_j,b_j]}(f) \le \varepsilon. \qquad \square$$

Here is the principle result regarding absolutely continuous functions.

5.6.4 Theorem. *Let μ be a finite measure on $\mathbb{R}$ with distribution function F. Then F is absolutely continuous iff $\mu \ll \lambda$. In this case, $F' = d\mu/d\lambda \in L^1(\lambda)$.*

Proof. Assume $\mu \ll \lambda$ and let $\varepsilon > 0$. By 5.3.2, there exists a $\delta > 0$ such that $\lambda(E) < \delta \Rightarrow \mu(E) < \varepsilon$. Thus if $(a_1, b_1), \ldots, (a_n, b_n)$ are disjoint with union E and $\sum_{j=1}^n (b_j - a_j) < \delta$, then

$$\sum_{j=1}^n \left[F(b_j) - F(a_j)\right] = \sum_{j=1}^n \mu(a_j, b_j] = \mu(E) < \varepsilon,$$

the last equality because $\mu\{b_j\} = 0$.

Conversely, assume that F is absolutely continuous and let $\lambda(E) = 0$. Given $\varepsilon > 0$, choose $\delta > 0$ as in the definition of absolutely continuity of F. By regularity, there exists an open set $U \supseteq E$ such that $\lambda(U) < \delta$. Now write U as a disjoint union of a sequence of open intervals (a_n, b_n). By absolute continuity of F, for all n

$$\sum_{j=1}^n \mu(a_j, b_j) = \sum_{j=1}^n [F(b_j) - F(a_j)] < \varepsilon.$$

Thus $\mu(E) \le \mu(U) \le \varepsilon$ and so $\mu(E) = 0$. The last assertion follows from 5.4.6. $\square$

Fundamental Theorems of Calculus

Theorem 5.6.4 implies a version of the fundamental theorem of calculus for Lebesgue integrable functions. For comparison, we first state the classical version:

5.6.5 Theorem. *Let $f : [a, b] \to \mathbb{C}$. The following are equivalent:*

(a) *f' exists and is continuous on $[a, b]$.*

(b) *There exists a continuous function $h : [a, b] \to \mathbb{C}$ such that*

$$f(x) = f(a) + \int_a^x h(t)\, dx \quad \textit{for all } x \in [a, b].$$

If these hold, then $f' = h$.

Note that condition (a) in the preceding theorem implies that f is absolutely continuous. Weakening this condition by requiring f to be *merely* absolutely continuous yields the following version of the fundamental theorem of calculus.

5.6.6 Theorem. *Let $f : [a, b] \to \mathbb{C}$. The following are equivalent:*

(a) *f is absolutely continuous.*

(b) *There exists a Lebesgue integrable $h : [a, b] \to \mathbb{C}$ such that*

$$f(x) = f(a) + \int_a^x h(t)\, dt \quad \text{for all } x \in [a, b].$$

If these hold, then f' exists and equals h λ-a.e.

Proof. Assume (a) holds. Using 5.6.3, we may suppose that f is a distribution function. Extend f to $\mathbb{R}$ by defining $f(x) = f(a)$ for $x < a$ and $f(x) = f(b)$ for $x > b$, obtaining an absolutely continuous distribution function on $\mathbb{R}$. Let μ be the corresponding finite measure. By 5.6.4, $\mu \ll \lambda$. Therefore, part (b) holds for the function $h = d\mu/\,d\lambda = f'$.

Now suppose that (b) holds. Then

$$\sum_{j=1}^{n} |f(b_j) - f(a_j)| \le \sum_{k=1}^{n} \int_{a_j}^{b_j} |h(t)|\, dt = \int_A |h(t)|\, dt, \quad A := \bigcup_j (a_j, b_j).$$

Since $|h|\lambda \ll \lambda$, f is absolutely continuous. □

5.6.7 Remark. The preceding theorem implies that if f is absolutely continuous and $f' = 0$ λ-a.e., then f = constant. On the other hand, the Cantor function $f \in BV[0, 1]$ satisfies $f' = 0$ a.e. but is nonconstant. Therefore, $AC[0, 1] \subsetneqq BV[0, 1]$. it follows that $AC(I) \subsetneqq BV(I)$ for all bounded (nondegenerate) intervals I. ◊

5.6.8 Example. For $\alpha > 0$, define a continuous function f on $[0, 1]$ by

$$f(x) := \begin{cases} x^\alpha \sin(1/x) & \text{if } 0 < x \le 1, \\ 0 & \text{if } x = 0. \end{cases}$$

We show that $f \in AC[0, 1]$ if $\alpha > 1$ and $f \notin BV[0, 1]$ if $\alpha \le 1$

Suppose $\alpha > 1$. Since

$$|f'(x)| = |\alpha x^{\alpha-1} \sin(1/x) - x^{\alpha-2} \cos(1/x)| \le \alpha x^{\alpha-1} + x^{\alpha-2}$$

and $\int_0^1 (x^{\alpha-1} + x^{\alpha-2})\, dx$ converges, f' is Lebesgue integrable. Moreover, for $a > 0$,

$$f(x) - f(a) = \int_a^x f'(t)\, dt.$$

Letting $a \to 0$ yields (b) of 5.6.6. Therefore, $f \in AC[0, 1]$.

Now let $\alpha \le 1$. Set

$$a_k := \frac{2}{(4k+1)\pi}, \quad b_k := \frac{1}{2k\pi}, \quad c := \frac{2^\alpha}{\pi^\alpha},$$

and note that

$$b_{k+1} < a_k < b_k, \quad f(b_k) = 0 \text{ and, } \quad f(a_k) = \frac{c}{(4k+1)^\alpha}.$$

For sufficiently small $\varepsilon > 0$, form the partition

$$\mathcal{P}_\varepsilon = \{\varepsilon < a_p < b_p < a_{p-1} < \cdots < a_k < b_k < \cdots < b_{q+1} < a_q < b_q < 1\}$$

of $[\varepsilon, 1]$, where p and q are, respectively, the largest and smallest integers satisfying the inequalities $\varepsilon < a_p < b_q < 1$, or equivalently

$$\frac{1}{2\pi} < q < p < \frac{2 - \pi\varepsilon}{4\pi\varepsilon}.$$

From $f(a_k) - f(b_k) = \dfrac{c}{(4k+1)^\alpha}$ we have

$$V_{[0,1]}(f) \geq V_{[\varepsilon,1]}(f) \geq V_{\mathcal{P}}(f) \geq c\sum_{k=q}^{p} \frac{1}{(4k+1)^\alpha}.$$

By choosing ε arbitrarily small, the upper limit p of the sum may be made arbitrarily large. Since $\sum_{k=1}^\infty (4k+1)^{-\alpha}$ diverges, $V_{[0,1]}(f) = \infty$. ◊

Exercises

5.49 Show that $f, g \in AC[a,b] \Rightarrow fg \in AC[a,b]$.

5.50 Let p, $q > 0$. Show that the function $f(x) := x^p \sin(x^{-q})$ $(x > 0)$, $f(0) = 0$, is absolutely continuous on $[0,1]$ iff $p > q$.

5.51 The Cantor function f is an example of a continuous nondecreasing function on $[0,1]$ with $f' = 0$ a.e. Extend the Cantor function to a nondecreasing function on $\mathbb{R}$ by defining $f(x) = 0$, $x \leq 0$, and $f(x) = 1$, $x \geq 1$. Define g on $[0,1]$ by $g(x) = \sum_{n=1}^\infty 2^{-n} f\big((x - a_n)/(b_n - a_n)\big)$, where the $[a_n, b_n]$ are the closed intervals in $[0,1]$ with rational endpoints. Show that g is continuous, *strictly* increasing, and $g' = 0$ a.e.

5.52 Let $f \in AC[c,d]$ and let $g : [a,b] \to [c,d]$ be strictly increasing with $g([a,b]) = [c,d]$. Show that $g \in AC[a,b] \Rightarrow f \circ g \in AC[a,b]$. Give an example of a strictly increasing function g for which $f \circ g \in BV[a,b] \setminus AC[a,b]$ for nontrivial $f \in AC[c,d]$.

5.53 Let $\varphi : [a,b] \to [c,d]$ be absolutely continuous and $\varphi' > 0$ a.e.

(a) Show that φ is 1-1 hence φ^{-1} exists and is continuous.

(b) Let $f : [c,d] \to \mathbb{R}$ be Lebesgue integrable. Prove the change of variables formula

$$\int_{[c,d]} f\, d\lambda = \int_{[a,b]} (f \circ \varphi)\varphi'\, d\lambda.$$

5.54 Let F be monotone nondecreasing on $[a,b]$. Show that

$$\int_{[a,x]} F'\, d\lambda \leq F(x+) - F(a+)$$

Chapter 6

Fourier Analysis on $\mathbb{R}^d$

The subject of this chapter plays an important role in many areas of science, technology, and mathematics, including quantum physics, image processing, probability theory, statistics, and differential equations. We begin with the notion of *convolution*, which is central to Fourier analysis.

6.1 Convolution of Functions

Definition and Basic Properties

Let $f,\ g \in L^1(\mathbb{R}^d, \lambda^d)$. The **convolution of** f **and** g is the function

$$(f * g)(x) = \int f(x-y)g(y)\,dy, \quad x \in \mathbb{R}^d.$$

The basic properties of convolution are summarized in the following proposition. Note that parts (a)–(e) of the proposition collectively assert that $L^1(\mathbb{R}^d)$ is a commutative Banach algebra under convolution.

6.1.1 Proposition. *Let $f,\ g,\ h \in L^1(\mathbb{R}^d)$, $c \in \mathbb{C}$, $\phi \in C_c^\infty(\mathbb{R}^d)$, and α a multi-index. Then convolution $f * g$ is well-defined, $f * g \in L^1(\mathbb{R}^d)$, and the following hold.*

- **(a)** $\|f * g\|_1 \le \|f\|_1 \|g\|_1$.
- **(b)** $f * g = g * f$.
- **(c)** $(f * g) * h = f * (g * h)$.
- **(d)** $f * (cg) = (cf) * g = c(f * g)$.
- **(e)** $f * (g + h) = f * g + f * h$.
- **(f)** $\partial^\alpha (f * \phi) = f * \partial^\alpha \phi$.

Proof. To see that convolution is well-defined, note first that since the function $(x, y) \to x - y$ is Borel measurable, the integrand is measurable in (x, y). Thus if $f, g \ge 0$, the integral exists for all x. The inequality $\|f * g\|_1 \le \|f\|_1 \|g\|_1$, proved next, shows that $f * g$ is finite a.e. and in L^1. Considering real and imaginary parts and then positive and negative parts, we see that $f * g \in L^1$ for every $f,\ g \in L^1$.

(a) By Fubini's theorem and translation invariance,

$$\int |(f * g)(x)|\,dx \le \iint |f(x-y)g(y)|\,dx\,dy = \iint |f(x)g(y)|\,dx\,dy = \|f\|_1 \|g\|_1.$$

(b) By translation and reflection invariance,

$$(f * g)(x) = \int f(x-y)g(y)\,dy = \int f(-y)g(x+y)\,dy = \int f(y)g(x-y)\,dy = (g * f)(x).$$

(c) By Fubini's theorem,

$$\begin{aligned}[f*(g*h)](x) &= \int f(x-y)(g*h)(y)\,dy = \iint f(x-y)g(y-z)h(z)\,dz\,dy \\ &= \iint f(x-y)g(y-z)h(z)\,dy\,dz = \iint f\big(x-z-y)\big)g(y)h(z)\,dy\,dz \\ &= \int f*g(x-z)h(z)\,dz = [(f*g)*h](x).\end{aligned}$$

Parts (d) and (e) are left as exercises.
(f) By 3.4.5 and the chain rule,

$$\partial^\alpha(f*\phi)(x) = \int \partial_x^\alpha\big[\phi(x-y)\big]f(y)\,dy = \int (\partial^\alpha\phi)(x-y)f(y)\,dy. \qquad \square$$

Approximate Identities

The Banach algebra $L^1(\mathbb{R}^d)$ does not possess an identity, that is, there is no function e such that $f*e = e*f$ for all $f \in L^1$ (Ex. 6.2). However, $L^1(\mathbb{R}^d)$ has an **approximate identity**, as described in the following.

6.1.2 Theorem. *Let $\phi \in L^1(\mathbb{R}^d)$ with $\int \phi\,d\lambda^d = 1$. For $n \in \mathbb{N}$ and $x \in \mathbb{R}^d$ define $\phi_n(x) := n^d\phi(nx)$. If $1 \le p < \infty$, then $f*\phi_n \overset{L^p}{\to} f$ for all $f \in L^p$. The same conclusion holds for $p = \infty$ if f is uniformly continuous and bounded.*

Proof. Let T_z denote translation by $-z$, that is, $T_z f(x) = f(x-z)$. By the dilation property of λ^d,

$$\begin{aligned}f*\phi_n(x) - f(x) &= n^d\int f(x-y)\phi(ny)\,dy - \int f(x)\phi(y)\,dy \\ &= \int \big[T_{y/n}f(x) - f(x)\big]\phi(y)\,dy. \qquad (\dagger)\end{aligned}$$

Therefore,

$$\begin{aligned}\|f*\phi_n - f\|_p &\le \left(\int\left[\int \left|T_{y/n}f(x) - f(x)\right|\left|\phi(y)\right|dy\right]^p dx\right)^{1/p} \\ &\le \int\left(\int \left|T_{y/n}f(x) - f(x)\right|^p dx\right)^{1/p}\left|\phi(y)\right|dy \\ &= \int \|T_{y/n}f - f\|_p|\phi(y)|\,dy,\end{aligned}$$

the second inequality by 4.1.5. Since $\|T_{y/n}f - f\|_p \to 0$ (4.2.3) and $\|T_{y/n}f - f\|_p|\phi(y)| \le 2\|f\|_p|\phi(y)|$, the dominated convergence theorem implies that $\|f*\phi_n - f\|_p \to 0$. This proves the first part of the theorem. The second part follows from (†), since by uniform continuity $\|T_{y/n}f - f\|_\infty \to 0$. $\square$

6.1.3 Remark. The function ϕ in the statement of the theorem may be taken to be C^∞ with support contained in a given compact interval. To see this, let $h : \mathbb{R}^d \to [0, +\infty)$ be a C^∞ function such that $h > 0$ on $(-\boldsymbol{a}, \boldsymbol{a})$, and $h = 0$ on $(-\boldsymbol{a}, \boldsymbol{a})^c$, where $\boldsymbol{a} = (1, \dots, 1)$ (0.14.4). Then $\phi := (\int h)^{-1}h$ is C^∞ with support contained in $[-\boldsymbol{a}, \boldsymbol{a}]$ and $\int \phi = 1$. As a consequence, given $\boldsymbol{\varepsilon} = (\varepsilon, \dots, \varepsilon)$, the support of ϕ_n is contained in $[-\boldsymbol{\varepsilon}, \boldsymbol{\varepsilon}]$ for all large n. For future reference we note that because the interval is symmetric, h may be taken to be even. (Take $a = -b$ in 0.14.2.) $\Diamond$

The following application of 6.1.2 asserts that L^p functions may be approximated by smooth functions with compact support.

6.1.4 Corollary. *Let $1 \le p < \infty$ and $f \in L^p(\mathbb{R}^d)$. Then for each $\varepsilon > 0$ there exists a function $\varphi \in C^\infty$ that vanishes outside a compact interval such that $\|f - \varphi\|_p < \varepsilon$. If f is continuous with compact support, then the same assertion holds for the case $p = \infty$.*

Proof. By 4.2.2 we may assume that f is continuous and vanishes outside an interval $[\boldsymbol{a}, \boldsymbol{b}]$. Let ϕ_n be as in 6.1.3 with support contained in $[-\varepsilon, \varepsilon]$. Then

$$\operatorname{supp}(f * \phi_n) \subseteq \operatorname{supp}(f) + \operatorname{supp}(\phi_n) \subseteq [\boldsymbol{a} - \varepsilon, \boldsymbol{b} + \varepsilon].$$

Moreover, by 6.1.1(f), $f * \phi_n \in C^\infty$. Since $\|f * \phi_n - f\|_p \to 0$, we need only take $\varphi = f * \phi_n$ for sufficiently large n to complete the argument. □

Exercises

6.1 Let $f, g \in L^1(\mathbb{R}^d)$. Show that $\int (f * g)(x)\,dx = \int f(x)\,dx \cdot \int g(x)\,dx$.

6.2 Show that there is no function $e \in L^1(\mathbb{R}^d)$ such that $f * e = f$ for all $f \in L^1$.

6.3 Let $a > 0$ and $f(x) = \mathbf{1}_{[-a,a]}$. Show that $f * f(x) = (2a - |x|)\mathbf{1}_{[-2a,2a]}$.

6.4 Let T_a denote translation by a. Show that $T_a(f * g) = (T_a f) * g = f * (T_a g)$.

6.5 Let $1 \le p < \infty$, q conjugate to p, $f \in L^p$ and $g \in L^q$. Prove:
(a) $\|f * g\|_\infty \le \|f\|_p \|g\|_q$. (b) $f * g$ is uniformly continuous. (c) $\lim_{|x|\to\infty} f * g(x) = 0$ $(p > 1)$.

6.6 Show that if f, $g \in L^1(\mathbb{R}^d)$, then $\operatorname{supp} f * g$ is contained in the closure K of $\operatorname{supp} f + \operatorname{supp} g$. In particular, the members of $C_c(\mathbb{R}^d) * C_c(\mathbb{R}^d)$ have compact support.

6.7 Let $f \in L^1(\mathbb{R}^d)$ and $g \in L^p(\mathbb{R}^d)$ $(1 \le p \le \infty)$. Prove that $f * g(x)$ exists for a.a x and that $\|f * g\|_p \le \|f\|_1 \|g\|_p$.

6.8 [↑6.7, 4.6] Let $p, q, r \in [1, \infty]$ such that $p^{-1} + q^{-1} = 1 + r^{-1}$, and let $f \in L^p(\mathbb{R}^d)$, $g \in L^q(\mathbb{R}^d)$. Prove that $\int |f(x-y)g(y)|\,dy < \infty$ for a.a. x and that $\|f * g\|_r \le \|f\|_p \|g\|_q$. ⟦Eliminate the special cases (1) $p = q/(q-1)$, $r = \infty$, (2) $q = 1$, $r = p$, and (3) $p = 1$, $r = q$. Then let p, q, r be finite and write $|f(x-y)g(y)| = |f(x-y)|^{1-p/r}|g(y)|^{1-q/r}|f(x-y)|^{p/r}|g(y)|^{q/r}$. ⟧

6.9 Let $T : \mathbb{R}^d \to \mathbb{R}^d$ be linear and nonsingular. Prove that $(f * g) \circ T = |\det T|(f \circ T) * (g \circ T)$. In particular, if T is orthogonal, then $(f * g) \circ T = (f \circ T) * (g \circ T)$.

6.2 The Fourier Transform

Definition and Basic Properties

The **Fourier transform** $\mathfrak{F}(f) = \widehat{f}$ of a function $f \in L^1(\mathbb{R}^d)$ is defined by

$$\mathfrak{F}(f)(\boldsymbol{\xi}) = \widehat{f}(\boldsymbol{\xi}) := \int e^{-2\pi i\,\boldsymbol{\xi}\cdot\boldsymbol{x}} f(\boldsymbol{x})\,d\boldsymbol{x}, \quad \boldsymbol{\xi} \in \mathbb{R}^d. \text{ [1]} \tag{6.1}$$

It is clear that $\mathfrak{F}$ is linear and

$$\|\mathfrak{F}(f)\|_\infty = \|\widehat{f}\|_\infty \le \|f\|_1. \tag{6.2}$$

[1]Some authors omit the factor 2π in the exponent of (6.1). Its presence, however, simplifies the inversion formula.

Additional properties of the transform are given in the next proposition. The following notation will be needed:

$$x^\alpha := x_1^{\alpha_1} \cdots x_d^{\alpha_d}, \text{ where } x = (x_1, \ldots, x_d) \text{ and } \alpha = (\alpha_1, \ldots, \alpha_d) \text{ is a multi-index.}$$

6.2.1 Proposition. *Let $f, g \in L^1(\mathbb{R}^d)$, $\phi \in C_c^\infty(\mathbb{R}^d)$, $T : \mathbb{R}^d \to \mathbb{R}^d$ linear and nonsingular with adjoint T^*, $T_a f(x) := f(a + x)$, and $D_r f(x) := f(rx)$. Then*

(a) $\widehat{f * g} = \widehat{f}\,\widehat{g}$.

(b) $\widehat{\partial^\alpha \phi}(\xi) = (2\pi i\, \xi)^\alpha \widehat{\phi}(\xi)$.

(c) $\widehat{f \circ T} = |\det T|^{-1} \widehat{f} \circ T^{*-1}$.

(d) $\widehat{T_a f}(\xi) = e^{2\pi i\, \xi \cdot a} \widehat{f}(\xi)$.

(e) $T_a \widehat{f} = \widehat{h}, \quad h(x) := e^{-2\pi i\, a \cdot x} f(x)$.

(f) $\widehat{D_r f} = r^{-d} D_{1/r} \widehat{f}$.

Proof. (a) By Fubini's theorem and translation invariance,

$$\widehat{f * g}(\xi) = \iint e^{-2\pi i\, \xi \cdot x} f(x - y) g(y)\, dy\, dx = \iint e^{-2\pi i\, \xi \cdot (x+y)} f(x) g(y)\, dx\, dy = \widehat{f}(\xi) \widehat{g}(\xi).$$

(b) An integration by parts yields

$$\widehat{\frac{\partial \phi}{\partial x_1}}(\xi) = \int \cdots \int e^{-2\pi\, i\, \xi \cdot x} \phi_{x_1}(x)\, dx_1 \cdots dx_d = 2\pi\, i\, \xi_1 \int e^{-2\pi\, i\, \xi \cdot x} \phi(x)\, dx = 2\pi\, i\, \xi_1 \widehat{\phi}(\xi).$$

(The constant term is absent because ϕ has compact support.) The analogous result holds for the remaining variables. The desired formula now follows by induction.

For (c), we apply the change of variable theorem:

$$\widehat{f \circ T}(\xi) = \int e^{-2\pi\, ix \cdot \xi} f(Tx)\, dx = \int e^{-2\pi\, i\, Tx \cdot T^{*-1} \xi} f(Tx)\, dx = |\det T|^{-1} \widehat{f}(T^{*-1} \xi).$$

Parts (d) – (f) are left as exercises (Ex. 6.11). □

6.2.2 Theorem (Riemann-Lebesgue Lemma). *The Fourier transform $\mathfrak{F}$ is a continuous linear mapping from the Banach space $L^1(\mathbb{R}^d)$ into the Banach space $C_0(\mathbb{R}^d)$.*

Proof. We have already noted that $\mathfrak{F}$ is linear. Let $f \in L^1$. Using the dominated convergence theorem one easily establishes that $\widehat{f}$ is continuous. Since $\|\widehat{f}\|_\infty \le \|f\|_1$, $\mathfrak{F}$ is a continuous linear mapping from $L^1(\mathbb{R}^d)$ into $C_b(\mathbb{R}^d)$. It remains to show that $\widehat{f} \in C_0(\mathbb{R}^d)$.

Suppose first that $f \in C_c^\infty(\mathbb{R}^d)$. Then $\widehat{\partial f / \partial x_j}$ is bounded, hence, by 6.2.1(b), there exists a constant C such that $|\xi_j|\, |\widehat{f}(\xi)| \le C$ for all j and ξ. This implies that $\widehat{f} \in C_0(\mathbb{R}^d)$.

In the general case, given $\varepsilon > 0$ choose $g \in C_c^\infty(\mathbb{R}^d)$ such that $\|f - g\|_1 < \varepsilon$ (6.1.4). By the preceding paragraph, $\widehat{g} \in C_0(\mathbb{R}^d)$. Since $\|\widehat{f} - \widehat{g}\|_\infty \le \|f - g\|_1$, we have $\|\widehat{f} - \widehat{g}\|_\infty < \varepsilon$. Therefore, $\widehat{f} \in C_0(\mathbb{R}^d)$. □

The Fourier Inversion Theorem

The **inverse Fourier transform** $\check{f}$ of a function $f \in L^1(\mathbb{R}^d)$ is defined by

$$\check{f}(\xi) := \int e^{2\pi i\, \xi \cdot x} f(x)\, dx, \quad \xi \in \mathbb{R}^d. \tag{6.3}$$

The next theorem describes one of the most important properties of the Fourier transform, one that is largely responsible for the utility of the transform. For the proof we need the following lemma.

6.2.3 Lemma. *Let $a > 0$ and $b \in \mathbb{R}$. Then*

$$\int \exp\left(ib\xi \cdot x - a|\xi|^2\right) d\xi = \left(\frac{\pi}{a}\right)^{d/2} \exp\left(-b^2|x|^2/4a\right).$$

Proof. Let $F(x)$ denote the left side of the equation. Consider first the case $d = 1$. Differentiating and then integrating by parts, we have

$$\begin{aligned} F'(x) &= ib \int \xi \exp\left(ib\xi x - a\xi^2\right) d\xi \\ &= -\frac{ib}{2a} \exp\left(ib\xi x - a\xi^2\right)\Big|_{-\infty}^{\infty} - \frac{b^2}{2a} x \int \exp\left(ib\xi x - a\xi^2\right) d\xi \\ &= \frac{-b^2}{2a} x F(x). \end{aligned}$$

It follows that the derivative of $F(x) \exp\left(b^2 x^2/4a\right)$ is zero and so

$$F(x) \exp\left(b^2 x^2/4a\right) = F(0) = \int_{-\infty}^{\infty} e^{-a\xi^2}\, d\xi = \sqrt{\frac{\pi}{a}},$$

the last equality from (3.16).

In the general case, apply Fubini's theorem and the case $d = 1$:

$$\begin{aligned} F(x) &= \int \prod_{k=1}^{d} \exp\left(ib\xi_k x_k - a x_k^2\right) d\xi = \prod_{k=1}^{d} \int \exp\left(ib\xi_k x_k - a x_k^2\right) d\xi_k \\ &= \prod_{k=1}^{d} \sqrt{\frac{\pi}{a}} \exp\left(-b^2 x_k^2/4a\right) = \left(\frac{\pi}{a}\right)^{d/2} \exp\left(-b^2|x|^2/4a\right). \end{aligned}$$ □

Here is the inversion theorem:

6.2.4 Theorem. *If f and $\widehat{f}$ are both in $L^1(\mathbb{R}^d)$, then $f = \check{\widehat{f}}$, that is,*

$$f(x) = \int e^{2\pi i \xi \cdot x} \widehat{f}(\xi)\, d\xi, \quad \textit{for a.a. } x \in \mathbb{R}^d.$$

Proof. We divide the proof into several steps.

(1) *Set $\phi(x) := \exp\left(-\pi|x|^2\right)$. Then $\int \phi = 1$.*

⟦For $d = 1$ this follows from (3.16). The general case follows from Fubini's theorem.⟧

(2) *For the function ϕ in (1) define $\phi_n(x) = n^d \phi(nx)$ as in 6.1.2. For $n \in \mathbb{N}$ and $x \in \mathbb{R}^d$, define $\psi_{n,x}(\xi) = \exp\left(2\pi i \xi \cdot x - \pi n^{-2}|\xi|^2\right)$. Then $\widehat{\psi}_{n,x}(y) = \phi_n(x - y)$.*

⟦Take $a = \pi/n^2$ and $b = 2\pi$ in 6.2.3 to obtain

$$\widehat{\psi}_{n,x}(y) = \int e^{2\pi i \xi \cdot (x-y) - \pi n^{-2}|\xi|^2}\, d\xi = n^d e^{-\pi n^2|x-y|^2} = \phi_n(x - y).⟧$$

(3) *For $g, h \in L^1(\lambda^d)$, $\int g\widehat{h} = \int h\widehat{g}$.*

⟦By Fubini's theorem, $\iint g(x)h(y)e^{-2\pi i x \cdot y}\, dx\, dy = \iint g(x)h(y)e^{-2\pi i x \cdot y}\, dy\, dx$.⟧

(4) $\displaystyle\int \psi_{n,x}(\xi)\widehat{f}(\xi)\,d\xi = f * \phi_n(x).$

⟦By (2) and (3), the left side of (4) is $\int \widehat{\psi}_{n,x}(y) f(y)\,dy = \int \phi_n(x-y) f(y)\,dy$, which is the right side.⟧

To complete the proof of the theorem, let $n \to \infty$ in (4). Since $\psi_{n,x}(\xi) \to \exp(2\pi i \xi \cdot x)$ and $\widehat{f} \in L^1$, the left side tends to $\check{\widehat{f}}(x)$ by the dominated convergence theorem. By 6.1.2 the right side tends in L^1 to f, hence a subsequence tends to $f(x)$ a.e. (4.3.3). Thus the two functions are equal a.e. □

Exercises

6.10 The indicator function $h = \mathbf{1}_{(0,\infty)}$ is called the *Heaviside function.* Let $f \in L^1(\mathbb{R})$ be differentiable with $f' \in L^1(\mathbb{R})$. Find $(f'h) * h$ and $h * h * \cdots * h$ (n factors).

6.11 Verify (d) – (f) of 6.2.1.

6.12 Let $a > 0$ and $f(x) = \mathbf{1}_{[-a,a]}$. Show that $\widehat{f}(\xi) = (\pi\xi)^{-1}\sin 2\pi a\xi$.

6.13 Let $f \in L^1(\lambda^d)$. Prove that f even (odd) $\Rightarrow \widehat{f}$ even (odd).

6.14 Let $f(x) = \pi e^{-2\pi b|x|}$, $b > 0$. Show that $\widehat{f}(\xi) = b/(b^2 + \xi^2)$.

6.15 Define $\widetilde{f}(x) = f(-x)$. Verify that $\widehat{\widetilde{f}} = \widetilde{\widehat{f}}$ and $\overline{\widehat{f}} = \widehat{\widetilde{\overline{f}}}(x)$.

6.16 Let $f \in L^1$. Show that for an orthogonal linear transformation T, $\widehat{f \circ T} = \widehat{f}$. Use this to show that if f is a radial function (§3.6), then so is $\widehat{f}$.

6.17 Let $f \in L^1(\mathbb{R}^d)$. Show that $\check{\overline{f}} = \overline{\widehat{f}}$.

6.18 [↑6.14] Let $g(x) = (1 + x^2)^{-1}$. Use the inversion formula to show that $g * g(x) = \pi/(4 + x^2)$.

6.3 Rapidly Decreasing Functions

These functions are particularly well behaved with respect to the Fourier inversion operation, a fact that will allow a relatively simple proof of the Plancherel theorem.

Definition and Basic Properties

For a function $\phi \in C^\infty(\mathbb{R}^d)$, define seminorms

$$p_{\alpha,\beta}(\phi) = \sup_{x\in\mathbb{R}^d} |x^\beta \partial^\alpha \phi(x)| \quad\text{and}\quad q_{\alpha,n}(\phi) = \sup_{x\in\mathbb{R}^d} \left(1 + |x|^2\right)^n |\partial^\alpha \phi(x)|, \tag{6.4}$$

where $n \in \mathbb{Z}^+$ and $\alpha = (\alpha_1, \ldots, \alpha_d)$ and $\beta = (\beta_1, \ldots, \beta_d)$ are multi-indices. The following proposition makes a connection between these seminorms.

6.3.1 Proposition. (a) *For each $n \in \mathbb{Z}^+$, there exists a finite set S of multi-indices and a constant $A > 0$, each depending only on n, such that*

$$q_{\alpha,n}(\phi) \le A \sum_{\beta\in S} p_{\alpha,\beta}(\phi) \quad \forall \text{ multi-index } \alpha \text{ and } \forall\ \phi \in C^\infty(\mathbb{R}^d).$$

(b) *For each multi-index β, there exists a constant $B > 0$ and $m \in \mathbb{N}$, each depending only on β, such that*

$$p_{\alpha,\beta}(\phi) \le B\, q_{\alpha,m}(\phi) \quad \forall \text{ multi-index } \alpha \text{ and } \forall\, \phi \in C^\infty(\mathbb{R}^d).$$

Proof. The function $y \to \sum_{k=1}^d |y_k|^{2n}$ has a positive minimum s on the unit sphere $|y| = 1$. Taking $y = x/|x|$ we see that $s^{-1}\sum_{k=1}^d |x_k|^{2n} \ge |x|^{2n}$ for all x. For a suitable constant A depending only on n we then have

$$\left(1+|x|^2\right)^n \le 2^n\left(1+|x|^{2n}\right) \le 2^n\left(1+s^{-1}\sum_{k=1}^d |x_k|^{2n}\right) \le A\left(1+\sum_{k=1}^d |x_k|^{2n}\right) = A\sum_{\beta\in S} |x^\beta|,$$

where S denotes the set of $d+1$ multi-indices $(0,0\ldots,0)$, $(0,\ldots,0,2n,0,\ldots,0)$. Multiplying by $|\partial^\alpha\phi(x)|$ and taking suprema yields the inequality in (a).

(b) For the multi-index β, set $m := \sum_{j=1}^d \beta_j$, $t_j := \beta_j/m$, and $t = (t_1,\ldots,t_d)$. By Ex. 4.28 and the Cauchy-Schwarz inequality,

$$\prod_{j=1}^d |x_j|^{\beta_j} = \left[\prod_{j=1}^d |x_j|^{t_j}\right]^m \le \left[\sum_{j=1}^d t_j|x_j|\right]^m \le \left[\sum_{j=1}^d t_j^2\right]^{m/2}\left[\sum_{j=1}^d |x_j|^2\right]^{m/2} \le d^{m/2}\left[\sum_{j=1}^d |x_j|^2\right]^{m/2}.$$

Therefore, $|x^\beta| \le d^{m/2}(1+|x|^2)^m$, proving (b). □

As a consequence we have

6.3.2 Corollary. *The following are equivalent for a function $\phi \in C^\infty(\mathbb{R}^d)$.*

(a) $p_{\alpha,\beta}(\phi) < \infty$ *for all multi-indices α and β.*

(b) $q_{\alpha,m}(\phi) < \infty$ *for all multi-indices α and all $m \in \mathbb{Z}^+$.*

A function ϕ that satisfies the equivalent conditions (a) and (b) of the corollary is called a **rapidly decreasing** or **Schwartz function**. For example, $x^\alpha \exp(-|x|^2)$ is rapidly decreasing. The collection of all rapidly decreasing functions is called the **Schwartz space** on $\mathbb{R}^d$ and is denoted by $\mathcal{S} = \mathcal{S}(\mathbb{R}^d)$. Clearly, the following inclusions hold:

$$C_c^\infty(\mathbb{R}^d) \subseteq \mathcal{S}(\mathbb{R}^d) \subseteq C_0(\mathbb{R}^d).$$

Moreover, from the sum and product rules for ∂^α it follows that $\mathcal{S}(\mathbb{R}^d)$ is an algebra and is closed under the operations ∂^α and multiplication by x^α.

6.3.3 Proposition. *Let $1 \le p < \infty$. Then in the L^p norm, $C_c^\infty(\mathbb{R}^d)$ is dense in $\mathcal{S}(\mathbb{R}^d)$ and $\mathcal{S}(\mathbb{R}^d)$ is dense in $L^p(\mathbb{R}^d)$.*

Proof. Let $\phi \in \mathcal{S}(\mathbb{R}^d)$. Choose $n > p/d$ and $C > 0$ so that $(1+|x|)^n|\phi(x)| \le C$ for all x. Then $|\phi(x)| \le C(1+|x|)^{-n}$, so $\phi \in L^p$ by 3.6.3. Thus $\mathcal{S}(\mathbb{R}^d) \subseteq L^p$. Since $C_c^\infty(\mathbb{R}^d) \subseteq \mathcal{S}(\mathbb{R}^d)$ and $C_c^\infty(\mathbb{R}^d)$ is dense in $L^p(\mathbb{R}^d)$ (6.1.4), the assertions follow. □

The following result will be needed in the proof of the Plancherel theorem below.

6.3.4 Theorem. $\widehat{\mathcal{S}} = \mathcal{S}$.

Proof. Let $\phi \in \mathcal{S}$. Let α and β be any multi-indices and set $c := -2\pi i$. Differentiating under the integral sign, we have

$$(c\xi)^\beta \partial^\alpha \widehat{\phi}(\xi) = \int (c\xi)^\beta \left(\partial_\xi^\alpha e^{c\,\xi\cdot x}\right) \phi(x)\, dx = \int (c\xi)^\beta e^{c\,\xi\cdot x} (cx)^\alpha \phi(x)\, dx$$
$$= \int \left(\partial_x^\beta e^{c\,\xi\cdot x}\right) (cx)^\alpha \phi(x)\, dx.$$

Performing a sequence of integration by parts on the last integral, we obtain

$$(c\xi)^\beta \partial^\alpha \widehat{\phi}(\xi) = \int e^{c\,\xi\cdot x} \psi(x)\, dx = \widehat{\psi}(\xi)$$

for some function ψ in $\mathcal{S}$. In particular, $\xi^\beta \partial^\alpha \widehat{\phi}(\xi)$ is bounded, hence $\widehat{\phi} \in \mathcal{S}$. Therefore, $\widehat{\mathcal{S}} \subseteq \mathcal{S}$.

For the reverse inclusion, set $\phi_1(x) := \widehat{\phi}(-x)$. By the first paragraph, $\phi_1 \in \mathcal{S}$, and by the inversion theorem

$$\widehat{\phi_1}(x) = \int e^{2\pi\, i\,\xi\cdot x} \widehat{\phi}(-\xi)\, d\xi = \int e^{-2\pi i\,\xi\cdot x} \widehat{\phi}(\xi)\, d\xi = \phi(x).$$

Therefore, $\phi \in \widehat{\mathcal{S}}$. □

The Plancherel Theorem

The following classical result is an easy consequence of 6.3.4.

6.3.5 Theorem (Plancherel). *There exists an linear mapping T from $L^2(\mathbb{R}^d)$ onto $L^2(\mathbb{R}^d)$ such that $Tf = \widehat{f}$ on $\mathcal{S}(\mathbb{R}^d)$ and $\|Tf\|_2 = \|f\|_2$ for all $f \in L^2(\mathbb{R}^d)$.*

Proof. Let $\phi \in \mathcal{S}$. By the inversion theorem and Fubini's theorem,

$$\|\phi\|_2^2 = \int \phi(x)\overline{\phi}(x)\, dx = \iint \exp\left(2\pi i x\cdot\xi\right) \widehat{\phi}(\xi)\overline{\phi}(x)\, d\xi\, dx$$
$$= \iint \overline{\exp\left(-2\pi i x\cdot\xi\right)\phi(x)}\, \widehat{\phi}(\xi)\, dx\, d\xi = \int \overline{\widehat{\phi}}(\xi)\widehat{\phi}(\xi)\, d\xi$$
$$= \|\widehat{\phi}\|_2^2.$$

Now define a linear mapping $T : \mathcal{S} \to \mathcal{S} = \widehat{\mathcal{S}}$ by $T\phi = \widehat{\phi}$. By what has just been proved, $\|T\phi\|_2 = \|\phi\|_2$. Given $f \in L^2$, choose a sequence (ϕ_n) in $\mathcal{S}$ such that $\|f - \phi_n\|_2 \to 0$. Then $\|T\phi_n - T\phi_m\|_2 = \|\phi_n - \phi_m\|_2 \to 0$ hence $(T\phi_n)$ is a Cauchy sequence in L^2 and therefore converges to a unique L^2 function Tf. It follows that T is linear and $\|Tf\|_2 = \|f\|_2$. □

Notation. For functions f in $L^2(\mathbb{R}^d)$ one frequently writes $\widehat{f}$ for Tf and $\check{f}$ for $T^{-1}f$, even though f may not lie in $L^1(\mathbb{R}^d)$. We then have $\check{\widehat{f}} = f = \widehat{\check{f}}$ in L^2.

6.3.6 Corollary. *For all $f, g \in L^2(\mathbb{R}^d)$, $\int fg = \int \widehat{f}\, \check{g}$ and $\int f\,\overline{g} = \int \widehat{f}\, \overline{\widehat{g}}$.*

Proof. By step (3) of proof the inversion theorem, $\int f\, g = \int f\, \widehat{\check{g}} = \int \widehat{f}\check{g}$ $(f, g \in \mathcal{S})$. Since $\mathcal{S}$ is dense in L^2, the first equality in the conclusion follows from the Plancherel theorem and the L^2 continuity of the map $(f, g) \to \int fg$. (Ex. 4.3). The second equality follows from the first by replacing g by $\overline{g}$ and using $\check{\overline{g}} = \overline{\widehat{g}}$ (Ex. 6.17). □

Exercises

6.19 Prove that a C_0-function f on $\mathbb{R}^d$ is uniformly continuous. Conclude that a Schwartz function is uniformly continuous.

6.20 [$\uparrow$6.12, 3.43] Use the Plancherel theorem to show that

$$\int_0^\infty \frac{\sin x}{x}\,dx = \int_0^\infty \left(\frac{\sin x}{x}\right)^2 dx = \frac{\pi}{2}.$$

6.21 Let $\phi \in \mathscr{S}$ and α a multi-index. Show that $\partial^\alpha \widehat{\phi} = \widehat{\psi}$, where $\psi(x) := (-2\pi i x)^\alpha \phi(x)$.

6.22 Let $f, g \in L^2(\lambda^d)$. Show that $f * g(x) = \int \widehat{f}(\xi)\widehat{g}(\xi)e^{2\pi i\,\xi\cdot x}\,d\xi$. [[Fix x and define $h(y) = \overline{g(x-y)}$. Then $(f*g)(x) = \int f(y)g(x-y)\,dx = \int \widehat{f}(\xi)\overline{\widehat{h}}(\xi)\,d\xi$.]]

6.23 Let $f, g \in \mathscr{S}$. Verify (a)–(e) below to conclude that $f * g \in \mathscr{S}$, that is, $q_{\alpha,n}(f*g) < \infty\ \forall\, n,\, \alpha$.

(a) $\partial^\alpha(f*g) = (\partial^\alpha f)*g = f*(\partial^\alpha g)$. This reduces the argument to the case $\alpha = 0$.

(b) For each n, there exists a constant C_n such that for all x,

$$|(f*g)(x)| \le C_n \int (1+|x-y|^2)^{-n}(1+|y|^2)^{-n}\,dy.$$

(c) Set $A := \{y : 2|y-x| \ge |x|\}$. Then

$$\int_A (1+|x-y|^2)^{-n}(1+|y|^2)^{-n}\,dy \le (1+\tfrac{1}{4}|x|^2)^{-n}\int_A (1+|y|^2)^{-n}\,dy.$$

(d) Set $B := \{y : 2|y-x| \le |x|\}$. Then

$$\int_B (1+|x-y|^2)^{-2n}(1+|y|^2)^{-2n}\,dy \le (1+\tfrac{1}{4}|x|^2)^{-n}\int_B (1+|y|^2)^{-n}\,dy,$$

(e) For sufficiently large n, there exists a constant D_n depending on d such that

$$|(f*g)(x)| \le D_n(1+|x|^2)^{-n} \quad \text{for all } x.$$

6.24 Let $\phi_n, \phi \in \mathscr{S}$ and $\psi_n := \phi_n - \phi$. Define convergence $\phi_n \to \phi$ in $\mathscr{S}$ to mean that $p_{\alpha,\beta}(\psi_n) \to 0$ for all multi-indices α and β, or equivalently, $q_{\alpha,m}(\psi_n) \to 0$ for all multi-indices α and all $m \in \mathbb{Z}^+$. Let $\phi_n \to \phi$ in $\mathscr{S}$. Show that $\widehat{\phi_n} \to \widehat{\phi}$ in $\mathscr{S}$ by verifying the following:

(a) $\partial^\alpha \phi_n \to \partial^\alpha \phi$ in L^p for all $1 \le p \le \infty$ and for all multi-indices α.

(b) For any $\psi \in \mathscr{S}$ and multi-indices α, β,

$$\begin{aligned}(-2\pi i\,\xi)^\beta \partial^\alpha \widehat{\psi}(\xi) &= \int (-2\pi i\,x)^\alpha \psi(x)\partial_x^\beta e^{-2\pi i\,x\cdot\xi}\,dx \\ &= (-1)^\beta \int e^{-2\pi i\,x\cdot\xi}\,\partial_x^\beta\left[(-2\pi i\,x)^\alpha\psi(x)\right]dx\end{aligned}$$

(c) For any $\psi \in \mathscr{S}$ and multi-indices α, β, there exists a constant C and a finite set F of multi-indices such that

$$\left\|\xi^\beta \partial^\alpha \widehat{\psi}(\xi)\right\|_\infty \le C \sum_{\alpha',\beta' \in F} \left\|x^{\beta'}\partial^{\alpha'}\psi(x)\right\|_1.$$

6.25 (*Heisenberg uncertainty principle*). The principle states that a nonzero function and its Fourier transform cannot both be *sharply localized.* The precise analytical statement takes the form

$$\int |x|^2|\phi(x)|^2\,dx \cdot \int |\xi|^2|\widehat{\phi}(\xi)|^2\,d\xi \ge \frac{\|\phi\|_2^4}{16\pi^2}$$

Establish this for $\phi \in \mathscr{S}(\mathbb{R})$ by verifying (a) and (b) and then using $\widehat{\phi'}(\xi) = (2\pi i\,\xi)\widehat{\phi}(\xi)$.

(a) $\int |\phi(x)|^2\,dx = -2\mathrm{Re}\int x\phi(x)\overline{\phi'(x)}\,dx$.

(b) $\|\phi\|_2^4 \le 4\int x^2|\phi(x)|^2\,dx \int |\phi'(x)|^2\,dx = 4\int x^2|\phi(x)|^2\,dx \int |\widehat{\phi'}(\xi)|^2\,d\xi$.

6.4 Fourier Analysis of Measures on $\mathbb{R}^d$

The notions of convolution and Fourier transform of functions have natural extensions to measures. These have important applications in probability theory, for example, in the proof of the central limit theorem (18.4.19).

Convolution of Measures

The **convolution of complex measures** μ **and** ν on $\mathfrak{B}(\mathbb{R}^d)$ is the complex measure $\mu * \nu$ defined by

$$(\mu*\nu)(E) = \int \mathbf{1}_E(x+y)\, d(\mu\otimes\nu)(x,y) = \int\int \mathbf{1}_E(x+y)\, d\mu(x)\, d\nu(y), \quad E \in \mathfrak{B}(\mathbb{R}^d). \quad (6.5)$$

Note that if $A : \mathbb{R}^d \times \mathbb{R}^d \to \mathbb{R}^d$ is the addition operator $A(x,y) := x+y$, then $\mu*\nu = A(\mu\otimes\nu)$, the image measure of $\mu\otimes\nu$ under A. Thus, by 3.2.15 and Fubini's theorem, for all suitable h

$$\int h(z)\, d(\mu*\nu)(z) = \int h(x+y)\, d(\mu\otimes\nu)(x,y) = \iint h(x+y)\, d\mu(x)\, d\nu(y). \quad (6.6)$$

A related notion is the **convolution** $f * \nu$ of $f \in L^1(\nu)$ and ν defined by

$$(f*\nu)(x) = \int f(x-y)\, d\nu(y), \quad x \in \mathbb{R}^d.$$

The following proposition gives the basic properties of measure convolution.

6.4.1 Proposition. *Let μ, ν, and η be complex measures on $\mathfrak{B}(\mathbb{R}^d)$, f, $g \in L^1(\nu)$, and $c \in \mathbb{C}$. Then*

- **(a)** $\mu * \nu = \nu * \mu$.
- **(b)** $(\mu * \nu) * \eta = \mu * (\nu * \eta)$.
- **(c)** $c(\mu * \nu) = (c\mu) * \nu) = \mu * (c\nu)$.
- **(d)** $\mu * (\nu + \eta) = \mu * \nu + \mu * \eta$.
- **(e)** $|\mu * \nu| \le |\mu| * |\nu|$.
- **(f)** $\|\mu * \nu\| \le \|\mu\|\,\|\nu\|$.
- **(g)** $c(f * \nu) = (cf) * \nu = f * (c\nu)$.
- **(h)** $(f+g) * \nu = f * \nu + g * \nu$.
- **(i)** $f * (\mu + \nu) = f * \mu + f * \nu$.
- **(j)** $\|f * \nu\|_1 \le \|f\|_1\,\|\nu\|$.

Proof. Parts (a)–(d) and (g)–(i) are exercises (6.26). For (e) we use (6.6) and 5.18:

$$\begin{aligned}|\mu*\nu|(E) &= \sup_{|f|\le 1}\left|\int_E f\, d(\mu*\nu)\right| \le \sup_{|f|\le 1}\int \mathbf{1}_E(x+y)|f(x+y)|\, d|\mu|(x)\, d|\nu|(y)\\ &\le \int \mathbf{1}_E(x+y)\, d|\mu|(x)\, d|\nu|(y) = (|\mu| * |\nu|)(E).\end{aligned}$$

For (f) we have $|\mu*\nu|(\mathbb{R}^d) \le |\mu| * |\nu|(\mathbb{R}^d) = \iint \mathbf{1}_{\mathbb{R}^d}(x+y)\, d|\mu|\, d|\nu| = |\mu|(\mathbb{R}^d)|\nu|(\mathbb{R}^d)$.

For (j) note first that $\iint |f(x-y)|\, d|\nu|(y)\, dx \le \iint |f(x)|\, dx\, d|\nu|(y) = \|f\|_1\, |\nu|(\mathbb{R}^d) < \infty$, hence the function $\int f(x-y)\, d\nu(y)$ is defined for a.a. x and is integrable. Moreover, this calculation together with (5.4) shows that $\|f * \nu\|_1 \le \|f\|_1\,\|\nu\|$. □

From 6.4.1 and 5.2.5 we have

6.4.2 Corollary. *The space $M(\mathbb{R}^d)$ of complex measures on $\mathfrak{B}(\mathbb{R}^d)$ is a commutative Banach algebra under convolution and the total variation norm.*

The Fourier-Stieltjes Transform

The **Fourier-Stieltjes transform** $\widehat{\mu}$ of a complex measure μ is defined by

$$\widehat{\mu}(\boldsymbol{\xi}) = \int e^{-2\pi i\, \boldsymbol{\xi}\cdot \boldsymbol{x}}\, d\mu(\boldsymbol{x}), \quad \boldsymbol{\xi} \in \mathbb{R}^d.$$

Note that for the measure $d\mu = f\, d\lambda \;\; (f \in L^1(\lambda^d))$ we have $\widehat{\mu} = \widehat{f}$, hence the Fourier-Stieltjes transform may be seen as a generalization of the Fourier transform.

The Fourier-Stieltjes transform enjoys many of the properties of the Fourier transform, a notable exception being the Riemann-Lebesgue lemma (consider δ_0). The following proposition summarizes the properties of the transform.

6.4.3 Proposition. *Let μ and ν be complex Borel measures on $\mathbb{R}^d$, $T : \mathbb{R}^d \to \mathbb{R}^d$ linear, $a, b \in \mathbb{C}$, and $\boldsymbol{\alpha} \in \mathbb{R}^d$. Then*

(a) *$\widehat{\mu}$ is continuous.* **(b)** $\|\widehat{\mu}\|_\infty \le \|\mu\|$.

(c) $\widehat{a\mu + b\nu} = a\,\widehat{\mu} + b\,\widehat{\nu}$. **(d)** $\widehat{\mu * \nu} = \widehat{\mu}\,\widehat{\nu}$.

(e) $\widehat{T(\mu)} = \widehat{\mu} \circ T^*$. **(f)** $\widehat{\mu}(\boldsymbol{\alpha} + \boldsymbol{\xi}) = \widehat{\nu}(\boldsymbol{\xi})$, *where* $d\nu(\boldsymbol{x}) := e^{-2\pi i(\boldsymbol{\alpha}\cdot\boldsymbol{x})}\, d\mu(\boldsymbol{x})$.

Proof. Part (a) follows from the dominated convergence theorem. Parts (b) and (c) are clear. For (d) we have

$$\widehat{\mu * \nu}(\boldsymbol{\xi}) = \int e^{-2\pi i\, \boldsymbol{x}\cdot\boldsymbol{\xi}}\, d(\mu * \nu)(\boldsymbol{x}) = \iint e^{-2\pi i\, (\boldsymbol{x}+\boldsymbol{y})\cdot\boldsymbol{\xi}}\, d\mu(\boldsymbol{x})d\nu(\boldsymbol{y}) = \widehat{\mu}(\boldsymbol{\xi})\widehat{\nu}(\boldsymbol{\xi}).$$

For (e),

$$\widehat{T\mu}(\boldsymbol{\xi}) = \int e^{-2\pi i\, \boldsymbol{x}\cdot\boldsymbol{\xi}}\, dT(\mu)(\boldsymbol{x}) = \int e^{-2\pi i\, T(\boldsymbol{x})\cdot\boldsymbol{\xi}}\, d\mu(\boldsymbol{x}) = \int e^{-2\pi i\, \boldsymbol{x}\cdot T^*\boldsymbol{\xi}}\, d\mu(\boldsymbol{x}) = \widehat{\mu}(T^*\boldsymbol{\xi}).$$

Part (f) is left as an exercise. □

The next result has important applications to probability distributions (§18.1).

6.4.4 Uniqueness Theorem for Fourier-Stieltjes Transforms. *Let μ and ν be finite measures on $\mathcal{B}(\mathbb{R}^d)$ such that $\widehat{\mu} = \widehat{\nu}$. Then $\mu = \nu$.*

Proof. Let $\phi \in \mathscr{S}$. By Fubini's theorem

$$\begin{aligned}\int \widehat{\phi}(\boldsymbol{\xi})\, d\mu(\boldsymbol{\xi}) &= \iint \exp(-2\pi i\, \boldsymbol{\xi}\cdot\boldsymbol{x})\phi(\boldsymbol{x})\, d\boldsymbol{x}\, d\mu(\boldsymbol{\xi}) = \int \widehat{\mu}(\boldsymbol{x})\phi(\boldsymbol{x})\, d\boldsymbol{x} = \int \widehat{\nu}(\boldsymbol{x})\phi(\boldsymbol{x})\, d\boldsymbol{x} \\ &= \int \widehat{\phi}(\boldsymbol{\xi})\, d\nu(\boldsymbol{\xi}).\end{aligned}$$

Since the Fourier transform $\mathscr{S} \to \mathscr{S}$ is surjective, $\int \phi\, d\mu = \int \phi\, d\nu$ for all $\phi \in \mathscr{S}$. Let $\boldsymbol{\varepsilon} = (\varepsilon, \ldots, \varepsilon)$ and $\boldsymbol{a}, \boldsymbol{b} \in \mathbb{R}^d$ with $a_j < b_j$ for all j. Choose a C^∞ function ϕ_ε so that $\mathbf{1}_{[\boldsymbol{a},\boldsymbol{b}]} \le \phi_\varepsilon \le \mathbf{1}_{(\boldsymbol{a}-\boldsymbol{\varepsilon},\boldsymbol{b}+\boldsymbol{\varepsilon})}$ (0.14.5). By dominated convergence, $\lim_{\varepsilon\to 0} \int \phi_\varepsilon\, d\mu = \mu[\boldsymbol{a}, \boldsymbol{b}]$ and similarly for ν. Therefore, $\mu[\boldsymbol{a}, \boldsymbol{b}] = \nu[\boldsymbol{a}, \boldsymbol{b}]$ for all $[\boldsymbol{a}, \boldsymbol{b}]$. By the uniqueness theorem for measures (1.6.8), $\mu = \nu$. □

Exercises

6.26 Verify parts (a)–(d) and (g)–(i) of 6.1.1.

6.27 Let ν be a complex measure on $\mathbb{R}^d$, $1 \le p \le \infty$, and $f \in L^p(\mathbb{R}^d)$. Show that $(f * \nu)(x)$ exists λ a.e., $f * \nu \in L^p(\mathbb{R}^d)$ and $\|f * \nu\|_p \le \|f\|_p \|\nu\|$. ⟦Let $1 \le p < \infty$. Consider first the case $f \ge 0$ and $\nu \ge 0$ and use Minkowski's inequality for integrals to show that $\|f * \nu\|_p \le \|f\|_p \|\nu\|$. Apply this to $|f|$ and $|\nu|$ in the general case.⟧

6.28 Let $f, g \in L^1$ and let ν a finite measure. Prove:

(a) $\delta_a * \nu = T_a(\nu)$ and $\delta_0 * \nu = \nu$, where $a \in \mathbb{R}^d$ and T_a is translation by a.

(b) $\delta_a * \delta_b = \delta_{a+b}$.

(c) $(f\lambda^d) * \nu = (f * \nu)\lambda^d$

(d) $(f\lambda^d) * (g\lambda^d) = (f * g)\lambda^d$.

6.29 Let $T : \mathbb{R}^d \to \mathbb{R}^d$ be linear, and μ, ν complex measures on $\mathcal{B}(\mathbb{R}^d)$. Show that $(T\mu) * (T\nu) = T(\mu * \nu)$.

6.30 Let μ_j, ν_j be finite measures on $\mathcal{B}(\mathbb{R}^d)$ with $\mu_j \ll \nu_j$ $(j = 1, 2)$. Show that $\mu_1 * \mu_2 \ll \nu_1 * \nu_2$.

6.31 Let μ and ν be finite measures on $\mathcal{B}(\mathbb{R}^p)$ and $\mathcal{B}(\mathbb{R}^q)$. Express $\widehat{\mu \otimes \nu}$ in terms of $\widehat{\mu}$ and $\widehat{\nu}$.

6.32 Verify 6.4.3(f).

6.33 Let μ be a complex measure on $\mathcal{B}(\mathbb{R}^d)$ and $f \in L^1(\mathbb{R}^d)$. Show that $\widehat{f * \mu} = \widehat{f} \cdot \widehat{\mu}$.

6.34 Let μ, $\nu \in M(\mathbb{R}^d)$ and $\mu \ll \lambda$. Show that $\mu * \nu \ll \lambda$ and find $d(\mu * \nu)/d\lambda$ in terms of $d\mu/d\lambda$.

Chapter 7

Measures on Locally Compact Spaces

In this chapter we describe a fundamental connection between topology and measure in the setting of locally compact Hausdorff spaces. Many of the results will be seen as generalizations of already established links between Borel measures on $\mathbb{R}^d$ and the Euclidean topology.

7.1 Radon Measures

Definition and Basic Properties

A **Radon measure** on a locally compact Hausdorff space X is a measure μ on $\mathcal{B}(X)$ with the following properties:

$$\begin{aligned}&\text{(a)}\ \ \mu(K) < \infty \quad \text{for all compact } K \subseteq X.\\ &\text{(b)}\ \ \mu(U) = \sup\{\mu(K) : K \subseteq U,\ K \text{ compact}\} \quad \text{for all open } U \subseteq X.\\ &\text{(c)}\ \ \mu(E) = \inf\{\mu(U) : U \supseteq E,\ U \text{ open}\} \quad \text{for all } E \in \mathcal{B}(X).\end{aligned} \tag{7.1}$$

Properties (b) and (c) assert, respectively, that μ is **inner regular on open sets** and **outer regular on Borel sets**. If μ is a Radon measure on X, we shall call the pair (X, μ) a **Radon measure space**.

If μ is a finite measure, then conditions (b) and (c) are equivalent to the assertion that for each Borel set E and each $\varepsilon > 0$ there exist a compact set K and an open set U such that $K \subseteq E \subseteq U$ and $\mu(U \setminus K) < \varepsilon$. It follows that if η is a measure with $\eta \leq \mu$, then η is a Radon measure.

A Radon measure that satisfies (b) for *every* Borel set U is said to be **regular**. For example, Lebesgue-Stieltjes measures on $\mathbb{R}^d$ are regular Radon measures (1.8.1). The following proposition shows that if μ is σ-finite, then a Radon measure is regular.

7.1.1 Proposition. *A Radon measure μ is inner regular on σ-finite sets E, that is,*

$$\mu(E) = \sup\{\mu(K) : K \subseteq E,\ K \textit{ compact}\}. \tag{7.2}$$

In particular, a finite Radon measure is regular.

Proof. Let s denote the supremum. Clearly, $s \leq \mu(E)$. For the reverse inequality, assume first that $\mu(E) < \infty$. Given $\varepsilon > 0$, choose an open set $U \supseteq E$ such that $\mu(U) < \mu(E) + \varepsilon$, and a compact set $K \subseteq U$ such that $\mu(K) > \mu(U) - \varepsilon$. Since $\mu(U \setminus E) = \mu(U) - \mu(E) < \varepsilon$, we may choose an open set $V \supseteq U \setminus E$ with $\mu(V) < \varepsilon$. Then $K \setminus V$ is compact, contained in U, and $s \geq \mu(K \setminus V) = \mu(K) - \mu(K \cap V) > \mu(U) - \mu(V) - \varepsilon \geq \mu(E) - 2\varepsilon$. Since ε was arbitrary, $s \geq \mu(E)$. Therefore, (7.2) holds for sets E of finite measure.

Now assume that $\mu(E) = \infty$ and E is σ-finite. Let $E_n \uparrow E$ with $\mu(E_n) < \infty$ for every n. Given $k \in \mathbb{N}$, choose n so that $\mu(E_n) > k$ and by the first paragraph choose a compact set $K \subseteq E_n$ such that $s \geq \mu(K) > k$. Since k is arbitrary, $s = \infty$. $\square$

Consequences of Regularity

The proof of following result is the same as that for special case $X = \mathbb{R}^d$ (4.2.2), since the proof of the latter uses only the properties (7.1) of λ^d.

7.1.2 Theorem. *Let (X, μ) be a Radon measure space and $f \in L^p(\mu)$ $(1 \le p < \infty)$. Then for each $\varepsilon > 0$ there exists $g \in C_c(X)$ such that $\|f - g\|_p < \varepsilon$.*

The following is an important application of the preceding theorem. The proof brings together several familiar results on convergence of sequences of functions as well as Tietze's extension theorem.

7.1.3 Lusin's Theorem. *Let (X, μ) be a Radon measure space and $f : X \to \mathbb{C}$ Borel measurable such that $\mu\{f \neq 0\} < \infty$. Then for each $\varepsilon > 0$ there exists $g \in C_c(X)$ such that $g = f$ except on a set of measure $< \varepsilon$. Moreover, if f is bounded, then g may be chosen so that $\|g\|_\infty \le \|f\|_\infty$.*

Proof. Set $E := \{f \neq 0\}$. Suppose first that f is bounded. Then $f \in L^1(\mu)$, hence by 7.1.2 there exists a sequence of continuous functions f_n with compact support such that $f_n \xrightarrow{L^1} f$. By 4.3.3, there exists a subsequence (f_{n_k}) that converges to f a.e. By Egoroff's theorem (2.4.5), there exists set $A \subseteq E$ with $\mu(E \setminus A) < \varepsilon/3$ such that $f_{n_k} \to f$ uniformly on A. In particular, f is continuous on A. By regularity, we may choose a compact set and an open set U such that $K \subseteq A \subseteq E \subseteq U$, $\mu(A \setminus K) < \varepsilon/3$, and $\mu(U \setminus E) < \varepsilon/3$, hence $\mu(U \setminus K) < \varepsilon$. By 0.12.8, there exists a continuous function F on X with compact support contained in U such that $F = f$ on K. Now define a continuous function $\phi : \mathbb{C} \to \mathbb{C}$ by

$$\phi(z) = \begin{cases} z & \text{if } |z| \le \|f\|_\infty, \\ \|f\|_\infty \operatorname{sgn} z & \text{if } |z| > \|f\|_\infty \end{cases}$$

and set $g := \phi \circ F$. Then $g = f$ on K, $g = f = 0$ on $U^c \subseteq E^c$, and $\{g \neq f\} \subseteq U \setminus K$. The function g satisfies the requirements.

In the unbounded case, set $E_n = \{0 < |f| < n\}$. Then $E_n \uparrow E$, hence we may choose n so that $\mu(E \setminus E_n) < \varepsilon/2$. Since $f_n := f\mathbf{1}_{E_n}$ is bounded, by the first paragraph there exists a continuous function g_n with compact support such that $\mu\{g_n \neq f_n\} < \varepsilon/2$. Since $\{f_n \neq f\}$ is contained in $E \setminus E_n$,

$$\{g_n \neq f\} \subseteq \{g_n \neq f_n\} \cup \big(\{g_n = f_n\} \cap \{f_n \neq f\}\big) \subseteq \{g_n \neq f_n\} \cup \big(E \setminus E_n\big),$$

hence $\mu\{g_n \neq f\} < \varepsilon$. □

The Space of Complex Radon Measures

A **signed Radon measure** on a locally compact Hausdorff space X is a signed Borel measure μ whose positive and negative variations $\mu^\pm$ are Radon measures. A **complex Radon measure** on X is a complex Borel measure μ whose real and imaginary parts $\mu_{r,i}$ are signed Radon measures. The collection of all complex Radon measures on X will be denoted by $M_{ra}(X)$. We show in this subsection that $M_{ra}(X)$ is a Banach space under the total variation norm. For this we prepare the following lemmas.

7.1.4 Lemma. *$M_{ra}(X)$ is a linear space over $\mathbb{C}$.*

Proof. Let $\mu_j \in M_{ra}(X)$, $c_j \in \mathbb{C}$, $j = 1, 2$. The proof that $c_1\mu_1 + c_2\mu_2 \in M_{ra}(X)$ is divided into the following steps:

(1) *If $\mu_j \geq 0$ and $c_j \geq 0$, then $c_1\mu_1 + c_2\mu_2 \in M_{ra}(X)$.*

⟦Let $E \in \mathcal{B}(X)$, $\varepsilon > 0$, K_j compact, and U_j open such that $K_j \subseteq E \subseteq U_j$ and $\mu_j(U_j \setminus K_j) < \varepsilon(c_1 + c_2)^{-1}$. Set $K = K_1 \cup K_2$ and $U = U_1 \cap U_2$. Then $K \subseteq E \subseteq U$ and $(c_1\mu_1 + c_2\mu_2)(U \setminus K) < \varepsilon$.⟧

(2) *If $\mu_j \geq 0$, then $\mu_1 - \mu_2 \in M_{ra}(X)$.*

⟦Set $\eta := \mu_1 - \mu_2$. By Ex. 5.7, $\mu_1 \geq \eta^+$ and $\mu_2 \geq \eta^-$. It follows that the measures $\eta^\pm$ are regular and so, by definition, η is regular.⟧

(3) *If μ_j are signed measures and $c_j \in \mathbb{R}$, then $c_1\mu_1 + c_2\mu_2 \in M_{ra}(X)$.*

⟦Write $\mu_j = \mu_j^+ - \mu_j^-$ and $c_j = c_j^+ - c_j^-$ and use (1) and (2).⟧

(4) *If μ_j are complex measures and $c_j \in \mathbb{C}$, then $c_1\mu_1 + c_2\mu_2 \in M_{ra}(X)$.*

⟦Write $\mu_j = \mu_{jr} + i\,\mu_{ji}$ and $c_j = \operatorname{Re} c_j + i \operatorname{Im} c_j$. Then $c_1\mu_1 + c_2\mu_2$ is of the form $a\nu + ib\eta$, where $a, b \in \mathbb{R}$ and ν, η are signed measures. By (3), $a\nu, b\eta \in M_{ra}(X)$.⟧ □

7.1.5 Lemma. $\mu \in M_{ra}(X)$ *iff* $|\mu| \in M_{ra}(X)$.

Proof. We may write $\mu = \mu_r^+ - \mu_r^- + i(\mu_i^+ - \mu_i^-)$, which, for notational convenience, we express as $\mu = \mu_1 - \mu_2 + i\,(\mu_3 - \mu_4)$. If $\mu \in M_{ra}(X)$, then by definition $\mu_j \in M_{ra}(X)$, hence given $E \in \mathcal{B}(X)$ and $\varepsilon > 0$ there exist compact sets K_j and open sets U_j such that $K_j \subseteq E \subseteq U_j$ and $\mu_j(U_j \setminus K_j) < \varepsilon$. Set $U = \bigcap_{j=1}^4 U_j$ and $K = \bigcup_{j=1}^4 K_j$ and let $E_1, \ldots, E_n$ be a measurable partition of $U \setminus K$. Then

$$\sum_{j=1}^n |\mu(E_j)| \leq \sum_{j=1}^n \sum_{i=1}^4 \mu_i(E_j) = \sum_{i=1}^4 \sum_{j=1}^n \mu_i(E_j) = \sum_{i=1}^4 \mu_i(U \setminus K) < 4\varepsilon.$$

Taking the supremum over all such partitions yields $|\mu|(U \setminus K) \leq 4\varepsilon$. Therefore, $|\mu| \in M_{ra}(X)$.

Conversely, let $|\mu| \in M_{ra}(X)$. The inequality $\mu_j(E) \leq |\mu|(E)$ implies that $\mu_j \in M_{ra}(X)$. By definition, $\mu \in M_{ra}(X)$. □

7.1.6 Theorem. *$M_{ra}(X)$ is a Banach space under the total variation norm.*

Proof. By 7.1.4, $M_{ra}(X)$ is a linear subspace of $M(X)$, the space of all complex Borel measures on X. Since the latter is complete (5.2.5), it suffices to show that $M_{ra}(X)$ is closed in $M(X)$. Let $\mu_n \in M_{ra}(X)$ and $\mu \in M(X)$ such that $\|\mu_n - \mu\| = |\mu_n - \mu|(X) \to 0$. Let $\varepsilon > 0$ and choose n so that $|\mu_n - \mu|(X) < \varepsilon$. Given $E \in \mathcal{B}(X)$, choose a compact set K and an open set U such that $K \subseteq E \subseteq U$ and $|\mu_n|(U \setminus K) < \varepsilon$ (7.1.5). Then

$$|\mu|(U \setminus K) \leq |\mu_n - \mu|(U \setminus K) + |\mu_n|(U \setminus K) < 2\varepsilon,$$

hence $\mu \in M_{ra}(X)$. □

The Support of a Radon Measure

Let (X, μ) be a Radon measure space and set $U = \bigcup\{V : V \text{ is open in } X \text{ and } \mu(V) = 0\}$. Then U is the largest open set such that $\mu(U) = 0$. To see this, let $K \subseteq U$ be compact and choose a finite cover $V_1, \ldots, V_n$ of K such that $\mu(V_j) = 0$. Then $\mu(K) = 0$, and since μ is inner regular, $\mu(U) = 0$.

The complement of U is called the **support of** μ and is denoted by $\operatorname{supp}\mu$:

$$\operatorname{supp}\mu = \left(\bigcup\{V : V \subseteq X \text{ open}, \mu(V) = 0\}\right)^c = \bigcap\{C : C \subseteq X \text{ closed}, \mu(C^c) = 0\}.$$

Thus $\operatorname{supp}(\mu)$ is the smallest closed set on which the measure μ is concentrated. Exercise 7.2 gives various properties of the support.

Exercises

7.1 Let μ be a regular Radon measure on a locally compact, Hausdorff space X and let $Y \subseteq X$ be closed, hence locally compact (0.12.1). Show that the restriction ν of μ to $\mathcal{B}(Y) = \mathcal{B}(X) \cap Y$ is a Radon measure on Y.

7.2 Let (X, μ) be a Radon measure space.

(a) Show that if V is open in X and $V \cap \operatorname{supp} \mu \neq \emptyset$, then $\mu(V) > 0$.

(b) Show that $\operatorname{supp}(\mu) = \bigcap \{f^{-1}(\{0\}) : f \in C_0(X),\ f \geq 0,\ \int f\, d\mu = 0\}$.

(c) Let f be nonnegative and continuous on $\mathbb{R}^d$. Show that $\operatorname{supp}(f\lambda) = \operatorname{supp}(f)$.

(d) Let μ_1 and μ_2 be Radon measures on X. Prove that

$$\operatorname{supp}(\mu_1 \vee \mu_2) = \operatorname{supp}(\mu_1) \cup \operatorname{supp}(\mu_2) \quad \text{and} \quad \operatorname{supp}(\mu_1 \wedge \mu_2) \subseteq \operatorname{supp}(\mu_1) \cap \operatorname{supp}(\mu_2)$$

and that the inclusion may be proper.

(e) Let $x \in X$, $c > 0$. Show that $\operatorname{supp}(\mu + c\delta_x) = \operatorname{supp}(\mu) \cup \{x\}$. Conclude that the support of $\sum_{j=1}^n c_j \delta_{x_j}$ $(c_j > 0)$ is $\{x_1, \ldots, x_n\}$.

7.3 Let μ be a Radon measure on X and $f \in L^1(X)$. Given $\varepsilon > 0$, show that there exists a compact set K such that $\int_{X \setminus K} |f|\, d\mu < \varepsilon$.

7.4 The **Baire σ-field** $\mathcal{B}_a(X)$ is the smallest σ-field relative to which each member of $C_c(X)$ is measurable. Show that $\mathcal{B}_a(X)$ is generated by the compact G_δ sets.

7.5 [↑1.85] (*Intermediate value property of measures*). Let μ be a regular Borel measure on X with the property that $\mu\{x\} = 0$ for all $x \in X$. Let E be a Borel set and $0 < c < \mu(E)$. Verify the following assertions to show that there exists a compact subset C of E such that $\mu(C) = c$.

(a) Let $\mathcal{A} := \{C : C \text{ is compact } C \subseteq E \text{ and } \mu(C) \geq c\}$. Then $\mathcal{A}$ is nonempty.

(b) Order $\mathcal{A}$ by reverse inclusion. If $\mathcal{C}$ is a chain in $\mathcal{A}$ and $B = \bigcap \mathcal{C}$, then $\mu(B) = \inf_{C \in \mathcal{C}} \mu(C)$. Conclude that $\mathcal{A}$ has a minimal element C. ⟦Argue by contradiction, using outer regularity on B and the finite intersection property for compact sets.⟧

(c) If $\mu(C) > c$ and $x \in C$, then there exists an open set $U \ni x$ such that $\mu(U) < \mu(C) - c$. Hence there exists a proper closed subset C_1 of C such that $\mu(C_1) > c$.

7.6 Let μ be a σ-finite Radon measure on X and ν a complex measure such that $\nu \ll \mu$. Show that ν is a Radon measure.

7.7 Let X and Y be locally compact Hausdorff spaces and $T : X \to Y$ continuous. Let μ be a regular Borel measure on X. Prove:

(a) $T\mu$ is inner regular on Borel subsets of Y. (b) If X and Y are compact, then $T\mu$ is regular.

7.2 The Riesz Representation Theorem

Let X be a locally compact Hausdorff space. A **positive linear functional** on $C_c(X)$ is a linear mapping $I : C_c(X) \to \mathbb{R}$ such that $f \geq 0 \Rightarrow I(f) \geq 0$. For example, if μ a Borel measure on X which is finite on compact sets, then $I(f) = \int f\, d\mu$ defines a positive linear functional on $C_c(X)$. Positive linear functionals of the latter type are said to be *represented by an integral.* The goal of this section is to prove that *every* positive linear functional on $C_c(X)$ may be so represented. As we shall see in §7.5 below as well as in Chapter 8, this result is one of several regarding concrete representations of linear functionals.

7.2.1 Riesz Representation Theorem. *Let I be a positive linear functional on $C_c(X)$. Then there exists a unique Radon measure μ on X such that*

$$I(f) = \int f\,d\mu \quad \textit{for all } f \in C_c(X). \tag{7.3}$$

Proof. The basic idea of the proof is to construct an outer measure from I and then use Carathéodory's theorem to obtain μ. This accomplished in the following steps, the first of which establishes uniqueness, for which regularity is crucial. The remaining steps establish existence.

(1) *Let μ be a Radon measure satisfying* (7.3). *For an open set $U \subseteq X$,*

$$\mu(U) = \sup\{I(f) : f \in C_U\}, \quad \textit{where} \quad C_U := \{f \in C_c(X) : \ 0 \le f \le \mathbf{1}_U\}.$$

In particular, the Radon measure satisfying (7.3) *is unique.*

⟦Denote the supremum by s. If $f \in C_U$, then $I(f) \le \int \mathbf{1}_U\,d\mu = \mu(U)$, hence $s \le \mu(U)$. For the reverse inequality, let K be a compact set contained in U and by Urysohn's lemma choose $f \in C_U$ such that $\mathbf{1}_K \le f \le \mathbf{1}_U$. Then $\mu(K) \le I(f) \le s$. By inner regularity, $\mu(U) \le s$. Two Radon measures satisfying (7.3) therefore agree on all open sets and hence, by outer regularity, are equal.⟧

(2) *For an open $U \subseteq X$, define $\mu(U)$ by the equation in (1). Then μ is monotone, $\mu(\emptyset) = 0$, and for any sequence of open sets U_n with union U*

$$\mu(U) \le \sum\nolimits_n \mu(U_n).$$

⟦The first two assertions are clear. For the inequality, let $f \in C_U$ and set $K := \operatorname{supp}(f)$. Since $K \subseteq U$, there exists a finite subcover $\{U_1, \ldots, U_p\}$ of K and nonnegative $f_i \in C_c(X)$ such that $\operatorname{supp}(f_i) \subseteq U_i$ and $\sum_{i=1}^p f_i = 1$ on K (0.14.1). Then $\sum_{j=1}^n f \cdot f_j = f$, and since $f \cdot f_j \in C_{U_j}$,

$$I(f) = \sum_{j=1}^n I(f \cdot f_j) \le \sum_{j=1}^n \mu(U_j) \le \sum_{j=1}^\infty \mu(U_j).$$

Taking the supremum over all $f \in C_U$ and applying (1) yields the desired inequality.⟧

(3) *For an arbitrary $E \subseteq X$, define $\mu^*(E)$ by*

$$\mu^*(E) = \inf\{\mu(U) : U \supseteq E,\ U \textit{ open}\}.$$

Then μ^* is an outer measure on X that agrees with μ on open sets.

⟦It is obvious that $\mu^*(U) = \mu(U)$ for open sets U. In particular, $\mu^*(\emptyset) = \mu(\emptyset) = 0$. It follows from (2) that

$$\mu^*(E) = \inf\Big\{\sum_n \mu(U_n) : E \subseteq \bigcup_n U_n,\ U_n \text{ open}\Big\}.$$

Thus by 1.5.1, μ^* is an outer measure.⟧

(4) *$\mathcal{B}(X) \subseteq \mathcal{M}(\mu^*)$, hence μ^* restricted to $\mathcal{B}(X)$ is a measure that extends μ. Denoting this extension by μ, we have by step (3)*

$$\mu(E) = \inf\{\mu(U) : U \supseteq E,\ U \textit{ open}\} \textit{ for every } E \in \mathcal{B}(X).$$

⟦For the inclusion, it suffices to show that $\mathcal{M}(\mu^*)$ contains all open sets U, that is,

$$\mu^*(E) \geq \mu^*(E \cap U) + \mu^*(E \cap U^c) \quad \text{for all } E \subseteq X. \tag{†}$$

For this we may assume that $\mu^*(E) < \infty$. Suppose first that E is open. Then $V := E \cap U$ is open, so given $\varepsilon > 0$ there exists $f \in C_V$ such that $I(f) > \mu(V) - \varepsilon$. Also, $W := E \setminus \text{supp}(f)$ is open, so there exists $g \in C_W$ such that $I(g) > \mu(W) - \varepsilon$. Since $f = 0$ on $V^c = E^c \cup U^c$ and $g = 0$ on $W^c = E^c \cup \text{supp}(f)$, $f + g \in C_E$. Therefore,

$$\mu(E) \geq I(f+g) = I(f) + I(g) > \mu(V) + \mu(W) - 2\varepsilon.$$

Since $\text{supp}(f) \subseteq U$, $W \supseteq E \setminus U$. We now have

$$\mu(E) \geq \mu(E \cap U) + \mu(E \setminus U) - 2\varepsilon \quad \text{for all } \varepsilon > 0.$$

this shows that (†) holds for open sets E.

Now let $E \subseteq X$ be arbitrary and let A be open with $E \subseteq A$ and $\mu(A) < \mu^*(E) + \varepsilon$. Then by the first part

$$\mu^*(E) + \varepsilon > \mu(A) \geq \mu^*(A \cap U) + \mu^*(A \cap U^c) \geq \mu^*(E \cap U) + \mu^*(E \cap U^c).$$

Therefore, (†) holds.⟧

(5) *If K is compact, then* $\mu(K) = \inf\{I(f) : f \in C_c(X),\ f \geq \mathbf{1}_K\}$.

⟦Let α denote the infimum. For an open set $U \supseteq K$ choose $f \in C_c(X)$ such that $\mathbf{1}_K \leq f \leq \mathbf{1}_U$. Then $I(f) \leq \mu(U)$, and taking the infimum over all such U and using (4) yields $I(f) \leq \mu(K)$. Therefore, $\alpha \leq \mu(K)$.

For the reverse inequality, let $f \in C_c(X)$ with $f \geq \mathbf{1}_K$, and for fixed $0 < r < 1$ set $U := \{f > r\}$. If $g \in C_U$, then $g = 0$ on $\{f \leq r\}$ and $g \leq 1 < r^{-1}f$ on $\{f > r\}$, so in each case $rg \leq f$. Therefore $rI(g) \leq I(f)$. Taking the supremum over all such g we have $r\mu(U) \leq I(f)$. Therefore, $r\mu(K) \leq I(f)$, and letting $r \to 1$ we obtain $\mu(K) \leq I(f)$. Taking the infimum over all such f yields $\mu(K) \leq \alpha$.⟧

(6) *For every open set $U \subseteq X$,* $\mu(U) = \sup\{\mu(K) : K \text{ compact and } K \subseteq U\}$.

⟦Let s denote the supremum. Clearly, $s \leq \mu(U)$. For the reverse inequality, let $r < \mu(U)$ and choose $g \in C_U$ such that $I(g) > r$. Let $K = \text{supp}(g)$ and $f \in C_c(X)$ with $f \geq \mathbf{1}_K$. Then $f \geq g$, so $I(f) \geq I(g)$. Taking the infimum over all such f we see from (5) that $\mu(K) \geq I(g)$. Therefore, $s \geq \mu(K) > r$, and since r was arbitrary, $s \geq \mu(U)$.⟧

(7) *If $f \in C_c(X)$ and $0 \leq f \leq 1$, then for each n there exist $f_j \in C_c(X)$ and compact sets $K_0 \supseteq K_1 \supseteq \cdots \supseteq K_n$ such that*

$$f = \sum_{j=1}^{n} f_j \quad \text{and} \quad n^{-1}\mathbf{1}_{K_j} \leq f_j \leq n^{-1}\mathbf{1}_{K_{j-1}}.$$

⟦Set $\varphi_0 = 0$, $K_0 = \text{supp}(f)$, and for $j = 1, \ldots, n$ define

$$\varphi_j(x) := \min\{f(x), j/n\},\ f_j := \varphi_j - \varphi_{j-1}, \text{ and } K_j := \{f \geq j/n\}.$$

The sets K_j are compact and decreasing, $\varphi_j = f$ on K_j^c, $\varphi_j = j/n$ on K_j, $\varphi_j = 0$ on K_0^c, and $\varphi_n = f$ on K_n. The assertions follow.⟧

(8) *Let f, f_j, and K_j be as in (7). Then for $j = 1, \ldots, n$,*

$$\frac{1}{n}\,\mu(K_j) \le \int f_j\,d\mu \le \frac{1}{n}\,\mu(K_{j-1}) \quad\text{and}\quad \frac{1}{n}\,\mu(K_j) \le I(f_j) \le \frac{1}{n}\,\mu(K_{j-1}).$$

⟦The first set of inequalities are an immediate consequence of (7). For the second set, observe that for any open set U containing K_{j-1}, $nf_j \in C_U$, hence $I(nf_j) \le \mu(U)$. Taking the infimum over U and applying (4) and (5) produces the second inequality.⟧

(9) *$I(f) = \int f\,d\mu$ for all $f \in C_c(X)$.*

⟦Let $f \in C_c(X)$. By considering positive and negative parts, we may assume that $f \ge 0$. Furthermore, dividing by $\|f\|_\infty$, we may also assume that $f \le 1$. Summing the inequalities in (8) and using $\sum_j f_j = f$, we obtain

$$\frac{1}{n}\sum_{j=1}^{n}\mu(K_j) \le \int f\,d\mu \le \frac{1}{n}\sum_{j=0}^{n-1}\mu(K_j) \quad\text{and}\quad \frac{1}{n}\sum_{j=1}^{n}\mu(K_j) \le I(f) \le \frac{1}{n}\sum_{j=0}^{n-1}\mu(K_j),$$

from which it follows that

$$\left|\int f\,d\mu - I(f)\right| \le \frac{1}{n}\left[\mu(K_0) - \mu(K_n)\right] \le \frac{1}{n}\,\mu(K_0).$$

Letting $n \to \infty$ yields the desired conclusion.⟧ □

The following result is immediate from step (1) of the preceding proof.

7.2.2 Corollary. *Let (X, μ) be a Radon measure space. Then for each open subset U of X,*

$$\mu(U) = \sup\left\{\int f\,d\mu : 0 \le f \le 1,\ \operatorname{supp}(f) \subseteq U\right\}.$$

Exercises

7.8 [↑7.1] Let μ be a regular Radon measure on X and Y a closed subset of X. For each $f \in C_c(Y)$, define a measurable function f_e on X by $f_e = f$ on Y and $f_e = 0$ on $X\setminus Y$. Then $I(f) = \int_X f_e\,d\mu$ defines a positive linear functional on $C_c(Y)$. Describe the corresponding Radon measure in terms of μ and justify your assertion.

7.9 Let I be a positive linear functional on $C_0(X)$ with corresponding Radon measure μ and let $U \subseteq X$ be open. Then U is a locally compact Hausdorff space. For $g \in C_0(U)$ define g_e by $g_e = g$ on U and $g_e(X \setminus U) = 0$.

(a) Show that $g_e \in C_0(X)$.

(b) Show that $J(g) = I(g_e)$ defines a positive linear functional on $C_0(U)$.

(c) What is the connection between μ and the Radon measure corresponding to J?

7.10 [↑7.7] Let X and Y be compact Hausdorff spaces and $T : X \to Y$ continuous. Given a positive linear functional I on $C_c(X)$ define a positive linear functional J on $C_c(Y)$ by $J(f) = I(f \circ T)$. Find a connection between the associated Radon measures and justify your assertions.

7.3 Products of Radon Measures

Finitely Many Measures

Let μ and ν be Borel measures on locally compact Hausdorff spaces X and Y, respectively. The product measure $\mu \otimes \nu$ is then defined on $\mathcal{B}(X) \otimes \mathcal{B}(Y)$. Denoting the collection of open sets on a topological space Z by $\mathcal{O}_Z$, we have

$$\mathcal{B}(X) \otimes \mathcal{B}(Y) = \sigma(\mathcal{O}_X) \otimes \sigma(\mathcal{O}_Y) = \sigma(\mathcal{O}_X \times \mathcal{O}_Y) \subseteq \sigma(\mathcal{O}_{X \times Y}) = \mathcal{B}(X \times Y), \qquad (7.4)$$

the second equality by 1.2.4. If X and Y are second countable with countable bases (U_n) and (V_n), respectively, then every open set in $X \times Y$ is a countable union of sets of the form $U_n \times V_m$, hence the inclusion in (7.4) is equality and so $\mathcal{B}(X) \otimes \mathcal{B}(Y) = \mathcal{B}(X \times Y)$. In general, however, the inclusion may be strict (see, for example, [20]), in which case $\mu \otimes \nu$ is not a Borel measure on $X \times Y$. In spite of this shortcoming, if μ and ν are Radon measures it is possible to extend $\mu \otimes \nu$ to a Borel measure on $X \times Y$. For this we need a preliminary result which is of some independent interest. The development is facilitated by the introduction of some standard notation.

Given functions g on X and h on Y, define the **tensor product** $g \otimes h$ of g and h on $X \times Y$ by

$$(g \otimes h)(x, y) = g(x)h(y), \quad x \in X, \ y \in Y.$$

If $\mathscr{G}$ and $\mathscr{H}$ are linear spaces of functions on X and Y, respectively, the **tensor product** $\mathscr{G} \otimes \mathscr{H}$ of $\mathscr{G}$ and $\mathscr{H}$ is the linear span of the set of all functions $g \otimes h$, $g \in \mathscr{G}$ and $h \in \mathscr{H}$.

7.3.1 Proposition. *$C_c(X) \otimes C_c(Y)$ is dense in $C_c(X \times Y)$ in the uniform norm.*

Proof. Let $\pi_X : X \times Y \to X$ and $\pi_Y : X \times Y \to Y$ denote the projection mappings. For $f \in C_c(X \times Y)$, the sets $K_X := \pi_X(\mathrm{supp}(f))$ and $K_Y := \pi_Y(\mathrm{supp}(f))$ are compact and $\mathrm{supp}(f) \subseteq K_X \times K_Y$. Choose open sets $U_X \subseteq X$ and $U_Y \subseteq Y$ with compact closure such that $K_X \subseteq U_X$ and $K_Y \subseteq U_Y$ and set $K := \mathrm{cl}\, U_X \times \mathrm{cl}\, V_Y$. By the Stone-Weierstrass theorem, $C(\mathrm{cl}\, U_X) \otimes C(\mathrm{cl}\, U_Y)$ is dense in $C(K)$, hence, given $\varepsilon > 0$, there exists a function

$$F := \sum_{i=1}^{n} g_i \otimes h_i \in C(\mathrm{cl}\, U_X) \otimes C(\mathrm{cl}\, U_Y)$$

such that $|F - f| < \varepsilon$ on K and so $|F| < \varepsilon$ on $K \setminus (K_X \times K_Y)$. By the Tietze extension theorem, g_i and h_i extend to members of $C_b(X)$ and $C_b(Y)$, respectively. By Urysohn's lemma, there exist $g \in C_c(X)$ and $h \in C_c(Y)$ with $\mathbf{1}_{K_X} \le g \le \mathbf{1}_{U_X}$ and $\mathbf{1}_{K_Y} \le h \le \mathbf{1}_{U_Y}$. Define $G \in C_c(X \times Y)$ by

$$G = (g \otimes h) \sum_{i=1}^{n} g_i \otimes h_i = \sum_{i=1}^{n} (g g_i) \otimes (h h_i).$$

Then $G = F$ on $K_X \times K_Y$, $G = f = 0$ on $(U_X \times U_Y)^c$, and $|G - f| = |G| \le |F| < \varepsilon$ on $(U_X \times U_Y) \setminus (K_X \times K_Y)$. Therefore, $\|G - f\|_\infty < \varepsilon$. □

It follows from 7.3.1 that members of $C_c(X \times Y)$ are $\mathcal{B}(X) \otimes \mathcal{B}(Y)$-measurable. Indeed, one need only show this for functions $g \otimes h \in C_c(X) \otimes C_c(Y)$, and such a function is $\mathcal{B}(X) \otimes \mathcal{B}(Y)$-measurable since it is the product of continuous functions $g \circ \pi_X$ and $h \circ \pi_Y$. With this observation we may now prove the following version of Fubini's theorem:

7.3.2 Theorem. *Let (X,μ) and (Y,ν) be Radon measure spaces. Then $C_c(X \times Y) \subseteq L^1(\mu \otimes \nu)$, and for all $f \in C_c(X \times Y)$*

$$\int f(x,y)\, d(\mu \otimes \nu)(x,y) = \iint f(x,y)\, d\mu(x)\, d\nu(y) = \iint f(x,y)\, d\nu(y)\, d\mu(x). \tag{7.5}$$

Proof. In the notation of the proof of 7.3.1, $f = 0$ off $K_X \times K_Y$, which has finite measure. Therefore, the inclusion holds and (7.5) is a consequence of Fubini's theorem applied to $K_X \times K_Y$. □

Now define a positive linear functional $I(f)$ on $C_c(X \times Y)$ by the common value in (7.5). The corresponding measure from the Riesz representation theorem is then defined on $\mathcal{B}(X \times Y)$ and is an extension of $\mu \otimes \nu$. We denote this measure by $\mu \overline{\otimes} \nu$. In summary:

7.3.3 Corollary. *There exists a unique Radon measure $\mu \overline{\otimes} \nu$ on $\mathcal{B}(X \times Y)$ whose restriction to $\mathcal{B}(X) \otimes \mathcal{B}(Y)$ is $\mu \otimes \nu$.*

Infinitely Many Measures

The preceding results extend in the obvious way to finitely many Radon measure spaces. More interestingly, the results extend to the case of *arbitrarily many* spaces. However, for this we require that the spaces be compact and the measures be probability measures. Here are the details:

Let $\mathfrak{I}$ be an arbitrary index set and for each $i \in \mathfrak{I}$ let X_i be a compact Hausdorff space. The product topological space $X := \prod_i X_i$ is compact by Tychonoff's theorem. In what follows we consider finite sequences

$$s := (i_1, \ldots, i_n), \quad i_j \in \mathfrak{I}, \quad i_j \neq i_k,$$

which we shall **index sequences.** For such a sequence, define

$$X_s := X_{i_1} \times \cdots \times X_{i_n} \quad \text{and} \quad \pi_s : X \to X_s, \quad \pi_s(x) := \big(x(i_1), \ldots, x(i_n)\big).$$

Denote by $C_s(X)$ the subalgebra of $C(X)$ consisting of all functions of the form $f \circ \pi_s$, $f \in C(X_s)$, and set $\mathcal{F} := \bigcup_s C_s(X)$, the collection of the functions in $C(X)$ that depend on only finitely many coordinates. Note that $s_1 \leq s_2 \Rightarrow C_{s_1}(X) \subseteq C_{s_2}(X)$, where the notation means that the terms of the sequence s_1 are contained in the sequence s_2. It follows that $\mathcal{F}$ is an algebra. Since $\mathcal{F}$ contains the constant functions and separates points of X, $\mathcal{F}$ is dense in $C(X)$ by the Stone-Weierstrass theorem. With these preliminaries out of the way, we may prove

7.3.4 Theorem. *For each $i \in \mathfrak{I}$ let X_i be a compact Hausdorff space and μ_i a Radon probability measure on X_i. Set $X := \prod_i X_i$. Then there exists a unique Radon measure μ on $\mathcal{B}(X)$ such that for all index sequences $s = (i_1, \ldots, i_n)$ the image measure $\pi_s(\mu)$ equals $\mu_{i_1} \overline{\otimes} \cdots \overline{\otimes} \mu_{i_n}$. Equivalently,*

$$\int_X f \circ \pi_s \, d\mu = \int_{X_s} f \, d(\mu_{i_1} \overline{\otimes} \cdots \overline{\otimes} \mu_{i_n}) \tag{7.6}$$

for all bounded Borel functions f on X_s.

Proof. For $g = f \circ \pi_s \in C_s(X)$, $f \in C(X_s)$, let $I_s(g)$ denote the right side of (7.6). Clearly, I_s is a well-defined, positive linear functional on $C_s(X)$ and

$$|I_s(g)| = |I_s(f \circ \pi_s)| \leq \|f\|_\infty = \|g\|_\infty \,.$$

Moreover, because the μ_i are probability measures, $s \le s' \Rightarrow I_s(g) = I_{s'}(g)$. Define a positive linear functional I on $\mathscr{F}$ by

$$I(g) = I_s(g), \quad g \in C_s(X).$$

If also $g \in C_{s'}(X)$, then $I_s(g) = I_{s\cup s'}(g) = I_{s'}(g)$, hence I is well-defined. Since $\mathscr{F}$ is dense in $C(X)$ and $|I(g)| \le \|g\|_\infty$, I has an extension to a positive linear functional on $C(X)$. Indeed, if (g_n) is a sequence in $\mathscr{F}$ and $g_n \to g \in C(X)$, then $\big(I(g_n)\big)$ is a Cauchy sequence in $\mathbb{C}$ hence converges to some $I(g) \in \mathbb{C}$, independent of the sequence (g_n), giving the desired extension. By the Riesz representation theorem, there exists a unique Radon probability measure μ on X such that $I(g) = \int_X g\,d\mu$, $g \in C(X)$, which implies (7.6) for continuous f.

It remains to show that $\pi_s(\mu) = \mu_{i_1}\overline{\otimes}\cdots\overline{\otimes}\,\mu_{i_n}$. Since these define equal positive linear functionals on $C(X_s)$ and since $\mu_{i_1}\overline{\otimes}\cdots\overline{\otimes}\,\mu_{i_n}$ is a Radon measure on X_s, it suffices by the uniqueness part of the Riesz representation theorem to show that $\pi_s(\mu)$ is a Radon measure on X_s. But since $\pi_s : X \to X_s$ is continuous, this follows directly from Ex. 7.7. □

Exercises

7.11 For each $n \in \mathbb{N}$, let X_n be a compact Hausdorff space, μ_n a Radon probability measure on X_n, $\pi_n : X \to X_n$ the projection map, and (X, μ) the product measure space. Show that the projection mappings $\pi_n : X \to X_n$ are *independent*, that is, if $n_1 < n_2 < \cdots < n_k$ and $B_j \in \mathcal{B}(X_{n_j})$, then

$$\mu\{\pi_{n_j} \in B_j,\ j = 1, \ldots, k\} = \prod_{j=1}^{k} \mu\{\pi_{n_j} \in B_j\}.$$

7.12 In the preceding exercise, assume that $X_1 = X_2 = \ldots$ and $\mu_1 = \mu_2 = \ldots$. For $E \in \mathcal{B}(X_1)$, define

$$T_E(x_1, x_2, \ldots) = \inf\{n \in \mathbb{N} : x_n \in E\},$$

where by convention $\inf \emptyset = \infty$. Show that T_E is Borel measurable and find $\mu\{T_E = n\}$, the probability that the first time a sequence enters E is at time n. The function T_E is an example of a *stopping time*. (See §18.5.)

7.13 [↑7.11] (Coin Toss) For each $n \in \mathbb{N}$, let $X_n = \{0, 1\}$ and define $\mu_n\{1\} = p$ and $\mu_n\{0\} = 1 - p$, where $0 < p < 1$. Set $S_n = \sum_{j=1}^{n} \pi_j$.

(a) Show that S_n is *binomially distributed*, that is,

$$\mu\{S_n = k\} = \binom{n}{k} p^k (1-p)^{n-k}, \quad k = 0, 1, \ldots, n.$$

(b) Show that $\int_X S_n\,d\mu = np$ and $\int_X S_n^2\,d\mu = np + n(n-1)p^2$.

(c) Show that $n^{-1}S_n \xrightarrow{\mu} p$. ⟦Use 3.21.⟧

Part (c) is a special case of the *weak law of large numbers*. (See §18.4.)

7.14 For $n \in \mathbb{N}$ let X_n be a compact Hausdorff space and set $Y_n = X_1 \times \cdots \times X_n$ and $X = \prod_{n=1}^{\infty} X_n$. Let μ_1 be a probability measure on $\mathcal{B}(X_1)$ and suppose for each $n > 2$ and $(x_1, \ldots, x_{n-1}) \in Y_{n-1}$ that there exists a probability measure $\mu_{n-1}(x_1, \ldots, x_{n-1}, \cdot) = \mu_{n-1}(x_1, \ldots, x_{n-1}, dx_n)$ on $\mathcal{B}(X_n)$ such that $\mu(x_1, \ldots, x_{n-1}, B)$ is Borel measurable in $(x_1, \ldots, x_{n-1})$ for each $B \in \mathcal{B}(X_n)$. Let $\Pi_n : X \to Y_n$ denote the projection mapping. Show that there exists a unique Radon probability measure μ on X such that for all bounded Borel functions f on Y_n,

$$\int_X f \circ \Pi_n\,d\mu = \int_{X_1} \cdots \int_{X_n} f(x_1, \ldots, x_n)\mu_{n-1}(x_1, \ldots, x_{n-1}, dx_n) \cdots \mu_1(dx_1).$$

7.4 Vague Convergence

Let X be a locally compact Hausdorff space. A sequence of complex measures μ_n on $\mathcal{B}(X)$ is said to **converge vaguely** to a complex measure μ, written $\mu_n \stackrel{v}{\to} \mu$, if

$$\lim_n \int f\,d\mu_n = \int f\,d\mu \tag{7.7}$$

for all $f \in C_0(X)$. For example, if $(\boldsymbol{x}_n)$ is a sequence in $\mathbb{R}^d$ and $\boldsymbol{x}_n \to \boldsymbol{x}$, then $\delta_{\boldsymbol{x}_n} \stackrel{v}{\to} \delta_{\boldsymbol{x}}$. Note that since the measures μ_n and μ may be identified with continuous linear functionals on $C_0(X)$, vague convergence is simply weak* sequential convergence in the dual of $C_0(X)$ (see §10.2).

Vague convergence does not necessarily imply that (7.7) holds for all $f \in C_b(X)$ (Ex. 7.16). Additional conditions are needed, as described in the next theorem.

7.4.1 Theorem. *Let $\mu, \mu_n \in M_{ra}(X)$ be nonnegative. Then (7.7) holds for all $f \in C_b(X)$ iff $\mu_n \stackrel{v}{\to} \mu$ and $\|\mu_n\| \to \|\mu\|$.*

Proof. The necessity is obvious. For the sufficiency, we may assume that $\|\mu\| > 0$. Choose $0 < \varepsilon < \|\mu\|$. By 7.2.2, there exists a $\phi \in C_c(X)$ with $0 \le \phi \le 1$ such that $\int \phi\,d\mu > \mu(X) - \varepsilon$. Let $f \in C_b(X)$. Since $\mu_n(X) \to \mu(X)$,

$$\overline{\lim_n}\left|\int f(1-\phi)\,d\mu_n\right| \le \|f\|_\infty \overline{\lim_n}\int (1-\phi)\,d\mu_n = \|f\|_\infty \int (1-\phi)\,d\mu \le \varepsilon\,\|f\|_\infty. \tag{†}$$

Since $\int f\phi\,d\mu_n \to \int f\phi\,d\mu$ (because $f\phi \in C_c(X)$), we see from the expansion

$$\int f\,d\mu_n - \int f\,d\mu = \int f(1-\phi)\,d\mu_n - \int f(1-\phi)\,d\mu + \int f\phi\,d\mu_n - \int f\phi\,d\mu$$

and (†) that

$$\overline{\lim_n}\left|\int f\,d\mu_n - \int f\,d\mu\right| \le 2\varepsilon\,\|f\|_\infty. \qquad \square$$

The following result gives a sufficient condition for vague convergence on $\mathcal{B}(\mathbb{R}^d)$ in terms of Fourier-Stieltjes transforms. It will be needed later in the proof of the central limit theorem.

7.4.2 Theorem. *Let $\mu, \mu_1, \mu_2, \ldots$ be complex measures on $\mathcal{B}(\mathbb{R}^d)$ such that $\sup_n \|\mu_n\| < \infty$ and $\widehat{\mu}_n \to \widehat{\mu}$ pointwise. Then $\mu_n \stackrel{v}{\to} \mu$.*

Proof. We use the Fourier inversion formula: For $\phi \in \mathcal{S}(\mathbb{R}^d)$ and any complex measure ν,

$$\int \phi(\boldsymbol{x})\,d\nu(\boldsymbol{x}) = \iint \widehat{\phi}(\boldsymbol{\xi})e^{2\pi\boldsymbol{\xi}\cdot\boldsymbol{x}}\,d\nu(\boldsymbol{x})\,d\boldsymbol{\xi} = \int \widehat{\phi}(\boldsymbol{\xi})\,\widehat{\nu}(-\boldsymbol{\xi})\,d\boldsymbol{\xi},$$

hence

$$\left|\int \phi(\boldsymbol{x})\,d\mu_n(\boldsymbol{x}) - \int \phi(\boldsymbol{x})\,d\mu(\boldsymbol{x})\right| \le \int |\widehat{\phi}(\boldsymbol{\xi})|\,|\widehat{\mu_n}(-\boldsymbol{\xi}) - \widehat{\mu}(-\boldsymbol{\xi})|\,d\boldsymbol{\xi}.$$

By hypothesis, the integrand on the right tends pointwise to 0. Since the integrand is dominated by the L^1 function $2|\widehat{\phi}(\boldsymbol{\xi})|$, $\int \phi(\boldsymbol{x})\,d\mu_n \to \int \phi(\boldsymbol{x})\,d\mu$. Now let $f \in C_c(\mathbb{R}^d)$ and choose $\phi \in C_c^\infty(\mathbb{R}^d)$ such that $\|f-\phi\|_\infty < \varepsilon$. (6.1.4). Then

$$\begin{aligned}\left|\int f\,d\mu_n - \int f\,d\mu\right| &\le \left|\int (f-\phi)\,d\mu_n\right| + \left|\int (f-\phi)\,d\mu\right| + \left|\int \phi\,d\mu_n - \int \phi\,d\mu\right| \\ &\le \|f-\phi\|_\infty\left(\|\mu_n\| + \|\mu\|\right) + \left|\int \phi\,d\mu_n - \int \phi\,d\mu\right|.\end{aligned}$$

Since $\int \phi\,d\mu_n \to \int \phi\,d\mu$, $\overline{\lim}_n \left|\int f\,d\mu_n - \int f\,d\mu\right| \le \varepsilon\big(\sup_n \|\mu_n\| + \|\mu\|\big)$. □

The following theorem characterizes vague convergence of finite measures on $\mathbb{R}$ in terms of convergence of the associated distribution functions.

7.4.3 Theorem. *Let $\mu, \mu_1, \mu_2, \ldots$ be finite measures on $\mathcal{B}(\mathbb{R})$ such that $\sup_n \|\mu_n\| < \infty$. Set $F(x) := \mu(-\infty, x]$ and $F_n(x) := \mu_n(-\infty, x]$. Then $\mu_n \xrightarrow{v} \mu$ iff $F_n(x) \to F(x)$ at each continuity point x of F.*

Proof. Suppose $F_n(x) \to F(x)$ at each continuity point x of F. Because F has only countably many discontinuities, $F_n \to F$ λ-a.e. Let $g \in C_c^1(\mathbb{R})$. Integrating by parts (§3.6) and applying the dominated convergence theorem, we have

$$\int g\,d\mu_n = \int g(x)\,dF_n(x) = -\int g'(x)F_n(x)dx \to -\int g'(x)F(x)dx = \int g\,d\mu.$$

Since an arbitrary $f \in C_c(\mathbb{R})$ may be uniformly approximated by functions $g \in C_c^1(\mathbb{R})$ (6.1.4) and since $\sup_n \|\mu_n\| < \infty$, it follows that $\mu_n \xrightarrow{v} \mu$.

For the converse, let x be a continuity point of F. Fix $k \in \mathbb{N}$ and $\delta > 0$ and construct a piecewise linear function $f \in C_c(\mathbb{R})$ such that $f = 1$ on the interval $[-k, x]$, and $f = 0$ on $(-\infty, -k-\delta] \cup [x+\delta, \infty)$. Given $\varepsilon > 0$, choose N so that $\int f\,d\mu_n \le \int f\,d\mu + \varepsilon$ for all $n > N$. For such n and all k

$$F_n(x) - F_n(-k) = \mu_n(-k, x] \le \int f\,d\mu_n \le \varepsilon + \int f\,d\mu \le \varepsilon + F(x+\delta) - F(-k-\delta),$$

hence

$$F_n(x) \le \varepsilon + F(x+\delta) - F(-k-\delta) + F_n(-k).$$

Letting $k \to \infty$ we have $F_n(x) \le \varepsilon + F(x+\delta)$ for all $n \ge N$ and so $\overline{\lim}_n F_n(x) \le F(x+\delta)$. Letting $\delta \to 0$ we then have $\overline{\lim}_n F_n(x) \le F(x)$. Similarly, by taking $g \in C_c(\mathbb{R})$ such that $g = 1$ on $[-k+\delta, x-\delta]$ and $g = 0$ on $(-\infty, k] \cup [x, \infty)$ and linear on the remaining intervals, we see that $\underline{\lim}_n F_n(x) \ge F(x)$. Therefore, $F_n(x) \to F(x)$. □

Exercises

7.15 Let X be locally compact and Hausdorff and μ, μ_n complex measures with $\sup_n \|\mu_n\| < \infty$. Show that the set $\mathscr{V}$ of all functions $f \in C_b(X)$ for which $\int f\,d\mu_n \to \int f\,d\mu$ is a closed linear subspace of $C_b(X)$ in the uniform norm. Conclude that if $\int f\,d\mu_n \to \int f\,d\mu$ for all $f \in C_c(X)$, then $\mu_n \xrightarrow{v} \mu$.

7.16 Show that the condition $\|\mu_n\| \to \|\mu\|$ in 7.4.1 cannot be removed.

7.17 Show that the convergence $F_n(x) \to F(x)$ in 7.4.3 need not hold at points x where F is discontinuous.

7.18 Consider the space $[0,1]$ with Lebesgue measure λ. Set $f_n = n\mathbf{1}_{[0,1/n]}$. Show that $f_n \cdot \lambda \xrightarrow{v} \delta_0$.

7.19 Let $\mu_n(E) = 2^{-n}\sum_{j=-\infty}^{\infty} \mathbf{1}_E(j/2^n)$, $E \in \mathcal{B}([a,b])$. Show that $\mu_n \xrightarrow{v} \lambda\big|_{[a,b]}$

7.20 [↑7.7] Let X and Y be compact Hausdorff spaces and $T : X \to Y$ continuous. Let μ_n and μ be Radon measures on X. Show that if $\mu_n \xrightarrow{v} \mu$, then $T(\mu_n) \xrightarrow{v} T(\mu)$.

7.21 Let $(X, \mathcal{F}, \mu)$ be a probability space and g_n, g real-valued, measurable functions on $\mathbb{R}$ such that $g_n \xrightarrow{\mu} g$. Show that $g_n(\mu) \xrightarrow{v} g(\mu)$. ⟦Let $f \in C_0(\mathbb{R})$, $\varepsilon > 0$, and set $E_n = \{|g_n - g)| \geq \delta\}$ for a suitable δ obtained from the uniform continuity of f.⟧ Show that the converse is false. ⟦ Consider the space $[0, 1)$ with Lebesgue measure λ. Set

$$A_n = [0, 1/2^n) \cup [2/2^n, 3/2^n) \cup [4/2^n, 5/2^n) \cup \cdots \cup [(2^n - 2)/2^n, (2^n - 1)/2^n)$$

and $r_n = \mathbf{1}_{A_n}$. (The functions r_n are called *Rademacher functions*.) Show that $r_n(\lambda) \xrightarrow{v} r_1(\lambda)$ but $r_n \overset{\lambda}{\not\to} f$ for any f.⟧

7.22 Let $\mu, \mu_1, \mu_2, \ldots$ be probability measures on $\mathcal{B}(\mathbb{R})$ such that $\mu_n \xrightarrow{v} \mu$. Let F, F_n be as in 7.4.3. Carry out the following steps to show that if F is continuous, then $F_n(x) \to F(x)$ uniformly on $\mathbb{R}$. Give an example to show that the continuity of F is needed here, that is, in general F_n need not converge uniformly to F on the set of continuity points of F.

(a) Given $\varepsilon > 0$, choose $a < b$ so that $F(a) < \varepsilon$ and $1 - F(b) < \varepsilon$. Then there exists a partition $\mathcal{P} = \{x_0 = a < x_1 < \cdots < x_k = b\}$ such that $|F(x_i) - F(x_{i-1})| < \varepsilon$ for all i.

(b) There exists N such that $|F_n(x_i) - F(x_i)| < \varepsilon$ for all $n \geq N$ and all i. Fix such an n.

(c) If $x \leq a$, then $0 \leq F(x) < \varepsilon$ and $0 \leq F_n(x) < 2\varepsilon$.

(d) If $x \geq b$, then $0 \leq 1 - F(x) < \varepsilon$ and $0 \leq 1 - F_n(x) < 2\varepsilon$.

(e) If $x \in [x_{i-1}, x_i]$, then $F(x_{i-1}) \leq F(x) < F(x_{i-1}) + \varepsilon$ and $F(x_{i-1}) - \varepsilon < F_n(x) \leq F(x_{i-1}) + 2\varepsilon$.

(f) Conclude that $|F_n(x) - F(x)| < 4\varepsilon$ for all x.

---

*7.5 The Daniell-Stone Representation Theorem

The Riesz representation theorem asserts that for a locally compact Hausdorff space X, a positive linear functional I on $C_c(X)$, may be represented as an integral against a Radon measure. In this section we consider the representation problem in more general setting, one which admits integral representations of positive linear functionals on a variety of other function spaces.

A (real) linear space $\mathscr{L}$ of real-valued functions on a set X is called a **Stone vector lattice** if

$$f, g \in \mathscr{L} \Rightarrow f \vee g,\ f \wedge g,\ f \wedge 1 \in \mathscr{L}.$$

The second requirement is actually redundant, since $f \wedge g = \big[(-f) \vee (-g)\big]$. The third requirement is also redundant if $\mathscr{L}$ contains the constant function 1; however, we do *not* assume this. Note that the conditions on $\mathscr{L}$ imply the additional properties

$$f \in \mathscr{L} \Rightarrow f^+,\ f^-,\ |f| \in \mathscr{L}.$$

Here are some examples: The set of real-valued measurable functions on a measurable space is a Stone vector lattice, as is the subspace of simple functions. If X a topological space, then $C_b(X)$, $C_0(X)$, and $C_c(X)$ are Stone vector lattices. If $(X, \mathcal{F}, \nu)$ is an arbitrary measure space, then $L^1(X, \mathcal{F}, \nu)$ is a Stone vector lattice.

A **Daniell-Stone integral** on a Stone vector lattice $\mathscr{L}$ is a linear functional $I : \mathscr{L} \to \mathbb{R}$ with the following properties:

- I is **positive**: $f \geq 0 \Rightarrow I(f) \geq 0$.
- I is **continuous from above**: $f_n \downarrow 0 \Rightarrow I(f_n) \to 0$.

Note that I must then have the additional properties

- $f \leq g \Rightarrow I(f) \leq I(g)$ • $f_n \downarrow f \Rightarrow I(f_n) \downarrow I(f)$ • $f_n \uparrow f \Rightarrow I(f_n) \uparrow I(f)$,

as may be seen by considering differences.

7.5.1 Example. Let X be a locally compact Hausdorff topological space. A positive linear functional I on $C_c(X)$ is a Daniell-Stone integral. Indeed, if $f_n \downarrow 0$, then $\operatorname{supp} f_n \subseteq K := \operatorname{supp} f_1$, hence by Dini's theorem the convergence is uniform on K. Choose a continuous function g with compact support such that $\mathbf{1}_K \leq g \leq 1$. Given $\varepsilon > 0$, choose N so that $f_n \leq \varepsilon / I(g)$ on K for all $n \geq N$. For such n, $f_n \leq \varepsilon g / I(g)$ on X, hence $I(f_n) \leq \varepsilon$. This shows that I is continuous from above and hence is a Daniell-Stone integral. ◊

The preceding example shows that 7.2.1 is a special case of the following general result:

7.5.2 Daniell-Stone Representation Theorem. *Let X be a nonempty set, I a Daniell-Stone integral on a Stone vector lattice $\mathscr{L}$ of functions on X, and let $\mathcal{F}$ denote the σ-field on X generated by $\mathscr{L}$, that is, by the sets $f^{-1}(B)$, $f \in \mathscr{L}$, $B \in \mathcal{B}(\mathbb{R})$. Then there exists a measure μ on $\mathcal{F}$ such that $I(f) = \int f \, d\mu$ for every $f \in \mathscr{L}$.*

Proof. We follow [26], where first a measure is constructed on the regions between the graphs of functions in $\mathscr{L}$. The proof is broken down into the following steps:

(1) *For functions $f, g : X \to \mathbb{R}$ with $f \leq g$, define*

$$(f, g] := \{(x, t) \in X \times \mathbb{R} : f(x) < t \leq g(x)\}.$$

Then for any $B \subseteq X$ and $c > 0$, $(0, c\mathbf{1}_B] = B \times (0, c]$.

⟦$(x, t) \in (0, c\mathbf{1}_B]$ iff $0 < t \leq c\mathbf{1}_B(x)$ iff $\mathbf{1}_B(x) = 1$ and $0 < t \leq c$ iff $(x, t) \in B \times (0, c]$.⟧

(2) *Let $\mathcal{H}$ denote the collection of all sets $(f, g]$ with f, $g \in \mathscr{L}$ and $f \leq g$. Then $\mathcal{H}$ is a semiring on $X \times \mathbb{R}$.*

⟦Clearly $\emptyset = (f, f] \in \mathcal{H}$. Let $(f_1, g_1]$, $(f_2, g_2] \in \mathcal{H}$. Then

$$(f_1, g_1] \cap (f_2, g_2] = \big(f_1 \vee f_2, (f_1 \vee f_2) \vee (g_1 \wedge g_2)\big],$$

hence $\mathcal{H}$ is a π system. Moreover,

$$(f_1, g_1] \setminus (f_2, g_2] = \big(f_1, f_1 \vee (g_1 \wedge f_2)\big] \cup \big(g_1 \wedge (g_2 \vee f_1), g_1\big].$$

Indeed, (x, t) in the left side iff $f_1(x) \leq t < g_1(x)$ and either $t < f_2(x)$ or $t \geq g_2(x)$, that is, iff (a) $f_1(x) \leq t < f_1(x) \vee \big(g_1(x) \wedge f_2(x)\big)$ or (b) $g_1(x) \wedge \big(g_2(x) \wedge f_1(x)\big) \leq t < g_1(x)$. Moreover, since (a) and (b) cannot occur simultaneously, the union is disjoint. Therefore $\mathcal{H}$ is a semiring.⟧

(3) *Define a set function ν on $\mathcal{H}$ by $\nu(f, g] = I(g - f)$. Then ν is a measure on $\mathcal{H}$ and hence, by 1.6.4, has an extension to $\sigma(\mathcal{H})$.*

⟦For countable additivity, let $(f, g] = \bigcup_{n=1}^{\infty}(f_n, g_n]$ (disjoint). Then for each $x \in X$, $(f(x), g(x)] = \bigcup_{n=1}^{\infty}(f_n(x), g_n(x)]$ (disjoint). Applying Lebesgue measure λ, we have $g(x) - f(x) = \sum_n \big[g_n(x) - f_n(x)\big]$. Since the partial sums of the series increase monotonically to $g - f$, $I(g - f) = \sum_n I(g_n - f_n)$, that is, $\nu(f, g] = \sum_n \nu(f_n, g_n]$.⟧

(4) *Let* $f \in \mathscr{L}$ *with* $f \geq 0$ *and* $c > 0$. *Then there exists a sequence of nonnegative functions* f_n *in* $\mathscr{L}$ *such that* $f_n \uparrow \mathbf{1}_{\{f>1\}}$, *hence* $(0, cf_n] \uparrow (0, c\mathbf{1}_{\{f>1\}}] = \{f > 1\} \times (0, c]$ *(by (1))*.

⟦Define $f_n = [n(f - f \wedge 1)] \wedge 1$. If $f(x) \leq 1$, then $f_n(x) = 0$ for all n. If $f(x) > 1$, then eventually $f_n(x) = 1$. Therefore, $f_n \uparrow \mathbf{1}_{\{f>1\}}$ ⟧

(5) $\sigma(\mathcal{H})$ *contains all sets of the form* $\{a < f \leq b\} \times (0, c]$, $f \in \mathscr{L}$, $0 < a < b$, $c > 0$.

⟦Since $a \geq 0$, the sets are unchanged when f is replaced by f^+, so we may assume that $f \geq 0$. By (4), $\{f > 1\} \times (0, c] \in \mathcal{H}$. Since

$$\{a < f \leq b\} \times (0, c] = \{1 < a^{-1}f\} \times (0, c] \setminus \left(\{1 < b^{-1}f\} \times (0, c]\right),$$

the assertion follows.⟧

(6) $\sigma(\mathcal{H})$ *contains all sets of the form* $B \times (0, c]$, $c > 0$, $B \in \mathcal{F}$.

⟦For fixed c, the collection of all $B \in \mathcal{F}$ for which $B \times (0, c] \in \sigma(\mathcal{H})$ is a σ-field containing the generators $\{a < f \leq b\}$ of $\mathcal{F}$, hence must equal $\mathcal{F}$.⟧

(7) *Define a measure* μ *on* $\mathcal{F}$ *by* $\mu(B) := \nu\big(B \times (0, 1]\big)$. *Then for* $f \in \mathscr{L}$ *and* $f \geq 0$,

$$\nu\{a < f \leq b\} \times (0, c] = c\mu\{a < f \leq b\}.$$

⟦Using the sequence in (4), we have

$$\begin{aligned}\nu\big(\{f > 1\} \times (0, c]\big) &= \lim_n \nu\big(0, cf_n\big] = \lim_n I(cf_n) = c \lim_n I(f_n) = c\nu\big(\{f > 1\} \times (0, 1]\big) \\ &= c\mu\{f > 1\}.\end{aligned}$$

Now apply the set identities in the proof of (5).⟧

(8) $I(f) = \int f \, d\mu$ *for every* $f \in \mathscr{L}$.

⟦We may take $f \geq 0$. Moreover, since $f \wedge n \uparrow f$, we may assume that f is bounded. Let $h = \sum_{i=1}^k c_i \mathbf{1}_{C_i}$ be a simple function in standard form, where $c_i > 0$, $C_i := \{a_i < f \leq b_i\}$, and $0 \leq a_i < b_i$. Since $(0, h]$ is a disjoint union of the sets $(0, c_i \mathbf{1}_{C_i}] = C_i \times (0, c_i]$, we have by (7)

$$\nu(0, h] = \sum_i \nu\big(C_i \times (0, c_i]\big) = \sum_i c_i \mu(C_i) = \int h \, d\mu.$$

By a minor modification of the proof of 2.3.1 (necessitated by the use of left open rather than right open intervals in the definition of C_i), there exists a sequence (h_n) of such simple functions such that $h_n \uparrow f$. Taking limits in $\nu(0, h_n] = \int h_n \, d\mu$ yields $I(f) = \nu(0, f] = \int f \, d\mu$.⟧ □

Part II

Functional Analysis

Chapter 8

Banach Spaces

Several examples of Banach spaces have played important roles in Part I of the text, notably L^p spaces and various spaces of continuous functions. In this chapter we develop the basic properties of general normed spaces. Additional properties are considered in Chapters 10 and 14.

8.1 General Properties of Normed Spaces

We remind the reader that a norm $\|\cdot\|$ on a linear space $\mathscr{X}$ over $\mathbb{K}$ has the following properties:

$$\textbf{(a)}\ \|\boldsymbol{x}\| \ge 0, \quad \textbf{(b)}\ \boldsymbol{x} \ne 0 \Rightarrow \|\boldsymbol{x}\| \ne 0, \quad \textbf{(c)}\ \|c\boldsymbol{x}\| = |c|\,\|\boldsymbol{x}\|, \quad \textbf{(d)}\ \|\boldsymbol{x}+\boldsymbol{y}\| \le \|\boldsymbol{x}\| + \|\boldsymbol{y}\|.$$

A seminorm has the same properties with the possible exception of (b). We also recall the following variations of the triangle inequality:

$$\left\|\sum_{j=1}^{n} \boldsymbol{x}_j\right\| \le \sum_{j=1}^{n} \|\boldsymbol{x}_j\| \quad \text{and} \quad \big|\, \|\boldsymbol{x}\| - \|\boldsymbol{y}\| \,\big| \le \|\boldsymbol{x}-\boldsymbol{y}\|. \tag{8.1}$$

For ease of reference, we list below the main examples of normed spaces discussed in the first part of the text together with some new ones. All are Banach spaces except (d) and (j). The sequence spaces (h) – (k) are special cases of the function spaces (a) – (e). We remind the reader that $\|\cdot\|_p$ is in general only a seminorm unless one adopts the convention (which we do) of identifying functions that are equal a.e.

8.1.1 Examples.

(a) $L^p(X, \mathfrak{F}, \mu) = \{f : X \to \mathbb{K} : f \text{ is } \mathfrak{F}\text{-measurable and } \|f\|_p < \infty\}$, where

$$\|f\|_p = \left(\int |f|^p \, d\mu\right)^{1/p} \ (1 \le p < \infty), \quad \|f\|_\infty = \sup\{t : \mu\{|f| > t\} > 0\}.$$

(b) $B(X)$ = the space of all bounded functions $f : X \to \mathbb{C}$ with norm $\|f\|_\infty = \sup|f(X)|$, where X is a nonempty set.

(c) $C_b(X)$ = the space of all bounded continuous functions $f : X \to \mathbb{C}$ with norm $\|\cdot\|_\infty$, where X is a topological space.

(d) $C_c(X) = \{f \in C_b(X) : \text{supp}(f) \text{ is compact}\}$ with norm $\|\cdot\|_\infty$, where X is a locally compact Hausdorff topological space.

(e) $C_0(X)$ = closure of $C_c(X)$ in $C_b(X)$, X a locally compact Hausdorff space.

(f) $M(X)$ = space of complex measures on a measurable space $(X, \mathfrak{F})$ with the total variation norm $\|\mu\| = |\mu|(X)$

(g) $M_{ra}(X)$ = space of complex Radon measures on $\mathcal{B}(X)$ with the total variation norm $\|\mu\| = |\mu|(X)$, where X is a locally compact Hausdorff space.
(h) $\ell^p = \ell^p(\mathbb{N}) := \{\boldsymbol{x} = (x_n) : \|\boldsymbol{x}\|_p < \infty\}$, where

$$\|\boldsymbol{x}\|_p = \left(\sum_{n=1}^{\infty} |x_n|^p\right)^{1/p} \quad (1 \le p < \infty), \quad \|\boldsymbol{x}\|_\infty = \sup_n |x_n|.$$

(i) $\ell^p(\mathbb{Z})$ = the space of all **bilateral sequences** $\boldsymbol{x} = (\ldots, x_{-1}, x_0, x_1, \ldots)$ such that $\|\boldsymbol{x}\|_p < \infty$, where

$$\|\boldsymbol{x}\|_p := \left(\sum_{n=-\infty}^{\infty} |x_n|^p\right)^{1/p} \quad (1 \le p < \infty), \quad \|\boldsymbol{x}\|_\infty := \sup_{n\in\mathbb{Z}} |x_n|.$$

(j) $\mathfrak{c}_{00} := \{\boldsymbol{x} = (x_n) : x_n = 0 \text{ for all but finitely many } n\}$, $\|\boldsymbol{x}\|_\infty := \sup_n |x_n|$.
(k) $\mathfrak{c}_0 := \{\boldsymbol{x} = (x_n) : \lim_n x_n = 0\}$, $\|\boldsymbol{x}\|_\infty := \sup_n |x_n|$.
(l) $\mathfrak{c} := \{\boldsymbol{x} = (x_n) : \lim_n x_n \text{ exists}\}$, $\|\boldsymbol{x}\|_\infty := \sup_n |x_n|$. ◊

The Topology and Geometry of Normed Spaces

Let $\mathcal{X}$ be a normed space. As noted in §0.4, the mapping $(\boldsymbol{x}, \boldsymbol{y}) \mapsto \|\boldsymbol{x} - \boldsymbol{y}\|$ is a metric. The metric topology of $\mathcal{X}$ is called the **norm topology**. The second inequality in (8.1) shows that $\|\boldsymbol{x}\|$ is a continuous function of $\boldsymbol{x}$. Using the sequential form of continuity, one may easily establish the continuity of vector space operations $(\boldsymbol{x}, \boldsymbol{y}) \mapsto \boldsymbol{x} + \boldsymbol{y}$ and $(c, \boldsymbol{x}) \mapsto c\boldsymbol{x}$. (Ex. 8.1). As a consequence, the closure of a subspace of $\mathcal{X}$ is a linear space and the closure of a convex set is convex. Moreover, for fixed $\boldsymbol{y}$ and $c \ne 0$, the mappings $\boldsymbol{x} \mapsto \boldsymbol{x} + \boldsymbol{y}$ and $\boldsymbol{x} \mapsto c\boldsymbol{x}$ are homeomorphisms, hence one has

$$\text{cl}(\boldsymbol{y} + A) = \boldsymbol{y} + \text{cl}(A), \ \text{cl}(cA) = c\,\text{cl}(A), \ \text{int}(\boldsymbol{y} + A) = \boldsymbol{y} + \text{int}(A) \text{ and } \text{int}(cA) = c\,\text{int}(A).$$

The open ball, closed ball, and sphere of radius r and center $\boldsymbol{x}$ in a normed space $\mathcal{X}$ take the forms

$$B_r(\boldsymbol{x}) := \{\boldsymbol{y} \in \mathcal{X} : \|\boldsymbol{x} - \boldsymbol{y}\| < r\}, \quad C_r(\boldsymbol{x}) := \{\boldsymbol{y} \in \mathcal{X} : \|\boldsymbol{x} - \boldsymbol{y}\| \le r\} \text{ and}$$
$$S_r(\boldsymbol{x}) := \{\boldsymbol{y} \in \mathcal{X} : \|\boldsymbol{x} - \boldsymbol{y}\| = r\}.$$

In case of ambiguity, we include the norm symbol in the notation, as in $B_r(\boldsymbol{x}, \|\cdot\|)$. We also use the simplified notation

$$B_r := B_r(\boldsymbol{0}), \quad C_r := C_r(\boldsymbol{0}), \quad \text{and} \quad S_r := S_r(\boldsymbol{0}).$$

The ball B_1 is called the **open unit ball** and C_1 is called the **closed unit ball**. The following relations are occasionally useful (Ex. 8.5):

$$B_r(\boldsymbol{x}) = \boldsymbol{x} + rB_1, \qquad C_r(\boldsymbol{x}) = \boldsymbol{x} + rC_1, \qquad S_r(\boldsymbol{x}) = \boldsymbol{x} + rS_1. \tag{8.2}$$

The reader may check that $C_r(\boldsymbol{x})$ is the closure of $B_r(\boldsymbol{x})$ and $B_r(\boldsymbol{x})$ is the interior of $C_r(\boldsymbol{x})$ (Ex. 8.3), properties *not* shared by general metric spaces (consider a discrete space). The balls $B_r(\boldsymbol{x})$ and $C_r(\boldsymbol{x})$ are easily seen to be convex; B_r and C_r have the additional property of being balanced (see §0.2).

Separable Spaces

A normed linear space is **separable** if it is separable in the metric topology. Such spaces are important in contexts where a metric is needed for the weak or weak* topologies discussed in Chapter 10.

8.1.2 Examples.

(a) For $1 \le p < \infty$, the space $L^p(\mathbb{R}^d)$ is separable. For example, the collection of all step functions $\sum_{i=1}^n a_i \mathbf{1}_{I_j}$, where $a_j \in \mathbb{Q}$ and I_j is a bounded open interval whose coordinate intervals have rational endpoints, is dense in L^p.

(b) The space $L^\infty(\mathbb{R}^d)$ is not separable. To see this for the case $d = 1$, let $f_t := \mathbf{1}_{(-\infty,t)}$ and note that the balls $B_{1/2}(f_t)$ are disjoint. Since there are uncountably many of these, $L^\infty(\mathbb{R})$ cannot contain a countable dense set.

(c) The space $C[a,b]$ is separable under the uniform norm. Indeed, by the Weierstrass approximation theorem, the set of polynomials on $[a,b]$ with rational coefficients is dense in $C[a,b]$. A similar argument shows that $C(X)$ is separable for any compact subset X of $\mathbb{R}^d$.

(d) The space $C_b(\mathbb{R})$ of bounded continuous functions on $\mathbb{R}$ is not separable in the uniform norm. The basic idea is a variation of the argument for L^∞: For each doubly infinite sequence $s = (\ldots, s_{-1}, s_0, s_1, \ldots,)$, where $s_n = 0$ or 1, define $f_s \in C_b(\mathbb{R})$ such that $f_s(n) = s_n$ and f_s is linear for $n \le x \le n+1$ $(n \in \mathbb{Z})$. Then $\|f_s - f_t\| = 1$ $(s \ne t)$, hence the balls $B_{1/2}(f_s)$ are disjoint. Since the set of all such sequences is uncountable, $C_b(\mathbb{R})$ cannot contain a countable dense set. ◊

(e) The **disk algebra** $A(\mathbb{D})$ is the algebra of continuous functions on the closed unit disk $\operatorname{cl}\mathbb{D}$ that are analytic on $\mathbb{D}$. We show that $A(\mathbb{D})$ is separable in the uniform norm by showing that the set of all polynomials $P(z)$ is dense in $A(\mathbb{D})$. To this end, let $0 < r < 1$ and note that if $f \in A(\mathbb{D})$, then $f_r(z) := f(rz)$ is analytic on the disk $r^{-1}\mathbb{D} \supseteq \operatorname{cl}\mathbb{D}$. The Taylor series $\sum_{k=0}^\infty c_k z^k$ for f_r therefore converges uniformly to f_r on $\operatorname{cl}\mathbb{D}$. Given $\varepsilon > 0$, choose a partial sum P_n of the series such that $|f_r(z) - P_n(z)| < \varepsilon$ for all $z \in \mathbb{D}$. Letting $r \to 1$, we obtain $|f(z) - P_n(z)| \le \varepsilon$ on $\mathbb{D}$.

Equivalent Norms

Two norms $\|\cdot\|$ and $|||\cdot|||$ on a vector space $\mathscr{X}$ are said to be **equivalent** if the associated metrics are equivalent, that is, if there exist positive real numbers a and b such that

$$\|x\| \le a\,|||x||| \quad\text{and}\quad |||x||| \le b\,\|x\| \quad \text{for all } x \in \mathscr{X}. \tag{8.3}$$

The notion of equivalence of norms is an equivalence relation on the collections of norms on $\mathscr{X}$ (Ex. 8.7).

The norms $\|\cdot\|_p$, $1 \le p \le \infty$, on $\mathbb{K}^d$ are easily seen to be equivalent. For an infinite dimensional example, let μ be a finite measure on $\mathscr{B}[0,1]$, g a positive continuous function on $[0,1]$, and $\nu = g\mu$. Then the L^1 norms of μ and ν are equivalent on $C[0,1]$. On the other hand, the norms $\|\cdot\|_1$ and $\|\cdot\|_\infty$ on $C[0,1]$ satisfy $\|\cdot\|_1 \le \|\cdot\|_\infty$ but are not equivalent (Ex. 8.8).

8.1.3 Proposition. *Let $\|\cdot\|$ and $|||\cdot|||$ be norms on a vector space $\mathscr{X}$. Each of the following statements implies the other two.*

(a) *$\|\cdot\|$ and $|||\cdot|||$ are equivalent.*

(b) *The corresponding metric topologies are equal.*

(c) *For any sequence (x_n) in $\mathscr{X}$, $\|x_n\| \to 0$ iff $|||x_n||| \to 0$.*

Proof. (a) ⇒ (b): If (8.3) holds, then $B_r(\boldsymbol{x}, \|\cdot\|) \subseteq B_{br}(\boldsymbol{x}, |||\cdot|||)$ and $B_r(\boldsymbol{x}, |||\cdot|||) \subseteq B_{ar}(\boldsymbol{x}, \|\cdot\|)$. Since open sets are unions of open balls, the topologies coincide.

(b) ⇒ (c): Let $\|\boldsymbol{x}_n\| \to 0$ and $\varepsilon > 0$. Since $B_\varepsilon(\mathbf{0}, |||\cdot|||)$ is open in the $\|\cdot\|$-topology, $B_\delta(\mathbf{0}, \|\cdot\|) \subseteq B_\varepsilon(\mathbf{0}, |||\cdot|||)$ for some $\delta > 0$. Therefore, $\boldsymbol{x}_n$ is eventually in $B_\varepsilon(\mathbf{0}, |||\cdot|||)$, which shows that $|||\boldsymbol{x}_n||| \to 0$. The converse is similar.

(c) ⇒ (a): We claim that if (c) holds, then there exists $m \in \mathbb{N}$ such that

$$B_{1/m}(\mathbf{0}, \|\cdot\|) \subseteq B_1(\mathbf{0}, |||\cdot|||). \tag{†}$$

Indeed, if the claim is false then for each n we could choose $\boldsymbol{x}_n \in B_{1/n}(\mathbf{0}, \|\cdot\|) \setminus B_1(\mathbf{0}, |||\cdot|||)$ to obtain $\|\boldsymbol{x}_n\| \to 0$ and $|||\boldsymbol{x}_n||| \geq 1$, contradicting (c). Now let $\boldsymbol{x} \neq \mathbf{0}$ and set $\boldsymbol{y} := (2m\|\boldsymbol{x}\|)^{-1}\boldsymbol{x}$. Then $\|\boldsymbol{y}\| < 1/m$, hence, by (†), $|||\boldsymbol{y}||| < 1$. Therefore, $|||\boldsymbol{x}||| < 2m\|\boldsymbol{x}\|$, which is the second inequality in (8.3) with $b = 2m$. The verification of the first inequality is similar. □

Finite Dimensional Spaces

Let $\mathscr{X}$ be a finite dimensional vector space. If $\boldsymbol{v}_1, \ldots, \boldsymbol{v}_d$ is a basis for $\mathscr{X}$ and $1 \leq p \leq \infty$, then a norm on $\mathscr{X}$ is given by

$$|||\boldsymbol{x}|||_p := \|\vec{x}\|_p, \quad \text{where } \boldsymbol{x} := x_1\boldsymbol{v}_1 + \cdots + x_d\boldsymbol{v}_d \text{ and } \vec{x} := (x_1, \ldots, x_d). \tag{8.4}$$

These norms are easily seen to be equivalent. A somewhat surprising result is the following:

8.1.4 Theorem. *All norms on a finite dimensional vector space $\mathscr{X}$ are equivalent.*

Proof. Let $\|\cdot\|$ be an arbitrary norm on $\mathscr{X}$. It suffices to show that $\|\cdot\|$ is equivalent to the complete norm $|||\cdot|||_2$ defined in (8.4).

One inequality in (8.3) is easy: In the notation of (8.4), we have, by the triangle and CBS inequalities,

$$\|\boldsymbol{x}\| \leq \sum_{k=1}^d \|\boldsymbol{v}_k\|\,|x_k| \leq \left(\sum_{k=1}^d \|\boldsymbol{v}_k\|^2\right)^{1/2} \left(\sum_{k=1}^d |x_k|^2\right)^{1/2} = a\,|||\boldsymbol{x}|||_2,$$

where $a := \left(\sum_{k=1}^d \|\boldsymbol{v}_k\|^2\right)^{1/2}$. For the other inequality, define a function $F : \mathbb{K}^d \to \mathbb{R}^+$ by $F(\vec{x}) = \|\boldsymbol{x}\|$. Then

$$|F(\vec{x}) - F(\vec{y})| \leq \|\boldsymbol{x} - \boldsymbol{y}\| \leq a\,\|\vec{x} - \vec{y}\|_2,$$

hence F is continuous. Moreover, if $\vec{x} \neq 0$ then, by linear independence, $\boldsymbol{x} \neq \mathbf{0}$. Thus F is positive on the compact Euclidean sphere $\{\vec{x} : \|\vec{x}\|_2 = 1\}$ and so has a positive minimum m there. For any $\vec{x} \neq \mathbf{0}$ we then have $\|\boldsymbol{x}/\|\vec{x}\|_2\| = F(\vec{x}/\|\vec{x}\|_2) \geq m$, hence $\|\boldsymbol{x}\| \geq m\|\vec{x}\|_2 = m|||\boldsymbol{x}|||_2$. □

Theorem 8.1.4 shows that in a finite dimensional normed space $\mathscr{X}$ one may always choose an equivalent norm relative to which $\mathscr{X}$ isometrically isomorphic to a Euclidean space $\mathbb{K}^d$. This implies that the metric properties of $\mathbb{K}^d$ carry over to $\mathscr{X}$. In particular,

8.1.5 Corollary. *A finite dimensional normed space is complete, its subspaces are closed, and its bounded sets are relatively compact.*

Interestingly, the last assertion of the corollary actually characterizes finite dimensional spaces: a normed space with a compact ball is finite dimensional. The proof depends on the following result, which guarantees the existence of vectors in a normed space that are "nearly orthogonal" to a given closed subspace.

8.1.6 Theorem (F. Riesz). *Let $\mathscr{Y}$ be a proper closed subspace of a normed space $\mathscr{X}$. Then for each $\varepsilon \in (0,1)$ there exists $x_\varepsilon \in \mathscr{X}$ such that*

$$\|x_\varepsilon\| = 1 \quad and \quad \inf\{\|x_\varepsilon - y\| : y \in \mathscr{Y}\} \geq 1 - \varepsilon. \tag{8.5}$$

Proof. Choose any $x \in \mathscr{X} \setminus \mathscr{Y}$ and set $d := \inf\{\|x - y\| : y \in \mathscr{Y}\}$. Since $\mathscr{Y}$ is closed, $d > 0$.

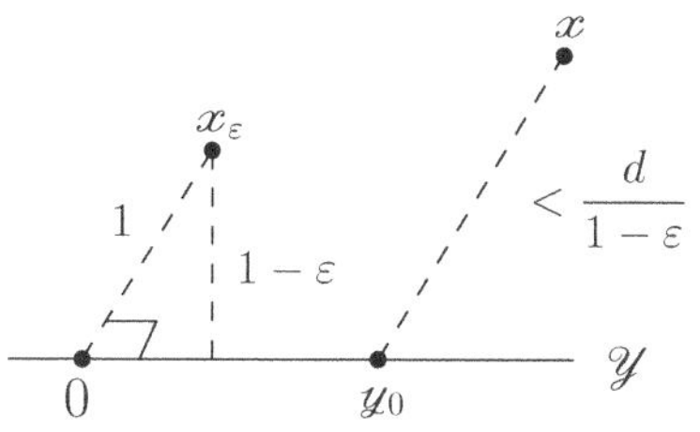

FIGURE 8.1: x_ε nearly orthogonal to $\mathscr{Y}$

Choose $y_0 \in \mathscr{Y}$ such that $\|x - y_0\| < d/(1-\varepsilon)$ and set

$$x_\varepsilon := \|x - y_0\|^{-1}(x - y_0).$$

Then for any $y \in \mathscr{Y}$ the vector $z := y_0 + \|x - y_0\|\, y$ is in $\mathscr{Y}$ and

$$y = \|x - y_0\|^{-1}(z - y_0),$$

hence

$$\|x_\varepsilon - y\| = \frac{1}{\|x - y_0\|}\|x - z\| \geq \frac{d}{\|x - y_0\|} \geq 1 - \varepsilon. \qquad \square$$

8.1.7 Theorem. *Let $\mathscr{X}$ be a normed space with $S_1 = \{x \in \mathscr{X} : \|x\| = 1\}$ compact. Then $\mathscr{X}$ is finite dimensional.*

Proof. Assume that $\mathscr{X}$ is infinite dimensional. Choose $x_1 \in \mathscr{X}$ with $\|x_1\| = 1$. Since the span of x_1 is a proper closed subspace of $\mathscr{X}$, by 8.1.6 there exists a vector x_2 with $\|x_2\| = 1$ such that $\|x_2 - y\| \geq 1/2$ for all $y \in \text{span}\{x_1\}$. Proceeding by induction, we obtain an infinite sequence (x_n) in S_1 such that

$$\|x_{n+1} - y\| \geq 1/2 \quad \text{for all } y \in \text{span}\{x_1, \ldots, x_n\}.$$

In particular, $\|x_m - x_n\| \geq 1/2$ for all $m \neq n$. On the other hand, the compactness of S_1 implies that (x_n) has a convergent subsequence. As these assertions are incompatible, $\mathscr{X}$ must be finite dimensional. $\square$

*Strictly Convex Spaces

A normed space is **strictly convex** if it satisfies the equivalent conditions in the following proposition, these conditions asserting in various ways that a sphere does not contain line segments.

8.1.8 Proposition. *Let $\mathscr{X}$ be a normed space. The following statements are equivalent:*

(a) $x \neq y$ *and* $\|x\| = \|y\| = 1 \Rightarrow \|x + y\| < 2$.

(b) $x \neq y$ *and* $\|x\| = \|y\| = 1 \Rightarrow \|(1-t)x + ty\| < 1$ *for all* $0 < t < 1$.

(c) $x \neq 0$ *and* $y \neq 0$ *and* $\|x + y\| = \|x\| + \|y\| \Rightarrow x = ty$ *for some* $t > 0$.

Proof. (a) ⇒ (b): Let $\boldsymbol{x} \neq \boldsymbol{y}$ and $\|\boldsymbol{x}\| = \|\boldsymbol{y}\| = 1$. By hypothesis, the inequality in (b) holds for $t = 1/2$. Now let $0 < t < 1/2$. Then $0 < 2t < 1$ and

$$t\boldsymbol{x} + (1-t)\boldsymbol{y} = t(\boldsymbol{x} + \boldsymbol{y}) + (1-2t)\boldsymbol{y},$$

hence

$$\|t\boldsymbol{x} + (1-t)\boldsymbol{y}\| \le t\|\boldsymbol{x} + \boldsymbol{y}\| + (1-2t)\,\|\boldsymbol{y}\| < 2t + (1-2t) = 1.$$

Thus the inequality in (b) holds for $0 < t < 1/2$. Similarly, if $1/2 < t < 1$, then $0 < 2t - 1 < 1$ and

$$t\boldsymbol{x} + (1-t)\boldsymbol{y} = (1-t)(\boldsymbol{x} + \boldsymbol{y}) + (2t-1)\boldsymbol{x},$$

hence

$$\|t\boldsymbol{x} + (1-t)\boldsymbol{y}\| \le (1-t)\|\boldsymbol{x} + \boldsymbol{y}\| + (2t-1)\,\|\boldsymbol{x}\| < 2(1-t) + 2t - 1 = 1.$$

Therefore the inequality in (b) holds for all $t \in (0, 1)$.

(b) ⇒ (c): Let $\boldsymbol{x} \neq \boldsymbol{0}$ and $\boldsymbol{y} \neq \boldsymbol{0}$ and $\|\boldsymbol{x} + \boldsymbol{y}\| = \|\boldsymbol{x}\| + \|\boldsymbol{y}\|$. Then

$$\frac{\|\boldsymbol{x}\|}{\|\boldsymbol{x} + \boldsymbol{y}\|} + \frac{\|\boldsymbol{y}\|}{\|\boldsymbol{x} + \boldsymbol{y}\|} = 1,$$

which forces $\|\boldsymbol{x}\|^{-1}\boldsymbol{x} = \|\boldsymbol{y}\|^{-1}\boldsymbol{y}$; otherwise, by (b) with $t := \|\boldsymbol{y}\|\,\|\boldsymbol{x} + \boldsymbol{y}\|^{-1}$,

$$1 = \left\|\frac{\boldsymbol{x} + \boldsymbol{y}}{\|\boldsymbol{x} + \boldsymbol{y}\|}\right\| = \left\|(1-t)\frac{\boldsymbol{x}}{\|\boldsymbol{x}\|} + t\frac{\boldsymbol{y}}{\|\boldsymbol{y}\|}\right\| < 1.$$

(c) ⇒ (a): Let $\|\boldsymbol{x}\| = \|\boldsymbol{y}\| = 1$, $\boldsymbol{x} \neq \boldsymbol{y}$. Suppose for a contradiction that $\|\boldsymbol{x} + \boldsymbol{y}\| = 2$. Then $2 = \|\boldsymbol{x} + \boldsymbol{y}\| \le \|\boldsymbol{x}\| + \|\boldsymbol{y}\| \le 2$, hence $\|\boldsymbol{x} + \boldsymbol{y}\| = \|\boldsymbol{x}\| + \|\boldsymbol{y}\|$ and so, by hypothesis, $\boldsymbol{x} = t\boldsymbol{y}$ for some $t > 0$. But then $t = 1$, contradicting that $\boldsymbol{x} \neq \boldsymbol{y}$. □

A Euclidean space is an example of a strictly convex space. By contrast, $\mathbb{R}^d$ with the norm $\|x\|_1 = \sum_{j=1}^{d} |x_j|$ or $\|x\|_\infty = \max_{1 \le j \le d} |x_j|$ is not strictly convex. A more interesting example is $L^p(X, \mathcal{F}, \mu)$. If $1 < p < \infty$, then L^p is strictly convex by the second part of 4.1.3. On the other hand, if $p = 1$ or ∞, then L^p is strictly convex only in trivial circumstances (Ex. 8.23).

We conclude this subsection with an application of strict convexity to approximation theory. Let $\mathcal{X}$ be a normed space and C a nonempty convex subset of $\mathcal{X}$. If $\boldsymbol{x} \in \mathcal{X}$, then an element $\boldsymbol{y}_0 \in C$ is called a **best approximation to $\boldsymbol{x}$ out of C** if

$$\|\boldsymbol{x} - \boldsymbol{y}_0\| = \inf\{\|\boldsymbol{x} - \boldsymbol{y}\| : \boldsymbol{y} \in C\}.$$

The relevance of this notion here is that if $\mathcal{X}$ is strictly convex then best approximations, if they exist, are unique. To see this, let α denote the infimum and suppose that $\|\boldsymbol{x} - \boldsymbol{z}_0\| = \alpha$ for some point $\boldsymbol{z}_0 \in C$ distinct from $\boldsymbol{y}_0$. Then $\boldsymbol{x} - \boldsymbol{z}_0 \neq \boldsymbol{x} - \boldsymbol{y}_0$, and since $\|\boldsymbol{x} - \boldsymbol{z}_0\| = \|\boldsymbol{x} - \boldsymbol{y}_0\| = \alpha$, we have, by strict convexity,

$$\left\|\boldsymbol{x} - \tfrac{1}{2}(\boldsymbol{z}_0 + \boldsymbol{y}_0)\right\| = \tfrac{1}{2}\,\|(\boldsymbol{x} - \boldsymbol{z}_0) + (\boldsymbol{x} - \boldsymbol{y}_0)\| < \alpha.$$

But this is impossible as $\frac{1}{2}(\boldsymbol{z}_0 + \boldsymbol{y}_0) \in C$. We have proved

8.1.9 Proposition. *Let C be a nonempty convex subset of a strictly convex space $\mathcal{X}$. Then each $\boldsymbol{x} \in \mathcal{X}$ has at most one best approximation out of C.*

Note that, as a special case, a nonempty convex subset of a strictly convex space $\mathscr{X}$ cannot have more than one member with smallest norm.

While strict convexity guarantees uniqueness of best approximations, it does not guarantee existence. For this additional conditions must be placed on $\mathscr{X}$. One such condition is *uniform convexity*, discussed in §10.4. For now, we offer the following more modest result, the proof of which is left to the reader as an exercise (8.25).

8.1.10 Proposition. *Let $\mathscr{X}$ be a normed space and $\mathscr{Y}$ a finite dimensional subspace of $\mathscr{X}$. Then for each $x \in \mathscr{X}$ there exists a best approximation to x out of $\mathscr{Y}$.*

For example, if $1 < p < \infty$ and $f \in L^p[0,1]$, then there exists a unique polynomial on $[0,1]$ of degree $\leq n$ that best approximates f in L^p norm out of all polynomials of degree $\leq n$.

Exercises

8.1 Prove that the operations of addition and scalar multiplication in a normed linear space are continuous.

8.2 Let $\mathscr{X}$ be a normed space and $x \neq 0 \in \mathscr{X}$. Show that if (c_n) is a sequence in $\mathbb{K}$ such that $c_n x \to y \in \mathscr{X}$, then $c := \lim_n c_n$ exists in $\mathbb{K}$ and $cx = y$.

8.3 Show that in a normed linear space, $C_r(x) = \operatorname{cl} B_r(x)$ and $B_r(x) = \operatorname{int} C_r(x)$.

8.4 Let C be a nonempty, closed subset C of a normed space $\mathscr{X}$ with the property $x, y \in C \Rightarrow \frac{1}{2}(x+y) \in C$. Show that C is convex. ⟦Consider the dyadic rationals.⟧

8.5 Verify the relations (8.2).

8.6 Let $\mathscr{Y}$ be a dense subspace of a normed linear space $\mathscr{X}$. Show that the open unit ball $B_1 \cap \mathscr{Y}$ of $\mathscr{Y}$ is dense in open unit ball B_1 of $\mathscr{X}$.

8.7 Prove that equivalence of norms is an equivalence relation.

8.8 Show that the norms $\|\cdot\|_1$ and $\|\cdot\|_\infty$ on $C[0,1]$ are not equivalent.

8.9 Let $M > 0$ and let (c_n) be a sequence in $\mathbb{C}$ such that $|\sum_{n\in F} c_n| \leq M$ for all finite $F \subseteq \mathbb{N}$. Show that $\sum_n |c_n| \leq 4M$, hence $(c_n) \in \ell^1(\mathbb{N})$.

8.10 Show that an infinite dimensional Banach space $\mathscr{X}$ has a nonclosed linear subspace. ⟦Use the Baire category theorem.⟧

8.11 Show that the linear space $D[a,b]$ of differentiable functions on $[a,b]$ is not complete in either the uniform norm or the L^1 norm.

8.12 Show that the space $\mathfrak{c}$ of all convergent sequences in $\mathbb{C}$ is a Banach space under the sup norm.

8.13 Let $0 < \alpha < 1$. A function $f \in C_b(\mathbb{R})$ is **Hölder continuous of order** α if

$$\|f\|_{0,\alpha} := \sup_{x\neq y} \frac{|f(x) - f(y)|}{|x-y|^\alpha} < \infty.$$

Show that the set $C_{0,\alpha}(\mathbb{R})$ of all such functions is a Banach space under the norm $|||f||| := \|f\|_{0,\alpha} + \|f\|_\infty$.

8.14 (a) Prove that $\|f\|_{bv} := |f(a)| + V_{[a,b]}(f)$ defines a norm on the space $BV[a,b]$ of functions of bounded variation (see §5.5) and that $BV[a,b]$ is a Banach space under this norm.

(b) Show that the space $AC[a,b]$ of absolutely continuous functions on $[a,b]$ is a closed subspace of the Banach space $BV[a,b]$.

(c) Show that the norms $\|\cdot\|_{bv}$ and $\|\cdot\|_\infty$ on $AC[a,b]$ are not equivalent.

8.15 Show that the spaces $C_c(\mathbb{R}^d)$ and $C_0(\mathbb{R}^d)$ are separable.

8.16 Prove that if $\mathscr{X}$ is a separable Banach space and $\mathscr{Y}$ is a closed subspace, then $\mathscr{Y}$ is separable.

8.17 Let $A \subseteq \mathbb{R}$ be Lebesgue measurable with $\lambda(A) > 0$. Show that $L^p(A)$ is infinite dimensional, $1 \le p \le \infty$.

8.18 Show that the L^p and L^q norms on C[0,1] are not equivalent if $1 \le p < q \le \infty$.

8.19 Let X and Y be Hausdorff topological spaces and let Z be a dense subset of X. Let $C_b(X)$ and $C_b(Y)$ have the supremum norms. Suppose that $f : X \times Y \to \mathbb{C}$ is bounded such that $f(\,\cdot\,,y) \in C_b(X)$ for all $y \in Y$, $f(z,\,\cdot\,) \in C_b(Y)$ for all $z \in Z$, and the set of mappings $f(Z,\,\cdot\,) := \{f(z,\,\cdot\,) : z \in Z\}$ is relatively compact in $C_b(Y)$. Prove the following:

(a) The collection of mappings $f(X,\,\cdot\,) = \{f(x,\,\cdot\,) : x \in X\}$ is relatively compact in $C_b(Y)$.

⟦Let $x \in X$ and $z_\alpha \in Z$ with $z_\alpha \to x$. Then $f(z_\beta,\,\cdot\,) \to g \in C_b(Y)$.⟧

(b) $f \in C_b(X \times Y)$. ⟦Let $(x_\alpha, y_\alpha) \to (x_0, y_0)$ in $X \times Y$ and use (a).⟧

(c) If Y is compact, then $x \to f(x,\,\cdot\,)$ maps X continuously into $C_b(Y)$. ⟦Argue by contradiction, using (b).⟧

8.20 Show that each of the conditions below is equivalent to strict convexity:

(i) $\boldsymbol{x} \ne \boldsymbol{y}$ and $\|\boldsymbol{x}\| = \|\boldsymbol{y}\| \Rightarrow \|\boldsymbol{x} + \boldsymbol{y}\| < \|\boldsymbol{x}\| + \|\boldsymbol{y}\|$.

(ii) $\boldsymbol{x} \ne \boldsymbol{y}$ and $\|\boldsymbol{x}\|, \|\boldsymbol{y}\| \le 1 \Rightarrow \|\boldsymbol{x} + \boldsymbol{y}\| < 2$.

(iii) $\boldsymbol{x} \ne \boldsymbol{y}$ and $\|\boldsymbol{x}\| = \|\boldsymbol{y}\| = 1 \Rightarrow \|(1-s)\boldsymbol{x} + s\boldsymbol{y}\| < 1$ for some $0 < s < 1$.

8.21 Prove the converse of 8.1.9: Let a normed space $\mathscr{X}$ have the property that for each closed convex subset C, every $\boldsymbol{x} \in \mathscr{X}$ has at most one best approximation out of C. Then $\mathscr{X}$ is strictly convex.

8.22 Show that for a locally compact Hausdorff space containing at least two points, $C_0(X)$ is not strictly convex.

8.23 Show that L^1 and L^∞ are not strictly convex except in trivial cases.

8.24 Let $\mathscr{X}$ be a normed space and C a nonempty subset of $\mathscr{X}$. Show that the set of best approximations to $\boldsymbol{x}$ out of C is convex.

8.25 Prove 8.1.10.

8.2 Bounded Linear Transformations

Throughout this section, $\mathscr{X}$, $\mathscr{Y}$, and $\mathscr{Z}$ denote normed spaces over $\mathbb{K}$.

Recall from §0.4 that a linear transformation $T : \mathscr{X} \to \mathscr{Y}$ is continuous iff it is bounded, that is, iff there exists a constant $M \ge 0$ such that $\|T\boldsymbol{x}\| \le M\|\boldsymbol{x}\|$ for all $\boldsymbol{x} \in \mathscr{X}$ (0.4.8). The set of all bounded linear transformations (operators) from $\mathscr{X}$ to $\mathscr{Y}$ is denoted by $\mathscr{B}(\mathscr{X}, \mathscr{Y})$:

$$\mathscr{B}(\mathscr{X}, \mathscr{Y}) := \big\{T : \mathscr{X} \to \mathscr{Y} : T \text{ is linear and bounded}\big\}.$$

Note that $\mathscr{B}(\mathscr{X}, \mathscr{Y})$ is a vector space under pointwise addition and scalar multiplication:

$$(S+T)(\boldsymbol{x}) := S\boldsymbol{x} + T\boldsymbol{x}, \quad (cT)\boldsymbol{x} := c(T\boldsymbol{x}), \quad S, T \in \mathscr{B}(\mathscr{X}, \mathscr{Y}), \quad c \in \mathbb{K}.$$

An operator $T \in \mathscr{B}(\mathscr{X}, \mathscr{Y})$ is said to be **invertible** if T^{-1} exists and is a member of $\mathscr{B}(\mathscr{Y}, \mathscr{X})$. In this case T is a topological isomorphism from $\mathscr{X}$ onto $\mathscr{Y}$. Recall that a special case is a surjective isometry, which satisfies $\|T\boldsymbol{x}\| = \|\boldsymbol{x}\|$ for all $\boldsymbol{x}$.

The Operator Norm

The **operator norm of** $T \in \mathscr{B}(\mathscr{X},\mathscr{Y})$ is defined by

$$\|T\| := \inf\{M : \|T\boldsymbol{x}\| \le M\|\boldsymbol{x}\| \text{ for all } \boldsymbol{x} \in \mathscr{X}\}. \tag{8.6}$$

The boundedness of T implies that the set on the right is nonempty, hence $\|T\| < \infty$. Taking a sequence of M's in (8.6) tending to $\|T\|$, we see that

$$\|T\boldsymbol{x}\| \le \|T\|\,\|\boldsymbol{x}\| \text{ for all } \boldsymbol{x} \in \mathscr{X}. \tag{8.7}$$

It follows that

$$\|T\| = \sup\{\|T\boldsymbol{x}\| : \|\boldsymbol{x}\| \le 1\}. \tag{8.8}$$

Indeed, denoting the supremum by s, we have $s \le \|T\|$ by (8.7). For the reverse inequality, let $\boldsymbol{x} \ne \boldsymbol{0}$ and set $\boldsymbol{y} := \boldsymbol{x}/\|\boldsymbol{x}\|$. Then $\|T\boldsymbol{y}\| \le s$, hence $\|T\boldsymbol{x}\| \le s\,\|\boldsymbol{x}\|$. By (8.6), $\|T\| \le s$.

8.2.1 Theorem. *$\|T\|$ is a norm on $\mathscr{B}(\mathscr{X},\mathscr{Y})$.*

Proof. If $\|T\| = 0$, then $T = 0$ by (8.7). The triangle inequality follows from (8.8) and the calculation $\|(S+T)\boldsymbol{x}\| \le \|S\boldsymbol{x}\| + \|T\boldsymbol{x}\| \le (\|S\| + \|T\|)\,\|\boldsymbol{x}\|$. Absolute homogeneity is an easy consequence of definition (8.6). □

8.2.2 Theorem. *If $\mathscr{Y}$ is complete, then $\mathscr{B}(\mathscr{X},\mathscr{Y})$ is complete in the norm $\|T\|$, that is, $\mathscr{B}(\mathscr{X},\mathscr{Y})$ is a Banach space.*

Proof. Let (T_n) be a Cauchy sequence in $\mathscr{B}(\mathscr{X},\mathscr{Y})$. In particular, $M := \sup_n \|T_n\| < \infty$. Since $\|T_n\boldsymbol{x} - T_m\boldsymbol{x}\| \le \|T_n - T_m\|\,\|\boldsymbol{x}\|$, $(T_n\boldsymbol{x})$ is a Cauchy sequence in $\mathscr{Y}$. Thus there exists a unique element $T\boldsymbol{x} \in \mathscr{Y}$ such that $T_n\boldsymbol{x} \to T\boldsymbol{x}$. The mapping $T : \mathscr{X} \to \mathscr{Y}$ is obviously linear, and the calculation $\|T\boldsymbol{x}\| = \lim_n \|T_n\boldsymbol{x}\| \le M\,\|\boldsymbol{x}\|$ shows that T is bounded. To see that $\|T_n - T\| \to 0$, let $\varepsilon > 0$ and choose N so that $\|T_n - T_m\| \le \varepsilon$ for all $m, n \ge N$. For such m and n and all $\boldsymbol{x}$ with $\|\boldsymbol{x}\| \le 1$, $\|T_n\boldsymbol{x} - T_m\boldsymbol{x}\| \le \varepsilon$. Letting $m \to \infty$ yields $\|(T_n - T)\boldsymbol{x}\| \le \varepsilon$. Thus $\|T_n - T\| \le \varepsilon$ for all $n \ge N$. □

8.2.3 Examples.

(a) (*Convolution operator*). For a fixed $g \in L^1(\mathbb{R}^d)$, the mapping $T_g f := f * g$ (§6.1) is a bounded linear operator on L^1 with $\|T_g\| = \|g\|_1$. The inequality $\|T_g\|_1 \le \|g\|_1$ follows from 6.1.1. For the reverse inequality, choose an approximate identity (ϕ_n) in L^1 (6.1.2). Then $\lim_n \|g * \phi_n - g\|_1 = 0$, hence $\|T_g\| \ge \|T_g\phi_n\|_1 \to \|g\|_1$.

(b) (*Integral operator*). Let X be a compact Hausdorff space, μ a Radon probability measure on X, and $k : X \times X \to \mathbb{C}$ a continuous function. Define K on $C(X)$ by

$$(Kf)(x) = \int k(x,y)f(y)\,d\mu(y), \quad f \in C(X).$$

Then K is a bounded linear operator with $\|K\| \le \|k\|_\infty$.

(c) (*Multiplication operator*). Let $(X, \mathscr{F}, \mu)$ be a σ-finite measure space and $\phi \in L^\infty(\mu)$. For $1 \le p < \infty$, define a linear operator $M_\phi : L^p \to L^p$ by $M_\phi f = \phi f$. We claim that $\|M_\phi\|_p = \|\phi\|_\infty$. The inequality $\|M_\phi\|_p \le \|\phi\|_\infty$ follows from

$$\|M_\phi f\|_p = \left(\int |\phi f|^p\right)^{1/p} \le \|\phi\|_\infty \left(\int |f|^p\right)^{1/p} = \|\phi\|_\infty \|f\|_p.$$

For the reverse inequality, recall that $\|\phi\|_\infty = \inf\{t > 0 : |\phi| \le t \text{ a.e.}\}$ (4.1.6). Thus for $0 < r < 1$, the set on which $|\phi| \ge r\,\|\phi\|_\infty$ has positive measure. Since X is σ-finite, the inequality holds on some set E of positive *finite* measure. Since $f := \mu(E)^{-1/p}\mathbf{1}_E$ has L^p norm equal to one, $\|M_\phi\|^p \ge \|f\phi\|_p^p = \mu(E)^{-1}\int_E |\phi|^p\,du \ge (r\,\|\phi\|_\infty)^p$. Letting $r \to 1$ shows that $\|M_\phi\| \ge \|\phi\|_\infty$. ◊

The Banach Algebra $\mathscr{B}(\mathscr{X})$

We use the notation $\mathscr{B}(\mathscr{X})$ for $\mathscr{B}(\mathscr{X},\mathscr{X})$. Note that operator composition in $\mathscr{B}(\mathscr{X})$ satisfies

$$(RS)T = R(ST) \quad R(aS+bT) = aRS + bRT, \quad \text{and} \quad (aS+bT)R = aSR + bTR.$$

Thus $\mathscr{B}(\mathscr{X})$ is an algebra under composition, with identity the identity operator I. Moreover, one has

$$\|ST\| \le \|S\|\,\|T\|, \quad T, S \in \mathscr{B}(\mathscr{X}). \tag{8.9}$$

(Ex. 8.29.) Thus

8.2.4 Proposition. *$\mathscr{B}(\mathscr{X})$ is a normed algebra under operator composition and a Banach algebra if $\mathscr{X}$ is complete.*

The Dual Space $\mathscr{X}'$

The space $\mathscr{B}(\mathscr{X},\mathbb{K})$ is called the **dual space of** $\mathscr{X}$ and is denoted by $\mathscr{X}'$. If $f \in \mathscr{X}'$ and $\boldsymbol{x} \in \mathscr{X}$, the value $f(\boldsymbol{x})$ is frequently written $\langle \boldsymbol{x}, f\rangle$, depending on notational requirements. Thus for a, $b \in \mathbb{K}$, $\boldsymbol{x}, \boldsymbol{y} \in \mathscr{X}$ and $f, g \in \mathscr{X}'$ we have the **duality relations**

$$\langle a\boldsymbol{x} + b\boldsymbol{y}, f\rangle = \langle a\boldsymbol{x}, f\rangle + \langle b\boldsymbol{y}, f\rangle \quad \text{and} \quad \langle \boldsymbol{x}, af + bg\rangle = \langle \boldsymbol{x}, af\rangle + \langle \boldsymbol{x}, bg\rangle.$$

By Theorem 8.2.2, $\mathscr{X}'$ is a Banach space under the norm

$$\|f\| = \sup\{|\langle \boldsymbol{x}, f\rangle| : \|\boldsymbol{x}\| \le 1\}.$$

Moreover, we have the inequality

$$|\langle \boldsymbol{x}, f\rangle| \le \|\boldsymbol{x}\|\,\|f\| \quad \boldsymbol{x} \in \mathscr{X}, \quad f \in \mathscr{X}'.$$

We denote the closed unit ball in $\mathscr{X}'$ by C_1':

$$C_1' := \{\boldsymbol{x}' \in \mathscr{X} : \|\boldsymbol{x}'\| \le 1\}.$$

Bilinear Transformations

Let $\mathscr{X}$, $\mathscr{Y}$, and $\mathscr{Z}$ be normed spaces over $\mathbb{K}$. A mapping $B : \mathscr{X} \times \mathscr{Y} \to \mathscr{Z}$ is said to be **bilinear** if $B(\boldsymbol{x}, \boldsymbol{y})$ is linear in $\boldsymbol{x}$ for each fixed $\boldsymbol{y}$ and linear in $\boldsymbol{y}$ for each fixed $\boldsymbol{x}$. B is said to be **bounded** if for some $M > 0$

$$\|B(\boldsymbol{x}, \boldsymbol{y})\| \le M\,\|\boldsymbol{x}\|\,\|\boldsymbol{y}\| \quad \text{for all } \boldsymbol{x} \in \mathscr{X} \text{ and } \boldsymbol{y} \in \mathscr{Y}. \tag{8.10}$$

The set $\mathscr{B}\mathscr{I}(\mathscr{X} \times \mathscr{Y}, \mathscr{Z})$ of all bounded bilinear mappings is easily seen to be a vector space under pointwise addition scalar multiplication. Defining $\|B\|$ to be the infimum of the constants M in (8.10), we have

8.2.5 Theorem. *$\mathscr{B}\mathscr{I}(\mathscr{X} \times \mathscr{Y}, \mathscr{Z})$ is a normed space and*

$$\|B\| = \sup\{\|B(\boldsymbol{x}, \boldsymbol{y})\| : \|\boldsymbol{x}\| \le 1, \ \|\boldsymbol{y}\| \le 1\}. \tag{8.11}$$

Moreover, if $\mathscr{Z}$ is complete, then $\mathscr{B}\mathscr{I}(\mathscr{X} \times \mathscr{Y}, \mathscr{Z})$ is complete.

Proof. The proof is similar to that of (8.2.2). For example, to verify (8.11) let s denote the supremum and let M be as in (8.10). If $\|\boldsymbol{x}\|$, $\|\boldsymbol{y}\| \le 1$, then $\|B(\boldsymbol{x}, \boldsymbol{y})\| \le M$, hence $s \le M$. Taking the infimum of the M's yields $s \le \|B\|$. Since $\left\|B(\|\boldsymbol{x}\|^{-1}\boldsymbol{x}, \|\boldsymbol{y}\|^{-1}\boldsymbol{y})\right\| \le s$ ($\boldsymbol{x}$, $\boldsymbol{y} \ne \boldsymbol{0}$), we have $\|B(\boldsymbol{x}, \boldsymbol{y})\| \le s\,\|\boldsymbol{x}\|\,\|\boldsymbol{y}\|$ and so $\|B\| \le s$ by (8.10). □

A bilinear transformation $B : \mathscr{X} \times \mathscr{X} \to \mathbb{K}$ is called a **bilinear form** on $\mathscr{X}$. For example, for f, $g \in \mathscr{X}'$, the mapping $f \otimes g : \mathscr{X} \times \mathscr{X} \to \mathbb{K}$ defined by $(f \otimes g)(\boldsymbol{x}, \boldsymbol{y}) = f(\boldsymbol{x})g(\boldsymbol{y})$ is a bilinear form with $\|f \otimes g\| = \|f\|\,\|g\|$. The mapping $(f, g) \to \int fg\,d\mu$ is a bilinear form on $L^2(\mu)$ with norm $\le \|f\|_2\,\|g\|_2$.

Exercises

8.26 Show that every infinite dimensional normed linear space has an unbounded operator. ⟦Extend a linearly independent sequence (x_n) to a basis and start by defining Tx_n. ⟧

8.27 Show that $\|T\| = \sup\{\|Tx\| : \|x\| = 1\}$.

8.28 Let $\mathscr{X}$ and $\mathscr{Y}$ be normed spaces and $T : \mathscr{X} \to \mathscr{Y}$ linear. Show that if $T(S_r)$ is bounded for some $r > 0$, then $T \in \mathscr{B}(\mathscr{X}, \mathscr{Y})$.

8.29 Prove (8.9).

8.30 Let $\mathscr{X}$ and $\mathscr{Y}$ be normed spaces over $\mathbb{K}$ and $T : \mathscr{X} \to \mathscr{Y}$ linear. Prove that T is bounded iff $\sum_{n=1}^{\infty} Tx_n$ converges for all sequences (x_n) for which $\sum_{n=1}^{\infty} x_n$ converges absolutely.

8.31 Let $\mathscr{X}$ and $\mathscr{Y}$ be normed spaces with $\mathscr{X}$ finite dimensional. Show that every linear transformation $T : \mathscr{X} \to \mathscr{Y}$ is continuous.

8.32 [↑3.2.16] (*Translation operator*). For $f : \mathbb{R}^d \to \mathbb{C}$, define $(T_x f)(y) = f(x + y)$. Show that T_x is an isometric isomorphism on both $C_b(\mathbb{R}^d)$ and $L^p(\lambda^d)$ $(1 \le p \le \infty)$. If we consider T_x on $C_b[0, \infty)$ and take $x > 0$, then $\|T_x\| = 1$ but T_x is not an isometry.

8.33 [↑3.2.16] (*Dilation operator*). For $f : \mathbb{R}^d :\to \mathbb{R}$, define $(D_r f)(x) = f(rx)$, $r \neq 0$. Show that D_r is an isometric isomorphism on $C_b(\mathbb{R}^d)$ and $|r|^{d/p} D_r$ is an isometric isomorphism on $L^p(\lambda^d)$.

8.34 (*Left and right shift operators*). Define T_ℓ and T_r on sequences $x = (x_n)$ by

$$T_\ell(x_1, x_2, \ldots) = (x_2, x_3, \ldots) \quad \text{and} \quad T_r(x_1, x_2, \ldots) = (0, x_1, x_2, \ldots).$$

Clearly $T_\ell \mathscr{X} \subseteq \mathscr{X}$ and $T_r \mathscr{X} \subseteq \mathscr{X}$ for the spaces ℓ^p, $\mathfrak{c}_0, \mathfrak{c}$. Show that in each case the operators have norm one and that T_r is an isometry but T_ℓ is not. Show also that $T_\ell T_r = I \neq T_r T_\ell$.

8.35 (*Evaluation functional*). For $x \in [0, 1]$ define the linear functional $\widehat{x}(f) = f(x)$, $f \in C[0, 1]$. Show that $\widehat{x}$ is continuous on $C[0, 1]$ in the uniform norm but not in the L^1 norm.

8.36 Let $\mathscr{P}[0, 1]$ denote the space of all polynomials on $[0, 1]$. Show that the derivative operator $Df = f'$ on $\mathscr{P}[0, 1]$ is unbounded in both the uniform norm and the L^1 norm.

8.37 Let $\mathscr{X}$ and $\mathscr{Y}$ be normed linear spaces such that $\mathscr{B}(\mathscr{X}, \mathscr{Y})$ is complete. Show that if $\mathscr{X}' \neq \{0\}$, then $\mathscr{Y}$ is complete.

8.38 Let $\mathscr{X}$ and $\mathscr{Z}$ be Banach spaces, $\mathscr{Y}$ a dense subspace of $\mathscr{X}$, and $T \in \mathscr{B}(\mathscr{Y}, \mathscr{Z})$. Show that T extends uniquely to a member of $\mathscr{B}(\mathscr{X}, \mathscr{Z})$ with the same norm.

8.39 Fix $a \in \ell^\infty$ and define $T : \ell^\infty \to \ell^\infty$ by $T(x) = (a_1 x_1, a_2 x_2, \ldots)$. (a) Find $\|T\|$. (b) Show that $\operatorname{ran}(T)$ need not be closed. (c) If T is 1-1, show that T^{-1} may not be bounded on $\operatorname{ran} T$.

8.40 For $x \in \mathfrak{c}$ define $L(x) = \lim_n x_n$. Show that $L \in \mathfrak{c}'$ and $\|L\| = 1$.

8.41 Let μ and ν be σ-finite measures on a measurable space $(X, \mathscr{F})$ such that $\mu \ll \nu$ and set $\varphi = d\mu/d\nu$. Show that the mapping $Tf = \varphi^{1/p} f$ is a linear isometry from $L^p(\mu)$ to $L^p(\nu)$. Show that T is surjective iff $\nu \ll \mu$.

8.42 Let $(X, \mathscr{F}, \mu)$ be a σ-finite measure space and $k : X \times X \to \mathbb{C}$ measurable such that the functions $F(x) := \int |k(x, y)|\, d\mu(y)$ and $G(y) := \int |k(x, y)|\, d\mu(x)$ are in L^∞. Let $1 < p < \infty$ and let q be conjugate to p. Show that the integral operator

$$Kf(x) := \int k(x, y) f(y)\, d\mu(y), \quad f \in L^p(\mu),$$

is bounded with norm $\le \|F\|_\infty^{1/p} \|G\|_\infty^{1/q}$.

8.43 Let $\mathscr{X}$ be strictly convex and $P \in \mathscr{B}(\mathscr{X})$ such that $P^2 = P$ and $\|P\| \le 1$. Suppose that for each $x \in \mathscr{X}$ there exists $T_x \in \mathscr{B}(\mathscr{X})$ with $\|T_x\| \le 1$ such that $T_x P x = x$. Show that P is the identity operator.

8.3 Concrete Representations of Dual Spaces

Dual spaces play a fundamental role in functional analysis and appear frequently in the development of the subject in the book. In this section we give several examples of concrete representations of dual spaces. Later sections treat the properties of general dual spaces.

The arguments in the first two examples make frequent use of the signum function $\operatorname{sgn} z$ defined in Chapter 0. We shall also need the special sequences

$$\boldsymbol{e}_n := (0, \ldots, 0, \overset{n}{1}, 0, \ldots) \quad \text{and} \quad \boldsymbol{e} := (1, 1, \ldots).$$

The Dual of $\mathfrak{c}_0$ is ℓ^1

For $\boldsymbol{x} = (x_1, x_2, \ldots) \in \ell^1$, define a linear map f_x by

$$f_x(\boldsymbol{y}) := \sum_{n=1}^{\infty} x_n y_n, \quad \boldsymbol{y} := (y_1, y_2, \ldots) \in \mathfrak{c}_0.$$

We show that the mapping $\boldsymbol{x} \to f_x$ is an isometric isomorphism of ℓ^1 onto $\mathfrak{c}_0'$ with inverse $f \to \boldsymbol{x}_f$, where

$$\boldsymbol{x}_f := \big(f(\boldsymbol{e}_1), f(\boldsymbol{e}_2), \ldots\big).$$

Clearly, $f_x \in \mathfrak{c}_0'$ with $\|f_x\| \le \|\boldsymbol{x}\|_1$. Now let $f \in \mathfrak{c}_0'$ be arbitrary and set $y_j := \operatorname{sgn} f(\boldsymbol{e}_j)$. Then

$$\boldsymbol{y}^{(n)} := \sum_{j=1}^{n} y_j \boldsymbol{e}_j = (y_1, \ldots, y_n, 0, 0, \ldots) \in \mathfrak{c}_0 \quad \text{and} \quad \|\boldsymbol{y}^{(n)}\|_\infty \le 1,$$

hence $\|f\| \ge |\langle \boldsymbol{y}^{(n)}, f\rangle| = \left|\sum_{j=1}^n y_j f(\boldsymbol{e}_j)\right| = \sum_{j=1}^n |f(\boldsymbol{e}_j)|$, which shows that $\|f\| \ge \|\boldsymbol{x}_f\|_1$. Moreover, if $\boldsymbol{z} = (z_1, z_2, \ldots) \in \mathfrak{c}_0$ and $\boldsymbol{z}^{(n)} := (z_1, \ldots, z_n, 0, \ldots)$, then $\left\|\boldsymbol{z}^{(n)} - \boldsymbol{z}\right\|_\infty \to 0$ and so $f(\boldsymbol{z}) = \lim_n f(\boldsymbol{z}^{(n)}) = \lim_n \sum_{k=1}^n z_k f(\boldsymbol{e}_k) = f_{x_f}(\boldsymbol{z})$. Therefore, $f = f_{x_f}$.

The Dual of $\mathfrak{c}$ is ℓ^1

For $\boldsymbol{x} = (x_1, x_2, \ldots) \in \ell^1$ define a linear map f_x by

$$f_x(\boldsymbol{y}) := x_1 \lim_n y_n + \sum_{n=1}^{\infty} x_{n+1} y_n, \quad \boldsymbol{y} := (y_1, y_2, \ldots) \in \mathfrak{c}$$

We show that the mapping $\boldsymbol{x} \to f_x$ is an isometric isomorphism of ℓ^1 onto $\mathfrak{c}'$ with inverse $f \to \boldsymbol{x}_f$, where

$$\boldsymbol{x}_f := \left(f(\boldsymbol{e}) - \sum_{n=1}^{\infty} f(\boldsymbol{e}_n), f(\boldsymbol{e}_1), f(\boldsymbol{e}_2), \ldots\right).$$

Clearly, $\boldsymbol{x} \to f_x$ is linear. Moreover, since $|\lim_n y_n| \le \|\boldsymbol{y}\|_\infty$, $\|f_x\| \le \|\boldsymbol{x}\|_1$. Now let $f \in \mathfrak{c}'$. As above, $\sum_{j=1}^{\infty} |f(\boldsymbol{e}_j)| < \infty$, hence $\boldsymbol{x}_f \in \ell^1$. Set

$$\boldsymbol{d}_n = (0, \ldots, \overset{n}{0}, 1, 1, \ldots) = \boldsymbol{e} - \sum_{j=1}^{n} \boldsymbol{e}_j \quad \text{and} \quad \boldsymbol{y}^{(n)} = \operatorname{sgn}(f(\boldsymbol{d}_n))\boldsymbol{d}_n + \sum_{j=1}^{n} \operatorname{sgn}(f(\boldsymbol{e}_j))\boldsymbol{e}_j.$$

Then $\boldsymbol{y}^{(n)} \in \mathfrak{c}$ and $\|\boldsymbol{y}^{(n)}\| \le 1$, and since $f(\boldsymbol{d}_n) \to f(\boldsymbol{e}) - \sum_{n=1}^{\infty} f(\boldsymbol{e}_n)$ we have

$$\|f\| \ge |\langle \boldsymbol{y}^{(n)}, f\rangle| = |f(\boldsymbol{d}_n)| + \sum_{j=1}^{n} |f(\boldsymbol{e}_j)| \to \|\boldsymbol{x}_f\|_1.$$

Finally, if $\boldsymbol{z} = (z_1, z_2, \ldots) \in \mathfrak{c}$, $\alpha := \lim_n z_n$, and $\boldsymbol{z}^{(n)} := (z_1, \ldots, z_n, \alpha, \alpha, \ldots)$, then $\left\|\boldsymbol{z}^{(n)} - \boldsymbol{z}\right\|_\infty \to 0$, hence

$$f(\boldsymbol{z}) = \lim_n f(\boldsymbol{z}^{(n)}) = \lim_n \left(\alpha f(\boldsymbol{d}_n) + \sum_{k=1}^{n} z_k f(\boldsymbol{e}_k)\right) = f_{x_f}(\boldsymbol{z}).$$

Therefore, $f = f_{x_f}$, completing the proof.

The Dual of L^p is L^q

Let $(X, \mathscr{F}, \mu)$ be a σ-finite measure space, $1 \le p < \infty$, and let q be conjugate to p. For $g \in L^q$ define φ_g on L^p by

$$\varphi_g(f) = \int fg\, d\mu, \quad f \in L^p.$$

We show that the mapping $g \mapsto \varphi_g$ is an isometric isomorphism from L^q onto $(L^p)'$.

Clearly, φ_g is linear, and by Hölder's inequality $\|\varphi_g\| \le \|g\|_q$. Now let $\varphi \in (L^p)'$ be arbitrary. It remains to find a $g \in L^q$ such that $\varphi = \varphi_g$ and $\|\varphi_g\| \ge \|g\|_q$.

Suppose that the existence of g has been established for finite measure spaces. For the σ-finite case, let $X_n \uparrow X$ with $0 < \mu(X_n) < \infty$ for all n. Consider the restriction φ_n of φ to the subspace L^p_n of L^p consisting of those functions that vanish outside X_n. We identify L^p_n with $L^p(X_n, \mathcal{F} \cap X_n, \mu)$ and make the analogous identification for L^q_n. For each n choose a function g_n in L^q_n such that

$$\varphi_n(f) = \int f g_n\, d\mu, \quad \varphi_n := \varphi\big|_{L^p_n}, \quad f \in L^p_n, \quad \|g_n\|_q = \|\varphi_n\| \le \|\varphi\|.$$

Since $L^p_n \subseteq L^p_{n+1}$, $g_{n+1} = g_n$ a.e. on X_n, hence we may define a measurable function g on X such that $g = g_n$ on X_n. Since $|g_n| \le |g_{n+1}|$ on X_{n+1}, $\|g_n\|_q \to \|g\|_q$ by the monotone convergence theorem, hence $\|g\|_q \le \|\varphi\|$. Furthermore, if $f \in L^p(\mu)$, then $\|f\mathbf{1}_{X_n} \to f\|_p$ by the dominated convergence theorem and so

$$\varphi(f) = \lim_n \varphi(f\mathbf{1}_{X_n}) = \lim_n \int f g_n\, d\mu = \lim_n \int f\mathbf{1}_{X_n} g\, d\mu = \int fg\, d\mu = \varphi_g(f).$$

Thus if the assertion holds for the finite case then it holds for the σ-finite case.

We now establish the existence of g for the case $\mu(X) < \infty$. To this end, define a set function ν on $\mathcal{F}$ by $\nu(E) = \langle \mathbf{1}_E, \varphi\rangle$. Then ν is countably additive. Indeed, if (E_n) is a disjoint sequence in $\mathcal{F}$ with union E, then

$$\int \left|\mathbf{1}_E - \sum_{j=1}^{n} \mathbf{1}_{E_j}\right|^p d\mu = \int \sum_{j>n} \mathbf{1}_{E_j}\, d\mu = \sum_{j>n} \mu(E_j) \to 0,$$

that is, $\sum_{j=1}^{n} \mathbf{1}_{E_j} \xrightarrow{L^p} \mathbf{1}_E$. Countable additivity now follows from the linearity and continuity of φ.

Next, observe that the inequality $|\nu(E)| \le \|\varphi\| \, \|\mathbf{1}_E\|_p$ implies that $\nu \ll \mu$. Thus, by the

Radon-Nikodym theorem, there exists a function $g \in L^1(\mu)$ such that $\langle \mathbf{1}_E, \varphi\rangle = \int_E g\, d\mu$ for all $E \in \mathcal{F}$. In particular, for all simple functions f, we have

$$\varphi(f) = \int fg\, d\mu. \tag{a}$$

We claim that $fg \in L^1$ and that (a) holds for all $f \in L^p(\mu)$. Define $f_n = \mathbf{1}_{E_n} f$, where $E_n = \{|f| \le n\}$. For each n choose a sequence of simple functions $(f_{n,k})_k$ such that $f_{n,k} \stackrel{a.e.}{\to} f_n$ and $|f_{n,k}| \le |f_n|$ for all k (2.3.1). By the dominated convergence theorem, $\lim_k \|f_{n,k} - f_n\|_p \to 0$, hence from (a)

$$\varphi(f_n) = \lim_k \varphi(f_{n,k}) = \lim_k \int f_{n,k} g = \int f_n g, \tag{b}$$

the last equality by the dominated convergence theorem, since $|f_{n,k} g| \le |ng|$. Now set $|g| = e^{i\theta} g$. Replacing f by $e^{i\theta}|f|$ in the above, we have

$$\|\varphi\|\,\|f\|_p \ge \|\varphi\|\,\|f_n\|_p = \|\varphi\|\,\|e^{i\theta}|f_n|\|_p \ge |\varphi(e^{i\theta}|f_n|)| = \left|\int g e^{i\theta}|f_n|\, d\mu\right| = \int |g f_n|\, d\mu.$$

Letting $n \to \infty$ and applying Fatou's lemma yields

$$\|\varphi\|\,\|f\|_p \ge \int |gf|\, d\mu, \tag{c}$$

hence $fg \in L^1$. Moreover, since $\|f - f_n\|_p^p = \int_{|f|>n} |f|\, d\mu \to 0$ we have $\varphi(f_n) \to \varphi(f)$. Using the dominated convergence theorem in (b), we see that (a) holds for all $f \in L^p(\mu)$.

We now show that $g \in L^q$ and that $\|g\|_q \le \|\varphi\|$, completing the argument. Suppose first that $q < \infty$. Define $g_n = g$ if $|g| \le n$ and $g_n = 0$ otherwise, so that

$$g_n \to g, \quad |g_n| \le |g|, \quad \text{and} \quad \|g\|_q \le \varliminf_n \|g_n\|_q, \tag{d}$$

the last by Fatou's lemma. By (c), $\|\varphi\|\,\|f\|_p \ge \int |gf|\, d\mu \ge \int |g_n f|\, d\mu$ ($f \in L^p$). Taking $f = |g_n|^{q/p}$ and applying Ex. 4.5 we have

$$\|\varphi\| \left(\int |g_n|^q\right)^{1/p} \ge \int |g_n| \cdot |g_n|^{q/p} = \left(\int |g_n|^q\right)^{1/p} \left(\int |g_n|^q\right)^{1/q},$$

hence $\|\varphi\| \ge \|g_n\|_q$. Therefore, by ($d$), $\|\varphi\| \ge \|g\|_q$.

Now suppose $q = \infty$. Set $A := \{|g| > \|\varphi\| + \varepsilon\}$ ($\varepsilon > 0$) and suppose that $\mu(A) > 0$. Define $f(x) = \mu(A)^{-1} \operatorname{sgn} g(x) \mathbf{1}_A(x)$. Then $\|f\|_1 = 1$ and $\int fg\, d\mu = \mu(A)^{-1} \int_A |g|\, d\mu > \|\varphi\| + \varepsilon$, contradicting ($d$). Therefore, $|g| \le \|\varphi\| + \varepsilon$ a.e. and so $\|g\|_\infty \le \|\varphi\|$, as required. ◊

The Dual of $C_0(X)$ is $M_{ra}(X)$

Let X be a locally compact Hausdorff space. For $\mu \in M_{ra}(X)$ define

$$\varphi_\mu(f) = \int f\, d\mu, \quad f \in C_0(X).$$

We show that the mapping $\mu \to \varphi_\mu$ is an isometry from $M_{ra}(X)$ onto $C_0(X)'$. This result is known as the *Riesz representation theorem.*

Clearly, φ_μ is linear and $|\varphi_\mu(f)| \le \|\mu\|\,\|f\|_\infty$, hence $\varphi_\mu \in C_0(X)'$ and $\|\varphi_\mu\| \le \|\mu\|$. To show equality, let $\mu = e^{i\theta}|\mu|$ be the polar decomposition of μ (5.3.6(b)). Since $|\mu|$ is a Radon measure (7.1.5), by Lusin's theorem (7.1.3) given $\varepsilon > 0$ there exists $g \in C_c(X)$ such that $|g| \le 1$ and $g = e^{-i\theta}$ on a set E with $|\mu|(E^c) < \varepsilon/2$. Then

$$\|\mu\| = \int e^{-i\theta}\,d\mu \le \left|\int g\,d\mu\right| + \left|\int_{E^c} (e^{-i\theta} - g)\,d\mu\right| \le |\varphi_\mu(g)| + 2|\mu|(E^c) \le \|\varphi_\mu\| + \varepsilon,$$

hence $\|\mu\| \le \|\varphi_\mu\|$.

It remains to show that if $\varphi \in C_0(X)'$ then $\varphi = \varphi_\mu$ for some $\mu \in M_{ra}(X)$. Assume first that φ is real-valued on the real linear space $C_0(X, \mathbb{R})$. Define φ^+ on $C_0\big(X, [0, \infty)\big)$ by

$$\varphi^+(f) := \sup\{\varphi(g) : g \in C_0(X, \mathbb{R}),\ \ 0 \le g \le f\}.$$

Clearly $\varphi^+(f) \ge 0$ and $\varphi^+(cf) = c\varphi^+(f)$ for $c \ge 0$. We claim that

$$\varphi^+(f_1 + f_2) = \varphi^+(f_1) + \varphi^+(f_2). \tag{$\dagger$}$$

Let $0 \le g_j \le f_j$. Then $0 \le g_1 + g_2 \le f_1 + f_2$, hence $\varphi(g_1) + \varphi(g_2) \le \varphi^+(f_1 + f_2)$. Taking suprema over g_1 and g_2 yields $\varphi^+(f_1) + \varphi^+(f_2) \le \varphi^+(f_1 + f_2)$. For the reverse inequality, let $g \le f_1 + f_2$ and set $g_1 = g \wedge f_1$ and $g_2 = g - g_1$. Then $0 \le g_i \le f_i$, hence $\varphi(g) = \varphi(g_1 + g_2) \le \varphi^+(f_1) + \varphi^+(f_2)$. Taking the supremum over g yields $\varphi(f_1 + f_2) \le \varphi^+(f_1) + \varphi^+(f_2)$.

Next, extend the definition of φ^+ to $C_0(X, \mathbb{R})$ by defining $\varphi^+(f) = \varphi^+(f^+) - \varphi^+(f^-)$. Using ($\dagger$) one shows by an argument entirely similar to the first part of the proof of 3.2.12 that φ^+ is linear on $C_0(X, \mathbb{R})$. Defining $\varphi^- = \varphi^+ - \varphi$ we now have $\varphi = \varphi^+ - \varphi^-$, where $\varphi^\pm$ are positive linear functionals on $C_0(X, \mathbb{R})$.

By 7.2.1, there exist unique Radon measures $\mu^\pm$ such that

$$\varphi^+(f) = \int f\,d\mu^+ \quad \text{and} \quad \varphi^+(f) = \int f\,d\mu^- \ \text{ for all } f \in C_c(X, \mathbb{R}).$$

Then $\mu^+ - \mu^-$ is a signed Radon measure on X such that

$$\varphi(f) = \varphi^+(f) - \varphi^-(f) = \int f\,d(\mu^+ - \mu^-), \quad f \in C_c(X, \mathbb{R}).$$

For a complex-valued φ, apply this result to $\operatorname{Re}\varphi$ and $\operatorname{Im}\varphi$ to obtain Radon measures μ_r, μ_i such that for all $f \in C_c(X, \mathbb{R})$,

$$\varphi(f) = \operatorname{Re}\varphi(f) + i\operatorname{Im}\varphi(f) = \int f\,d\mu_r + i\int f\,d\mu_i = \int f\,d(\mu_r + i\mu_i). \tag{$\ddagger$}$$

By considering real and imaginary parts of $f \in C_c(X)$ we see that ($\ddagger$) holds for $f \in C_c(X)$. Since $C_c(X)$ is dense in $C_0(X)$, ($\ddagger$) holds for all $f \in C_0(X)$. Therefore, $\varphi = \varphi_\mu$ with $\mu = \mu_r + i\mu_i$. ◊

Exercises

8.44 Show that the dual of $\mathfrak{c}_{00}$ is ℓ^1.

8.45 Let ba($\mathbb{N}$) denote the linear space of finitely additive, complex set functions μ on $\mathbb{N}$ with the total variation norm $\|\mu\|$. (The latter is defined exactly as in the case of complex measures.) If $g = \sum_{j=1}^n a_j \mathbf{1}_{E_j}$ is a simple function in standard form, then $\int g\,d\mu$ may be defined as in § 3.1. Moreover, (a) and (b) of 3.1.1 (linearity) hold since only finite additivity is used in the proof. Verify the following to show that the dual of $\ell^\infty(\mathbb{N})$ is ba($\mathbb{N}$).

(a) For $\mu \in$ ba($\mathbb{N}$), $\varphi_\mu(g) := \int g\,d\mu$ is a bounded linear functional on the subspace of ℓ^∞

consisting of simple functions. Therefore, φ_μ extends to a bounded linear functional on ℓ^∞ such that $|\varphi_\mu(f)| \le \|\mu\| \, \|f\|_\infty$.

(b) $\|\varphi_\mu\| = \|\mu\|$ ⟦Let $E_1, \ldots E_n$ be a partition of $\mathbb{N}$ with $\sum_{j=1}^n |\mu(E_j)| > \|\mu\| - \varepsilon$.⟧

(c) The mapping $\mu \to \varphi_\mu$ is an isometric isomorphism from $\mathrm{ba}(\mathbb{N})$ onto the dual of ℓ^∞.

8.46 Give the space $C^k[0,1]$ of k-times continuously differentiable functions on $[0,1]$ the norm

$$|||\, f \,||| = \sum_{j=0}^{k} \|f^{(j)}\|_\infty.$$

(a) Show that $C^k[0,1]$ is a Banach space.

(b) Show that for any $a \in [0,1]$, an equivalent norm is $\sum_{j=0}^{k-1} |f^{(j)}(a)| + \|f^{(k)}\|_\infty$.

(c) Show that the dual of $C^k[0,1]$ consists of all function of the form

$$\varphi_{a,\mu}(f) := a \cdot \vec{f}(a) + \int f^{(k)}\, d\mu,$$

where μ is a complex Radon measure on $[0,1]$, $a = (a_0, \ldots, a_{k-1}) \in \mathbb{R}^k$, and $\vec{f}(a) = (f(a), f'(a), \ldots, f^{(k-1)}(a))$.

(d) Show that the mapping $S : (a, \mu) \to \varphi_{a,\mu}$ is a topological isomorphism from the product space $\mathbb{K}^k \times M_{ra}[0,1]$ onto the dual of $C^k[0,1]$.

8.4 Some Constructions

In this section we describe several standard ways of constructing new normed spaces from given spaces.

Product Spaces

Let $\mathscr{X}$ and $\mathscr{Y}$ be normed linear spaces over $\mathbb{K}$. The **product vector space** is the set $\mathscr{X} \times \mathscr{Y}$ together with the operations

$$(\boldsymbol{x}_1, \boldsymbol{y}_1) + (\boldsymbol{x}_2, \boldsymbol{y}_2) := (\boldsymbol{x}_1 + \boldsymbol{x}_2, \boldsymbol{y}_1 + \boldsymbol{y}_2), \quad c(\boldsymbol{x}, \boldsymbol{y}) := (c\boldsymbol{x}, c\boldsymbol{y}).$$

There is no canonical norm for $\mathscr{X} \times \mathscr{Y}$; however, the following equivalent norms are frequently used:

$$\|(\boldsymbol{x}, \boldsymbol{y})\|_1 := \|\boldsymbol{x}\| + \|\boldsymbol{y}\|, \quad \|(\boldsymbol{x}, \boldsymbol{y})\|_2 := \sqrt{\|\boldsymbol{x}\|^2 + \|\boldsymbol{y}\|^2}, \quad \|(\boldsymbol{x}, \boldsymbol{y})\|_\infty := \max\{\|\boldsymbol{x}\|, \|\boldsymbol{y}\|\}.$$

Each of these norms induces the product topology on $\mathscr{X} \times \mathscr{Y}$. More generally, we have the following result, which may be seen as a direct consequence of 8.1.3. The proof is left as an exercise (8.47).

8.4.1 Proposition. *All norms on $\mathscr{X} \times \mathscr{Y}$ that generate the product topology are equivalent.*

The **projection maps** $P_X : \mathscr{X} \times \mathscr{Y} \to \mathscr{X}$ and $P_Y : \mathscr{X} \times \mathscr{Y} \to \mathscr{Y}$ are defined by

$$P_X(\boldsymbol{x}, \boldsymbol{y}) = \boldsymbol{x} \quad \text{and} \quad P_Y(\boldsymbol{x}, \boldsymbol{y}) = \boldsymbol{y}.$$

These are clearly linear and continuous in the product topology. The straightforward proof of following proposition is left to the reader.

8.4.2 Proposition. *Let $\mathscr{X}$, $\mathscr{Z}$, and $\mathscr{Y}$ be normed spaces.*

(a) *$\mathscr{X} \times \mathscr{Y}$ is complete iff $\mathscr{X}$ and $\mathscr{Y}$ are complete.*

(b) *The projection mappings are open.*

(c) *If $T : \mathscr{Z} \to \mathscr{X} \times \mathscr{Y}$ is linear, then T is bounded iff $P_X T$ and $P_Y T$ are bounded.*

The preceding discussion may be generalized to a product $\mathscr{X}_1 \times \cdots \times \mathscr{X}_n$ of finitely many normed linear spaces. The analogs of 8.4.1 and 8.4.2 are easily seen to hold in this setting. The details are left to the reader.

Direct Sums

Let $\mathscr{Z}$ be a vector space over $\mathbb{K}$ with subspaces $\mathscr{X}$ and $\mathscr{Y}$. Then $\mathscr{Z}$ is said to be the **algebraic direct sum** of $\mathscr{X}$ and $\mathscr{Y}$ if the following conditions hold:

$$\mathscr{Z} = \mathscr{X} + \mathscr{Y} := \{\boldsymbol{x} + \boldsymbol{y} : \boldsymbol{x} \in \mathscr{X},\ \boldsymbol{y} \in \mathscr{Y}\} \text{ and } \mathscr{X} \cap \mathscr{Y} = \{0\}. \tag{8.12}$$

In this case we write

$$\mathscr{Z} = \mathscr{X} \oplus \mathscr{Y}.$$

Conditions (8.12) are equivalent to the property that every member of $\mathscr{Z}$ is uniquely expressible as a sum $\boldsymbol{x} + \boldsymbol{y}$, $\boldsymbol{x} \in \mathscr{X}$, $\boldsymbol{y} \in \mathscr{Y}$. This in turn is equivalent to $\mathscr{X} \times \mathscr{Y}$ being algebraically isomorphic to $\mathscr{Z}$ under the mapping

$$(\boldsymbol{x}, \boldsymbol{y}) \mapsto \boldsymbol{x} + \boldsymbol{y} : \mathscr{X} \times \mathscr{Y} \to \mathscr{Z}. \tag{8.13}$$

Projection mappings P_X and P_Y on $\mathscr{X} \oplus \mathscr{Y}$ are defined by analogy with product spaces:

$$P_X(\boldsymbol{x} + \boldsymbol{y}) = \boldsymbol{x} \text{ and } P_Y(\boldsymbol{x} + \boldsymbol{y}) = \boldsymbol{y}.$$

Uniqueness of representation implies that the mappings are well-defined. Moreover, the mappings are easily seen to be linear. The identities

$$P_X + P_Y = I, \quad P_X^2 = P_X, \quad P_Y^2 = P_Y, \text{ and } P_X P_Y = P_Y P_X = 0 \tag{8.14}$$

follow easily from the definitions. Furthermore,

$$\mathscr{X} = \operatorname{ran} P_X = \ker P_Y \text{ and } \mathscr{Y} = \operatorname{ran} P_Y = \ker P_X. \tag{8.15}$$

Conversely, if P_X and P_Y are linear mappings on $\mathscr{Z}$ that satisfy (8.14), then $\mathscr{Z} = \mathscr{X} \oplus \mathscr{Y}$, where $\mathscr{X} := \operatorname{ran} P_X$ and $\mathscr{Y} := \ker P_X$. Indeed, any $\boldsymbol{z} \in \mathscr{Z}$ may be written $\boldsymbol{z} = P\boldsymbol{z} + (\boldsymbol{z} - P\boldsymbol{z})$, where $P\boldsymbol{z} \in \mathscr{X}$ and $\boldsymbol{z} - P\boldsymbol{z} \in \mathscr{Y}$. Since $\boldsymbol{z} \in \mathscr{X} \cap \mathscr{Y} \Rightarrow \boldsymbol{z} = P\boldsymbol{z} = 0$, the assertion follows.

Now let $\mathscr{Z}$ be a normed space with subspaces $\mathscr{X}$ and $\mathscr{Y}$ and let $\mathscr{X} \times \mathscr{Y}$ have a norm that generates the product topology. The algebraic isomorphism $(\boldsymbol{x}, \boldsymbol{y}) \mapsto \boldsymbol{x} + \boldsymbol{y}$ is obviously continuous. If the inverse $\boldsymbol{x} + \boldsymbol{y} \mapsto (\boldsymbol{x}, \boldsymbol{y})$ is also continuous, that is, if $(\boldsymbol{x}, \boldsymbol{y}) \mapsto \boldsymbol{x} + \boldsymbol{y}$ is a *topological* isomorphism, then $\mathscr{Z}$ is said to be the **topological direct sum** of $\mathscr{Y}$ and $\mathscr{Z}$.

8.4.3 Proposition. *Let $\mathscr{Z}$ be a normed space that is the algebraic direct sum of subspaces $\mathscr{X}$ and $\mathscr{Y}$. Then $\mathscr{Z}$ is a topological direct sum iff P_X (equivalently P_Y) is continuous.*

Proof. The map P_X is the composition of the algebraic isomorphism $\boldsymbol{x} + \boldsymbol{y} \mapsto (\boldsymbol{x}, \boldsymbol{y})$ with the continuous map $(\boldsymbol{x}, \boldsymbol{y}) \mapsto \boldsymbol{x}$. Hence if $(\boldsymbol{x}, \boldsymbol{y}) \mapsto \boldsymbol{x} + \boldsymbol{y}$ is a topological isomorphism, then P_X is continuous. Conversely, if P_X is continuous, then so is $P_Y = I - P_X$, hence $\boldsymbol{x}_n + \boldsymbol{y}_n \to \boldsymbol{x} + \boldsymbol{y} \Rightarrow \boldsymbol{x}_n \to \boldsymbol{x}$ and $\boldsymbol{y}_n \to \boldsymbol{y} \Rightarrow (\boldsymbol{x}_n, \boldsymbol{y}_n) \to (\boldsymbol{x}, \boldsymbol{y})$. Therefore, $(\boldsymbol{x}, \boldsymbol{y}) \mapsto \boldsymbol{x} + \boldsymbol{y}$ is a topological isomorphism. □

The notion of a normed linear space $\mathscr{X}$ as an algebraic direct sum $\mathscr{X}_1 \oplus \cdots \oplus \mathscr{X}_n$ of subspaces $\mathscr{X}_j$ of $\mathscr{X}$ is defined analogously. The requirement here is that every member of $\mathscr{X}$ be uniquely expressible as a sum $x_1 + \cdots + x_n$. The associated projection mappings P_j, defined by $P_j(x_1 + \cdots + x_n) = x_j$, satisfy

$$\sum_{j=1}^{n} P_j = I, \quad P_j^2 = P_j \text{ and } P_i P_j = 0, \quad i \neq j.$$

Topological direct sums are defined as above by requiring that the map $x_1 + \cdots + x_n \mapsto (x_1, \cdots, x_n)$ be continuous. The proof of the following proposition is a straightforward modification of that of 8.4.3.

8.4.4 Proposition. *Let $\mathscr{X}$ be a normed space which is the algebraic direct sum of subspaces $\mathscr{X}_1, \ldots, \mathscr{X}_n$. Then $\mathscr{X}$ is the topological direct sum iff the projection mappings P_j are continuous.*

Quotient Spaces

Recall that if $\mathscr{Y}$ is a subspace of a linear space $\mathscr{X}$, then $\mathscr{X}/\mathscr{Y}$ is the linear space of all equivalence classes $x + \mathscr{Y}$ with the operations

$$(x_1 + \mathscr{Y}) + (x_2 + \mathscr{Y}) = (x_1 + x_2) + \mathscr{Y} \text{ and } c(x + \mathscr{Y}) = cx + \mathscr{Y}.$$

Relative to these operations the quotient map

$$Q : \mathscr{X} \to \mathscr{X}/\mathscr{Y}, \quad Qx := x + \mathscr{Y},$$

is linear with kernel $\mathscr{Y}$. We show in this subsection that if $\mathscr{X}$ is a normed space, then the quotient space has a natural norm, called the **quotient norm**, with respect to which Q is continuous and open.

8.4.5 Theorem. *Let $\mathscr{Y}$ be a closed linear subspace of a normed space $\mathscr{X}$. Then*

$$\|Qx\| = \|x + \mathscr{Y}\| := \inf\{\|x + y\| : y \in \mathscr{Y}\} \tag{8.16}$$

defines a norm on $\mathscr{X}/\mathscr{Y}$. Moreover, if $\mathscr{X}$ is complete, then so is $\mathscr{X}/\mathscr{Y}$.

Proof. Since $\mathbf{0} \in \mathscr{Y}$, $\|Q(\mathbf{0})\| = 0$. Let x, x_1, $x_2 \in \mathscr{X}$ and $c \in \mathbb{K}$. If $c \neq 0$, then

$$\|c(x + \mathscr{Y})\| = \inf\{\|cx + y\| : y \in \mathscr{Y}\} = |c| \inf\{\|x + c^{-1}y\| : y \in \mathscr{Y}\} = |c|\, \|x + \mathscr{Y}\|.$$

For the triangle inequality, note that for any $y_1, y_2 \in \mathscr{Y}$,

$$\|(x_1 + \mathscr{Y}) + (x_2 + \mathscr{Y})\| \leq \|x_1 + x_2 + y_1 + y_2\| \leq \|x_1 + y_1\| + \|x_2 + y_2\|.$$

Taking infima over y_1 and y_2 yields $\|(x_1 + \mathscr{Y}) + (x_2 + \mathscr{Y})\| \leq \|x_1 + \mathscr{Y})\| + \|x_2 + \mathscr{Y}\|$. For positivity, assume that $\|Qx\| = 0$. Then there exists a sequence (y_n) in $\mathscr{Y}$ such that $\|x + y_n\| \to 0$. Since $\mathscr{Y}$ is closed, $x \in \mathscr{Y}$, hence $x + \mathscr{Y} = \mathscr{Y}$, that is, $Qx = 0$. Therefore, (8.16) defines a norm.

Now assume that $\mathscr{X}$ is complete. To show that $\mathscr{X}/\mathscr{Y}$ is complete, we use 0.4.3: Let (x_n) be a sequence in $\mathscr{X}$ such that $\sum_{n=1}^{\infty} \|x_n + \mathscr{Y}\| < \infty$. For each n choose $y_n \in \mathscr{Y}$ such that $\|x_n + y_n\| < \|x_n + \mathscr{Y}\| + 1/2^n$. Then $\sum_{n=1}^{\infty} \|x_n + y_n\| < \infty$, so the sequence of partial sums $\sum_{j=1}^{n}(x_j + y_j)$ converges to some $x \in \mathscr{X}$. Since

$$\left\| \sum_{j=1}^{n} (x_j + \mathscr{Y}) - (x + \mathscr{Y}) \right\| \leq \left\| \sum_{j=1}^{n} (x_j + y_j) - x \right\|,$$

the series $\sum_{n=1}^{\infty}(x_n + \mathscr{Y})$ converges to $x + \mathscr{Y}$. □

8.4.6 Theorem. *Let $\mathscr{Y}$ be a closed linear subspace of a normed space $\mathscr{X}$ and let $\mathscr{X}/\mathscr{Y}$ have the quotient norm. Then the quotient map Q has the following properties:*

(a) *Q is a bounded linear operator. If $\mathscr{Y} \neq \mathscr{X}$, then $\|Q\| = 1$.*

(b) *Q is an open mapping.*

(c) *If $\mathscr{Z}$ is a normed space and $T : \mathscr{X}/\mathscr{Y} \to \mathscr{Z}$ a linear mapping such that TQ is bounded, then T is bounded.*

Proof. (a) By (8.16), $\|Q\boldsymbol{x}\| \le \|\boldsymbol{x}\|$, hence $\|Q\| \le 1$. If $\mathscr{Y} \neq \mathscr{X}$, then $\|Q\boldsymbol{x}\| = 1$ for some $\boldsymbol{x} \in \mathscr{X}$ and so for each $r > 1$ there exists $\boldsymbol{y}_r \in \mathscr{Y}$ such that $\|\boldsymbol{x} + \boldsymbol{y}_r\| < r$. Then $\|Q\|\,\|\boldsymbol{x} + \boldsymbol{y}_r\| \ge \|Q(\boldsymbol{x} + \boldsymbol{y}_r)\| = \|Q\boldsymbol{x}\| = 1$, hence $\|Q\| \ge 1/\|\boldsymbol{x} + \boldsymbol{y}_r\| > 1/r$. Since r was arbitrary, $\|Q\| \ge 1$.

(b) Note first that $Q\big(B_r(\mathbf{0})\big) = B_r\big(Q(\mathbf{0})\big)$ (Ex. 8.57). Since an open set is a union of open balls and since translations of open balls are open balls, it follows that Q is open.

(c) If TQ is continuous and V is open in $\mathscr{Z}$, then $U := Q^{-1}\big(T^{-1}(V)\big)$ is open in $\mathscr{X}/\mathscr{Y}$ and so $T^{-1}(V) = Q(U)$ is open in $\mathscr{X}/\mathscr{Y}$ by (b). □

The machinery of quotient spaces allows a simple proof of the following result.

8.4.7 Proposition. *Let $\mathscr{X}$ be a normed space, $\mathscr{Y}$ a closed subspace of $\mathscr{X}$, and $\mathscr{F}$ a finite dimensional subspace of $\mathscr{X}$. Then $\mathscr{Y} + \mathscr{F}$ is a closed subspace of $\mathscr{X}$.*

Proof. Since $Q(\mathscr{F})$ is a finite dimensional subspace of $\mathscr{X}/\mathscr{Y}$, it is closed. Therefore, $\mathscr{Y} + \mathscr{F} = Q^{-1}\big(Q(\mathscr{F})\big)$ is closed. □

Exercises

8.47 Prove 8.4.1.

8.48 Prove 8.4.2.

8.49 Prove that $\mathfrak{c}$ is the topological direct sum $\mathbb{K}\boldsymbol{e} \oplus \mathfrak{c}_0$, where $\boldsymbol{e} = (1, 1, \ldots)$.

8.50 Let $(\mathscr{X}_n)$ be a sequence of normed spaces. The **product vector space** $\mathscr{X} := \prod_n \mathscr{X}_n$ is the collection of all sequences $(\boldsymbol{x}_1, \boldsymbol{x}_2, \ldots)$ $(\boldsymbol{x}_n \in \mathscr{X}_n)$, with coordinate-wise addition and scalar multiplication. Show that there exists a norm on $\mathscr{X}$ that induces the product topology of $\mathscr{X}$ iff $\mathscr{X}_n = \{0\}$ for all sufficiently large n.

8.51 [↑ 8.50] Let $(\mathscr{X}_n)$ be a sequence of normed spaces and let $\prod_n \mathscr{X}_n$ have the product vector space structure. For $\boldsymbol{x} = (\boldsymbol{x}_1, \boldsymbol{x}_2, \ldots)$ define $\|\boldsymbol{x}\|_\infty := \sup_n \|\boldsymbol{x}_n\|$. Show that $\mathscr{X} := \{\boldsymbol{x} : \|\boldsymbol{x}\| < \infty\}$ is a normed space under $\|\cdot\|_\infty$. Show also that $\mathscr{X}$ is complete iff each space $\mathscr{X}_n$ is complete.

8.52 Let $\mathscr{Z}$ be a linear space with $\mathscr{Z} = \mathscr{X} \oplus \mathscr{Y}$ and let $T : \mathscr{Z} \to \mathscr{Z}$ be linear. Prove that $TP_X = P_X T$ iff $T\mathscr{X} \subseteq \mathscr{X}$ and $T\mathscr{Y} \subseteq \mathscr{Y}$.

8.53 Let $\mathscr{Z}$ be a normed space that is a topological direct sum of closed subspaces $\mathscr{X}$ and $\mathscr{Y}$. Show that $\mathscr{Z}/\mathscr{X}$ is topologically isomorphic to $\mathscr{Y}$.

8.54 Let $\mathscr{X}$ denote any of the sequence spaces $\mathfrak{c}_0$, $\mathfrak{c}$, ℓ^1, or ℓ^∞. Show that $\mathscr{X} = \mathscr{X}_1 \oplus \mathscr{X}_2$, where the summands are closed subspaces of $\mathscr{X}$ isometrically isomorphic to $\mathscr{X}$.

8.55 Let $\mathscr{X}$ be a normed space, $\mathscr{Y}$ a complete (hence closed) subspace of $\mathscr{X}$. Show that if $\mathscr{X}/\mathscr{Y}$ is complete, then $\mathscr{X}$ is complete.

8.56 Let $\|\cdot\|$ be a seminorm on a linear space $\mathscr{X}$. The notions of sequential convergence and Cauchy sequence still make sense in this setting, except that limits, if they exist, may not be unique. The device for handling this situation is as follows: Let $\mathscr{Y} = \{y \in \mathscr{X} : \|y\| = 0\}$. Show that $\mathscr{Y}$ is a linear subspace of $\mathscr{X}$ and that $\|x + \mathscr{Y}\| = \|x\|$ defines a norm on $\mathscr{Z} := \mathscr{X}/\mathscr{Y}$. Show also that if the seminorm has the property that every Cauchy sequence converges in $\mathscr{X}$ then $\mathscr{Z}$ is complete.

8.57 Let $\mathscr{Y}$ be a closed linear subspace of a normed space $\mathscr{X}$ and $Q : \mathscr{X} \to \mathscr{X}/\mathscr{Y}$ the quotient map. Show that $Q(B_r(0)) = B_r(Q(0))$.

8.58 Let X be a noncompact locally compact Hausdorff space and let $X_\infty := X \cup \{\infty\}$ be the one-point compactification of X (§ 0.12). Show that $C_0(X)$ is isometrically isomorphic to the space $\{f \in C(X_\infty) : f(\infty) = 0\}$ and that $C(X_\infty)$ is topologically isomorphic to the direct product $C_0(X) \times \mathbb{C}$.

8.59 Let $\mathscr{X}$ and $\mathscr{Y}$ be normed spaces with dual spaces $\mathscr{X}'$ and $\mathscr{Y}'$, and let $\mathscr{X} \times \mathscr{Y}$ and $\mathscr{X}' \times \mathscr{Y}'$ have the $\|\cdot\|_2$-norms. Given $z' \in (\mathscr{X} \times \mathscr{Y})'$, define $Sz' \in \mathscr{X}'$ and $Tz' \in \mathscr{Y}'$ by $\langle x, Sz'\rangle = \langle (x, 0), z'\rangle$ and $\langle y, Tz'\rangle = \langle (0, y), z\rangle$. Show that the mapping $Rz' := (Sz', Tz')$ is an isometric isomorphism from $(\mathscr{X} \times \mathscr{Y})'$ onto $\mathscr{X}' \times \mathscr{Y}'$ such that $\langle (x, y), z'\rangle = \langle x, Sz'\rangle + \langle y, Tz'\rangle$.

⟦For $\|z'\| \le \|Rz\|$ use the Cauchy-Schwartz inequality in $\mathbb{K}^2$. For the reverse inequality let $\varepsilon > 0$, and find $x \in \mathscr{X}$ and $y \in \mathscr{Y}$ with norm one such that $\|Sz'\| \le |\langle x, Sz'\rangle| + \varepsilon$ and $\|Tz'\| \le |\langle y, Tz'\rangle| + \varepsilon$. Choose $|a| = |b| = 1$ such that $|\langle x, Sz'\rangle| = a\langle x, Sz'\rangle$ and $|\langle y, Tz'\rangle| = b\langle y, Tz'\rangle$ and consider $\langle (a\|Sz'\|\, x, b\|Tz'\|\, y), z'\rangle$. ⟧

8.5 Hahn-Banach Extension Theorems

The Hahn-Banach theorem in its various forms guarantees the existence of a rich supply of continuous linear functionals. The versions described in this section treat the problem of extending a linear functional while preserving a certain crucial inequality. Geometric versions in the form of separation theorems are considered in §9.3.

The version of the Hahn-Banach theorem for real linear spaces is based on the following notion: A **Minkowski functional** on a real linear space $\mathscr{X}$ is a function $p : \mathscr{X} \to \mathbb{R}$ satisfying

- **subadditivity:** $p(x + y) \le p(x) + p(y)$,
- **positive homogeneity:** $p(tx) = p(tx)$, $t > 0$.

Clearly, every seminorm is a Minkowski functional. The function $p(f) = \sup f(X)$ on $B(X, \mathbb{R})$ is an example of a Minkowski functional that is not a seminorm.

The Real Hahn-Banach Theorem

8.5.1 Theorem. *Let $\mathscr{X}$ be real linear space, p a Minkowski functional on $\mathscr{X}$, and g a real-valued linear functional on subspace of $\mathscr{Y}$ of $\mathscr{X}$ such that $g \le p$ on $\mathscr{Y}$. Then there exists a real-valued linear functional f on $\mathscr{X}$ such that $f = g$ on $\mathscr{Y}$ and $f \le p$ on $\mathscr{X}$.*

Proof. Let $x_0 \notin \mathscr{Y}$. We show first that g extends linearly to a function f on $\mathscr{Z} := \mathbb{K}x_0 \oplus \mathscr{Y}$ such that $f \le p$ on $\mathscr{Z}$. Since the representation $z = cx_0 + y$ is unique, for any fixed $a \in \mathbb{R}$ the function

$$f(cx_0 + y) := ac + g(y), \quad x \in \mathscr{X},$$

is a well-defined linear functional on $\mathscr{Z}$ that extends g. We claim that a may be chosen so that

$$f(cx_0 + y) \le p(cx_0 + y) \text{ for all } c \in \mathbb{R} \text{ and } y \in \mathscr{Y}. \qquad (\dagger)$$

The inequality obviously holds if $c = 0$. If $c \neq 0$, then, by positive homogeneity, (†) is equivalent to

$$a + g(\mathcal{y}/c) \le p(\mathcal{x}_0 + \mathcal{y}/c) \text{ if } c > 0 \quad \text{and} \quad -a + g(-\mathcal{y}/c) \le p(-\mathcal{x}_0 - \mathcal{y}/c) \text{ if } c < 0.$$

These inequalities are clearly implied by the conditions

$$a + g(\mathcal{y}) \le p(\mathcal{x}_0 + \mathcal{y}) \text{ and } -a + g(\tilde{\mathcal{y}}) \le p(-\mathcal{x}_0 + \tilde{\mathcal{y}}), \quad \mathcal{y}, \tilde{\mathcal{y}} \in \mathcal{Y},$$

or, equivalently,

$$-p(-\mathcal{x}_0 + \tilde{\mathcal{y}}) + g(\tilde{\mathcal{y}}) \le a \le p(\mathcal{x}_0 + \mathcal{y}) - g(\mathcal{y}), \quad \mathcal{y}, \tilde{\mathcal{y}} \in \mathcal{Y}.$$

Such a choice of a is possible if

$$\sup_{\tilde{\mathcal{y}} \in \mathcal{Y}} \big[-p(-\mathcal{x}_0 + \tilde{\mathcal{y}}) + g(\tilde{\mathcal{y}}) \big] \le \inf_{\mathcal{y} \in \mathcal{Y}} \big[p(\mathcal{x}_0 + \mathcal{y}) - g(\mathcal{y}) \big].$$

But this inequality holds by virtue of the calculations

$$g(\mathcal{y}) + g(\tilde{\mathcal{y}}) = g(\mathcal{y} + \tilde{\mathcal{y}}) \le p(\mathcal{y} + \tilde{\mathcal{y}}) = p(\mathcal{x}_0 + \mathcal{y} - \mathcal{x}_0 + \tilde{\mathcal{y}}) \le p(\mathcal{x}_0 + \mathcal{y}) + p(-\mathcal{x}_0 + \tilde{\mathcal{y}}).$$

Thus (†) holds, which shows that g has the required extension to $\mathcal{Z}$.

Now consider the collection $\mathscr{E}$ of all real linear extensions f of g for which $f \le p$ on $\operatorname{dom}(f)$. For two such functions, write $f_1 \preceq f_2$ if f_2 is an extension of f_1, that is, $\operatorname{dom}(f_1) \subseteq \operatorname{dom}(f_2)$ and $f_1 = f_2$ on $\operatorname{dom}(f_1)$. Then $\preceq$ is a partial order on $\mathscr{E}$ such that every chain has an upper bound. By Zorn's lemma, there exists a maximal extension $f \in \mathscr{E}$. From the first part of the proof and maximality, $\operatorname{dom} f = \mathcal{X}$. Thus f is the desired extension of g. □

The Complex Hahn-Banach Theorem

For the seminorm version of the Hahn-Banach theorem, we need the following lemma.

8.5.2 Lemma (Bohnenblust-Sobczyk). *Let $\mathcal{X}$ be complex linear space and f a linear functional on $\mathcal{X}$. Then the real and imaginary parts f_r and f_i of f are real linear functionals on $\mathcal{X}$, considered as a linear space over $\mathbb{R}$, and $f_i(\mathcal{x}) = -f_r(i\mathcal{x})$. Conversely, if f_r is a real linear functional on $\mathcal{X}$, then the equation*

$$f(\mathcal{x}) = f_r(\mathcal{x}) - i f_r(i\mathcal{x}), \quad \mathcal{x} \in \mathcal{X}, \tag{8.17}$$

defines a complex linear functional on $\mathcal{X}$.

Proof. That f_r and f_i are real linear functionals is clear. Moreover, for $\mathcal{x} \in \mathcal{X}$,

$$f_r(i\mathcal{x}) + i f_i(i\mathcal{x}) = f(i\mathcal{x}) = i f(\mathcal{x}) = i\big(f_r(\mathcal{x}) + i f_i(\mathcal{x})\big) = -f_i(\mathcal{x}) + i f_r(\mathcal{x}),$$

hence $f_r(i\mathcal{x}) = -f_i(\mathcal{x})$.

Conversely, if f_r is a real linear functional on $\mathcal{X}$, then the functional f defined by (8.17) is additive and satisfies $f(t\mathcal{x}) = t f(\mathcal{x})$ for $t \in \mathbb{R}$. The equality

$$f(i\mathcal{x}) = f_r(i\mathcal{x}) - i f_r(-\mathcal{x}) = f_r(i\mathcal{x}) + i f_r(\mathcal{x}) = i f(\mathcal{x})$$

then implies that f is a complex linear functional. □

8.5.3 Theorem. *Let $\mathcal{X}$ be a real or complex linear space and p a seminorm on $\mathcal{X}$. Let $\mathcal{Y}$ be a subspace of $\mathcal{X}$ and g a linear functional on $\mathcal{Y}$ such that $|g| \le p$ on $\mathcal{Y}$. Then there exists a linear functional f on $\mathcal{X}$ such that $f = g$ on $\mathcal{Y}$ and $|f| \le p$ on $\mathcal{X}$.*

Proof. The real case follows from 8.5.1, so we may assume that $\mathscr{X}$ is a complex linear space. By the lemma, $g_r := \operatorname{Re} g$ and $g_i := \operatorname{Im} g$ are real linear functionals on $\mathscr{Y}$. Since $g_r \leq p$ on $\mathscr{Y}$, there exists a real linear extension f_r of g_r such that $f_r \leq p$ on $\mathscr{X}$. Define f as in (8.17). By the lemma,

$$f(\boldsymbol{y}) = f_r(\boldsymbol{y}) - i f_r(i\boldsymbol{y}) = g_r(\boldsymbol{y}) - i g_r(i\boldsymbol{y}) = g(\boldsymbol{y}), \quad \boldsymbol{y} \in \mathscr{Y},$$

hence f is an extension of g. Writing $f(\boldsymbol{x}) = |f(\boldsymbol{x})|e^{-i\theta}$, we have

$$|f(\boldsymbol{x})| = f(\boldsymbol{x})e^{i\theta} = f(e^{i\theta}\boldsymbol{x}) = f_r(e^{i\theta}\boldsymbol{x}) \leq p(e^{i\theta}\boldsymbol{x}) = |e^{i\theta}|p(\boldsymbol{x}) = p(\boldsymbol{x}). \qquad \square$$

The Hahn-Banach Theorem for Normed Spaces

8.5.4 Theorem. *Let $\mathscr{Y}$ be a subspace of a normed space $\mathscr{X}$. If $g \in \mathscr{Y}'$, then there exists an $f \in \mathscr{X}'$ that extends g such that $\|f\| = \|g\|$.*

Proof. Define a seminorm $p(\boldsymbol{x}) := \|g\| \|\boldsymbol{x}\|$. Then $|g| \leq p$ on $\mathscr{Y}$, hence g has a linear extension f to $\mathscr{X}$ such that $|f| \leq p$. It follows that $f \in \mathscr{X}'$ and $\|f\| = \|g\|$. $\square$

8.5.5 Corollary. *Let $\mathscr{Y}$ be a closed subspace of a normed space $\mathscr{X}$ and $\boldsymbol{x}_0 \notin \mathscr{Y}$. Then there exists $f \in \mathscr{X}'$ such that $\|f\| = 1$, $f(\mathscr{Y}) = 0$, and $f(\boldsymbol{x}_0) = \inf\{\|\boldsymbol{x}_0 + \boldsymbol{y}\| : \boldsymbol{y} \in \mathscr{Y}\}$.*

Proof. Define a linear functional g on the subspace $\mathscr{Y}_0 := \mathbb{K}\boldsymbol{x}_0 \oplus \mathscr{Y}$ by $g(c\boldsymbol{x}_0 + \boldsymbol{y}) = cd$, where d is the above infimum. Since

$$|g(c\boldsymbol{x}_0 + \boldsymbol{y})| = |c|d \leq |c| \left\|\boldsymbol{x}_0 + c^{-1}\boldsymbol{y}\right\| = \|c\boldsymbol{x}_0 + \boldsymbol{y}\|,$$

$\|g\| \leq 1$. Now choose a sequence $(\boldsymbol{y}_n)$ in $\mathscr{Y}$ such that $\|\boldsymbol{x}_0 + \boldsymbol{y}_n\| \to d$. Then

$$d = g(\boldsymbol{x}_0 + \boldsymbol{y}_n) \leq \|g\| \|\boldsymbol{x}_0 + \boldsymbol{y}_n\| \to \|g\| d,$$

hence, since $d > 0$ (because $\mathscr{Y}$ is closed), $\|g\| \geq 1$ and so $\|g\| = 1$. An application of 8.5.4 completes the argument. $\square$

The second part of the next corollary asserts that $\mathscr{X}'$ **separates points of** $\mathscr{X}$.

8.5.6 Corollary. *For any $\boldsymbol{x}_0 \neq 0$ in a normed space $\mathscr{X}$, there exists $f \in \mathscr{X}'$ such that $\|f\| = 1$ and $f(\boldsymbol{x}_0) = \|\boldsymbol{x}_0\|$. In particular, if $\boldsymbol{x}_1 \neq \boldsymbol{x}_2$ then there exists $f \in \mathscr{X}'$ such that $f(\boldsymbol{x}_1) \neq f(\boldsymbol{x}_2)$,*

Proof. For the first part, take $\mathscr{Y} = \{0\}$ in 8.5.5. For the second part take $\boldsymbol{x}_0 = \boldsymbol{x}_1 - \boldsymbol{x}_2$. $\square$

8.5.7 Corollary. *Let $\mathscr{X}$ be a normed space. If $\mathscr{X}'$ is separable, then $\mathscr{X}$ is separable.*

Proof. Let (f_n) be dense in $\mathscr{X}'$. For each n, choose $\boldsymbol{x}_n \in \mathscr{X}$ such that $\|\boldsymbol{x}_n\| = 1$ and $|\langle \boldsymbol{x}_n, f_n\rangle| \geq \|f_n\|/2$, and set $\mathscr{Y} = \operatorname{cl}\operatorname{span}\{\boldsymbol{x}_1, \boldsymbol{x}_2, \ldots\}$. We claim that $\mathscr{Y} = \mathscr{X}$. If not, then by 8.5.5 we may choose $f \in \mathscr{X}'$ with $\|f\| = 1$ and $f(\mathscr{Y}) = \{0\}$. But then

$$1 - \|f - f_n\| \leq \|f_n\| \leq 2|\langle \boldsymbol{x}_n, f_n\rangle| = 2|\langle \boldsymbol{x}_n, f - f_n\rangle| \leq 2\|f - f_n\|,$$

hence $\|f - f_n\| \geq 1/3$ for all n, contradicting that (f_n) is dense in $\mathscr{X}'$. $\square$

8.5.8 Corollary. *Let $\mathscr{X}$ be a normed space. Then for each $\boldsymbol{x} \in \mathscr{X}$,*

$$\|\boldsymbol{x}\| = \sup\{|\langle \boldsymbol{x}, f\rangle| : f \in \mathscr{X}', \ \|f\| \leq 1\}.$$

Proof. Let s denote the supremum. Since $|\langle \boldsymbol{x}, f\rangle| \leq \|\boldsymbol{x}\| \|f\|$, $s \leq \|\boldsymbol{x}\|$. By 8.5.6, there exists $f \in \mathscr{X}'$ such that $\|f\| = 1$ and $\langle \boldsymbol{x}, f\rangle = \|\boldsymbol{x}\|$. Therefore, $s \geq \|\boldsymbol{x}\|$. $\square$

The Bidual of a Normed Space

The **bidual** $\mathcal{X}''$ of a normed space $\mathcal{X}$ is the dual of the dual: $\mathcal{X}'' := (\mathcal{X}')'$. Given $\boldsymbol{x} \in \mathcal{X}$, the **evaluation functional** corresponding to $\boldsymbol{x} = (x_n)$ is the linear functional $\widehat{\boldsymbol{x}}$ defined by

$$\langle f, \widehat{\boldsymbol{x}} \rangle = \langle \boldsymbol{x}, f \rangle, \quad f \in \mathcal{X}'.$$

The collection of all evaluation functionals is denoted by $\widehat{\mathcal{X}}$. Corollary 8.5.8 asserts that $\widehat{\mathcal{X}} \subseteq \mathcal{X}''$ and $\|\widehat{\boldsymbol{x}}\| = \|\boldsymbol{x}\|$. For example, from §8.3 we see that the bidual of $\mathfrak{c}_0$ may be identified with ℓ^∞. To find $\widehat{\mathfrak{c}_0}$ in this identification, note that for $\boldsymbol{x} = (x_n) \in \mathfrak{c}_0$, $\widehat{\boldsymbol{x}}$ is the mapping

$$\langle \boldsymbol{y}, \widehat{\boldsymbol{x}} \rangle = \langle \boldsymbol{x}, \boldsymbol{y} \rangle = \sum_{n=1}^{\infty} x_n y_n, \quad \boldsymbol{y} \in \ell^1 = \mathfrak{c}_0'.$$

Now recall that in the identification of ℓ^∞ with $(\ell^1)'$, a sequence (x_n) in ℓ^∞ is identified with the linear functional on ℓ^1 defined precisely by the above equation. Thus we see that $\widehat{\mathfrak{c}_0}$ may be identified with the subspace $\mathfrak{c}_0$ of ℓ^∞.

8.5.9 Theorem. *Let $\mathcal{X}$ be a normed space. Then the mapping $\boldsymbol{x} \to \widehat{\boldsymbol{x}}$ is a linear isometry into the bidual of $\mathcal{X}$. Thus the closure of $\widehat{\mathcal{X}}$ in $\mathcal{X}''$ is a concrete realization of the completion of $\mathcal{X}$.*

*Invariant Versions of the Hahn-Banach Theorem

A **semigroup of operators** on a vector space $\mathcal{X}$ is a set $\mathcal{S}$ of linear operators $S : \mathcal{X} \to \mathcal{X}$ that is closed under composition. A subspace $\mathcal{Y}$ of $\mathcal{X}$ is said to be **$\mathcal{S}$-invariant** if $S\mathcal{Y} \subseteq \mathcal{Y}$ for all $S \in \mathcal{S}$. A function G on an $\mathcal{S}$-invariant subspace $\mathcal{Y}$ is said to be **$\mathcal{S}$-invariant** if $G(S\boldsymbol{y}) = G(\boldsymbol{y})$ for all $\boldsymbol{y} \in \mathcal{Y}$ and $S \in \mathcal{S}$. The following versions of the Hahn-Banach theorem, due to Agnew and Morse, address the problem of extending linear functionals that are invariant under the action of a semigroup of operators.

8.5.10 Theorem. *Let $\mathcal{X}$ be a real vector space, $\mathcal{S}$ a commutative semigroup of operators on $\mathcal{X}$, and p a Minkowski functional on $\mathcal{X}$ such that $p(S\boldsymbol{x}) \le p(\boldsymbol{x})$ for all $\boldsymbol{x} \in \mathcal{X}$ and $S \in \mathcal{S}$. Let $\mathcal{Y}$ be an $\mathcal{S}$-invariant subspace of $\mathcal{X}$ and G an $\mathcal{S}$-invariant, real-valued, linear functional on $\mathcal{Y}$ such that $G \le p$ on $\mathcal{Y}$. Then G extends to a real-valued $\mathcal{S}$-invariant linear functional F on $\mathcal{X}$ such that $F \le p$ on $\mathcal{X}$.*

Proof. We may assume that $\mathcal{S}$ contains the identity operator I. Let $\operatorname{co}\mathcal{S}$ denote the set of convex combinations of members of $\mathcal{S}$:

$$\operatorname{co}\mathcal{S} = \left\{ \sum_{j=1}^{n} t_j S_j : S_j \in \mathcal{S},\ t_j \ge 0, \text{ and } \sum_{j=1}^{n} t_j = 1 \right\}.$$

Define $q(\boldsymbol{x})$ on $\mathcal{X}$ by

$$q(\boldsymbol{x}) := \inf\left\{ p(T\boldsymbol{x}) : T \in \operatorname{co}\mathcal{S} \right\}.$$

By linearity of T, q is a Minkowski functional on $\mathcal{X}$. Since $G(\boldsymbol{y}) = G(T\boldsymbol{y}) \le p(T\boldsymbol{y})$ for all $T \in \operatorname{co}\mathcal{S}$ and $\boldsymbol{y} \in \mathcal{Y}$, $G \le q$ on $\mathcal{Y}$. By the Hahn-Banach theorem, G has a linear extension F on $\mathcal{X}$ such that $F \le q$. It remains to show that F is $\mathcal{S}$-invariant.

Fix $S \in \mathcal{S}$ and for each n define $T_n = \frac{1}{n}\sum_{j=0}^{n-1} S^j$ where $S^0 := I$. Then

$$T_n(I - S) = \frac{1}{n}\left[\sum_{j=0}^{n-1} S^j - \sum_{j=1}^{n} S^j \right] = \frac{1}{n}(I - S^n),$$

hence for all x

$$F(x) - F(Sx) \le q((I - S)x) \le p(T_n(I - S)x) = \tfrac{1}{n}p(x - S^n x) \le \tfrac{1}{n}[p(x) + p(-x)].$$

Letting $n \to \infty$ we see that $F(x) - F(Sx) \le 0$. Replacing x by $-x$ gives the reverse inequality. Therefore, F is $\mathcal{S}$-invariant. □

The proof of the following seminorm version of 8.5.10 may be modelled along the lines of the proof of 8.5.3. The details are left to the reader.

8.5.11 Theorem. *Let $\mathcal{X}$ be a real or complex linear space, $\mathcal{S}$ a commutative semigroup of operators on $\mathcal{X}$, and p seminorm on $\mathcal{X}$ such that $p(Sx) \le p(x)$ for all $x \in \mathcal{X}$ and $S \in \mathcal{S}$. Let $\mathcal{Y}$ be an $\mathcal{S}$-invariant subspace of $\mathcal{X}$ and G an $\mathcal{S}$-invariant linear functional on $\mathcal{Y}$ such that $|G| \le p$ on $\mathcal{Y}$. Then G extends to an $\mathcal{S}$-invariant linear functional F on $\mathcal{X}$ such that $|F| \le p$ on $\mathcal{X}$.*

Exercises

8.60 Show that the converse of 8.5.7 is false.

8.61 Let $\mathcal{X}$ be a normed space. Show that $\widehat{\mathcal{X}}$ is closed in $\mathcal{X}''$ iff $\mathcal{X}$ is a Banach space.

8.62 Let $x_1, \ldots, x_n$ be linearly independent vectors in a normed space $\mathcal{X}$. Show that $\mathcal{X}'$ has at least n linearly independent vectors.

8.63 Show that if $\mathcal{X}$ is strictly convex, then for each $f \in \mathcal{X}'$ with $\|f\| > 0$ there is at most one x such that $\|x\| = 1$ and $f(x) = \|f\|$.

8.64 (a) Show that for each $n \in \mathbb{N}$ there exists a probability measure μ_n on $[0,1]$ such that $\int_0^1 x^k \, d\mu(x) = k + 1$ for all integers $0 \le k \le n$. Can this hold for all $k \ge 0$? (b) Show that there exists a probability measure μ on $[0,1]$ such that $\int_0^1 x^k \, d\mu(x) = (k+1)^{-1}$ for all integers $k \ge 0$.

8.65 (Krein) Let X be a set and $\mathcal{G} \subseteq \mathcal{F}$ linear spaces of real-valued functions f on X such that for each $f \in \mathcal{F}$ there exists $g \in \mathcal{G}$ with $g \ge f$. Let I be a positive linear functional on $\mathcal{G}$, that is, $I(g) \ge 0$ whenever $g \in \mathcal{G}$ and $g \ge 0$, Show that I extends to a positive linear functional on $\mathcal{F}$. ⟦Consider $p(f) := \inf\{I(g) : g \in \mathcal{G} \text{ and } g \ge f\}$.⟧

8.66 Show that a finite dimensional subspace $\mathcal{F}$ of a normed space $\mathcal{X}$ is **complemented**, that is, there exists a closed subspace $\mathcal{Y}$ of $\mathcal{X}$ such that $\mathcal{X} = \mathcal{F} \oplus \mathcal{Y}$. ⟦Let $\{x_1, \ldots, x_d\}$ be a basis for $\mathcal{F}$ and define suitable $x'_j \in \mathcal{X}'$.⟧

---

*8.6 Applications of the Hahn-Banach Theorem

The Moment Problem

Given a sequence (c_n) of real numbers, the classical version of the moment problem asks when there exists a real-valued function F of bounded variation on $[0,1]$ such that

$$\int_0^1 t^n \, dF(t) = c_n, \quad n = 0, 1, \ldots$$

If $c_n \in [0,1]$ and $c_0 = 1$, then the problem can be stated in probabilistic terms: When does there exist a probability measure on $[0,1]$ with given moments c_n? Note that by the Stone-Weierstrass theorem, the solution, if one exists, is unique.

Since the integral $\int_0^1 g\,dF$ defines a continuous linear functional on $C[0,1]$, the moment problem may be stated somewhat more abstractly as follows: Given a sequence (c_n), when does there exist a continuous linear functional F on $C[0,1]$ such that $\langle t^n, F\rangle = c_n$ for all n? This suggests that the problem may be cast in a broader context, where $C[0,1]$ is replaced by an arbitrary normed space $\mathscr{X}$ and the functions t^n are replaced by members of $\mathscr{X}$. Here is the precise statement of the general moment problem, the resolution of which is a consequence of the Hahn-Banach theorem.

8.6.1 Theorem. *Let $\mathscr{X}$ be a normed space, $\mathfrak{I}$ an arbitrary index set, $\{\boldsymbol{x}_i :\in \mathfrak{I}\} \subseteq \mathscr{X}$, and $\{c_i : i \in \mathfrak{I}\} \subseteq \mathbb{K}$. Then the following statements are equivalent:*

(a) *There exists $\boldsymbol{x}' \in \mathscr{X}'$ such that $\langle \boldsymbol{x}_i, \boldsymbol{x}'\rangle = c_i$ for all $i \in \mathfrak{I}$.*

(b) *There exists $M > 0$ such that for all finite subsets $\mathfrak{I}_0 \subseteq \mathfrak{I}$ and all $t_i \in \mathbb{K}$,*

$$\Big|\sum_{i\in\mathfrak{I}_0} t_j c_j\Big| \le M \Big\|\sum_{j\in\mathfrak{I}_0} t_j \boldsymbol{x}_j\Big\|.$$

Proof. (a) $\Rightarrow$ (b): If such an $\boldsymbol{x}'$ exists, then

$$\Big|\sum_{j\in\mathfrak{I}_0} t_j c_j\Big| = \Big|\sum_{j\in\mathfrak{I}_0} \langle t_j \boldsymbol{x}_j, \boldsymbol{x}'\rangle\Big| \le \|\boldsymbol{x}'\| \Big\|\sum_{j\in\mathfrak{I}_0} t_j \boldsymbol{x}_j\Big\|.$$

(b) $\Rightarrow$ (a): Let $\mathscr{Y}$ be the linear span of the set $\{\boldsymbol{x}_i : i \in \mathfrak{I}\}$. A typical member of $\mathscr{Y}$ may be written $\sum_{j\in\mathfrak{I}_0} t_j \boldsymbol{x}_j$, where $\mathfrak{I}_0 \subseteq \mathfrak{I}$ is finite. Define a mapping $\boldsymbol{x}'$ on $\mathscr{Y}$ by

$$\boldsymbol{x}'\Big(\sum_{j\in\mathfrak{I}_0} t_j \boldsymbol{x}_j\Big) = \sum_{j\in\mathfrak{I}_0} t_j c_j.$$

The inequality in (b) implies that $\boldsymbol{x}'$ is well-defined. Moreover, $\boldsymbol{x}'$ is linear and (b) shows that $\|\boldsymbol{x}'\| \le M$. Therefore, by 8.5.4, $\boldsymbol{x}'$ has an extension to a member of $\mathscr{X}'$. □

Invariant Means

Let S be a nonempty set. A **mean** on $B(S,\mathbb{R})$ is a linear functional m such that

$$\inf\{f(s) : s \in S\} \le m(f) \le \sup\{f(s) : s \in S\} \quad \text{for all } f \in B(S,\mathbb{R}).$$

Now let S be a semigroup. A mean m on $B(S,\mathbb{R})$ is **invariant** if

$$m(R_s f) = m(f) = m(L_s f) \quad \text{for all } s \in S \text{ and } f \in B(S,\mathbb{R}),$$

where R_s and L_s are the **right** and **left translation operators** on $B(S,\mathbb{R})$ defined by

$$R_s f(t) = f(ts) \quad\text{and}\quad L_s f(t) = f(st), \quad s,t \in S.$$

8.6.2 Theorem. *If S is a commutative semigroup, then $B(S,\mathbb{R})$ has an invariant mean.*

Proof. Take p to be the functional $p(f) = \sup f$ on $B(S,\mathbb{R})$ and m the identity function on the space of constant function on S and apply 8.5.10. □

In certain circumstances the commutativity hypothesis may be removed,[1] but not generally, as the following example shows.

[1]For example, every finite group has an invariant mean. (See Chapter 16.)

8.6.3 Example. Let S be the free group on two generators a and b. Thus S consists of an identity 1 and all concatenations of the symbols a, b, a^{-1}, and b^{-1}, these concatenations called *words*. A word may be reduced to a unique expression of the form $s_1^{\varepsilon_1} s_2^{\varepsilon_2} \ldots s_n^{\varepsilon_n}$ where the ε_j are integers and $a^1 := a$, $b^1 := b$. (Any pairs aa^{-1} etc. are omitted.) Assume $B(S, \mathbb{R})$ has an invariant mean m. Let B denote the subset of S consisting of the identity and all reduced words starting with b. Since the sets $a^j B$ are disjoint, $1 \geq \sum_{j=1}^{n} \mathbf{1}_{a^j B}$, hence

$$1 = m(1) \geq \sum_{j=1}^{n} m(\mathbf{1}_{a^j B}) = nm(\mathbf{1}_B) \quad \text{for all } n$$

and so $m(\mathbf{1}_B) = 0$. Now set

$$f = \sum_{n \in \mathbb{Z} \setminus \{0\}} \mathbf{1}_{a^n B}.$$

Since $0 \leq L_{b^{-1}} f \leq \mathbf{1}_B$ we have $0 \leq m(f) = m(L_{b^{-1}} f) \leq m(\mathbf{1}_B) = 0$, hence $m(f) = 0$. On the other hand, from $f + \mathbf{1}_B = 1$ we obtain the contradictory statement $m(f) = 1$. Thus $B(S, \mathbb{R})$ cannot have an invariant mean. ◊

<u>Banach Limits</u>

A **Banach limit** on $B\big((0, \infty), \mathbb{R}\big)$ is linear functional, typically denoted by $\mathrm{Lim}_{t \to \infty} f(t)$, with the following properties:

(a) $\mathrm{Lim}_{t \to \infty} f(t) = \lim_{t \to \infty} f(t)$ whenever the limit on the right exists.

(b) $\mathrm{Lim}_{t \to \infty} f(t + s) = \mathrm{Lim}_{t \to \infty} f(t)$ for all $f \in B\big((0, \infty), \mathbb{R}\big)$ and $s \in (0, \infty)$.

(c) $\underline{\lim}_{t \to \infty} f(t) \leq \mathrm{Lim}_{t \to \infty} f(t) \leq \overline{\lim}_{t \to \infty} f(t)$ for all $f \in B\big((0, \infty), \mathbb{R}\big)$.

8.6.4 Theorem. *Banach limits exist.*

Proof. Define a Minkowski functional p on $B\big((0, \infty), \mathbb{R}\big)$ by $p(f) = \overline{\lim}_{t \to \infty} f(t)$. Let $\mathscr{F}$ be the subspace of all functions $f : (0, \infty) \to \mathbb{R}$ such that the limit $L(f) := \lim_{t \to \infty} f(t)$ exists in $\mathbb{R}$. If T_s is the translation operator $T_s f(t) = f(s + t)$, then $T_s \mathscr{F} \subseteq \mathscr{F}$, $L(T_s f) = L(f) = p(f)$ for all $f \in \mathscr{F}$, and $p(T_s f) = p(f)$ for all $f \in B\big((0, \infty), \mathbb{R}\big)$. An application of 8.5.10 yields the desired functional. □

The reader may easily formulate the analogous notion of Banach limit on $\ell^\infty(\mathbb{N})$, replacing $\lim_{t \to \infty} f(t)$ by $\lim_n x_n$, etc.

<u>Invariant Set Functions</u>

It follows From Ex. 3.72 and the material in §1.9 that there is no translation invariant measure μ on $\mathcal{P}(\mathbb{R})$ with the property $\mu[a, b] = b - a$ for all $a \leq b$, the countable additivity requirement being essentially responsible for this state of affairs. One may then reasonably ask if there exists a translation invariant *finitely additive* set function μ on $\mathcal{P}(\mathbb{R})$ with the aforementioned length property. Banach has answered this question in the affirmative:

8.6.5 Theorem (Banach). *There exists a finitely additive, translation invariant set function μ on $\mathcal{P}(\mathbb{R})$ such that $\mu[a, b] = b - a$ for all $a \leq b$.*

Proof. Let $\mathscr{X}$ denote the real linear space of all bounded, real-valued functions on $\mathbb{R}$ with period one. For $f \in \mathscr{X}$, set $p(f) = \|f\|_\infty$. Clearly, $p(T_t f) = f$, where T_t denotes the translation operator $T_t f(s) = f(s + t)$. Let $\mathscr{Y}$ be the subspace of continuous functions and define a positive linear functional G on $\mathscr{Y}$ by $G(f) = \int_0^1 f(t)\, dt$. Then $G \leq p$, and by periodicity G is translation invariant. By 8.5.10, G extends to a translation invariant positive linear functional F on $\mathscr{X}$. Now define $\mu(E) := F(\mathbf{1}_E)$, $E \subseteq \mathbb{R}$. □

Exercises

8.67 Show that a Banach limit on $B(0,\infty)$ or $\ell^\infty(\mathbb{N})$ is continuous.

8.68 Let $a_j \in \mathbb{R}$. Find the Banach limit of the sequence $\boldsymbol{x} = (a_1, \ldots a_m, a_1, \ldots a_m, \ldots)$.

8.69 Show that there exists a continuous linear functional $f \to \mathrm{Lim}_{t\to 0} f(t)$ on $B := B\big((-1,1),\mathbb{R}\big)$ with the following properties:

(a) $\mathrm{Lim}_{t\to 0} f(t) = \lim_{t\to 0} f(t)$ whenever the limit on the right exists.

(b) $\mathrm{Lim}_{t\to 0} f(rt) = \mathrm{Lim}_{t\to 0} f(t)$ for all $f \in B$ and $0 < r < 1$.

(c) $\underline{\lim}_{t\to 0} f(t) \le \mathrm{Lim}_{t\to 0} f(t) \le \overline{\lim}_{t\to 0} f(t)$ for all $f \in B$.

8.70 Show that there exists a finitely additive, translation invariant measure μ on $\mathcal{P}(\mathbb{R})$ such that $\mu(E) = \lambda(E)$ for every bounded, Lebesgue measurable set $E \subseteq \mathbb{R}$.

8.7 Baire Category in Banach Spaces

In this section we prove three basic results which, together with the Hahn-Banach theorem, form the core of functional analysis.

The Uniform Boundedness Principle

The following theorem asserts that under suitable conditions a family of bounded linear transformations that is pointwise bounded is uniformly bounded on bounded sets. The proof depends on the Baire category theorem (0.3.12).

8.7.1 Uniform Boundedness Principle. *Let $\mathcal{X}$ and $\mathcal{Y}$ be Banach spaces and let $\mathcal{T}$ be a subset of $\mathcal{B}(\mathcal{X},\mathcal{Y})$ such that $\sup_{T\in\mathcal{T}} \|T\boldsymbol{x}\| < \infty$ for each $\boldsymbol{x} \in \mathcal{X}$. Then $\sup_{T\in\mathcal{T}} \|T\| < \infty$.*

Proof. The set $X_n := \{\boldsymbol{x} \in \mathcal{X} : \|T\boldsymbol{x}\| \le n \ \forall\ T \in \mathcal{T}\}$ is closed and, by hypothesis, $\mathcal{X} = \bigcup_n X_n$. By Baire's theorem, some X_n contains a closed ball $C(\boldsymbol{x}_0, r)$. Thus

$$\|T\boldsymbol{y}\| \le n \text{ for all } T \in \mathcal{T} \text{ and all } \boldsymbol{y} \in \mathcal{X} \text{ with } \|\boldsymbol{y} - \boldsymbol{x}_0\| \le r.$$

Now let $\boldsymbol{x} \ne 0 \in \mathcal{X}$ with $\|\boldsymbol{x}\| \le 1$ and set $\boldsymbol{y} := \boldsymbol{x}_0 + r\,\|\boldsymbol{x}\|^{-1}\boldsymbol{x}$. Then $\|\boldsymbol{y} - \boldsymbol{x}_0\| \le r$ and $\boldsymbol{x} = r^{-1}\|\boldsymbol{x}\|(\boldsymbol{y} - \boldsymbol{x}_0)$, hence $\|T\boldsymbol{x}\| \le r^{-1}\|\boldsymbol{x}\|\big(\|T\boldsymbol{y}\| + \|T\boldsymbol{x}_0\|\big) \le r^{-1}\big[n + \sup_{S\in\mathcal{T}}\|S\boldsymbol{x}_0\|\big]$. □

The following application will have important consequences later.

8.7.2 Banach-Steinhaus Theorem. *Let $\mathcal{X}$ and $\mathcal{Y}$ be Banach spaces and let (T_n) be a sequence in $\mathcal{B}(\mathcal{X},\mathcal{Y})$. Then $\lim_n T_n\boldsymbol{x}$ exists in $\mathcal{Y}$ for all $\boldsymbol{x} \in \mathcal{X}$ iff the following conditions hold:*

(a) $\sup_n \|T_n\| < \infty$ *and*

(b) $\lim_n T_n\boldsymbol{u}$ *exists in $\mathcal{Y}$ for all $\boldsymbol{u}$ in a dense subset D of $\mathcal{X}$.*

Moreover, if (a) *and* (b) *hold, then the pointwise limit $T := \lim_n T_n$ is a member of $\mathcal{B}(\mathcal{X},\mathcal{Y})$ and $\|T\| \le \underline{\lim}_n \|T_n\| \le \sup_n \|T_n\|$.*

Proof. If $T\boldsymbol{x} := \lim_n T_n\boldsymbol{x}$ exists for all $\boldsymbol{x} \in \mathscr{X}$, then T is linear and $\sup_n \|T_n\boldsymbol{x}\| < \infty$ for all $\boldsymbol{x}$. Therefore, by the uniform boundedness theorem, $\sup_n \|T_n\| < \infty$. From $\|T\boldsymbol{x}\| = \lim_n \|T_n\boldsymbol{x}\| \leq \|\boldsymbol{x}\| \varliminf_n \|T_n\|$ we have $\|T\| \leq \varliminf_n \|T_n\|$.

Now assume (a) and (b) hold and set $s = \sup_n \|T_n\|$. For $\boldsymbol{x} \in \mathscr{X}$ and $\varepsilon > 0$, choose $\boldsymbol{u} \in D$ such that $\|\boldsymbol{x} - \boldsymbol{u}\| < \varepsilon/s$. Then

$$\|T_n\boldsymbol{x} - T_m\boldsymbol{x}\| \leq \|T_n(\boldsymbol{x} - \boldsymbol{u})\| + \|T_n\boldsymbol{u} - T_m\boldsymbol{u}\| + \|T_m(\boldsymbol{u} - \boldsymbol{x})\| \leq 2\varepsilon + \|T_n\boldsymbol{u} - T_m\boldsymbol{u}\|.$$

The expression on the right is $< 3\varepsilon$ for all sufficiently large m and n, hence $(T_n\boldsymbol{x})$ is a Cauchy sequence. Since $\mathscr{Y}$ is complete, $(T_n\boldsymbol{x})$ converges in $\mathscr{Y}$. □

The Open Mapping Theorem

Let $\mathscr{X}$ and $\mathscr{Y}$ be normed spaces. by definition, a mapping $T : \mathscr{X} \to \mathscr{Y}$ is open iff for each $\boldsymbol{x} \in \mathscr{X}$ and $r > 0$ the image $T\big(B_r(\boldsymbol{x})\big)$ contains an open ball $B_s(T\boldsymbol{x})$. If T is linear, then one has the following simplification:

8.7.3 Proposition. *A linear mapping $T : \mathscr{X} \to \mathscr{Y}$ is open iff $T(B_1) \supseteq B_t$ for some $t > 0$, in which case T is surjective.*

Proof. The necessity is clear. For the sufficiency, let $T(B_1) \supseteq B_t$ for some t. For any $r > 0$ and $\boldsymbol{x} \in \mathscr{X}$, $B_r(\boldsymbol{x}) = \boldsymbol{x} + rB_1$, hence by linearity

$$T\big(B_r(\boldsymbol{x})\big) = T\boldsymbol{x} + rT(B_1) \supseteq T\boldsymbol{x} + rB_t = B_{rt}(T\boldsymbol{x}).$$ □

Finally, for any $\boldsymbol{y} \neq 0 \in \mathscr{Y}$, $t(2\|\boldsymbol{y}\|)^{-1}\boldsymbol{y} \in B_t$, hence $\boldsymbol{y} \in \operatorname{ran} T$.

The proof of the open mapping theorem rests on the following lemma.

8.7.4 Lemma. *Let $\mathscr{X}$ and $\mathscr{Y}$ be Banach spaces, $T \in \mathscr{B}(\mathscr{X}, \mathscr{Y})$, and $\varepsilon > 0$. Suppose that $B_\varepsilon \subseteq \operatorname{cl} T(B_1)$. Then $B_{\varepsilon/2} \subseteq T(B_1)$.*

Proof. Let $\boldsymbol{y} \in B_\varepsilon$. By hypothesis, we have $\|\boldsymbol{y} - \boldsymbol{y}_1\| < \varepsilon/2$ for some $\boldsymbol{y}_1 \in T(B_1)$. Then $\boldsymbol{y} - \boldsymbol{y}_1 \in B_{\varepsilon/2} \subseteq \operatorname{cl} T(B_{1/2})$, hence $\|\boldsymbol{y} - \boldsymbol{y}_1 - \boldsymbol{y}_2\| < \varepsilon/4$ for some $\boldsymbol{y}_2 \in T(B_{1/2})$. By induction, we obtain a sequence $(\boldsymbol{x}_n)$ in $\mathscr{X}$ such that

$$\|\boldsymbol{y} - T(\boldsymbol{x}_1 + \cdots + \boldsymbol{x}_n)\| < \varepsilon/2^n, \quad \text{where} \quad \boldsymbol{x}_n \in B_{1/2^{n-1}}.$$

The sums $\boldsymbol{x}_1 + \cdots + \boldsymbol{x}_n$ form a Cauchy sequence with limit $\boldsymbol{x}$, say. By continuity, $T\boldsymbol{x} = \boldsymbol{y}$. Since

$$\|\boldsymbol{x}\| = \lim_n \left\| \sum_{k=1}^n \boldsymbol{x}_k \right\| \leq \sum_{k=1}^\infty 1/2^{k-1} = 2,$$

$\boldsymbol{y} \in T(B_2)$. Therefore, $B_\varepsilon \subseteq T(B_2)$, hence $B_{\varepsilon/2} \subseteq T(B_1)$. □

8.7.5 Open Mapping Theorem. *If $\mathscr{X}$ and $\mathscr{Y}$ are Banach spaces and $T \in \mathscr{B}(\mathscr{X}, \mathscr{Y})$ is surjective, then T is an open mapping.*

Proof. By 8.7.3 and 8.7.4, it suffices to show that $\operatorname{cl} T(B_1) \supseteq B_\varepsilon$ for some ε. Now, by surjectivity, $\mathscr{Y} = \bigcup_{n=1}^\infty T(B_n)$, hence, by the Baire category theorem, for some n the closure $\operatorname{cl} T(B_n) = n \operatorname{cl} T(B_1)$ contains an open ball. Thus for some $\boldsymbol{y}_0 \in \mathscr{Y}$ and $\varepsilon > 0$,

$$\operatorname{cl} T(B_1) \supseteq B_\varepsilon(\boldsymbol{y}_0) = \boldsymbol{y}_0 + B_\varepsilon, \text{ and } \operatorname{cl} T(B_1) \supseteq -B_\varepsilon(\boldsymbol{y}_0) = -\boldsymbol{y}_0 + B_\varepsilon,$$

the second inclusion following from the first because $-B_r = B_r$. Thus for $\boldsymbol{y} \in B_\varepsilon$ we have $\boldsymbol{y} \pm \boldsymbol{y}_0 \in \operatorname{cl} T(B_1)$, so by convexity $\boldsymbol{y} = \frac{1}{2}(\boldsymbol{y} + \boldsymbol{y}_0) + \frac{1}{2}(\boldsymbol{y} - \boldsymbol{y}_0) \in \operatorname{cl} T(B_1)$, as required. □

The following fundamental result is immediate.

8.7.6 Banach Isomorphism Theorem. *Let $\mathscr{X}$ and $\mathscr{Y}$ be Banach spaces and $T : \mathscr{X} \to \mathscr{Y}$ a continuous algebraic isomorphism onto $\mathscr{Y}$. Then T is a topological isomorphism.*

8.7.7 Corollary. *Let $\mathscr{Z}$ be a Banach space that is an algebraic direct sum of closed subspaces $\mathscr{X}$ and $\mathscr{Y}$. Then $\mathscr{Z}$ is a topological direct sum of $\mathscr{X}$ and $\mathscr{Y}$.*

Proof. Since the algebraic isomorphism $(\boldsymbol{x}, \boldsymbol{y}) \to \boldsymbol{x} + \boldsymbol{y}$ is continuous, the assertion follows from the Banach isomorphism theorem. □

8.7.8 Corollary. *Let $\mathscr{Z}$ be a Banach space and $P : \mathscr{Z} \to \mathscr{Z}$ linear such that $P^2 = P$. If* $\operatorname{ran} P$, $\ker P$ *are closed, then P is continuous and $\mathscr{Z}$ is the topological direct sum of* $\operatorname{ran} P$ *and* $\ker P$.

Proof. As noted earlier, $\mathscr{Z}$ is the algebraic direct sum of $\operatorname{ran} P$ and $\ker P$. By the preceding corollary, the sum is topological, hence P is continuous (8.4.3). □

8.7.9 Corollary. *Let $\mathscr{X}$ and $\mathscr{Y}$ be Banach spaces, let $T \in \mathscr{B}(\mathscr{X}, \mathscr{Y})$ be surjective, and let $Q : \mathscr{X} \to \mathscr{X}/\ker T$ be the quotient map. Then there exists a topological isomorphism $S : \mathscr{X}/\ker T \to \mathscr{Y}$ such that $SQ = T$.*

Proof. Define S by the equation $SQ = T$. Since $\ker Q = \ker T$, S is well-defined and bijective. Moreover, S is clearly linear, and since SQ is continuous so is S (8.4.6(c)). Banach's isomorphism theorem now implies that S is a topological isomorphism. □

8.7.10 Corollary (Sard Quotient Theorem). *Let $\mathscr{X}$, $\mathscr{Y}$, and $\mathscr{Z}$ be Banach spaces and let $T_{XY} \in \mathscr{B}(\mathscr{X}, \mathscr{Y})$ and $T_{XZ} \in \mathscr{B}(\mathscr{X}, \mathscr{Z})$ with T_{XY} is surjective. Then there exists $T_{YZ} \in \mathscr{B}(\mathscr{Y}, \mathscr{Z})$ such that $T_{XZ} = T_{YZ}T_{XY}$ iff* $\ker T_{XY} \subseteq \ker T_{XZ}$.

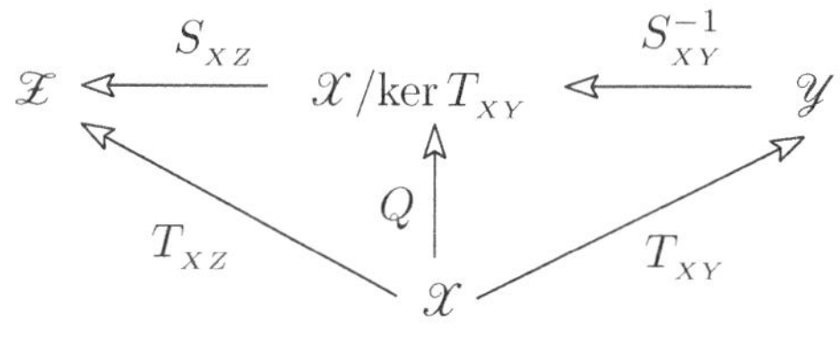

FIGURE 8.2: Sard Quotient Theorem

Proof. The necessity is obvious. For the sufficiency use 8.7.9 to obtain a topological isomorphism $S_{XY} : \mathscr{X}/\ker T_{XY} \to \mathscr{Y}$ such that $S_{XY}Q = T_{XY}$, where $Q : \mathscr{X} \to \mathscr{X}/\ker T_{XY}$ is the quotient map. Since $\ker Q = \ker T_{XY} \subseteq \ker T_{XZ}$, we may define $S_{XZ} \in \mathscr{B}(\mathscr{X}/\ker T_{XY}, \mathscr{Z})$ so that $S_{XZ}Q = T_{XZ}$. Since $Q = S_{XY}^{-1}T_{XY}$, we have $T_{XZ} = S_{XZ}S_{XY}^{-1}T_{XY}$. Therefore, $S_{XZ}S_{XY}^{-1}$ is the desired map T_{YZ}. □

8.7.11 Example. Let X be a compact Hausdorff space and $Y \subseteq X$ closed. We show that $C(Y)$ is isometrically isomorphic to $C(X)/\mathscr{Y}$, where $\mathscr{Y} = \{g \in C(X) : g(Y) = 0\}$.

Define a bounded linear map $T : C(X) \to C(Y)$ by $Tf = f\big|_Y$. Then $\ker T = \mathscr{Y}$, and T is surjective by Tietze's extension theorem. Let $Q : C(X) \to C(X)/\mathscr{Y}$ denote the quotient map. By 8.7.9, there exists a topological isomorphism $S \in \mathscr{B}(C(X)/\ker T, \mathscr{Y})$ such that $SQ = T$. It remains to show S is an isometry, that is,

$$\inf\left\{ \|f + g\|_\infty : g \in \mathscr{Y} \right\} = \left\| f\big|_Y \right\|_\infty, \quad f \in C(X).$$

Let α denote the infimum. For any $g \in \mathscr{Y}$ and $y \in Y$, $|f(y)| = |f(y) + g(y)| \le \|f + g\|_\infty$, hence $\|f|_Y\|_\infty \le \alpha$. For the reverse inequality, let $\varepsilon > 0$ and set

$$U := \{x \in X : |f(x)| < \|f|_Y\|_\infty + \varepsilon\}.$$

Then U is open and contains the compact set Y, hence there exists continuous function h such that $0 \le h \le 1$, $h = 0$ on Y, and $h = 1$ on U^c. Setting $g = -fh$ we have $g \in \mathscr{Y}$ and

$$|f(x) + g(x)| = |f(x)|\,|1 - h(x)| \le \|f|_Y\|_\infty + \varepsilon, \quad x \in \mathscr{X}.$$

Therefore $\alpha \le \|f + g\|_\infty \le \|f|_Y\|_\infty + \varepsilon$. Since ε was arbitrary, $\alpha \le \|f|_Y\|_\infty$. ◊

The Closed Graph Theorem

Let $\mathscr{X}$ and $\mathscr{Y}$ be Banach spaces and let $\mathscr{X} \times \mathscr{Y}$ have the product topology and vector space structure. For a mapping $T : \mathscr{X} \to \mathscr{Y}$, set $G_T := \{(\boldsymbol{x}, T\boldsymbol{x}) : \boldsymbol{x} \in \mathscr{X}\}$. Note that G_T is closed iff the following condition holds:

$$(\boldsymbol{x}_n) \subseteq \mathscr{X},\ (\boldsymbol{x}, \boldsymbol{y}) \in \mathscr{X} \times \mathscr{Y} \text{ and } (\boldsymbol{x}_n, T\boldsymbol{x}_n) \to (\boldsymbol{x}, \boldsymbol{y}) \Rightarrow T\boldsymbol{x} = \boldsymbol{y}. \tag{8.18}$$

In particular, if T is continuous, then G_T is closed. The converse holds for linear maps:

8.7.12 Closed Graph Theorem. *Let $\mathscr{X}$ and $\mathscr{Y}$ be Banach spaces and let $T : \mathscr{X} \to \mathscr{Y}$ be a linear map such that G_T is closed in $\mathscr{X} \times \mathscr{Y}$. Then T is continuous.*

Proof. Give $\mathscr{X} \times \mathscr{Y}$ the norm $\|(\boldsymbol{x}, \boldsymbol{y})\| = \max\{\|\boldsymbol{x}\|, \|\boldsymbol{y}\|\}$, which generates the product topology (see §8.4). Since T is linear, G_T is a linear subspace of $\mathscr{X} \times \mathscr{Y}$. Define projection mappings $P_X : G_T \to \mathscr{X}$ and $P_Y : G_T \to \mathscr{Y}$ by $P_X(\boldsymbol{x}, T\boldsymbol{x}) = \boldsymbol{x}$ and $P_Y(\boldsymbol{x}, T\boldsymbol{x}) = T\boldsymbol{x}$. These maps are clearly linear and $P_X(\boldsymbol{x}, T\boldsymbol{x})$ is trivially continuous. Moreover, because T is closed, P_Y is continuous. Since P_X is a bijection, $P_X^{-1} : \mathscr{X} \to G_T$ is continuous by the Banach isomorphism theorem. Thus $T = P_Y P_X^{-1}$ is continuous. □

The following corollary is sometimes called the *two norm theorem*.

8.7.13 Corollary. *Let $\mathscr{X}$ be a Banach space with respect to norms $\|\boldsymbol{x}\|$ and $|||\,\boldsymbol{x}\,|||$. Suppose there exists a constant c such that $|||\,\boldsymbol{x}\,||| \le c\,\|\boldsymbol{x}\|$ for all $\boldsymbol{x}$. Then the norms are equivalent.*

Proof. We show that the identity map $I : (\mathscr{X}, |||\cdot|||) \to (\mathscr{X}, \|\cdot\|)$ is continuous. It will follow that $\|\boldsymbol{x}\| = \|I\boldsymbol{x}\| \le \|I\|\,|||\,\boldsymbol{x}\,|||$, proving the corollary.

Let $(\boldsymbol{x}_n, I\boldsymbol{x}_n) = (\boldsymbol{x}_n, \boldsymbol{x}_n) \to (\boldsymbol{x}, \boldsymbol{y})$ in $(\mathscr{X}, |||\cdot|||) \times (\mathscr{X}, \|\cdot\|)$, so $|||\,\boldsymbol{x}_n - \boldsymbol{x}\,||| \to 0$ and $\|\boldsymbol{x}_n - \boldsymbol{y}\| \to 0$. Since $|||\,\boldsymbol{x}_n - \boldsymbol{y}\,||| \le c\,\|\boldsymbol{x}_n - \boldsymbol{y}\|$, we also have $|||\,\boldsymbol{x}_n - \boldsymbol{y}\,||| \to 0$. Therefore, $\boldsymbol{x} = \boldsymbol{y}$. By the closed graph theorem, I is continuous. □

Exercises

8.71 Define linear functionals $f_n(\boldsymbol{x}) = \sum_{j=1}^n x_j$ on $\mathfrak{c}_{00}$. Show that $\sup_n |f_n(\boldsymbol{x})| < \infty$ for all $\boldsymbol{x}$, yet $\sup_n \|f_n\| = \infty$. Conclude that the completeness of $\mathscr{X}$ and $\mathscr{Y}$ in 8.7.1 is essential.

8.72 Let $\mathscr{X}$ be a normed space and $A \subseteq \mathscr{X}$ such that $\sup\{|f(\boldsymbol{x})| : \boldsymbol{x} \in A\} < \infty$ is bounded for every $f \in \mathscr{X}'$. Prove that $\sup\{\|\boldsymbol{x}\| : \boldsymbol{x} \in A\} < \infty$. Thus **weak boundedness** implies norm boundedness.

8.73 Let $\mathscr{X}$, $\mathscr{Y}$ be Banach spaces and $T : \mathscr{X} \to \mathscr{Y}$ linear such that T is **weakly continuous**, that is, $f \circ T$ is continuous for each $f \in \mathscr{Y}'$. Show that T is continuous.

8.74 [↑8.2] Let $\mathscr{X}$, $\mathscr{Y}$, $\mathscr{Z}$ be Banach spaces and let $B : \mathscr{X} \times \mathscr{Y} \to \mathscr{Z}$ be bilinear and separately continuous, that is, continuous in $\boldsymbol{x}$ for each $\boldsymbol{y}$ and continuous in $\boldsymbol{y}$ for each $\boldsymbol{x}$. Show that B is bounded.

8.75 Let $\mathscr{X}$, $\mathscr{Y}$ be Banach spaces and $T \in \mathscr{B}(\mathscr{X},\mathscr{Y})$ injective. Prove: T^{-1} is continuous on $\operatorname{ran}(T)$ iff $\operatorname{ran}(T)$ is closed.

8.76 [↓10.2.11] Let $\mathscr{X}$, $\mathscr{Y}$ be Banach spaces and $T \in \mathscr{B}(\mathscr{X},\mathscr{Y})$ surjective. Show that there exists $c > 0$ such that for each $\boldsymbol{x}$ there exists $\boldsymbol{x}_1$ with $T\boldsymbol{x}_1 = T\boldsymbol{x}$ and $\|\boldsymbol{x}_1\| \le c\,\|T\boldsymbol{x}\|$. ⟦Use 8.7.9.⟧

8.77 Let $\mathscr{X}$ and $\mathscr{Y}$ be Banach spaces and $T : \mathscr{X} \to \mathscr{Y}$ linear. Suppose T has the property that $\boldsymbol{x}_n \to \mathbf{0}$ and $T\boldsymbol{x}_n \to \boldsymbol{y} \Rightarrow \boldsymbol{y} = \mathbf{0}$. Prove that T is continuous.

8.78 Let $(X,\mathscr{F},\mu)$ be a measure space and $T : L^1(\mu) \to L^1(\mu)$ linear with the property that if (f_n) is a sequence in L^1 with $f_n \overset{\text{a.e}}{\to} 0$, then $Tf_n \overset{\text{a.e}}{\to} 0$. Show that T is bounded.

8.79 Let $\mathscr{X}$ be a Banach space, $T \in \mathscr{B}(\mathscr{X})$ injective, and $S : \mathscr{X} \to \mathscr{X}$ linear with TS is continuous. Prove that S is continuous.

8.80 Let $C^1[0,1]$ and $C[0,1]$ have the sup norms. Show that the linear map $D : C^1[0,1] \to C[0,1]$, $Df = f'$, has a closed graph but is unbounded. Thus the completeness hypothesis in the closed graph theorem is essential.

8.81 Let $\|\cdot\|$ be a complete norm on $C[0,1]$ such that the evaluation maps $\widehat{x}(f) = f(x)$, $x \in [0,1] \cap \mathbb{Q}$ are continuous on $C[0,1]$. Show that $\|\cdot\|$ and $\|\cdot\|_\infty$ are equivalent.

8.82 Let $(X,\mathscr{F},\mu)$ be a measure space and g measurable such that $fg \in L^1$ for all $f \in L^1$. Show that the linear mapping $T : f \to fg$ on L^1 is continuous and that $g \in L^\infty$.

8.83 Let $(X,\mathfrak{F},\mu)$ be a measure space and $E \in \mathfrak{F}$. Let $\mathscr{Y} = \{g \in L^1 : g(E) = 0\}$. Show that $L^1(X,\mathfrak{F},\mu)/\mathscr{Y}$ is isometrically isomorphic to $L^1(E,\mathfrak{F}\cap E,\nu)$, where $\nu = \mu\big|_{\mathfrak{F}\cap E}$.

---

*8.8 Applications

Divergent Fourier Series

Let $f : \mathbb{R} \to \mathbb{C}$ be a periodic function with period 2π. The **Fourier series** of f is the formal series

$$f(t) \sim \sum_{k=-\infty}^{\infty} c_k e^{ikt}, \quad \text{where } c_k := \frac{1}{2\pi}\int_0^{2\pi} e^{-ikx} f(x)\,dx.$$

The L^2 convergence of Fourier series is discussed in §11.3. Deeper questions center around pointwise convergence. In the current subsection, we merely demonstrate the existence of a continuous f for which the above series diverges at $t = 0$. The proof here does not actually construct such a function; however, a concrete example was given by Fejer [19].

Let $\mathscr{X}$ be the space of all continuous functions $f : \mathbb{R} \to \mathbb{C}$ with period 2π. Then $\mathscr{X}$ is a Banach space under the sup norm. For $n \in \mathbb{N}$, define $F_n \in \mathscr{X}'$ by

$$F_n(f) = \sum_{k=-n}^{n} \frac{1}{2\pi}\int_0^{2\pi} e^{-ikx} f(x)\,dx,$$

which is the nth partial sum of the Fourier series for f evaluated at $t = 0$. We show that $\lim_n \|F_n\| = \infty$. It will then follow from the uniform boundedness principle that $\sup_n |F_n(f)| = \infty$ for some $f \in \mathscr{X}$, as claimed.

As a first step, we express F_n in terms of the **Dirichlet kernel**

$$D_n(t) = \begin{cases} \dfrac{\sin\left(n+\frac{1}{2}\right)t}{2\sin\frac{1}{2}t} & \text{if } t \notin 2\pi\mathbb{Z}, \\ n+\frac{1}{2} & \text{otherwise.} \end{cases} \tag{†}$$

Expand the numerator in (†) as

$$\sin\left(n+\tfrac{1}{2}\right)t = \sin\tfrac{1}{2}t + \sum_{k=1}^{n}\left[\sin\left(k+\tfrac{1}{2}\right)t - \sin\left(k-\tfrac{1}{2}\right)t\right] = \sin\tfrac{1}{2}t + 2\sum_{k=1}^{n}\cos kt\,\sin\tfrac{1}{2}t.$$

Since $2\cos\theta = e^{i\theta} + e^{-i\theta}$, we then have

$$D_n(t) = \frac{1}{2} + \sum_{k=1}^{n}\cos kt = \frac{1}{2}\sum_{k=-n}^{n} e^{ikt}.$$

The nth partial sum of the Fourier series for f may now be written

$$\sum_{k=-n}^{n}\left(\frac{1}{2\pi}\int_0^{2\pi} e^{-ikx} f(x)\,dx\right)e^{ikt} = \frac{1}{\pi}\int_0^{2\pi} f(x)D_n(t-x)\,dx.$$

Setting $t=0$ and noting that D_n is an even function, we have

$$F_n(f) = \frac{1}{\pi}\int_0^{2\pi} f(x)D_n(x)\,dx, \quad f \in \mathscr{X}.$$

We claim that

$$\|F_n\| = \frac{1}{\pi}\int_0^{2\pi} |D_n(x)|\,dx. \tag{‡}$$

The inequality $\|F_n\| \le \pi^{-1}\int_0^{2\pi}|D_n(x)|\,dx$ is clear. For the reverse inequality, define $g(x)=1$ if $D_n(x)>0$ and $g(x)=-1$ if $D_n(x)\le 0$, so that $|D_n(x)| = g(x)D_n(x)$. Since D_n changes sign at only finitely many points in $[0,2\pi]$, given $\varepsilon>0$ there exists $f\in\mathscr{X}$ with norm one such that $\int_0^{2\pi}|f(x)-g(x)|\,dx < \varepsilon\pi/\|D_n\|_\infty$. Therefore,

$$\begin{aligned}
\frac{1}{\pi}\int_0^{2\pi}|D_n(x)|\,dx &\le |F_n(f)| + \left|F_n(f) - \frac{1}{\pi}\int_0^{2\pi}|D_n(x)|\,dx\right| \\
&= |F_n(f)| + \left|\frac{1}{\pi}\int_0^{2\pi}[f(x)-g(x)]D_n(x)\,dx\right| \\
&< \|F_n\| + \varepsilon,
\end{aligned}$$

verifying (‡).

Finally, from (†), (‡), and the inequality $2|\sin\frac{1}{2}x| \le |x|$ we have

$$\begin{aligned}
\frac{1}{\pi}\|F_n\| &= \int_0^{2\pi}|D_n(x)|\,dx \ge \int_0^{2\pi}\frac{\left|\sin\left(n+\frac{1}{2}\right)x\right|}{x}\,dx = \int_0^{(2n+1)\pi}\frac{|\sin t|}{t}\,dt \\
&= \sum_{k=1}^{2n+1}\int_{(k-1)\pi}^{k\pi}\frac{|\sin t|}{t}\,dt \ge \sum_{k=1}^{2n+1}\frac{1}{k\pi}\int_{(k-1)\pi}^{k\pi}|\sin t|\,dt \\
&= \sum_{k=1}^{2n+1}\frac{2}{k\pi}.
\end{aligned}$$

Thus $\|F_n\|\to\infty$, as required.

Vector-Valued Analytic Functions

Let $\mathscr{X}$ be a complex Banach space, U an open subset of $\mathbb{C}$, and $f : U \to \mathscr{X}$. Then f is said to be **strongly analytic on** U if the limit

$$\lim_{z \to z_0} \frac{f(z) - f(z_0)}{z - z_0}$$

exists in the norm topology of $\mathscr{X}$ for each $z_0 \in U$. Analogously, f is said to be **weakly analytic on** U if the limit holds in the weak topology, that is, if $\boldsymbol{x}' \circ f$ is analytic on U for each $\boldsymbol{x}' \in \mathscr{X}'$. Clearly, strong analyticity implies weak analyticity. The following theorem is the converse.

8.8.1 Theorem. *A weakly analytic function f is strongly analytic.*

Proof. Let $z_0 \in U$. Since $\mathscr{X}$ is complete, it suffices to prove the Cauchy property

$$\lim_{z,w \to z_0} \|g(z) - g(w)\| = 0, \quad \text{where } g(z) = \frac{f(z) - f(z_0)}{z - z_0}.$$

Let $C : z = z_0 + re^{it}$ $(0 \le t \le 2\pi)$ with r so small that C and its interior are contained in U. By continuity, $\sup_{z \in C} |(\boldsymbol{x}' \circ f)(z)| < \infty$, hence, by the uniform boundedness principle applied to the mappings $T_z : \boldsymbol{x}' \to (\boldsymbol{x}' \circ f)(z)$, there exists a constant $M > 0$ such that $\sup_{z \in \gamma} |(\boldsymbol{x}' \circ f)(z)| \le M$ for all $\boldsymbol{x}' \in \mathscr{X}'$ with $\|\boldsymbol{x}'\| \le 1$. Fix such an $\boldsymbol{x}'$ and set $h = \boldsymbol{x}' \circ f$. By the Cauchy integral formula,

$$h(z) = \frac{1}{2\pi i} \int_C \frac{h(\xi)}{\xi - z} d\,\xi, \qquad |z - z_0| < r,$$

hence

$$\frac{h(z) - h(z_0)}{z - z_0} = \frac{1}{2\pi i(z - z_0)} \int_C \left[\frac{h(\xi)}{\xi - z} - \frac{h(\xi)}{\xi - z_0} \right] d\xi = \frac{1}{2\pi i} \int_C \frac{h(\xi)}{(\xi - z)(\xi - z_0)} d\,\xi.$$

Now let $|z - z_0| < r/2$ and $|w - w_0| < r/2$. Then for $\xi \in C$, $|\xi - z| \ge r/2$ and $|\xi - w| \ge r/2$, so from the calculations

$$\langle g(z) - g(w), \boldsymbol{x}' \rangle = \frac{h(z) - h(z_0)}{z - z_0} - \frac{h(w) - h(z_0)}{w - z_0} = \frac{1}{2\pi i} \int_C h(\xi) \frac{z - w}{(\xi - z)(\xi - w)(\xi - z_0)} d\xi$$

we see that $|\langle g(z) - g(w), \boldsymbol{x}' \rangle| \le 4Mr^{-2}|z - w|$. Therefore, $\|g(z) - g(w)\| \le 4Mr^{-2}|z - w|$, verifying the Cauchy property. □

Summability

Let A be an infinite matrix with entries $a_{mn} \in \mathbb{C}$. Then A maps sequences $\boldsymbol{x} = (x_1, x_2, \ldots)$ onto sequences $\boldsymbol{y} = A\boldsymbol{x}$ with mth term the series $y_m := \sum_{n=1}^{\infty} a_{mn} x_n$ (which may or may not converge). We denote the limit of a sequence $\boldsymbol{x}$, if it exists, by $\lim \boldsymbol{x}$. The following theorem characterizes those matrices that preserve limits. It asserts that the *summability property* $\lim A\boldsymbol{x} = \lim \boldsymbol{x}$ holds iff the $\ell^1(\mathbb{N})$ norms of the rows of A are uniformly bounded, the columns are members of $\mathfrak{c}_0$, and the row sums tend to one.

8.8.2 Theorem (Silverman-Toeplitz). *$A\boldsymbol{x} \in \mathfrak{c}$ and $\lim A\boldsymbol{x} = \lim \boldsymbol{x}$ for all $\boldsymbol{x} \in \mathfrak{c}$ iff the following conditions are satisfied:*

$$\text{(a)}\ \sup_m \sum_{n=1}^{\infty} |a_{mn}| < \infty, \qquad \text{(b)}\ \lim_m a_{mn} = 0\ \forall\, n \in \mathbb{N}, \qquad \text{(c)}\ \lim_m \sum_{n=1}^{\infty} a_{mn} = 1.$$

Proof. (*Sufficiency*) Let $\boldsymbol{x} \in \mathfrak{c}$ and $x = \lim \boldsymbol{x}$. By (a), the series $y_m := \sum_{n=1}^{\infty} a_{mn}x_n$ is absolutely convergent for each m. Now write

$$y_m = \sum_{n=1}^{\infty} a_{mn}(x_n - x) + x\sum_{n=1}^{\infty} a_{mn}.$$

By (c), the second term on the right has limit x as $m \to \infty$. Therefore, to show that $y_m \to x$ it suffices to show that the first term on the right tends to zero as $m \to \infty$. Let s denote the supremum in (a). Given $\varepsilon > 0$, choose N so that $|x_n - x| < \varepsilon/s$ for all $n > N$. Then

$$\left|\sum_{n=1}^{\infty} a_{mn}(x_n - x)\right| \le \sum_{n=1}^{N} |a_{mn}|\,|x_n - x| + \sum_{n>N} |a_{mn}|\,|x_n - x| \le \sum_{n=1}^{N} |a_{mn}|\,|x_n - x| + \varepsilon,$$

hence, by (b), $\overline{\lim}_m \left|\sum_{n=1}^{\infty} a_{mn}(x_n - x)\right| \le \varepsilon$, which implies the desired conclusion.

(*Necessity*) Fix n and let $\boldsymbol{x} := (0, \ldots, 0, \overset{n}{1}, 0, \ldots)$. Then $A\boldsymbol{x} = (a_{1n}, a_{2n}, \ldots)$ and $\lim \boldsymbol{x} = 0$, hence $\lim_{m\to\infty} a_{mn} = \lim A\boldsymbol{x} = 0$, proving (b). For (c), take $\boldsymbol{x} = (1, 1 \ldots)$ and argue similarly.

To prove (a), we show first that $\sum_{n=1}^{\infty} |a_{mn}| < \infty$. If this is not the case, then there exists a strictly increasing sequence of indices n_k such that $\sum_{j=n_k+1}^{n_{k+1}} |a_{mj}| > k$. Define

$$x_j = 0, \ 1 \le j \le n_1 \quad x_j = k^{-1}\operatorname{sgn} a_{mj}, \ n_k + 1 \le j \le n_{k+1}, \ k = 1, 2, \ldots.$$

Then $x_j \to 0$, hence $\boldsymbol{x} = (x_j) \in \mathfrak{c}$ and $\lim A\boldsymbol{x} = 0$. On the other hand,

$$(A\boldsymbol{x})_m = \left|\sum_{j=1}^{\infty} a_{mj}x_j\right| = \sum_{k=1}^{\infty}\sum_{j=n_k+1}^{n_{k+1}} \frac{|a_{mj}|}{k} = \infty.$$

This verifies the claim.

Now define a sequence of linear functionals $\boldsymbol{x}'_m$ on $\mathfrak{c}$ by

$$\langle \boldsymbol{x}, \boldsymbol{x}'_m \rangle = \sum_{n=1}^{\infty} a_{mn}x_n = (A\boldsymbol{x})_m.$$

By the above, $\boldsymbol{x}'_m \in \mathfrak{c}'$. Since $\lim_m \langle \boldsymbol{x}, \boldsymbol{x}'_m \rangle$ exists for each $\boldsymbol{x} \in \mathfrak{c}$, $\sup_m \|\boldsymbol{x}'_m\| < \infty$ by the uniform boundedness principle. Since $\|\boldsymbol{x}'_m\| = \sum_{n=1}^{\infty} |a_{mn}|$ (§8.3), (a) holds. □

Schauder Bases

A sequence $(\boldsymbol{e}_n)$ in a normed space $\mathcal{X}$ is said to be a **Schauder basis** or simply a **basis** for $\mathcal{X}$ if $\|\boldsymbol{e}_n\| = 1$ for all n and if each $\boldsymbol{x} \in \mathcal{X}$ can be represented uniquely as a series $\sum_{k=1}^{\infty} c_k\boldsymbol{e}_k$, that is, there exist unique scalars $c_k \in \mathbb{K}$ such that

$$\lim_n \left\|\boldsymbol{x} - \sum_{k=1}^{n} c_k\boldsymbol{e}_k\right\| = 0.$$

For example, the sequences $\boldsymbol{e}_n := (0, \ldots, 0, \overset{n}{1}, 0 \ldots)$ form a basis for each of the spaces $\mathfrak{c}_0$ and ℓ^p, $1 \le p < \infty$. In $\mathfrak{c}$, one must augment this set by $\boldsymbol{e} = (1, 1, \ldots)$ (Ex. 8.85, 8.86).

The uniqueness of the representation implies that the coefficients c_k depend linearly on $\boldsymbol{x}$. Thus we may write

$$\boldsymbol{x} = \sum_{k=1}^{\infty} c_k(\boldsymbol{x})\boldsymbol{e}_k, \quad c_k(\boldsymbol{x}_j) = \delta_{jk}. \tag{8.19}$$

The functions c_k are called **coordinate functionals.**

A normed space with a basis is clearly separable. Schauder conjectured that the converse holds: every separable Banach space has a basis. Later, Enflo disproved the conjecture by exhibiting a separable Banach space with no basis. For details on Schauder bases, see [44]. Our goal in this subsection is simply to prove the following noteworthy result.

8.8.3 Theorem (Banach)**.** *In a Banach space $\mathscr{X}$, the coordinate functionals of a basis are continuous.*

Proof. Define $|||\, \boldsymbol{x} \,||| = \sup_n \left\| \sum_{j=1}^n c_j(\boldsymbol{x})\boldsymbol{e}_j \right\|$. It is easy to check that $|||\cdot|||$ is a norm. Moreover, since $|||\, \boldsymbol{x} \,||| \geq \| \sum_{j=1}^n c_j(\boldsymbol{x})\boldsymbol{e}_j \| \to \|\boldsymbol{x}\|$, we have $\|\boldsymbol{x}\| \leq |||\, \boldsymbol{x} \,|||$. We claim that $\mathscr{X}$ is complete in the new norm. To this end, let $(\boldsymbol{x}_n)$ be a $|||\cdot|||$-Cauchy sequence. Given $\varepsilon > 0$, choose N such that for all $m, n \geq N$, $|||\, \boldsymbol{x}_m - \boldsymbol{x}_n \,||| < \varepsilon$. Thus

$$\left\| \sum_{j=q}^{p} c_j(\boldsymbol{x}_m - \boldsymbol{x}_n)\boldsymbol{e}_j \right\| < \varepsilon, \quad p \geq q \geq 1, \quad m, n \geq N. \tag{$\dagger$}$$

In particular, for each j the sequence $\left(c_j(\boldsymbol{x}_n)\boldsymbol{e}_j\right)_n$ is $\|\cdot\|$-Cauchy and so $\|\cdot\|$-converges. It follows that the limit $\alpha_j := \lim_n c_j(\boldsymbol{x}_n)$ exists in $\mathbb{K}$ (Ex. 8.2). Moreover, for $n \geq N$ and all sufficiently large q,

$$\begin{aligned}\left\| \sum_{j=q}^{p} \alpha_j \boldsymbol{e}_j \right\| &\leq \left\| \sum_{j=q}^{p} \left(\alpha_j - c_j(\boldsymbol{x}_n)\right)\boldsymbol{e}_j \right\| + \left\| \sum_{j=q}^{p} c_j(\boldsymbol{x}_n - \boldsymbol{x}_N)\boldsymbol{e}_j \right\| + \left\| \sum_{j=q}^{p} c_j(\boldsymbol{x}_N)\boldsymbol{e}_j \right\| \\ &\leq \left\| \sum_{j=q}^{p} \left(\alpha_j - c_j(\boldsymbol{x}_n)\right)\boldsymbol{e}_j \right\| + 2\varepsilon.\end{aligned}$$

Letting $n \to \infty$, we obtain the inequality $\left\| \sum_{j=q}^{p} \alpha_j \boldsymbol{e}_j \right\| \leq 2\varepsilon$. Thus the partial sums of the series $\sum_{j=1}^{\infty} \alpha_j \boldsymbol{e}_j$ form a Cauchy sequence, so the series converges in $(\mathscr{X}, \|\cdot\|)$. Set $\boldsymbol{x} := \sum_{j=1}^{\infty} \alpha_j \boldsymbol{e}_j$, so that $c_j(\boldsymbol{x}) = \alpha_j = \lim_m c_j(\boldsymbol{x}_m)$. Letting $m \to \infty$ in ($\dagger$) with $q = 1$ we have

$$\left\| \sum_{j=1}^{p} c_j(\boldsymbol{x} - \boldsymbol{x}_n)\boldsymbol{e}_j \right\| \leq \varepsilon, \quad n \geq N,\ p \geq 1.$$

Taking the supremum over all p, we have $|||\, \boldsymbol{x} - \boldsymbol{x}_n \,||| \leq \varepsilon$, proving that $\boldsymbol{x}_n \to \boldsymbol{x}$ in $(\mathscr{X}, |||\cdot|||)$. Therefore, $(\mathscr{X}, |||\cdot|||)$ is complete.

It now follows from 8.7.13 that $\|\cdot\|$ and $|||\cdot|||$ are equivalent. Since $|c_k(\boldsymbol{x})| = \|c_k(\boldsymbol{x})\boldsymbol{e}_k\| \leq |||\, \boldsymbol{x} \,|||$, c_k is $|||\cdot|||$-continuous, hence also $\|\cdot\|$-continuous. □

Exercises

8.84 Show that the matrix $A = [a_{ij}]$, where $a_{ij} = 1/i$ for $j \leq i$ and $a_{ij} = 0$ otherwise, satisfies (a), (b), and (c) of 8.8.2. Conclude that $\lim_n (x_1 + \cdots + x_n)/n = \lim_n x_n$.

8.85 Show that $\{\boldsymbol{e}_n = (0, \ldots, 0, \overset{n}{1}, 0, \ldots) : n \in \mathbb{N}\}$ is a basis for $\mathfrak{c}_0$ and ℓ^p, $1 \leq p < \infty$, but not for ℓ^∞.

8.86 Show that the vectors $\boldsymbol{e}_n = (0, \ldots, 0, \overset{n}{1}, 0, \ldots)$ together with $\boldsymbol{e} = (1, 1 \ldots)$ form a basis for $\mathfrak{c}$.

8.87 Let $\boldsymbol{d}_1 = (1, 0, \ldots)$ and $\boldsymbol{d}_n = (1, 0 \ldots, 0, \overset{n}{1/n}, 0, \ldots)$, $n \geq 2$.

(a) Show that $(\boldsymbol{d}_n)$ is a basis for $\mathfrak{c}_{00}$.

⟦Consider $c_1(\boldsymbol{x}) := x_1 - \sum_{k=2}^{\infty} k x_k$ and $c_n(\boldsymbol{x}) := n x_n$ $(n \geq 2)$.⟧

(b) Show that $\|\boldsymbol{d}_n - \boldsymbol{d}_1\| \to 0$ but $c_1(\boldsymbol{d}_n - \boldsymbol{d}_1) \not\to 0$. Conclude that completeness of the normed space $\mathscr{X}$ in 8.8.3 is essential.

8.9 The Dual Operator

Definition and Properties

Let $\mathscr{X}$ and $\mathscr{Y}$ be normed spaces. The **dual** of an operator $T \in \mathscr{B}(\mathscr{X}, \mathscr{Y})$ is the mapping $T' : \mathscr{Y}' \to \mathscr{X}'$ defined by

$$\langle \boldsymbol{x}, T'f\rangle = \langle T\boldsymbol{x}, f\rangle, \quad \boldsymbol{x} \in \mathscr{X}, \quad f \in \mathscr{Y}'.$$

Clearly, T' is linear. Moreover, by definition of the norm in $\mathscr{X}'$,

$$\|T'f\| = \sup\{|\langle \boldsymbol{x}, T'f\rangle| : \|\boldsymbol{x}\| \le 1\} = \sup\{|\langle T\boldsymbol{x}, f\rangle| : \|\boldsymbol{x}\| \le 1\} \le \|T\|\,\|f\|,$$

hence $T' \in \mathscr{B}(\mathscr{Y}', \mathscr{X}')$ with $\|T'\| \le \|T\|$. On the other hand, by 8.5.8,

$$\|T\boldsymbol{x}\| = \sup\{|\,\langle T\boldsymbol{x}, f\rangle\,| : \|f\| \le 1\} = \sup\{|\,\langle \boldsymbol{x}, T'f\rangle\,| : \|f\| \le 1\} \le \|T'\|\,\|\boldsymbol{x}\|,$$

hence $\|T'\| \le \|T\|$. We have proved

8.9.1 Proposition. *$T' \in \mathscr{B}(\mathscr{Y}', \mathscr{X}')$ and $\|T'\| = \|T\|$.*

The elementary algebraic properties of the dual operation are given in the following proposition. The proof is left as an exercise.

8.9.2 Proposition. *Let $\mathscr{X}$, $\mathscr{Y}$, and $\mathscr{Z}$ be normed spaces, T, $U \in \mathscr{B}(\mathscr{X}, \mathscr{Y})$, $S \in \mathscr{B}(\mathscr{Y}, \mathscr{Z})$, and $c \in \mathbb{K}$. Then*

$$(T+U)' = T' + U', \quad (cT)' = cT', \quad \text{and} \quad (ST)' = T'S'.$$

Annihilators

Let $\mathscr{X}$ be a normed space. The **annihilators** $A^\perp$ and ${}^\perp B$ of subsets $A \subseteq \mathscr{X}$ and $B \subseteq \mathscr{X}'$ are defined by

$$A^\perp := \{\boldsymbol{x}' \in \mathscr{X}' : \langle \boldsymbol{x}, \boldsymbol{x}'\rangle = 0 \;\forall\; \boldsymbol{x} \in A\} \;\text{ and }\; {}^\perp B := \{\boldsymbol{x} \in \mathscr{X} : \langle \boldsymbol{x}, \boldsymbol{x}'\rangle = 0 \;\forall\; \boldsymbol{x}' \in B\}.$$

Clearly, $A^\perp$ and ${}^\perp B$ are closed linear subspaces of $\mathscr{X}'$ and $\mathscr{X}$, respectively. Moreover, it is easily established that

$$A \subseteq {}^\perp(A^\perp), \quad B \subseteq ({}^\perp B)^\perp, \quad \mathscr{X}^\perp = \{0\}, \quad \text{and} \quad {}^\perp\mathscr{X}' = \{0\},$$

the last property by 8.5.6.

8.9.3 Proposition. *${}^\perp(A^\perp)$ is the closed linear span of $A \subseteq \mathscr{X}$. Thus the linear span of A is dense in $\mathscr{X}$ iff $A^\perp = \{0\}$.*

Proof. Let $\mathscr{Z}$ denote the closed linear span of A. Since ${}^\perp(A^\perp)$ is closed. linear, and contains A, it must contain $\mathscr{Z}$. Let $f \in \mathscr{X}'$ with $f(\mathscr{Z}) = \{0\}$. Then $f \in A^\perp$, hence $f = 0$ on ${}^\perp(A^\perp)$. By 8.5.5, $\mathscr{Z} = {}^\perp(A^\perp)$. The last assertion follows from the obvious fact that ${}^\perp\{0\} = \mathscr{X}$. □

The proof of the next proposition is an exercise for the reader (8.96).

8.9.4 Proposition. *Let $\mathscr{X}$ and $\mathscr{Y}$ be normed spaces and $T \in \mathscr{B}(\mathscr{X}, \mathscr{Y})$. Then*

$$\ker T' = \big[\operatorname{ran} T\big]^\perp \quad \text{and} \quad \ker T = {}^\perp[\operatorname{ran} T'].$$

The following lemma is sometimes useful in establishing invertibility of an operator.

8.9.5 Lemma. *Let $\mathscr{X}$ and $\mathscr{Y}$ be Banach spaces and $T \in \mathscr{B}(\mathscr{X}, \mathscr{Y})$. Then T is injective and* $\operatorname{ran} T$ *is closed iff there exists $a > 0$ such that $\|T\boldsymbol{x}\| \geq a\|\boldsymbol{x}\|$ for all $\boldsymbol{x}$. In this case $T^{-1} \in \mathscr{B}(\operatorname{ran} T, \mathscr{X})$.*

Proof. If T is injective and $\operatorname{ran} T$ is closed, then, by the Banach isomorphism theorem, $T^{-1} \in \mathscr{B}(\operatorname{ran} T, \mathscr{X})$ and $\|\boldsymbol{x}\| = \|T^{-1}T\boldsymbol{x}\| \leq \|T^{-1}\|\,\|T\boldsymbol{x}\|$ for all $\boldsymbol{x} \in \mathscr{X}$. Conversely, assume the inequality holds. Then T is obviously injective. If $T\boldsymbol{x}_n \to \boldsymbol{y}$, then the inequality shows that $(\boldsymbol{x}_n)$ is a Cauchy sequence and so converges to some $\boldsymbol{x} \in \mathscr{X}$. Therefore, $\boldsymbol{y} = T\boldsymbol{x} \in \operatorname{ran} T$, hence $\operatorname{ran} T$ is closed. □

8.9.6 Proposition. *Let $\mathscr{X}$ and $\mathscr{Y}$ be Banach spaces and $T \in \mathscr{B}(\mathscr{X}, \mathscr{Y})$. Then T is invertible iff T' is invertible, in which case $(T^{-1})' = (T')^{-1}$.*

Proof. Assume that T' is invertible and set $c := \|(T')^{-1}\|$. Then $c\,\|T'\boldsymbol{y}'\| \geq \|\boldsymbol{y}'\|$ for all $\boldsymbol{y}'$, hence

$$\begin{aligned} c\,\|T\boldsymbol{x}\| &= \sup\{|\langle \boldsymbol{x}, T'(c\boldsymbol{y}')\rangle| : \|\boldsymbol{y}'\| \leq 1\} = \sup\{|\langle \boldsymbol{x}, T'\boldsymbol{y}'\rangle| : \|\boldsymbol{y}'\| \leq c\} \\ &\geq \sup\{|\langle \boldsymbol{x}, T'\boldsymbol{y}'\rangle| : \|T'\boldsymbol{y}'\| \leq 1\} = \|\boldsymbol{x}\|\,. \end{aligned}$$

By the lemma, $\operatorname{ran} T$ is closed and $T^{-1} \in \mathscr{B}(\operatorname{ran} T, \mathscr{X})$. Since $\left[\operatorname{ran}\, T\right]^{\perp} = \ker\, T' = \{0\}$, $\operatorname{ran}\, T$ is also dense in $\mathscr{Y}$. Therefore, $\operatorname{ran}\, T = \mathscr{Y}$, hence T is invertible. Since

$$T'(T^{-1})' = (T^{-1}T)' = I \quad \text{and} \quad (T^{-1})'T' = (TT^{-1})' = I,$$

$(T^{-1})' = (T')^{-1}$. A similar argument shows that if T is invertible, then T' is invertible. □

Duals of Quotient Spaces and Subspaces

The following theorem uses the quotient map to identify the dual of $\mathscr{X}/\mathscr{Y}$ with $\mathscr{Y}^{\perp}$.

8.9.7 Theorem. *Let $\mathscr{X}$ be a normed space, $\mathscr{Y}$ a closed subspace, and $Q : \mathscr{X} \to \mathscr{X}/\mathscr{Y}$ the quotient map. Then $Q' : (\mathscr{X}/\mathscr{Y})' \to \mathscr{X}'$ is an isometry onto $\mathscr{Y}^{\perp}$.*

Proof. We claim that the range of Q' is $\mathscr{Y}^{\perp}$. Indeed, if $\psi \in (\mathscr{X}/\mathscr{Y})'$ and $\boldsymbol{y} \in \mathscr{Y}$, then $\langle \boldsymbol{y}, Q'\psi\rangle = \langle Q\boldsymbol{y}, \psi\rangle = 0$, hence $Q'\psi \in \mathscr{Y}^{\perp}$. Conversely, if $f \in \mathscr{Y}^{\perp}$, then the equation $\langle Q\boldsymbol{x}, \psi\rangle = \langle \boldsymbol{x}, f\rangle$ defines $\psi \in (\mathscr{X}/\mathscr{Y})'$ with $Q'\psi = f$.

Now, since $\|Q\| \leq 1$,

$$\|Q'\psi\| = \sup\{|\,\langle \boldsymbol{x}, Q'\psi\rangle\,| : \|\boldsymbol{x}\| \leq 1\} = \sup\{|\,\langle Q\boldsymbol{x}, \psi\rangle\,| : \|\boldsymbol{x}\| \leq 1\} \leq \|\psi\|\,.$$

To see that $\|\psi\| \leq \|Q'\psi\|$, let $0 < r < 1$. Since Q is surjective, we may choose $Q\boldsymbol{x}$ with norm one such that $|\langle Q\boldsymbol{x}, \psi\rangle| > r\,\|\psi\|$. Since $\|Q\boldsymbol{x}\| < r^{-1}$ we may choose $\boldsymbol{y} \in \mathscr{Y}$ with $\|\boldsymbol{x} + \boldsymbol{y}\| < r^{-1}$. Then

$$r\,\|\psi\| < |\langle Q\boldsymbol{x}, \psi\rangle| = |\langle Q(\boldsymbol{x} + \boldsymbol{y}), \psi\rangle| = |\langle \boldsymbol{x} + \boldsymbol{y}, Q'\psi\rangle| < r^{-1}\,\|Q'\psi\|\,,$$

and letting $r \to 1$ yields $\|\psi\| \leq \|Q'\psi\|$. □

The next result identifies the dual of a subspace with a quotient space. An analogous result is given in Ex. 8.97.

8.9.8 Theorem. *Let $\mathscr{X}$ be a normed space, $\mathscr{Y}$ a closed subspace, and $Q : \mathscr{X}' \to \mathscr{X}'/\mathscr{Y}^{\perp}$ the quotient map. Then there exists a linear isometry T from $\mathscr{X}'/\mathscr{Y}^{\perp}$ onto $\mathscr{Y}'$ such that $TQ : \mathscr{X}' \to \mathscr{Y}'$ is the restriction mapping $f \to f\big|_{\mathscr{Y}}$.*

Proof. Let $I : \mathscr{Y} \hookrightarrow \mathscr{X}$ denote the inclusion map. Then $I' : \mathscr{X}' \to \mathscr{Y}'$ is the restriction mapping $f \to f|_{\mathscr{Y}}$, which has kernel $\mathscr{Y}^\perp$ and which is surjective by 8.5.4. By 8.7.9, there exists a topological isomorphism T from $\mathscr{X}'/\mathscr{Y}^\perp$ onto $\mathscr{Y}'$ such that $TQ = I'$. To show that T is an isometry, let $f \in \mathscr{X}'$ and $g \in \mathscr{Y}^\perp$. Then

$$\|TQf\| = \|f|_{\mathscr{Y}}\| = \sup\{|\langle \boldsymbol{y}, f\rangle| : \|\boldsymbol{y}\| \leq 1\} = \sup\{|\langle \boldsymbol{y}, f+g\rangle| : \|\boldsymbol{y}\| \leq 1\} \leq \|f+g\|.$$

Taking the infimum on g yields $\|TQf\| \leq \|Qf\|$. On the other hand, given $h \in \mathscr{Y}'$ there exists an $f \in \mathscr{X}'$ such that $I'f = h$ and $\|f\| = \|h\|$ (8.5.4), so $\|TQf\| = \|I'f\| = \|f\| \geq \|Qf\|$. □

Exercises

8.88 Prove 8.9.2.

8.89 Let $\mathscr{X}$, $\mathscr{Y}$ be normed linear spaces and $T \in \mathscr{B}(\mathscr{X}, \mathscr{Y})$. Prove that $\|T\boldsymbol{x}\| = \|T''\widehat{\boldsymbol{x}}\|$.

8.90 Let $\mathscr{X}$ and $\mathscr{Y}$ be normed spaces and $T \in \mathscr{B}(\mathscr{X}, \mathscr{Y})$. Prove that T is an isometry onto $\mathscr{Y}$ iff T' is an isometry onto $\mathscr{X}'$.

8.91 [↑8.34] Let T_r and T_ℓ be the right and left shift operators on $\mathfrak{c}_0$. Identify $\mathfrak{c}_0'$ with ℓ^1 as in §8.3. Find T_r' and $T_\ell{}'$

8.92 Find the dual of the multiplication map M_ϕ of 8.2.3(c) for the case $1 < p < \infty$.

8.93 Let X be a compact Hausdorff space and $\varphi : X \to X$ continuous. Define T on $C(X)$ by $Tf = f \circ \varphi$. Find T' by identifying $C(X)'$ with $M_{ra}(X)$.

8.94 Define $T : L^1(0, \infty) \to C_0(0, \infty)$ by $(Tf)(x) = \displaystyle\int_x^\infty f(t)\,dt$. Find the dual of T.

8.95 Let $1 \leq p < \infty$, $r \neq 0$, and let $D_r : L^p(\lambda^d) \to L^p(\lambda^d)$ be the dilation operator $D_r f(\boldsymbol{x}) = f(r\boldsymbol{x})$. Find D_r' by identifying the dual of L^p with L^q $(p^{-1} + q^{-1} = 1)$.

8.96 Prove 8.9.4.

8.97 Let $\mathscr{Y}$ be a closed subspace of a normed space $\mathscr{X}$. Prove that $\mathscr{X}/\mathscr{Y}$ is isometrically isomorphic to $(\mathscr{Y}^\perp)'$.

8.98 Show that there is a norm one projection of $\mathscr{X}'''$ onto $\mathscr{X}'$ (identified with $\widehat{\mathscr{X}'}$).

8.10 Compact Operators

Throughout this section, $\mathscr{X}$, $\mathscr{Y}$, and $\mathscr{Z}$ denote Banach spaces over $\mathbb{K}$.

In this section we describe the basic properties of compact operators on Banach spaces. A detailed analysis of compact operators on Hilbert spaces is given in Chapter 12.

An operator $T \in \mathscr{B}(\mathscr{X}, \mathscr{Y})$ is said to be **compact** if T maps bounded sets onto relatively compact sets. Equivalently, T is compact iff for any bounded sequence $(\boldsymbol{x}_n)$ in $\mathscr{X}$ the image sequence $(T\boldsymbol{x}_n)$ has a convergent subsequence in $\mathscr{Y}$. The collection of all compact operators in $\mathscr{B}(\mathscr{X}, \mathscr{Y})$ is denoted by $\mathscr{B}_0(\mathscr{X}, \mathscr{Y})$:

$$\mathscr{B}_0(\mathscr{X}, \mathscr{Y}) = \{T \in \mathscr{B}(\mathscr{X}, \mathscr{Y}) : T \text{ is compact}\}.$$

We write $\mathscr{B}_0(\mathscr{X})$ for $\mathscr{B}_0(\mathscr{X}, \mathscr{X})$.

A simple yet fundamental example of a compact operator $T \in \mathscr{B}(\mathscr{X},\mathscr{Y})$ is an operator with finite dimensional range. Such an operator is said to be of **finite rank** and may be expressed in the form

$$T\boldsymbol{x} = \sum_{j=1}^{n} \langle \boldsymbol{x}, \boldsymbol{x}'_j \rangle \boldsymbol{y}_j,$$

where $\boldsymbol{y}_1,\ldots,\boldsymbol{y}_n$ is a basis for $\operatorname{ran} T$ and $\boldsymbol{x}'_j \in \mathscr{X}$. The collection of all operators of finite rank is denoted by $\mathscr{B}_{00}(\mathscr{X},\mathscr{Y})$:

$$\mathscr{B}_{00}(\mathscr{X},\mathscr{Y}) = \{T \in \mathscr{B}(\mathscr{X},\mathscr{Y}) : T \text{ has finite rank}\}.$$

We write $\mathscr{B}_{00}(\mathscr{X})$ for $\mathscr{B}_{00}(\mathscr{X},\mathscr{X})$.

The proof of the following proposition is an exercise for the reader.

8.10.1 Proposition. *$\mathscr{B}_{00}(\mathscr{X},\mathscr{Y})$ and $\mathscr{B}_0(\mathscr{X},\mathscr{Y})$ are linear subspaces of $\mathscr{B}(\mathscr{X},\mathscr{Y})$. Moreover, in the obvious notation,*

$$\mathscr{B}_0(\mathscr{X},\mathscr{Y}) \circ \mathscr{B}(\mathscr{Z},\mathscr{X}) \subseteq \mathscr{B}_0(\mathscr{Z},\mathscr{Y}) \quad \textit{and} \quad \mathscr{B}(\mathscr{Y},\mathscr{Z}) \circ \mathscr{B}_0(\mathscr{X},\mathscr{Y}) \subseteq \mathscr{B}_0(\mathscr{X},\mathscr{Z}),$$

with the analogous inclusions holding for $\mathscr{B}_{00}$. In particular, $\mathscr{B}_0(\mathscr{X})$ and $\mathscr{B}_{00}(\mathscr{X})$ are ideals in the Banach algebra $\mathscr{B}(\mathscr{X})$.

8.10.2 Theorem. *$\mathscr{B}_0(\mathscr{X},\mathscr{Y})$ is operator-norm closed in $\mathscr{B}(\mathscr{X},\mathscr{Y})$.*

Proof. Let $T \in \mathscr{B}(\mathscr{X},\mathscr{Y})$ and $T_n \in \mathscr{B}_0(\mathscr{X},\mathscr{Y})$ with $\|T_n - T\| \to 0$. Given $\varepsilon > 0$ choose n such that $\|T_n - T\| < \varepsilon$, and let $\boldsymbol{x}_1,\ldots,\boldsymbol{x}_m \in C_1$ so that $T_n(C_1) \subseteq \bigcup_{j=1}^m B_\varepsilon(T_n\boldsymbol{x}_j)$. Then for each $\boldsymbol{x} \in C_1$ there exists j such that $\|T_n\boldsymbol{x} - T_n\boldsymbol{x}_j\| < \varepsilon$, hence

$$\|T\boldsymbol{x} - T\boldsymbol{x}_j\| \le \|T\boldsymbol{x} - T_n\boldsymbol{x}\| + \|T_n\boldsymbol{x} - T_n\boldsymbol{x}_j\| + \|T_n\boldsymbol{x}_j - T\boldsymbol{x}_j\| < 3\varepsilon.$$

Therefore, $T(C_1) \subseteq \bigcup_{j=1}^m B_{3\varepsilon}(T\boldsymbol{x}_j)$. Since ε was arbitrary, $T(C_1)$ is totally bounded and hence relatively compact. □

Here is one of several similar examples that are prototypical compact operators.

8.10.3 Example. Let X be a compact Hausdorff topological space, μ a Borel probability measure on X, and $k \in L^2(X \times X)$. Define

$$(Kf)(x) = \int_X k(x,y) f(y)\, d\mu(y), \quad f \in L^2(\mu).$$

An application of the CBS inequality shows that $\|Kf\|_2 \le \|k\|_2 \|f\|_2$, hence K is a bounded linear operator on $L^2(\mu)$ with $\|K\| \le \|k\|_2$. The operator K is called an **integral operator with kernel** k. We show that K is compact.

First, assume that k is continuous. Then the collection of functions $F := \{k(\cdot,y) : y \in X\}$ is compact in $C(X)$, hence given $\varepsilon > 0$ there exist $y_j \in X$ such that $F \subseteq \bigcup_{j=1}^n B_\varepsilon(k(\cdot,y_j))$. Let

$$A_j = \{y \in X : \|k(\cdot,y) - k(\cdot,y_j)\|_\infty < \varepsilon\}, \quad B_1 := A_1, \text{ and } B_j := A_j \cap A_1^c \cap \cdots \cap A_{j-1}^c.$$

Then X is the disjoint union of the sets B_j. Define

$$Tf(x) = \sum_{j=1}^{n} k(x,y_j) \int_{B_j} f(y)\, d\mu(y), \quad f \in L^2(\mu).$$

Then T has finite rank, and for all f with $\|f\|_2 \le 1$ and all $x \in X$,

$$|Kf(x) - Tf(x)| \le \sum_{j=1}^{n} \int_{B_j} |k(x,y) - k(x,y_j)|\,|f(y)|\,d\mu(y) \le \varepsilon \int |f| \le \varepsilon.$$

It follows that $\|K - T\| < \varepsilon$. Therefore, K can be approximated by finite rank operators and so is compact.

In the general case, let k_n be a sequence in $C(X \times X)$ such that $\|k_n - k\|_2 \to 0$ (7.1.2). By the preceding paragraph, the corresponding integral operators K_n are compact. From $\|K - K_n\| \le \|k - k_n\|_2 \to 0$ we see that K is compact. ◊

8.10.4 Theorem. *Let $T \in \mathscr{B}(\mathscr{X}, \mathscr{Y})$. Then T is compact iff T' is compact.*

Proof. Suppose T is compact. We show that if $(f_n) \subseteq \mathscr{Y}'$ with $s := \sup_n \|f_n\| < \infty$, then $(T'f_n)$ has a convergent subsequence. Since T is compact, $\operatorname{ran} T$ has a countable dense set, say $(\boldsymbol{y}_n)$ (Ex. 8.101). A standard diagonal argument shows that (f_n) has a subsequence (g_n) such that $\lim_n g_n(\boldsymbol{y}_k)$ exists for all k. We claim that $\lim_n g_n(\boldsymbol{y})$ exists for all $\boldsymbol{y} \in \operatorname{cl} \operatorname{ran} T$. Indeed, for such $\boldsymbol{y}$ and any m, n, k,

$$\begin{aligned}|g_m(\boldsymbol{y}) - g_n(\boldsymbol{y})| &\le |g_m(\boldsymbol{y}) - g_m(\boldsymbol{y}_k)| + |g_m(\boldsymbol{y}_k) - g_n(\boldsymbol{y}_k)| + |g_n(\boldsymbol{y}_k) - g_n(\boldsymbol{y})| \\ &\le 2C\|\boldsymbol{y} - \boldsymbol{y}_k\| + |g_m(\boldsymbol{y}_k) - g_n(\boldsymbol{y}_k)|,\end{aligned}$$

and since $\boldsymbol{y}$ may be approximated by a $\boldsymbol{y}_k$ we see that $(g_n(\boldsymbol{y}))_n$ is a Cauchy sequence, verifying the claim.

Now let $g(\boldsymbol{y}) := \lim_n g_n(\boldsymbol{y})$ ($\boldsymbol{y} \in \operatorname{cl} \operatorname{ran} T$). Clearly, g is linear and $|g(\boldsymbol{y})| = \lim_n |g_n(\boldsymbol{y})| \le s\|\boldsymbol{y}\|$, hence g is continuous on $\operatorname{cl} \operatorname{ran} T$. Therefore, $g \circ T \in \mathscr{X}'$, and for any $\boldsymbol{x} \in \mathscr{X}$

$$\lim_n \langle \boldsymbol{x}, T'g_n\rangle = \lim_n \langle T\boldsymbol{x}, g_n\rangle = \langle \boldsymbol{x}, g \circ T\rangle.$$

We claim that $\|T'g_n - g \circ T\| \to 0$. Suppose the claim is false. Then there exists $\varepsilon > 0$ such that $\|T'g_n - g \circ T\| \ge \varepsilon$ for infinitely many n, say for $n \in S$. For each $n \in S$ choose $\boldsymbol{x}_n$ with norm one such that

$$|g_n(T\boldsymbol{x}_n) - g(T\boldsymbol{x}_n)| = |\langle \boldsymbol{x}_n, T'g_n - g \circ T\rangle| \ge \varepsilon/2. \tag{†}$$

Since T is compact, there exists a strictly increasing sequence $(n_k)_k$ in S and $\boldsymbol{y} \in \mathscr{Y}$ such that $T\boldsymbol{x}_{n_k} \to \boldsymbol{y}$. Since $\sup_n \|g_n\| < \infty$, $g_{n_k}(T\boldsymbol{x}_{n_k}) \to g(\boldsymbol{y})$. But this contradicts (†). Therefore, $\|T'g_n - g \circ T\| \to 0$, hence T' is compact. The proof that T' compact $\Rightarrow T$ compact is left as an exercise (8.103). □

*Fredholm Alternative for Compact Operators

Let A be an $n \times n$ matrix. A standard argument shows that one of the following holds:

(i) The system of equations $Ax = 0$ has a nonzero solution in $\mathbb{K}^n$.

(ii) The system $Ax = y$ has a unique solution for each $y \in \mathbb{K}^n$.

In this subsection we prove an infinite dimensional version of this result using the following lemmas.

8.10.5 Lemma. *Let $T \in \mathscr{B}(\mathscr{X})$ be compact, and for each $\boldsymbol{x} \in \mathscr{X}$ let $d(\boldsymbol{x})$ denote the distance from $\boldsymbol{x}$ to $\ker(I - T)$. Then there exists $M > 0$ such that $d(\boldsymbol{x}) \le M\,\|(I - T)\boldsymbol{x}\|$ for all $\boldsymbol{x}$.*

Proof. We may assume $(I - T)\boldsymbol{x} \neq \mathbf{0}$. If the conclusion of the lemma is false, then the ratio $d(\boldsymbol{x})/\|(I-T)\boldsymbol{x}\|$ is unbounded, so there exists a sequence $(\boldsymbol{x}_n)$ such that $d(\boldsymbol{x}_n)/\|(I-T)\boldsymbol{x}_n\| \to \infty$. Since $\ker(I-T)$ is closed, $d(\boldsymbol{x}_n) = \|\boldsymbol{x}_n - \boldsymbol{u}_n\|$ for some $\boldsymbol{u}_n \in \ker(I-T)$. Set $\boldsymbol{y}_n = d(\boldsymbol{x}_n)^{-1}(\boldsymbol{x}_n - \boldsymbol{u}_n)$. We then have

$$(I-T)\boldsymbol{y}_n = \frac{1}{d(\boldsymbol{x}_n)}(I-T)\boldsymbol{x}_n \to 0. \tag{$\dagger$}$$

Since $\|\boldsymbol{y}_n\| = 1$ and T is compact, some subsequence of $(T\boldsymbol{y}_n)$ converges, say $T\boldsymbol{y}_{n_k} \to \boldsymbol{y}$. By ($\dagger$), $\boldsymbol{y}_{n_k} \to \boldsymbol{y}$, hence $(I-T)\boldsymbol{y} = \mathbf{0}$ and so $\boldsymbol{u}_{n_k} + d(\boldsymbol{x}_{n_k})\boldsymbol{y} \in \ker(I-T)$. But then

$$\|\boldsymbol{y}_{n_k} - \boldsymbol{y}\| = \frac{1}{d(\boldsymbol{x}_{n_k})}\|\boldsymbol{x}_{n_k} - (\boldsymbol{u}_{n_k} + d(\boldsymbol{x}_{n_k}))\boldsymbol{y}\| \geq 1,$$

contradicting that $\boldsymbol{y}_{n_k} \to \boldsymbol{y}$. □

8.10.6 Lemma. *Let $T \in \mathscr{B}(\mathcal{X})$ be compact. Then* $\operatorname{ran}(I-T')$ *is closed.*

Proof. Set $S := I-T$. Since $\operatorname{ran} S' \subseteq \operatorname{cl}\operatorname{ran} S' = (\ker S)^\perp$, it suffices to show that $(\ker S)^\perp \subseteq \operatorname{ran} S'$. Let $f \in (\ker S)^\perp$ and define a linear map g on $\operatorname{ran} S$ by $g(S\boldsymbol{x}) = f(\boldsymbol{x})$. By 8.10.5, there exists $M > 0$ and for each $\boldsymbol{x} \in \mathcal{X}$ a member $\widetilde{\boldsymbol{x}} \in \ker S$ such that $d(\boldsymbol{x}) = \|\boldsymbol{x} - \widetilde{\boldsymbol{x}}\| \leq M\|S\boldsymbol{x}\|$. Therefore,

$$|g(S\boldsymbol{x})| = |g(S(\boldsymbol{x} - \widetilde{\boldsymbol{x}}))| = |f(\boldsymbol{x} - \widetilde{\boldsymbol{x}})| \leq \|f\|\,\|\boldsymbol{x} - \widetilde{\boldsymbol{x}}\| \leq M\|f\|\,\|S\boldsymbol{x}\|,$$

hence g is continuous on $\operatorname{ran} S$ and so has a continuous extension to $\operatorname{cl}\operatorname{ran} S$. By the Hahn-Banach theorem, g extends to a member h of $\mathcal{X}'$. Then for all $\boldsymbol{x}$,

$$\langle \boldsymbol{x}, f\rangle = \langle S\boldsymbol{x}, g\rangle = \langle S\boldsymbol{x}, h\rangle = \langle \boldsymbol{x}, S'h\rangle,$$

hence $f = S'h \in \operatorname{ran} S'$. □

We may now prove

8.10.7 Theorem (Fredholm). *Let $T \in \mathscr{B}_0(\mathcal{X})$ and $\lambda \neq 0$. Then $\lambda I - T$ is surjective iff $\lambda I - T$ is injective. Thus one of the following holds:*

(i) *The equation $T\boldsymbol{x} - \lambda\boldsymbol{x} = 0$ has a nonzero solution.*
(ii) *The equation $T\boldsymbol{x} - \lambda\boldsymbol{x} = \boldsymbol{y}$ has a unique solution for any $\boldsymbol{y} \in \mathcal{X}$.*

Proof. Since $\lambda I - T = \lambda(I - \lambda^{-1}T)$ and $\lambda^{-1}T$ is compact, we may take $\lambda = 1$. Set $S := I-T$. Suppose that S is surjective but not injective. Then $S\boldsymbol{x}_1 = 0$ for some $\boldsymbol{x}_1 \neq 0$. We claim that the containment $\ker(S^{n-1}) \subseteq \ker(S^n)$ is proper. Indeed, since S is surjective, there exists x_2 such that $S\boldsymbol{x}_2 = \boldsymbol{x}_1$, and in general there exists a vector x_n such that $S\boldsymbol{x}_n = \boldsymbol{x}_{n-1}$. Then $S^n x_n = S^{n-1}x_{n-1} = \cdots = Sx_1 = 0$ and $S^{n-1}x_n = S^{n-2}x_{n-1} = \cdots = Sx_2 = x_1 \neq 0$, so $x_n \in \ker(S^n) \setminus \ker(S^{n-1})$, verifying the claim. By 8.1.6 there exists $\boldsymbol{y}_n \in \ker(S^n) \setminus \ker(S^{n-1})$ such that

$$\|\boldsymbol{y}_n\| = 1 \quad\text{and}\quad \inf\{\|\boldsymbol{y}_n - \boldsymbol{y}\| : \boldsymbol{y} \in \ker(S^{n-1})\} \geq 1/2.$$

Now write

$$T\boldsymbol{y}_n - T\boldsymbol{y}_m = (I-S)\boldsymbol{y}_n - (I-S)\boldsymbol{y}_m = \boldsymbol{y}_n + [S\boldsymbol{y}_m - \boldsymbol{y}_m - S\boldsymbol{y}_n].$$

The term in square brackets is in $\ker(S^{n-1})$ for all $n > m$, hence $\|T\boldsymbol{y}_n - T\boldsymbol{y}_m\| \geq 1/2$. But then $(T\boldsymbol{y}_n)$ has no convergent subsequence, contradicting that T is compact. Therefore, S is injective.

Conversely, assume that S is injective. We claim that $\operatorname{ran}(S)$ is closed. To verify this, we use the following simple observation regarding sequences $(\boldsymbol{x}_n)$ in $\mathscr{X}$:

$$T\boldsymbol{x}_n \to \boldsymbol{u} \text{ and } S\boldsymbol{x}_n \to \boldsymbol{v} \Rightarrow \boldsymbol{x}_n \to \boldsymbol{u} + \boldsymbol{v} \Rightarrow S(\boldsymbol{u} + \boldsymbol{v}) = \boldsymbol{v}. \tag{†}$$

Now let $S\boldsymbol{x}_n \to \boldsymbol{v}$. We consider two cases:

Case 1. $(\boldsymbol{x}_n)$ has a bounded subsequence. By compactness of T, there exists a subsequence $(\boldsymbol{x}_{n_k})$ of $(\boldsymbol{x}_n)$ such that $T\boldsymbol{x}_{n_k} \to \boldsymbol{u}$ for some $\boldsymbol{u} \in \mathscr{X}$. Applying (†) we then have $\boldsymbol{v} = S(\boldsymbol{u} + v) \in \operatorname{ran} S$, as required.

Case 2. $(\boldsymbol{x}_n)$ has no bounded subsequence. Then $\|\boldsymbol{x}_n\| \to \infty$, and setting $\boldsymbol{y}_n = \|\boldsymbol{x}_n\|^{-1} \boldsymbol{x}_n$ we have $\|\boldsymbol{y}_n\| = 1$ and $S\boldsymbol{y}_n \to 0$. Going to a subsequence if necessary, we may assume that $T\boldsymbol{y}_n \to \boldsymbol{u}$ for some $\boldsymbol{u}$. Applying (†) with $\boldsymbol{v} = 0$, we then have $S\boldsymbol{u} = 0$ and $\boldsymbol{y}_n \to \boldsymbol{u}$. But the latter implies that $\boldsymbol{u} \neq 0$, which contradicts the injectivity of S. Therefore, Case 2 is not possible, verifying the claim.

Now, since S is injective, $\operatorname{cl}\operatorname{ran}(S') = \ker(S)^{\perp} = \mathscr{X}$. But by 8.10.6, $\operatorname{ran}(S')$ is closed. Therefore, S' is surjective. By the necessity of the theorem applied to S', recalling that T' is compact, S' is injective. Thus, since $\operatorname{ran}(S)$ is closed, $\operatorname{ran}(S) = {}^{\perp}\ker(S') = \mathscr{X}$, that is, S is surjective, completing the proof. □

The Fredholm alternative is typically applied to integral equations in L^2 of the form $\lambda f - Kf = g$, where K is defined as in 8.10.3. If $|\lambda| > \|K\|$ it may be shown that $(\lambda I - K)^{-1}$ exists and may be expanded into an operator norm convergent series:

$$(\lambda I - K)^{-1} = \lambda^{-1} I + \lambda^{-2} K + \lambda^{-3} K^2 + \cdots$$

(see 13.1.3). The solution to the equation $\lambda f - Kf = g$ is therefore given by

$$f = \lambda^{-1} g + \lambda^{-2} K g + \lambda^{-3} K^2 g + \cdots .$$

Exercises

8.99 Let $\mathscr{X}$ be a Banach space, $(\boldsymbol{x}_n)$, $(\boldsymbol{x}'_n)$ sequences in $\mathscr{X}$ and $\mathscr{X}'$ with norm ≤ 1, and $(c_n) \subseteq \ell^1(\mathbb{N})$. Show that the operator $T\boldsymbol{x} = \sum_{k=1}^{\infty} c_k \langle \boldsymbol{x}, \boldsymbol{x}'_k \rangle \boldsymbol{x}_k$ is compact.

8.100 Prove 8.10.1.

8.101 Prove that if $T \in \mathscr{B}(\mathscr{X}, \mathscr{Y})$ is compact, then T has separable range.

8.102 Let $\mathscr{X}$ be a normed space, $\lambda \neq 0$, and $T \in \mathscr{B}(\mathscr{X})$ compact. Prove that $\ker(\lambda - T)^m$ is finite dimensional for all $m \in \mathbb{N}$.

8.103 Let $\mathscr{X}$ and $\mathscr{Y}$ be Banach spaces and $T \in \mathscr{B}(\mathscr{X}, \mathscr{Y})$. Prove: If T' is compact, then T is compact.

Chapter 9

Locally Convex Spaces

A locally convex linear space $\mathcal{X}$ is a generalization of a normed space, the topology on $\mathcal{X}$ given by a family of seminorms rather than a single norm. These spaces occur in a variety of contexts, including operator theory and distributions. In the present chapter we develop the properties of locally convex spaces to a sufficient extent that will allow the discussion of weak and weak* topologies in the next chapter and the material on distributions in Chapter 15 to be seen from a general vantage point. Additional properties of locally convex spaces as well as applications are considered in Chapter 14.

9.1 General Properties

Let $\mathcal{X}$ be a vector space over $\mathbb{K}$. A **vector topology** on $\mathcal{X}$ is a topology relative to which the vector space operations

$$(\boldsymbol{x}, \boldsymbol{y}) \mapsto \boldsymbol{x} + \boldsymbol{y} : \mathcal{X} \times \mathcal{X} \to \mathcal{X} \quad \text{and} \quad (c, \boldsymbol{x}) \mapsto c\boldsymbol{x} : \mathbb{K} \times \mathcal{X} \to \mathcal{X}$$

are continuous. A vector space $\mathcal{X}$ with a Hausdorff vector topology $\mathcal{T}$ is called a **topological vector space** (TVS) and is denoted by $\mathcal{X}_{\mathcal{T}}$. We omit the subscript when there is no possibility of confusion. Since translation $\boldsymbol{x} \mapsto \boldsymbol{x} + \boldsymbol{y}$ in a TVS is a homeomorphism, a neighborhood base at $\boldsymbol{x}$ is of the form $\boldsymbol{x} + \mathcal{U}_0$, where $\mathcal{U}_0$ is a neighborhood base at zero. It follows that if $\mathcal{X}$ and $\mathcal{Y}$ are topological vector spaces, then a linear map $T : \mathcal{X} \to \mathcal{Y}$ is continuous iff it is continuous at zero.

A **locally convex topology** is a vector topology with a neighborhood base at zero consisting of open convex sets. A **locally convex space** (LCS) is vector space with a Hausdorff locally convex topology. Every normed space is a LCS since the balls B_r are convex. Additional examples are given in the exercises and in later chapters.

Geometry and Topology

The continuity of vector operations in a TVS implies that the closure of a subspace is a linear space, the closure of a convex set is convex, and the closure of a balanced set is balanced. The **closed convex hull** of a subset A of a TVS $\mathcal{X}$ is the intersection of all closed convex subsets of $\mathcal{X}$ containing A. It may also be characterized as the closure of the convex hull of A. Similarly, the **closed, convex, balanced hull** of A is the intersection of all closed, convex, balanced subsets of $\mathcal{X}$ containing A and may be realized as the closure of the convex balanced hull of A (Ex. 9.3.)

The open, balanced neighborhoods of zero in TVS form a neighborhood base at zero, as do the open, convex, balanced neighborhoods of zero in LCS. To establish these facts we need the following lemma:

9.1.1 Lemma. *Let $\mathcal{X}$ be a TVS and E a subset with nonempty interior. If E is balanced (convex), then* $\operatorname{int} E$ *is balanced (convex).*

Proof. Let E be balanced and $0 < |c| \leq 1$. Since $\boldsymbol{x} \to c\boldsymbol{x}$ is a homeomorphism, $c \operatorname{int} E = \operatorname{int} cE \subseteq \operatorname{int} E$. Therefore, $\operatorname{int} E$ is balanced.

Now let E be convex and let $\boldsymbol{x}, \boldsymbol{y} \in \operatorname{int} E$. Choose a neighborhood U of zero such that $\boldsymbol{x} + U \subseteq E$ and $\boldsymbol{y} + U \subseteq E$. For arbitrary $\boldsymbol{u} \in U$, the vectors $\boldsymbol{x} + \boldsymbol{u}$ and $\boldsymbol{y} + \boldsymbol{u}$ are in E, hence, by convexity,

$$t\boldsymbol{x} + (1-t)\boldsymbol{y} + \boldsymbol{u} = t(\boldsymbol{x} + \boldsymbol{u}) + (1-t)(\boldsymbol{y} + \boldsymbol{u}) \in E.$$

Thus $t\boldsymbol{x} + (1-t)\boldsymbol{y} + U \subseteq E$ and so $t\boldsymbol{x} + (1-t)\boldsymbol{y} \in \operatorname{int} E$. Therefore, $\operatorname{int} E$ is convex. □

9.1.2 Proposition. *A TVS has a neighborhood base at zero of open (closed) balanced sets. A LCS has a neighborhood base at zero of open (closed) convex balanced sets.*

Proof. Let W be any neighborhood of zero. By continuity of scalar multiplication at zero, there exists an open neighborhood V of zero and $\delta > 0$ such that $cV \subseteq W$ for all $|c| \leq \delta$. Then $\bigcup_{|c| \leq \delta} cV$ is a balanced, open neighborhood of zero contained in W. Thus a TVS has a neighborhood base at zero of open balanced sets.

By the first paragraph and continuity of addition at $(\mathbf{0}, \mathbf{0})$, we may choose a balanced neighborhood U of zero so that $U + U \subseteq W$. If $\boldsymbol{x} \in \operatorname{cl} U$, then $(\boldsymbol{x} + U) \cap U \neq \emptyset$, hence there exists $\boldsymbol{u}_1, \boldsymbol{u}_2 \in U$ such that $\boldsymbol{x} + \boldsymbol{u}_1 = \boldsymbol{u}_2$. Since U is balanced, $-\boldsymbol{u}_1 \in U$, hence $\boldsymbol{x} = \boldsymbol{u}_2 - \boldsymbol{u}_1 \in U + U \subseteq W$. Therefore, $\operatorname{cl} U \subseteq W$. This shows that a TVS has a neighborhood base at zero of closed balanced sets.

For a LCS, we may take W to be convex. Let V be a balanced, open neighborhood of zero contained in W. The convex hull $\operatorname{co} V$ is still balanced and $\operatorname{co} V \subseteq W$. Since $V \subseteq \operatorname{co} V$ and V is open, $V \subseteq \operatorname{int} \operatorname{co} V$. In particular, $\mathbf{0} \in \operatorname{int} \operatorname{co} V$. Therefore, $\operatorname{int} \operatorname{co} V$ is an open, convex, balanced neighborhood of zero contained in W. This shows that a LCS has a neighborhood base at zero of open, convex, balanced sets. An argument similar to that in the preceding paragraph shows that a LCS has a neighborhood base at zero of closed, convex, balanced sets. □

Seminormed Spaces

Let $\mathscr{P}$ be a family of seminorms on a vector space $\mathscr{X}$. The initial topology induced by the collection of all functions of the form $\boldsymbol{z} \mapsto p(\boldsymbol{z} - \boldsymbol{y})$, where $p \in \mathscr{P}$ and $\boldsymbol{y} \in \mathscr{X}$, is called the **seminorm topology generated by** $\mathscr{P}$. The space $\mathscr{X}$ with this topology is called a **seminormed space**. A neighborhood base at $\boldsymbol{x}$ for a seminorm topology consists of finite intersections of sets of the form

$$\{\boldsymbol{z} \in \mathscr{X} : |p(\boldsymbol{z} - \boldsymbol{y}) - p(\boldsymbol{x} - \boldsymbol{y})| < \varepsilon\}, \quad p \in \mathscr{P}, \quad \boldsymbol{y} \in \mathscr{X}, \quad \varepsilon > 0.$$

Taking $\boldsymbol{y} = \boldsymbol{x}$ produces the smaller collection of sets

$$\{\boldsymbol{z} \in \mathscr{X} : p(\boldsymbol{z} - \boldsymbol{x}) < \varepsilon\}, \quad p \in \mathscr{P}, \quad \varepsilon > 0.$$

It is sufficient to consider these, as the inequality $|p(\boldsymbol{z} - \boldsymbol{y}) - p(\boldsymbol{x} - \boldsymbol{y})| \leq p(\boldsymbol{z} - \boldsymbol{x})$ implies that the smaller collection produces an equivalent neighborhood base at $\boldsymbol{x}$. Thus a neighborhood base at $\boldsymbol{x}$ consist of sets of the form

$$\left\{\boldsymbol{z} \in \mathscr{X} : \max_{1 \leq j \leq n} p_j(\boldsymbol{z} - \boldsymbol{x}) < \varepsilon\right\} = \boldsymbol{x} + U, \quad U := \left\{\boldsymbol{y} \in \mathscr{X} : \max_{1 \leq j \leq n} p_j(\boldsymbol{y}) < \varepsilon\right\}, \tag{9.1}$$

where $p_j \in \mathscr{P}$ and $\varepsilon > 0$. In particular, a net $(\boldsymbol{x}_\alpha)$ converges to $\boldsymbol{x}$ in this topology iff $p(\boldsymbol{x}_\alpha - \boldsymbol{x}) \to 0$ for all $p \in \mathscr{P}$. It follows easily from properties of seminorms that the seminorm topology on $\mathscr{X}$ is a vector topology.

A family $\mathscr{P}$ of seminorms on a vector space $\mathcal{X}$ is said to be **separating** if

$$\bigcap_{p\in\mathscr{P}}\{\boldsymbol{x} : p(\boldsymbol{x}) = 0\} = \{\boldsymbol{0}\};$$

equivalently, $\boldsymbol{x} \neq \boldsymbol{y} \Rightarrow p(\boldsymbol{x} - \boldsymbol{y}) > 0$ for some $p \in \mathscr{P}$. Setting $\varepsilon = p(\boldsymbol{x} - \boldsymbol{y})$ and defining $V := \{\boldsymbol{z} : p(\boldsymbol{z}) < \varepsilon/2\}$ we see that the neighborhoods $\boldsymbol{x} + V$ and $\boldsymbol{y} + V$ are disjoint, hence the seminorm topology of a separating family $\mathscr{P}$ is Hausdorff. Conversely, if the topology is Hausdorff, then for each $\boldsymbol{x} \neq \boldsymbol{0} \in \mathcal{X}$ there exists a set U as in (9.1) such that $\boldsymbol{x} \notin U$, hence $p_j(\boldsymbol{x}) \geq \varepsilon$ for some j. We have proved

9.1.3 Proposition. *A seminormed space is Hausdorff iff the family of seminorms is separating.*

Now let $\mathcal{X}$ have the seminorm topology generated by a separating family $\mathscr{P}$ of seminorms. The triangle inequality $|p(\boldsymbol{x}) - p(\boldsymbol{y})| \leq p(\boldsymbol{x} - \boldsymbol{y})$ implies that a seminorm is continuous. Conversely, if q is *any* continuous seminorm on $\mathcal{X}$, then the set $\{x : q(\boldsymbol{x}) < \varepsilon\}$ is open in $\mathcal{X}$, hence the enhanced family of seminorms $\mathscr{P} \cup \{q\}$ generates no new open sets and therefore produces the same topology as the original family. This observation is occasionally useful in reducing the complexity of some arguments. For example, the basic neighborhood of zero U in (9.1) may be described in terms of a single continuous seminorm $p(\boldsymbol{x}) = \varepsilon^{-1}\max_j p_j(\boldsymbol{x})$ as $U = \{\boldsymbol{x} : p(\boldsymbol{x}) < 1\}$.

We show next that every Hausdorff seminormed space is a locally convex space and vice versa. For this we need the following notion: For a neighborhood of zero U in a TVS $\mathcal{X}$, define the **Minkowski functional of** U by

$$p_{U}(\boldsymbol{x}) := \inf\{t > 0 : \boldsymbol{x} \in tU\}. \tag{9.2}$$

By continuity of scalar multiplication, $s\boldsymbol{x} \in U$ for sufficiently small $s > 0$, hence $p_U(\boldsymbol{x}) < \infty$. The following result is the key to establishing the connection between locally convex spaces and seminormed spaces.

9.1.4 Proposition. *Let U be an open, convex, balanced neighborhood of zero in a TVS $\mathcal{X}$. Then p_U is a Minkowski functional that is continuous in the topology of $\mathcal{X}$. Moreover,*

$$U = \{\boldsymbol{x} : p_U(\boldsymbol{x}) < 1\}. \tag{9.3}$$

Proof. To verify the subadditivity property, let $\boldsymbol{x} \in sU$ and $\boldsymbol{y} \in tU$ $(s, t > 0)$. By convexity of U,

$$\frac{1}{s+t}(\boldsymbol{x} + \boldsymbol{y}) = \frac{s}{s+t}(s^{-1}\boldsymbol{x}) + \frac{t}{s+t}(t^{-1}\boldsymbol{y}) \in U,$$

hence $p_U(\boldsymbol{x} + \boldsymbol{y}) \leq s + t$. Since s and t were arbitrary, $p_U(\boldsymbol{x} + \boldsymbol{y}) \leq p_U(\boldsymbol{x}) + p_U(\boldsymbol{y})$.

For positive homogeneity, let $c \in \mathbb{F}$, $c \neq 0$. Since U is balanced, $c^{-1}U = |c^{-1}|U$, hence

$$\begin{aligned} p_U(c\boldsymbol{x}) &= \inf\{t > 0 : c\boldsymbol{x} \in tU\} = \inf\{t > 0 : \boldsymbol{x} \in tc^{-1}U\} = \inf\{|c|s > 0 : \boldsymbol{x} \in sU\} \\ &= |c|p_U(\boldsymbol{x}). \end{aligned}$$

Therefore, p_U is a Minkowski functional.

To see that p_U is continuous, let $\boldsymbol{x}_\alpha \to \boldsymbol{x}$ in $\mathcal{X}$ and $\varepsilon > 0$. Then $\varepsilon^{-1}(\boldsymbol{x} - \boldsymbol{x}_\alpha) \to 0$ hence, eventually, $\varepsilon^{-1}(\boldsymbol{x} - \boldsymbol{x}_\alpha) \in U$ and so $|p_U(\boldsymbol{x}) - p_U(\boldsymbol{x}_\alpha)| \leq p_U(\boldsymbol{x} - \boldsymbol{x}_\alpha) \leq \varepsilon$.

To establish (9.3), let $x \in U$ and note that by continuity of scalar multiplication there exists $t > 1$ such that $t\boldsymbol{x} \in U$, hence $p_U(\boldsymbol{x}) \leq 1/t < 1$. Therefore, $U \subseteq \{\boldsymbol{x} : p_U(\boldsymbol{x}) < 1\}$. For the reverse inclusion, let $p_U(\boldsymbol{x}) < 1$ and choose t so that $\boldsymbol{x} \in tU$ and $p_U(\boldsymbol{x}) < t < 1$. By the balanced property, $\boldsymbol{x} = t(\boldsymbol{x}/t) \in U$. □

9.1.5 Theorem. *A Hausdorff seminormed space $\mathscr{X}$ is locally convex. Conversely, every LCS is a Hausdorff seminormed space.*

Proof. Let $\mathscr{X}$ be a Hausdorff seminormed space, and let $\boldsymbol{x}_\alpha \to \boldsymbol{x}$, $\boldsymbol{y}_\alpha \to \boldsymbol{y}$ in $\mathscr{X}$ and $c_\alpha \to c$ in $\mathbb{K}$. Then

$$p(\boldsymbol{x}_\alpha + \boldsymbol{y}_\alpha - (\boldsymbol{x}+\boldsymbol{y})) \le p(\boldsymbol{x}_\alpha - \boldsymbol{x}) + p(\boldsymbol{y}_\alpha - \boldsymbol{y}) \to 0 \text{ and}$$
$$p(c_\alpha \boldsymbol{x}_\alpha - c\boldsymbol{x}) \le |c_\alpha| p(\boldsymbol{x}_\alpha - \boldsymbol{x}) + |c_\alpha - c| p(\boldsymbol{x}) \to 0,$$

hence $\mathscr{X}$ is a TVS. As the sets U in (9.1) are open and convex, $\mathscr{X}$ is a LCS. Conversely, let $\mathscr{X}$ be a LCS. By (9.3), the seminorms p_U, where U runs through a base of open, convex, balanced neighborhoods of zero, generate the given locally convex topology. □

In view of Theorem 9.1.5, we may (and shall) abandon the phrase "seminormed space" and use instead the LCS terminology.

Fréchet Spaces

If $\mathscr{X}$ is a LCS with a countable generating class (p_n) of seminorms (or, equivalently, a countable basis of open convex neighborhoods of zero), then

$$d(\boldsymbol{x}, \boldsymbol{y}) := \sum_{n=1}^{\infty} 2^{-n} \frac{p_n(\boldsymbol{x}-\boldsymbol{y})}{1 + p_n(\boldsymbol{x}-\boldsymbol{y})} \tag{9.4}$$

defines a metric for the locally convex topology of $\mathscr{X}$, as is readily verified. If $\mathscr{X}$ is complete in this metric, then $\mathscr{X}$ is called a **Fréchet space**. The metric d is not derived from a norm, since homogeneity fails (dramatically). We shall call d the **standard metric** for $\mathscr{X}$. Clearly every Banach space is a Fréchet space. Here are some nontrivial examples:

9.1.6 Examples.

(a) *The space $C(U)$.* Let $U \subseteq \mathbb{R}^d$ be open. Define compact subsets of U by

$$K_n := \{x \in \mathbb{R}^d : |x| \le n,\ d(x, U^c) \ge 1/n\} \quad n \in \mathbb{N}.$$

Then $K_n \subseteq \operatorname{int} K_{n+1}$ and $K_n \uparrow U$. Now define seminorms p_n on $C(U)$ by

$$p_n(f) = \sup\{|f(x)| : x \in K_n\}.$$

Since the sets $\{x : d(x, U^c) > 1/n\}$ form an increasing open cover of U, every compact set is contained in some K_n. Thus convergence in the locally convex topology generated by the seminorms p_n is uniform convergence on compact subsets of U, also called **local uniform convergence**. Since each space $C(K_n)$ is complete, $C(U)$ is a Fréchet space.

(b) *The space $H(U)$.* For $U \subseteq \mathbb{R}^2$, the space $H(U)$ of analytic (holomorphic) functions is a closed subspace of $C(U)$ in (a), since the property of analyticity is conveyed by local uniform convergence. Therefore $H(U)$ is also a Fréchet space.

(c) *The space $C^\infty(U)$.* Let U and (K_n) be as in (a). Define a countable family of seminorms $p_{m,\alpha}$ on $C^\infty(U)$ by

$$p_{m,\alpha}(f) = \sup\{|\partial^\alpha f(x)| : x \in K_m\},$$

where $\alpha = (\alpha_1, \ldots, \alpha_d)$ $(\alpha_j \in \mathbb{Z}^+)$, is a multi-index. A sequence (f_n) converges to zero in the locally convex topology generated by these seminorms iff $\partial^\alpha f_n \to 0$ locally uniformly for all α. To see that $C^\infty(U)$ is a Fréchet space, let (ϕ_n) be a Cauchy sequence with respect to the standard metric, so that

$$\lim_{m,n} \sup_{x \in K_j} |\partial^\alpha \phi_n(x) - \partial^\alpha \phi_m(x)| = 0 \quad \forall\, j \text{ and } \forall \text{ multi-index } \alpha.$$

Since $C(K_j)$ is complete and $K_j \uparrow U$, for each multi-index-α there exists $\phi^\alpha \in C(U)$ such that $\partial^\alpha \phi_n \to \phi^\alpha$ uniformly on each compact subset of U. Set $\phi = \phi^{(0,\dots,0)}$, so $\phi_n \to \phi$ locally uniformly. Letting $n \to \infty$ in

$$\phi_n(x_1, \dots, x_d) = \int_0^{x_1} \partial^{(1,0\dots,0)} \phi_n(t_1, x_2 \dots, x_d)\, dt_1,$$

we obtain

$$\phi(x_1, \dots, x_d) = \int_0^{x_1} \phi^{(1,0,\dots,0)}(t_1, x_2 \dots, x_d)\, dt_1.$$

This shows that $\partial^{(1,0,\dots,0)}\phi(x_1, \dots, x_d)$ exists and equals $\phi^{(1,0,\dots,0)}(x_1, \dots, x_d)$. In a similar manner, it may be shown that $\partial^\alpha \phi(x_1, \dots, x_d)$ exists and equals $\phi^\alpha(x_1, \dots, x_d)$ for all multi-indices α. Therefore, $C^\infty(U)$ is complete.

For later reference we note that the space $C_c^\infty(U)$ is dense in $C^\infty(U)$. Indeed, by Urysohn's lemma for C^∞ functions, for each n there exists a function $\phi_n \in C_c^\infty(U)$ such that $\phi_n = 1$ on K_n. For any $f \in C^\infty(U)$ we then have $\phi_n f \in C_c^\infty(U)$ and $\phi_n f = f$ on K_n, hence for $n > m$ and all α, $p_{m,\alpha}(\phi_n f - f) = 0$.

(d) *Schwartz space.* The space $\mathcal{S}$ of rapidly decreasing functions is a Fréchet space under the countable family of norms $q_{\alpha,m}$ defined by

$$q_{\alpha,m}(\phi) = \sup_{x \in \mathbb{R}^d} (1 + |x|)^m |\partial^\alpha \phi(x)|.$$

The proof that $\mathcal{S}$ is complete with respect to the standard metric is similar to that of (b). By 6.3.2, the same Fréchet topology is obtained by using the countable family of norms

$$p_{\alpha,\beta}(\phi) = \sup_{x \in \mathbb{R}^d} |x^\alpha \partial^\beta \phi(x)|. \qquad \lozenge$$

There are metrizable TVS that are not locally convex and hence not Fréchet spaces. Here is one such example:

9.1.7 Example. Let $(X, \mathcal{F}, \mu)$ be a finite measure space and let $L^0 = L^0(X, \mathcal{F}, \mu)$ denote the linear space of measurable functions $f : X \to \mathbb{K}$. Then

$$d(f, g) = \int \frac{|f - g|}{1 + |f - g|}\, d\mu$$

defines a metric on L^0 (where, as usual, we identify functions equal a.e.). Convergence in this metric is simply convergence in measure (Ex. 3.22). The inequalities

$$\begin{aligned} \mu\{|(f_n + g_n) - (f + g)| \ge 2\varepsilon|\} &\le \mu\{|f - f_n| \ge \varepsilon|\} + \mu\{|g - g_n| \ge \varepsilon|\} \quad \text{and} \\ \mu\{|c_n f_n - cf| \ge 2\varepsilon|\} &\le \mu\{|c_n f_n - c_n f| \ge \varepsilon|\} + \mu\{|c_n f - cf| \ge \varepsilon|\} \\ &\le \mu\{|f_n - f| \ge (|c_n| + 1)^{-1}\varepsilon|\} + \mu\{|f| \ge |c_n - c|^{-1}\varepsilon|\} \end{aligned}$$

then imply that L^0 is a TVS under the usual pointwise operations.

Now consider the measure space $([0, 1], \mathcal{B}[0, 1], \lambda)$. If L^0 were locally convex, then the open ball $B_{1/2}(0)$ would contain an open convex neighborhood of zero, which in turn would contain an open ball $B_r(0)$, whose convex hull is then contained in $B_{1/2}(0)$. For each $n > 1/r$, let $f_j = \mathbf{1}_{[(j-1)/n, j/n)}$, $1 \le j \le n$. Then

$$d(f_j, 0) = \int \frac{|f_j|}{1 + |f_j|}\, d\lambda = \frac{1}{2n} < r,$$

hence the convex combination $f := (1/n) \sum_{j=1}^n f_j$ is in $B_{1/2}(0)$. But $f = 1/n$ a.e. and so

$$d(f, 0) = \int \frac{1/n}{1 + 1/n}\, d\lambda = \frac{n}{n+1} > \frac{1}{2} \quad (n > 1). \qquad \lozenge$$

Exercises

9.1 Let $\mathscr{X}$ be a TVS and A, $B \subseteq \mathscr{X}$ with A compact and B closed. Show that $A + B$ is closed.

9.2 Let $\mathscr{U}$ be a neighborhood base at zero in TVS $\mathscr{X}$ and A, $B \subseteq \mathscr{X}$. Prove:

(a) $\operatorname{cl} A = \bigcap_{U \in \mathscr{U}}(U + A)$.

(b) $A + \operatorname{int} B \subseteq \operatorname{int}(A + B)$.

(c) If A is compact, B is closed, and $A \cap B = \emptyset$, then $(A+U) \cap (B+U) = \emptyset$ for some neighborhood of zero.

9.3 Show that the closed convex hull of subset A of a TVS $\mathscr{X}$ is the closure of $\operatorname{co} A$ and that the closed, convex, balanced hull of A is the closure of $\operatorname{cobal} A$.

9.4 Let $\mathscr{X}$ be a linear space. Show that if U is balanced and $|a| \le |b|$, then $aU \subseteq bU$.

9.5 Let U and V be open, convex, balanced neighborhoods of zero in a TVS $\mathscr{X}$. Show that $p_{U \cap V} = \max\{p_U, p_V\}$.

9.6 A subset E of a TVS is **bounded** if for each neighborhood V of zero there exists $t > 0$ such that $E \subseteq tV$. Verify the following

(a) If $E_1, \ldots, E_n$ are bounded, then $E_1 + \cdots + E_n$, $\bigcup_{j=1}^n E_j$, cE_1, and $\operatorname{cl} E_1$ are bounded.

(b) Every compact set K is bounded.

(c) E is bounded iff $x_n \in E$, $t_n \in \mathbb{K}$ and $t_n \to 0 \Rightarrow t_n x_n \to 0$.

(d) In a LCS, E is bounded iff $p(E)$ is bounded for every continuous seminorm p.

9.7 (Kolmogorov). Let $\mathscr{X}$ be a TVS with a bounded, convex, balanced neighborhood U of zero. Show that p_U is a norm that gives the original topology of $\mathscr{X}$. ⟦For positivity, let $x \ne 0$ and V a balanced neighborhood of zero that does contain x. If $U \subseteq tV$, then $p_U(x) \ge 1/t$. For the equality of topologies, consider suitable nets.⟧

9.8 Let $\mathscr{X}$ be a LCS generated by a family of seminorms $\mathscr{P}$ and let $\mathscr{Y}$ a linear subspace of $\mathscr{X}$. Prove that the relative topology of $\mathscr{Y}$ is the locally convex topology τ induced by the seminorms $p\big|_{\mathscr{Y}}$ $(p \in \mathscr{P})$.

9.9 Let p and q be seminorms on a vector space $\mathscr{X}$ such that $\{x : p(x) < 1\} = \{x : q(x) < 1\}$. Show that $p = q$.

9.10 Let $\mathscr{X}$ be a TVS and p a seminorm on $\mathscr{X}$ such that the set $\{x \in \mathscr{X} : p(x) < 1\}$ is open. Show that p is continuous.

9.11 Let $\mathscr{X}$ be a vector space with locally convex topologies τ_1 and τ_2. Show that $\tau_1 \le \tau_2$ iff every τ_1-continuous seminorm is τ_2-continuous.

9.2 Continuous Linear Functionals

Continuity on Topological Vector Spaces

As in the case of a normed space, the space of all continuous linear functionals on a TVS $\mathscr{X}$ is called the **dual space of** $\mathscr{X}$ and is denoted by $\mathscr{X}'$. We continue to use both notations $\langle x, f\rangle$ and $f(x)$ ($x \in \mathscr{X}$ and $f \in \mathscr{X}'$). The next two propositions summarize the general properties of linear functionals on $\mathscr{X}$. The first asserts that the kernel of a nontrivial (that is, not identically zero) linear functional f has co-dimension one, hence the equation $f = t$ ($t \in \mathbb{R}$) defines a *hyperplane*.

9.2.1 Proposition. *Let f be a nontrivial linear functional on a TVS $\mathscr{X}$. Then f is an open map, and for each $\boldsymbol{x}_0 \notin \ker f$*

$$\mathscr{X} = \mathbb{K}\,\boldsymbol{x}_0 \oplus \ker f. \tag{9.5}$$

Proof. For the first part, it suffices to prove that $f(U)$ is open for every open neighborhood of U of zero. By 9.1.2, we may take U to be balanced. Choose $\boldsymbol{x}$ such that $f(\boldsymbol{x}) = 1$. By continuity of scalar multiplication at zero, for sufficiently small $\varepsilon > 0$ we have $\varepsilon\boldsymbol{x} \in U$. Since U is balanced, $|c| < \varepsilon \Rightarrow c\boldsymbol{x} \in U \Rightarrow c = f(c\boldsymbol{x}) \in f(U)$. Therefore, $f(U)$ contains the open disk $B_\varepsilon(0)$ in $\mathbb{K}$ and so is open.

To verify (9.5), for $\boldsymbol{x} \in \mathscr{X}$ write

$$\boldsymbol{x} = \frac{f(\boldsymbol{x})}{f(\boldsymbol{x}_0)}\,\boldsymbol{x}_0 + \left(\boldsymbol{x} - \frac{f(\boldsymbol{x})}{f(\boldsymbol{x}_0)}\,\boldsymbol{x}_0\right)$$

and observe that the second term is in $\ker f$. Therefore, $\mathscr{X} = \mathbb{K}\,\boldsymbol{x}_0 + \ker f$. The sum is direct since if $c\boldsymbol{x}_0 \in \ker f$, then $0 = f(c\boldsymbol{x}_0) = cf(\boldsymbol{x}_0)$, hence $c = 0$. □

9.2.2 Proposition. *Let f be a linear functional on a TVS $\mathscr{X}$. The following statements are equivalent:*

(a) *f is continuous.*

(b) *$\ker f$ is closed.*

(c) *f is bounded on some neighborhood of zero.*

Proof. That (a) $\Rightarrow$ (b) is clear. For (b) $\Rightarrow$ (c) we may assume that f is not identically zero. Let $\boldsymbol{x} \notin \ker f$ and choose a neighborhood U of $\mathbf{0}$ such that $(U + \boldsymbol{x}) \cap \ker f = \emptyset$. By 9.1.2, we may assume that U is balanced. We claim that f is bounded on U. If not, then for any $c \in \mathbb{K}$ there exists $\boldsymbol{u} \in U$ such that $|f(\boldsymbol{u})| > |c|$. Setting $a := c/f(\boldsymbol{u})$ we have $|a| < 1$ and so $c = f(a\boldsymbol{u}) \in f(U)$. Thus $f(U) = \mathbb{K}$, and in particular $f(\boldsymbol{u}) = -f(\boldsymbol{x})$ for some $\boldsymbol{u} \in U$. But this contradicts $(U + \boldsymbol{x}) \cap \ker f = \emptyset$. Therefore, $f(U)$ must be bounded.

To prove (c) $\Rightarrow$ (a), let $|f(\boldsymbol{u})| < r$ for all $\boldsymbol{u}$ in a neighborhood U of zero. If $\boldsymbol{x}_\alpha \to 0$ and $\varepsilon > 0$, then eventually $(r/\varepsilon)\boldsymbol{x}_\alpha \in U$ and so $|f(\boldsymbol{x}_\alpha)| < \varepsilon$. Therefore, f is continuous. □

It is possible for a TVS not to have *any* nontrivial continuous linear functionals, as the following example demonstrates.

9.2.3 Example. We show that the space $L^0[0,1]$ of Example 9.1.7 has no nontrivial continuous linear functionals. Let F be such a functional and choose $f \in L^0[0,1]$ such that $F(f) \neq 0$. Next, choose whichever of the functions $f\mathbf{1}_{[0,1/2)}$ or $f\mathbf{1}_{[1/2,1]}$, call it f_1, has the property $F(f_1) \neq 0$, and note that $\lambda\{f_1 \neq 0\} \leq 1/2$. By induction, we obtain a sequence (f_n) such that $\alpha_n := F(f_n) \neq 0$ and $\lambda\{f_n \neq 0\} \leq 1/2^n$. Set $g_n := \alpha_n^{-1} f_n$. Then

$$d(g_n, 0) = \int \frac{|f_n|}{|\alpha_n| + |f_n|}\,d\lambda = \int_{|f_n| \neq 0} \frac{|f_n|}{|\alpha_n| + |f_n|}\,d\lambda \leq \lambda\{f_n \neq 0\} \to 0,$$

hence $g_n \to 0$ yet $F(g_n) = 1$ for all n. ◊

We shall see in the next section that, unlike the TVS case, a LCS always has a rich supply of continuous linear functionals.

Continuity on Locally Convex Spaces

Continuity in a LCS may be conveniently characterized in terms of seminorms.

9.2.4 Proposition. *Let $\mathscr{X}$ be a LCS with generating family $\mathscr{P}$ of seminorms and let f be a linear functional on $\mathscr{X}$. The following statements are equivalent:*

(a) *f is continuous.*

(b) *There exist $p_1, \ldots, p_n \in \mathscr{P}$ and $M > 0$ such that $|f(\boldsymbol{x})| \leq M \max_j p_j(\boldsymbol{x})$ for all $\boldsymbol{x}$.*

(c) *There exists a continuous seminorm q on $\mathscr{X}$ such that $|f(\boldsymbol{x})| \leq q(\boldsymbol{x})$ for all $\boldsymbol{x}$.*

Proof. (a) $\Rightarrow$ (b): By continuity of f at zero, there exists a basic neighborhood U of zero as in (9.1) such that $|f(\boldsymbol{u})| < 1$ for all $\boldsymbol{u} \in U$. Set $p(\boldsymbol{x}) = \max_j p_j(\boldsymbol{x})$. For any $\boldsymbol{x} \in \mathscr{X}$ and $\delta > 0$, $\varepsilon\boldsymbol{x}/(p(\boldsymbol{x}) + \delta) \in U$ hence $|f(\boldsymbol{x})| < \varepsilon^{-1}(p(\boldsymbol{x}) + \delta)$. Letting $\delta \to 0$ yields (b) with $M = 1/\varepsilon$.

(b) $\Rightarrow$ (c): Take $q = M \max_j p_j(\boldsymbol{x})$.

(c) $\Rightarrow$ (a): If $\boldsymbol{x}_\alpha \to 0$, then $q(\boldsymbol{x}_\alpha) \to 0$, hence $f(\boldsymbol{x}_\alpha) \to 0$. □

Continuity on Finite Dimensional Spaces

The following theorem shows that Hausdorff vector topologies on finite dimensional spaces are unique.

9.2.5 Theorem. *Let $\mathscr{X}$ be a finite dimensional TVS with dimension d. Then $\mathscr{X}$ is topologically isomorphic to $\mathbb{K}^d$.*

Proof. Let $\boldsymbol{v}_1, \ldots, \boldsymbol{v}_d$ be a basis for $\mathscr{X}$. We show that the algebraic isomorphism $T(x_1, \ldots, x_d) = x_1\boldsymbol{v}_1 + \cdots + x_d\boldsymbol{v}_d$ is a homeomorphism. By continuity of the vector operations, T is continuous. We show that there exists a neighborhood U of zero in $\mathscr{X}$ such that $U \subseteq T(B_1)$, where B_1 is the open unit ball in Euclidean space $\mathbb{K}^d$, which will prove that T is an open map and hence is a topological isomorphism.

Since the unit sphere S_1 in $\mathbb{K}^d$ is compact, $T(S_1)$ is compact in $\mathscr{X}$. Let $\mathscr{U}$ be the collection of all closed, balanced neighborhoods of zero in $\mathscr{X}$. Since $\mathscr{X}$ is Hausdorff $\bigcap \mathscr{U} = \{0\}$, hence $\bigcap_{U \in \mathscr{U}} U \cap T(S_1) = \emptyset$. Since $T(S_1)$ is compact it cannot have the finite intersection property, hence there exist $U_1, \ldots, U_n$ such that $U := U_1 \cap \ldots \cap U_n$ does not meet $T(S_1)$. Therefore, $U \subseteq T(S_1^c)$. Since U is balanced, $U \subseteq T(B_1)$, as required. □

Since every linear functional on Euclidean space is continuous, we have

9.2.6 Corollary. *A linear functional on a finite dimensional TVS is continuous.*

Exercises

9.12 Let $\mathscr{X}$ be a real TVS and f a linear functional on $\mathscr{X}$ such that $\{f \leq t\}$ is closed for some t. Show that f continuous.

9.13 Let $\mathscr{X}$ be a TVS and f a linear functional on $\mathscr{X}$ such that $\ker f$ is not dense in $\mathscr{X}$. Show that f is continuous.

9.14 Let $\mathscr{X}$ and $\mathscr{Y}$ be locally convex spaces. Show that a linear transformation $T : \mathscr{X} \to \mathscr{Y}$ is continuous iff $p \circ T$ is continuous for every continuous seminorm p on $\mathscr{Y}$.

9.15 Let $\mathscr{X}$ be a TVS, $f \in \mathscr{X}'$ a nontrivial real linear functional, and $t \in \mathbb{R}$. Let $C = \{\boldsymbol{x} : f(\boldsymbol{x}) \leq t\}$ and $U = \{\boldsymbol{x} : f(\boldsymbol{x}) < t\}$. Show that $\operatorname{cl} U = C$ and $\operatorname{int} C = U$.

9.3 Hahn-Banach Separation Theorems

The theorems in this section consider the problem of separating disjoint convex sets by hyperplanes. The proofs rely on the Minkowski functional of a neighborhood of zero, defined in §9.1.

Weak Separation in a TVS

9.3.1 Theorem. *Let A and B be nonempty, disjoint, convex subsets of a TVS $\mathscr{X}$. If A is open, then there exist $f \in \mathscr{X}'$ and $t \in \mathbb{R}$ such that*

$$\operatorname{Re} f(x) < t \le \operatorname{Re} f(y) \quad \textit{for all } x \in A \textit{ and } y \in B. \tag{9.6}$$

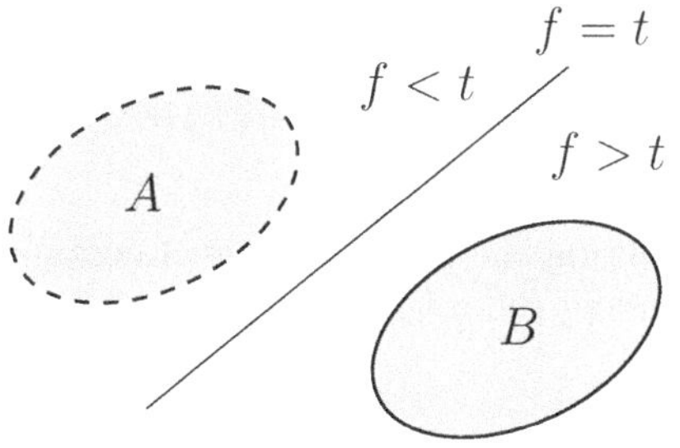

FIGURE 9.1: Separation by a hyperplane.

Proof. Suppose first that $\mathbb{K} = \mathbb{R}$. Fix $x_0 \in A$ and $y_0 \in B$ and let $z_0 := y_0 - x_0$. The set $U := A - B + z_0$ is convex, contains zero, and is open, the last property because U is a union of the open sets $A - y + z_0$ $(y \in B)$. Let p be the Minkowski functional of U. Since A and B are disjoint, $z_0 \notin U$, hence $p(z_0) \ge 1$ by (9.3). Define g on the one-dimensional space $\mathscr{Y} := \mathbb{R}z_0$ by $g(cz_0) = c$. Then $g \le p$ on $\mathscr{Y}$, hence g extends to a linear functional f on $\mathscr{X}$ with $f \le p$. (8.5.1). Since $p < 1$ on U, $-\varepsilon < f < \varepsilon$ on the open set $-\varepsilon U \cap \varepsilon U$, hence f is continuous at zero and therefore everywhere. If $x \in A$ and $y \in B$, then $x - y + z_0 \in U$, hence

$$f(x) - f(y) = f(x - y + z_0) - 1 \le p(x - y + z_0) - 1 < 0$$

and so $f(x) < f(y)$. Since convex sets are connected, $f(A)$ and $f(B)$ are disjoint intervals in $\mathbb{R}$, hence $f(A)$ lies to the left of $f(B)$. Moreover, since A is open and f is nontrivial, $f(A)$ is open. Therefore, we may take t in (9.6) to be the right endpoint of $f(A)$.

For the case $\mathbb{K} = \mathbb{C}$, apply the first part to $\mathscr{X}$ as a real linear space to obtain a real linear functional f_r that satisfies $f_r(x) < t \le f_r(y)$ for all $x \in A$ and $y \in B$. Then $f(x) := f_r(x) - if_r(ix)$ defines a complex linear functional satisfying (9.6). □

Strict Separation in a LCS

In Theorem 9.3.1 it is possible that the hyperplane $f = t$ intersects B. The next theorem asserts that under suitable conditions one actually has strict separation.

9.3.2 Theorem. *Let A and B be nonempty disjoint closed convex subsets of a LCS $\mathscr{X}$. If A or B is compact, then there exists $f \in \mathscr{X}'$ such that*

$$\sup\{\operatorname{Re} f(x) : x \in A\} < \inf\{\operatorname{Re} f(y) : y \in B\}. \tag{9.7}$$

Proof. Suppose A is compact. Let $\mathscr{U}_0$ be a neighborhood base at zero of open convex sets. We claim that there exists $U \in \mathscr{U}_0$ such that $(U + A) \cap B = \emptyset$. Assuming this and noting that $C := U + A$ is open and convex, we may choose by 9.3.1 $f \in \mathscr{X}'$ and $t \in \mathbb{R}$ such that

$$\operatorname{Re} f(\boldsymbol{x}) < t \le \operatorname{Re} f(\boldsymbol{y}) \text{ for all } \boldsymbol{x} \in C \text{ and } \boldsymbol{y} \in B.$$

Since A is a compact subset of C,

$$\sup\{\operatorname{Re} f(\boldsymbol{x}) : \boldsymbol{x} \in A\} < t \le \operatorname{Re} f(\boldsymbol{y}) \text{ for all } \boldsymbol{y} \in B,$$

proving (9.7).

To verify the claim, for each $\boldsymbol{x} \in A \subseteq B^c$ choose $V_{\boldsymbol{x}} \in \mathscr{U}_0$ such that $\boldsymbol{x} + V_{\boldsymbol{x}} \subseteq B^c$. Next, choose $U_{\boldsymbol{x}} \in \mathscr{U}_0$ so that $U_{\boldsymbol{x}} + U_{\boldsymbol{x}} \subseteq V_{\boldsymbol{x}}$. This is possible by continuity of addition at $(\boldsymbol{0}, \boldsymbol{0})$. Then the sets $\boldsymbol{x} + U_{\boldsymbol{x}} + U_{\boldsymbol{x}}$ and B are disjoint. Moreover, by compactness, there exist $\boldsymbol{x}_1, \ldots, \boldsymbol{x}_n \in A$ such that $A \subseteq \bigcup_{j=1}^n (\boldsymbol{x}_j + U_{\boldsymbol{x}_j})$. Setting $U = \bigcap_{j=1}^n U_{\boldsymbol{x}_j}$, we have

$$A + U \subseteq \bigcup_{j=1}^n (\boldsymbol{x}_j + U_{\boldsymbol{x}_j} + U) \subseteq \bigcup_{j=1}^n (\boldsymbol{x}_j + U_{\boldsymbol{x}_j} + U_{\boldsymbol{x}_j}) \subseteq B^c,$$

verifying the claim and completing the proof for case A compact.

If B is compact, then reversing the roles of A and B yields

$$\sup\{\operatorname{Re} f(\boldsymbol{x}) : \boldsymbol{x} \in B\} < \inf\{\operatorname{Re} f(\boldsymbol{y}) : \boldsymbol{y} \in A\}.$$

Equation (9.7) then holds with f replaced by $-f$. □

Here is an important variant of the preceding theorem.

9.3.3 Theorem. *Let A and B be nonempty, disjoint, closed, convex subsets of a LCS $\mathscr{X}$. If A is balanced and either A or B is compact, then there exists $f \in \mathscr{X}'$ such that*

$$\sup\{|f(\boldsymbol{x})| : \boldsymbol{x} \in A\} < \inf\{|f(\boldsymbol{y})| : \boldsymbol{y} \in B\}. \tag{9.8}$$

Proof. Let f be as in 9.3.2 and choose t with

$$\sup\{\operatorname{Re} f(\boldsymbol{x}) : \boldsymbol{x} \in A\} < t < \inf\{\operatorname{Re} f(\boldsymbol{y}) : \boldsymbol{y} \in B\}.$$

For $\boldsymbol{x} \in A$, write $|f(\boldsymbol{x})| = e^{i\theta} f(\boldsymbol{x}) = f(e^{i\theta}\boldsymbol{x}) = \operatorname{Re} f(e^{i\theta}\boldsymbol{x})$. Since $e^{i\theta}\boldsymbol{x} \in A$ we have $\sup\{|f(\boldsymbol{x})| : \boldsymbol{x} \in A\} < t < \operatorname{Re} f(\boldsymbol{y}) \le |f(\boldsymbol{y})|$ for all $\boldsymbol{y} \in B$, verifying (9.8). □

Some Consequences of the Separation Theorems

The following are generalizations of results proved earlier for normed spaces.

9.3.4 Theorem. *Let $\mathscr{Y}$ be a closed subspace of a LCS $\mathscr{X}$ and let $\boldsymbol{x}_0 \notin \mathscr{Y}$. Then there exists an $f \in \mathscr{X}'$ such that $f(\boldsymbol{x}_0) = 1$ and $f(\mathscr{Y}) = 0$. In particular, if $\boldsymbol{x}_1 \ne \boldsymbol{x}_2$ then there exists an $f \in \mathscr{X}'$ such that $f(\boldsymbol{x}_1) \ne f(\boldsymbol{x}_2)$, that is, $\mathscr{X}'$ separates points of $\mathscr{X}$.*

Proof. Take $A = \mathscr{Y}$ and $B = \{\boldsymbol{x}_0\}$ in 9.3.2 to obtain $g \in \mathscr{X}'$ with

$$\sup\{\operatorname{Re} g(\boldsymbol{y}) : \boldsymbol{y} \in \mathscr{Y}\} < \operatorname{Re} g(\boldsymbol{x}_0).$$

But because $\mathscr{Y}$ is a linear space, $\operatorname{Re} g(\mathscr{Y})$ cannot be bounded above unless $\operatorname{Re} g(\mathscr{Y}) = \{0\}$, which then implies that $g(\boldsymbol{x}_0) \ne 0$. Since $\operatorname{Im} g(\boldsymbol{y}) = -\operatorname{Re} g(i\boldsymbol{y})$, we have $g(\mathscr{Y}) = \{0\}$. Now take $f = g/g(\boldsymbol{x}_0)$. □

9.3.5 Corollary. *Let $\mathscr{Y}$ be a subspace of $\mathscr{X}$ and let g be a continuous linear functional on $\mathscr{Y}$. Then there exists an $f \in \mathscr{X}'$ such that $f = g$ on $\mathscr{Y}$.*

Proof. We may assume that g is not identically zero. Choose $\boldsymbol{y}_1 \in \mathscr{Y}$ such that $g(\boldsymbol{y}_1) = 1$. By 9.3.4 applied to the closure of $\mathscr{Y}_0 := \ker g$ in $\mathscr{X}$, there exists $f \in \mathscr{X}'$ such that $f(\boldsymbol{y}_1) = 1$ and $f(\mathscr{Y}_0) = 0$. If $\boldsymbol{y} \in \mathscr{Y}$, then $\boldsymbol{y} - g(\boldsymbol{y})\boldsymbol{y}_1 \in \mathscr{Y}_0$ and so

$$f(\boldsymbol{y}) - g(\boldsymbol{y}) = f(\boldsymbol{y}) - g(\boldsymbol{y})f(\boldsymbol{y}_1) = f\big(\boldsymbol{y} - g(\boldsymbol{y})\boldsymbol{y}_1\big) = 0.$$

Therefore, f extends g. □

9.3.6 Corollary. *A finite dimensional subspace $\mathscr{Y}$ of a LCS $\mathscr{X}$ is closed.*

Proof. Let $\boldsymbol{y}_1, \ldots, \boldsymbol{y}_d$ be a basis for $\mathscr{Y}$. Then $\boldsymbol{y} = \sum_{j=1}^{d} g_j(\boldsymbol{y})\boldsymbol{y}_j$ ($\boldsymbol{y} \in \mathscr{Y}$), where g_j is a linear functional on $\mathscr{Y}$. By 9.2.6, g_j is continuous and so has a continuous extension $f_j \in \mathscr{X}'$. Therefore, if $\boldsymbol{y}_\alpha \in \mathscr{Y}$ and $\boldsymbol{y}_\alpha \to \boldsymbol{x} \in \mathscr{X}$ we have

$$\boldsymbol{x} = \lim_i \sum_j f_j(\boldsymbol{y}_\alpha)\boldsymbol{y}_j = \sum_j f_j(\boldsymbol{x})\boldsymbol{y}_j \in \mathscr{Y}.$$ □

9.3.7 Corollary. *A LCS $\mathscr{X}$ is finite dimensional iff it has a compact neighborhood of zero.*

Proof. If $\mathscr{X}$ is finite dimensional, then $\mathscr{X}$ is topologically isomorphic to $\mathbb{K}^d$ (9.2.5), proving the necessity.

For the sufficiency, let V be a neighborhood of zero in $\mathscr{X}$ with compact closure. Then there exists a finite subset F of $\mathscr{X}$ such that

$$\operatorname{cl} V \subseteq \bigcup_{\boldsymbol{x} \in F} (\boldsymbol{x} + \tfrac{1}{2}V) = F + \tfrac{1}{2}V.$$

Let $\mathscr{Y}$ be the finite dimensional subspace of $\mathscr{X}$ spanned by F. We claim that

$$V \subseteq \mathscr{Y} + \tfrac{1}{2^n}V \quad \text{for all } n. \tag{†}$$

This is clear for $n = 1$. If the assertion holds for n, then

$$\tfrac{1}{2}V \subseteq \tfrac{1}{2}\mathscr{Y} + \tfrac{1}{2^{n+1}}V = \mathscr{Y} + \tfrac{1}{2^{n+1}}V,$$

hence

$$V \subseteq \mathscr{Y} + \tfrac{1}{2}V \subseteq \mathscr{Y} + \mathscr{Y} + \tfrac{1}{2^{n+1}}V = \mathscr{Y} + \tfrac{1}{2^{n+1}}V,$$

verifying the claim.

Now let p be any continuous seminorm on $\mathscr{X}$. Since $\operatorname{cl} V$ is compact, $s := \sup p(V) < \infty$. Let $\boldsymbol{v} \in V$. Using (†), we may write $\boldsymbol{v} = \boldsymbol{y}_n + 2^{-n}\boldsymbol{w}_n$ for some $\boldsymbol{y}_n \in \mathscr{Y}$ and $\boldsymbol{w}_n \in V$. Then $p(\boldsymbol{v} - \boldsymbol{y}_n) = 2^{-n}p(\boldsymbol{w}_n) \leq 2^{-n}s$, so for sufficiently large n, $p(\boldsymbol{v} - \boldsymbol{y}_n) < 1$. Since p was arbitrary, $\boldsymbol{v} \in \operatorname{cl}\mathscr{Y}$. Since $\mathscr{Y}$ is finite dimensional, $\operatorname{cl}\mathscr{Y} = \mathscr{Y}$ (9.3.6). We have shown that $V \subseteq \mathscr{Y}$. Now, for any $\boldsymbol{x} \in \mathscr{X}$, eventually $\boldsymbol{x}/n \in V$ and so $\boldsymbol{x} \subset \mathscr{Y}$. Therefore, $\mathscr{X} = \mathscr{Y}$, hence $\mathscr{X}$ is finite dimensional. □

9.3.8 Corollary. *Let $\mathscr{X}$ be a LCS that is the algebraic direct sum of subspaces $\mathscr{Y}$ and $\mathscr{Z}$ with $\mathscr{Y}$ finite dimensional. Then the direct sum is topological, that is, the linear isomorphism $(\boldsymbol{y}, \boldsymbol{z}) \mapsto \boldsymbol{y} + \boldsymbol{z}$ is a homeomorphism in the product topology.*

Proof. Let $\{\boldsymbol{y}_1, \ldots, \boldsymbol{y}_n\}$ be a basis for $\mathscr{Y}$. By 9.3.4 there exist $f_1, \ldots, f_n \in \mathscr{X}'$ such that $f_i(\boldsymbol{x}_j) = \delta_{ij}$. Define $P\boldsymbol{x} = \sum_{j=1}^{n} \langle \boldsymbol{x}, f_j \rangle \boldsymbol{x}_j$. Then P is a continuous projection and the continuity of $\boldsymbol{y} + \boldsymbol{z} \to (\boldsymbol{y}, \boldsymbol{z})$ follows as in 8.4.3 (using nets). □

The Bipolar Theorem

Let $\mathscr{X} = \mathscr{X}_\tau$ be a LCS. The **polars** of $A \subseteq \mathscr{X}$ and $B \subseteq \mathscr{X}'$ are defined by

$$A^0 = \{f \in \mathscr{X}' : |f(x)| \le 1 \ \forall\ x \in A\} \text{ and } {}^0B = \{x \in \mathscr{X} : |f(x)| \le 1 \ \forall\ f \in B\}.$$

It is easy to check that each polar is convex and balanced.

Polars are related to annihilators, introduced in §8.9. Indeed, if A and B are linear subspaces, then the two constructions coincide (Ex. 9.17). Here is the polar analog of 8.9.3.

9.3.9 Theorem. *If $A \subseteq \mathscr{X}$, then ${}^0A^0$ is the τ-closed, convex, balanced hull of A.*

Proof. Let C denote the closed convex balanced hull of A. Since $A \subseteq {}^0A^0$ and ${}^0A^0$ is closed, convex, and balanced, $C \subseteq {}^0A^0$. For the reverse inclusion, let $y \in C^c$ and choose $f \in \mathscr{X}$ and $t \in \mathbb{R}$ so that $\sup\{|f(x)| : x \in C\} < t < |f(y)|$ (9.3.3). Set $g := f/t$. Then $\sup\{|g(x)| : x \in A\} < 1 < |g(y)|$, hence $y \notin {}^0A^0$. □

Exercises

9.16 A **half-space** in $\mathbb{R}^d$ is a set of the form $\{x \in \mathbb{R}^d : a_1x_1 + \cdots + a_dx_d \le a\}$. Show that a closed, convex subset C of $\mathbb{R}^d$ is the intersection of all half-spaces that contain it.

9.17 Let $\mathscr{X}$ be a LCS, $A \subseteq \mathscr{X}$, and $B \subseteq \mathscr{X}'$. The **annihilators** of A and B are defined as for normed spaces by

$$A^\perp = \{f \in \mathscr{X}' : \langle x, f\rangle = 0 \ \forall\ x \in A\} \text{ and } {}^\perp B = \{x \in \mathscr{X} : \langle x, f\rangle = 0 \ \forall\ f \in B\}.$$

Prove that if $\mathscr{Y}$ is a subspace of $\mathscr{X}$ and $\mathscr{Z}$ is a subspace of $\mathscr{X}'$, then $\mathscr{Y}^\perp = \mathscr{Y}^0$ and ${}^\perp\mathscr{Z} = {}^0\mathscr{Z}$.

9.18 Show that $\mathscr{Y}$ is dense in $\mathscr{X}$ iff $\mathscr{Y}^\perp = \{0\}$.

9.19 Show that if A and B are open in 9.3.1, then there exists $t \in \mathbb{R}$ such that $\mathrm{Re} f(x) < t < \mathrm{Re} f(y)$ for all $x \in A$ and $y \in B$.

9.20 Let A, B, and A_i $(i \in \mathfrak{I})$ be subsets of a LCS $\mathscr{X}$. Prove that

(a) $A \subseteq B \Rightarrow B^0 \subseteq A^0$. (b) $\left(\bigcup_{i\in\mathfrak{I}} A_i\right)^0 = \bigcap_{i\in\mathfrak{I}} A_i^0$. (c) $(cA)^0 = c^{-1}A^0$, $c \ne 0 \in \mathbb{K}$.

*9.4 Some Constructions

Product Spaces

Let $\{\mathscr{X}_i\}$ be a family of TVS and let $\mathscr{X}$ denote the product vector space $\prod_{i\in\mathfrak{I}} \mathscr{X}_i$. If $\pi_i : \mathscr{X} \to \mathscr{X}_i$ denotes the projection map, then a net (f_α) converges to f in the product topology iff $\pi_i(f_\alpha) \to \pi_i(f)$ for each i. Since the projections are linear, $\mathscr{X}$ is easily seen to be a TVS in the product topology.

Now assume that each $\mathscr{X}_i$ is a locally convex space. Consider seminorms on $\mathscr{X}$ of the form $p_F(f) = \max\{p_i(f(i)) : i \in F\}$, where p_i is a continuous seminorm on $\mathscr{X}_i$ and $F \subseteq \mathfrak{I}$ is finite. By the first paragraph, the family (p_F) generates a locally convex topology which is the product topology.

Quotient Spaces

The following results generalize theorems in §8.4 on quotients of normed spaces.

9.4.1 Theorem. *Let $\mathscr{X}$ be a TVS, $\mathscr{Y}$ a closed subspace of $\mathscr{X}$, and $\mathscr{X}/\mathscr{Y}$ the algebraic quotient space with quotient map $Q : \mathscr{X} \to \mathscr{X}/\mathscr{Y}$. Then $\mathscr{X}/\mathscr{Y}$ is a TVS in the quotient topology and Q is an open map. Moreover, if $\mathscr{X}$ is locally convex (Fréchet), then $\mathscr{X}/\mathscr{Y}$ is locally convex (Fréchet).*

Proof. Recall that the quotient topology on $\mathscr{X}/\mathscr{Y}$ is the strongest topology relative to which Q is continuous; equivalently, W is open in $\mathscr{X}/\mathscr{Y}$ iff $Q^{-1}(W)$ is open in $\mathscr{X}$. Now, if U is open in $\mathscr{X}$, then

$$Q^{-1}(Q(U)) = U + \mathscr{Y} = \bigcup_{y \in \mathscr{Y}} U + y,$$

which is open in $\mathscr{X}$. Therefore, $Q(U)$ is open in $\mathscr{X}/\mathscr{Y}$, hence Q is an open map.

To see that the quotient topology is Hausdorff, suppose that $Q(\boldsymbol{x}_1) \neq Q(\boldsymbol{x}_2)$, so that $\boldsymbol{x}_1 - \boldsymbol{x}_2$ is in the open set $\mathscr{Y}^c$. Choose a neighborhood of zero in $\mathscr{X}$ such that $\boldsymbol{x}_1 - \boldsymbol{x}_2 + U \subseteq \mathscr{Y}^c$. By continuity of the vector difference operation, there exists a neighborhood V of zero such that $V - V \subseteq U$. Then $Q(\boldsymbol{x}_1 + V)$ and $Q(\boldsymbol{x}_2 + V)$ are disjoint neighborhoods $Q(\boldsymbol{x}_1)$ and $Q(\boldsymbol{x}_2)$, respectively.

To see that the quotient topology is a vector topology, consider nets $Q(\boldsymbol{x}_\alpha) \to Q(\boldsymbol{x})$ and $Q(\boldsymbol{y}_\alpha) \to Q(\boldsymbol{y})$. A typical neighborhood of $Q(\boldsymbol{x}) + Q(\boldsymbol{y}) = Q(\boldsymbol{x} + \boldsymbol{y})$ in $\mathscr{X}/\mathscr{Y}$ is of the form $\boldsymbol{x} + \boldsymbol{y} + Q(U)$, where U is a neighborhood of zero in $\mathscr{X}$. By continuity of addition, there exists a neighborhood V of zero in $\mathscr{X}$ such that $V + V \subseteq U$. Then, eventually,

$$Q(\boldsymbol{x}_\alpha + \boldsymbol{y}_\alpha) = Q(\boldsymbol{x}_\alpha) + Q(\boldsymbol{y}_\alpha) \in Q(\boldsymbol{x}) + Q(V) + Q(\boldsymbol{x}) + Q(V) \subseteq Q(\boldsymbol{x} + \boldsymbol{y}) + Q(U),$$

which shows that $Q(\boldsymbol{x}_\alpha + \boldsymbol{y}_\alpha) \to Q(\boldsymbol{x} + \boldsymbol{y})$. Therefore, vector addition in $\mathscr{Z}$ is continuous. A similar argument shows that scalar multiplication is continuous.

For the last assertion, note that if $\mathscr{U}$ is a basis of open, convex neighborhoods of zero in $\mathscr{X}$, then $Q(\mathscr{U})$ is a basis of open, convex neighborhoods of zero in $\mathscr{X}/\mathscr{Y}$. □

9.4.2 Corollary. *Let $\mathscr{X}$ and $\mathscr{Z}$ be TVS and $T : \mathscr{X} \to \mathscr{Z}$ linear, continuous, surjective, and open. Then $\mathscr{X}/\ker T$ is topologically isomorphic to $\mathscr{Z}$.*

Proof. Let Q denote the quotient map. Then $S : Q(\boldsymbol{x}) \mapsto T(\boldsymbol{x})$ from $\mathscr{X}/\ker T$ to $\mathscr{Z}$ is well defined, linear, bijective, and $SQ = T$. By definition of the final topology, S is continuous. Since Q and T are open maps, S is open. Therefore, S is a topological isomorphism. □

9.4.3 Corollary. *Let $\mathscr{X}$ be a LCS and $f_j \in \mathscr{X}'$, $j = 1, \ldots, n$. Let $\mathscr{Y} = \bigcap_{j=1}^n \ker f_j$. Then $\mathscr{X}$ is topologically isomorphic to a direct product $\mathscr{Y} \times \mathbb{K}^d$ for some $d \leq n$.*

Proof. Let $T(\boldsymbol{x}) = (f_1(\boldsymbol{x}), \ldots, f_n(\boldsymbol{x}))$. Then $\ker T = \mathscr{Y}$ and T maps $\mathscr{X}$ onto a d-dimensional subspace of $\mathbb{K}^n$, which we may identify with $\mathbb{K}^d$. Choose $\boldsymbol{x}_0 \in \mathscr{X}$ such that $T(\boldsymbol{x}_0) = (1, 1, \ldots, 1)$ and let U be any convex balanced neighborhood of zero. Then $(-\varepsilon, \varepsilon)\boldsymbol{x}_0 \subseteq U$ for sufficiently small $\varepsilon > 0$, hence $T(U)$ contains the d-dimensional open neighborhood of zero $T\big((-\varepsilon, \varepsilon)\boldsymbol{x}_0\big) = (-\varepsilon, \varepsilon) \times \cdots \times (-\varepsilon, \varepsilon)$. Therefore, $T : \mathscr{X} :\to \mathbb{K}^d$ is open. By 9.4.2, $\mathscr{X}/\mathscr{Y}$ is topologically isomorphic to $\mathbb{K}^d$ under a mapping S with $SQ = T$. In particular, there exist $\boldsymbol{x}_1, \ldots, \boldsymbol{x}_d \in \mathscr{X}$ such that $SQ(\boldsymbol{x}_j) = T\boldsymbol{x}_j = (0, \ldots, 0, 1, 0 \ldots, 0)$, 1 in the jth position. Let P_j denote the jth projection map on $\mathbb{K}^d$ and define

$$R\boldsymbol{x} = \left(\boldsymbol{x} - \sum_{j=1}^d P_j(T(\boldsymbol{x}))\boldsymbol{x}_j, T\boldsymbol{x}\right).$$

Then R is a linear isomorphism of $\mathscr{X}$ onto $\mathscr{Y} \times \mathbb{K}^d$. Moreover, it is easy to see that $\boldsymbol{x}_\alpha \to 0$ iff $R\boldsymbol{x}_\alpha \to 0$. Therefore, R is the desired topological isomorphism. □

Strict Inductive Limits

Let $\mathscr{X}$ be a vector space and $(\mathscr{X}_n)$ a sequence of subspaces with union $\mathscr{X}$ such that $\mathscr{X}_n \subsetneq \mathscr{X}_{n+1}$ for all n. Let $\mathscr{X}_n$ have a locally convex topology $\mathcal{T}_n$ such that the relative topology on $\mathscr{X}_n$ induced by $\mathcal{T}_{n+1}$ is $\mathcal{T}_n$. The sequence $(\mathscr{X}_n, \mathcal{T}_n)$ is called a **strict inductive system for** $\mathscr{X}$. Such a system gives rise to a locally convex topology $\mathcal{T}$ on $\mathscr{X}$ which will have important applications in Chapter 15 on distributions. For the construction of $\mathcal{T}$ we need the following lemmas.

9.4.4 Lemma. *Let $\mathscr{Y}$ be a LCS and $\mathscr{Z}$ a linear subspace of $\mathscr{Y}$. If U is a convex, balanced, open neighborhood of $\mathscr{Z}$ in the relative topology induced by $\mathscr{Y}$, then there exists a convex, balanced, open neighborhood V of zero in $\mathscr{Y}$ such that $V \cap \mathscr{Z} = U$. Moreover, if $\boldsymbol{y} \notin \operatorname{cl}\mathscr{Z}$, then V may be chosen to exclude $\boldsymbol{y}$.*

Proof. Choose a convex, balanced, open neighborhood W of zero in $\mathscr{Y}$ such that $W \cap \mathscr{Z} \subseteq U$, and let V be the convex hull of $U \cup W$. Since U and W are balanced, it is readily verified that V is balanced. Moreover, $V \cap \mathscr{Z} \supseteq U$. To show equality, note that, by the convexity of U and W, a member $\boldsymbol{z}$ of $V \cap \mathscr{Z}$ may be written as $\boldsymbol{z} = t\boldsymbol{u} + (1-t)\boldsymbol{w}$, where $\boldsymbol{u} \in U$, $\boldsymbol{w} \in W$, and $0 \le t \le 1$. If $t = 1$ then $\boldsymbol{z} = \boldsymbol{u} \in U$. If $t < 1$ then $\boldsymbol{w} = (1-t)^{-1}(\boldsymbol{z} - t\boldsymbol{u}) \in \mathscr{Z}$, hence $\boldsymbol{w} \in U$ and so, again, $\boldsymbol{z} \in U$. Therefore, $V \cap \mathscr{Z} = U$.

To see that V is open, note first that for each $t \in [0,1)$, the set $tU + (1-t)W$ is open in $\mathscr{Y}$, as it is the sum of two sets, one of which is open in $\mathscr{Y}$. By the observation in the first paragraph, $V = \bigcup_{t\in[0,1]}[tU + (1-t)W]$. We claim that, in fact, $V = \bigcup_{t\in[0,1)}[tU + (1-t)W]$. For the verification, we show that each $\boldsymbol{u} \in U$ is in the set $tU + (1-t)W$ for some $0 < t < 1$. To this end choose $r > 0$ so that $r\boldsymbol{u} \in W$ and write

$$\boldsymbol{u} = t\{t^{-1}[1 - r + tr]\boldsymbol{u}\} + (1-t)r\boldsymbol{u}, \quad 0 < t < 1.$$

The braced expression is in $\mathscr{Z}$ and tends to $\boldsymbol{u}$ as $t \to 1$. Since U is open in $\mathscr{Z}$, the braced expression is in U for some $t < 1$. Since $r\boldsymbol{u} \in W$, we see that $\boldsymbol{u} \in tU + (1-t)W$, verifying the claim and proving that V is open.

For the last assertion, choose W so that $(\boldsymbol{y} + W) \cap \mathscr{Z} = \emptyset$. If $\boldsymbol{y} = t\boldsymbol{u} + (1-t)\boldsymbol{w}$, where $\boldsymbol{u} \in U$, $\boldsymbol{w} \in W$, and $0 \le t \le 1$, then the vector $\boldsymbol{y} - (1-t)\boldsymbol{w} = t\boldsymbol{u}$ is simultaneously in $\boldsymbol{y} + W$ and U, impossible since $U \subseteq \mathscr{Z}$. □

9.4.5 Lemma. *Every continuous seminorm q on $\mathscr{Z}$ extends to a continuous seminorm p on $\mathscr{Y}$. Moreover, if $\boldsymbol{y} \notin \operatorname{cl}\mathscr{Z}$, then p may be chosen so that $p(\boldsymbol{y}) > 1$.*

Proof. Take $U := \{\boldsymbol{z} \in \mathscr{Z} : q(\boldsymbol{z}) < 1\}$ in 9.4.4 and set $p = p_V$, the Minkowski functional of the set V of 9.4.4. Since V is open, p is continuous (9.1.4). Since $V \cap \mathscr{Z} = U$,

$$\{\boldsymbol{z} \in \mathscr{Z} : p(\boldsymbol{z}) < 1\} = \{\boldsymbol{z} \in \mathscr{Z} : q(\boldsymbol{z}) < 1\},$$

hence $p\big|_{\mathscr{Z}} = q$ (Ex. 9.9). If $\boldsymbol{y} \notin \operatorname{cl}\mathscr{Z}$ and V is chosen so that $\boldsymbol{y} \notin V$, then $p(\boldsymbol{y}) \ge 1$. □

Here is the main result of the subsection:

9.4.6 Theorem. *Let $(\mathscr{X}_n, \mathcal{T}_n)$ be a strict inductive system for $\mathscr{X}$.*

(a) *There exists a strongest locally convex topology $\mathcal{T}$ on $\mathscr{X}$ such that the relative topology on $\mathscr{X}_n$ induced by $\mathcal{T}$ is $\mathcal{T}_n$.*

(b) *A seminorm p on $\mathscr{X}$ is $\mathcal{T}$-continuous iff its restriction to $\mathscr{X}_n$ is $\mathcal{T}$-continuous for each n.*

(c) *$\mathcal{T}$ is Hausdorff iff each $\mathcal{T}_n$ is Hausdorff.*

Proof. Let $\mathscr{P}$ denote the family of all seminorms on $\mathscr{X}$ with property that the restriction $p|_{\mathscr{X}_n}$ is a continuous seminorm on $\mathscr{X}_n$. The identically zero seminorm obviously has this property, hence $\mathscr{P}$ is nonempty. Let τ denote the locally convex topology on $\mathscr{X}$ generated by $\mathscr{P}$ and let τ_n' denote the relative topology on $\mathscr{X}_n$ induced by τ. By Ex. 9.8, τ_n' is generated by the collection $\mathscr{P}|_{\mathscr{X}_n}$. Since, by definition, the seminorms in $\mathscr{P}|_{\mathscr{X}_n}$ are τ_n-continuous, $\tau_n' \leq \tau_n$ (Ex. 9.11). To show that $\tau_n \leq \tau_n'$, it suffices to show that every τ_n-continuous seminorm p_n may be extended to a τ-continuous seminorm p on $\mathscr{X}$. Indeed, it will then follow that p_n is continuous in the relative topology, implying the inequality. To construct the extension, we use 9.4.5. By induction, for each $m \geq n$ there exists a τ_{m+1}-continuous seminorm p_{m+1} on $\mathscr{X}_{m+1}$ such that $p_{m+1}|_{\mathscr{X}_m} = p_m$. Define p on $\mathscr{X} = \bigcup_{m \geq n} \mathscr{X}_m$ so that $p = p_m$ on each $\mathscr{X}_m$. Then p is a well-defined seminorm on $\mathscr{X}$ and by construction $p \in \mathscr{P}$.

Now let σ be a locally convex topology with property that the relative topology on $\mathscr{X}_n$ induced by σ is τ_n. If q is a σ-continuous seminorm on $\mathscr{X}$, then $U := \{\boldsymbol{x} \in \mathscr{X} : q(\boldsymbol{x}) < 1\}$ is σ-open hence $U \cap \mathscr{X}_n$ is τ_n-open, which implies that $q|_{\mathscr{X}_n}$ is τ_n-continuous (Ex. 9.10). Thus $q \in \mathscr{P}$, hence $\sigma \leq \tau$.

It remains to verify (d). Assume that $\mathscr{X}_n$ is Hausdorff for all n. Let $\boldsymbol{x} \in \mathscr{X}$ and $\boldsymbol{x} \neq \mathbf{0}$. Then $\boldsymbol{x} \in \mathscr{X}_n$ for some n, hence there exists a continuous seminorm p_n with $p_n(\boldsymbol{x}) \neq 0$. By the preceding, p_n extends to a τ-continuous seminorm p on $\mathscr{X}$. Since $p(\boldsymbol{y}) \neq 0$, τ is Hausdorff. The converse is similar. □

The space $\mathscr{X}$ with the topology τ is called the **inductive limit** of the system $(\mathscr{X}_n, \tau_n)$.

9.4.7 Corollary. *Let each $\mathscr{X}_n$ be a Fréchet space. Then the inductive limit topology τ has the following properties:*

(a) *A sequence $(\boldsymbol{x}_n)$ τ-converges to $\boldsymbol{x}$ in $\mathscr{X}$ iff there exists a k such that $(\boldsymbol{x}_n) \subseteq \mathscr{X}_k$ and $\boldsymbol{x}_n \to \boldsymbol{x}$ in the topology τ_k.*

(b) *If T is a linear mapping from $\mathscr{X}$ to a LCS $\mathscr{Y}$, then T is τ-continuous iff for each k the restriction of T to $\mathscr{X}_k$ is τ_k-continuous. In particular, τ-continuity and τ-sequential continuity of linear maps on $\mathscr{X}$ are equivalent.*

Proof. (a) The sufficiency is clear. For the necessity, we may take $\boldsymbol{x} = 0$. Suppose, for a contradiction, that the necessity is false. Thus for each k, $\boldsymbol{x}_n \notin \mathscr{X}_k$ for infinitely many n. Set $\mathscr{Y}_1 = \mathscr{X}_1$ and choose $\boldsymbol{x}_{n_1} \notin \mathscr{Y}_1$. Next, choose $j > 1$ such that $\boldsymbol{x}_{n_1} \in \mathscr{X}_j$ and set $\mathscr{Y}_2 = \mathscr{X}_j$. Continuing in this manner, we obtain a subsequence $(\boldsymbol{y}_k := \boldsymbol{x}_{n_k})$ of $(\boldsymbol{x}_n)$ and a subsequence $(\mathscr{Y}_k)$ of $(\mathscr{X}_n)$ such that $\mathscr{Y}_k \uparrow \mathscr{X}$ and $\boldsymbol{y}_k \in \mathscr{Y}_{k+1} \setminus \mathscr{Y}_k$. It is easy to see that the inductive limit of $(\mathscr{Y}_k)$ is the same as that of $(\mathscr{X}_n)$ (Ex. 9.21). Now let p_1 be a continuous seminorm on $\mathscr{Y}_1$ such that $p_1(\boldsymbol{y}_1) = 1$. By the construction in the proof of 9.4.6, there exists a continuous seminorm p on $\mathscr{X}$ that extends p_1 such that $p|_{\mathscr{Y}_k}$ is a continuous seminorm on $\mathscr{Y}_k$ for each k. Incorporating the second assertion of 9.4.5 into this construction shows that p may be chosen so that $p(\boldsymbol{y}_k) \geq 1$ for all k. Then $(\boldsymbol{y}_k)$ cannot converge to zero in $\mathscr{X}$.

(b) The necessity is clear. For the sufficiency, let q be any continuous seminorm on $\mathscr{Y}$. Then $p := q \circ T$ is a seminorm on $\mathscr{X}$. Since $T|_{\mathscr{X}_n}$ is continuous, $p|_{\mathscr{X}_n}$ is continuous, so p is continuous on $\mathscr{X}$. Therefore, T is continuous (Ex. 9.14). □

Exercises

9.21 Let $(\mathscr{X}_n, \tau_n)$ be a strict inductive system for $\mathscr{X}$ and let (n_k) be a strictly increasing sequence of positive integers. Set $\mathscr{Y}_k = \mathscr{X}_{n_k}$ and $\sigma_k = \tau_{n_k}$. Show that the inductive limit of $(\mathscr{Y}_k, \sigma_k)$ is the same as that of $(\mathscr{X}_n, \tau_n)$.

9.22 Show that $\mathscr{X}$ is not a Fréchet space. ⟦Assume the contrary. Choose $\boldsymbol{x}_n \in \mathscr{X}_{n+1} \setminus \mathscr{X}_n$ and $\varepsilon_n > 0$ so that $d(\varepsilon_n \boldsymbol{x}_n, \mathbf{0}) < 1/n$ and apply (a) of 9.4.7.⟧

Chapter 10

Weak Topologies on Normed Spaces

In this chapter we consider two important locally convex topologies: the weak topology on a normed space $\mathscr{X}$ and the weak* topology on its dual $\mathscr{X}'$. The chapter relies on some of the material developed in Sections 9.1–9.3.

10.1 The Weak Topology

Definition and General Properties

The **weak topology** on a LCS $\mathscr{X} = \mathscr{X}_\tau$ is the initial topology induced by the family of functions $\mathscr{X}'$ (see §0.6). We denote this topology by w and the space $\mathscr{X}$ with the weak topology by $\mathscr{X}_w$. Net convergence in $\mathscr{X}_w$ is described by

$$\boldsymbol{x}_\alpha \overset{w}{\to} \boldsymbol{x} \text{ iff } f(\boldsymbol{x}_\alpha) \to f(\boldsymbol{x}) \text{ for every } f \in \mathscr{X}'.$$

In particular, if $\boldsymbol{x}_\alpha \to \boldsymbol{x}$ and $\boldsymbol{y}_\alpha \to \boldsymbol{y}$ in $\mathscr{X}$ and $c_\alpha \to c$ in $\mathbb{K}$, then, by applying continuous linear functionals, we have

$$\boldsymbol{x}_\alpha + \boldsymbol{y}_\alpha \to \boldsymbol{x} + \boldsymbol{y} \text{ and } c_\alpha \boldsymbol{x}_\alpha \to c\boldsymbol{x}.$$

It follows that w is a vector topology. By 0.6.4, a neighborhood base at zero is given by the open, convex, balanced sets

$$U(f_1, \ldots, f_k; \varepsilon) := \{\boldsymbol{y} : |f_j(\boldsymbol{y})| < \varepsilon,\ j = 1, \ldots, k\},\quad f_j \in \mathscr{X}',\ \varepsilon > 0. \tag{10.1}$$

Thus X_w is a LCS with generating seminorms $p_f(\boldsymbol{x}) = |f(\boldsymbol{x})|$. (The separating property is a consequence of 9.3.4.)

By definition of initial topologies, every member of X'_τ is w continuous, and since $w \leq \tau$, every member of X'_w is τ continuous. Thus

$$\mathscr{X}_w{}' = \mathscr{X}_\tau{}' \text{ and } (\mathscr{X}_w)_w = \mathscr{X}_w.$$

For the remainder of the chapter, we shall be mainly concerned with the weak topology on normed spaces rather than on general LCS. (We return to the general case in later chapters.) For ease and uniformity of notation, we frequently denote the norm topology on $\mathscr{X}$ by s (for *strong topology*). The following result shows that for infinite dimensional normed spaces it is always the case that $w < s$.

10.1.1 Proposition. *If $\mathscr{X}$ is a normed space, then $w = s$ iff $\mathscr{X}$ is finite dimensional.*

Proof. Assume $w = s$. Then $U := \{\boldsymbol{x} : \|\boldsymbol{x}\| < 1\}$ is w-open and hence contains a neighborhood of 0 of the form $U_0 := U(f_1, \ldots, f_n; \varepsilon)$, as in (10.1). We then have $\bigcap_{j=1}^n \ker f_j \subseteq U_0 \subseteq U$, and since U is norm bounded, $\bigcap_{j=1}^{n} \ker f_j = \{0\}$. The linear map $\boldsymbol{x} \mapsto (f_1(\boldsymbol{x}), \ldots, f_n(\boldsymbol{x}))$ from $\mathscr{X}$ to $\mathbb{K}^n$ is therefore 1-1 and so $\mathscr{X}$ is finite dimensional.

Conversely, assume that $\mathscr{X}$ is finite dimensional. We may then identify $\mathscr{X}$ with Euclidean space $\mathbb{K}^d$ for some d. Since the open ball B_n with center $\mathbf{0}$ and radius n has compact closure, the weak and norm topologies agree on B_n (0.8.5). Thus if U is norm open, then $U \cap B_n$ is open in the weak topology for every n and so $U = \bigcup_n U \cap B_n$ is weakly open. □

Weak Sequential Convergence

While sequences generally do not have the utility of nets, they have the advantage of being easier to work with. The following theorem gives necessary and sufficient conditions for a sequence to converge weakly. It is an immediate consequence of the Banach-Steinhaus theorem applied to the functionals $\widehat{x}_n$ on the Banach space $\mathscr{X}'$.

10.1.2 Theorem. *Let $\mathscr{X}$ be a normed space and $(\boldsymbol{x}_n)$ a sequence in $\mathscr{X}$. Then $\boldsymbol{x}_n \xrightarrow{w} \boldsymbol{x} \in \mathscr{X}$ iff the following conditions hold:*

(a) $\sup_n \|\boldsymbol{x}_n\| < \infty$;

(b) *$\langle \boldsymbol{x}_n, \boldsymbol{x}' \rangle \to \langle \boldsymbol{x}, \boldsymbol{x}' \rangle$ for all $\boldsymbol{x}'$ in some subset D of $\mathscr{X}'$ with* $\operatorname{cl}\operatorname{span} D = \mathscr{X}'$.

If these hold, then $\|\boldsymbol{x}\| \le \underline{\lim}_n \|\boldsymbol{x}_n\|$.

10.1.3 Corollary. *Let $(X, \mathcal{F}, \mu)$ be a σ-finite measure space. Then a sequence (f_n) in L^1 converges weakly to some f in L^1 iff the following conditions hold:*

(a) $\sup_n \|f_n\|_1 < \infty$ *and*

(b) *$\lim_n \int_E f_n \, d\mu$ exists for all $E \in \mathcal{F}$.*

In particular, $\boldsymbol{x}_n \xrightarrow{w} \boldsymbol{x}$ in ℓ^1 iff $\sup_n \|\boldsymbol{x}_n\|_1 < \infty$ and $\boldsymbol{x}_n(j) \to \boldsymbol{x}(j)$ for each j.

Proof. The necessity is clear. For the sufficiency, define complex measures

$$\nu_n(E) = \int_E f_n(x) \, d\mu(x), \quad E \in \mathcal{F}.$$

By the Vitali-Hahn-Saks theorem (5.2.4), $\nu(E) := \lim_n \nu_n(E)$ defines a complex measure on $\mathcal{F}$. Moreover, $\nu \ll \mu$, hence $d\nu = f d\mu$ for some $f \in L^1(\mu)$. Thus

$$\lim_n \int f_n \mathbf{1}_E \, d\mu = \nu(E) = \int f \mathbf{1}_E \, d\mu, \quad E \in \mathcal{F}.$$

Taking D in 10.1.2 to be the collection of measurable indicator functions we see that (f_n) converges weakly to f in L^1. □

Proposition 10.1.1, together with Corollary 0.5.6, imply that in every infinite dimensional normed space there are nets that converge weakly but not strongly. The same assertion cannot be made for sequences:

10.1.4 Theorem (Schur). *A weakly convergent sequence in $\ell^1(\mathbb{N})$ converges in norm. Thus the notions of weak and norm sequential convergence in $\ell^1(\mathbb{N})$ coincide.*

Proof. Suppose the assertion is false. Then there exists a sequence $(\boldsymbol{x}_n) \in \ell^1$ and $\varepsilon > 0$ such that $\boldsymbol{x}_n \xrightarrow{w} \mathbf{0}$ and $\|\boldsymbol{x}_n\| \ge 5\varepsilon$ for all n. We construct a subsequence $\boldsymbol{x}_{n_k}$ and a member $\boldsymbol{y}$ of the dual space ℓ^∞ such that $|\langle \boldsymbol{x}_{n_k}, \boldsymbol{y} \rangle| \ge \varepsilon$ for all k, producing the desired contradiction.

Since $\boldsymbol{x}_n \xrightarrow{w} \mathbf{0}$, $\boldsymbol{x}_n(j) \to 0$ for all j. Set $m_0 = n_0 = 1$. Let n_1 be an integer $> n_0$

such that $\sum_{j=1}^{m_0} |\boldsymbol{x}_{n_1}(j)| = |\boldsymbol{x}_{n_1}(m_0)| < \varepsilon$, and let m_1 be an integer $> m_0$ such that $\sum_{j=m_1+1}^{\infty} |\boldsymbol{x}_{n_1}(j)| < \varepsilon$. Next, let n_2 be an integer $> n_1$ such that $\sum_{j=1}^{m_1} |\boldsymbol{x}_{n_2}(j)| < \varepsilon$, and let m_2 be an integer $> m_1$ such that $\sum_{j=m_2+1}^{\infty} |\boldsymbol{x}_{n_2}(j)| < \varepsilon$. In this way we construct strictly increasing sequences (m_k) and (n_k) such that

$$\sum_{j=1}^{m_{k-1}} |\boldsymbol{x}_{n_k}(j)| < \varepsilon \text{ and } \sum_{j=m_k+1}^{\infty} |\boldsymbol{x}_{n_k}(j)| < \varepsilon \ \forall\, k. \tag{$\dagger$}$$

Now define $\boldsymbol{y} \in \ell^\infty$ by $\boldsymbol{y}(j) = \operatorname{sgn} \boldsymbol{x}_{n_k}(j)$ $(m_{k-1} < j \le m_k \in \mathbb{N})$. Fix k and set

$$\alpha_j := \boldsymbol{x}_{n_k}(j)\boldsymbol{y}(j) - |\boldsymbol{x}_{n_k}(j)|.$$

Then $\alpha_j = 0$ for $m_{k-1} < j \le m_k$, hence

$$\sum_{j=1}^{\infty} \alpha_j = \sum_{j=1}^{m_{k-1}} \alpha_j + \sum_{j=m_k+1}^{\infty} \alpha_j. \tag{$\ddagger$}$$

Since $|\boldsymbol{y}(j)| \le 1$, we have $|\alpha_j| \le 2|\boldsymbol{x}_{n_k}(j)|$ and so from ($\dagger$) and ($\ddagger$) $\left|\sum_{j=1}^{\infty} \alpha_j\right| \le 4\varepsilon$. Therefore, by definition of α_j,

$$|\langle \boldsymbol{x}_{n_k}, \boldsymbol{y}\rangle| = \left|\sum_{j=1}^{\infty} \boldsymbol{x}_{n_k}(j)\boldsymbol{y}(j)\right| \ge \sum_{j=1}^{\infty} |\boldsymbol{x}_{n_k}(j)| - \left|\sum_{j=1}^{\infty} \alpha_j\right| \ge \varepsilon,$$

as required. □

Combining the last theorem with 10.1.3, we obtain

10.1.5 Corollary. *A bounded sequence $(\boldsymbol{x}_n)$ in $\ell^1(\mathbb{N})$ converges in norm to $\boldsymbol{x}$ iff $\boldsymbol{x}_n(j) \to \boldsymbol{x}(j)$ for each j.*

Note that Theorem 10.1.4 does not hold in ℓ^p for $1 < p < \infty$ (see Ex. 10.1).

Convexity and Closure in the Weak Topology

Since $w \le s$, every weakly closed subset of a normed space is norm closed. On the other hand, 10.1.1 shows that in every infinite dimensional normed space there are norm closed sets that are not weakly closed. Thus the notions of strong and weak closures in normed spaces are generally distinct. However, for *convex* sets the two closures coincide. We prove this in the general setting of a LCS:

10.1.6 Theorem. *Let $\mathcal{X} = \mathcal{X}_\tau$ be a LCS and C a convex subset of $\mathcal{X}$. Then $\operatorname{cl}_w C = \operatorname{cl}_\tau C$. Thus every τ-closed convex set is weakly closed.*

Proof. Obviously, $\operatorname{cl}_\tau C \subseteq \operatorname{cl}_w C$. Now let $\boldsymbol{x}_0 \in (\operatorname{cl}_s C)^c$. By 9.3.2, there exist $f \in \mathcal{X}'$ and $t \in \mathbb{R}$ such that $\operatorname{Re} f(\boldsymbol{x}_0) < t < \inf\{\operatorname{Re} f(\boldsymbol{y}) : \boldsymbol{y} \in \operatorname{cl}_\tau C\}$. The weak neighborhood $\{\boldsymbol{x} : \operatorname{Re} f(\boldsymbol{x}) < t\}$ of $\boldsymbol{x}_0$ is therefore disjoint from C and so $\boldsymbol{x}_0 \in (\operatorname{cl}_w C)^c$. □

10.1.7 Corollary (Mazur's Theorem). *Let $\boldsymbol{x}_n \xrightarrow{w} \boldsymbol{x}$ in a normed space $\mathcal{X}$. Then there exists a sequence of convex combinations of members of $(\boldsymbol{x}_n)$ that converges in norm to $\boldsymbol{x}$.*

Proof. Let C denote the set of all convex combinations of members of the sequence $(\boldsymbol{x}_n)$. Then $\boldsymbol{z}_n := n^{-1}\sum_{j=1}^{n} \boldsymbol{x}_j \in C$ and $\boldsymbol{z}_n \xrightarrow{w} \boldsymbol{x}$, hence by the theorem $\boldsymbol{x}$ is in the norm closure of C, verifying the assertion. □

*Application: Weak Bases

A sequence $(\boldsymbol{e}_n)$ in a normed linear space $\mathscr{X}$ is a **weak basis** if $\|\boldsymbol{e}_n\| = 1$ and for each $\boldsymbol{x} \in \mathscr{X}$ there exists a unique sequence (c_n) in $\mathbb{K}$ such that the sequence of partial sums $\sum_{j=1}^{n} c_n \boldsymbol{e}_n$ converges weakly to $\boldsymbol{x}$. By analogy with the strong case, we may then write

$$\boldsymbol{x} = w\text{-}\sum_{j=1}^{\infty} c_j(\boldsymbol{x})\boldsymbol{e}_j,$$

where the c_j are linear functionals satisfying $c_j(\boldsymbol{e}_i) = \delta_{ij}$. We show that if $\mathscr{X}$ is a Banach space, then a weak basis is a Schauder basis.

Let $\mathscr{X}^\infty$ denote the linear space of all functions $f = (f(1), f(2), \ldots) : \mathbb{N} \to \mathscr{X}$ such that $\|f\|_\infty := \sup_n \|f(n)\| < \infty$. The space $\mathscr{X}^\infty$ is easily seen to be a Banach space under this norm. Define a linear map $T : \mathscr{X} \to \mathscr{X}^\infty$ by

$$T\boldsymbol{x} = (S_1\boldsymbol{x}, S_2\boldsymbol{x}, \ldots), \quad \text{where} \quad S_m\boldsymbol{x} := \sum_{j=1}^{m} c_j(\boldsymbol{x})\boldsymbol{e}_j \xrightarrow{w} \boldsymbol{x}.$$

Note that $\|T\boldsymbol{x}\|_\infty = \sup_n \|S_n\boldsymbol{x}\|$. We use the closed graph theorem to show that T is continuous. Let $\boldsymbol{x}_n \to \boldsymbol{x}$ in $\mathscr{X}$ and $T\boldsymbol{x}_n \to f$ in $\mathscr{X}^\infty$. In particular, we have the coordinate-wise convergence

$$\lim_n S_m\boldsymbol{x}_n = f(m) \quad \text{for each } m. \tag{†}$$

We claim that

$$f(m) = \sum_{j=1}^{m} \alpha_j\boldsymbol{e}_j \quad \text{for some } \alpha_j \in \mathbb{C}. \tag{‡}$$

For $m = 1$, we have $c_1(\boldsymbol{x}_n)\boldsymbol{e}_1 = S_1\boldsymbol{x}_n \to f(1)$, hence $f(1) = \alpha_1\boldsymbol{e}_1$ for some $\alpha_1 \in \mathbb{C}$ (Ex. 8.2). If the assertion holds for m, then, since $c_{m+1}(\boldsymbol{x}_n)\boldsymbol{e}_{m+1} = S_{m+1}(\boldsymbol{x}_n) - S_m(\boldsymbol{x}_n)$ converges,

$$f(m+1) = \lim_n S_{m+1}\boldsymbol{x}_n = \lim_n \left[S_m\boldsymbol{x}_n + c_{m+1}(\boldsymbol{x}_n)\boldsymbol{e}_{m+1}\right] = \sum_{j=1}^{m+1} \alpha_j\boldsymbol{e}_j$$

for some α_{m+1}. Therefore, the claim holds by induction. Now let $\boldsymbol{x}' \in \mathscr{X}'$. From (†),

$$\begin{aligned}\lim_m \langle f(m), \boldsymbol{x}'\rangle &= \lim_m \lim_n \langle S_m\boldsymbol{x}_n, \boldsymbol{x}'\rangle = \lim_n \lim_m \langle S_m\boldsymbol{x}_n, \boldsymbol{x}'\rangle = \lim_n \langle \boldsymbol{x}_n, \boldsymbol{x}'\rangle \\ &= \langle \boldsymbol{x}, \boldsymbol{x}'\rangle.\end{aligned}$$

The interchange of limits is justified because the convergence in (†) is uniform in m. Therefore, $\boldsymbol{x} = w\text{-}\sum_{j=1}^{\infty} \alpha_j\boldsymbol{e}_j$, hence, by uniqueness, $\alpha_j = c_j(\boldsymbol{x})$ for all j. By (‡) we then have $f(m) = S_m\boldsymbol{x}$ for all m, that is, $f = T\boldsymbol{x}$, proving continuity of T.

It now follows that $\|S_m\boldsymbol{x} - \boldsymbol{x}\| \to 0$, as required. Indeed, since $\boldsymbol{x}$ is in the weak closure of the span of $(\boldsymbol{e}_n)$ and since the weak closure is the same as the norm closure (10.1.6), given $\varepsilon > 0$ we may choose $\boldsymbol{y}_n := \sum_{j=1}^{n} a_j\boldsymbol{e}_j$ such that $\|\boldsymbol{x} - \boldsymbol{y}_n\| < \varepsilon(\|T\| + 1)^{-1}$. But for $m \geq n$, $S_m\boldsymbol{y}_n = \boldsymbol{y}_n$ since $c_j(\boldsymbol{e}_i) = \delta_{ij}$. Therefore, for such m,

$$\|S_m\boldsymbol{x} - \boldsymbol{x}\| \leq \|S_m(\boldsymbol{x} - \boldsymbol{y}_n)\| + \|\boldsymbol{x} - \boldsymbol{y}_n\| \leq \|T\|\,\|\boldsymbol{x} - \boldsymbol{y}_n\| + \|\boldsymbol{x} - \boldsymbol{y}_n\| \leq \varepsilon.$$

Exercises

10.1 Find a sequence (x_n) in $\mathfrak{c}_0$ that converges weakly but not strongly to zero. Do the same for ℓ^p, $1 < p < \infty$.

10.2 (von Neumann). Let $1 < p < \infty$. For each pair $m, n \in \mathbb{N}$ with $1 \le m < n$, define $x_{m,n} \in \ell^p$ by $x_{m,n} := (0, \ldots 0, \overset{m}{1}, 0 \ldots, 0, \overset{n}{m}, 0, \ldots)$. Let A be the set of all $x_{m,n}$. Show that zero is in the weak closure of A in $\ell^p(\mathbb{N})$, but no sequence in A converges in norm to zero.

10.3 Show that the sequence of functions $x_n(t) = t^n$ in $(C[0,1], \|\cdot\|_\infty)$ converges weakly but not strongly.

10.4 Show that $x_n \xrightarrow{w} x$ in $\mathfrak{c}$ iff the following hold:

(a) $\sup_n \|x_n\| < \infty$, (b) $x_n(j) \to x(j) \ \forall\, j$, and (c) $\lim_n \lim_j x_n(j) = \lim_j x(j)$.

10.5 Let $(X, \mathcal{F}, \mu)$ be a σ-finite measure space and $1 < p < \infty$. Show that a sequence (f_n) in L^p converges weakly to $f \in L^p$ iff the following hold:

(a) $\sup \|f_n\|_\infty < \infty$, (b) $\int_E f_n \, d\mu \to \int_E f \, d\mu \ \forall\, E \in \mathcal{F}$ with $\mu(E) < \infty$.

10.6 Let X be a locally compact Hausdorff space. Show that $f_n \xrightarrow{w} f$ in $C_0(X)$ iff $\sup \|f_n\|_\infty < \infty$ and $f_n \to f$ pointwise on X.

10.7 Let $\mathcal{X}$ and $\mathcal{Y}$ be Banach spaces and $T : \mathcal{X} \to \mathcal{Y}$ linear. Show that T is norm continuous iff T is weak-weak continuous.

10.8 Let $\mathcal{X}$ be a normed space. Show that if C is weakly compact, then $\{cx : x \in C, \ |c| \le r\}$ is weakly compact.

10.9 Let $\mathcal{X}$ be a Banach space. Prove the following:

(a) If $\mathcal{X}$ is infinite dimensional, then every weak neighborhood U of zero is unbounded.

(b) The weak topology of a normed space $\mathcal{X}$ is metrizable iff $\mathcal{X}$ is finite dimensional. ⟦Consider $\{x \in \mathcal{X} : d(x, 0) < 1/n\}$ and use the uniform boundedness principle.⟧

10.10 A sequence (x_n) in a normed space $\mathcal{X}$ is said to be **weakly Cauchy** if $(\langle x_n, x' \rangle)$ is Cauchy in $\mathbb{K}$ for all $x' \in \mathcal{X}'$. The space $\mathcal{X}$ is **weakly sequentially complete** if every weakly Cauchy sequence (x_n) in $\mathcal{X}$ converges weakly to a member of $\mathcal{X}$. Prove:

(a) A weakly Cauchy sequence is norm bounded.

(b) ℓ^1 is weakly sequentially complete.

(c) $\mathfrak{c}_0$ is not weakly sequentially complete.

(d) $C[0,1]$ with the uniform norm is not weakly sequentially complete.

10.11 Let X be compact and (f_n) a bounded sequence in $C(X)$ that converges pointwise to $f \in C(X)$. Show that there exists a sequence of convex combinations of members of (f_n) that converges in the uniform norm to $\mathcal{X}$.

10.12 Prove that in an infinite dimensional normed space the weak closure of S_1 is C_1. ⟦Suppose there exists $x_0 \in C_1 \setminus \mathrm{cl}_w S_1$. Choose an open, convex, weak neighborhood U of zero such that $V := U + x_0$ does not meet S_1.⟧

10.13 Prove the following result on *compact convergence* of bounded nets: A bounded net (x_α) converges weakly to x_0 in a normed space $\mathcal{X}$ iff $\langle x_\alpha, f \rangle \to \langle x_0, f \rangle$ uniformly in f on compact subsets of $\mathcal{X}'$.

⟦For each norm compact $K \subseteq \mathcal{X}'$ and $\varepsilon > 0$, define

$$U(K; \varepsilon) := \left\{ x \in X : \sup_{f \in K} |\langle x, f \rangle| < \varepsilon \right\}.$$

If B a bounded subset of $\mathcal{X}$, then, for each $x_0 \in B$, the sets $(x_0 + U(K; \varepsilon)) \cap B$ form a neighborhood base of x_0 in the relative weak topology of B.⟧

10.2 The Weak* Topology

Definition and General Properties

Let $\mathscr{X}$ be a normed space. The **weak* topology** on $\mathscr{X}'$, denoted by w^*, is the initial topology with respect to the family of functions $\widehat{\mathscr{X}}$. Since these functions are weakly continuous on $\mathscr{X}'$, $w^* \le w :=$ the weak topology on $\mathscr{X}'$. Net convergence in the weak* topology is characterized by

$$f_\alpha \overset{w^*}{\to} f \text{ iff } \langle \boldsymbol{x}, f_\alpha\rangle \to \langle \boldsymbol{x}, f\rangle \text{ for every } \boldsymbol{x} \in \mathscr{X}.$$

It follows that w^* is a vector topology. By 0.6.4, a neighborhood base at zero is given by the open, convex, balanced sets

$$U(\boldsymbol{x}_1, \ldots, \boldsymbol{x}_k; \varepsilon) := \big\{f \in \mathscr{X}' : |f(\boldsymbol{x}_j)| < \varepsilon,\ j = 1, \ldots, k\big\}. \tag{10.2}$$

Therefore, $\mathscr{X}'_{w^*}$ is a LCS with generating seminorms $p_{\boldsymbol{x}}(f) = |f(\boldsymbol{x})|$.

The Dual of $\mathscr{X}'_{w^*}$

10.2.1 Proposition. *Let $\mathscr{X}$ be a normed space. A w^*-continuous linear functional φ on $\mathscr{X}'$ is of the form $\widehat{\boldsymbol{x}}$ for some $\boldsymbol{x} \in \mathscr{X}$. Thus $(\mathscr{X}'_{w^*})' = \widehat{\mathscr{X}}$.*

Proof. By definition of the w^*-topology, there exist $\boldsymbol{x}_j \in X$ and $\varepsilon > 0$ such that $|\varphi(f)| < 1$ for all $f \in U := U(\boldsymbol{x}_1, \ldots, \boldsymbol{x}_k; \varepsilon)$. In particular, if $f(\boldsymbol{x}_j) = 0$ for all j then $nf \in U$ for all $n \in \mathbb{N}$, hence $\varphi(f) = 0$. Thus $\bigcap_j \ker \widehat{\boldsymbol{x}_j} \subseteq \ker \varphi$, which implies that φ is a linear combination of the $\widehat{\boldsymbol{x}_j}$, say $\varphi = \sum_{j=1}^k c_j \widehat{\boldsymbol{x}_j}$ (0.2.3). Therefore, $\varphi = \widehat{\boldsymbol{x}}$, where $\boldsymbol{x} = \sum_{j=1}^k c_j \boldsymbol{x}_j$. □

10.2.2 Corollary. *The space $\widehat{\mathscr{X}}$ is dense in the weak* topology of $\mathscr{X}''$.*

Proof. Suppose the assertion is false. Choose $\boldsymbol{x}'' \in \mathscr{X}''$ not in the w^*-closure of $\widehat{\mathscr{X}}$ in $\mathscr{X}''$. By 9.3.3, there exists a weak* continuous linear functional F on $\mathscr{X}''$ such that

$$\sup\{|F(\widehat{\boldsymbol{x}})| : \boldsymbol{x} \in \mathscr{X}\} < |F(\boldsymbol{x}'')|.$$

By the proposition, $F = \widehat{f}$ for some $f \in \mathscr{X}'$. Therefore, the preceding inequality becomes

$$\sup\{|f(\boldsymbol{x})| : \boldsymbol{x} \in \mathscr{X}\} < |\boldsymbol{x}''(f)|.$$

But the left side is unbounded unless $f = 0$, in which case $\boldsymbol{x}''(f) = 0$, impossible. □

10.2.3 Corollary. *Let $A \subseteq \mathscr{X}$ be finite. Then $A^{00} \subseteq \widehat{\mathscr{X}}$.*

Proof. Let $\psi \in A^{00}$. We show that ψ is weak* continuous on $\mathscr{X}'$. Let $f_\alpha \overset{w^*}{\to} 0$ in $\mathscr{X}'$ and $\varepsilon > 0$. Then $\varepsilon^{-1} f_\alpha(\boldsymbol{x}) = f_\alpha(\varepsilon^{-1}\boldsymbol{x}) \to 0$ for each $\boldsymbol{x} \in \mathscr{X}$, and since A is finite, there exists α_0 such that

$$\sup\{\varepsilon^{-1}|f_\alpha(\boldsymbol{x})| : \boldsymbol{x} \in A\} \le 1 \text{ for all } \alpha \ge \alpha_0.$$

For such α, $\varepsilon^{-1} f_\alpha \in A^0$, hence $|\psi(f_\alpha)| < \varepsilon$. Therefore, $\psi(f_\alpha) \to 0$. □

The Banach-Alaoglu Theorem

The next theorem implies that $w^* < s$ for every infinite dimensional normed space.

10.2.4 Banach-Alaoglu Theorem. *Let $\mathscr{X}$ be a normed space. Then the norm closed unit ball C_1' in $\mathscr{X}'$ is weak* compact.*

Proof. The proof uses Tychonoff's theorem (0.8.9). For each $x \in \mathscr{X}$, let K_x denote the compact set $\{z \in \mathbb{K} : |z| \le \|x\|\}$ and let $K := \prod_{x\in\mathscr{X}} K_x$. Then K is the collection of all functions $f : \mathscr{X} \to \mathbb{K}$, linear or not, such that $|f(x)| \le \|x\|$. In particular, $C_1' \subseteq K$. Since the product topology of K is the topology of pointwise convergence on $\mathscr{X}$, the relative topology on C_1' from K is precisely the weak* topology. Now let (f_α) be a net in C_1' that converges pointwise to a member f of K. Then f is linear and since $|f(x)| \le \|x\|$ for all x, $f \in C_1'$. Therefore, C_1' is closed in K. Since K is compact in the topology of pointwise convergence, C_1' is w^*-compact. □

10.2.5 Corollary (Banach). *If $\mathscr{X}$ is separable, then the norm closed unit ball C_1' in $\mathscr{X}'$ is metrizable in the weak* topology and is weak* sequentially compact.*

Proof. Let (x_n) be a sequence with dense span in $\mathscr{X}$. Then

$$d(f,g) = \sum_{k=1}^{\infty} 2^{-k}|f(x_k) - g(x_k)|$$

defines a metric on C_1' such that $d(f_\alpha, f) \to 0$ iff $f_\alpha \overset{w^*}{\to} f$. Therefore, the metric and w^*-topologies agree on C_1', and the conclusion follows from the theorem. □

*Application: Means on Function Spaces

Let S be a set and $\mathscr{F}$ a norm-closed, conjugate-closed, linear subspace of $B(S)$ that contains the constant functions. A **mean** on $\mathscr{F}$ is a linear functional m on $\mathscr{F}$ such that

$$\inf_{s\in S} \operatorname{Re} f(s) \le m(\operatorname{Re} f) \le \sup_{s\in S} \operatorname{Re} f(s), \quad f \in \mathscr{F}. \tag{10.3}$$

(See 8.6.) The set of means on $\mathscr{F}$ is denoted by $M(\mathscr{F})$.

10.2.6 Proposition. *A mean m on $\mathscr{F}$ has the following properties:*

(a) $m(1) = 1$.

(b) *m is positive, that is, $f \ge 0 \Rightarrow m(f) \ge 0$.*

(c) $m(\operatorname{Re} f) = \operatorname{Re} m(f)$, $m(\operatorname{Im} f) = \operatorname{Im} m(f)$, *and* $m(\overline{f}) = \overline{m(f)}$.

(d) *m is a bounded linear functional with norm one.*

(e) *$m(f)$ is in the closure of $\operatorname{co} f(S)$ in $\mathbb{C}$.*

Conversely, a linear functional m on $\mathscr{F}$ that satisfies (a) and (b) is a mean.

Proof. Parts (a) and (b) are clear from 10.3.

(c) If $f = g + ih$, where $g = \operatorname{Re} f = (f + \overline{f})/2$ and $h = \operatorname{Im} f = (f - \overline{f})/2i$, then $m(f) = m(g) + im(h)$, and $m(\overline{f}) = m(g - ih) = m(g) - im(h)$. Since (10.3) implies that $m(g)$ and $m(h)$ are real, (c) follows.

(d) Let $|m(f)| = e^{i\theta}m(f) = m(e^{i\theta}f)$. Since the last expression is real, we have from (c)

$$|m(f)| = m\big(\mathrm{Re}\,(e^{i\theta}f)\big) \le \sup_{s\in S} \mathrm{Re}\, e^{i\theta}f(s) \le \|f\|_\infty,$$

which shows that $\|m\| \le 1$. Since $m(1) = 1$, $\|m\| = 1$.

(e) Let H be any closed half-space in $\mathbb{C} = \mathbb{R}^2$ containing $f(S)$, say $H = \{(x,y) : ax+by \le c\}$. If $f = g + ih$, then $ag + bh \le c$, hence by positivity, $am(g) + bm(h) \le m(c) = c$ and so $m(f) = m(g) + im(h) \in H$. Since the closed convex hull of $f(S)$ is the intersection of all half spaces containing $f(S)$ (Ex. 9.16), part (e) follows.

The last assertion of the proposition is clear. □

10.2.7 Theorem. *$M(\mathscr{F})$ is convex and w^*-compact and is the w^*-closed convex hull of δ_S, where $\delta_s(f) := f(s)$.*

Proof. It is easy to see that a convex combination of means is a mean. If (m_α) is a net in $M(\mathscr{F})$ and $m_\alpha \xrightarrow{w^*} m$ in $\mathscr{F}'$, then (10.3) holds for each m_α and so must hold for m. Therefore, $M(\mathscr{F})$ is w^*-closed, hence, by Alaoglu's theorem, is w^*-compact. Let C denote the w^*-closed convex hull of δ_S. If $m \in M(\mathscr{F}) \setminus C$, then by the separation theorem there exists $f \in \mathscr{F}$ such that

$$\sup \mathrm{Re}\, f(S) = \sup\{\mathrm{Re}\,\langle f, \delta_s\rangle : s \in S\} < \mathrm{Re}\, m(f),$$

contradicting 10.3. Therefore, $M(\mathscr{F}) = C$. □

The proof of following corollary is an exercise (Ex. 10.27).

10.2.8 Corollary. *Let $\mathscr{G}$ be a norm-closed, conjugate-closed subspace of $\mathscr{F}$ that contains the constant functions. Then every mean on $\mathscr{G}$ extends to a mean on $\mathscr{F}$.*

Weak* Continuity

Here is a significant extension of 10.2.1 for Banach spaces:

10.2.9 Theorem. *Let $\mathscr{X}$ be a Banach space. If φ is a linear functional on $\mathscr{X}'$ whose restriction to the closed unit ball C_1' is w^*-continuous, then $\varphi = \widehat{x}$ for some $x \in \mathscr{X}$.*

Proof. Fix $n \in \mathbb{N}$. By hypothesis, the set

$$U := \{f \in C_1' : |\langle f, \varphi\rangle| < 1/n\}$$

is a w^*-neighborhood of zero in C_1', hence there exists a weak* neighborhood V of zero in $\mathscr{X}'$ such that $V \cap C_1' \subseteq U$. We may assume that

$$V := \{f \in \mathscr{X}' : |f(x)| \le 1,\ x \in A\} = A^0, \quad \text{where } A \subseteq \mathscr{X} \text{ is finite.}$$

Note that by the bipolar theorem, ${}^0V^0 = V$. We claim that

$$n\varphi \in V^0 + C_1'', \quad n \in \mathbb{N}, \tag{α}$$

where C_1'' is the closed unit ball in $\mathscr{X}''$. To see this, note first that $V^0 + C_1''$ convex and balanced, and since V^0 is weak* closed and C_1'' is weak* compact, $V^0 + C_1''$ is weak* closed in $\mathscr{X}''$ (Ex. 9.1). If $n\varphi \notin V^0 + C_1''$, then by 9.3.3 and 10.2.1 there exists $f \in \mathscr{X}'$ such that

$$\sup\{|\widehat{f}(\psi + \phi)| : \psi \in V^0,\ \phi \in C_1''\} < 1 < |\widehat{f}(n\varphi)| = n|\varphi(f)|. \tag{β}$$

In particular, $\|f\| = \|\widehat{f}\| = \sup\{|\widehat{f}(\phi)| : \phi \in C_1''\} < 1$ and $\sup\{|\widehat{f}(\psi)| : \psi \in V^0\} < 1$. These inequalities show that $f \in C_1' \cap {}^0V^0 = C_1' \cap V \subseteq U$. But then $|\varphi(nf)| < 1$, contradicting (β). Therefore, (α) holds.

By 10.2.3, $V^0 = A^{00} \subseteq \widehat{\mathscr{X}}$, hence from ($\alpha$) there exists $\widehat{x}_n \in V^0$ and $\phi_n \in C_1''$ such that $n\varphi = \widehat{x}_n + \phi_n$ ($n \in \mathbb{N}$). It follows that

$$|\varphi(f) - f(x_n/n)| = |\phi_n(f/n)| \leq 1/n \quad \forall\, f \in C_1' \text{ and } n \in \mathbb{N}. \qquad (\gamma)$$

By the triangle inequality,

$$\|x_n/n) - x_m/m\| = \sup\left\{\left|f(x_n/n) - f(x_m/m)\right| : f \in C_1'\right\} \leq 1/n + 1/m \quad \forall\, n, m,$$

hence (x_n/n) is a Cauchy sequence. Setting $x := \lim_n x_n/n$, we have $\varphi = \widehat{x}$ by (γ). □

10.2.10 Corollary. *Let $\mathscr{X}$ and $\mathscr{Y}$ be Banach spaces and $T : \mathscr{X}' \to \mathscr{Y}$ linear. If T restricted to the closed unit ball C_1' of $\mathscr{X}'$ is weak*-weak continuous, then T is weak*-weak continuous on $\mathscr{X}'$.*

Proof. For each $y' \in \mathscr{Y}'$, the map $x' \mapsto \langle Tx', y'\rangle$ is w^*-continuous on C_1', hence there exists x depending on y' such that $\langle Tx', y'\rangle = \langle x, x'\rangle$ for all $x' \in \mathscr{X}'$. Thus $\langle Tx', y'\rangle$ is weak* continuous in x'. □

*The Closed Range Theorem

Let $\mathscr{X}$ and $\mathscr{Y}$ be normed spaces and $T \in \mathscr{B}(\mathscr{X}, \mathscr{Y})$. Recall that

$$\left[\operatorname{ran} T\right]^{\perp} = \ker T' \quad \text{and} \quad {}^{\perp}[\operatorname{ran} T'] = \ker T.$$

By the bipolar theorem applied to $\operatorname{ran} T \subseteq \mathscr{X}_s$ and $\operatorname{ran} T' \subseteq \mathscr{X}'_{w^*}$ we then have

$$\operatorname{cl}_w \operatorname{ran} T = \operatorname{cl} \operatorname{ran} T = {}^{\perp}[\ker T'] \quad \text{and} \quad \operatorname{cl}_{w^*} \operatorname{ran} T' = [\ker T]^{\perp}.$$

With this context we may now prove

10.2.11 Theorem (Banach). *Let $\mathscr{X}$ and $\mathscr{Y}$ be Banach spaces and $T \in \mathscr{B}(\mathscr{X}, \mathscr{Y})$. The following are equivalent:*

(a) $\operatorname{ran} T$ *is norm closed.* **(b)** $\operatorname{ran} T'$ *is w^* closed.* **(c)** $\operatorname{ran} T'$ *is norm closed.*

Proof. (a) ⇒ (b): By the preceding, it suffices to show that $[\ker T]^{\perp} \subseteq \operatorname{ran} T'$. To this end let $x' \in [\ker T]^{\perp}$ and define g on $\operatorname{ran} T$ by $g(Tx) = \langle x, x'\rangle$. Then g is well-defined and linear. We claim that g is continuous. For the verification, we use Ex. 8.76, which asserts that for some $c > 0$ and each $y \in \operatorname{ran} T$, the inequality $\|x\| \leq c\,\|y\|$ holds for some x with $Tx = y$. Let $y_n \in \mathscr{X}$ such that $y_n \to 0$, and choose $\|x_n\| \leq c\,\|y_n\|$ such that $Tx_n = y_n$. Then $x_n \to 0$, hence $g(y_n) = \langle x_n, x'\rangle \to 0$, establishing the claim. By the Hahn-Banach theorem there exists $y' \in \mathscr{Y}'$ that extends g, that is, $\langle Tx, y'\rangle = \langle x, x'\rangle$ for all x. It follows that $T'y' = x'$, hence $x' \in \operatorname{ran} T'$.

(c) ⇒ (a): Let $S : \mathscr{X} \to \mathscr{Z} := \operatorname{cl} \operatorname{ran} T$ be the mapping T but with the indicated new codomain. Let $I : \mathscr{Z} \hookrightarrow \mathscr{Y}$ denote the inclusion map, so that $T = IS$ and the dual map $I' : \mathscr{Y}' \to \mathscr{Z}'$ is the restriction mapping. By the Hahn-Banach theorem, I' is surjective. It follows that $\operatorname{ran} S' = S'(\mathscr{Z}') = S'(I'(\mathscr{Y}')) = T'(\mathscr{Y}') = \operatorname{ran} T'$, hence $\operatorname{ran} S'$ is closed. Moreover, if $z' \in \mathscr{Z}'$ and $S'z' = 0$ then $z' = 0$ on $\operatorname{ran} S = \operatorname{ran} T$ hence $z' = 0$ on $\mathscr{Z}$. Therefore, S' is 1-1 and so $S' : \mathscr{Z}' \to \operatorname{ran} S'$ is invertible. Thus there exists $\varepsilon > 0$ such that $\|S'z'\| \geq \varepsilon\,\|z'\|$ for all $z' \in \mathscr{Z}'$. We claim that in the space $\mathscr{Z}$, $B_\varepsilon \subseteq \operatorname{cl} S(B_1)$ $(= \operatorname{cl} S(C_1))$; it will follow from 8.7.4 that S is surjective, hence $\operatorname{ran} T = S(\mathscr{X}) = \mathscr{Z} = \operatorname{cl} \operatorname{ran} T$, completing the proof. To verify the claim, let $z \in \mathcal{Z} \setminus \operatorname{cl} S(C_1)$. By 9.3.3, there exists $z' \in \mathcal{Z}'$ with norm one such that

$$\sup\{|\langle Sx, z'\rangle| : \|x\| \le 1\} < |\langle z, z'\rangle|.$$

The right side is $\le \|z\|$ and the left side equals $\|x\| \le 1\} = \|S'z'\| \ge \varepsilon \|z'\| = \varepsilon$. Therefore, $\|z\| > \varepsilon$, as required. □

10.2.12 Corollary. *T is surjective iff* $\operatorname{ran} T'$ *is closed and T' is injective. In this case, T' has a continuous inverse* $(T')^{-1} : \operatorname{ran} T' \to \mathcal{Y}'$.

Proof. (*Necessity*). By the theorem, $\operatorname{ran} T'$ is closed. Moreover, $\ker T' = \left[\operatorname{ran} T\right]^{\perp} = \{0\}$, hence T' is injective. By the open mapping theorem, $(T')^{-1} : \operatorname{ran} T' \to \mathcal{Y}'$ is continuous.

(*Sufficiency*). By the theorem, $\operatorname{ran} T$ is closed. Thus $\operatorname{ran} T = {}^{\perp}[\ker T'] = {}^{\perp}\{0\} = \mathcal{X}$. □

10.2.13 Corollary. *T' is surjective iff* $\operatorname{ran} T$ *is closed and T is injective. In this case, T has a continuous inverse* $T^{-1} : \operatorname{ran} T \to \mathcal{Y}$.

Proof. (*Necessity*). By the theorem, $\operatorname{ran} T$ is closed. Moreover, $\ker T = {}^{\perp}[\operatorname{ran} T'] = \{0\}$. Therefore, T is injective and so has a continuous inverse $T^{-1} : \operatorname{ran} T \to \mathcal{Y}$.

(*Sufficiency*). By the theorem, $\operatorname{ran} T'$ is w^*-closed. Thus $\operatorname{ran} T' = [\ker T]^{\perp} = \{0\}^{\perp} = \mathcal{X}'$. □

Exercises

10.14 [↑9.6] Show that the set $E := \{n\boldsymbol{e}_n : n \in \mathbb{N}\}$ is bounded in the weak* topology of $\ell^1(\mathbb{N}) = \mathfrak{c}'_{00}$ but is not norm bounded.

10.15 Find an example of a Banach space $\mathcal{X}$ for which the unit sphere in $\mathcal{X}'$ is not weak* compact.

10.16 Show that $\ell^1(\mathbb{N})$ has two distinct weak* topologies.

10.17 Formulate and prove the weak* analog of Ex. 10.13.

10.18 Let $\mathcal{X}$ be a normed space. Show that $w^* = s$ iff $\mathcal{X}$ is finite dimensional.

10.19 [↑10.9] Show that the weak* topology of $\mathcal{X}'$ is metrizable iff $\mathcal{X}$ is finite dimensional.

10.20 Let $\mathcal{X}$ and $\mathcal{Y}$ be Banach spaces with $\mathcal{Y}$ separable and let $T \in \mathscr{B}(\mathcal{X}, \mathcal{Y})$. Prove that T' is compact iff T' carries weak* convergent sequences (y'_n) in $\mathcal{Y}'$ onto norm convergent sequences (Ty'_n) in $\mathcal{X}'$.

10.21 Let $\mathcal{X}$ be a normed space and $E \subseteq \mathcal{X}'$. Show that E is weak* dense in $\mathcal{X}'$ iff for every $x \ne 0$ there exists $x' \in E$ such that $\langle x, x'\rangle \ne 0$.

10.22 Let $\mathcal{Z}$ be a subspace of $\mathcal{X}'$. Show that $({}^{\perp}\mathcal{Z})^{\perp}$ is the w^*-closure of $\mathcal{Z}$.

10.23 Show that the dual space of a Banach space w^***-sequentially complete**, that is, if $(\langle x_n, x'\rangle)$ is Cauchy in $\mathbb{K}$ for all $x' \in \mathcal{X}'$, then (x_n) in $\mathcal{X}$ weak* converges to a member of $\mathcal{X}$. Give an example to show that the assertion is generally false if $\mathcal{X}$ is not complete.

10.24 Let $\mathcal{X}$ and $\mathcal{Y}$ be Banach spaces. Prove that a linear map $T : \mathcal{Y}' \to \mathcal{X}'$ is w^*-w^* continuous iff $T = S'$ for some $S \in \mathscr{B}(\mathcal{X}, \mathcal{Y})$. Thus w^*-w^* continuity implies s-s continuity.

10.25 [↑§7.4] Prove the analog of 10.1.2 for weak* sequential convergence. Conclude the following:

Let X be compact and Hausdorff and let (μ_n) be a sequence in $M(X)$. Then (μ_n) converges in the weak* topology iff $\sup_n \|\mu_n\| < \infty$ and $\lim_n \mu_n(E)$ exists for every $E \in \mathscr{F}$.

10.26 Let X be a locally compact Hausdorff space and μ a Radon measure on X. Prove that $C_0(X)$ is weak* dense in $L^\infty(\mu)$.

10.27 Prove 10.2.8.

10.28 Show that $\mathscr{F} := \{f\,d\lambda : f \in L^1[a,b]\}$ is a norm-closed, non-weak* closed subspace of $M[a,b]$.

10.29 Let X be a locally compact Hausdorff space, $\{f_i : i \in \mathfrak{I}\} \subseteq C_0(X)$, and $\{c_i : i \in \mathfrak{I}\} \subseteq \mathbb{C}$. Suppose for each finite set $F \subseteq \mathfrak{I}$ there exits $\mu_F \in M_{ra}(X)$ with $\|\mu_F\| \leq 1$ such that $\int f_i\,d\mu_F = c_i$ for all $i \in F$. Prove that there exists $\mu \in M_{ra}(X)$ with $\|\mu\| \leq 1$ such that $\int f_i\,d\mu = c_i$ for all $i \in \mathfrak{I}$. Formulate more generally.

10.30 Let S be a nonempty set and $\mathscr{F}$ a conjugate closed, norm closed subspace of $B(S)$. Show that the convex balanced hull of $\{\delta_s : s \in S\}$ is weak* -dense in the closed unit ball C_1' of $\mathscr{F}'$.

10.31 Let X be a compact Hausdorff space and $\mathcal{P}(X)$ the space of probability measures on X. Identifying $\mathcal{P}(X)$ with a subset of $C(X)'$, show that $\mathcal{P}(X)$ is the w^*-closed convex hull of the set δ_X of all Dirac measures on X.

10.3 Reflexive Spaces

A normed space $\mathscr{X}$ is said to be **reflexive** if $\widehat{\mathscr{X}} = \mathscr{X}''$. The mapping $x \to \widehat{x}$ is then a linear isometry from $\mathscr{X}$ onto $\mathscr{X}''$ (§8.5). Identifying $\mathscr{X}$ with $\mathscr{X}''$ under this isometry, we see that a reflexive space is a dual space and hence is complete. Moreover, by Alaoglu's theorem, the ball C_1 in a reflexive space is $\mathscr{X}$ is weakly compact.[1] Note that the property of reflexivity is invariant under a change to an equivalent norm. This is a consequence of the fact that dual spaces are defined topologically and hence remain the same under such a change.

Examples and Basic Properties

Every finite dimensional space $\mathscr{X}$ is reflexive since $\widehat{\mathscr{X}}$ and $\mathscr{X}''$ have the same dimension. The spaces L^p $(1 < p < \infty)$ are reflexive, as can be seen by identifying $(L^p)'$ with L^q and $(L^q)'$ with L^p, where q is conjugate to p. The space L^1 is not reflexive unless it is finite dimensional. This may be seen as a simple consequence of a later result on extreme points that implies in the infinite dimensional case that L^1 is not a dual space (see 14.4.7(b)). The spaces $\mathfrak{c}_0$ and $\mathfrak{c}$ are not reflexive, as their bidual is ℓ^∞ (§8.3). The spaces $C(X)$, X compact, and L^∞ are not reflexive unless they are finite dimensional (Ex. 10.42, 10.43).

The next theorem shows that the property of reflexivity is either common to both $\mathscr{X}$ and $\mathscr{X}'$ or to neither. The proof is a simple consequence of the following general result.

10.3.1 Lemma. *In any normed space $\mathscr{X}$, $\mathscr{X}''' = \widehat{\mathscr{X}'} \oplus \widehat{\mathscr{X}}^\perp$.*

Proof. Let $\varphi \in \mathscr{X}'''$ and define $f \in \mathscr{X}'$ by $\langle x, f\rangle = \langle \widehat{x}, \varphi\rangle$. Noting that $\langle \widehat{x}, \widehat{f}\rangle = \langle x, f\rangle$, we see that $\varphi - \widehat{f} \in \widehat{\mathscr{X}}^\perp$ and so $\mathscr{X}''' = \widehat{\mathscr{X}'} + \widehat{\mathscr{X}}^\perp$. To see that the sum is direct, let φ be in the intersection of the spaces, so that $\varphi = \widehat{f}$, $f \in \mathscr{X}'$ and $\widehat{f} \in \widehat{\mathscr{X}}^\perp$. Then $\langle x, f\rangle = \langle \widehat{x}, \widehat{f}\rangle = 0$ for all x, hence $f = 0$ and therefore $\varphi = 0$. □

10.3.2 Theorem. *$\mathscr{X}$ is reflexive iff $\mathscr{X}'$ is reflexive.*

Proof. By the lemma, $\mathscr{X}'$ is reflexive iff $\widehat{\mathscr{X}}^\perp = \{0\}$ iff $\widehat{\mathscr{X}} = \mathscr{X}''$. □

10.3.3 Theorem. *If $\mathscr{X}$ is reflexive and $\mathscr{Y}$ is a closed subspace of $\mathscr{X}$, then $\mathscr{Y}$ is reflexive.*

[1]This property actually characterizes reflexivity. See 10.3.5.

Proof. Let $\varphi \in \mathscr{Y}''$ and let $T : \mathscr{X}' \to \mathscr{Y}'$ be the restriction map $Tf = f\big|_{\mathscr{Y}}$. Since $T'\varphi \in \mathscr{X}''$, $T'\varphi = \widehat{x}$ for some $x \in \mathscr{X}$. We claim that $x \in \mathscr{Y}$. Indeed, if $f \in \mathscr{X}'$ and $f(\mathscr{Y}) = \{0\}$, then $Tf = 0$ and so

$$\langle x, f\rangle = \langle f, \widehat{x}\rangle = \langle f, T'\varphi\rangle = \langle Tf, \varphi\rangle = \langle 0, \varphi\rangle = 0.$$

An application of 8.5.6 verifies the claim.

For all $f \in \mathscr{X}'$, we now have

$$\langle Tf, \varphi\rangle = \langle f, T'\varphi\rangle = \langle f, \widehat{x}\rangle = \langle x, f\rangle = \langle x, Tf\rangle = \langle Tf, \widehat{x}\rangle .$$

Since T is surjective (8.5.4), $\varphi = \widehat{x}$. □

Weak Compactness and Reflexivity

As noted earlier, if $\mathscr{X}$ is reflexive then the closed unit ball C_1 of $\mathscr{X}$ is weakly compact. The converse also holds. For the proof we need the following result.

10.3.4 Lemma (Goldstine Theorem). *Let $\mathscr{X}$ be a normed space. The image $\widehat{C_1}$ of C_1 under the canonical imbedding $x \to \widehat{x}$ is w^*-dense in the closed unit ball C_1'' of $\mathscr{X}''$.*

Proof. Let K denote the w^*-closure of $\widehat{C_1}$ in $\mathscr{X}''$. Since $\widehat{C_1} \subseteq C_1''$ and C_1'' is w^*-closed, $K \subseteq C_1''$. Suppose there exists $\varphi \in C_1'' \setminus K$. By 9.3.2 and 10.2.1, we may choose $f \in \mathscr{X}'$ such that

$$\sup\{|\langle \widehat{x}, \widehat{f}\rangle| : \|x\| \le 1\} < |\langle \varphi, \widehat{f}\rangle|.$$

But the left side is $\|f\|$ while the right side is $\le \|f\|$. Therefore, $K = C_1''$. □

10.3.5 Theorem. *Let $\mathscr{X}$ be a normed space. Then $\mathscr{X}$ is reflexive iff C_1 is weakly compact.*

Proof. We have already noted the necessity. For the sufficiency, if C_1 is weakly compact, then, by definition of the w^*-topology, $\widehat{C_1}$ is weak*-compact in $\mathscr{X}''$. By the lemma, $\widehat{C_1} = C_1''$, hence $\widehat{\mathscr{X}} = \mathscr{X}''$. □

Theorem 10.3.5 holds if weak compactness is replaced by weak sequential compactness. We state and prove the necessity here; the sufficiency is a consequence of the Eberlein-Šmulian theorem (§14.1).

10.3.6 Theorem. *The closed unit ball C_1 in a reflexive Banach space $\mathscr{X}$ is weakly sequentially compact.*

Proof. Let $x_n \in \mathscr{X}$ with $\|x_n\| \le 1$ for all n. The closed linear span $\mathscr{Y}$ of (x_n) is separable and by 10.3.3 is also reflexive. Thus $\widehat{\mathscr{Y}} = \mathscr{Y}''$ is separable and therefore so is $\mathscr{Y}'$ (8.5.8). By 10.2.5, the closed unit ball in $\mathscr{Y}''$ is weak* sequentially compact. By reflexivity, this is simply the assertion that the closed unit ball in $\mathscr{Y}$ is weakly sequentially compact. Thus (x_n) has a weakly convergent subsequence. □

Exercises

10.32 Let $\mathscr{X}$ and $\mathscr{Y}$ be reflexive spaces. Show that $T \in \mathscr{B}(\mathscr{X}, \mathscr{Y})$ is compact iff $T(C_1)$ is compact.

10.33 Let $\mathscr{X}$ be a normed linear space. Show that the weak and weak* topologies on $\mathscr{X}'$ are equal iff $\mathscr{X}$ is reflexive.

10.34 Let $\mathscr{X}$ be a Banach space and $T : \mathscr{X} \to \mathscr{X}'' : x \to \widehat{x}$ the canonical embedding. Show that $\mathscr{X}$ is reflexive iff the adjoint $T' : \mathscr{X}''' \to \mathscr{X}'$ is 1-1.

10.35 Let $\mathcal{Y}$ be a subspace of a normed space $\mathcal{X}$. Prove that $\widehat{\mathcal{Y}} \subseteq \mathcal{Y}^{\perp\perp}$ and that equality holds iff $\mathcal{Y}$ is reflexive. ⟦If $\mathcal{Y}$ is reflexive and $F \in \mathcal{Y}^{\perp\perp}$, define G on $\mathcal{Y}'$ by $G(g) = F(\widetilde{g})$, where $\widetilde{g}$ is an extension of g to $\mathcal{X}$ with $\|\widetilde{g}\| = \|g\|$.⟧

10.36 Prove that $\mathcal{X}$ is reflexive iff every norm closed subspace of $\mathcal{X}'$ is weak* closed.

10.37 Show that the weak* analog of 10.1.6 holds in $\mathcal{X}'$ iff $\mathcal{X}$ is reflexive.

10.38 Let $\mathcal{X}$ be reflexive and $A \subseteq \mathcal{X}'$. Show that $({}^{\perp}A)^{\perp}$ is the norm closed linear span of A.

10.39 [↑10.10] Show that a reflexive space $\mathcal{X}$ is weakly sequentially complete.

10.40 Use 10.3.6 to show that ℓ^1 is not reflexive. Conclude that $\mathfrak{c}_0$ and $\mathfrak{c}$ are not reflexive.

10.41 Let $\mathcal{X}$ be reflexive and $x' \in \mathcal{X}'$. Prove that $\|x'\| = \langle x, x' \rangle$ for some x with $\|x\| = 1$. (R.C. James showed that every space with this property is reflexive.) Give an example of a Banach space for which the assertion is false.

10.42 Argue as follows to show that $L^1[0,1]$ is not reflexive: Let $A_n = \big(1/(n+1), 1/n\big)$, $n \in \mathbb{N}$, and define $T : \ell^1 \to L^1[0,1]$ by $Tx = \sum_{n=1}^{\infty} x_n [\lambda(A_n)]^{-1} \mathbf{1}_{A_n}$. Show that T is an isometric isomorphism into L^1.

10.43 Show that $M_{ra}[0,1]$, and therefore $C[0,1]$, is not reflexive by using the following argument: Consider first the version of ℓ^1 that consists of real sequences and define $T : \ell^1 \to M_{ra}[0,1]$ by $Tx = \sum x_n \delta_{1/n}$. Show that T is an isometry. Then consider the complex case.

*10.4 Uniformly Convex Spaces

Definition and General Properties

A normed linear space $\mathcal{X}$ is said to be **uniformly convex** if for each $\varepsilon \in (0,2)$ there exists $\delta > 0$ such that

$$\|x\| \le 1,\ \|y\| \le 1, \text{ and } \|x - y\| \ge \varepsilon \Rightarrow \left\|\tfrac{1}{2}(x+y)\right\| \le 1 - \delta. \tag{10.4}$$

Geometrically, this says that the midpoints of line segments in the closed unit ball with lengths bounded away from zero are uniformly distant from the surface. For ease of reference we let $P(x,y)$ denote antecedent of the implication 10.4 and $Q(x,y)$ the consequent.

A normed space that satisfies the *parallelogram law* [2]

$$\|x+y\|^2 + \|x-y\|^2 = 2\|x\|^2 + 2\|y\|^2$$

is uniformly convex. Indeed, if $P(x,y)$ holds, then $\|x+y\|^2 \le 4 - \varepsilon^2$, hence $Q(x,y)$ holds for $\delta := 1 - \frac{1}{2}\sqrt{4-\varepsilon^2}$. In particular, $L^2(X, \mathcal{F}, \mu)$ is uniformly convex. More generally, Clarkson has shown that $L^p(X, \mathcal{F}, \mu)$ is uniformly convex for $1 < p < \infty$ [8]. (See Ex. 10.44 for the case $p \ge 2$.)

Here is a useful sequential characterization of uniform convexity:

10.4.1 Proposition. *A normed linear space $\mathcal{X}$ is uniformly convex iff for any sequences (x_n) and (y_n) in C_1 with $\|\frac{1}{2}(x_n + y_n)\| \to 1$ it follows that $\|x_n - y_n\| \to 0$.*

[2]Spaces that satisfy the parallelogram law are called *inner product spaces*. These are discussed in detail in the next chapter.

Proof. Let $\mathcal{X}$ be uniformly convex with sequences in C_1 such that $\|\boldsymbol{x}_n - \boldsymbol{y}_n\| \not\to 0$. Then there exist $\varepsilon > 0$ such that $\|\boldsymbol{x}_n - \boldsymbol{y}_n\| \geq \varepsilon$ for infinitely many n. By uniform convexity, there exists $\delta > 0$ such that $\|\frac{1}{2}(\boldsymbol{x}_n + \boldsymbol{y}_n)\| \leq 1 - \delta$ for infinitely many n. Therefore, $\|\frac{1}{2}(\boldsymbol{x}_n + \boldsymbol{y}_n)\| \not\to 1$.

Now suppose that $\mathcal{X}$ is not uniformly convex. Then there exists an $\varepsilon \in (0,2)$ and sequences $(\boldsymbol{x}_n)$ and $(\boldsymbol{y}_n)$ with $\|\boldsymbol{x}_n\| \leq 1$, $\|\boldsymbol{y}_n\| \leq 1$, $\|\boldsymbol{x}_n - \boldsymbol{y}_n\| \geq \varepsilon$, and $\left\|\frac{1}{2}(\boldsymbol{x}_n + \boldsymbol{y}_n)\right\| > 1 - 1/n$. It follows that $\left\|\frac{1}{2}(\boldsymbol{x}_n + \boldsymbol{y}_n)\right\| \to 1$. Since $\|\boldsymbol{x}_n - \boldsymbol{y}_n\| \not\to 0$, the sequential criterion fails. □

Connections with Strict Convexity

It follows directly from Ex. 8.20(ii) that a uniformly convex space is strictly convex. The following example shows that the two notions are distinct.

10.4.2 Example. (*A strictly convex non-uniformly convex Banach space*). Consider the linear space ℓ^1 with the norm $\|\boldsymbol{x}\| := \|\boldsymbol{x}\|_1 + \|\boldsymbol{x}\|_2$. Since $\|\boldsymbol{x}\|_1 \leq \|\boldsymbol{x}\|_2$, $\|\cdot\|$ is equivalent to $\|\cdot\|_1$. Recalling that $(\ell^1, \|\cdot\|_1)$ is not reflexive and that reflexivity is preserved under a change to an equivalent norm, we see that $(\ell^1, \|\cdot\|)$ is not reflexive. Thus by 10.4.6 below, $(\ell^1, \|\cdot\|)$ is not uniformly convex.

To see that $(\ell^1, \|\cdot\|)$ is strictly convex let $\boldsymbol{x} \neq \boldsymbol{y}$ with $\|\boldsymbol{x}\| = \|\boldsymbol{y}\| = 1$. We show that $\|\boldsymbol{x} + \boldsymbol{y}\| < 2$. Set $a_j = |x_j|$, $b_j = |y_j|$, $\boldsymbol{a} = (a_1, a_2, \ldots)$, and $\boldsymbol{b} = (b_1, b_2, \ldots)$. Then $\|\boldsymbol{a}\| = \|\boldsymbol{x}\| = 1$, $\|\boldsymbol{b}\| = \|\boldsymbol{y}\| = 1$, and $\|\boldsymbol{x} + \boldsymbol{y}\| \leq \|\boldsymbol{a} + \boldsymbol{b}\|$. We consider two cases.

Suppose that $\boldsymbol{a} \neq \boldsymbol{b}$. Since $\|\boldsymbol{a}\|_1 = 1 - \|\boldsymbol{a}\|_2$, $\|\boldsymbol{b}\|_1 = 1 - \|\boldsymbol{b}\|_2$, and $a_j, b_j \geq 0$, we have

$$\|\boldsymbol{a} + \boldsymbol{b}\| = \|\boldsymbol{a}\|_1 + \|\boldsymbol{b}\|_1 + \|\boldsymbol{a} + \boldsymbol{b}\|_2 = 2 - \|\boldsymbol{a}\|_2 - \|\boldsymbol{b}\|_2 + \|\boldsymbol{a} + \boldsymbol{b}\|_2 < 2,$$

the last inequality by the strict convexity of ℓ^2. Therefore, $\|\boldsymbol{x} + \boldsymbol{y}\| < 2$.

Now suppose $\boldsymbol{a} = \boldsymbol{b}$, that is, $|x_j| = |y_j|$ for all j. Since $x_k \neq y_k$ for some k, it follows that $|x_k + y_k| < |x_k| + |y_k|$. Therefore $\|\boldsymbol{x} + \boldsymbol{y}\| < \|\boldsymbol{a} + \boldsymbol{b}\| \leq 2$. ♢

In the finite dimensional case, the two notions are equivalent:

10.4.3 Proposition. *A finite dimensional strictly convex space is uniformly convex.*

Proof. Suppose $\mathcal{X}$ is not uniformly convex. By 10.4.1, we may choose sequences $(\boldsymbol{x}_n)$ and $(\boldsymbol{y}_n)$ in C_1 such that $\|\frac{1}{2}(\boldsymbol{x}_n + \boldsymbol{y}_n)\| \to 1$ and $\|\boldsymbol{x}_n - \boldsymbol{y}_n\| \not\to 0$. By compactness of C_1 we may assume that $\boldsymbol{x}_n \to \boldsymbol{x}$ and $\boldsymbol{y}_n \to \boldsymbol{y}$. Then $\|\frac{1}{2}(\boldsymbol{x} + \boldsymbol{y})\| = 1$ and $\boldsymbol{x} \neq \boldsymbol{y}$, so $\mathcal{X}$ is not strictly convex. □

Weak and Strong Convergence In Uniformly Convex Spaces

The next theorem makes an important connection between weak and strong sequential convergence in uniformly convex spaces. For the proof we need

10.4.4 Lemma. *Let $\mathcal{X}$ be uniformly convex. If $(\boldsymbol{y}_n)$ is a sequence in C_1 such that $\lim_{m,n\to\infty} \left\|\frac{1}{2}(\boldsymbol{y}_n + \boldsymbol{y}_m)\right\| = 1$, then $(\boldsymbol{y}_n)$ is Cauchy.*

Proof. If $(\boldsymbol{y}_n)$ is not Cauchy, then there exists an $\varepsilon > 0$ and strictly increasing sequences of indices n_k and m_k such that $\|\boldsymbol{y}_{n_k} - \boldsymbol{y}_{m_k}\| \geq \varepsilon$ for all k. But this contradicts 10.4.1, since $\|\frac{1}{2}(\boldsymbol{y}_{n_k} + \boldsymbol{y}_{m_k})\| \to 1$. □

10.4.5 Theorem (Radon-Riesz). *Let $(\boldsymbol{x}_n)$ be a sequence in a uniformly convex Banach space $\mathcal{X}$. Then $\boldsymbol{x}_n \xrightarrow{s} \boldsymbol{x} \in \mathcal{X}$ iff $\boldsymbol{x}_n \xrightarrow{w} \boldsymbol{x}$ and $\|\boldsymbol{x}_n\| \to \|\boldsymbol{x}\|$.*

Proof. The necessity is clear. For the sufficiency we may take $\boldsymbol{x} \neq \boldsymbol{0}$, since otherwise there is nothing to prove. Then $\|\boldsymbol{x}_n\| \neq 0$ for all large n, hence we may assume that $\|\boldsymbol{x}_n\| \neq 0$

for *all* n. Set $y_n = \|x_n\|^{-1} x_n$ and $y = \|x\|^{-1} x$ and note that the hypotheses imply that $y_n \xrightarrow{w} y$. Thus for any $f \in \mathcal{X}'$,

$$\lim_{m,n\to\infty} f\left(\tfrac{1}{2}(y_m + y_n)\right) = f(y). \tag{†}$$

Choose f so that $f(y) = \|f\| = 1$ (8.5.6). Then $f\left(\frac{1}{2}(y_m+y_n)\right) \le \left\|\frac{1}{2}(y_m + y_n)\right\| \le 1$, hence from (†), $\lim_{m,n\to\infty} \left\|\frac{1}{2}(y_m + y_n)\right\| = 1$. By the lemma, (y_n) is Cauchy and so converges in norm to y. From the inequality

$$\|x_n - x\| = \big\| \|x_n\| y_n - \|x\| y \big\| \le \big|\|x_n\| - \|x\|\big| \, \|y_n\| + \|x\| \, \|y_n - y\|$$

we see that $x_n \xrightarrow{s} x$. □

Connection with Reflexivity

Here is the main result of the section. It is remarkable in the sense that uniform convexity is a purely geometric notion, defined exclusively as a property of the norm, while reflexivity may be viewed, via say 10.3.5, as being a topological property.

10.4.6 Theorem (Milman-Pettis). *A uniformly convex Banach space is reflexive.*

Proof. We follow [39]. Suppose for a contradiction that $\mathcal{X}$ is uniformly convex but not reflexive. Let C_1 and C_1'' denote the closed unit balls in $\mathcal{X}$ and $\mathcal{X}''$, respectively. By assumption, $\widehat{C_1}$ is a proper subset of C_1'', hence since $\widehat{C_1}$ is closed there exists $x'' \in C_1''$ at a positive distance from $\widehat{C_1}$, that is,

$$\|x'' - \widehat{x}\| \ge 2\varepsilon \quad \text{for all } x \in C_1. \tag{†}$$

Define

$$\delta = \delta(\varepsilon) := \inf\left\{1 - \|\tfrac{1}{2}(x + y)\| : x,\, y \in C_1,\, \|x - y\| \ge \varepsilon\right\}.$$

It follows easily from 10.4.1 that $\delta(\varepsilon) > 0$. Since $\|x''\| = 1$, we may choose $f \in C_1'$ such that $|\langle f, x''\rangle - 1| < \delta/2$. Thus x'' is in the weak* open set

$$V := \{y'' \in \mathcal{X}'' : |\langle f, y''\rangle - 1| < \delta/2\}.$$

By 10.3.4, x'' is in the weak* closure of $V \cap \widehat{C}_1$ (approximate $\langle f, x''\rangle$ by $\langle f, \widehat{x}\rangle$.) Now, for any $\widehat{x}$ and $\widehat{y}$ in $V \cap \widehat{C}_1$,

$$|\langle x + y, f\rangle| = |2 + (\langle f, \widehat{x}\rangle - 1) + (\langle f, \widehat{y}\rangle - 1)| \ge 2 - \delta,$$

hence $\|x + y\| \ge 2 - \delta$ and so $1 - \|\frac{1}{2}(x + y)\| \le \frac{1}{2}\delta < \delta$. From the definition of δ, $\|x - y\| < \varepsilon$. Thus $V \cap \widehat{C_1} \subseteq \widehat{x} + \varepsilon C_1''$. Since $\widehat{x} + \varepsilon C_1''$ is weak* closed and since x'' is in the weak* closure of $V \cap \widehat{C_1}$, we conclude that $x'' \in \widehat{x} + \varepsilon C_1''$. But this contradicts (†). □

Exercises

10.44 Verify steps (a)–(d) below and then use (d) to show that L^p is uniformly convex for $p \ge 2$.

(a) For $c > 0$, the function $f(p) = (1 + c^p)^{1/p}$ is strictly decreasing in p on $[2, \infty)$.

(b) $(a^p + b^p)^{1/p} \le (a^2 + b^2)^{1/2}$ $(a \ge 0,\, b \ge 0,\, p \ge 2)$.

(c) $c^2 + d^2 \le 2^{1-2/p}(c^p + d^p)^{2/p}$ $(c \ge 0,\, d \ge 0)$. ⟦For $p > 2$, use Hölder's inequality.⟧

(d) $|s + t|^p + |s - t|^p \le 2^{p-1}(|s|^p + |t|^p)$ $(s, t \in \mathbb{R},\, p \ge 2)$.

10.45 Let $\mathcal{X}$ be a uniformly convex Banach space and $f \ne 0 \in \mathcal{X}'$. Show that there exists a unique $x \in S_1$ such that $\|f\| = f(x)$. ⟦It suffices to consider the case $\|f\| = 1$. Let $x_n \in S_1$ and $f(x_n) \to 1$. Use 10.4.4 to show that (x_n) is Cauchy. (One may also use 10.41.) For uniqueness, suppose also that $f(y) = 1$ for some $y \in S_1$ with $\|x - y\| > 0$ and consider $\frac{1}{2}(x + y)$.⟧

Chapter 11

Hilbert Spaces

A Hilbert space is a Banach space whose norm is derived from an inner product. This feature endows Hilbert spaces with rich geometric structure that accounts for the broad applicability of the subject to areas such as harmonic analysis, differential equations, and quantum mechanics. In this chapter we examine the structure of Hilbert spaces. The next chapter treats operators on these spaces.

11.1 General Principles

Sesquilinear Forms

Let $\mathscr{X}$ and $\mathscr{Y}$ be vector spaces over $\mathbb{K}$. A **sesquilinear**[1] **functional** on $\mathscr{X} \times \mathscr{Y}$ is a mapping $B : \mathscr{X} \times \mathscr{Y} \to \mathbb{K}$ that is linear in the first variable and *conjugate* linear in the second; that is, for all $\boldsymbol{x}$, $\boldsymbol{x}_j$, $\boldsymbol{y}$, $\boldsymbol{y}_j \in \mathscr{Y}$, and $c_j \in \mathbb{K}$,

$$\begin{aligned} B(c_1\boldsymbol{x}_1 + c_2\boldsymbol{x}_2, \boldsymbol{y}) &= c_1 B(\boldsymbol{x}_1, \boldsymbol{y}) + c_2 B(\boldsymbol{x}_2, \boldsymbol{y}), \quad \text{and} \\ B(\boldsymbol{x}, c_1\boldsymbol{y}_1 + c_2\boldsymbol{y}_2) &= \overline{c_1} B(\boldsymbol{x}, \boldsymbol{y}_1) + \overline{c_2} B(\boldsymbol{x}, \boldsymbol{y}_2). \end{aligned}$$

Of course, if $\mathbb{K} = \mathbb{R}$, then B is simply a bilinear functional. The set of sesquilinear functionals is easily seen to a linear space over $\mathbb{K}$ under the usual pointwise operations.

A sesquilinear functional B on $\mathscr{X} \times \mathscr{X}$ is called a **sesquilinear form on** $\mathscr{X}$. Such a mapping is said to be **Hermitian** or **self-adjoint**[2] if $\overline{B(\boldsymbol{x}, \boldsymbol{y})} = B(\boldsymbol{y}, \boldsymbol{x})$, and **positive** if $B(\boldsymbol{x}, \boldsymbol{x}) \geq 0$. The set of Hermitian sesquilinear forms is easily seen to a real linear subspace of the vector space of all sesquilinear forms.

The following result is useful in reducing some proofs involving sesquilinear forms to simpler arguments.

11.1.1 Theorem (Polarization Identities). *Let B be a sesquilinear form on $\mathscr{X}$.*

(a) *If $\mathbb{K} = \mathbb{R}$, then $4B(\boldsymbol{x}, \boldsymbol{y}) = B(\boldsymbol{x} + \boldsymbol{y}, \boldsymbol{x} + \boldsymbol{y}) - B(\boldsymbol{x} - \boldsymbol{y}, \boldsymbol{x} - \boldsymbol{y})$.*

(b) *If $\mathbb{K} = \mathbb{C}$, then $4B(\boldsymbol{x}, \boldsymbol{y}) = \sum_{k=0}^{3} i^k B(\boldsymbol{x} + i^k\boldsymbol{y}, \boldsymbol{x} + i^k\boldsymbol{y})$.*

Proof. For (b) we have

$$B(\boldsymbol{x} + i^k\boldsymbol{y}, \boldsymbol{x} + i^k\boldsymbol{y}) = B(\boldsymbol{x}, \boldsymbol{x}) + i^k B(\boldsymbol{y}, \boldsymbol{x}) + \overline{i^k} B(\boldsymbol{x}, \boldsymbol{y}) + +B(\boldsymbol{y}, \boldsymbol{y}).$$

Multiplying by i^k and summing, we see that

$$\sum_{k=0}^{3} i^k B(\boldsymbol{x} + i^k\boldsymbol{y}, \boldsymbol{x} + i^k\boldsymbol{y}) = 4B(\boldsymbol{x}, \boldsymbol{y}) + \sum_{k=0}^{3} \left[i^k B(\boldsymbol{x}, \boldsymbol{x}) + (-1)^k B(\boldsymbol{y}, \boldsymbol{x}) + i^k B(\boldsymbol{y}, \boldsymbol{y})\right].$$

[1] From the Latin "semis qui" meaning one and a half.
[2] For $\mathbb{K} = \mathbb{R}$, the term **symmetric** is also used.

Since $\sum_{k=0}^{3} i^k = 0$ and $\sum_{k=0}^{3}(-1)^k = 0$, the desired formula follows. □

11.1.2 Corollary. *If $\mathbb{K} = \mathbb{C}$, then a sesquilinear form on $\mathscr{X}$ is Hermitian iff $B(\boldsymbol{x}, \boldsymbol{x})$ is real for all $\boldsymbol{x}$.*

Proof. The necessity is clear. For the sufficiency, factor out i^k to write the general term of the sum in (b) as $i^k B(\bar{i}^{\,k}\boldsymbol{x}+\boldsymbol{y}, \bar{i}^{\,k}\boldsymbol{x}+\boldsymbol{y})$. Taking conjugates in (b) and using the hypothesis we have

$$4\overline{B(\boldsymbol{x},\boldsymbol{y})} = \sum_{k=0}^{3} \bar{i}^{\,k} B(\bar{i}^{\,k}\boldsymbol{x}+\boldsymbol{y}, \bar{i}^{\,k}\boldsymbol{x}+\boldsymbol{y}) = \sum_{k=0}^{3} i^k B(i^k\boldsymbol{x}+\boldsymbol{y}, i^k\boldsymbol{x}+\boldsymbol{y}) = 4B(\boldsymbol{y},\boldsymbol{x}). \quad \square$$

Semi-Inner-Product Spaces

For the remainder of the chapter, we use the notation $(\cdot\,|\,\cdot)$ for positive Hermitian sesquilinear forms. A vector space $\mathscr{X}$ over $\mathbb{K}$ equipped with such a form is called a **semi-inner-product space**. Define an associated function $\|\cdot\| : \mathscr{X} \to [0,\infty)$ by

$$\|\boldsymbol{x}\| = \sqrt{(\boldsymbol{x}\,|\,\boldsymbol{x})}, \quad \boldsymbol{x} \in \mathscr{X}. \tag{11.1}$$

The polarization identities may then be written as

$$4\,(\boldsymbol{x}\,|\,\boldsymbol{y}) = \|\boldsymbol{x}+\boldsymbol{y}\|^2 - \|\boldsymbol{x}-\boldsymbol{y}\|^2 \ (\mathbb{K}=\mathbb{R}) \ \text{ and } \ 4\,(\boldsymbol{x}\,|\,\boldsymbol{y}) = \sum_{k=0}^{3} i^k \left\|\boldsymbol{x}+i^k\boldsymbol{y}\right\|^2 \ (\mathbb{K}=\mathbb{C}).$$

Moreover, a direct calculation yields the **parallelogram law**

$$\|\boldsymbol{x}+\boldsymbol{y}\|^2 + \|\boldsymbol{x}-\boldsymbol{y}\|^2 = 2\,\|\boldsymbol{x}\|^2 + 2\,\|\boldsymbol{y}\|^2. \tag{11.2}$$

The following inequality is one of the most important tools in Hilbert space theory. As we shall see, it is the essential ingredient in frequent calculations involving operators on Hilbert spaces.

11.1.3 CBS Inequality (Cauchy, Bunyakovsky, Schwarz). *Let $\mathscr{X}$ be a semi-inner-product space. Then*

$$|\,(\boldsymbol{x}\,|\,\boldsymbol{y})\,| \le \|\boldsymbol{x}\|\,\|\boldsymbol{y}\|, \quad \boldsymbol{x},\boldsymbol{y} \in \mathscr{X}. \tag{11.3}$$

Equality holds iff $\|t\boldsymbol{x}+\alpha\boldsymbol{y}\| = 0$ for some $t \in \mathbb{R}$ and $\alpha \in \mathbb{K}$.

Proof. We may assume $(\boldsymbol{x}\,|\,\boldsymbol{y}) \ne 0$. Set $\alpha = (\boldsymbol{x}\,|\,\boldsymbol{y})/|\,(\boldsymbol{x}\,|\,\boldsymbol{y})\,|$. For any $t \in \mathbb{R}$,

$$\begin{aligned} 0 \le \|t\boldsymbol{x}+\alpha\boldsymbol{y}\| &= (t\boldsymbol{x}+\alpha\boldsymbol{y}\,|\,t\boldsymbol{x}+\alpha\boldsymbol{y}) = t^2\,\|\boldsymbol{x}\|^2 + t\bar{\alpha}\,(\boldsymbol{x}\,|\,\boldsymbol{y}) + t\alpha\,(\boldsymbol{y}\,|\,\boldsymbol{x}) + |\alpha|^2|\,\|\boldsymbol{y}\|^2 \\ &= t^2\,\|\boldsymbol{x}\|^2 + 2t|\,(\boldsymbol{x}\,|\,\boldsymbol{y})\,| + \|\boldsymbol{y}\|^2 =: at^2+bt+c. \end{aligned} \tag{†}$$

Since the quadratic in (†) is never negative, it has at most one zero, hence the discriminant $b^2 - 4ac = 4|\,(\boldsymbol{x}\,|\,\boldsymbol{y})\,|^2 - 4\,\|\boldsymbol{x}\|\,\|\boldsymbol{y}\|$ cannot be positive. This establishes (11.3).

Equality holds in (11.3) iff the discriminant $b^2 - 4ac$ is zero. Letting t be the unique zero of at^2+bt+c, we see that $\|t\boldsymbol{x}+\alpha\boldsymbol{y}\| = 0$. □

11.1.4 Corollary. *$\|\cdot\|$ is a seminorm.*

Proof. The only property that is not immediately evident is the triangle inequality. For this we use the CBS inequality as follows:

$$\begin{aligned} \|\boldsymbol{x}+\boldsymbol{y}\|^2 &= (\boldsymbol{x}+\boldsymbol{y}\,|\,\boldsymbol{x}+\boldsymbol{y}) = \|\boldsymbol{x}\|^2 + \|\boldsymbol{y}\|^2 + 2\mathrm{Re}\,(\boldsymbol{x}\,|\,\boldsymbol{y}) \le \|\boldsymbol{x}\|^2 + \|\boldsymbol{y}\|^2 + 2|\,(\boldsymbol{x}\,|\,\boldsymbol{y})\,| \\ &\le \|\boldsymbol{x}\|^2 + \|\boldsymbol{y}\|^2 + 2\,\|\boldsymbol{x}\|\,\|\boldsymbol{y}\| = \big(\,\|\boldsymbol{x}\| + \|\boldsymbol{y}\|\,\big)^2 \end{aligned}$$

Taking square roots yields the desired inequality. □

11.1.5 Corollary. $\|x\| = \sup\{|\,(x\,|\,y)\,| : \|y\| \le 1\}$.

Proof. Let s denote the supremum. By the CBS inequality, $s \le \|x\|$, hence the assertion holds $\|x\| = 0$. If $\|x\| \ne 0$, take $y = \|x\|^{-1} x$ in the definition of s to get $\|x\| = |\,(x\,|\,y)\,| \le s$. □

Inner Product Spaces. Hilbert Spaces

A positive Hermitian sesquilinear form on $\mathscr{X}$ whose associated seminorm (11.1) is a norm is called an **inner product**. A vector space equipped with an inner product is called an **inner product space**. An inner product space that is complete with respect to the induced norm is called a **Hilbert space**.

Note that in an inner product space, the CBS inequality implies that $(x\,|\,y)$ is continuous in (x, y) in the norm topology, as may be seen by writing

$$|\,(x\,|\,y) - (a\,|\,b)\,| \le |\,(x - a\,|\,y)\,| + |\,(a\,|\,y - b)\,|$$

A norm on a vector space $\mathscr{X}$ is an **inner product norm** if there is an inner product that induces the norm via (11.1). Such a norm must satisfy the parallelogram law (11.2). The converse holds:

11.1.6 Theorem. *A norm $\|\cdot\|$ on a vector space $\mathscr{X}$ over $\mathbb{K}$ that satisfies the parallelogram law is an inner product norm.*

Proof. (For the case $\mathbb{K} = \mathbb{C}$): Define $(x\,|\,y)$ by

$$(x\,|\,y) = \frac{1}{4}\sum_{k=1}^{4} i^k \,\|x + i^k y\|^2. \tag{†}$$

Then

$$\begin{aligned} 4(x\,|\,y) &= \|x + y\|^2 - \|x - y\|^2 + i\left(\|x + iy\|^2 - \|x - iy\|^2\right) \\ &= \|y + x\|^2 - \|y - x\|^2 - i\left(\|y + ix\|^2 - \|y - ix\|^2\right) \\ &= 4\overline{(y\,|\,x)}. \end{aligned}$$

In particular, from the first equality we have

$$\begin{gathered} (-x\,|\,x) = -\,(x\,|\,x), \quad (ix\,|\,x) = i\,(x\,|\,x) \quad \text{and} \\ (x\,|\,x) = \tfrac{1}{4}\left[\|2x\|^2 + i\left(\|(1+i)x\|^2 - \|(1-i)x\|^2\right)\right] = \|x\|^2. \end{gathered}$$

To prove the additive property of inner products, use (†) to write

$$4\,(x \pm y\,|\,z) = \|x \pm y + z\|^2 - \|x \pm y - z\|^2 + i\left(\|x \pm y + iz\|^2 - \|x \pm y - iz\|^2\right)$$

and then add to obtain

$$\begin{aligned} &4\,(x + y\,|\,z) + 4\,(x - y\,|\,z) \\ &= \left[\|(x + z) + y\|^2 + \|(x + z) - y\|^2\right] - \left[\|(x - z) + y\|^2 + \|(x - z) - y\|^2\right] \\ &+ i\left[\|(x + iz) + y\|^2 + \|(x + iz) - y\|^2\right] - i\left[\|(x - iz) + y\|^2 + \|(x - iz) - y\|^2\right]. \end{aligned}$$

Applying the parallelogram identity to each bracketed expression reduces the right side to

$$2\left[\|x + z\|^2 - \|x - z\|^2\right] + 2i\left[\|x + iz\|^2 - \|x - iz\|^2\right] = 8\,(x\,|\,z).$$

We now have

$$(\mathscr{x} + \mathscr{y} \,|\, \mathscr{z}) + (\mathscr{x} - \mathscr{y} \,|\, \mathscr{z}) = 2\,(\mathscr{x} \,|\, \mathscr{z}). \tag{‡}$$

Taking $\mathscr{x} = \mathscr{y}$ and noting that $(0 \,|\, \mathscr{z}) = 0$, we have $(2\mathscr{y} \,|\, \mathscr{z}) = 2\,(\mathscr{y} \,|\, \mathscr{z})$ hence $2\left(\frac{1}{2}\mathscr{y} \,|\, \mathscr{z}\right) = (\mathscr{y} \,|\, \mathscr{z})$ for all $\mathscr{y}$. Setting $\mathscr{x} + \mathscr{y} = \mathscr{u}$ and $\mathscr{x} - \mathscr{y} = \mathscr{v}$ in (‡) yields

$$(\mathscr{u} \,|\, \mathscr{z}) + (\mathscr{v} \,|\, \mathscr{z}) = 2\left(\tfrac{1}{2}(\mathscr{u} + \mathscr{v}) \,|\, \mathscr{z}\right) = (\mathscr{u} + \mathscr{v} \,|\, \mathscr{z}).$$

This, together with $(\mathscr{y} \,|\, \mathscr{x}) = \overline{(\mathscr{x} \,|\, \mathscr{y})}$, proves the biadditivity of $(\cdot \,|\, \cdot)$.

By induction, $(n\mathscr{y} \,|\, \mathscr{z}) = n\,(\mathscr{y} \,|\, \mathscr{z})$ for all $n \in \mathbb{N}$, and since $(-\mathscr{y} \,|\, \mathscr{z}) = -\,(\mathscr{y} \,|\, \mathscr{z})$, the equality holds for all integers n. Replacing $\mathscr{y}$ by $(1/n)\mathscr{y}$ yields $((1/n)\mathscr{y} \,|\, \mathscr{z}) = (1/n)\,(\mathscr{y} \,|\, \mathscr{z})$, hence

$$((m/n)\mathscr{y} \,|\, \mathscr{z}) = m\,((1/n)\mathscr{y} \,|\, \mathscr{z}) = (m/n)\,(\mathscr{y} \,|\, \mathscr{z}) \quad \text{for all } \ m, n \in \mathbb{Z},\ n \neq 0.$$

Noting that $(\mathscr{y} \,|\, \mathscr{z})$ is continuous in $\mathscr{y}$ we see that $(a\mathscr{y} \,|\, \mathscr{z}) = a\,(\mathscr{y} \,|\, \mathscr{z})$ for all real a. Combining this with $(i\,\mathscr{y} \,|\, \mathscr{z}) = i\,(\mathscr{y} \,|\, \mathscr{z})$ we obtain $(i\,b\mathscr{y} \,|\, \mathscr{z}) = i\,b\,(\mathscr{y} \,|\, \mathscr{z})$ and so $((a + i\,b)\mathscr{y} \,|\, \mathscr{z}) = (a + i\,b)\,(\mathscr{y} \,|\, \mathscr{z})$, for all $a,\ b \in \mathbb{R}$, completing the proof. □

11.1.7 Corollary. *Let $\mathscr{X}$ be a Banach space and $\mathscr{Y}$ a dense subspace of $\mathscr{X}$ with an inner product that induces the norm on $\mathscr{Y}$. Then $\mathscr{X}$ is a Hilbert space. In particular, the norm completion of an inner product space is a Hilbert space.*

Proof. The norm of $\mathscr{X}$ satisfies the parallelogram law on the dense subspace $\mathscr{Y}$. Since the norm and vector operations are continuous, the law holds on $\mathscr{X}$. □

11.1.8 Examples.

(a) The space $\mathbb{K}^d$ is a Hilbert space with respect to the **Euclidean inner product**

$$(x \,|\, y) = \sum_{j=1}^{d} x_j \overline{y}_j, \quad x := (x_1, \ldots, x_d), \ \ y := (y_1, \ldots, y_d).$$

The associated norm is the Euclidean norm on $\mathbb{K}^d$. Note that the parallelogram law fails for the norms $\|\cdot\|_1$ and $\|\cdot\|_\infty$ on $\mathbb{K}^d$, hence these are not inner product norms.

(b) Let $A := [a_{ij}]_{d\times d}$ be a matrix with entries in $\mathbb{K}^d$ that satisfies

- $a_{ji} = \overline{a_{ij}}$ (A is *Hermitian* or *self-adjoint*), and
- $\displaystyle\sum_{j=1}^{d} a_{ij} x_i \overline{x_j} > 0$ for all $(x_1, \ldots, x_d) \neq \mathbf{0}$ (A is *positive definite*).

Then

$$(x \,|\, y) := \sum_{i=1}^{d}\sum_{j=1}^{d} a_{ij} x_j \overline{y_j}, \quad x := (x_1, \ldots, x_d), \ \ y := (y_1, \ldots, y_d),$$

defines an inner product on $\mathbb{K}^d$. One obtains the Euclidean inner product of (a) by taking A to be the identity matrix.

(c) The **trace** $\operatorname{tr}(A)$ of a square matrix A is the sum of the diagonal elements of A. Clearly, $\operatorname{tr}(\cdot)$ is linear and $\operatorname{tr}(A^*) = \overline{\operatorname{tr}(A)}$, where A^* is the conjugate transpose of A. Let $M_{mn} = M_{mn}(\mathbb{C})$ denote the vector space over $\mathbb{C}$ of $m \times n$ complex matrices. For $A, B \in M_{mn}$, $(A \,|\, B) = \operatorname{tr}(B^* A)$ defines an inner product on M_{mn} called the **trace inner product**.

(d) The space $L^2(X, \mathcal{F}, \mu)$ with the L^2 norm is a Hilbert space under the inner product

$$(f \,|\, g) = \int f\overline{g}\, d\mu, \quad f, g \in L^2.$$

(As usual, we identify functions equal a.e.) In particular, ℓ^2 is a Hilbert space. On the other hand, for $p \neq 2$ the L^p norm is not induced by an inner product, since the parallelogram law fails (Ex. 11.3).

(e) Let U be open in $\mathbb{C}$ and let $A^2(U)$ be the space of functions in $L^2(U)$ that are analytic on U. Then $A^2(U)$ is closed in the L^2-norm and hence is a Hilbert space. To see this, we first establish the formula

$$f(z) = \frac{1}{\pi r^2} \int_{C_r(z)} f(w)\, d\lambda^2(w) \qquad z \in U, \;\; C_r(z) \subseteq U, \;\; f \in A^2(U). \tag{†}$$

In fact, by radial integration (§3.6) we have

$$\int_{C_r(z)} f(w)\, d\lambda^2(w) = \int_{C_r(0)} f(w+z)\, d\lambda^2(w) = \int_0^r t \int_0^{2\pi} f(te^{i\theta} + z)\, d\theta\, dt,$$

and if $f(w) = \sum_{n=0}^{\infty} c_n (w-z)^n$ we can evaluate the inner integral by integrating term by term:

$$\int_0^{2\pi} f(te^{i\theta} + z)\, d\theta = \sum_{n=0}^{\infty} c_n t^n \int_0^{2\pi} e^{in\theta}\, d\theta = 2\pi c_0 = 2\pi f(z).$$

Thus

$$\int_{C_r(z)} f(w)\, d\lambda^2(w) = 2\pi f(z) \int_0^r t\, dt = \pi r^2 f(z),$$

which is (†).

Now let K be a compact subset of U and let $r := \frac{1}{2}\mathrm{dist}(K, U^c)$. If $z \in K$, then $C_r(z) \subseteq U$, hence from (†) and the CBS inequality,

$$|f(z)| \leq \frac{1}{\pi r^2} \int_{C_r(z)} |f(w)|\, d\lambda^2(w) \leq \frac{1}{\sqrt{\pi} r} \left(\int_{C_r(z)} |f(w)|^2\, d\lambda^2(w) \right)^{1/2} \leq \frac{1}{\sqrt{\pi} r} \|f\|_2 \,.$$

By considering a finite cover of K by disks $C_r(z)$, we see from the above inequality that if $f_n \in A^2(U)$ and $\|f_m - f_n\|_2 \to 0$, then (f_n) is uniformly Cauchy on compact subsets of U and therefore converges uniformly (and in L^2) to a continuous function f. Thus f is analytic and so $A^2(U)$ is closed in $L^2(U)$. ◊

Isomorphisms of Hilbert Spaces

Hereafter, we use the notation $\mathcal{H}$ and $\mathcal{K}$ for Hilbert spaces. If there is a possibility of ambiguity, we use a subscript for the inner product, as in $(\cdot \,|\, \cdot)_{\mathcal{H}}$. An **isomorphism** of $\mathcal{H}$ and $\mathcal{K}$ (both over $\mathbb{K}$) is a bijective linear mapping $T : \mathcal{H} \to \mathcal{K}$ that preserves inner products:

$$(T\boldsymbol{x} \,|\, T\boldsymbol{y}) = (\boldsymbol{x} \,|\, \boldsymbol{y}), \quad \boldsymbol{x}, \boldsymbol{y} \in \mathcal{H}. \tag{11.4}$$

If an isomorphism exists, then $\mathcal{H}$ and $\mathcal{K}$ are said to be **isomorphic** (as Hilbert spaces). Thus isomorphic Hilbert spaces are "structurally" identical.

Taking $\boldsymbol{x} = \boldsymbol{y}$ in (11.4), we see that an isomorphism of Hilbert spaces is an isometry. The converse is true: every isometry of $\mathcal{H}$ onto $\mathcal{K}$ is a Hilbert space isomorphism. This follows directly from the polarization identity.

Exercises

11.1 Verify the parallelogram law.

11.2 Show that the uniform norm on $C[0, 1]$ is not an inner product norm.

11.3 Show that for $1 \le p < \infty$, the L^p norm on $C[0, 1]$ is not an inner product norm unless $p = 2$.

11.4 (Apollonius' identity). Prove that in an inner product space,

$$\|\boldsymbol{x} - \boldsymbol{z}\|^2 + \|\boldsymbol{y} - \boldsymbol{z}\|^2 = \tfrac{1}{2}\|\boldsymbol{x} - \boldsymbol{y}\|^2 + 2\left\|\tfrac{1}{2}(\boldsymbol{x} + \boldsymbol{y}) - \boldsymbol{z}\right\|^2.$$

11.5 Let $(\mathscr{X}, (\cdot\,|\,\cdot))$ be a semi-inner-product space with associated seminorm $\|\cdot\|$. By Ex. 8.56, $\mathscr{Y} := \{\boldsymbol{x} : \|\boldsymbol{x}\| = 0\}$ is a subspace of $\mathscr{X}$. Let $Q : \mathscr{X} \to \mathscr{X}/\mathscr{Y}$ denote the quotient map. Show that $\langle Q\boldsymbol{x} \mid Q\boldsymbol{y}\rangle := (\boldsymbol{x}\,|\,\boldsymbol{y})$ is a well-defined inner product on $\mathscr{X}/\mathscr{Y}$.

11.6 Let $\mathscr{H}$ denote the linear space of absolutely continuous functions f on $[0, 1]$ such that $f(0) = 0$ and $f' \in L^2[0, 1]$. Show that $(f\,|\,g) = \int f'(t)\overline{g'(t)}\,dt$ defines an inner product on $\mathscr{H}$ relative to which $\mathscr{H}$ is a Hilbert space.

11.7 Prove directly (without using uniform convexity) that an inner product space is strictly convex.

11.8 Let $U : \mathscr{H} \to \mathscr{K}$ be a bijection that preserves inner products: $(U\boldsymbol{x}\,|\,U\boldsymbol{y}) = (\boldsymbol{x}\,|\,\boldsymbol{y})$ for all $\boldsymbol{x}, \boldsymbol{y} \in \mathscr{H}$. Show that U must be linear.

11.2 Orthogonality

Throughout this section, $\mathscr{H}$ and $\mathscr{K}$ denote Hilbert spaces over $\mathbb{K}$.

The central feature of a Hilbert space that accounts for its rich structure is the concept of orthogonality. This leads to the notions of orthogonal complement and orthonormal bases, considered in this section and the next.

Orthogonal Complements

Vectors $\boldsymbol{x}$ and $\boldsymbol{y}$ in $\mathscr{H}$ are said to be **orthogonal**, written $\boldsymbol{x} \perp \boldsymbol{y}$, if $(\boldsymbol{x}\,|\,\boldsymbol{y}) = 0$. The following result on orthogonality generalizes the classical Pythagorean theorem.

11.2.1 Proposition. *Let $\boldsymbol{x}, \boldsymbol{y} \in \mathscr{H}$.*

(a) *If $\mathbb{K} = \mathbb{R}$, then $\boldsymbol{x} \perp \boldsymbol{y}$ iff $\|\boldsymbol{x} + \boldsymbol{y}\|^2 = \|\boldsymbol{x}\|^2 + \|\boldsymbol{y}\|^2$.*

(b) *If $\mathbb{K} = \mathbb{C}$, then $\boldsymbol{x} \perp \boldsymbol{y}$ iff $\|\alpha\boldsymbol{x} + \beta\boldsymbol{y}\|^2 = \|\boldsymbol{x}\|^2 + \|\boldsymbol{y}\|^2$ for all $\alpha, \beta \in \mathbb{T}$.*

Proof. Part (a) follows from the expansion

$$\|\boldsymbol{x} + \boldsymbol{y}\|^2 = (\boldsymbol{x} + \boldsymbol{y}\,|\,\boldsymbol{x} + \boldsymbol{y}) = \|\boldsymbol{x}\|^2 + \|\boldsymbol{y}\|^2 + 2\,(\boldsymbol{x}\,|\,\boldsymbol{y}).$$

For part (b), a similar calculation yields

$$\|\alpha\boldsymbol{x} + \beta\boldsymbol{y}\|^2 = (\alpha\boldsymbol{x} + \beta\boldsymbol{y}\,|\,\alpha\boldsymbol{x} + \beta\boldsymbol{y}) = \|\boldsymbol{x}\|^2 + \|\boldsymbol{y}\|^2 + \alpha\overline{\beta}\,(\boldsymbol{x}\,|\,\boldsymbol{y}) + \overline{\alpha}\beta\overline{(\boldsymbol{x}\,|\,\boldsymbol{y})}$$

for all $\alpha, \beta \in \mathbb{T}$. Thus the norm identity in (b) is equivalent to

$$\alpha\overline{\beta}\,(\boldsymbol{x}\,|\,\boldsymbol{y}) + \overline{\alpha}\beta\overline{(\boldsymbol{x}\,|\,\boldsymbol{y})} = 0 \quad \text{for all } \alpha, \beta \in \mathbb{T}. \tag{†}$$

This is trivially satisfied if $\boldsymbol{x} \perp \boldsymbol{y}$. Conversely, if (†) holds take $\alpha = \beta = 1$ to obtain $(\boldsymbol{x} \,|\, \boldsymbol{y}) + \overline{(\boldsymbol{x} \,|\, \boldsymbol{y})} = 0$, and take $\alpha = i$, $\beta = 1$ to obtain $i\big((\boldsymbol{x} \,|\, \boldsymbol{y}) - \overline{(\boldsymbol{x} \,|\, \boldsymbol{y})}\big) = 0$. Conclude that $(\boldsymbol{x} \,|\, \boldsymbol{y}) = 0$. □

By induction, we have

11.2.2 Corollary. *If $\boldsymbol{x}_j \in \mathscr{H}$ and $\boldsymbol{x}_j \perp \boldsymbol{x}_k$ for $j \neq k$, then $\left\| \sum_{j=1}^n \boldsymbol{x}_j \right\|^2 = \sum_{j=1}^n \|\boldsymbol{x}_j\|^2$.*

If $\boldsymbol{x} \perp \boldsymbol{y}$ for all $\boldsymbol{y}$ in a subset S of $\mathscr{H}$, then $\boldsymbol{x}$ is said to be **orthogonal to** S, written $\boldsymbol{x} \perp S$. The collection of all vectors orthogonal to S is called the **orthogonal complement of** S and is denoted by $S^\perp$:

$$S^\perp := \{\boldsymbol{x} \in \mathscr{H} : (\boldsymbol{x} \,|\, \boldsymbol{y}) = 0 \;\forall\; \boldsymbol{y} \in S\}.$$

A direct argument shows that $S^\perp$ is a closed linear subspace of $\mathscr{H}$.

The theorem below regarding orthogonal complements is of fundamental importance in Hilbert space theory. For the proof we need the following lemma.

11.2.3 Lemma. *A nonempty, closed, convex subset K of $\mathscr{H}$ has a unique member of smallest norm.*

Proof. Let $d := \inf\{\|\boldsymbol{x}\| : \boldsymbol{x} \in K\}$. We claim that for $\boldsymbol{x}, \boldsymbol{y} \in K$,

$$\|\boldsymbol{x} - \boldsymbol{y}\|^2 \leq 2\|\boldsymbol{x}\|^2 + 2\|\boldsymbol{y}\|^2 - 4d^2. \tag{†}$$

Indeed, since K is convex, $\frac{1}{2}(\boldsymbol{x} + \boldsymbol{y}) \in K$, hence

$$4d^2 \leq (\boldsymbol{x} + \boldsymbol{y} \,|\, \boldsymbol{x} + \boldsymbol{y}) = \|\boldsymbol{x}\|^2 + \|\boldsymbol{y}\|^2 + 2\mathrm{Re}\,(\boldsymbol{x} \,|\, \boldsymbol{y}),$$

and so

$$\|\boldsymbol{x} - \boldsymbol{y}\|^2 = \|\boldsymbol{x}\|^2 + \|\boldsymbol{y}\|^2 - 2\mathrm{Re}\,(\boldsymbol{x} \,|\, \boldsymbol{y}) \leq 2\|\boldsymbol{x}\|^2 + 2\|\boldsymbol{y}\|^2 - 4d^2.$$

Now let $\boldsymbol{x}_n \in K$ and $\|\boldsymbol{x}_n\| \to d$. Then $2\|\boldsymbol{x}_n\|^2 + 2\|\boldsymbol{x}_m\|^2 \to 2d^2$, hence from (†), $\|\boldsymbol{x}_n - \boldsymbol{x}_m\|^2 \to 0$. The limit $\boldsymbol{x} := \lim_n \boldsymbol{x}_n$ is then a member of K with smallest norm. If $\boldsymbol{y} \in K$ also has smallest norm, then by (†), $\|\boldsymbol{x} - \boldsymbol{y}\|^2 \leq 2d^2 + 2d^2 - 4d^2 = 0$, hence $\boldsymbol{x} = \boldsymbol{y}$. □

11.2.4 Theorem. *If $\mathscr{M}$ is a closed subspace of $\mathscr{H}$, then $\mathscr{H} = \mathscr{M} \oplus \mathscr{M}^\perp$. Moreover, if $\boldsymbol{x} = \boldsymbol{m} + \boldsymbol{m}^\perp$, then $\boldsymbol{m}$ is the unique member of $\mathscr{M}$ closest to $\boldsymbol{x}$.*

Proof. For a fixed $\boldsymbol{x} \in \mathscr{H}$, there exists, by the lemma, a unique member $\boldsymbol{y}$ of $\boldsymbol{x} + \mathscr{M}$ such that

$$\|\boldsymbol{y}\| \leq \|\boldsymbol{x} + \boldsymbol{m}\| \quad \text{for all } \boldsymbol{m} \in \mathscr{M}.$$

We show that $\boldsymbol{y} \in \mathscr{M}^\perp$. Let $\boldsymbol{m} \in \mathscr{M}$. Since $\boldsymbol{y} + t\boldsymbol{m} = \boldsymbol{x} + (\boldsymbol{y} - \boldsymbol{x} + t\boldsymbol{m}) \in \boldsymbol{x} + \mathscr{M}$, the function

$$f(t) := \|\boldsymbol{y} + t\boldsymbol{m}\|^2 = \|\boldsymbol{y}\|^2 + 2t\,\mathrm{Re}\,(\boldsymbol{y} \,|\, \boldsymbol{m}) + t^2\|\boldsymbol{m}\|^2$$

has minimum value $\|\boldsymbol{y}\|^2 = f(0)$. It follows that $f'(0) = 0$ and so $\mathrm{Re}\,(\boldsymbol{y} \,|\, \boldsymbol{m}) = 0$. Replacing $\boldsymbol{m}$ by $i\boldsymbol{m}$ yields $\mathrm{Im}\,(\boldsymbol{y} \,|\, \boldsymbol{m}) = 0$. Therefore, $(\boldsymbol{y} \,|\, \boldsymbol{m}) = 0$, hence $\boldsymbol{y} \in \mathscr{M}^\perp$.

We may now write $\boldsymbol{x} = (\boldsymbol{x} - \boldsymbol{y}) + \boldsymbol{y}$, which shows that $\mathscr{H} = \mathscr{M} + \mathscr{M}^\perp$. The sum is direct since $\boldsymbol{z} \in \mathscr{M} \cap \mathscr{M}^\perp \Rightarrow (\boldsymbol{z} \,|\, \boldsymbol{z}) = 0$. Since

$$\|(\boldsymbol{x} - \boldsymbol{y}) - \boldsymbol{x}\| = \|\boldsymbol{y}\| \leq \|\boldsymbol{m} - \boldsymbol{x}\| \quad \text{for all } \boldsymbol{m} \in \mathscr{M},$$

$\boldsymbol{x} - \boldsymbol{y}$ is the unique member of $\mathscr{M}$ closest to $\boldsymbol{x}$. □

11.2.5 Corollary. *If $\mathscr{M}$ is a closed subspace of $\mathscr{H}$, then $\mathscr{M}^{\perp\perp} = \mathscr{M}$.*

Proof. Let $x \in \mathscr{M}^{\perp\perp}$ and write $x = y + z$, $y \in \mathscr{M}$, $z \in \mathscr{M}^{\perp}$. Then

$$0 = (x \,|\, z) = (y \,|\, z) + (z \,|\, z) = (z \,|\, z),$$

hence $z = 0$ and so $x = y \in \mathscr{M}$. Therefore, $\mathscr{M}^{\perp\perp} \subseteq \mathscr{M}$. The reverse inclusion is clear. □

The Riesz Representation Theorem

For $y \in \mathscr{H}$ define a mapping f_y on $\mathscr{H}$ by

$$f_y(x) := (x \,|\, y)$$

Then f_y is a linear functional with $\|f_y\| = \|y\|$ (11.1.5). Furthermore, $f_{ay+bz} = \bar{a} f_y + \bar{b} f_z$. Thus the map $y \to f_y$ is a conjugate linear isometry from $\mathscr{H}$ into $\mathscr{H}'$. The next theorem asserts that the mapping is surjective.

11.2.6 Riesz Representation Theorem. *Every $f \in \mathscr{H}'$ is of the form f_y for some $y \in \mathscr{H}$.*

Proof. We may assume that f is not the zero functional. Then $\mathscr{H} = \ker f \oplus (\ker f)^{\perp}$ where $(\ker f)^{\perp}$ has dimension one. Choose $z \in (\ker f)^{\perp}$ with $f(z) = 1$ and set $a = 1/\|z\|^2$. For $x \in \mathscr{H}$ we may write $x = u + cz$, where $u \in \ker f$, hence

$$f(x) = cf(z) = c = (cz \,|\, az) = (u + cz \,|\, az) = f_{az}(x).$$

Therefore, $f = f_{az}$. □

Recall that a net (x_α) in a normed space $\mathscr{X}$ converges weakly to $x \in \mathscr{X}$ if $\langle x_\alpha, f\rangle \to \langle x, f\rangle$ for all $f \in \mathscr{X}'$. Thus from the Riesz representation theorem we have

11.2.7 Corollary. *A net (x_α) in $\mathscr{H}$ converges weakly to x iff $(x_\alpha \,|\, y) \to (x \,|\, y)$ $\forall\, y \in \mathscr{H}$.*

11.2.8 Corollary. *A Hilbert space is reflexive.*

Proof. The dual space $\mathscr{H}'$ is a Hilbert space under the inner product

$$(f_x \,|\, f_y) = (y \,|\, x).$$

(The transposition of the elements on the right side is necessary to compensate for the conjugate linearity of the mapping $x \to f_x$.) Let $\varphi \in \mathscr{H}''$. By the Riesz representation theorem applied to $\mathscr{H}'$, there exists f_y such that for all x,

$$\varphi(f_x) = (f_x \,|\, f_y) = (y \,|\, x) = f_x(y) = \widehat{y}(f_x).$$

Therefore, $\varphi = \widehat{y}$ and so $\mathscr{H}'' = \widehat{\mathscr{H}}$. □

Exercises

11.9 Let S be a subset of a Hilbert space $\mathscr{H}$. Use 11.2.4 to prove that $S^{\perp\perp} = \operatorname{cl}\operatorname{span}(S)$.

11.10 Let $\mathscr{X}$ and $\mathscr{Y}$ be subspaces of $\mathscr{H}$. Show that $(\mathscr{X} + \mathscr{Y})^{\perp} = \mathscr{X}^{\perp} \cap \mathscr{Y}^{\perp}$.

11.11 Prove that $x \perp y$ iff $\|x\| \le \|x + cy\|$ for all $c \in \mathbb{C}$.

11.12 Show that in a Hilbert space, if $x_n \xrightarrow{w} x$ and $\|x_n\| \to \|x\|$ then $\|x_n - x\| \to 0$. Find an example of a Banach space for which the assertion is false.

11.13 A function $f \in L^2[-1,1]$ is **odd** (**even**) if $f(-t) = -f(t)$ ($f(-t) = f(t)$) for all a.a. $t \in [-1,1]$. Let $\mathscr{O}$ ($\mathscr{E}$) denote the linear space of odd (even) functions. Show that each space is the orthogonal complement of the other and that $L^2[-1,1] = \mathscr{O} \oplus \mathscr{E}$.

11.14 For each linear functional F on $\ell^2(\mathbb{N})$, find a function g such that $F(f) = (f \,|\, g)$ for all f.
(a) $F(f) = \sum_{j=1}^m f(j)$. (b) $F(f) = f(2) - f(1)$. (c) $F(f) = \sum_{n=1}^\infty 2^{-n}[f(n) - f(n+1)]$.

11.15 Find an example of an inner product space $\mathscr{X}$ and a continuous linear functional $f \in \mathscr{X}'$ such that no vector $y \in \mathscr{X}$ exists for which $f(x) = (x \,|\, y)$ for all $x \in \mathscr{X}$.

11.16 Let $\mathscr{H}$ be the Hilbert space defined in Ex. 11.6. Let F be the evaluation functional $F(f) = f(1/2)$. Find a function g such that $F(f) = (f \,|\, g)$ for all $f \in \mathscr{H}$.

11.17 Let $T \in \mathscr{B}(\mathscr{H})$ be weak-norm continuous. Show that $\operatorname{ran} T$ is finite dimensional. [[There exist $x_j \in \mathscr{H}$ and $\varepsilon > 0$ such that $|(x \,|\, x_j)| < \varepsilon$ $(1 \le j \le n)$ implies $\|Tx\| < 1$.]]

11.18 [↑8.34] Let T_r and T_ℓ denote the right and left shift operators on ℓ^2 and let $x \in \mathscr{H}$. Compute the weak limits $\lim_n T_\ell^n x$ and $\lim_n T_r^n x$

11.3 Orthonormal Bases

Throughout this section, $\mathscr{H}$ and $\mathscr{K}$ denote Hilbert spaces over $\mathbb{K}$.

A subset $\mathscr{E}$ of $\mathscr{H}$ is said to be **orthonormal** if for all $e, f \in \mathscr{E}$

$$(e \,|\, f) = \begin{cases} 1 & \text{if } e = f, \\ 0 & \text{otherwise.} \end{cases}$$

The scalars $(x \,|\, e)$ are called the **Fourier coefficients of x with respect to $\mathscr{E}$**. An **orthonormal basis** (or, simply, **basis**) is an orthonormal set whose span is dense in $\mathscr{H}$. We discuss the role orthonormal bases in Fourier series on $\mathbb{R}$ below and in a more general setting in Chapter 16.

11.3.1 Proposition. *An orthonormal set $\mathscr{E}$ is a basis iff $\mathscr{E}^\perp = \{0\}$.*

Proof. Let $\mathscr{M} := \operatorname{cl}\operatorname{span}(\mathscr{E})$. Since $\mathscr{H} = \mathscr{M} \oplus \mathscr{M}^\perp$ and $\mathscr{M}^\perp = \mathscr{E}^\perp$, the assertion follows. □

11.3.2 Proposition. *Every orthonormal subset $\mathscr{F}$ of $\mathscr{H}$ is contained in an orthonormal basis. In particular, every (nontrivial) Hilbert space has an orthonormal basis.*

Proof. Order the family $\mathfrak{E}$ of all orthonormal subsets of $\mathscr{H}$ containing $\mathscr{F}$ upward by inclusion. The union of a chain in $\mathfrak{E}$ is clearly orthonormal, hence is an upper bound for the chain. By Zorn's lemma, $\mathfrak{E}$ has a maximal element. If $x \in \mathscr{E}^\perp$ and $x \ne 0$, then $\mathscr{E} \cup \{\|x\|^{-1} x\}$ is an orthonormal set properly containing $\mathscr{E}$, contradicting maximality. Therefore, $\mathscr{E}$ is an orthonormal basis, proving the first assertion of the theorem. The second assertion follows from the first by taking $\mathscr{F}$ to consist of a single vector of norm one. □

The next results depend on the material in §0.4 on unordered sums.

11.3.3 Theorem (Bessel's Inequality). *Let $\mathscr{E}$ be an orthonormal set and $x \in \mathscr{H}$. Then*

$$\sum_{e \in \mathscr{E}} |(x \,|\, e)|^2 \le \|x\|^2,$$

where at most countably many of the terms in the sum are nonzero.

Proof. Let $F \subseteq \mathscr{E}$ be finite and set $y = \sum_{e \in F} (x \mid e)\, e$. By orthonormality and sesquilinearity,

$$(y \mid y) = \sum_{e \in F} (x \mid e) \overline{(x \mid e)} = (x \mid y),$$

hence $(x - y \mid y) = 0$. Thus

$$\|x\|^2 = \|x - y + y\|^2 = \|x - y\|^2 + \|y\|^2 \geq \|y\|^2 = \sum_{e \in F} |(x \mid e)|^2.$$

Since F was arbitrary, the assertions follow from 0.4.5. □

11.3.4 Theorem. *Let $\mathscr{E}$ be an orthonormal set in $\mathscr{H}$. The following are equivalent:*

(a) *$\mathscr{E}$ is a basis.*

(b) *For each $x \in \mathscr{H}$, $x = \sum_{e \in \mathscr{E}} (x \mid e)\, e$* (Fourier expansion of x).

(c) *For each pair $x, y \in \mathscr{H}$, $(x \mid y) = \sum_{e \in \mathscr{E}} (x \mid e)\, (e \mid y)$.*

(d) *For each $x \in \mathscr{H}$, $\|x\|^2 = \sum_{e \in \mathscr{E}} |(x \mid e)|^2$* (Parseval's identity).

In (b) – (d), at most countably many of the Fourier coefficients $(x \mid e)$ are nonzero.

Proof. (a) $\Rightarrow$ (b): Denote the nonzero Fourier coefficients of x by $(x \mid e_n)$. We show that

$$x = \sum_{n=1}^{\infty} (x \mid e_n)\, e_n.$$

By the Pythagorean relation,

$$\left\| \sum_{k=n}^{m} (x \mid e_k)\, e_k \right\|^2 = \sum_{k=n}^{m} |(x \mid e_k)|^2, \quad m > n,$$

which, by Bessel's inequality, tends to 0 as $n \to \infty$. Therefore, the sequence of partial sums $\sum_{k=1}^{n} (x \mid e_k)\, e_k$ is Cauchy and so converges to some y. It remains to show that $y = x$. Now, for any $e \in \mathscr{E}$, by continuity of the inner product we have

$$(y \mid e) = \lim_n \sum_{k=1}^{n} (x \mid e_k)\, (e_k \mid e).$$

If $e = e_m$ for some m, then the right side is $(x \mid e_m)$. If $e \neq e_m$ for all m, then both $(x \mid e)$ and $(y \mid e)$ are zero. Thus $(x - y \mid e) = 0$ for all $e \in \mathscr{E}$. Since $\mathscr{E}$ is a basis, $x = y$.

(b) $\Rightarrow$ (c): Using a common sequence (e_n) for x and y we have

$$x = \sum_{n=1}^{\infty} (x \mid e_n)\, e_n \quad \text{and} \quad y = \sum_{n=1}^{\infty} (y \mid e_n)\, e_n.$$

Now observe that

$$\left(\sum_{j=1}^{n} (x \mid e_j)\, e_j \,\middle|\, \sum_{k=1}^{n} (y \mid e_k)\, e_k \right) = \sum_{j,k} (x \mid e_j)\, (e_k \mid y)\, (e_j \mid e_k) = \sum_{j=1}^{n} (x \mid e_j)\, (e_j \mid y).$$

Letting $n \to \infty$ and using the continuity of the inner product yields (c).

(d) $\Rightarrow$ (a): Then $\|x\|^2 = 0$ for every $x \in \mathscr{E}^{\perp}$, hence $\mathscr{E}$ is a basis. □

The Dimension of a Hilbert Space

The notion of dimension rests on the following result.

11.3.5 Proposition. *All bases in a Hilbert space have the same cardinality.*

Proof. We may assume that $\mathscr{H}$ is not finite dimensional. Let $\mathscr{E}$ and $\mathscr{F}$ be bases with cardinality $|\mathscr{E}|$ and $|\mathscr{F}|$, respectively. It suffices to show that $|\mathscr{E}| \leq |\mathscr{F}|$. For each $f \in \mathscr{F}$, define $E_f = \{e \in \mathscr{E} : (e \,|\, f) \neq 0\}$. Then $\mathscr{E} = \bigcup_{f \in \mathscr{F}} E_f$, and since each E_f is countable, the cardinality of the union is $\leq |\mathscr{F}| \cdot \aleph_0 = |\mathscr{F}|$. Therefore, $|\mathscr{E}| \leq |\mathscr{F}|$. □

The cardinality of a basis in a Hilbert space $\mathscr{H}$ is called the **dimension of** $\mathscr{H}$. The following corollary shows that Hilbert spaces with the same dimension are "structurally identical."

11.3.6 Corollary. *$\mathscr{H}$ and $\mathscr{K}$ have the same dimension iff they are isomorphic as Hilbert spaces.*

Proof. Let $\mathscr{E}$ and $\mathscr{F}$ be bases for $\mathscr{H}$ and $\mathscr{K}$, respectively. If a Hilbert space isomorphism $T : \mathscr{H} \to \mathscr{K}$ exists, then $T(\mathscr{E})$ is an orthonormal set in $\mathscr{K}$, hence $|\mathscr{E}| = |T(\mathscr{E})| \leq |\mathscr{F}|$. But T^{-1} is also a Hilbert space isomorphism, hence $|\mathscr{F}| \leq |\mathscr{E}|$. Therefore, $|\mathscr{E}| = |\mathscr{F}|$.

Conversely, let $\mathscr{E}$ and $\mathscr{F}$ have the same cardinality and let $\Psi : \mathscr{E} \to \mathscr{F}$ be any bijection. For $x \in \mathscr{H}$, define $T : \mathscr{H} \to \mathscr{K}$ by

$$Tx = \sum_{e \in \mathscr{E}} (x \,|\, e)\, \Psi(e).$$

By Bessel's inequality, Tx is well-defined, and at most countably many terms are nonzero. By sesquilinearity and continuity of the inner product,

$$(Tx \,|\, Ty) = \left(\sum_{e \in \mathscr{E}} (x \,|\, e)\, \Psi(e) \,\middle|\, \sum_{\tilde{e} \in \mathscr{E}} (y \,|\, \tilde{e})\, \Psi(\tilde{e}) \right) = \sum_{e \in \mathscr{E}} (x \,|\, e)\,(e \,|\, y) = (x \,|\, y). \quad \square$$

The Gram-Schmidt Process

The members of an orthonormal set are easily seen to be linearly independent. Indeed, if the vectors $e_1, \ldots, e_n$ are orthonormal and $c_1 e_1 + \cdots + c_n e_n = 0$, then taking inner products of both sides of the equation with e_j shows that $c_j = 0$. The following proposition is a converse of sorts: it allows a finite set of linearly independent vectors to be replaced by an orthonormal set without changing the span. The technique in the proof is known as the **Gram-Schmidt process.**

11.3.7 Proposition. *Let $\mathscr{H}$ be an inner product space and $A := \{x_1, x_2, \ldots\}$ a linearly independent set. Then there exist an orthonormal set $E := \{e_1, e_2, \ldots\}$ such that*

$$\operatorname{span}\{x_1, \ldots, x_n\} = \operatorname{span}\{e_1, \ldots, e_n\} \quad \textit{for all } n.$$

Proof. Set $A_n := \{x_1, \ldots, x_n\}$. We construct E inductively. Define $e_1 = x_1 / \|x_1\|$. Assume that the desired set $E_n := \{e_1, \ldots, e_n\}$ has been constructed. Then $x_{n+1} \notin \operatorname{span} E_n$, so the vector

$$y_{n+1} := x_{n+1} - \sum_{j=1}^{n} (x_{n+1} \,|\, e_j)\, e_j$$

is not zero. Define $e_{n+1} = y_{n+1} / \|y_{n+1}\|$. Then $e_{n+1} \in \operatorname{span} A_{n+1}$, $(e_{n+1} \,|\, e_k) = 0$ for $k \leq n$, and $\operatorname{span} A_{n+1} = \operatorname{span}\{e_1, \ldots, e_{n+1}\}$. □

For example, applying the Gram-Schmidt process to the set of monomials $x_n = t^n$ $(n \geq 0)$ on $[-1, 1]$ yields the **Legendre polynomials**

$$e_n(t) := \left(\frac{2n+1}{2}\right)^{1/2} \frac{1}{2^n n!} \frac{d^n}{dt^n}(t^2 - 1)^n, \quad n = 0, 1, \ldots.$$

For this and other interesting examples of orthonormal bases on $L^2[a, b]$, the reader is referred to [28].

Most infinite dimensional Hilbert spaces one encounters in applications are separable. Analysis of such spaces is somewhat easier because of the following result:

11.3.8 Proposition. *If a Hilbert space $\mathscr{H}$ is separable, then it has a countable basis.*

Proof. We may assume that $\mathscr{H}$ is not finite dimensional. Let (x_n) be a dense sequence of nonzero vectors in $\mathscr{H}$. If x_2 is a multiple of x_1, we may remove it without changing the span of (x_n). Likewise, if x_n is a linear combination of its predecessors, then it may be removed without affecting the span. By induction, we obtain a linear independent subsequence (y_n) of (x_n) with span $(y_n) =$ span (x_n). The Gram-Schmidt process may be applied to (y_n) to obtain an orthonormal sequence (e_n) such that span $(e_n) =$ span $(y_n) =$ span (x_n). If $x \perp e_n$ for all n, then $x \perp x_n$ for all n, and since (x_n) is dense in $\mathscr{H}$, $x = 0$. Therefore, (e_n) is a basis. □

For example, the vectors $e_n = (0, \ldots, 0, 1, 0, \ldots)$ $(n \in \mathbb{N})$ form an orthonormal basis in ℓ^2. It follows from 11.3.8 that every separable Hilbert space is isomorphic to ℓ^2. This fact, however, does not necessarily lead to simplifications in the study of separable Hilbert spaces, as the isomorphism may obscure certain essential properties of concrete Hilbert spaces such as $L^2[0, 1]$. Nonetheless, it is of some interest to know that, structurally, all separable Hilbert spaces are "like" ℓ^2.

Fourier Series

We show that the functions

$$e_n(t) = e^{2\pi i n t}, \quad t \in [0, 1], \quad n \in \mathbb{Z}, \tag{11.5}$$

form an orthonormal basis for $L^2[0, 1]$ with respect to Lebesgue measure. The calculation

$$\int_0^1 e^{2\pi i n t} e^{2\pi i m t}\, dt = \int_0^1 e^{2\pi i(n-m)t}\, dt$$

shows that $(e_n)_n$ is an orthonormal set. Let $\mathscr{A}$ denote the algebra of continuous functions $f : [0, 1] \to \mathbb{C}$ with $f(0) = f(1)$. Since $C[0, 1]$ is dense in $L^2[0, 1]$, a simple linearization argument shows that the same is true for $\mathscr{A}$. For each $f \in \mathscr{A}$ define $F_f : \mathbb{T} \to \mathbb{C}$ by

$$F_f(e^{2\pi i t}) = f(t), \quad t \in [0, 1].$$

By the periodicity of f, the function F_f is well-defined and continuous. Moreover,

$$F_\tau(z) = \sum_{k=-n}^{n} c_k z^k \ (z \in \mathbb{T}), \quad \text{where } \tau(t) := \sum_{k=-n}^{n} c_k e^{2\pi i n t}.$$

Let $\mathscr{T}$ denote the collection of all such functions τ. By the Stone-Weierstrass theorem, F_τ is uniformly dense in $C(\mathbb{T})$. It follows that $\mathscr{T}$ is dense in $\mathscr{A}$ in the uniform norm and is therefore dense in $L^2[0, 1]$. Thus $(e_n)_{n \in \mathbb{Z}}$ is a basis, as claimed.

From 11.3.4 we see that every $f \in L^2[0,1]$ has a **Fourier series expansion**

$$f = \sum_{n=-\infty}^{\infty} \widehat{f}(n)\boldsymbol{e}_n, \quad \widehat{f}(n) := (f \mid \boldsymbol{e}_n) = \int_0^1 f(t)e^{-2\pi int}\,dt, \tag{11.6}$$

where convergence is in $L^2[0,1]$. The function $\widehat{f}$ is called the **Fourier transform** of f. The convergence of the series in (11.6) implies that $\lim_n \widehat{f}(n) = 0$, which is the classical *Riemann-Lebesgue lemma.*

The following is an interesting application to the Fourier transform of a rapidly decreasing function on $\mathbb{R}$ (see §6.3).

11.3.9 Theorem (Poisson Summation Formula). *Let φ be a rapidly decreasing function on $\mathbb{R}$ with Fourier transform $\widehat{\varphi}$. Then*

$$\sum_{n=-\infty}^{\infty} \varphi(n) = \sum_{n=-\infty}^{\infty} \widehat{\varphi}(n).$$

Proof. Define

$$f(t) := \sum_{n=-\infty}^{\infty} \varphi(t+n), \quad t \in \mathbb{R}.$$

The rapidly decreasing property of φ implies that the series, as well as all derived series, converge absolutely and locally uniformly. Thus f is a C^∞ function. Moreover, $\sum_{n=-\infty}^{\infty} \widehat{\varphi}(n)$ converges because $\widehat{\varphi}$ is also rapidly decreasing. Since $f(t+1) = f(t)$ for all t, we may consider $f \in L^2[0,1]$. Multiplying (†) by $e^{-2\pi imt}$ and integrating term by term, we have

$$\begin{aligned}\widehat{f}(m) &= \sum_{n=-\infty}^{\infty} \int_0^1 \varphi(t+n)e^{-2\pi imt}\,dt = \sum_{n=-\infty}^{\infty} \int_n^{n+1} \varphi(t)e^{-2\pi im(t-n)}\,dt \\ &= \sum_{n=-\infty}^{\infty} \int_n^{n+1} \varphi(t)e^{-2\pi imt}\,dt = \int_{-\infty}^{\infty} \varphi(t)e^{-2\pi imt}\,dt \\ &= \widehat{\varphi}(m),\end{aligned}$$

Thus for a.a. t,

$$\sum_{n=-\infty}^{\infty} \varphi(t+n) = f(t) = \sum_{n=-\infty}^{\infty} \widehat{f}(n)\boldsymbol{e}_n(t) = \sum_{n=-\infty}^{\infty} \widehat{\varphi}(n)e^{2\pi int}.$$

Since both series are continuous in t, the equation holds for all t. Setting $t = 0$ yields the desired equality. □

Exercises

11.19 Let $(\boldsymbol{e}_n)_n$ be an orthonormal basis and fix $\boldsymbol{y} \in \mathscr{H}$. Show that the minimum value of the function $\boldsymbol{x} \to \|\boldsymbol{x} - \boldsymbol{y}\|$ for $\boldsymbol{x} \in \operatorname{span}\{\boldsymbol{e}_1, \ldots, \boldsymbol{e}_m\}$ occurs when $\boldsymbol{x} = \sum_{j=1}^{m} (\boldsymbol{y} \mid \boldsymbol{e}_j)\,\boldsymbol{e}_j$.

11.20 Show that the sequence $1, z, z^2, \ldots$ is orthogonal in $L^2(\mathbb{D}, \lambda^2)$. Is the normalized sequence $\left(z^n \|z^n\|_2^{-1}\right)$ a basis?

11.21 A **Hamel basis** for a vector space is a linearly independent set that spans the space. Let $\mathscr{H}$ be an infinite dimensional Hilbert space. Show that an orthonormal basis cannot be a Hamel basis. Show that a Hamel basis in $\mathscr{H}$ is uncountable.

11.22 Show that in a Hilbert space, $\boldsymbol{x}_n \xrightarrow{w} \mathbf{0}$ iff $\sup_n \|\boldsymbol{x}_n\| < \infty$ and $(\boldsymbol{x}_n \,|\, \boldsymbol{e}) \to 0$ for every $\boldsymbol{e}$ in an orthonormal basis.

11.23 (Wirtinger's inequality). Let $f \in C^1[0,a]$ with $f(0) = f(a) = 0$. Show that $\pi \|f\|_2 \le a \|f'\|_2$. ⟦Extend f to $[-a,a]$ as an odd function. Use Parseval's identity on $f \in L^2[-1,1]$ with the basis $\frac{1}{\sqrt{2a}} e^{ibnt}\,dt$ $(b := 2\pi/a)$ and integrate $\widehat{f}(n)$ by parts.⟧

11.24 Show that the Fourier transform is a linear isometry from $L^2[0,1]$ onto $\ell^2(\mathbb{Z})$.

11.25 Let $(X, \mathcal{F}, \mu)$ be σ-finite and $\phi \in L^\infty(\mu)$. Show that the range of the multiplication mapping $M_\phi f := f\phi$ on $L^2(\mu)$ is closed iff $\phi = \mathbf{1}_E$ for some $E \in \mathcal{F}$.

11.26 Let $(X, \mathcal{F}, \mu)$ be σ-finite and $\phi \in L^\infty(\mu)$. Show that $\phi^{-1} \in L^\infty$ iff $\sup_n \|f_n\|_2 < \infty$ for any sequence (f_n) in L^2 for which (ϕf_n) converges in L^2. ⟦For the sufficiency, suppose $\phi^{-1} \notin L^\infty$. Choose $A_n \in \mathcal{F}$ such that $A_n \subseteq \{|\phi| < 1/n^2\}$ and $0 < \mu(A_n) < \infty$ (how?) and set $f_n = \sum_{k=1}^n \mu(A_k)^{-1/2} \mathbf{1}_{A_k}$.⟧

11.4 The Hilbert Space Adjoint

Throughout this section, $\mathscr{H}$ and $\mathscr{K}$ denote complex Hilbert spaces.

The Hilbert space adjoint of an operator $T \in \mathscr{B}(\mathscr{H})$ is closely related to the Banach space dual operator T', the essential difference being that the former acts on $\mathscr{H}$ while the latter acts on $\mathscr{H}'$. The existence of an adjoint operation in $\mathscr{B}(\mathscr{H})$ accounts to a large extent for the rich structure of $\mathscr{B}(\mathscr{H})$ and its various subalgebras, this structure absent in the Banach space case. For the construction of the adjoint we need the following notion.

Bounded Sesquilinear Functionals

A sesquilinear functional B on $\mathscr{H} \times \mathscr{K}$ is said to be **bounded** if

$$\|B\| := \sup\{|B(\boldsymbol{x}, \boldsymbol{y})| : \|\boldsymbol{x}\| \le 1,\ \|\boldsymbol{y}\| \le 1\} < \infty. \tag{11.7}$$

For example, if $T \in \mathscr{B}(\mathscr{K}, \mathscr{H})$, then

$$B_T(\boldsymbol{x}, \boldsymbol{y}) := (\boldsymbol{x} \,|\, T\boldsymbol{y})_{\mathscr{H}}, \quad \boldsymbol{x} \in \mathscr{H},\ \boldsymbol{y} \in \mathscr{K},$$

defines a bounded sesquilinear functional on $\mathscr{H} \times \mathscr{K}$ with

$$\|B_T\| = \sup\{|(\boldsymbol{x} \,|\, T\boldsymbol{y})_{\mathscr{H}}| : \|\boldsymbol{x}\| \le 1,\ \|\boldsymbol{y}\| \le 1\} = \|T\|,$$

the last equality from 11.1.5. One easily checks that (11.7) defines a norm on the linear space $\mathscr{S}(\mathscr{H} \times \mathscr{K})$ of all bounded sesquilinear functionals on $\mathscr{H} \times \mathscr{K}$ and that $\mathscr{S}(\mathscr{H} \times \mathscr{K})$ is complete in this norm (Ex. 11.27). Moreover, the mapping $T \to B_T$ is a conjugate linear isometric isomorphism from $\mathscr{B}(\mathscr{K}, \mathscr{H})$ into $\mathscr{S}(\mathscr{H} \times \mathscr{K})$. The following theorem shows that the mapping is surjective.

11.4.1 Theorem. *If B is a bounded sesquilinear functional on $\mathscr{H} \times \mathscr{K}$, then $B = B_T$ for some $T \in \mathscr{B}(\mathscr{K}, \mathscr{H})$.*

Proof. Fix $\mathbf{y} \in \mathscr{K}$. Since $B(\cdot, \mathbf{y}) \in \mathscr{H}'$, by the Riesz representation theorem there exists a unique vector $T\mathbf{y} \in \mathscr{H}$ such that

$$(\mathbf{x} \mid T\mathbf{y})_{\mathscr{H}} = B(\mathbf{x}, \mathbf{y}) \text{ for all } \mathbf{x} \in \mathscr{H}.$$

For each $\mathbf{x}$, the right side is conjugate linear in $\mathbf{y}$, so T is linear. Moreover, since $\|B\| < \infty$, T is bounded. □

The Lax-Milgram Theorem

The following consequence of 11.4.1 has important applications in the theory of partial differential equations (see §15.6).

11.4.2 Theorem (Lax-Milgram). *Let B be a bounded sesquilinear form on $\mathscr{H}$ such that for some $c > 0$*

$$B(\mathbf{x}, \mathbf{x}) \geq c \|\mathbf{x}\|^2 \quad \textit{for all} \quad \mathbf{x} \in \mathscr{H}.$$

Then for each continuous linear functional f on $\mathscr{H}$ there exists a unique $\mathbf{y} \in \mathscr{H}$ such that $B(\mathbf{x}, \mathbf{y}) = f(\mathbf{x})$ for all $\mathbf{x} \in \mathscr{H}$.

Proof. By 11.4.1, there exists $T \in \mathscr{B}(\mathscr{H})$ such that $B(\mathbf{x}, \mathbf{y}) = (\mathbf{x} \mid T\mathbf{y})$. Also, by the Riesz representation theorem, there exists $\mathbf{u} \in \mathscr{H}$ such that $f(\mathbf{x}) = (\mathbf{x} \mid \mathbf{u})$ for all $\mathbf{x} \in \mathscr{H}$. To complete the proof we must therefore find a $\mathbf{y} \in \mathscr{H}$ such that $(\mathbf{x} \mid T\mathbf{y}) = (\mathbf{x} \mid \mathbf{u})$ for all $\mathbf{x} \in \mathscr{H}$, that is, $T\mathbf{y} = \mathbf{u}$. Now, from $c\|\mathbf{x}\|^2 \leq B(\mathbf{x}, \mathbf{x}) = (\mathbf{x} \mid T\mathbf{x}) \leq \|\mathbf{x}\| \|T\mathbf{x}\|$ we have $\|T\mathbf{x}\| \geq c\|\mathbf{x}\|$, hence $\operatorname{ran} T$ is closed and T has a continuous inverse on $\operatorname{ran} T$. If $\mathbf{x} \in (\operatorname{ran} T)^{\perp}$, then $c\|\mathbf{x}\|^2 \leq B(\mathbf{x}, \mathbf{x}) = (\mathbf{x} \mid T\mathbf{x}) = 0$, hence $\mathbf{x} = 0$. Therefore, $\operatorname{ran} T = \mathscr{H}$ and so T is invertible. Now take $\mathbf{y} = T^{-1}\mathbf{u}$. □

Definition and Properties of the Adjoint

11.4.3 Theorem. *Let $T \in \mathscr{B}(\mathscr{H}, \mathscr{K})$. Then there exists a unique operator $T^* \in \mathscr{B}(\mathscr{K}, \mathscr{H})$ such that*

$$(T\mathbf{x} \mid \mathbf{y})_{\mathscr{K}} = (\mathbf{x} \mid T^*\mathbf{y})_{\mathscr{H}} \quad \textit{for all} \quad \mathbf{x} \in \mathscr{H}, \quad \mathbf{y} \in \mathscr{K}. \tag{11.8}$$

Proof. Take $B(\mathbf{x}, \mathbf{y}) = (T\mathbf{x} \mid \mathbf{y})_{\mathscr{K}}$ in 11.4.1, so that $B = B_{T^*}$ for some T^*. □

The operator $T^* : \mathscr{K} \to \mathscr{H}$ is called the **adjoint of** T. The operation $T \mapsto T^*$ on the Banach algebra $\mathscr{B}(\mathscr{H})$ is an example of an **involution**. The properties of the involution operation are summarized in the next theorem. Note that the last assertion of theorem implies that the involution operation is continuous in the operator norm.

11.4.4 Theorem. *Let $S, T \in \mathscr{B}(\mathscr{H})$ and $z \in \mathbb{C}$. Then*

$$(S+T)^* = S^*+T^*, \ \ (zT^*) = \overline{z}T^*, \ \ (ST)^* = T^*S^*, \ T^{**} = T, \ \textit{and} \ \|T^*T\| = \|T\|^2 = \|T^*\|^2.$$

Proof. The verification of the first three equalities is an exercise for the reader. That $T^{**} = T$ follows from

$$(T\mathbf{x} \mid \mathbf{y}) = (\mathbf{x} \mid T^*\mathbf{y}) = \overline{(T^*\mathbf{y} \mid \mathbf{x})} = \overline{(\mathbf{y} \mid T^{**}\mathbf{x})} = (T^{**}\mathbf{x} \mid \mathbf{y}), \quad \mathbf{x}, \mathbf{y} \in \mathscr{H}.$$

The norm equality is proved as follows: By the CBS inequality we have

$$\|T\mathbf{x}\|^2 = (T\mathbf{x} \mid T\mathbf{x}) = (\mathbf{x} \mid T^*T\mathbf{x}) \leq \|\mathbf{x}\| \|T^*T\mathbf{x}\| \leq \|T^*T\|, \quad \|\mathbf{x}\| \leq 1.$$

Taking the sup over all such $\mathbf{x}$ we obtain

$$\|T\|^2 \leq \|T^*T\| \leq \|T^*\| \|T\|. \tag{†}$$

Thus $\|T\| \leq \|T^*\|$. Replacing T by T^* and using $T^{**} = T$ we have $\|T^*\| \leq \|T\|$. Therefore, $\|T^*\| = \|T\|$. The inequalities in (†) are then equalities, giving the desired conclusion. □

The following is the Hilbert adjoint analog of 8.9.2. The proof is an exercise for the reader.

11.4.5 Proposition. *Let* $T \in \mathscr{B}(\mathscr{H})$. *Then* $\ker T^* = \left[\operatorname{ran} T\right]^{\perp}$ *and* $\ker T = [\operatorname{ran} T^*]^{\perp}$.

$\mathscr{B}(\mathscr{H})$ as a C^*-algebra

The properties in the conclusion of 11.4.4 assert that $\mathscr{B}(\mathscr{H})$ is a **C^*-algebra**. A norm closed subalgebra $\mathscr{C}$ of $\mathscr{B}(\mathscr{H})$ that is closed under the operation of involution is called a **C^*-subalgebra** of $\mathscr{B}(\mathscr{H})$. For example, if $T \in \mathscr{B}(\mathscr{H})$ and $TT^* = T^*T$, then the closure in $\mathscr{B}(\mathscr{H})$ of the set of all polynomials in T, T^* is a commutative C^*-algebra (see §13.1).

The following concept will occasionally be needed: The **commutant** of a subset $\mathscr{S}$ of $\mathscr{B}(\mathscr{H})$ is the set

$$\mathscr{S}' := \{T \in \mathscr{B}(\mathscr{H}) : TS = ST \ \forall\ S \in \mathscr{S}\}.$$

The notation is in conflict with that for dual spaces, but this should not be a problem, as context will indicate the intended meaning. The **bicommutant** of $\mathscr{S}$ is defined by $\mathscr{S}'' := (\mathscr{S}')'$, that is, the commutant of the commutant. The proof of following is an exercise (11.35).

11.4.6 Proposition. *The commutant of* $\mathscr{S} \subseteq \mathscr{H}$ *is a* C^**-subalgebra of* $\mathscr{B}(\mathscr{H})$ *containing the identity operator. Moreover,* $\mathscr{S} \subseteq \mathscr{S}''$.

Exercises

11.27 Prove that (11.7) defines a complete norm on $\mathscr{S}(\mathscr{H})$.

11.28 Let $S, T \in \mathscr{B}(\mathscr{H})$. Prove the *polarization identity for operators*

$$T^*S = \frac{1}{4}\sum_{i=0}^{3} i^k (S + i^k T)^*(S + i^k T).$$

11.29 Let $\mathscr{H}$ be a Hilbert space. Verify the following relations:

(a) $(\operatorname{ran} T)^{\perp} = \ker T^*$. (b) $(\operatorname{ran} T^*)^{\perp} = \ker T$. (c) $(\ker T)^{\perp} = \operatorname{cl}(\operatorname{ran} T^*)$.
(d) $(\ker T^*)^{\perp} = \operatorname{cl}(\operatorname{ran} T)$.

Conclude that T^* is injective iff $\operatorname{ran} T$ is dense in $\mathscr{H}$, and T is injective iff $\operatorname{ran} T^*$ is dense in $\mathscr{H}$.

11.30 Let $T \in \mathscr{B}(\mathscr{H})$. Suppose there exist $a, b > 0$ such that $\|Tx\| \geq a\,\|x\|$ and $\|T^*x\| \geq b\,\|x\|$ for all x. Show that T is invertible.

11.31 Let $\mathscr{H}$ be a Hilbert space, $T \in \mathscr{B}(\mathscr{H})$, and $\mathscr{M}$ a closed subspace $\mathscr{M}$. Then $\mathscr{M}$ is said to be **invariant under** T if $T\mathscr{M} \subseteq \mathscr{M}$. If both $\mathscr{M}$ and $\mathscr{M}^{\perp}$ are T-invariant, then $\mathscr{M}$ is said to **reduce** T. Let P be the orthogonal projection onto $\mathscr{M}$. Prove:

(a) $\mathscr{M}$ is T-invariant iff $\mathscr{M}^{\perp}$ is T^*-invariant. (b) $\mathscr{M}$ is T-invariant iff $PTP = TP$.

(c) $\mathscr{M}$ reduces T iff $\mathscr{M}^{\perp}$ reduces T^*. (d) $\mathscr{M}$ reduces T iff $PT = TP$.

11.32 Show that for $T \in \mathscr{B}(\mathscr{H})$, $\ker T^*T = \ker T$.

11.33 [↑8.34] Find the adjoints of the left and right shift operators T_ℓ and T_r on ℓ^2.

11.34 Show that $M_\phi^* = M_{\overline{\phi}}$ for the multiplication operator M_ϕ on $L^2(X, \mathfrak{F}, \mu)$, where $\phi \in L^\infty$.

11.35 Prove 11.4.6.

Chapter 12

Operator Theory

The special structure of Hilbert spaces allows the construction of classes of operators that have no analogs in general Banach spaces. In this chapter we discuss the main properties of these operators and consider as well various algebras of operators on Hilbert spaces.

Throughout the chapter, $\mathscr{H}$ and $\mathscr{K}$ denote complex Hilbert spaces.

12.1 Classes of Operators

Normal Operators

An operator $T \in \mathscr{B}(\mathscr{H})$ is said to be **normal** if $TT^* = T^*T$. For example, a normal operator in Euclidean space $\mathbb{C}^d$ is a linear transformation whose matrix commutes with the conjugate transpose. For an infinite dimensional example, consider a multiplication operator M_ϕ on $L^2(X, \mathfrak{F}, \mu)$, where $\phi \in L^\infty$. Since $M_\phi^* = M_{\overline{\phi}}$ (Ex. 11.34), we have $M_\phi M_\phi^* = M_{|\phi|^2} = M_\phi^* M_\phi$. Here is an important characterization of normal operators in terms of norms.

12.1.1 Proposition. *Let $T \in \mathscr{B}(\mathscr{H})$. Then T is normal iff $\|T\boldsymbol{x}\| = \|T^*\boldsymbol{x}\|$ for all $\boldsymbol{x} \in \mathscr{H}$.*

Proof. If T is normal, then $\|T\boldsymbol{x}\|^2 = (T\boldsymbol{x} \,|\, T\boldsymbol{x}) = (T^*T\boldsymbol{x} \,|\, \boldsymbol{x}) = (TT^*\boldsymbol{x} \,|\, \boldsymbol{x}) = \|T^*\boldsymbol{x}\|^2$. Conversely, if $\|T\boldsymbol{x}\| = \|T^*\boldsymbol{x}\|$ for all $\boldsymbol{x}$, then $(T\boldsymbol{x} \,|\, T\boldsymbol{x}) = (T^*\boldsymbol{x} \,|\, T^*\boldsymbol{x})$, hence, by the polarization identity, $(T\boldsymbol{x} \,|\, T\boldsymbol{y}) = (T^*\boldsymbol{x} \,|\, T^*\boldsymbol{y})$. Therefore, for all $\boldsymbol{x}$ and $\boldsymbol{y}$,

$$(T^*T\boldsymbol{x} \,|\, \boldsymbol{y}) = (T\boldsymbol{x} \,|\, T\boldsymbol{y}) = (T^*\boldsymbol{x} \,|\, T^*\boldsymbol{y}) = (TT^*\boldsymbol{x} \,|\, \boldsymbol{y}),$$

which shows that $T^*T = TT^*$. □

12.1.2 Corollary. *If $T \in \mathscr{B}(\mathscr{H})$ is normal, then $\|T^2\| = \|T\|^2$.*

Proof. By the proposition, $\|T^2\boldsymbol{x}\| = \|T(T\boldsymbol{x})\| = \|T^*(T\boldsymbol{x})\|$. Taking the supremum over all $\boldsymbol{x}$ with $\|\boldsymbol{x}\| = 1$, we obtain $\|T^2\| = \|T^*T\| = \|T\|^2$, the last equality by 11.4.4. □

Self-Adjoint Operators

An operator $T \in \mathscr{B}(\mathscr{H})$ is said to be **self-adjoint** if $T^* = T$. For example, a multiplication operator M_ϕ is self-adjoint iff ϕ is real-valued. Clearly, every self-adjoint operator is normal. On the other hand, the operator iI is normal but not self-adjoint.

It is clear that the sum of self-adjoint operators is self-adjoint. The product of self-adjoint operators S, T need not be self-adjoint. Indeed, the equality $(ST)^* = T^*S^* = TS$ shows that ST is self-adjoint iff $ST = TS$.

For any $S \in \mathscr{B}(\mathscr{H})$, the operators S^*S, SS^*, $S + S^*$ and $i(S - S^*)$ are self-adjoint. These examples suggests that self-adjoint operators may be viewed as the analogs of real numbers in the complex number system, the adjoint operation being the analog of conjugation. The following proposition strengthens this analogy. The proof is left as an exercise for the reader (12.2).

12.1.3 Proposition. *For $T \in \mathscr{B}(\mathscr{H})$, define*

$$\operatorname{Re} T = \tfrac{1}{2}(T + T^*) \quad \text{and} \quad \operatorname{Im} T = \tfrac{1}{2i}(T - T^*).$$

Then $\operatorname{Re} T$ and $\operatorname{Im} T$ are self-adjoint and $T = \operatorname{Re} T + i \operatorname{Im} T$. Moreover, the decomposition is unique. That is, if $T = A + iB$, where A and B are self-adjoint, then $A = \operatorname{Re} T$ and $B = \operatorname{Im} T$.

The operators $\operatorname{Re} T$ and $\operatorname{Im} T$ in the statement of the proposition are called the **real** and **imaginary parts** of T.

Proposition 12.1.5 below gives a characterization of self-adjointness in terms of the inner product. First, we prove

12.1.4 Lemma. *Let $T \in \mathscr{B}(\mathscr{H})$ such that $(T\boldsymbol{x} \,|\, \boldsymbol{x}) = 0$ for all $\boldsymbol{x}$. Then $T = 0$.*

Proof. For any $\boldsymbol{x}, \boldsymbol{y} \in \mathscr{H}$ and $c \in \mathbb{C}$,

$$\begin{aligned} 0 = (T(c\boldsymbol{x} + \boldsymbol{y}) \,|\, c\boldsymbol{x} + \boldsymbol{y}) &= |c|^2 (T\boldsymbol{x} \,|\, \boldsymbol{x}) + (T\boldsymbol{y} \,|\, \boldsymbol{y}) + c\,(T\boldsymbol{x} \,|\, \boldsymbol{y}) + \bar{c}\,(T\boldsymbol{y} \,|\, \boldsymbol{x}) \\ &= c\,(T\boldsymbol{x} \,|\, \boldsymbol{y}) + \bar{c}\,(T\boldsymbol{y} \,|\, \boldsymbol{x}). \end{aligned}$$

Taking $c = 1$ and $c = i$ yields $(T\boldsymbol{x} \,|\, \boldsymbol{y}) + (T\boldsymbol{y} \,|\, \boldsymbol{x}) = 0$ and $(T\boldsymbol{x} \,|\, \boldsymbol{y}) - (T\boldsymbol{y} \,|\, \boldsymbol{x}) = 0$, respectively. Adding the last two equations and taking $\boldsymbol{y} = T\boldsymbol{x}$, we see that $(T\boldsymbol{x} \,|\, T\boldsymbol{x}) = 0$. Therefore, $T\boldsymbol{x} = 0$. □

12.1.5 Proposition. *An operator T is self-adjoint iff $(T\boldsymbol{x} \,|\, \boldsymbol{x})$ is real for all $\boldsymbol{x} \in \mathscr{H}$.*

Proof. From the calculation $(T\boldsymbol{x} \,|\, \boldsymbol{x}) = (\boldsymbol{x} \,|\, T^*\boldsymbol{x}) = \overline{(T^*\boldsymbol{x} \,|\, \boldsymbol{x})}$ we see that $(T\boldsymbol{x} \,|\, \boldsymbol{x})$ is real iff $((T - T^*)\boldsymbol{x} \,|\, \boldsymbol{x}) = 0$. The conclusion now follows from the lemma. □

Corollary 11.1.5 asserts that for any $T \in \mathscr{B}(\mathscr{H})$, $\|T\| = \sup\{|\,(T\boldsymbol{x} \,|\, \boldsymbol{y})\,| : \|\boldsymbol{x}\|, \|\boldsymbol{y}\| \le 1\}$. For self-adjoint operators, there is a considerable simplification:

12.1.6 Theorem (Rayleigh). *Let $T \in \mathscr{B}(\mathscr{H})$ be self-adjoint. Then*

$$\|T\| = \sup\{|\,(T\boldsymbol{x} \,|\, \boldsymbol{x})\,| : \|\boldsymbol{x}\| \le 1\}.$$

Proof. Let s denote the supremum. Obviously, $s \le \|T\|$. For the reverse inequality, let $\boldsymbol{x}, \boldsymbol{y} \in C_1$. Since $(T\boldsymbol{y} \,|\, \boldsymbol{x}) = (\boldsymbol{y} \,|\, T\boldsymbol{x}) = \overline{(T\boldsymbol{x} \,|\, \boldsymbol{y})}$, we have

$$(T(\boldsymbol{x} + \boldsymbol{y}) \,|\, \boldsymbol{x} + \boldsymbol{y}) - (T(\boldsymbol{x} - \boldsymbol{y}) \,|\, \boldsymbol{x} - \boldsymbol{y}) = 2\,(T\boldsymbol{x} \,|\, \boldsymbol{y}) + 2\,(T\boldsymbol{y} \,|\, \boldsymbol{x}) = 4\operatorname{Re}\,(T\boldsymbol{x} \,|\, \boldsymbol{y}).$$

By definition of s, the left side of the equation is not bigger than

$$s\,\|\boldsymbol{x} + \boldsymbol{y}\|^2 + s\,\|\boldsymbol{x} - \boldsymbol{y}\|^2 = 2s(\|\boldsymbol{x}\|^2 + \|\boldsymbol{y}\|^2) \le 4s.$$

Therefore, $\operatorname{Re}\,(T\boldsymbol{x} \,|\, c\boldsymbol{y}) \le s$ for all $\boldsymbol{x}$ and $\boldsymbol{y}$ with norm ≤ 1 and all with $|c| = 1$. Choosing c so that $\operatorname{Re}\,(T\boldsymbol{x} \,|\, c\boldsymbol{y}) = |\,(T\boldsymbol{x} \,|\, \boldsymbol{y})\,|$, we have $|\,(T\boldsymbol{x} \,|\, \boldsymbol{y})\,| \le s$. Taking the supremum over all $\boldsymbol{x}, \boldsymbol{y} \in C_1$ shows that $\|T\| \le s$. □

We give an application of Rayleigh's theorem in 12.1.9. The theorem actually holds more generally for normal operators (13.2.10), but the proof is considerably deeper, depending on notions of spectral theory.

Positive Operators

An operator T is said to be **positive**, written $T \geq 0$, if $(T\boldsymbol{x} \mid \boldsymbol{x}) \geq 0$ for all $\boldsymbol{x} \in \mathscr{H}$. Thus a positive operator is self-adjoint; the converse is trivially false.

If $S \in \mathscr{B}(\mathscr{H})$, then S^*S and SS^* are clearly positive. The next theorem shows that all positive operators are of this form. The theorem reinforces the analogies between self-adjoint operators and real numbers and between positive operators and nonnegative real numbers. A direct proof of the theorem may be given now, but we prefer to wait until §13.6 when the machinery for a simpler proof will be available.

12.1.7 Theorem. *Let $T \in \mathscr{B}(\mathscr{H})$.*

(a) *If T is positive, then T has a unique positive square root, that is, a unique positive operator $T^{1/2}$ that satisfies $(T^{1/2})^2 = T$. Moreover, if T is invertible, then T^{-1} is positive, $T^{1/2}$ is invertible, and $(T^{1/2})^{-1} = (T^{-1})^{1/2}$.*

(b) *If T is self-adjoint, then there exists a unique pair of positive operators T^+ and T^- such that $T = T^+ - T^-$ and $T^+T^- = T^-T^+ = 0$.*

(c) *The operators $T^{1/2}$ in (a) and $T^{\pm}$ in (b) are members of the bicommutant $\{T\}''$ of T.*

12.1.8 Corollary. *The operator $|T| := (T^*T)^{1/2}$ is the unique positive operator $|T|$ with the property $\|T\boldsymbol{x}\| = \|\,|T|\boldsymbol{x}\,\|$ for all $\boldsymbol{x} \in \mathscr{H}$. Moreover, $|T| = T^+ + T^-$.*

Proof. For the norm equality we have

$$\|T\boldsymbol{x}\|^2 = (T\boldsymbol{x} \mid T\boldsymbol{x}) = (T^*T\boldsymbol{x} \mid \boldsymbol{x}) = \big(|T|^2\boldsymbol{x} \mid \boldsymbol{x}\big) = (|T|\boldsymbol{x} \mid |T|\boldsymbol{x}) = \|\,|T|\boldsymbol{x}\|^2 .$$

If also $S \geq 0$ and $\|S\boldsymbol{x}\| = \|T\boldsymbol{x}\|$ for all $\boldsymbol{x}$, then

$$\big(S^2\boldsymbol{x} \mid \boldsymbol{x}\big) = (S\boldsymbol{x} \mid S\boldsymbol{x}) = \|S\boldsymbol{x}\|^2 = \|T\boldsymbol{x}\|^2 = (T^*T\boldsymbol{x} \mid \boldsymbol{x}) .$$

By 12.1.4, $S^2 = T^*T$, hence, by uniqueness of the positive square root, $S = (T^*T)^{1/2} = |T|$. The last assertion of the corollary is left as an exercise (12.4). □

For self-adjoint operators $S, T \in \mathscr{B}(\mathscr{H})$ we write $S \geq T$ if $S - T$ is positive, that is, $(S\boldsymbol{x} \mid \boldsymbol{x}) \geq (T\boldsymbol{x} \mid \boldsymbol{x})$ for all $\boldsymbol{x} \in \mathscr{H}$. The relation is a partial order on the set of all self-adjoint operators on $\mathscr{H}$ (Ex. 12.8). The following theorem asserts the existence of a least upper bound for an increasing sequence of self-adjoint operators bounded above.

12.1.9 Theorem. *Let R and T_n be self-adjoint operators such that $T_n \leq T_{n+1} \leq R$ for all n. Then there exists self-adjoint operator T such that*

(a) *$T_n \leq T$ for all n.*

(b) *If S is self-adjoint and $T_n \leq S$ for all n, then $T \leq S$.*

(c) *$\|T_n\boldsymbol{x} - T\boldsymbol{x}\| \to 0$ for all $\boldsymbol{x} \in \mathscr{H}$.*

Proof. Replacing T_n by $T_n - T_1$ and R by $R - T_1$, we may assume that $T_n \geq 0$ for all n. By hypothesis, the sequence $\big((T_n\boldsymbol{x} \mid \boldsymbol{x})\big)$ is increasing and bounded, hence has a finite limit $B(\boldsymbol{x}, \boldsymbol{x}) \geq 0$. The polarization identity

$$(\boldsymbol{x} \mid T_n\boldsymbol{y}) = (T_n\boldsymbol{x} \mid \boldsymbol{y}) = \frac{1}{4}\sum_{k=1}^{4} i^k \left(T_n\boldsymbol{x} + i^k\boldsymbol{y} \mid T_n\boldsymbol{x} + i^k\boldsymbol{y}\right)$$

shows that $(T_n\boldsymbol{x} \mid \boldsymbol{y})$ converges to a function $B(\boldsymbol{x}, \boldsymbol{y})$ satisfying

$$B(\boldsymbol{y}, \boldsymbol{x}) = B(\boldsymbol{x}, \boldsymbol{y}) = \frac{1}{4}\sum_{k=0}^{3} i^k B(\boldsymbol{x} + i^k\boldsymbol{y}, \boldsymbol{x} + i^k\boldsymbol{y}).$$

By Rayleigh's theorem, $\|T_n\| = \sup\{(T_n\boldsymbol{x} \mid \boldsymbol{x}) : \|\boldsymbol{x}\| \le 1\} \le \sup\{(R\boldsymbol{x} \mid \boldsymbol{x}) : \|\boldsymbol{x}\| \le 1\} = \|R\|$, hence $c := \sup_n \|T_n\| < \infty$. Since $B(\boldsymbol{x}, \boldsymbol{x}) = \lim (T_n\boldsymbol{x} \mid \boldsymbol{x}) \le c\,\|\boldsymbol{x}\|^2$, we have for $\|\boldsymbol{x}\|, \|\boldsymbol{y}\| \le 1$,

$$|B(\boldsymbol{x}, \boldsymbol{y})| \le \frac{1}{4}\sum_{k=0}^{3} |B(\boldsymbol{x} + i^k\boldsymbol{y}, \boldsymbol{x} + i^k\boldsymbol{y})| \le c\|\boldsymbol{x} + i^k\boldsymbol{y}\|^2 \le 4c < \infty.$$

By 11.4.1, there exists a self-adjoint operator T such that $B(\boldsymbol{x}, \boldsymbol{y}) = (T\boldsymbol{x} \mid \boldsymbol{y})$ for all $\boldsymbol{x}, \boldsymbol{y} \in \mathscr{H}$. Thus $(T_n\boldsymbol{x} \mid \boldsymbol{y}) \to (T\boldsymbol{x} \mid \boldsymbol{y})$ and $(T_n\boldsymbol{x} \mid \boldsymbol{x}) \uparrow (T\boldsymbol{x} \mid \boldsymbol{x})$ for all $\boldsymbol{x}, \boldsymbol{y}$. In particular, T satisfies (a) and (b). Since $S_n := T - T_n \ge 0$, by the CBS inequality applied to the positive sesquilinear form $(S_n\boldsymbol{x} \mid \boldsymbol{y})$ we have, for any pair of unit vectors $\boldsymbol{x}, \boldsymbol{y}$,

$$|(S_n\boldsymbol{x} \mid \boldsymbol{y})|^2 \le (S_n\boldsymbol{x} \mid \boldsymbol{x})(S_n\boldsymbol{y} \mid \boldsymbol{y}) \le (S_n\boldsymbol{x} \mid \boldsymbol{x})\|S_n\| \le (S_n\boldsymbol{x} \mid \boldsymbol{x})(c + \|T\|).$$

Taking the sup over all such $\boldsymbol{y}$ yields $\|S_n\boldsymbol{x}\|^2 \le (S_n\boldsymbol{x} \mid \boldsymbol{x})(c + \|T\|)$. Since $(S_n\boldsymbol{x} \mid \boldsymbol{x}) \to 0$ we see that $\|S_n\boldsymbol{x}\| \to 0$, proving (c). □

Orthogonal Projections and Idempotents

Let $\mathscr{M}$ be a closed subspace of a Hilbert space $\mathscr{H}$, so that $\mathscr{H} = \mathscr{M} \oplus \mathscr{M}^\perp$. The projection P onto $\mathscr{M}$ is called the **orthogonal projection** of $\mathscr{H}$ onto $\mathscr{M}$. Thus, in the obvious notation, $P(\boldsymbol{m} + \boldsymbol{m}^\perp) = \boldsymbol{m}$. By 8.7.8 and the last assertion of 11.2.4, we have

12.1.10 Proposition. *Let $\mathscr{M}$ be a closed subspace of $\mathscr{H}$. Then the orthogonal projection $P : \mathscr{H} \to \mathscr{M}$ is continuous, $P\boldsymbol{x}$ is the unique member of $\mathscr{M}$ nearest $\boldsymbol{x}$, $\operatorname{ran} P = \mathscr{M}$, and $\ker P = \mathscr{M}^\perp$.*

An operator $T \in \mathscr{B}(\mathscr{H})$ is called an **idempotent** if $T^2 = T$. For such an operator, $\operatorname{ran} T = \{\boldsymbol{x} : T\boldsymbol{x} = \boldsymbol{x}\}$. Obviously, projection mappings are idempotents; the converse is false (Ex. 12.19).

12.1.11 Proposition. *Let $T \in \mathscr{B}(\mathscr{H})$ be a nonzero idempotent. Then $\operatorname{ran} T$ is closed and $\|T\| \ge 1$.*

Proof. The first assertion is clear. For the second use $\|T\| = \|T^2\| \le \|T\|\,\|T\|$. □

Here is a characterization of orthogonal projections in terms of idempotents.

12.1.12 Proposition. *An operator $P \ne 0 \in \mathscr{B}(\mathscr{H})$ is an orthogonal projection iff P is a self-adjoint idempotent. In this case, $P \ge 0$ and $\|P\| = 1$.*

Proof. Let P be an idempotent and $\mathscr{M} := \operatorname{ran} P = \{\boldsymbol{x} : P\boldsymbol{x} = \boldsymbol{x}\}$. If P is an orthogonal projection and $\boldsymbol{x} = \boldsymbol{m} + \boldsymbol{m}^\perp$, $\boldsymbol{y} = \boldsymbol{n} + \boldsymbol{n}^\perp$ $(\boldsymbol{m}, \boldsymbol{n} \in \mathscr{M})$, then

$$(P\boldsymbol{x} \mid \boldsymbol{y}) = \left(\boldsymbol{m} \mid \boldsymbol{n} + \boldsymbol{n}^\perp\right) = (\boldsymbol{m} \mid \boldsymbol{n}) = \left(\boldsymbol{m} + \boldsymbol{m}^\perp \mid \boldsymbol{n}\right) = (\boldsymbol{x} \mid P\boldsymbol{y}),$$

so P is self-adjoint and hence positive as well: $(P\boldsymbol{x} \mid \boldsymbol{x}) = \left(P^2\boldsymbol{x} \mid \boldsymbol{x}\right) = (P\boldsymbol{x} \mid P\boldsymbol{x}) \ge 0$. Also, $\|\boldsymbol{x}\|^2 = \|P\boldsymbol{x}\|^2 + \|(I - P)\boldsymbol{x}\|^2 \ge \|P\boldsymbol{x}\|^2$, hence $\|P\| \le 1$. By 12.1.11, $\|P\| = 1$.

Conversely, assume P is self-adjoint and positive. Since $P\boldsymbol{m}^\perp \in \mathscr{M}$, we have

$$0 = \left(P\boldsymbol{m}^\perp \mid \boldsymbol{m}^\perp\right) = \left(P^2\boldsymbol{m}^\perp \mid \boldsymbol{m}^\perp\right) = \left(P\boldsymbol{m}^\perp \mid P\boldsymbol{m}^\perp\right).$$

Therefore $P\boldsymbol{m}^\perp = 0$ and so $P(\boldsymbol{m} + \boldsymbol{m}^\perp) = P\boldsymbol{m} = \boldsymbol{m}$, as required. □

Here are some additional characterizations of orthogonal projections:

12.1.13 Proposition. *Let $P \neq 0$ be an idempotent. Then P is an orthogonal projection iff any one of the following holds.*

(a) *P is positive.*	**(b)** *P is self-adjoint.*	**(c)** *P is normal.*
(d) $\ker P = (\operatorname{ran} P)^{\perp}$.	**(e)** $(\ker P)^{\perp} = \operatorname{ran} P$.	**(f)** $\|P\| = 1$.

Proof. If P is an orthogonal projection, then (a) – (f) obviously hold. If (c) holds, then from 12.1.1, $Px = 0$ iff $P^*x = 0$, hence $\ker P = \ker P^* = (\operatorname{ran} P)^{\perp}$. Taking orthogonal complements yields (e). Therefore, (c) implies (d) and (e). Conversely, if (e) holds, then we have the orthogonal decomposition $\mathscr{H} = \operatorname{ran} P \oplus \ker P$, hence P is an orthogonal projection. It follows that (a) – (e) are equivalent and imply that P is an orthogonal projection.

Finally, we show that if $\|P\| \leq 1$, then (e) holds. Let $x \in (\ker P)^{\perp}$. Since $x - Px \in \ker P$,

$$\|x\|^2 = (x - Px + Px \mid x) = (Px \mid x) \leq \|Px\| \, \|x\| \leq \|x\|^2,$$

hence $\|x\|^2 = \|Px\|^2 = (Px \mid x)$. Therefore,

$$\|x - Px\|^2 = \|x\|^2 + \|Px\|^2 - 2\mathrm{Re}\,(Px \mid x) = 0$$

and so $x \in \operatorname{ran} P$. Thus $(\ker P)^{\perp} \subseteq \operatorname{ran} P$. For the reverse inclusion, let $x \in \operatorname{ran} P$ and write $x = y + z$, where $y \in \ker P$ and $z \in (\ker P)^{\perp} \subseteq \operatorname{ran} P$. Then $x = Px = Py + Pz = Pz = z$ hence x in $(\ker P)^{\perp}$. □

Unitary Operators

An operator $U \in \mathscr{B}(\mathscr{H})$ is said to be **unitary** if

$$U^*U = UU^* = I. \tag{12.1}$$

Thus a unitary operator is an invertible normal operator with $U^{-1} = U^*$.

12.1.14 Proposition. *An operator $U \in \mathscr{B}(\mathscr{H})$ is unitary iff it is a surjective isometry. In this case,*

$$(Ux \mid Uy) = (x \mid y) \quad \text{for all } \; x, y \in \mathscr{H}. \tag{12.2}$$

Proof. The necessity is clear. Conversely, if U is a surjective isometry, then (12.2) holds by the polarization identity, hence $U^*U = I$. Therefore, $U^* = U^{-1}$, hence $UU^* = I$. □

For example, the translation operator and the Fourier transform are unitary operators on $L^2(\mathbb{R}^d)$. The right shift on $\ell^2(\mathbb{N})$ is an isometry that is not unitary.

Note that the operator αI is unitary iff $|\alpha| = 1$. This suggests that the set of unitary operators is the analog of the subset $\mathbb{T}$ of $\mathbb{C}$. The next proposition reinforces this analogy.

12.1.15 Proposition. *The set $\mathscr{U}$ of all unitary operators in $\mathscr{B}(\mathscr{H})$ is a group under composition.*

Proof. If $U \in \mathscr{U}$, then $(U^{-1})^*U^{-1} = U^{**}U^* = UU^* = I$ and similarly $U^{-1}(U^{-1})^* = I$, hence $U^{-1} \in \mathscr{U}$. If $V \in \mathscr{U}$, then $(UV)^*(UV) = V^*U^*UV = V^*IV = I$ hence $UV \in \mathscr{U}$. □

Here is an application of unitary operators due to von Neumann. We give a generalization in Corollary 17.6.9.

12.1.16 Mean Ergodic Theorem. *Let $U \in \mathscr{B}(\mathscr{H})$ be unitary and let $P : \mathscr{H} \to \mathscr{M}$ be the orthogonal projection from $\mathscr{H}$ to $\mathscr{M} := \{\boldsymbol{m} \in \mathscr{H} : U\boldsymbol{m} = \boldsymbol{m}\}$. Then for every $\boldsymbol{x} \in \mathscr{H}$,*

$$\lim_n S_n\boldsymbol{x} = P\boldsymbol{x}, \quad \textit{where} \quad S_n := \frac{1}{n}\sum_{k=0}^{n-1} U^k. \tag{12.3}$$

Proof. (F. Riesz). The set $\mathscr{K}$ of all $\boldsymbol{x}$ for which (12.3) holds is clearly a linear space containing $\mathscr{M}$. We claim that $U\boldsymbol{x} - \boldsymbol{x} \in \mathscr{K}$ for all $\boldsymbol{x} \in \mathscr{H}$. Indeed, the calculation

$$(U\boldsymbol{x} - \boldsymbol{x} \,|\, \boldsymbol{m}) = (\boldsymbol{x} \,|\, U^{-1}\boldsymbol{m}) - (\boldsymbol{x} \,|\, \boldsymbol{m}) = (\boldsymbol{x} \,|\, \boldsymbol{m}) - (\boldsymbol{x} \,|\, \boldsymbol{m}) = 0$$

shows that $U\boldsymbol{x} - \boldsymbol{x} \perp \mathscr{M}$, hence $P(U\boldsymbol{x} - \boldsymbol{x}) = 0$, and because $\|U^n\| \le \|U\|^n \le 1$ we also have

$$\lim_n S_n(U\boldsymbol{x} - \boldsymbol{x}) = \lim_n \tfrac{1}{n}(U^n\boldsymbol{x} - \boldsymbol{x}) = 0,$$

verifying the claim.

Now let $\boldsymbol{x} \in \mathscr{K}^\perp$. By the first paragraph, $(U\boldsymbol{x} - \boldsymbol{x} \,|\, \boldsymbol{x}) = 0$, hence, by Ex. 12.5, $\|U\boldsymbol{x} - \boldsymbol{x}\|^2 = -2\mathrm{Re}\,(U\boldsymbol{x} - \boldsymbol{x} \,|\, \boldsymbol{x}) = 0$. Thus $\boldsymbol{x} \in \mathscr{M}$, and since also $\boldsymbol{x} \in \mathscr{M}^\perp$ we conclude that $\boldsymbol{x} = 0$. Therefore, $\mathscr{K}^\perp = \{0\}$, hence $\mathscr{K}$ is dense in $\mathscr{H}$. Since $\sup_n \|S_n\| \le 1$ and $T\boldsymbol{x} := \lim_n S_n\boldsymbol{x}$ exists for all $\boldsymbol{x}$ in a dense subset of $\mathscr{H}$, the limit exists for all $\boldsymbol{x} \in \mathscr{H}$ and defines a bounded linear operator T. Since $T = P$ on $\mathscr{K}$, $T = P$ on $\mathscr{H}$. □

*Partial Isometries

An operator $U \in \mathscr{B}(\mathscr{H})$ is a **partial isometry** if U is an isometry on $(\ker U)^\perp$. Thus U is a partial isometry iff there exists a closed subspace $\mathscr{K}$ of $\mathscr{H}$ such that U is an isometry on $\mathscr{K}$ and $U = 0$ on $\mathscr{K}^\perp$. If U is a partial isometry, then $(\ker U)^\perp$ is called the **initial space** of U and $\operatorname{ran} U$ the **final space**.

The following proposition characterizes a partial isometry in terms of the orthogonal projection onto the initial space.

12.1.17 Proposition. *Let $U \in \mathscr{B}(\mathscr{H})$ and set $P = U^*U$. Then U is a partial isometry iff P is an orthogonal projection. In this case* $(\ker U)^\perp = \operatorname{ran} P$.

Proof. In the proof we make frequent use of the identity

$$\|U\boldsymbol{x}\|^2 = (U\boldsymbol{x} \,|\, U\boldsymbol{x}) = (P\boldsymbol{x} \,|\, \boldsymbol{x}), \quad \boldsymbol{x} \in \mathscr{H}. \tag{†}$$

Let U be a partial isometry. Then $\|U\| \le 1$, and since $\|U^*\| = \|U\|$ we have $\|P\| \le \|U^*\|\,\|U\| \le 1$. Now, for $\boldsymbol{x} \in (\ker U)^\perp$, $\|\boldsymbol{x}\|^2 = \|U\boldsymbol{x}\|^2 = (P\boldsymbol{x} \,|\, \boldsymbol{x}) \le \|P\boldsymbol{x}\|\,\|\boldsymbol{x}\| \le \|\boldsymbol{x}\|^2$, hence $\|P\boldsymbol{x}\|^2 = (P\boldsymbol{x} \,|\, \boldsymbol{x}) = \|\boldsymbol{x}\|^2$ and so

$$\|P\boldsymbol{x} - \boldsymbol{x}\|^2 = (P\boldsymbol{x} - \boldsymbol{x} \,|\, P\boldsymbol{x} - \boldsymbol{x}) = \|P\boldsymbol{x}\|^2 + \|\boldsymbol{x}\|^2 - (P\boldsymbol{x} \,|\, \boldsymbol{x}) - (\boldsymbol{x} \,|\, P\boldsymbol{x}) = 0.$$

Therefore, P is the identity on $(\ker U)^\perp$. Since $P = 0$ on $\ker U$, P is the orthogonal projection onto $(\ker U)^\perp$.

Conversely, assume that P is an orthogonal projection. By (†), $\|U\boldsymbol{x}\|^2 = \|\boldsymbol{x}\|^2$ if $\boldsymbol{x} \in \operatorname{ran} P$, and $\|U\boldsymbol{x}\|^2 = 0$ if $\boldsymbol{x} \in (\operatorname{ran} P)^\perp = \ker P$. Therefore, U is a partial isometry with initial space $\operatorname{ran} P$. □

For example, the left shift $T_\ell(x_1, x_2, \ldots) = (x_2, x_3, \ldots)$ on $\ell^2(\mathbb{N})$ is a partial isometry with final space ℓ^2 and initial space consisting of all vectors of the form $(0, x_2, x_3, \ldots)$. The orthogonal projection P is $T_\ell^* T_\ell \boldsymbol{x} = T_r T_\ell \boldsymbol{x} = (0, x_2, x_3, \ldots)$.

12.1.18 Corollary. *An operator U on $\mathscr{H}$ is a partial isometry iff U^* is a partial isometry.*

Proof. If U is a partial isometry and $P = U^*U$, then $I - P$ is a projection onto $(\operatorname{ran} P)^\perp = \ker P = \ker U$, hence $U(I-P) = 0$. Therefore, $(UU^*)^2 - UU^* = U(P-I)U^* = 0$ and so UU^* is a self-adjoint idempotent, that is, an orthogonal projection. The converse follows from $U^{**} = U$. □

The next result is a generalization of the polar decomposition $z = e^{i\theta}|z|$ of a complex number.

12.1.19 Polar Decomposition Theorem. *Let $T \in \mathscr{B}(\mathscr{H})$. Then there exists a unique partial isometry $U \in \mathscr{B}(\mathscr{H})$ such that $T = U|T|$ and $\ker U = \ker T$.*

Proof. Define U on $\operatorname{ran}|T|$ by $U|T|\boldsymbol{x} = T\boldsymbol{x}$. Since $\|T\boldsymbol{x}\|^2 = \||T|\boldsymbol{x}\|^2$, U is a well-defined isometry on $\operatorname{ran}|T|$. As such, it has an extension to a linear isometry on $\operatorname{cl}\operatorname{ran}|T| = (\ker |T|)^\perp$. Extend U to all of $\mathscr{H}$ by defining U to be zero on $\ker |T|$. Since

$$(\operatorname{ran}|T|)^\perp = \ker |T| = \ker |T|^2 = \ker T^*T = \ker T,$$

we see that $U|T|\boldsymbol{x} = T\boldsymbol{x}$ on $\ker |T|$, hence $T = U|T|$.

If also $T = V|T|$, where V is a partial isometry and $\ker V = \ker T$, then $V = U$ on $\operatorname{ran}|T|$ and $V = 0 = U$ on $\ker |T| = (\operatorname{ran}|T|)^\perp$, hence $V = U$. □

12.1.20 Corollary. *Every invertible operator $T \in \mathscr{B}(\mathscr{H})$ may be written uniquely as $T = U|T|$, where U is unitary.*

Exercises

12.1 Show that if T is self-adjoint (normal), then T^n is self-adjoint (normal) ($n \in \mathbb{N}$).

12.2 Prove 12.1.3.

12.3 Let $S, T \in \mathscr{B}(\mathscr{H})$ be normal and $T^*S = ST^*$. Prove that $S + T$ and ST are normal.

12.4 Let $T \in \mathscr{B}(\mathscr{H})$ be self-adjoint. Show that $|T| = T^+ + T^-$.

12.5 Let $U \in \mathscr{B}(\mathscr{H})$. Prove that U is an isometry iff $\|U\boldsymbol{x} - \boldsymbol{x}\|^2 = 2\mathrm{Re}\,(\boldsymbol{x} - U\boldsymbol{x} \mid \boldsymbol{x})$.

12.6 Show that if $T_n \in \mathscr{B}(\mathscr{H})$ is normal for all n and $T_n \to T \in \mathscr{B}(\mathscr{H})$, then T is normal.

12.7 Show that if S, T and ST are self-adjoint, then $ST = TS$.

12.8 Show that the relation $T \le S$ iff $S - T \ge 0$ is a partial order on the set of all self-adjoint operators on $\mathscr{H}$.

12.9 [↑ 12.8] Prove that $|S+T|^2 \le 2|S|^2 + 2|T|^2$. ⟦Consider $(S \pm T)^*(S \pm T)$.⟧

12.10 Show that $|cT| = |c||T|$ ($c \in \mathbb{C}$).

12.11 Prove that $T^- = (-T)^+$.

12.12 Let $S, T \in \mathscr{B}(\mathscr{H})$ with $S \ge 0$ and $T \ge 0$. Show that $ST \ge 0$ iff $ST = TS$. Show that one then has $(ST)^{1/2} = S^{1/2}T^{1/2}$.

12.13 Show that $T \in \mathscr{B}(\mathscr{H})$ is normal iff the real and imaginary parts of T commute.

12.14 Let $T \in \mathscr{B}(\mathscr{H})$. Show that $I + T^*T$ is invertible and $\left\|(I + T^*T)^{-1}\right\| \le 1$.

12.15 [↑ 12.8] Let $T \in \mathscr{B}(\mathscr{H})$ and $0 \le T \le I$. Show that $0 \le T^2 \le T$.

12.16 Let $S, T \in \mathcal{B}(\mathcal{H})$ with $S \geq 0$ and T self-adjoint. Show that $TST \geq 0$.

12.17 (*Hellinger-Toeplitz*). Let $T : \mathcal{H} \to \mathcal{H}$ be linear and satisfy $(Tx \mid y) = (x \mid Ty)$ for all x, y. Show that $T \in \mathcal{B}(\mathcal{H})$. ⟦$T(C_1)$ is weakly bounded.⟧

12.18 Let P be an orthogonal projection. Prove: If $\|Px\| = \|x\|$, then $Px = x$.

12.19 Give an example of an idempotent in a Hilbert space that is not an orthogonal projection.

12.20 Let P, Q be orthogonal projections. Prove that the following are equivalent:

(a) PQ is a orthogonal projection. (b) QP is a orthogonal projection. (c) $PQ = QP$.

Show that if these hold, then $\operatorname{ran}(PQ) = \operatorname{ran} P \cap \operatorname{ran} Q$.

12.21 Let P, Q be orthogonal projections. Prove that the following are equivalent:

(a) $PQ = 0$. (b) $QP = 0$. (c) $\operatorname{ran} P \perp \operatorname{ran} Q$.

Show that if these hold, then $P+Q$ is an orthogonal projection and $\operatorname{ran}(P+Q) = \operatorname{ran} P \oplus \operatorname{ran} Q$.

12.22 Let P and Q be orthogonal projections on a Hilbert space $\mathcal{H}$. Show that $Q - P$ is an orthogonal projection iff any of the following holds:

(a) $Q - P$ is positive. (b) $\operatorname{ran} P \subseteq \operatorname{ran} Q$. (c) $PQ = P$. (d) $QP = P$.

Show that if these hold, then $\operatorname{ran}(Q - P) = \operatorname{ran} Q \cap \ker P$.

⟦For (a) ⇒ (b), let $x \in \operatorname{ran} P$ and use 12.18 to show that $Qx = x$.⟧

12.23 Let $\mathcal{M}$ be a closed subspace of $\mathcal{H}$ and $x \in \mathcal{H}$. Prove that

$$d(x, \mathcal{M}) = \sup\{|(x \mid y)| : y \in \mathcal{M}^\perp, \ \|y\| = 1\}$$

12.24 Let $T \in \mathcal{B}(\mathcal{H})$ be normal. Show that if T is invertible, then so is $|T|$ and $|T|^{-1} = |T^{-1}|$

12.25 (a) Let $T \in \mathcal{B}(\mathcal{H})$ be self-adjoint. Show that $T + iI$ is invertible.

(b) Define the **Cayley transform** U of T by $U = (T - iI)(T + iI)^{-1}$. Show that U is unitary.

(c) Let U be a unitary operator on $\mathcal{H}$ such that $I - U$ is invertible. Show that the operator $T := i(I + U)(I - U)^{-1}$ is self-adjoint and that U is its Cayley transform.

12.26 Let ϕ, $\phi^{-1} \in L^\infty(X, \mathcal{F}, \mu)$. Find the polar decomposition of the multiplication operator M_ϕ.

12.27 Let $T = U|T|$ be the polar decomposition of $T \in \mathcal{B}(\mathcal{H})$. Show that

(a) $U^*U|T| = |T|$, $U^*T = |T|$, and $UU^*T = T$. (b) $U|T|U^* = |T^*|$ (use uniqueness of $|T^*|$).

(c) T is normal iff $|T^*| = |T|$. (d) T is normal iff $U|T| = |T|U$ and $UU^* = U^*U$.

12.2 Compact Operators and Operators of Finite Rank

Recall that an operator on a Banach space is compact if it maps bounded sets onto relatively compact sets (§8.10). In a Hilbert space the property has an alternate formulation. Let $T \in \mathcal{B}(\mathcal{H}, \mathcal{K})$. Since C_1 is weakly compact and T is weak-weak continuous, $T(C_1)$ is weakly compact and therefore norm closed. It follows that T is compact iff $T(C_1)$ is norm compact.

Referring to §8.10 we see that the collection $\mathcal{B}_0(\mathcal{H}, \mathcal{K})$ of compact operators in $\mathcal{B}(\mathcal{H}, \mathcal{K})$ a closed linear space and that the set $\mathcal{B}_{00}(\mathcal{H}, \mathcal{K})$ of operators of finite rank is a linear subspace of $\mathcal{B}_0(\mathcal{H}, \mathcal{K})$. Moreover, both $\mathcal{B}_0(\mathcal{H})$ and $\mathcal{B}_{00}(\mathcal{H})$ are ideals in the Banach algebra $\mathcal{B}(\mathcal{H})$. In this section we show that a compact operator in $\mathcal{B}(\mathcal{H}, \mathcal{K})$ may be approximated in norm by members of $\mathcal{B}_{00}(\mathcal{H}, \mathcal{K})$.

Rank One Operators

For $x \in \mathscr{H}$ and $y \in \mathscr{K}$, define the **rank one** operator $x \otimes y \in \mathscr{B}_{00}(\mathscr{H}, \mathscr{K})$ by

$$(x \otimes y)\widetilde{x} = (\widetilde{x} \mid x)\, y, \quad \widetilde{x} \in \mathscr{H}. \tag{12.4}$$

It is easy to check that $x \otimes y$ is linear in y, conjugate linear in x and that $\|x \otimes y\| = \|x\|\,\|y\|$. Moreover,

$$T \circ (x \otimes y) = x \otimes (Ty) \text{ and } (x \otimes y) \circ S = (S^* x) \otimes y, \quad T \in \mathscr{B}(\mathscr{K}),\ S \in \mathscr{B}(\mathscr{H}). \tag{12.5}$$

Clearly, every linear combination of rank one operators is of finite rank. Conversely, every $T \in \mathscr{B}_{00}(\mathscr{H}, \mathscr{K})$ may be written

$$T = \sum_{j=1}^{n} x_j \otimes y_j \tag{12.6}$$

for suitable $x_j \in \mathscr{H}$ and $y_j \in \mathscr{K}$. Indeed, if $\{y_1, \ldots, y_n\}$ is an orthonormal basis for the finite dimensional space $T(\mathscr{H})$, then for all $x \in \mathscr{H}$,

$$Tx = \sum_{j=1}^{n} (Tx \mid y_j)\, y_j = \sum_{j=1}^{n} (x \mid T^* y_j)\, y_j = \sum_{j=1}^{n} \big[(T^* y_j) \otimes y_j\big] x,$$

hence (12.6) holds with $x_j = T^* y_j$.

An Approximation Theorem

Here is the main result of the section.

12.2.1 Theorem. *$\mathscr{B}_0(\mathscr{H}, \mathscr{K})$ is the operator norm closure of $\mathscr{B}_{00}(\mathscr{H}, \mathscr{K})$.*

Proof. We show that an arbitrary operator $T \in \mathscr{B}_0(\mathscr{H}, \mathscr{K})$ is the limit of a sequence of operators of finite rank. Since $\operatorname{cl} \operatorname{ran} T$ is separable (Ex. 8.101), it has a countable orthonormal basis (e_n). For each n define a finite rank operator

$$P_n := \sum_{k=1}^{n} (T^* e_k) \otimes e_k.$$

Since

$$Tx = \sum_{k=1}^{\infty} (Tx \mid e_k)\, e_k \quad \text{and} \quad P_n x = \sum_{k=1}^{n} (Tx \mid e_k)\, e_k,$$

by Parseval's identity and Bessel's inequality

$$\|(T - P_n)x\|^2 = \sum_{k>n} |\,(Tx \mid e_k)\,|^2 \to 0 \quad \text{and} \quad \|P_n x\|^2 = \sum_{k=1}^{n} |\,(Tx \mid e_k)\,|^2 \le \|Tx\|^2.$$

These facts, together with the compactness of $T(C_1)$, imply that $\|P_n - T\| \to 0$. Indeed, given $\varepsilon > 0$, choose $x_1, \ldots, x_m \in C_1$ such that $T(C_1) \subseteq \bigcup_{j=1}^{m} B_\varepsilon(Ty_j)$. Let $x \in C_1$ and choose j so that $\|Tx - Tx_j\| < \varepsilon$. Then

$$\|(T - P_n)x\| \le \|Tx - Tx_j\| + \|Tx_j - P_n x_j\| + \|P_n(x_j - x)\| \le 2\varepsilon + \|Tx_j - P_n x_j\|$$

and so $\|T - P_n\| \le 2\varepsilon + \max_j \|Ty_j - P_n y_j\|$. Therefore, $\overline{\lim}_n \|(T - P_n)\| \le 2\varepsilon$. □

For the case $\mathscr{K} = \mathscr{H}$, we have the following version:

12.2.2 Theorem. *Let $T \in \mathscr{B}_0(\mathscr{H})$. Then there exists a net (P_α) of projections of finite rank such that $\|P_\alpha T - T\| \to 0$.*

Proof. Let $\mathscr{E}$ be an orthonormal basis for $\mathscr{H}$, and for each finite set $\alpha \subseteq \mathscr{E}$ let P_α denote the projection of $\mathscr{H}$ onto $\operatorname{span}\alpha$. Then (P_α) is a net, where the indices are directed upward by inclusion. Set $Q_\alpha := P_\alpha - I$. For each $\boldsymbol{x} = \sum_{\boldsymbol{e}\in\mathscr{E}} (\boldsymbol{x} \mid \boldsymbol{e})\, \boldsymbol{e}$ we have, by Parseval's identity,

$$\|Q_\alpha \boldsymbol{x}\|^2 = \sum_{\boldsymbol{e}\in\mathscr{E}\setminus\alpha} |(\boldsymbol{x} \mid \boldsymbol{e})|^2 \to 0. \tag{†}$$

If it is not the case that $\|P_\alpha T - T\| \to 0$, then there exists an $\varepsilon > 0$, a subnet (Q_β), and a net $(\boldsymbol{x}_\beta)$ of unit vectors with $\|Q_\beta T\boldsymbol{x}_\beta\| \geq \varepsilon$ for all β. Since T is compact we may assume that $T\boldsymbol{x}_\beta \to \boldsymbol{y}$ for some $\boldsymbol{y}$. But then

$$\varepsilon \leq \|Q_\beta T\boldsymbol{x}_\beta\| \leq \|Q_\beta(T\boldsymbol{x}_\beta - \boldsymbol{y})\| + \|Q_\beta \boldsymbol{y}\| \leq \|T\boldsymbol{x}_\beta - \boldsymbol{y}\| + \|Q_\beta \boldsymbol{y}\| \to 0,$$

impossible. Thus $\|Q_\alpha T\| \to 0$. □

A subset $\mathscr{A}$ of $\mathscr{B}(\mathscr{K})$ is said to be **self-adjoint** if $T \in \mathscr{A} \Rightarrow T^* \in \mathscr{A}$.

12.2.3 Proposition. *$\mathscr{B}_0(\mathscr{H})$ and $\mathscr{B}_{00}(\mathscr{H})$ are self-adjoint.*

Proof. If $T \in \mathscr{B}_{00}(\mathscr{H})$, then $\mathscr{K} = \operatorname{ran} T \oplus \ker T^*$, hence $T^*(\mathscr{K}) = T^*(\operatorname{ran} T)$, which is finite dimensional. Therefore $T^* \in \mathscr{B}_{00}(\mathscr{H})$. Now let $T \in \mathscr{B}_0(\mathscr{H})$ and let (T_n) be a sequence in $\mathscr{B}_{00}(\mathscr{H})$ with $\|T_n - T\| \to 0$. Since $\|T_n^* - T^*\| = \|T_n - T\|$, we have $\|T_n^* - T^*\| \to 0$. Since $T_n^* \in \mathscr{B}_{00}(\mathscr{H})$, it follows that T^* is compact. □

12.2.4 Corollary. *$\mathscr{B}_0(\mathscr{H})$ is a C^*-subalgebra.*

Exercises

12.28 Let $T \in \mathscr{B}(\mathscr{H})$. Show that the commutant of $\mathscr{B}_{00}(\mathscr{H})$ is $\mathbb{C}I$, hence $\mathscr{B}_{00}(\mathscr{H})'' = \mathscr{B}(\mathscr{H})$.

12.29 Prove that $T \in \mathscr{B}(\mathscr{H})$ is compact iff $x_n \xrightarrow{w} 0 \Rightarrow \|Tx_n\| \to 0$. Show that this is false in $\ell^1(\mathbb{N})$.

12.30 Show that $T \in \mathscr{B}(\mathscr{H})$ is compact iff the following condition holds:

$$\boldsymbol{x}_n \xrightarrow{w} \boldsymbol{x} \text{ and } \boldsymbol{y}_n \xrightarrow{w} \boldsymbol{y} \Rightarrow (T\boldsymbol{x}_n \mid \boldsymbol{y}_n) \to (T\boldsymbol{x} \mid \boldsymbol{y}).$$

12.31 Let $\phi \in \mathfrak{C}_0$. Show that the multiplication operator M_ϕ on $\ell^2(\mathbb{N})$ is compact. Show that the analogous assertion for $\phi \in \mathfrak{C}$ is false.

12.32 Let $S, T \in \mathscr{B}(\mathscr{H})$. Prove: $S^*S \leq T^*T$ and T compact $\Rightarrow S$ compact.

12.33 Let $T \in \mathscr{B}(\mathscr{H})$. Show that T^*T compact $\Rightarrow T$ compact.

12.34 Let $\phi \in L^\infty(0,1)$. Show that if the multiplication operator M_ϕ on $L^2(0,1)$ is compact then $\phi = 0$ a.e. Find an example of a measure space $(X, \mathscr{F}, \mu)$ for which the assertion is false in $L^2(X, \mathscr{F}, \mu)$.

12.35 Prove that $\|\boldsymbol{x} \otimes \boldsymbol{y}\| = \|\boldsymbol{x}\|\,\|\boldsymbol{y}\|$.

12.36 Verify the assertions in (12.5).

12.37 Show that T is compact (has finite rank) iff $|T|$ is compact (has finite rank). ⟦Use a polar decomposition.⟧

12.3 The Spectral Theorem for Compact Normal Operators

Eigenvalues and Eigenvectors

An **eigenvalue** of $T \in \mathscr{B}(\mathscr{H})$ is a complex number α such that $\ker(\alpha I - T) \neq \{0\}$. The subspace $\ker(\alpha I - T)$ of $\mathscr{H}$ is called the **eigenspace of T corresponding to α**. A nonzero member of the eigenspace is called an **eigenvector.** The spectral theorem for finite dimensional spaces asserts that a normal operator T may be decomposed into a finite sum $T = \sum_j \alpha_j P_{\alpha_j}$, where the α_j are the eigenvalues of T and the operators P_{α_j} are the projections onto the mutually orthogonal eigenspaces with $\sum_j P_{\alpha_j} = I$. In this section, we prove an infinite dimensional version of this result for compact normal operators.[1]

12.3.1 Proposition. *Let $T \in \mathscr{B}(\mathscr{H})$ be normal.*

(a) *α is an eigenvalue of T iff $\overline{\alpha}$ is an eigenvalue of T^*.*

(b) *If α and β are distinct eigenvalues, then the eigenspaces $\ker(\alpha I - T)$ and $\ker(\beta I - T)$ are mutually orthogonal.*

Proof. (a) Since $\alpha I - T$ is normal with adjoint $\overline{\alpha} I - T^*$, we have $\|(\alpha I - T)\boldsymbol{x}\| = \|(\overline{\alpha} I - T^*)\boldsymbol{x}\|$. Therefore, $(\alpha I - T)\boldsymbol{x} = \boldsymbol{0}$ iff $(\overline{\alpha} I - T^*)\boldsymbol{x} = \boldsymbol{0}$.

(b) Let $T\boldsymbol{x} = \alpha\boldsymbol{x}$ and $T\boldsymbol{y} = \beta\boldsymbol{y}$, where $\boldsymbol{x}, \boldsymbol{y} \neq \boldsymbol{0}$. Then

$$\alpha\,(\boldsymbol{x} \,|\, \boldsymbol{y}) = (T\boldsymbol{x} \,|\, \boldsymbol{y}) = (\boldsymbol{x} \,|\, T^*\boldsymbol{y}) = \big(\boldsymbol{x} \,|\, \overline{\beta}\boldsymbol{y}\big) = \beta\,(\boldsymbol{x} \,|\, \boldsymbol{y}).$$

Since $\alpha \neq \beta$, $(\boldsymbol{x} \,|\, \boldsymbol{y}) = 0$. □

Diagonalizable Operators

An operator $T \in \mathscr{B}(\mathscr{H})$ is said to be **diagonalizable** if there exists an orthonormal basis $\{\boldsymbol{e}_i : i \in \mathfrak{I}\}$ of $\mathscr{H}$ and a bounded set of complex numbers $\{\alpha_i : i \in \mathfrak{I}\}$ such that

$$T\boldsymbol{x} = \sum_i \alpha_i\,(\boldsymbol{x} \,|\, \boldsymbol{e}_i)\,\boldsymbol{e}_i \quad \text{for all } \boldsymbol{x} \in \mathscr{H}. \tag{12.7}$$

Since $\alpha_i = (T\boldsymbol{e}_i \,|\, \boldsymbol{e}_i)$ we may write

$$T\boldsymbol{x} = \sum_i (T\boldsymbol{e}_i \,|\, \boldsymbol{e}_i)\,(\boldsymbol{x} \,|\, \boldsymbol{e}_i)\,\boldsymbol{e}_i = \sum_i (T\boldsymbol{e}_i \,|\, \boldsymbol{e}_i)\,(\boldsymbol{e}_i \otimes \boldsymbol{e}_i)\boldsymbol{x},$$

or simply

$$T = \sum_{i \in \mathfrak{I}} (T\boldsymbol{e}_i \,|\, \boldsymbol{e}_i)\,\boldsymbol{e}_i \otimes \boldsymbol{e}_i.$$

From (12.7) we see that $T\boldsymbol{x} = \alpha\boldsymbol{x}$ iff $\sum_i (\alpha_i - \alpha)\,(\boldsymbol{x} \,|\, \boldsymbol{e}_i)\,\boldsymbol{e}_i = 0$ iff $\sum_i |\alpha_i - \alpha|^2 |\,(\boldsymbol{x} \,|\, \boldsymbol{e}_i)\,|^2 = 0$. Thus the eigenvalues of T are the numbers α_i. Moreover, since $\boldsymbol{x}$ is an eigenvector corresponding to α iff $(\boldsymbol{x}, \boldsymbol{e}_i) = 0$ for all i with $\alpha_i \neq \alpha$, we see that the eigenspace corresponding to α is the span of those $\boldsymbol{e}_i$ for which $\alpha_i = \alpha$. Thus

$$\boldsymbol{x} = \sum_{i:\alpha_i = \alpha} (\boldsymbol{x} \,|\, \boldsymbol{e}_i)\,\boldsymbol{e}_i, \quad \boldsymbol{x} \in \ker(\alpha I - T). \tag{12.8}$$

The next two propositions give the basic properties of diagonalizable operators.

[1] We remove the compactness requirement in §13.6.

12.3.2 Proposition. *If T is diagonalizable relative to $\{\boldsymbol{e}_i : i \in \mathfrak{I}\}$ and $\{\alpha_i,\ i \in \mathfrak{I}\}$, then T^* is diagonalizable relative to $\{\boldsymbol{e}_i : i \in \mathfrak{I}\}$ and $\{\overline{\alpha}_i,\ i \in \mathfrak{I}\}$. In this case we have*

$$T^*\boldsymbol{x} = \sum_i \overline{\alpha}_i\,(\boldsymbol{x}\,|\,\boldsymbol{e}_i)\,\boldsymbol{e}_i \quad \textit{for all } \boldsymbol{x} \in \mathscr{H}. \tag{12.9}$$

Proof. Writing $\boldsymbol{x} = \sum_i (\boldsymbol{x}\,|\,\boldsymbol{e}_i)\,\boldsymbol{e}_i$ and $\boldsymbol{y} = \sum_j (\boldsymbol{y}\,|\,\boldsymbol{e}_j)\,\boldsymbol{e}_j$, we have from (12.7)

$$(\boldsymbol{x}\,|\,T^*\boldsymbol{y}) = (T\boldsymbol{x}\,|\,\boldsymbol{y}) = \sum_{i,j} \alpha_i\,(\boldsymbol{x}\,|\,\boldsymbol{e}_i)\,\overline{(\boldsymbol{y}\,|\,\boldsymbol{e}_j)}\,(\boldsymbol{e}_i\,|\,\boldsymbol{e}_j) = \Big(\boldsymbol{x}\,\Big|\,\sum_j \overline{\alpha}_j\,(\boldsymbol{y}\,|\,\boldsymbol{e}_j)\,\boldsymbol{e}_j\Big),$$

hence $T^*\boldsymbol{y} = \sum_j \overline{\alpha}_j\,(\boldsymbol{y}\,|\,\boldsymbol{e}_j)\,\boldsymbol{e}_j$. □

12.3.3 Proposition. *A diagonalizable operator T is normal. Moreover, T is self-adjoint iff α_i is real for every i, and T is positive iff $\alpha_i \ge 0$ for all i.*

Proof. From (12.7) and (12.9),

$$T^*T\boldsymbol{x} = \sum_i \alpha_i\,(\boldsymbol{x}\,|\,\boldsymbol{e}_i)\,T^*\boldsymbol{e}_i = \sum_i \alpha_i\overline{\alpha}_i\,(\boldsymbol{x}\,|\,\boldsymbol{e}_i)\,\boldsymbol{e}_i = TT^*\boldsymbol{x},$$

which shows that T is normal. Equations (12.7) and (12.9) also prove the self-adjoint part of the theorem. For the positivity part, use $(T\boldsymbol{x}\,|\,\boldsymbol{x}) = \sum_i \alpha_i|\,(\boldsymbol{x}\,|\,\boldsymbol{e}_i)\,|^2$, and $(T\boldsymbol{e}_i\,|\,\boldsymbol{e}_i) = \alpha_i$. □

For the spectral decomposition of compact diagonalizable operators, the following terminology will be convenient. A set A of complex numbers is said to **vanish at infinity** if $\{c \in A : |c| \ge \varepsilon\}$ is finite for all $\varepsilon > 0$. By taking $\varepsilon = 1/n$ we see that all but countably many members of such a set A are zero.

12.3.4 Lemma. *A diagonalizable operator T is compact iff the set $A := \{\alpha_i : i \in \mathfrak{I}\}$ of eigenvalues of T vanishes at infinity. In this case, the finite rank operator*

$$T_\varepsilon\boldsymbol{x} := \sum_{|\alpha_i| \ge \varepsilon} \alpha_i\,(\boldsymbol{x}\,|\,\boldsymbol{e}_i)\,\boldsymbol{e}_i$$

converges in operator norm to T as $\varepsilon \to 0$.

Proof. Suppose $A_\varepsilon := \{i \in \mathfrak{I} : |\alpha_i| \ge \varepsilon\}$ is infinite. Then there exists a sequence $(\boldsymbol{e}_n)$ of distinct members of $\{\boldsymbol{e}_i : i \in \mathfrak{I}\}$ such that $\|T\boldsymbol{e}_n\| = |\alpha_n| \ge \varepsilon$. Since $\boldsymbol{e}_n \xrightarrow{w} 0$, T cannot be compact.

Conversely, suppose that A_ε is finite for all ε. Then each operator $T_\varepsilon\boldsymbol{x}$ has finite rank. Moreover, by Bessel's inequality,

$$\|T\boldsymbol{x} - T_\varepsilon\boldsymbol{x}\|^2 = \sum_{|\alpha_i| < \varepsilon} \|\alpha_i\,(\boldsymbol{x}\,|\,\boldsymbol{e}_i)\,\boldsymbol{e}_i\|^2 \le \varepsilon^2\,\|\boldsymbol{x}\|^2.$$

Therefore, $\|T - T_\varepsilon\| \le \varepsilon$, and since ε was arbitrary, $T \in \operatorname{cl}\mathscr{B}_{00}(\mathscr{H}) = \mathscr{B}_0(\mathscr{H})$. □

Here is the main result regarding compact diagonalizable operators.

12.3.5 Theorem. *A compact diagonalizable operator T has only countably many distinct eigenvalues. If the nonzero distinct eigenvalues are denoted by λ_n and if P_n denotes the projection of $\mathscr{H}$ onto the eigenspace* $\ker(\lambda_n - T)$*, then*

$$T = \sum_n \lambda_n P_n, \quad P_n = \sum_{\alpha_i = \lambda_n} \boldsymbol{e}_i \otimes \boldsymbol{e}_i, \quad \textit{and } I = \sum_n P_n, \tag{12.10}$$

the first equation holding in the operator norm and the second and third holding pointwise in the norm topology of $\mathscr{H}$. Moreover, the sequence $(|\lambda_n|)$ may be taken to be decreasing, hence in the infinite case $|\lambda_n| \downarrow 0$.

Proof. The first assertion follows from the preceding lemma. For the proof of 12.10, we consider only the case where the sequence (λ_n) is infinite. Collecting together the terms in the expansion (12.7) corresponding to the same α_i, we have

$$Tx = \sum_n \lambda_n P_n x, \quad P_n x = \sum_{\alpha_i = \lambda_n} (x \mid e_i)\, e_i = \sum_{\alpha_i = \lambda_n} (e_i \otimes e_i) x \ \text{ and } \ x = \sum_n P_n x.$$

Since (λ_n) vanishes at infinity, given $0 < \varepsilon \le |\lambda_1|$ we may choose the smallest $n = n(\varepsilon)$ for which $|\lambda_k| < \varepsilon$ for all $k > n$. For $k \le n$ we then have $|\alpha_i| \ge \varepsilon$ for all α_i coinciding with λ_k. Thus $\sum_{k=1}^n \lambda_k P_k$ is the operator T_ε in the lemma. Since $n(\varepsilon)$ increases as ε decreases, the lemma implies that $T = \sum_n \lambda_n P_n$ holds in the operator norm. By considering the finite sets $\{|\lambda_n| \ge 1\} \subseteq \{|\lambda_n| \ge 1/2\} \subseteq \cdots$, we may arrange the sequence (λ_n) so that $|\lambda_{n+1}| \le |\lambda_n|$ for all n. □

The **multiplicity of an eigenvalue** λ_n is the dimension of $\operatorname{ran} P_n$, where P_n is the projection of the theorem.

The Spectral Theorem

For the main result of the section, we shall need some aspects of spectral theory, a subject developed fully and in a more general setting in the next chapter.

The **spectrum** of $T \in \mathscr{B}(\mathscr{H})$ is the set

$$\sigma(T) = \{\lambda \in \mathbb{C} : \lambda I - T \text{ is not invertible in } \mathscr{B}(\mathscr{H})\}.$$

The following result is proved in Chapter 13. It will be used here to prove the existence of eigenvalues for a compact normal operator, the essential ingredient in the proof of the spectral theorem.

12.3.6 Lemma. *Let $T \in \mathscr{B}(\mathscr{H})$. Then $\sigma(T)$ is nonempty and bounded. Moreover*

$$\sup\{|\lambda| : \lambda \in \sigma(T)\} = \lim_n \|T^n\|^{1/n}. \tag{12.11}$$

We shall need two more lemmas:

12.3.7 Lemma. *Let V be a normal operator on $\mathscr{H}$. Then V is invertible iff there exists $c > 0$ such that $\|Vx\| \ge c\,\|x\|$ for all x.*

Proof. The necessity follows by taking $c = \left\|V^{-1}\right\|^{-1}$. For the sufficiency, note that the inequality implies that V is injective and $\operatorname{ran} V$ is closed. Since V is normal, $\|V^* x\| = \|Vx\|$, hence V^* is also injective and so $\operatorname{ran} V = \ker(V^*)^\perp = \mathscr{H}$. □

12.3.8 Lemma. *A compact normal operator T has an eigenvalue.*

Proof. We may assume that $T \ne 0$. Noting that powers of a normal operator are normal (Ex. 12.1), by iterating the equality $\|T^2\| = \|T\|^2$ we see that $\|T^n\| = \|T\|^n$ for $n = 2^k$. Therefore, the limit in (12.11) is simply $\|T\|$, and since $\|T\| > 0$, $\sigma(T)$ must have a nonzero member λ. Set $S := \lambda^{-1} T$. Then S is compact and normal, and since $\lambda I - T$ is not invertible neither is $I - S$. By 12.3.7 applied to $V = I - S$, for each n there exists an x_n with unit norm such that $\|(I - S)\boldsymbol{x}_n\| \le 1/n$. By compactness of S we may take a convergent subsequence $S\boldsymbol{x}_{n_k} \to \boldsymbol{y}$. We then have

$$\boldsymbol{x}_{n_k} = (I - S)\boldsymbol{x}_{n_k} + S\boldsymbol{x}_{n_k} \to \boldsymbol{y}$$

and so $S\boldsymbol{y} = \boldsymbol{y}$, that is, $T\boldsymbol{y} = \lambda\boldsymbol{y}$, and $\|\boldsymbol{y}\| = 1$. □

Here is the main result of the section:

12.3.9 Theorem. *Let $T \in \mathscr{B}(\mathscr{H})$ be compact and normal. Then T is diagonalizable.*

Proof. Let $\mathfrak{O}$ denote the family of all orthonormal sets whose members are eigenvectors of T. By 12.3.8, $\mathfrak{O} \ne \emptyset$. A standard Zorn's lemma argument shows that $\mathfrak{O}$ has a maximal member, that is, an orthonormal set $\mathscr{E}$ of eigenvectors that is not properly contained in a larger such set. Let $\mathscr{K}$ denote the closed linear span of $\mathscr{E}$ and observe that $T(\mathscr{K}) \subseteq \mathscr{K}$. Also, by 12.3.2, $T^*(\mathscr{K}) \subseteq \mathscr{K}$, hence $T(\mathscr{K}^\perp) \subseteq \mathscr{K}^\perp$. Since T is diagonalizable on $\mathscr{K}$ it therefore suffices to show that $\mathscr{K}^\perp = \{0\}$.

Suppose that $\mathscr{K}^\perp \ne \{0\}$. We consider two cases: If $T(\mathscr{K}^\perp) = 0$, then every unit vector in $\mathscr{K}^\perp$ is an eigenvector with eigenvalue zero. If $T(\mathscr{K}^\perp) \ne 0$, then, by 12.3.8, T has an eigenvector in $\mathscr{K}^\perp$. Each outcome contradicts the maximality of $\mathscr{E}$, hence $\mathscr{K}^\perp = \{0\}$. □

The following application of the spectral theorem will be needed in the discussion of Hilbert-Schmidt integral operators in the next section.

12.3.10 Corollary. *If $T \in \mathscr{B}_0(\mathscr{H}, \mathscr{K})$ is not the zero operator, then there exist orthonormal (possibly finite) sequences $(\boldsymbol{x}_n) \subseteq \mathscr{H}$, $(\boldsymbol{y}_n) \subseteq \mathscr{K}$, and $(\alpha_n) \subseteq (0, \infty)$ such that in the operator norm*

$$T = \sum_n \alpha_n(\boldsymbol{x}_n \otimes \boldsymbol{y}_n). \tag{12.12}$$

If the sequences are infinite, then $\alpha_n \downarrow 0$.

Proof. By the spectral theorem applied to $T^*T \in \mathscr{B}_0(\mathscr{H})$, there exists an orthonormal sequence $(\boldsymbol{x}_n)$ of eigenvectors of T^*T and a decreasing sequence of corresponding eigenvalues $\beta_n > 0$ such that

$$T^*T\boldsymbol{x} = \sum_{n=1}^{\infty} \beta_n\,(\boldsymbol{x}\,|\,\boldsymbol{x}_n)\,\boldsymbol{x}_n, \quad \boldsymbol{x} \in \mathscr{H}. \tag{†}$$

We assume that (β_n) is an infinite sequence (hence $\beta_n \downarrow 0$); otherwise the sum in (†) is finite and the notation in the remainder of the proof may be adjusted accordingly.

Set $\alpha_n = \sqrt{\beta_n}$ and $\boldsymbol{y}_n = \alpha_n^{-1}T\boldsymbol{x}_n$. The calculation

$$\alpha_m\alpha_n\,(\boldsymbol{y}_m\,|\,\boldsymbol{y}_n) = (T\boldsymbol{x}_m\,|\,T\boldsymbol{x}_n) = (T^*T\boldsymbol{x}_m\,|\,\boldsymbol{x}_n) = \alpha_m^2\,(\boldsymbol{x}_m\,|\,\boldsymbol{x}_n)$$

implies that $(\boldsymbol{y}_n)$ is orthonormal, hence it remains to show that (12.12) holds.

Now, by Bessel's inequality,

$$\left\|\sum_{k=n}^{m} \alpha_k\,(\boldsymbol{x}\,|\,\boldsymbol{x}_k)\,\boldsymbol{y}_k\right\|^2 = \sum_{k=n}^{m} \alpha_k^2|\,(\boldsymbol{x}\,|\,\boldsymbol{x}_k)\,|^2 \le \beta_n\,\|\boldsymbol{x}\|^2 \le \beta_n, \quad \|\boldsymbol{x}\| \le 1,$$

hence the operators $\sum_{k=n}^{m} \alpha_k\boldsymbol{x}_k \otimes \boldsymbol{y}_k$ form a Cauchy sequence in $\mathscr{B}_{00}(\mathscr{H}, \mathscr{K})$. Let

$$S := \lim_n \sum_{k=1}^{n} \alpha_k\boldsymbol{x}_k \otimes \boldsymbol{y}_k \text{ (operator norm convergence)},$$

which is the operator on the right in (12.12). Since $S\boldsymbol{x}_m = \alpha_m\boldsymbol{y}_m = T\boldsymbol{x}_m$, $T = S$ on the closed linear span $\mathscr{X}$ of $(\boldsymbol{x}_n)$. Moreover, $T = S$ on $\mathscr{X}^\perp$. Indeed, S is obviously zero on $\mathscr{X}^\perp$, and (†) implies that $T^*T = 0$ on $\mathscr{X}^\perp$ so that $T = 0$ on $\mathscr{X}^\perp$ as well. This verifies (12.12). □

Exercises

12.38 Let $(X, \mathscr{F}, \mu)$ be a σ-finite measure space and let $\phi \in L^\infty(X, \mathscr{F}, \mu)$. Show that λ is an eigenvalue of the multiplication operator M_ϕ on L^2 iff $\phi = \lambda$ on a set of positive measure.

12.39 Find the eigenvalues of the left shift operator T_ℓ on ℓ^2. Show that the right shift operator T_r has no eigenvalues.

12.40 Let $f,\, g \in L^2[0,1]$ and extend f periodically to $\mathbb{R}$ so that the convolution operator $T_g f := f * g$ is defined on $[0,1]$:

$$(T_g f)(x) = \int_0^1 f(x-y)g(y)\,dy.$$

Referring to (11.5), show that e_n is an eigenvector of T_g with eigenvalue $\hat{g}(n)$.

12.41 Show that the operator T on $L^2[0,1]$ defined by $(Tf)(t) = tf(t)$ is self-adjoint with no eigenvalues.

12.42 Show that the operator T on $C[0,1]$ defined by $Tf(x) = \int_0^x f(t)\,dt$ does not have an eigenvalue.

12.43 Let $T \in \mathscr{B}(\mathscr{H})$ be self-adjoint, $\lambda \in \mathbb{C}$, and let P be the projection of $\mathscr{H}$ onto $\ker(\lambda - T)$. Show that $S \in \mathscr{B}(\mathscr{H})$ and $ST = TS \Rightarrow SP = PS$. Conclude in (12.10) that for T self-adjoint, $ST = TS \Rightarrow SP_n = P_nS$ for all n. (By 13.6.2, these assertions hold for normal T.)

---

*12.4 Hilbert-Schmidt Operators

In this section, $\mathscr{H}$, $\mathscr{K}$, and $\mathscr{L}$ denote Hilbert spaces over $\mathbb{C}$, and $\mathscr{E}$ and $\mathscr{F}$ are orthonormal bases for $\mathscr{H}$ and $\mathscr{K}$, respectively.

The Hilbert-Schmidt Norm

The **Hilbert-Schmidt norm** of an operator $T \in \mathscr{B}(\mathscr{H}, \mathscr{K})$ is defined by

$$\|T\|_2 := \left(\sum_{e \in \mathscr{E}} \|Te\|^2\right)^{1/2}.$$

Note that by Parseval's equality and an interchange of summations,

$$\sum_{e\in\mathscr{E}} \|Te\|^2 = \sum_{e\in\mathscr{E}} \sum_{f\in\mathscr{F}} |(Te \mid f)|^2 = \sum_{e\in\mathscr{E}} \sum_{f\in\mathscr{F}} |(e \mid T^* f)|^2 = \sum_{f\in\mathscr{F}} \|T^* f\|^2.$$

Thus the definition of $\|T\|_2$ is independent of the choice of the orthonormal basis and $\|T\|_2 = \|T^*\|_2$. If $\|T\|_2 < \infty$, then T is called a **Hilbert-Schmidt operator**. The set of all Hilbert-Schmidt operators is denoted by $\mathscr{B}_2(\mathscr{H}, \mathscr{K})$. It is easy to check that $\mathscr{B}_2(\mathscr{H}, \mathscr{K})$ is a linear space and $\|T\|_2$ is a norm. For example, the triangle inequality $\|T + S\|_2 \le \|T\|_2 + \|S\|_2$ follows easily from the CBS inequality in $\mathscr{H}$ and the triangle inequality in $\ell^2(\mathbb{N})$.

The following proposition makes important connections between the operator norm and the Hilbert-Schmidt norm.

12.4.1 Proposition. *If $S \in \mathscr{B}(\mathscr{L}, \mathscr{H})$, $T \in \mathscr{B}_2(\mathscr{H}, \mathscr{K})$, and $R \in \mathscr{B}(\mathscr{K}, \mathscr{L})$, then*

$$\|T\| \le \|T\|_2, \quad \|TS\|_2 \le \|S\|\,\|T\|_2, \quad \text{and} \quad \|RT\|_2 \le \|R\|\,\|T\|_2,$$

hence $TS \in \mathscr{B}_2(\mathscr{L}, \mathscr{K})$ and $RT \in \mathscr{B}_2(\mathscr{H}, \mathscr{L})$. In particular, $\mathscr{B}_2(\mathscr{H})$ is an ideal in $\mathscr{B}(\mathscr{H})$.

Proof. By Parsevals's identity and the CBS inequality, for $x \in \mathscr{H}$ with $\|x\| \leq 1$,

$$\|Tx\|^2 = \sum_{f} |(Tx \,|\, f)|^2 = \sum_{f} |(x \,|\, T^*f)|^2 \leq \sum_{f} \|T^*f\|^2 = \|T^*\|_2^2 = \|T\|_2^2,$$

proving the first inequality. For the remaining inequalities we have

$$\|RT\|_2^2 = \sum_{e} \|RTe\|^2 \leq \|R\|^2 \sum_{e} \|Te\|^2 = \|R\|^2 \|T\|_2^2, \quad \text{and}$$
$$\|TS\|_2^2 = \|(S^*T^*)^*\|_2^2 = \|S^*T^*\|_2^2 \leq \|S^*\|^2 \|T^*\|_2^2 = \|S\|^2 \|T\|_2^2. \qquad \square$$

12.4.2 Theorem. *The inclusions $\mathscr{B}_{00}(\mathscr{H},\mathscr{K}) \subseteq \mathscr{B}_2(\mathscr{H},\mathscr{K}) \subseteq \mathscr{B}_0(\mathscr{H},\mathscr{K})$ hold. Moreover, under the Hilbert-Schmidt norm, $\mathscr{B}_2(\mathscr{H},\mathscr{K})$ is a Banach space and $\mathscr{B}_{00}(\mathscr{H},\mathscr{K})$ is dense in $\mathscr{B}_2(\mathscr{H},\mathscr{K})$.*

Proof. To show that $\mathscr{B}_2(\mathscr{H},\mathscr{K})$ is complete, let (T_n) be a Cauchy sequence in $\mathscr{B}_2(\mathscr{H},\mathscr{K})$ with respect to $\|\cdot\|_2$. Then (T_n) is Cauchy with respect to the operator norm, hence there exists $T \in \mathscr{B}(\mathscr{H},\mathscr{K})$ such that $\|T_n - T\| \to 0$. Given $\varepsilon > 0$, choose N so that $\|T_m - T_n\|_2 < \varepsilon$ for all $m, n \geq N$. For such n and any finite $E \subseteq \mathscr{E}$,

$$\sum_{e \in E} \|(T - T_n)e\|^2 = \lim_m \sum_{e \in E} \|(T_m - T_n)e\|^2 \leq \overline{\lim_m} \|T_m - T_n\|_2^2 \leq \varepsilon.$$

Since E was arbitrary, $\|T - T_n\|_2^2 \leq \varepsilon$. Therefore, $T = T - T_n + T_n \in \mathscr{B}_2(\mathscr{H},\mathscr{K})$ and $T_n \to T$ in the Hilbert-Schmidt norm, proving that $\mathscr{B}_2(\mathscr{H},\mathscr{K})$ is a Banach space.

Now let $T \in \mathscr{B}_{00}(\mathscr{H},\mathscr{K})$ and choose an orthonormal basis $\{f_1, \ldots, f_n\}$ in $\operatorname{ran} T$. Then by Parseval's identity,

$$\sum_{e \in \mathscr{E}} \|Te\|^2 = \sum_{e \in \mathscr{E}} \sum_{j=1}^{n} |(Te \,|\, f_j)|^2 = \sum_{j=1}^{n} \sum_{e \in \mathscr{E}} |(e \,|\, T^*f_j)|^2 = \sum_{j=1}^{n} \|T^*f_j\|^2 < \infty.$$

Therefore, $\mathscr{B}_{00}(\mathscr{H},\mathscr{K}) \subseteq \mathscr{B}_2(\mathscr{H},\mathscr{K})$.

If $T \in \mathscr{B}_2(\mathscr{H},\mathscr{K})$, then there exists a sequence $(e_n) \subseteq \mathscr{E}$ such that $Te = 0$ for all $e \in \mathscr{E} \setminus (e_n)$. If $T_n \in \mathscr{B}_{00}(\mathscr{H},\mathscr{K})$ is defined so that

$$T_n e_j = Te_j,\ 1 \leq j \leq n, \text{ and } T_n e = 0 \text{ for } e \in \mathscr{E} \setminus \{e_1, \ldots, e_n\},$$

then $\lim_n \|T - T_n\|_2^2 = \lim_n \sum_{j>n} \|Te_j\|^2 = 0$, hence also $\lim_n \|T - T_n\| = 0$ and so $T \in \mathscr{B}_0(\mathscr{H},\mathscr{K})$. $\square$

The Hilbert-Schmidt Inner Product

Let $S, T \in \mathscr{B}_2(\mathscr{H},\mathscr{K})$. The **Hilbert-Schmidt inner product** of S and T is defined by

$$(S \,|\, T) := \sum_{e \in \mathscr{E}} (Se \,|\, Te) = \sum_{e \in \mathscr{E}} (T^*Se \,|\, e). \tag{12.13}$$

12.4.3 Proposition. *$(S \mid T)$ is independent of the orthonormal basis and is a well-defined inner product on $\mathscr{B}_2(\mathscr{H},\mathscr{K})$ with associated norm $\|\cdot\|_2$. Moreover,*

$$(x \otimes y \,|\, u \otimes v) = \overline{(x \,|\, u)}\,(y \,|\, v), \quad x, u \in \mathscr{H}, \quad y, v \in \mathscr{K}. \tag{12.14}$$

Proof. From the polarization identity $4(S\boldsymbol{e} \mid T\boldsymbol{e}) = \sum_{k=0}^{3} i^k \left\|S\boldsymbol{e} + i^k T\boldsymbol{e}\right\|^2$, we have

$$4(S \mid T) = \sum_{k=0}^{3} i^k \sum_{\boldsymbol{e}\in\mathscr{E}} \big((S + i^k T)\boldsymbol{e} \mid (S + i^k T)\boldsymbol{e}\big) = \sum_{k=0}^{3} i^k \|S + i^k T\|_2^2,$$

which shows that the series in (12.13) converges absolutely and that the definition of $(S \mid T)$ is independent of the basis. The proof that $(S \mid T)$ is an inner product is straightforward.

For the verification of (12.14), note that the left side is

$$\sum_{\boldsymbol{e}\in\mathscr{E}} \big((\boldsymbol{x} \otimes \boldsymbol{y})\boldsymbol{e} \mid (\boldsymbol{u} \otimes \boldsymbol{v})\boldsymbol{e}\big) = \sum_{\boldsymbol{e}\in\mathscr{E}} \big((\boldsymbol{e} \mid \boldsymbol{x})\,\boldsymbol{y} \mid (\boldsymbol{e} \mid \boldsymbol{u})\,\boldsymbol{v}\big) = (\boldsymbol{y} \mid \boldsymbol{v})\,(\boldsymbol{u} \mid \boldsymbol{x}). \qquad \square$$

12.4.4 Proposition. *The set* $\mathscr{G} := \{\boldsymbol{e} \otimes \boldsymbol{f} : \boldsymbol{e} \in \mathscr{E},\ \boldsymbol{f} \in \mathscr{F}\}$ *is an orthonormal basis for* $\mathscr{B}_2(\mathscr{H}, \mathscr{K})$.

Proof. By (12.14), $\mathscr{G}$ is orthonormal. Let $T \in \mathscr{G}^\perp$ so that $(T \mid \boldsymbol{e} \otimes \boldsymbol{f}) = 0$ for all $\boldsymbol{e}$ and $\boldsymbol{f}$. By sesquilinearity and joint continuity of the mapping $(\boldsymbol{y}, \boldsymbol{x}) \mapsto \boldsymbol{x} \otimes \boldsymbol{y}$,

$$\boldsymbol{x} \otimes \boldsymbol{y} = \sum_{\boldsymbol{e},\boldsymbol{f}} (\boldsymbol{x} \mid \boldsymbol{e})\,(\boldsymbol{e} \mid \boldsymbol{y})\,\boldsymbol{e} \otimes \boldsymbol{f},$$

hence $(T \mid \boldsymbol{x} \otimes \boldsymbol{y}) = 0$ for all $\boldsymbol{x}$ and $\boldsymbol{y}$. Since $\mathscr{B}_{00}(\mathscr{H}, \mathscr{K})$ is dense in $\mathscr{B}_2(\mathscr{H}, \mathscr{K})$, $(T \mid T) = 0$, hence $T = 0$. Therefore, $\mathscr{G}$ is a basis. $\square$

12.4.5 Example. Let $(X, \mathcal{F}, \mu)$ and $(Y, \mathcal{G}, \nu)$ be measure spaces such that $L^2(\mu)$ and $L^2(\nu)$ are separable with orthonormal bases (ϕ_n) and (ψ_n), respectively. We show that $\mathscr{B}_2\big(L^2(\nu), L^2(\mu)\big)$ and $L^2(\mu \otimes \nu)$ are isomorphic as Hilbert spaces under a mapping U such that $U(f \otimes g) = f \odot g$, where $(f \odot g)(x, y) = \overline{f(x)}g(y)$.

The calculation

$$\begin{aligned}
(\phi_m \odot \psi_n \mid \phi_j \odot \psi_k) &= \int_{X\times Y} (\phi_m \odot \psi_n)\,\overline{(\phi_j \odot \psi_k)} = \int_{X\times Y} \overline{\phi_m(x)}\psi_n(y)\phi_j(x)\overline{\psi_k(y)}\, d\mu(x)\, d\nu(y) \\
&= \int_X \overline{\phi_m}\phi_j \cdot \int_Y \psi_n \overline{\psi_k} = \overline{(\phi_m \mid \phi_j)}\,(\psi_n \mid \psi_k) \\
&= (\phi_m \otimes \psi_n \mid \phi_j \otimes \psi_k)
\end{aligned}$$

shows that the image $(\phi_m \odot \psi_n)_{m,n}$ under U of the orthonormal basis $(\phi_m \otimes \psi_n)_{m,n}$ is an orthonormal set. It remains then to show that the set is complete.

Let $f \in L^2(\mu \otimes \nu)$ such that for all m, n,

$$\begin{aligned}
0 = (\phi_m \odot \psi_n \mid f) &= \iint \overline{\phi_m(x)}\psi_n(y)\,\overline{f(x, y)}\, d\mu(x)\, d\nu(y) \\
&= \int \psi_n(y) \left[\int \phi_m(x)\, f(x, y)\, d\mu(x)\right]^{-} d\nu(y). \qquad (\dagger)
\end{aligned}$$

By Fubini's theorem,

$$\iint |f(x, y)|^2\, d\mu(x)\, d\nu(y) = \|f\|_2^2 < \infty,$$

hence $f(\cdot, y) \in L^2(\mu)$ for a.a. y. For such y, by the CBS inequality

$$\left|\int \phi_m(x) f(x, y)\, d\mu(x)\right| \le \int |\phi_m(x)|\, |f(x, y)|\, d\mu(x) \le \|\phi_m\|_2\, \|f(\cdot, y)\|_2 < \infty.$$

Thus the inner integral in (†) is an L^2 function of y and so must be zero, by the completeness of $(\psi_n)_n$. Using the completeness of $(\phi_m)_m$, we conclude that $f = 0$ a.e. Therefore, $(\phi_m \odot \psi_n)_{m,n}$ is complete. ◊

The Hilbert-Schmidt Operator $A \otimes B$

The following construction is closely related to the tensor product of operators, a notion that occurs frequently in physics, notably in quantum mechanics and electrodynamics. We shall use the construction developed here in a crucial part of the proof of the Peter-Weyl Theorem (16.5.18).

12.4.6 Theorem. *Given $A \in \mathscr{B}(\mathscr{H})$ and $B \in \mathscr{B}(\mathscr{K})$, there exists a unique bounded linear operator $A \otimes B$ on the Hilbert space $\mathscr{B}_2(\mathscr{H}, \mathscr{K})$ such that*

$$(A \otimes B)(\boldsymbol{x} \otimes \boldsymbol{y}) = (A\boldsymbol{x}) \otimes (B\boldsymbol{y}), \quad \boldsymbol{x} \in \mathscr{H}, \ \boldsymbol{y} \in \mathscr{K}. \tag{12.15}$$

Proof. The mapping $A \otimes B$ is defined by

$$A \otimes B : T \to BTA^* : \mathscr{B}_2(\mathscr{H}, \mathscr{K}) \to \mathscr{B}_2(\mathscr{H}, \mathscr{K}).$$

Clearly $A \otimes B$ is linear in T and since $\|BTA^*\|_2 \le \|B\| \, \|T\|_2 \, \|A^*\|$ (12.4.1), we see that $A \otimes B$ is bounded with $\|A \otimes B\|_2 \le \|A\| \, \|B\|$. By (12.5),

$$(A \otimes B)(\boldsymbol{x} \otimes \boldsymbol{y}) = B(\boldsymbol{x} \otimes \boldsymbol{y})A^* = (A\boldsymbol{x}) \otimes (B\boldsymbol{y}), \quad \boldsymbol{x} \in \mathscr{H}, \ \boldsymbol{y} \in \mathscr{K}.$$

For uniqueness, simply note that a pair of bounded linear operators on $\mathscr{B}_2(\mathscr{H}, \mathscr{K})$ that agree on the set $\{\boldsymbol{x} \otimes \boldsymbol{y} : \boldsymbol{x} \in \mathscr{H}, \ \boldsymbol{y} \in \mathscr{K}\}$ must in fact be equal, since the span of this set is dense in $\mathscr{B}_2(\mathscr{H}, \mathscr{K})$ (12.4.2). □

12.4.7 Proposition. *The following properties hold:*

- **(a)** *$(A, B) \to A \otimes B$ is sesquilinear.*
- **(b)** *$(A \otimes B)(C \otimes D) = (AC) \otimes (BD)$.*
- **(c)** *$(A \otimes B)^* = A^* \otimes B^*$.*
- **(d)** *$\|A \otimes B\|_2 = \|A\| \, \|B\|$.*
- **(e)** *$A \otimes B$ is invertible iff both A and B are invertible, and then $(A \otimes B)^{-1} = A^{-1} \otimes B^{-1}$.*
- **(f)** *If A and B are unitary, then $A \otimes B$ is unitary.*

Proof. Parts (a)–(c) follow from uniqueness and the properties of rank one operators. For example,

$$\begin{aligned}
[(A_1 + A_2) \otimes B](\boldsymbol{x} \otimes \boldsymbol{y}) &= \big[(A_1 + A_2)\boldsymbol{x}\big] \otimes (B\boldsymbol{y}) = (A_1\boldsymbol{x}) \otimes (B\boldsymbol{y}) + (A_2\boldsymbol{x}) \otimes (B\boldsymbol{y}) \\
&= (A_1 \otimes B)(\boldsymbol{x} \otimes \boldsymbol{y}) + (A_2 \otimes B)(\boldsymbol{x} \otimes \boldsymbol{y}), \\
(A \otimes B)(C \otimes D)(\boldsymbol{x} \otimes \boldsymbol{y}) &= (A \otimes B)(C\boldsymbol{x} \otimes D\boldsymbol{y}) = (AC\boldsymbol{x}) \otimes (BD\boldsymbol{y}) \\
&= [(AC) \otimes (BD)](\boldsymbol{x} \otimes \boldsymbol{y}) \quad \text{and} \\
\big((A \otimes B)^*(\boldsymbol{x} \otimes \boldsymbol{y}) \mid \boldsymbol{u} \otimes \boldsymbol{v}\big) &= \big(\boldsymbol{x} \otimes \boldsymbol{y} \mid (A\boldsymbol{u}) \otimes (B\boldsymbol{v})\big) = \overline{(\boldsymbol{x} \mid A\boldsymbol{u})} \, (\boldsymbol{y} \mid B\boldsymbol{v}) \\
&= \overline{(A^*\boldsymbol{x} \mid \boldsymbol{u})} \, (B^*\boldsymbol{y} \mid \boldsymbol{v}) \\
&= \big((A^*\boldsymbol{x}) \otimes (B^*\boldsymbol{y}) \mid \boldsymbol{u} \otimes \boldsymbol{v}\big).
\end{aligned}$$

For (d), we have already shown that $\|A \otimes B\|_2 \le \|A\| \, \|B\|$. For the reverse inequality, let $\|\boldsymbol{x}\| = \|\boldsymbol{y}\| = 1$. Then $\|\boldsymbol{x} \otimes \boldsymbol{y}\| = \|\boldsymbol{x}\| \, \|\boldsymbol{y}\| = 1$, hence

$$\|A \otimes B\|_2 \ge \|A \otimes B\| \ge \|(A \otimes B)(\boldsymbol{x} \otimes \boldsymbol{y})\| = \|A\boldsymbol{x} \otimes B\boldsymbol{y}\| = \|A\boldsymbol{x}\| \, \|B\boldsymbol{y}\|.$$

Taking the supremum over all such $\boldsymbol{x}$ and $\boldsymbol{y}$ yields $\|A \otimes B\|_2 \geq \|A\|\,\|B\|$.

For (e), if A and B are invertible, then $(A \otimes B)(A^{-1} \otimes B^{-1}) = AA^{-1} \otimes BB^{-1} = I \otimes I$, which is the identity operator in $\mathscr{B}_2(\mathscr{H}, \mathscr{K})$. Conversely, suppose that $A \otimes B$ is invertible. Then

$$I \otimes I = (A \otimes B)^{-1}(A \otimes B) = (A \otimes B)^{-1}(I \otimes B)(A \otimes I)$$

and

$$I \otimes I = (A \otimes B)(A \otimes B)^{-1} = (A \otimes I)(I \otimes B)(A \otimes B)^{-1},$$

hence $A \otimes I$ is invertible. Thus there exists $c > 0$ such that

$$\|A\boldsymbol{x}\|\,\|\boldsymbol{y}\| = \|(A \otimes I)(\boldsymbol{x} \otimes \boldsymbol{y})\| \geq c\,\|\boldsymbol{x} \otimes \boldsymbol{y}\| = \|\boldsymbol{x}\|\,\|\boldsymbol{y}\|$$

Taking $\boldsymbol{y} \neq \mathbf{0}$ we see that $\|A\boldsymbol{x}\| \geq c\,\|\boldsymbol{x}\|$ for all $\boldsymbol{x}$, which implies that A is injective. Since $(A \otimes B)^*$ is invertible and $(A \otimes B)^* = A^* \otimes B^*$, the same argument applied to $A^* \otimes B^*$ shows that A^* is injective. Therefore, A is surjective and so is invertible. Similarly, B is invertible.

Finally, if A and B are unitary, then $A^*A = I$ and $B^*B = I$, hence $(A^* \otimes B^*)(A \otimes B) = A^*A \otimes B^*B = I \otimes I$ and so $A \otimes B$ is unitary, proving (f). □

Note that the converse of (f) is false. (Take $A = (1/2)I$ and $B = 2I$.)

Hilbert-Schmidt Integral Operators

The prototype of the Hilbert-Schmidt operator is the Hilbert-Schmidt integral operator, which has applications in integral and differential equations. (See, for example, [9].)

Let $(X, \mathfrak{F}, \mu)$ be a separable, σ-finite measure space and let $k \in L^2(\mu \otimes \mu)$. The **Hilbert-Schmidt integral operator K with kernel k** is defined by

$$Kf(x) = \int k(x,y)f(y)\,d\mu(y), \quad f \in L^2(\mu). \tag{12.16}$$

By Ex. 12.44, K is bounded with $\|K\| \leq \|k\|_2$. We show in this subsection that K is a Hilbert-Schmidt operator.

First, we show that K is compact. Let $(\phi_n)_n$ be an orthonormal basis for $L^2(\mu)$ and define $\phi_n \odot \phi_m$ on $X \times X$ by

$$(\phi_n \odot \phi_m)(x,y) = \phi_n(x)\overline{\phi_m(y)}.$$

This is a slight variation of the definition given in (12.4.5), but still gives an orthonormal basis for $L^2(\mu \otimes \mu)$. Thus we have the Fourier expansion

$$k = \sum_{m,n=1}^{\infty} (k \,|\, \phi_n \odot \phi_m)\, \phi_n \odot \phi_m.$$

Moreover,

$$(K\phi_m \,|\, \phi_n) = \iint k(x,y)\phi_m(y)\overline{\phi_n(x)}\,d\mu(y)\,d\mu(x) = (k \,|\, \phi_n \odot \phi_m),$$

from which it follows that

$$\sum_{m,n=1}^{\infty} |(K\phi_m \,|\, \phi_n)|^2 < \infty. \tag{†}$$

Now let P_n denote the orthogonal projection of $L^2(\mu)$ onto the span of $\{\phi_1, \ldots, \phi_n\}$ and set $K_n = KP_n + P_nK - P_nKP_n$. Then K_n has finite rank, hence to show K is compact it suffices to show that $K_n \to K$ in operator norm. For $f \in L^2(\mu)$ and $c_k := (f \mid \phi_k)$ we have

$$\begin{aligned} f &= \sum_{k=1}^{\infty} c_k\phi_k, \quad P_nf = \sum_{k=1}^{n} c_k\phi_k, \\ Kf &= \sum_{k=1}^{\infty} c_kK\phi_k, \quad KP_nf = \sum_{k=1}^{n} c_kK\phi_k, \\ P_nKf &= \sum_{k=1}^{n} (Kf \mid \phi_k)\,\phi_k = \sum_{k=1}^{n}\sum_{i=1}^{\infty} c_i\,(K\phi_i \mid \phi_k)\,\phi_k, \quad \text{and} \\ P_nKP_nf &= \sum_{k=1}^{n} (KP_nf \mid \phi_k)\,\phi_k = \sum_{k=1}^{n}\sum_{i=1}^{n} c_i\,(K\phi_i \mid \phi_k)\,\phi_k. \end{aligned}$$

Thus

$$Kf - K_nf = (K - KP_n)f + (P_nKP_n - P_nK)f = \sum_{i>n} c_iK\phi_i - \sum_{k=1}^{n}\sum_{i>n} c_i\,(K\phi_i \mid \phi_k)\,\phi_k,$$

so for each j

$$(Kf - K_nf \mid \phi_j) = \sum_{i>n} c_i\,(K\phi_i \mid \phi_j) - \sum_{k=1}^{n}\sum_{i>n} c_i\,(K\phi_i \mid \phi_k)\,(\phi_k \mid \phi_j)\,.$$

The right side is zero if $j \le n$ and equals $\sum_{i>n} c_i\,(K\phi_i \mid \phi_j)$ otherwise. By the CBS inequality in $\ell^2(\mathbb{N})$ and, by Bessel's inequality, for $j > n$ we have

$$|\,(Kf - K_nf \mid \phi_j)\,|^2 \le \sum_{i>n} |c_i|^2 \sum_{i>n} |\,(K\phi_i \mid \phi_j)\,|^2 \le \|f\|_2^2 \sum_{i>n} |\,(K\phi_i \mid \phi_j)\,|^2,$$

hence

$$\|Kf - K_nf\|_2^2 = \sum_{j>n} |\,(Kf - K_nf \mid \phi_j)\,|^2 \le \|f\|_2^2 \sum_{j>n}\sum_{i>n} |\,(K\phi_i \mid \phi_j)\,|^2.$$

By (†), the term on the right tends to 0 as $n \to 0$, hence $\|K_n - K\| \to 0$. This shows that K is compact.

To show that K is a Hilbert-Schmidt operator we use 12.3.10, which guarantees the existence of orthonormal sequences (ϑ_n) and (ψ_n) in $L^2(X)$ and $\lambda_n \downarrow 0$ such that

$$Kf = \sum_n \lambda_n\,(f \mid \vartheta_n)\,\psi_n, \quad f \in L^2(X).$$

Now, for fixed x $\lambda_n\overline{\psi_n(x)} = \overline{K\vartheta_n(x)} = \int \overline{k(x,y)\vartheta_n(y)}\,d\mu(y)$, the integral being a Fourier coefficient of the function $\overline{k(x,\cdot)}$ with respect to the basis (ϑ_n). By Bessel's inequality,

$$\sum_n |K\vartheta_n(x)|^2 = \sum_n |\lambda_n\psi_n(x)|^2 \le \|k(x,\cdot)\|_2^2 = \int |k(x,y)|^2\,d\mu(y).$$

Since this holds for a.a. x, integrating we obtain

$$\sum_n \|K\vartheta_n\|_2^2 \le \iint |k(x,y)|^2\,d\mu(y)\,d\mu(x) = \|k\|_2^2 < \infty.$$

Therefore, K is a Hilbert-Schmidt operator.

Exercises

12.44 Let K be as in (12.16).

(a) Show that K is bounded with $\|K\| \le \|k\|_2$.

(b) Compute the adjoint of K. When is K self-adjoint?

(c) Let L be the Hilbert-Schmidt operator with kernel ℓ. Find the kernel of LK. Give a condition on the kernels of K and L that implies $LK = KL$.

(d) Use (b) and (c) to give a sufficient condition on k for K to be normal.

(e) Show that the **Volterra operator** $(Kf)(t) = \int_0^t f(s)\,ds$ $(t \in [0,1])$ is a Hilbert-Schmidt integral operator on $L^2[0,1]$.

12.45 Show that the Hilbert-Schmidt inner product satisfies $(T^* \mid S^*) = (S \mid T)$.

12.46 Show that $T \in \mathscr{B}_2(\mathscr{H})$ iff $|T| \in \mathscr{B}_2(\mathscr{H})$, in which case $\| |T| \|_2 = \| T \|_2$.

12.47 Let T be normal and $\mathscr{E}$ an orthonormal basis such that $T = \sum_{e\in\mathscr{E}} \alpha_e (e \otimes e)$. Show that $\|T\|_2 = \sum_{e\in\mathscr{E}} |\alpha_e|^2$.

---

*12.5 Trace Class Operators

The Trace Norm

Let $\mathscr{H}$ be a complex Hilbert space with orthonormal basis $\mathscr{E}$. The **trace norm** of an operator $T \in \mathscr{B}(\mathscr{H})$ is defined by

$$\|T\|_1 := \sum_{e\in\mathscr{E}} (|T|e \mid e). \tag{12.17}$$

If $\|T\|_1 < \infty$, then T is said to be of **trace class**. The set of all trace class operators is denoted by $\mathscr{B}_1(\mathscr{H})$. The calculation

$$\|T\|_1 = \sum_{e\in\mathscr{E}} (|T|^{1/2}e \mid |T|^{1/2}e) = \sum_{e\in\mathscr{E}} \| |T|^{1/2}e \|^2 = \| |T|^{1/2} \|_2^2$$

shows that $\|T\|_1$ is independent of the choice of orthonormal basis and that $T \in \mathscr{B}_1(\mathscr{H})$ iff $|T|^{1/2} \in \mathscr{B}_2(\mathscr{H})$.

We show below that $\mathscr{B}_1(\mathscr{H})$ is a linear space and that $\|\cdot\|_1$ is indeed a norm on $\mathscr{B}_1(\mathscr{H})$. First, we establish some preliminary results.

12.5.1 Proposition. *$T \in \mathscr{B}_1(\mathscr{H})$ iff any one (hence both) of the following conditions holds:*

(a) *$T = AB$ for some $A, B \in \mathscr{B}_2(\mathscr{H})$.*

(b) *$|T| = AB$ for some $A, B \in \mathscr{B}_2(\mathscr{H})$.*

Proof. Let $T \in \mathscr{B}_1(\mathscr{H})$. Using the polar decomposition $U|T|$ of T we have $T = AB$ with $A = U|T|^{1/2}$ and $B = |T|^{1/2}$. Since $|T|^{1/2} \in \mathscr{B}_2(\mathscr{H})$ and $\mathscr{B}_2(\mathscr{H})$ is an ideal, (a) holds. Since $|T| = |T|^{1/2}|T|^{1/2}$, (b) also holds.

Conversely, if (b) holds, then

$$\|T\|_1 = \sum_{e\in\mathscr{E}} (Be \mid A^*e) \le \sum_{e\in\mathscr{E}} \|Be\|\,\|A^*e\| \le \Big(\sum_{e\in\mathscr{E}} \|Be\|^2\Big)^{1/2} \Big(\sum_{e\in\mathscr{E}} \|A^*e\|^2\Big)^{1/2}$$
$$= \|B\|_2\,\|A^*\|_2,$$

hence $T \in \mathscr{B}_1(\mathscr{H})$.

Finally, if (a) holds, then using the polar decomposition of T again we have $|T| = U^*T = (U^*A)B$ (12.27), which gives (b). □

12.5.2 Corollary. $\mathscr{B}_{00}(\mathscr{H}) \subseteq \mathscr{B}_1(\mathscr{H}) \subseteq \mathscr{B}_2(\mathscr{H}) \subseteq \mathscr{B}_0(\mathscr{H})$.

Proof. The second inclusion follows from the proposition and the fact that $\mathscr{B}_2(\mathscr{H})$ is an algebra. For the first inclusion, let $T \in \mathscr{B}_{00}(\mathscr{H})$ and let $T = U|T|$ be the polar decomposition of T. From $U^*T = |T|$ we see that $\operatorname{ran}|T|$ is finite dimensional. Thus we may choose an orthonormal basis $\mathscr{E}$ for $\mathscr{H}$ so that some finite subset F is an orthonormal basis for $\operatorname{ran}|T|$. Since $e \perp \operatorname{ran}|T|$ for $e \in \mathscr{E} \setminus F$, the sum in (12.17) is finite and so $T \in \mathscr{B}_1(\mathscr{H})$. □

12.5.3 Theorem. *$\mathscr{B}_1(\mathscr{H})$ is a self-adjoint ideal of $\mathscr{B}(\mathscr{H})$ and the trace norm is a norm.*

Proof. Absolute homogeneity of $\|\cdot\|_1$ follows from Ex. 12.10. For the triangle inequality, let $S, T \in \mathscr{B}_1(\mathscr{H})$ and let $S = U|S|$, $T = V|T|$, and $S+T = W|S+T|$ be the polar decompositions. Then

$$|S| = U^*S,\ |T| = V^*T \text{ and } |S+T| = W^*(S+T) = W^*U|S| + W^*V|T|,$$

hence for any e in $\mathscr{E}$,

$$(|S+T|e \mid e) = ((W^*U|S|)e \mid e) + ((W^*V|T|) \mid e)\,e = (|S|e \mid U^*We) + (|T|e \mid V^*We)$$
$$= \big(|S|^{1/2}e \mid |S|^{1/2}U^*We\big) + \big(|T|^{1/2}e \mid |T|^{1/2}V^*We\big).$$

Thus by the CBS inequality,

$$(|S+T|e \mid e) \le \big\||S|^{1/2}e\big\|\,\big\||S|^{1/2}U^*We\big\| + \big\||T|^{1/2}e\big\|\,\big\||T|^{1/2}V^*We\big\|.$$

Summing over a finite subset F of $\mathscr{E}$ we then have

$$\sum_{e\in F} (|S+T|e \mid e) \le \sum_{e\in F} \big\||S|^{1/2}e\big\|\,\big\||S|^{1/2}U^*We\big\| + \sum_{e\in F} \big\||T|^{1/2}e\big\|\,\big\||T|^{1/2}V^*We\big\|. \quad (\dagger)$$

Applying the CBS inequality in ℓ^2 to the first sum in (†), we have

$$\sum_{e\in F} \big\||S|^{1/2}e\big\|\,\big\||S|^{1/2}U^*We\big\| \le \Big(\sum_{e\in F} \big\||S|^{1/2}e\big\|^2\Big)^{1/2} \Big(\sum_{e\in F} \big\||S|^{1/2}U^*We\big\|^2\Big)^{1/2}$$
$$\le \Big(\sum_{e\in\mathscr{E}} \big\||S|^{1/2}e\big\|^2\Big)^{1/2} \Big(\sum_{e\in\mathscr{E}} \big\||S|^{1/2}U^*We\big\|^2\Big)^{1/2}$$
$$= \big\||S|^{1/2}\big\|_2\,\big\||S|^{1/2}U^*W\big\|_2.$$

By 12.4.1 and the fact that U and W are partial isometries,

$$\big\||S|^{1/2}U^*W\big\|_2 \le \big\||S|^{1/2}\big\|_2 \|U^*W\| \le \big\||S|^{1/2}\big\|_2.$$

Similarly,

$$\left\| |T|^{1/2} V^* W \right\|_2 \le \left\| |T|^{1/2} \right\|_2 \|V^* W\| \le \left\| |T|^{1/2} \right\|_2.$$

Since F was arbitrary, we obtain from (†) the triangle inequality

$$\|S + T\|_1 \le \left\| |S|^{1/2} \right\|_2 + \left\| |T|^{1/2} \right\|_2 = \|S\|_1 + \|T\|_1 .$$

In particular, $\mathscr{B}_1(\mathscr{H})$ is a linear space.

Now let $\|T\|_1 = 0$. Since $|T|$ is compact and normal it is diagonalizable, hence there exists an orthonormal basis $\mathscr{F}$ such that

$$|T|\boldsymbol{x} = \sum_{\boldsymbol{f}} \alpha_{\boldsymbol{f}} \, (\boldsymbol{x} \,|\, \boldsymbol{f}) \, \boldsymbol{f} \quad \text{for all } \boldsymbol{x} \in \mathscr{H},$$

where the $\alpha_{\boldsymbol{f}}$ are the eigenvalues of $|T|$ with corresponding eigenvectors $\boldsymbol{f}$. Since $\sum_{\boldsymbol{f} \in \mathscr{F}} (|T|\boldsymbol{f} \,|\, \boldsymbol{f}) = \|T\|_1 = 0$ and the terms $\alpha_{\boldsymbol{f}} = (|T|\boldsymbol{f} \,|\, \boldsymbol{f})$ are nonnegative, $\alpha_{\boldsymbol{f}} = 0$ for all $\boldsymbol{f}$. Therefore, $|T| = 0$ and so $T = 0$.

To show that $\mathscr{B}_1(\mathscr{H})$ is an ideal in $\mathscr{B}(\mathscr{H})$, let $T \in \mathscr{B}_1(\mathscr{H})$ and $S \in \mathscr{B}(\mathscr{H})$. By 12.5.1, $T = AB$ for some $A, B \in \mathscr{B}_2(\mathscr{H})$, hence $TS = A(BS)$. Thus T is a product of members of $\mathscr{B}_2(\mathscr{H})$, hence $\mathscr{B}_1(\mathscr{H})\mathscr{B}(\mathscr{H}) \subseteq \mathscr{B}_1(\mathscr{H})$. Similarly $\mathscr{B}(\mathscr{H})\mathscr{B}_1(\mathscr{H}) \subseteq \mathscr{B}_1(\mathscr{H})$. Therefore, $\mathscr{B}_1(\mathscr{H})$ is an ideal of $\mathscr{B}(\mathscr{H})$. Since $T^* = B^*A^*$ and $A^*, B^* \in \mathscr{B}_2(\mathscr{H})$, $T^* \in \mathscr{B}_1(\mathscr{H})$. Therefore, $\mathscr{B}_1(\mathscr{H})$ is self-adjoint. □

The Trace

The **trace** $\operatorname{tr} T$ of $T \in \mathscr{B}_1(\mathscr{H})$ is defined in terms of the orthonormal basis $\mathscr{E}$ by

$$\operatorname{tr} T := \sum_{\boldsymbol{e} \in \mathscr{E}} (T\boldsymbol{e} \,|\, \boldsymbol{e}). \tag{12.18}$$

The following proposition shows that $\operatorname{tr} T$ is well-defined and independent of the basis.

12.5.4 Proposition. *For $T \in \mathscr{B}_1(\mathscr{H})$, the sum $\sum_{\boldsymbol{e} \in \mathscr{E}} (T\boldsymbol{e} \,|\, \boldsymbol{e})$ converges absolutely. Moreover,*

$$\operatorname{tr}(B^*A) = (A \,|\, B) \quad A, B \in \mathscr{B}_2(\mathscr{H}), \tag{12.19}$$

where the right side is the Hilbert-Schmidt inner product of A and B.

Proof. By 12.5.1, $T = B^*A$, where $A, B \in \mathscr{B}_2(\mathscr{H})$. Then

$$|(T\boldsymbol{e} \,|\, \boldsymbol{e})| = |(A\boldsymbol{e} \,|\, B\boldsymbol{e})| \le \|A\boldsymbol{e}\| \, \|B\boldsymbol{e}\| \le \tfrac{1}{2} \|A\boldsymbol{e}\|^2 + \tfrac{1}{2} \|B\boldsymbol{e}\|^2 .$$

Summing over $\boldsymbol{e} \in \mathscr{E}$ we have

$$\sum_{\boldsymbol{e} \in \mathscr{E}} |(T\boldsymbol{e} \,|\, \boldsymbol{e})| \le \tfrac{1}{2} \sum_{\boldsymbol{e} \in \mathscr{E}} \|A\boldsymbol{e}\|^2 + \tfrac{1}{2} \sum_{\boldsymbol{e} \in \mathscr{E}} \|B\boldsymbol{e}\|^2 = \tfrac{1}{2} \|A\|_2^2 + \tfrac{1}{2} \|B\|_2^2 < \infty.$$

This proves the first assertion of the proposition. The second assertion follows directly from the definition of the trace and the Hilbert-Schmidt inner product. □

Here are additional noteworthy properties of the trace and the trace norm.

12.5.5 Theorem. *Let $T \in \mathscr{B}_1(\mathscr{H})$, $S \in \mathscr{B}(\mathscr{H})$. Then*

(a) $\operatorname{tr}(\cdot)$ *is a linear functional on $\mathscr{B}_1(\mathscr{H})$ and is positive, that is, $T \ge 0 \Rightarrow \operatorname{tr} T \ge 0$.*

(b) $\operatorname{tr} T^* = \overline{\operatorname{tr}} T \quad (=: \overline{\operatorname{tr} T})$.

(c) $\operatorname{tr}(ST) = \operatorname{tr}(TS)$.

(d) $|\operatorname{tr}(ST)| \le \|S\|\,\|T\|_1$. *In particular,* $\operatorname{tr}(\cdot)$ *is* $\|\cdot\|_1$*-continuous.*

(e) $\|T^*\|_1 = \|T\|_1$.

(f) $\|ST\|_1 \le \|S\|\,\|T\|_1$ *and* $\|TS\|_1 \le \|S\|\,\|T\|_1$.

(g) $\|T\| \le \|T\|_1$, *hence if* $S \in \mathscr{B}_1(\mathscr{H})$, *then* $\|ST\|_1 \le \|S\|_1\,\|T\|_1$.

(h) *If T is normal, so that* $T = \sum_{e\in\mathscr{E}} \alpha_e(\boldsymbol{e}\otimes\boldsymbol{e})$ $(T\boldsymbol{e} = \alpha_e\boldsymbol{e})$, *then* $\|T\|_1 = \sum_{e\in\mathscr{E}} |\alpha_e|$.

Proof. Part (a) is clear. By 12.5.3, the left side of (b) is defined. A simple calculation shows that the equality holds.

For (c), note first that by Ex. 12.45, (12.19), and (b),

$$\operatorname{tr}(D^*C) = (C\,|\,D) = (D^*\,|\,C^*) = \operatorname{tr}(CD^*) = \operatorname{tr}[(DC^*)^*] = \overline{\operatorname{tr}}(DC^*),\quad C,\,D \in \mathscr{B}_2(\mathscr{H}).$$

Now write $T = B^*A$ for some $A,\,B \in \mathscr{B}_2(\mathscr{H})$ (12.5.1). Then

$$\operatorname{tr}(ST) = \operatorname{tr}\big((SB^*)A\big) = \overline{\operatorname{tr}}\big((BS^*)A^*\big) = \overline{\operatorname{tr}}\big(B(S^*A^*)\big) = \operatorname{tr}\big(B^*(AS)\big) = \operatorname{tr}\big(TS\big).$$

For (d), let $T = U|T|$ be the polar decomposition of T. For any $\boldsymbol{e} \in \mathscr{E}$,

$$|\,(ST\boldsymbol{e}\,|\,\boldsymbol{e})\,| = |\,(|T|\boldsymbol{e}\,|\,U^*S^*\boldsymbol{e})\,| = \big|(|T|^{1/2}\boldsymbol{e}\,\big|\,|T|^{1/2}U^*S^*\boldsymbol{e})\big| \le \|\,|T|^{1/2}\boldsymbol{e}\|\,\|\,|T|^{1/2}U^*S^*\boldsymbol{e}\|,$$

and summing over a finite subset F of $\mathscr{E}$ we have

$$\begin{aligned}\sum_{e\in F} |\,(ST\boldsymbol{e}\,|\,\boldsymbol{e})\,| &\le \sum_{e\in F} \|\,|T|^{1/2}\boldsymbol{e}\|\,\|\,|T|^{1/2}U^*S^*\boldsymbol{e}\| \\ &\le \Big(\sum_{e\in F} \|\,|T|^{1/2}\boldsymbol{e}\|^2\Big)^{1/2}\Big(\sum_{e\in F} \|\,|T|^{1/2}U^*S^*\boldsymbol{e}\|^2\Big)^{1/2}.\end{aligned}$$

Since F was arbitrary, we see that

$$|\operatorname{tr}(ST)| \le \big\||T|^{1/2}\big\|_2\,\big\||T|^{1/2}U^*S^*\big\|_2 \le \big\||T|^{1/2}\big\|_2^2\,\|U^*S^*\|,$$

the last inequality from 12.4.1. Finally, $\big\||T|^{1/2}\big\|_2^2 = \|T\|_1$ and $\|U^*S^*\| \le \|U^*\|\,\|S^*\| \le \|S^*\|$, completing the proof of (d).

For (e), (f), (g), and (h) let $T = U|T|$ be the polar decomposition of T. From (c) and the equalities $U|T|U^* = |T^*|$ and $U^*U|T| = |T|$ (Ex. 12.27) we have

$$\|T^*\|_1 = \operatorname{tr}|T^*| = \operatorname{tr}(U|T|U^*) = \operatorname{tr}(U^*U|T|) = \operatorname{tr}|T| = \|T\|_1,$$

proving (e).

For (f), let $ST = V|ST|$ be the polar decompositions of ST. Then $|ST| = V^*ST = V^*SU|T|$, hence, by (d), $\|ST\|_1 = \operatorname{tr}(V^*SU|T|) \le \|V^*SU\|\,\||T|\|_1 \le \|S\|\,\||T|\|_1$. Using this result and (e) we also have $\|TS\|_1 = \|S^*T^*\|_1 \le \|S^*\|\,\|T^*\|_1 = \|S\|\,\|T\|_1$.

To prove (g), note that since $|T|$ is compact and normal it is diagonalizable:

$$|T|\boldsymbol{x} = \sum_{e\in\mathscr{E}} \alpha_e\,(\boldsymbol{x}\,|\,\boldsymbol{e})\,\boldsymbol{e},\quad |T|\boldsymbol{e} = \alpha_e\boldsymbol{e},\ \alpha_e \ge 0.$$

Then for $\|\boldsymbol{x}\| \le 1$,

$$(|T|\boldsymbol{x}\,|\,\boldsymbol{x}) = \sum_{e\in\mathscr{E}} \alpha_e|\,(\boldsymbol{x}\,|\,\boldsymbol{e})\,|^2 \le \sum_{e\in\mathscr{E}} \alpha_e,$$

and using 12.1.6 we have

$$\|T\| = \|U|T|\| \le \|\,|T|\,\| \le \sum_{e\in\mathscr{E}} \alpha_e = \sum_{e\in\mathscr{E}} (|T|e \mid e) = \|T\|_1 .$$

The verification of (h) is left as an exercise (12.49). □

12.5.6 Theorem. *$\mathscr{B}_1(\mathscr{H})$ is a Banach algebra in the trace norm.*

Proof. By 12.5.5(g), $\mathscr{B}_1(\mathscr{H})$ is a normed algebra. To show completeness, let (T_n) be a Cauchy sequence in $\mathscr{B}_1(\mathscr{H})$ with respect to $\|\cdot\|_1$. Then (T_n) is Cauchy with respect to the operator norm (12.5.5(g)), hence there exists $T \in \mathscr{B}(\mathscr{H})$ such that $\|T_n - T\| \to 0$. Given $\varepsilon > 0$, choose N so that $\|T_m - T_n\|_1 < \varepsilon$ for all $m, n \ge N$. Let $F \subseteq \mathscr{E}$ be finite and let P be the projection of $\mathscr{H}$ onto the span of F. For a fixed $n \ge N$, let $T - T_n = U|T - T_n|$ be the polar decomposition of $T - T_n$. Then

$$\begin{aligned}\sum_{e\in F} (|T - T_n|e \mid e) &= \sum_{e\in F} (U^*(T - T_n)e \mid e) = \lim_m \Big|\sum_{e\in F} (U^*(T_m - T_n)e \mid e)\Big| \\ &= \lim_m \Big|\sum_{e\in\mathscr{E}} (U^*(T_m - T_n)e \mid Pe)\Big| = \lim_m |\mathrm{tr}(PU^*(T_m - T_n))| \\ &\le \overline{\lim_m}\, \|T_m - T_n\|_1 \le \varepsilon,\end{aligned}$$

the inequality by 12.5.5(d). Since F was arbitrary, $\|T - T_n\|_1 = \mathrm{tr}(|T - T_n|) \le \varepsilon$. □

The Dual Spaces $\mathscr{B}_0(\mathscr{H})'$ and $\mathscr{B}_1(\mathscr{H})'$

Define $\Psi_A(T) := \mathrm{tr}(TA)$, where either

$$(1)\ A \in \mathscr{B}_1(\mathscr{H}) \text{ and } T \in \mathscr{B}_0(\mathscr{H}) \quad\text{or}\quad (2)\ A \in \mathscr{B}(\mathscr{H}) \text{ and } T \in \mathscr{B}_1(\mathscr{H}).$$

In (1), $|\Psi_A(T)| \le \|A\|_1 \|T\|$ and in (2) $|\Psi_A(T)| \le \|T\|_1 \|A\|$ (12.5.5). Thus we obtain linear mappings

$$\begin{aligned}&(1')\quad \Psi : A \to \Psi_A : \mathscr{B}_1(\mathscr{H}) \to \mathscr{B}_0(\mathscr{H})', \quad \|\Psi_A\| \le \|A\|_1, \text{ and}\\ &(2')\quad \Psi : A \to \Psi_A : \mathscr{B}(\mathscr{H}) \to \mathscr{B}_1(\mathscr{H})', \quad \|\Psi_A\| \le \|A\|.\end{aligned}$$

The next theorem uses the map Ψ to identify $\mathscr{B}_1(\mathscr{H})$ with $\mathscr{B}_0(\mathscr{H})'$ and $\mathscr{B}(\mathscr{H})$ with $\mathscr{B}_1(\mathscr{H})'$.

12.5.7 Theorem. *The mappings Ψ in $(1')$ and $(2')$ are isometric isomorphisms.*

Proof. We follow the treatment in [36]. For $(1')$ we need to prove that $\|A\|_1 \le \|\Psi_A\|$ and that every member ψ of $\mathscr{B}_0(\mathscr{H})'$ is of the form Ψ_A for some $A \in \mathscr{B}_1(\mathscr{H})$.

Since $\|S\| \le \|S\|_2$ for $S \in \mathscr{B}_2(\mathscr{H}) \subseteq \mathscr{B}_0(\mathscr{H})$, ψ restricted to $\mathscr{B}_2(\mathscr{H})$ is a member of $\mathscr{B}_2(\mathscr{H})'$. By the Riesz representation theorem, there exists a T in the Hilbert space $\mathscr{B}_2(\mathscr{H})$ such that $\psi(\cdot) = (\cdot, T)$. Set $A := T^*$, so that $\psi(S) = \mathrm{tr}(AS)$ for all $S \in \mathscr{B}_2(\mathscr{H})$. It remains to show that $A \in \mathscr{B}_1(\mathscr{H})$. For this let $A = U|A|$ be the polar decomposition of A. If $F \subseteq \mathscr{E}$ is finite and P is the projection of $\mathscr{H}$ onto the span of F, then

$$\sum_{e\in F} (|A|e \mid e) = \sum_{e\in F} (U^*Ae \mid e) = \sum_{e\in\mathscr{E}} (PU^*Ae \mid e) = \mathrm{tr}(PU^*A) = \mathrm{tr}(APU^*) = \psi(PU^*).$$

Since $|\psi(PU^*)| \le \|\psi\|\,\|PU^*\| \le \|\psi\|$, we have $\sum_{e\in F} (|A|e \mid e) \le \|\psi\|$ for all finite F. Therefore $\|A\|_1 < \infty$, completing the proof of the first part of the theorem.

For (2'), we need to prove that $\|A\| \le \|\Psi_A\|$ and that every member ψ of $\mathscr{B}_1(\mathscr{H})'$ is of the form Ψ_A for some $A \in \mathscr{B}(\mathscr{H})$. Now, for any $\boldsymbol{x}$, $\boldsymbol{y} \in \mathscr{H}$, by direct calculation we have

$$(\boldsymbol{y} \otimes \boldsymbol{x})^* = \boldsymbol{x} \otimes \boldsymbol{y} \quad \text{and} \quad (\boldsymbol{x} \otimes \boldsymbol{y})(\boldsymbol{y} \otimes \boldsymbol{x}) = \|\boldsymbol{x}\|^2 (\boldsymbol{y} \otimes \boldsymbol{y}), \tag{†}$$

hence

$$|\boldsymbol{y} \otimes \boldsymbol{x}| = \left[(\boldsymbol{y} \otimes \boldsymbol{x})^*(\boldsymbol{y} \otimes \boldsymbol{x})\right]^{1/2} = \left[\|\boldsymbol{x}\|^2 (\boldsymbol{y} \otimes \boldsymbol{y})\right]^{1/2}. \tag{‡}$$

The operator $P := \|\boldsymbol{y}\| \|\boldsymbol{x}\| \left(\|\boldsymbol{y}\|^{-1} \boldsymbol{y} \otimes \|\boldsymbol{y}\|^{-1} \boldsymbol{y}\right)$ is positive, and it is easily verified that $P^2 = \|\boldsymbol{x}\|^2 (\boldsymbol{y} \otimes \boldsymbol{y})$. Thus by uniqueness of positive square roots, $P = \left[\|\boldsymbol{x}\|^2 (\boldsymbol{y} \otimes \boldsymbol{y})\right]^{1/2}$ and so by (‡)

$$|\boldsymbol{y} \otimes \boldsymbol{x}| = \|\boldsymbol{y}\| \|\boldsymbol{x}\| \left(\|\boldsymbol{y}\|^{-1} \boldsymbol{y} \otimes \|\boldsymbol{y}\|^{-1} \boldsymbol{y}\right).$$

Now define a sesquilinear map $B(\boldsymbol{x}, \boldsymbol{y}) = \psi(\boldsymbol{y} \otimes \boldsymbol{x})$. We then have

$$|B(\boldsymbol{x}, \boldsymbol{y})| \le \|\psi\| \|\boldsymbol{y} \otimes \boldsymbol{x}\|_1 = \|\psi\| \operatorname{tr} |\boldsymbol{y} \otimes \boldsymbol{x}| = \|\boldsymbol{x}\| \|\boldsymbol{y}\|^{-1} \|\psi\| \operatorname{tr}(\boldsymbol{y} \otimes \boldsymbol{y}) = \|\boldsymbol{x}\| \|\boldsymbol{y}\| \|\psi\|,$$

the last equality from the calculation

$$\operatorname{tr}(\boldsymbol{u} \otimes \boldsymbol{v}) = \sum_{\boldsymbol{e} \in \mathscr{E}} ((\boldsymbol{u} \otimes \boldsymbol{v})\boldsymbol{e} \mid \boldsymbol{e}) = \sum_{\boldsymbol{e} \in \mathscr{E}} (\boldsymbol{e} \mid \boldsymbol{u}) (\boldsymbol{v} \mid \boldsymbol{e}) = (\boldsymbol{v} \mid \boldsymbol{u}).$$

Therefore, B is bounded with $\|B\| \le \|\psi\|$. By 11.4.1, there exists an operator $S \in \mathscr{B}(\mathscr{H})$ with $\|S\| = \|B\| \le \|\psi\|$ such that

$$\psi(\boldsymbol{y} \otimes \boldsymbol{x}) = B(\boldsymbol{x}, \boldsymbol{y}) = (\boldsymbol{x} \mid S\boldsymbol{y}) \quad \boldsymbol{x}, \boldsymbol{y} \in \mathscr{H}.$$

Now, if $T \in \mathscr{B}_1(\mathscr{H})$ is self-adjoint, then T may be expressed as in 12.5.5(h), hence setting $A = S^*$ we have $\|A\| \le \|\psi\|$ and

$$\psi(T) = \sum_{\boldsymbol{e} \in \mathscr{E}} \alpha_{\boldsymbol{e}} \psi(\boldsymbol{e} \otimes \boldsymbol{e}) = \sum_{\boldsymbol{e} \in \mathscr{E}} \alpha_{\boldsymbol{e}} (\boldsymbol{e} \mid S\boldsymbol{e}) = \sum_{\boldsymbol{e} \in \mathscr{E}} (T\boldsymbol{e} \mid S\boldsymbol{e}) = \sum_{\boldsymbol{e} \in \mathscr{E}} (AT\boldsymbol{e} \mid \boldsymbol{e}) = \Psi_A(T),$$

the last equality from 12.5.5(c). Since every operator is a linear combination of self-adjoint operators, $\psi = \Psi_A$. □

Exercises

12.48 If $\mathscr{H}$ is finite dimensional, show that $\operatorname{tr}(T)$ is the sum of the diagonal elements of the matrix of T relative to any basis.

12.49 Verify part (h) of 12.5.5.

12.50 Verify the equations in (†) in the proof of 12.5.7.

Chapter 13

Banach Algebras

13.1 Introduction

In this chapter we develop the essential properties of commutative Banach algebras. The main goal is the Gelfand representation theorem, which asserts that such an algebra may be represented as the algebra of continuous functions on some topological space. Applications to operator theory, including the spectral theorem for normal operators, are given in §13.6.

Definitions and Examples

Recall that a Banach algebra is an algebra $\mathscr{A}$ over $\mathbb{C}$ which is a Banach space relative to a norm that satisfies

$$\|xy\| \le \|x\|\,\|y\| \quad \text{for all } x,\, y \in \mathscr{A}.$$

The inequality implies that multiplication in a Banach algebra is jointly continuous (Ex. 13.1).

An **involution** on a Banach algebra $\mathscr{A}$ is a mapping $x \mapsto x^*$ on $\mathscr{A}$ with the properties

$$(x+y)^* = x^* + y^*, \quad (cx)^* = \bar{c}x^*, \quad (xy)^* = y^*x^*, \quad x = x^{**}, \text{ and } \|x^*\| = \|x^*\|.$$

Note that if $\mathscr{A}$ is unital with identity e, then

$$e^* = ee^* = (ee^*)^* = e^{**} = e.$$

A Banach algebra with an involution is called a **Banach $*$-algebra**. A Banach $*$-algebra whose norm satisfies

$$\|xx^*\| = \|x\|^2$$

is called **C^*-algebra**. Note that the identity e in a unital C^*-algebra $\mathscr{A}$ satisfies $\|e\|^2 = \|ee^*\| = \|e\|$, hence if $\mathscr{A}$ is nontrivial then $\|e\| = 1$.

A **homomorphism** from a Banach algebra $\mathscr{A}$ into a Banach algebra $\mathscr{B}$ is bounded linear transformation $\varphi : \mathscr{A} \to \mathscr{B}$ such that

$$\varphi(xy) = \varphi(x)\varphi(y) \quad \text{for all } x, y \in \mathscr{A}.$$

If $\mathscr{A}$ and $\mathscr{B}$ are Banach $*$-algebras and $\varphi(x^*) = \varphi(x)^*$ for all $x \in \mathscr{A}$, then φ is called a $*$-homomorphism.

Recall that an ideal $\mathscr{J}$ of a Banach algebra $\mathscr{A}$ is a linear subspace such that $xy,\ yx \in \mathscr{A}$ for all $x \in \mathscr{A}$ and $y \in \mathscr{J}$. If $\mathscr{J} \ne \mathscr{A}$, then $\mathscr{J}$ is called a **proper ideal**. If $\mathscr{J}$ is closed, then $\mathscr{A}/\mathscr{J}$ is a Banach algebra under multiplication $(x+\mathscr{J})(y+\mathscr{J}) = xy+\mathscr{J}$, and the quotient map is an algebra homomorphism (Ex. 13.7). Quotient algebras will be of considerable importance later in connection with maximal ideals and characters of a Banach algebra.

We have seen several examples of Banach algebras and C^*-algebras throughout the text. For convenience, we include some of these in the following list.

13.1.1 Examples.

(a) If $\mathscr{X}$ is a (nontrivial) Banach space, then $\mathscr{B}(\mathscr{X})$ is a unital, noncommutative Banach algebra under the operator norm and with respect to operator composition.

(b) If $\mathscr{H}$ is a Hilbert space, then $\mathscr{B}(\mathscr{H})$ is a C^*-algebra, where involution is the adjoint operation. The spaces $\mathscr{B}_{00}(\mathscr{H})$, $\mathscr{B}_0(\mathscr{H})$, $\mathscr{B}_1(\mathscr{H})$, and $\mathscr{B}_2(\mathscr{H})$ are ideals of $\mathscr{B}(\mathscr{H})$.

(c) If X is a set, then $B(X)$ is a unital, commutative C^*-algebra with involution $f \to \overline{f}$.

(d) If X is a topological space, then $C_b(X)$ is a unital, commutative C^*-subalgebra of $B(X)$.

(e) If X is a noncompact, locally compact, Hausdorff topological space, then $C_0(X)$ is a non-unital C^*-subalgebra of $C_b(X)$.

(f) *ℓ^1 group algebra.* The space $\ell^1(\mathbb{Z})$ of all bilateral sequences $\boldsymbol{x} = (\ldots, x_{-1}, x_0, x_1, \ldots)$ is a commutative Banach $*$-algebra under the norm $\|\boldsymbol{x}\|_1 := \sum_{k=-\infty}^{\infty} |x_k| < \infty$ with convolution product $\boldsymbol{x} * \boldsymbol{y}$ and involution $\boldsymbol{x}^*$ defined by

$$(\boldsymbol{x} * \boldsymbol{y})(n) = \sum_{k=-\infty}^{\infty} x_{n-k} y_k, \quad \text{and} \quad \boldsymbol{x}^*(n) = \overline{\boldsymbol{x}^*(-n)}.$$

Moreover, $\ell^1(\mathbb{Z})$ has identity $\boldsymbol{e}_0 := (\ldots, 0, \overset{0}{1}, 0, \ldots)$. In general $\|\boldsymbol{x}^* * \boldsymbol{x}^*\| \neq \|\boldsymbol{x}\|^2$, hence $\ell^1(\mathbb{Z})$ not a C^*-algebra (Ex. 13.2).

(g) *L^1 group algebra.* The space $L^1(\mathbb{R}^d)$ is a commutative, non-unital Banach $*$-algebra under convolution $f * g$ and involution f^* defined by

$$f * g(\boldsymbol{x}) = \int f(\boldsymbol{x} - \boldsymbol{y}) g(\boldsymbol{y})\, d\boldsymbol{y}, \quad \text{and} \quad f^*(x) = \overline{f(-x)}.$$

(h) *Measure algebra.* The space $M(\mathbb{R}^d)$ of complex Borel measures on $\mathbb{R}^d$ with the total variation norm is a commutative Banach algebra under convolution.

The Group of Invertible Elements

A member $\boldsymbol{x}$ of a unital Banach algebra $\mathscr{A}$ is said to be **invertible** if there exists an element $\boldsymbol{y} \in \mathscr{A}$, called the **inverse of** $\boldsymbol{x}$, such that $\boldsymbol{x}\boldsymbol{y} = \boldsymbol{y}\boldsymbol{x} = \boldsymbol{e}$. In this case, $\boldsymbol{y}$ is unique and is denoted by $\boldsymbol{x}^{-1}$. If $\mathscr{A}$ is a $*$-algebra and $\boldsymbol{x} \in \mathscr{A}$ is invertible, then $\boldsymbol{x}^*$ is invertible and $(\boldsymbol{x}^*)^{-1} = (\boldsymbol{x}^{-1})^*$, as may be seen from the calculation $\boldsymbol{e} = \boldsymbol{e}^* = (\boldsymbol{x}\boldsymbol{x}^{-1})^* = (\boldsymbol{x}^{-1})^*\boldsymbol{x}^*$.

The set of invertible elements $G = G_{\mathscr{A}}$ of $\mathscr{A}$ is easily seen to be a group with identity $\boldsymbol{e}$. The following theorem shows that members of $\mathscr{A}$ sufficiently near $\boldsymbol{e}$ are invertible, a fact that is of critical importance in spectral theory.

13.1.2 Theorem (C. Neumann). *If $\|\boldsymbol{e} - \boldsymbol{x}\| < 1$, then $\boldsymbol{x} \in G$ and $\boldsymbol{x}^{-1} = \sum_{n=0}^{\infty} (\boldsymbol{e} - \boldsymbol{x})^n$.*

Proof. Let $\boldsymbol{y} := \boldsymbol{e} - \boldsymbol{x}$. By induction, $\|\boldsymbol{y}^n\| \leq \|\boldsymbol{y}\|^n$. Since $\|\boldsymbol{y}\| < 1$, the series $\boldsymbol{s} := \sum_{n=0}^{\infty} \boldsymbol{y}^n$, where $\boldsymbol{y}^0 := \boldsymbol{e}$, converges absolutely and hence converges. Since the partial sum $\boldsymbol{s}_n := \sum_{k=0}^{n} \boldsymbol{y}^k$ satisfies

$$\boldsymbol{s}_n(\boldsymbol{e} - \boldsymbol{y}) = \sum_{k=0}^{n} \boldsymbol{y}^k - \sum_{k=1}^{n+1} \boldsymbol{y}^k = \boldsymbol{e} - \boldsymbol{y}^{n+1}$$

and since $\|\boldsymbol{y}^{n+1}\| \leq \|\boldsymbol{y}\|^{n+1} \to 0$, we see that $\boldsymbol{s}\boldsymbol{x} = \boldsymbol{s}(\boldsymbol{e} - \boldsymbol{y}) = \lim_n \boldsymbol{s}_n(\boldsymbol{e} - \boldsymbol{y}) = \boldsymbol{e}$. Similarly, $\boldsymbol{x}\boldsymbol{s} = \boldsymbol{e}$. Therefore, $\boldsymbol{s} = \boldsymbol{x}^{-1}$. □

13.1.3 Corollary. *If $x \in \mathcal{A}$ and $z \in \mathbb{C}$ with $|z| > \|x\|$, then $ze - x$ is invertible and*

$$(ze - x)^{-1} = \sum_{n=0}^{\infty} z^{-n-1} x^n.$$

Proof. Since $e - (e - z^{-1}x) = z^{-1}x$ has norm less than one, $e - z^{-1}x$ is invertible and $(e - z^{-1}x)^{-1} = \sum_{n=0}^{\infty} z^{-n} x^n$. Multiplying by z^{-1} yields the desired expansion. □

13.1.4 Theorem. *The group G of invertible elements in $\mathcal{A}$ is open and the map $x \to x^{-1}$ on G is continuous.*

Proof. Let $x_0 \in G$ and set $r = \|x_0^{-1}\|^{-1}$. Then G contains the open ball $B_r(x_0)$. Indeed, if $\|x - x_0\| < r$, then

$$\|xx_0^{-1} - e\| = \|(x - x_0)x_0^{-1}\| \le \|x - x_0\| \, \|x_0^{-1}\| < 1,$$

hence xx_0^{-1} is invertible. Denoting the inverse by y and setting $a = x_0^{-1}y$, we see that $xa = xx_0^{-1}y = e$. A similar argument produces an element b such that $bx = e$. Thus x is invertible, verifying the claim and proving that G is open.

To show continuity of the inverse at e, let $x_n \to e$ in G. By 13.1.2, for sufficiently large n we have $x_n^{-1} = \sum_{k=0}^{\infty}(e - x_n)^k$, hence for $0 < \varepsilon < 1$ and $\|e - x_n\| < \varepsilon$,

$$\|x_n^{-1} - e\| \le \sum_{k=1}^{\infty} \|e - x_n\|^k \le \frac{\varepsilon}{1-\varepsilon}.$$

Therefore $\overline{\lim}_n \|x_n^{-1} - e\| \le \varepsilon(1-\varepsilon)^{-1}$ and letting $\varepsilon \to 0$ shows that $\lim_n x_n^{-1} = e$.

In the general case, let $x_n \to x$ in G. Then $x_n x^{-1} \to e$, hence, by the preceding paragraph, $xx_n^{-1} = (x_n x^{-1})^{-1} \to e$ and so $x_n^{-1} \to x^{-1}$. □

The Cauchy Product of Series

The **Cauchy product** of series $\sum_{n=0}^{\infty} a_n$ and $\sum_{n=0}^{\infty} b_n$ in a Banach algebra $\mathcal{A}$ is the series

$$\sum_{n=0}^{\infty} c_n, \quad \text{where } c_n = \sum_{k=0}^{n} a_k b_{n-k}.$$

The following result for numerical series is due to Mertens. The proof in the general setting of Banach algebras is the same.

13.1.5 Proposition. *If the series $A := \sum_{n=0}^{\infty} a_n$ and $B := \sum_{n=0}^{\infty} b_n$ converge in $\mathcal{A}$ and at least one of the series converges absolutely, then the Cauchy product C converges and $C = AB$.*

Proof. Assume that $\sum_{n=0}^{\infty} a_n$ converges absolutely. Let

$$A_n = \sum_{k=0}^{n} a_k, \quad B_n = \sum_{k=0}^{n} b_k, \quad C_n = \sum_{k=0}^{n} c_k, \quad \text{and} \quad \alpha = \sum_{n=0}^{\infty} \|a_n\|.$$

Then

$$\begin{aligned}
C_n &= a_0b_0 + (a_0b_1 + a_1b_0) + \cdots + (a_0b_n + a_1b_{n-1} + \cdots + a_nb_0) \\
&= a_0B_n + a_1B_{n-1} + \cdots + a_nB_0 \\
&= a_0(B_n - B + B) + a_1(B_{n-1} - B + B) + \cdots + a_n(B_0 - B + B) \\
&= a_0(B_n - B) + a_1(B_{n-1} - B) + \cdots + a_n(B_0 - B) + A_nB.
\end{aligned}$$

Thus to show that $C_n \to AB$ it suffices to verify that

$$X_n := \boldsymbol{a}_0(B_n - B) + \boldsymbol{a}_1(B_{n-1} - B) + \cdots + \boldsymbol{a}_n(B_0 - B) \to 0.$$

Given $\varepsilon > 0$, choose N such that

$$\|B_n - B\| < \varepsilon/2\alpha \quad \text{for all } n > N. \tag{†}$$

Since $\|\boldsymbol{a}_n\| \to 0$, we may choose $N' > N$ so that for all $n > N'$

$$\|\boldsymbol{a}_n(B_0 - B) + \boldsymbol{a}_{n-1}(B_1 - B) + \cdots + \boldsymbol{a}_{n-N}(B_N - B)\| < \varepsilon/2. \tag{‡}$$

For such n, from (†) and (‡) we have

$$\begin{aligned}\|X_n\| &\le \|\boldsymbol{a}_n(B_0 - B) + \boldsymbol{a}_{n-1}(B_1 - B) + \cdots + \boldsymbol{a}_{n-N}(B_N - B)\| \\ &\quad + \|\boldsymbol{a}_{n-N-1}\|\,\|B_{N+1} - B\| + \|\boldsymbol{a}_{n-N-2}\|\,\|B_{N+2} - B\| + \cdots + \|\boldsymbol{a}_0\|\,\|B_n - B\| \\ &< \varepsilon. \end{aligned}$$ □

Exercises

13.1 Verify that multiplication in a Banach algebra is jointly continuous.

13.2 Verify that $\ell^1(\mathbb{Z})$ is a Banach $*$-algebra but not a C^*-algebra.

13.3 (*Banach algebra generated by $\boldsymbol{x}$ and $\boldsymbol{e}$*). Let $\mathscr{A}$ be a commutative unital Banach algebra and let $\boldsymbol{x} \in \mathscr{A}$. Show that the intersection $\mathscr{B}$ of all closed subalgebras of $\mathscr{A}$ containing $\boldsymbol{x}$ and $\boldsymbol{e}$ is the closure of the set $\mathscr{P}$ of all polynomials $\sum_{j=0}^n c_j \boldsymbol{x}^j$ in $\boldsymbol{x}$, where $\boldsymbol{x}^0 := \boldsymbol{e}$.

13.4 The **commutant** of a nonempty subset E of a unital Banach algebra $\mathscr{A}$ is the set $E' := \{\boldsymbol{x} : \boldsymbol{x}\boldsymbol{y} = \boldsymbol{y}\boldsymbol{x}\ \forall\ \boldsymbol{y} \in E\}$. The **bicommutant** E'' of E is the commutant of the commutant: $E'' = (E')'$. Show that E' is a closed unital subalgebra of $\mathscr{A}$. Show also that if $\boldsymbol{x} \in G_{\mathscr{A}}$, then $\boldsymbol{x}^{-1} \in \{\boldsymbol{x}\}''$.

13.5 Let X, Y be topological spaces and $\tau : Y \to X$ a continuous function. Show that $\varphi(f) := f \circ \tau$ defines a $*$-homomorphism from $C_b(X)$ into $C_b(Y)$.

13.6 When is the dilation operator $(D_r f)(x) = f(rx)$ $(r > 0)$, a homomorphism on the group algebra $L^1(\mathbb{R}^d)$?

13.7 Let $\mathscr{A}$ be a Banach algebra and $\mathscr{I}$ a closed ideal in $\mathscr{C}$. Show that the Banach space $\mathscr{A}/\mathscr{I}$ is a Banach algebra under multiplication $(\boldsymbol{x} + \mathscr{I})(\boldsymbol{y} + \mathscr{I}) = \boldsymbol{x}\boldsymbol{y} + \mathscr{I}$ and that the quotient map Q is a homomorphism. Show also that if $\mathscr{A}$ is a Banach $*$-algebra and $\mathscr{I}$ is closed under involution, then $\mathscr{A}/\mathscr{I}$ is a Banach $*$algebra under involution $(\boldsymbol{x} + \mathscr{I})^* = \boldsymbol{x}^* + \mathscr{I}$ and Q is a $*$-homomorphism.

13.8 [↓13.3.4] Let $\mathscr{I}$ be a proper ideal of a unital Banach algebra. Show that $\mathrm{cl}\,\mathscr{I}$ is a proper ideal.

13.9 [↑8.46] Show that the space $C^n[0,1]$ of n-times continuously differentiable functions on $[0,1]$ is a Banach algebra with the norm $\|f\| = \sum_{k=0}^n \|f^{(k)}\|_\infty$.

13.10 Let X be a compact Hausdorff space, μ a probability Radon measure on X and $k : X \times X \to \mathbb{C}$ continuous and never zero. Define K on $L^1(\mu)$ by $Kf(x) = f(x) + \lambda \int_X k(x,y) f(y)\, d\mu(y)$, where $|\lambda| < \|k\|_\infty^{-1}$. Prove that $K \in \mathscr{B}(L^1(\mu))$ and is invertible. If $Kf = g$ show that for each $x \in X$ there exists a Borel measure μ_x on X such that $f(x) = \int_X g\, d\mu_x$.

13.11 Let $\mathscr{A}$ and $\mathscr{B}$ be unital Banach algebras and $\Phi : \mathscr{A} \to \mathscr{B}$ a homomorphism that maps identity onto identity. Show that $\Phi(G_{\mathscr{A}}) \subseteq G_{\mathscr{B}}$.

13.12 Let $\mathscr{A}$ denote the Banach algebra of bounded linear operators on $L^2[0,1]$, set $\boldsymbol{e}_n(t) = e^{2\pi int}$, and define $T \in \mathscr{A}$ so that $T\boldsymbol{e}_n = \boldsymbol{e}_{n+1}$, that is, $T\boldsymbol{x} = \sum_{n=-\infty}^{\infty} (\boldsymbol{x} \mid \boldsymbol{e}_n)\, \boldsymbol{e}_{n+1}$. Let $\mathscr{B}$ be the Banach algebra generated by T and I. Show that $T \in G_{\mathscr{A}} \setminus G_{\mathscr{B}}$.

13.13 Let $\mathscr{A}$ be a unital Banach algebra and $(\boldsymbol{x}_n) \subseteq G_{\mathscr{A}}$ such that $\boldsymbol{x}_n \to \boldsymbol{x} \notin G_{\mathscr{A}}$. Show that $\|\boldsymbol{x}_n^{-1}\| \to \infty$.

13.14 [↑8.1.2] (*Disk algebra*). Let $A(\mathbb{D})$ denote algebra of all bounded continuous functions on the closed unit disk $\text{cl}(\mathbb{D})$ that are analytic on $\mathbb{D}$. Show that $A(\mathbb{D})$ is a unital commutative C^*-algebra with respect to the sup norm and involution $f^*(z) = \overline{f(\overline{z})}$.

13.15 [↑6.4, 7.1.6] Show that the set of all measures $\mu \in M(\mathbb{R}^d)$ with $\mu \ll \lambda$ is an ideal in $M(\mathbb{R}^d)$.

13.16 (*Arens multiplication*). Let $\mathscr{A}$ be a Banach algebra. For f in the dual space $\mathscr{A}'$ and $\boldsymbol{x} \in \mathscr{A}$, define ${}_{\boldsymbol{x}}f \in \mathscr{A}'$ by ${}_{\boldsymbol{x}}f(\boldsymbol{y}) = f(\boldsymbol{x}\boldsymbol{y})$. Next, for F, G in the bidual $\mathscr{A}''$ and $f \in \mathscr{A}'$ define $Gf \in \mathscr{A}'$ by $Gf(\boldsymbol{x}) = G({}_{\boldsymbol{x}}f)$ and $FG \in \mathscr{A}''$ by $FG(f) = F(Gf)$. Show that $\mathscr{A}''$ is a Banach algebra under the multiplication $(F,G) \mapsto FG$ and that the canonical embedding $\boldsymbol{x} \mapsto \widehat{\boldsymbol{x}}$ is a homomorphism.

13.17 (*ℓ^1 semigroup algebra*). Show that $\ell(\mathbb{Z}^+)$ is a non-unital commutative Banach algebra under **convolution** $(f * g)(n) = \sum_{j+k=n} f(j)g(k)$. Determine whether the shift operators T_r and T_ℓ are homomorphisms on $\ell(\mathbb{Z}^+)$.

13.2 Spectral Theory

In this section, $\mathscr{A}$ denotes a unital Banach algebra.

The Spectrum of an Element

Let $\boldsymbol{x} \in \mathscr{A}$. The following definitions are used throughout spectral theory:

- The **spectrum of** $\boldsymbol{x}$: $\sigma(\boldsymbol{x}) = \{z \in \mathbb{C} : z\boldsymbol{e} - \boldsymbol{x} \text{ is not invertible}\}$.
- The **spectral radius of** $\boldsymbol{x}$: $r(\boldsymbol{x}) = \sup\{|z| : z \in \sigma(\boldsymbol{x})\}$.
- The **resolvent set of** $\boldsymbol{x}$: $\rho(\boldsymbol{x}) = \mathbb{C} \setminus \sigma(\boldsymbol{x}) = \{z \in \mathbb{C} : z\boldsymbol{e} - \boldsymbol{x} \text{ is invertible}\}$.

13.2.1 Proposition. *$\sigma(\boldsymbol{x})$ is compact and $r(\boldsymbol{x}) \le \|\boldsymbol{x}\|$. In particular,*

$$\{z \in \mathbb{C} : |z| > \|\boldsymbol{x}\|\} \subseteq \{z \in \mathbb{C} : |z| > r(\boldsymbol{x}\} \subseteq \rho(\boldsymbol{x}).$$

Proof. By 13.1.3, if $z\boldsymbol{e} - \boldsymbol{x}$ is not invertible, then $|z| \le \|\boldsymbol{x}\|$. Therefore, $\sigma(\boldsymbol{x})$ is bounded and $r(\boldsymbol{x}) \le \|\boldsymbol{x}\|$. Since the mapping $f(z) = z\boldsymbol{e} - \boldsymbol{x}$ is continuous and $\rho(\boldsymbol{x}) = f^{-1}(G_{\mathscr{A}})$ is open, $\sigma(\boldsymbol{x})$ is closed. □

The following lemma will be used to prove the key property that $\sigma(\boldsymbol{x}) \ne \emptyset$.

13.2.2 Lemma. *Let $\boldsymbol{x} \in \mathscr{A}$ and $\varphi \in \mathscr{A}'$. Define f on the open set $\rho(\boldsymbol{x})$ by*

$$f(z) = \langle (z\boldsymbol{e} - \boldsymbol{x})^{-1}, \varphi \rangle. \tag{13.1}$$

Then f is analytic on $\rho(\boldsymbol{x})$ and $f'(z) = -\langle [z\boldsymbol{e} - \boldsymbol{x}]^{-2}, \varphi \rangle$.

Proof. Fix $z \in \rho(\boldsymbol{x})$. For h sufficiently small, take $\boldsymbol{a} = (z+h)\boldsymbol{e} - \boldsymbol{x}$ and $\boldsymbol{b} = z\boldsymbol{e} - \boldsymbol{x}$ in the identity $\boldsymbol{a}^{-1} - \boldsymbol{b}^{-1} = \boldsymbol{a}^{-1}(\boldsymbol{b} - \boldsymbol{a})\boldsymbol{b}^{-1}$ to obtain

$$\big[(z+h)\boldsymbol{e} - \boldsymbol{x}\big]^{-1} - \big[z\boldsymbol{e} - \boldsymbol{x}\big]^{-1} = \big[(z+h)\boldsymbol{e} - \boldsymbol{x}\big]^{-1}(-h\boldsymbol{e})\big[z\boldsymbol{e} - \boldsymbol{x}\big]^{-1},$$

from which follows

$$\frac{f(z+h) - f(z)}{h} = -\Big\langle \big[(z+h)\boldsymbol{e} - \boldsymbol{x}\big]^{-1}\big[z\boldsymbol{e} - \boldsymbol{x}\big]^{-1}, \varphi \Big\rangle.$$

Letting $h \to 0$ and using the continuity of the inversion map we see that $f'(z)$ exists and equals $-\big\langle [z\boldsymbol{e} - \boldsymbol{x}]^{-2}, \varphi \big\rangle$. Since the latter is continuous in $z \in \rho(\boldsymbol{x})$, f is analytic. □

13.2.3 Theorem. *$\sigma(\boldsymbol{x})$ is nonempty.*

Proof. If $\sigma(\boldsymbol{x}) = \emptyset$, then the mapping f in (13.1) is entire. Moreover, for $|z| > \|\boldsymbol{x}\|$ and $\|\varphi\| \le 1$,

$$|f(z)| = |\langle (z\boldsymbol{e} - \boldsymbol{x})^{-1}, \varphi \rangle| \le \|\varphi\| \, \|(z\boldsymbol{e} - \boldsymbol{x})^{-1}\| \le \sum_{n=0}^{\infty} |z|^{-n-1} \|\boldsymbol{x}\|^n = \frac{1}{|z| - \|\boldsymbol{x}\|},$$

hence $\lim_{|z| \to \infty} |f(z)| = 0$. By Liouville's theorem, f is identically zero. Since φ was arbitrary, $(z\boldsymbol{e} - \boldsymbol{x})^{-1}$ is zero for all z, impossible. □

13.2.4 Theorem (Gelfand-Mazur). *If $\mathscr{A}$ is a division algebra (that is, every nonzero element in $\mathscr{A}$ is invertible), then $\mathscr{A} = \mathbb{C}\boldsymbol{e}$.*

Proof. Let $\boldsymbol{x} \in \mathscr{A}$ and $z \in \sigma(\boldsymbol{x})$. Then $z\boldsymbol{e} - \boldsymbol{x}$ is not invertible and so equals 0. □

The Spectral Radius Formula

The theorem in this subsection gives an important formula for $r(\boldsymbol{x})$. The formula has already played a key role in the proof of the spectral theorem for compact normal operators (§12.3) and will figure prominently in the proof of the Gelfand representation theorem later. To establish the formula we need

13.2.5 Lemma. *Let $g(z) = \sum_{n=0}^{\infty} a_n z^n$ be analytic in a region that contains the disk $D_r := \{z \in \mathbb{C} : |z| \le r\}$. For $\boldsymbol{x} \in \mathscr{X}$, set $g(\boldsymbol{x}) := \sum_{n=0}^{\infty} a_n \boldsymbol{x}^n$. Then*

$$g(\sigma(\boldsymbol{x}) \cap D_r) \subseteq \sigma(g(\boldsymbol{x})). \tag{13.2}$$

Proof. Note first that for $\|\boldsymbol{x}\| \le r$ the series $g(\boldsymbol{x})$ is absolutely convergent, hence converges. Now let $|z| \le r$. From the identity

$$(z^n\boldsymbol{e} - \boldsymbol{x}^n) = (z\boldsymbol{e} - \boldsymbol{x})\boldsymbol{y}_n, \quad \boldsymbol{y}_n := z^{n-1}\boldsymbol{e} + z^{n-2}\boldsymbol{x} + \cdots + z\boldsymbol{x}^{n-2} + \boldsymbol{x}^{n-1},$$

we have

$$g(z)\boldsymbol{e} - g(\boldsymbol{x}) = \sum_{n=1}^{\infty} a_n(z^n\boldsymbol{e} - \boldsymbol{x}^n) = (z\boldsymbol{e} - \boldsymbol{x})\sum_{n=1}^{\infty} a_n\boldsymbol{y}_n.$$

Since $\|\boldsymbol{y}_n\| \le nr^{n-1}$, the series on the right converges to some $\boldsymbol{y} \in \mathscr{A}$ which commutes with $(z\boldsymbol{e} - \boldsymbol{x})$, that is,

$$g(z)\boldsymbol{e} - g(\boldsymbol{x}) = (z\boldsymbol{e} - \boldsymbol{x})\boldsymbol{y} = \boldsymbol{y}(z\boldsymbol{e} - \boldsymbol{x}).$$

Thus if $g(z)\boldsymbol{e} - g(\boldsymbol{x})$ is invertible, then so is $z\boldsymbol{e} - \boldsymbol{x}$, verifying (13.2). □

13.2.6 Theorem. $r(x) = \lim_n \|x^n\|^{1/n}$.

Proof. By 13.2.5, $z \in \sigma(x) \Rightarrow z^n \in \sigma(x^n) \Rightarrow |z^n| \le \|x^n\| \Rightarrow |z| \le \|x^n\|^{1/n}$. Therefore, $r(x) \le \underline{\lim}_n \|x^n\|^{1/n}$.

To see that $\overline{\lim}_n \|x^n\|^{1/n} \le r(x)$, note first that if $|z| > \|x\|$, then the function f in (13.1) with $\|\varphi\| \le 1$ is well-defined and $f(z) = \sum_{k=0}^{\infty} \langle x^k, \varphi \rangle z^{-k-1}$. By 13.2.2, $f(z)$ is analytic on the larger set $|z| > r(x)$. It follows that the preceding Laurent series expansion for f is valid for $|z| > r(x)$ and converges uniformly on $|z| \ge r$ for any $r > r(x)$. Multiplying the series expansion by z^{n+1} and integrating term by term along the contour $z = re^{i\theta}$ yields

$$\int_0^{2\pi} r^{n+1} e^{i(n+1)\theta} f(re^{i\theta})\, d\theta = \sum_{k=0}^{\infty} \langle x^k, \varphi \rangle r^{n-k} \int_0^{2\pi} e^{i(n-k)\theta}\, d\theta = 2\pi \langle x^n, \varphi \rangle .$$

Now set $s := \sup_\theta \|(re^{i\theta} e - x)^{-1}\|$. Noting from (13.1) that $|f(z)| \le \left\|(ze - x)^{-1}\right\|$, we have

$$|\langle x^n, \varphi \rangle| = \frac{1}{2\pi} \left| \int_0^{2\pi} r^{n+1} e^{i(n+1)\theta} f(re^{i\theta})\, d\theta \right| \le r^{n+1} \sup_\theta |f(re^{i\theta})| \le r^{n+1} s.$$

Since φ was arbitrary, $\|x^n\| \le r^{n+1} s$. Thus $\overline{\lim}_n \|x^n\|^{1/n} \le r$, and since $r > r(x)$ was arbitrary, $\overline{\lim}_n \|x^n\|^{1/n} \le r(x)$, as required. □

Normal Elements in a C^*-Algebra $\mathscr{A}$

The following are generalizations of definitions given earlier for operators on a Hilbert space. An element x in $\mathscr{A}$ is said to be

(a) **normal** if $x^* x = x x^*$.

(b) **self-adjoint** if $x^* = x$.

(c) **positive**, written $x \ge 0$, if $x = y^* y$ for some $y \in \mathscr{A}$.

(d) **unitary** if $\mathscr{A}^*$ is unital and $x^* x = x x^* = e$.

(e) a **projection** if $x^2 = x = x^*$.

Note that for a unitary element x, $1 = \|e\| = \|x x^*\| = \|x\|^2$, hence $\|x\| = 1$.

If $\mathscr{A}$ is a unital C^*-algebra and $x \in \mathscr{A}$ is normal, we denote by $\mathscr{C}^*(x)$ the unital commutative C^*-algebra generated by x, x^*, and e. Thus $\mathscr{C}^*(x)$ is the closure in $\mathscr{A}$ of the algebra of polynomials in x and x^*.

13.2.7 Proposition. *If x is self-adjoint, then the series*

$$\exp(ix) := \sum_{n=0}^{\infty} \frac{1}{n!} (ix)^n$$

converges and is unitary.

Proof. The series clearly converges absolutely, hence converges. Set $u := \exp(ix)$. By continuity of involution,

$$u^* := \sum_{n=0}^{\infty} \frac{1}{n!} (-ix^*)^n = \sum_{n=0}^{\infty} \frac{1}{n!} (-ix)^n .$$

Let $\sum_{n=0}^{\infty} \boldsymbol{v}_n$ be the Cauchy product of series for $\boldsymbol{u}$ and $\boldsymbol{u}^*$. Then $\boldsymbol{v}_0 = \boldsymbol{e}$, and for $n \geq 1$

$$\boldsymbol{v}_n = \sum_{k=0}^{n} \frac{1}{k!}(i\boldsymbol{x})^k \frac{1}{(n-k)!}(-i\boldsymbol{x})^{n-k} = \frac{(i\boldsymbol{x})^n}{n!} \sum_{k=0}^{n} \binom{n}{k}(-1)^{n-k} = 0.$$

Therefore, $\boldsymbol{u}\boldsymbol{u}^* = \boldsymbol{u}^*\boldsymbol{u}^* = \sum_{n=0}^{\infty} \boldsymbol{v}_n = \boldsymbol{e}$. □

13.2.8 Theorem. *Let $\mathscr{A}$ be a unital C^*-algebra. The following hold for a member $\boldsymbol{x}$ of $\mathscr{A}$.*

(a) $\sigma(\boldsymbol{x}^*) = \overline{\sigma(\boldsymbol{x})}$.

(b) *If $\boldsymbol{x}$ is unitary, then $\sigma(\boldsymbol{x}) \subseteq \mathbb{T}$.*

(c) *If $\boldsymbol{x}$ is self-adjoint, then $\sigma(\boldsymbol{x}) \subseteq \mathbb{R}$.*

Proof. (a) $z \in \sigma(\boldsymbol{x}^*)$ iff $z\boldsymbol{e} - \boldsymbol{x}^*$ is not invertible iff $\overline{z}\boldsymbol{e} - \boldsymbol{x} = (z\boldsymbol{e} - \boldsymbol{x}^*)^*$ is not invertible iff $\overline{z} \in \sigma(\boldsymbol{x})$.

(b) If $z \in \sigma(\boldsymbol{x})$, then $\overline{z} \in \sigma(\boldsymbol{x}^*)$ by (a). Since $\boldsymbol{x}^* = \boldsymbol{x}^{-1}$, $\overline{z}^{-1} \in \sigma(\boldsymbol{x})$ by Ex. 13.21. Therefore, $|z|$ and $|z^{-1}|$ are both $\leq \|\boldsymbol{x}\| = 1$ and so $|z| = 1$.

(c) If $\boldsymbol{x}$ is self-adjoint, then $\exp(i\boldsymbol{x})$ is unitary (13.2.7), hence $\sigma(\exp(i\boldsymbol{x})) \subseteq \mathbb{T}$ by (b). Now let $z \in \sigma(\boldsymbol{x})$. By 13.2.5, $e^{iz} \in \sigma(\exp(i\boldsymbol{x}))$. Since $|e^{iz}| = 1$, $z \in \mathbb{R}$. □

For a normal element $\boldsymbol{x}$, the converses of (b) and (c) hold (Ex. 13.37). Moreover, if $\boldsymbol{x}$ is self-adjoint, then $\boldsymbol{x} \geq 0$ iff $\sigma(\boldsymbol{x}) \subseteq \mathbb{R}^+$ (Ex. 13.51). The proofs use the functional calculus developed in §13.6.

13.2.9 Proposition. *If $\boldsymbol{x} \in \mathscr{A}$ is normal, then $\|\boldsymbol{x}\| = r(\boldsymbol{x})$.*

Proof. If $\boldsymbol{x}$ is self-adjoint, then $\|\boldsymbol{x}^2\| = \|\boldsymbol{x}\|^2$; iterating yields $\|\boldsymbol{x}^{2^n}\| = \|\boldsymbol{x}\|^{2^n}$. In the general case, apply this result to the self-adjoint element $\boldsymbol{x}^*\boldsymbol{x}$ using $\|\boldsymbol{x}^*\boldsymbol{x}\| = \|\boldsymbol{x}\|^2$ to obtain

$$\|\boldsymbol{x}\|^{2^{n+1}} = \|\boldsymbol{x}^*\boldsymbol{x}\|^{2^n} = \|(\boldsymbol{x}^*\boldsymbol{x})^{2^n}\| = \|(\boldsymbol{x}^{2^n})^*\boldsymbol{x}^{2^n}\| = \|\boldsymbol{x}^{2^n}\|^2.$$

The assertion now follows from 13.2.6. □

Here is an application of 13.2.9 to normal operators. The formula for the special case of a self-adjoint operator was proved in 12.1.6.

13.2.10 Corollary. *Let $\mathscr{H}$ be a complex Hilbert and $T \in \mathscr{B}(\mathscr{H})$ normal. Then*

$$\|T\| = \sup\{|\,(T\boldsymbol{x}\,|\,\boldsymbol{x})\,| : \|\boldsymbol{x}\| = 1\}.$$

Proof. Let s denote the supremum. By 13.2.9, we may choose $\lambda \in \sigma(T)$ such that $|\lambda| = \|T\|$. By 12.3.7, there exists a sequence $(\boldsymbol{x}_n)$ with unit norm such that $\|T\boldsymbol{x}_n - \lambda\boldsymbol{x}_n\| \to 0$. Then

$$(T\boldsymbol{x}_n\,|\,\boldsymbol{x}_n) = (T\boldsymbol{x}_n - \lambda\boldsymbol{x}_n\,|\,\boldsymbol{x}_n) + (\lambda\boldsymbol{x}_n\,|\,\boldsymbol{x}_n) = (T\boldsymbol{x}_n - \lambda\boldsymbol{x}_n\,|\,\boldsymbol{x}_n) + \lambda \to \lambda.$$

Therefore, $s \geq |\,(T\boldsymbol{x}_n\,|\,\boldsymbol{x}_n)\,| \to |\lambda| = \|T\| \geq s$. □

Exercises

13.18 [↓13.6.1] Let $\mathscr{A}$ and $\mathscr{B}$ be unital Banach algebras and $\Phi : \mathscr{A} \to \mathscr{B}$ a homomorphism that maps the identity onto the identity. Show that $\sigma(\Phi(x)) \subseteq \sigma(x)$.

13.19 Let $\mathscr{A}$ and $\mathscr{B}$ be unital C^* algebras and $\Phi : \mathscr{A} \to \mathscr{B}$ a $*$-homomorphism that maps identity onto identity. Show that $\|\Phi(x)\| \le \|x\|$ and hence that Φ is continuous. ⟦Consider $r(x^*x)$ and $r(\Phi(x^*)\Phi(x))$.⟧

13.20 Let $\mathscr{A}$ be the finite dimensional algebra of upper triangular matrices

$$x = \begin{bmatrix} x_{11} & x_{12} & \cdots & x_{1n} \\ 0 & x_{22} & \cdots & x_{2n} \\ \vdots & \vdots & \ddots & \vdots \\ 0 & 0 & \cdots & x_{nn} \end{bmatrix}$$

Show that $\sigma(x) = \{x_{11}, x_{22}, \ldots, x_{nn}\}$.

13.21 Let $\mathscr{A}$ be a unital algebra and $x \in \mathscr{A}$ invertible. Show that $\sigma(x^{-1}) = \{z : z^{-1} \in \sigma(x)\}$.

13.22 Let X be a nonempty set and $f \in B(X)$. Show that $\sigma(f) = \operatorname{cl} f(S)$.

13.23 Let U be an open subset of $\mathbb{C}$ and let $\mathscr{A}$ be the Banach algebra of all bounded analytic functions on U with the sup norm. Show that for any $f \in \mathscr{A}$, $\sigma(f) = \operatorname{cl} f(U)$.

13.24 Define T on $L^2[0,1]$ by $(Tf)(x) = xf(x)$. Find $\sigma(T)$.

13.25 [↑8.34] Find the spectrum of the left shift and right shift operators on ℓ^2.

13.26 Let $\mathscr{A}$ be a unital Banach algebra and x, $y \in \mathscr{A}$. Prove:

(a) $e - xy$ is invertible iff $e - yx$ is invertible. ⟦If $z := (e - xy)^{-1}$, consider $e + yzx$.⟧

(b) $\sigma(xy) \setminus \{0\} = \sigma(yx) \setminus \{0\}$.

(c) $r(xy) = r(yx)$.

(d) For the shift operators on $\ell^2(\mathbb{N})$, $\sigma(T_r T_\ell) \neq \sigma(T_\ell T_r)$.

13.27 Let x, $y \in \mathscr{A}$ such that $xy = yx$. Show that $r(xy) \le r(x)r(y)$ and that equality holds if $x = y$.

13.28 (*Resolvent identity*). The **resolvent function** of a member x of a unital Banach algebra is the function $R(z) = (ze - x)^{-1}$, $z \in \rho(x)$. Verify that $R(z) - R(w) = (w - z)R(z)R(w)$.

13.29 Consider the Banach algebra $C^1[0,1]$ of Ex. 13.9. Let $f(x) = x$. Show that $r(f) = 1 < \|f\|$.

13.30 Let $\mathscr{A}$ be a unital C^*-algebra and $\mathscr{B}$ a closed C^*-subalgebra of $\mathscr{A}$ containing the identity. Let $x \in \mathscr{B}$. Obviously, $\sigma_{\mathscr{A}}(x) \subseteq \sigma_{\mathscr{B}}(x)$, hence $\rho_{\mathscr{B}}(x) \subseteq \rho_{\mathscr{A}}(x)$. Carry out the following steps to prove that $\sigma_{\mathscr{B}}(x) \subseteq \sigma_{\mathscr{A}}(x)$ and hence that $\sigma_{\mathscr{A}}(x) = \sigma_{\mathscr{B}}(x)$.

(a) If $U \subseteq V$ are open subsets of $\mathbb{C}$ and $V \cap \operatorname{bd}(U) = \emptyset$, then every component of U is a component of V. ⟦If U' is a component of U, then $\operatorname{bd} U' \subseteq \operatorname{bd} U$. ⟧

(b) If z is a boundary point of the open set $\rho_{\mathscr{B}}(x)$ and $z_n \in \rho_{\mathscr{B}}(x)$ with $z_n \to z$, then $\|(z_n e - x)^{-1}\| \to \infty$. ⟦Use Ex. 13.13.⟧

(c) $\rho_{\mathscr{A}}(x) \cap \operatorname{bd} \rho_{\mathscr{B}}(x) = \emptyset$.

(d) $\sigma_{\mathscr{B}}(x)$ is the union of $\sigma_{\mathscr{A}}(x)$ and certain bounded components of $\rho_{\mathscr{A}}(x)$.

(e) If x is self-adjoint, then $\sigma_{\mathscr{A}}(x) = \sigma_{\mathscr{B}}(x)$. ⟦$\rho_{\mathscr{A}}(x)$ is connected.⟧

(f) If x is invertible in $\mathscr{A}$, it is invertible in $\mathscr{B}$. ⟦x^*x is invertible in $\mathscr{B}$.⟧

(g) $\sigma_{\mathscr{A}}(x) = \sigma_{\mathscr{B}}(x)$.

13.3 The Spectrum of an Algebra

In this section, $\mathscr{A}$ denotes a commutative, unital Banach algebra.

Characters

A **character** of $\mathscr{A}$ is a homomorphism χ from $\mathscr{A}$ into $\mathbb{C}$ that is not identically zero. Thus $\chi(\boldsymbol{e}) \neq 0$, and it follows from the calculation $\chi(\boldsymbol{e}) = \chi(\boldsymbol{e}^2) = \chi(\boldsymbol{e})^2$ that $\chi(\boldsymbol{e}) = 1$. The collection of all characters of $\mathscr{A}$ is called the **spectrum** or **character space** of $\mathscr{A}$ and is denoted by $\sigma(\mathscr{A})$. For example, if X is a topological space and $x \in X$, then the mapping $f \mapsto f(x)$ is a character of the Banach algebra $C_b(X)$.

13.3.1 Proposition. *If χ is a character, then χ is continuous and $\|\chi\| \leq 1$.*

Proof. Let $\boldsymbol{x} \in \mathscr{A}$ and suppose that $|\chi(\boldsymbol{x})| > \|\boldsymbol{x}\|$. Set $\alpha = 1/\chi(\boldsymbol{x})$. Then $\|\alpha\boldsymbol{x}\| < 1$, so $\boldsymbol{e} - \alpha\boldsymbol{x}$ is invertible. Denote the inverse by $\boldsymbol{y}$, so that $\boldsymbol{y} - \alpha\boldsymbol{y}\boldsymbol{x} = \boldsymbol{y}(\boldsymbol{e} - \alpha\boldsymbol{x}) = \boldsymbol{e}$. But then $1 = \chi(\boldsymbol{e}) = \chi(\boldsymbol{y}) - \alpha\chi(\boldsymbol{y})\chi(\boldsymbol{x}) = \chi(\boldsymbol{y}) - \chi(\boldsymbol{y}) = 0$. Therefore, $|\chi(\boldsymbol{x})| \leq \|\boldsymbol{x}\|$, hence $\|\chi\| \leq 1$. □

The preceding proposition shows that $\sigma(\mathscr{A})$ is a subset of the closed unit ball of $\mathscr{A}'$. As such it inherits the weak* topology of $\mathscr{A}'$, also called the **Gelfand topology** of $\sigma(\mathscr{A})$.

13.3.2 Example. *The spectrum of $C(X)$.* Let X be a compact Hausdorff space. For $x \in X$ let $\widehat{x}$ denote the character $\widehat{x}(f) = f(x)$, $f \in C(X)$. We show that the mapping $x \to \widehat{x}$ is a homeomorphism onto the spectrum $\Sigma := \sigma(C(X))$ of $C(X)$.

The mapping $x \to \widehat{x}$ is obviously continuous in the weak* topology of $C(X)'$. Moreover, since the functions in $C(X)$ separate points (Urysohn's lemma), the mapping is 1-1. It remains to verify surjectivity.

Let $\chi \in \Sigma$. We claim that there exists $x_0 \in X$ such that $g(x_0) = 0$ for all $g \in \ker \chi$. If this is not the case, then for each $x \in X$ there exists $g_x \in \ker \chi$ such that $g_x(x) \neq 0$. By continuity, there exists an open neighborhood U_x of x such that $g_x \neq 0$ on U_x. By compactness of X, there exist $x_1, \ldots, x_n \in X$ such that $X = U_{x_1} \cup \cdots \cup U_{x_n}$. Set $g_j = g_{x_j}$. The function $g := \sum_{j=1}^n g_j \overline{g}_j = \sum_{j=1}^n |g_j|^2$ is then positive on X and hence invertible in $C(X)$. On the other hand, $\chi(g) = \sum_{j=1}^n \chi(g_j)\overline{\chi}(g_j) = 0$, impossible for an invertible element. This verifies the claim.

Now let $f \in C(X)$. Then $h := f - \chi(f) \cdot 1 \in \ker \chi$, hence $h(x_0) = 0$ and so $\widehat{x}_0(f) = \chi(f)$. Therefore, the mapping $x \to \widehat{x}$ is surjective. ◊

Maximal Ideals

A **maximal ideal** of $\mathscr{A}$ is a proper ideal that is not contained in a larger proper ideal. Here is an interesting and illuminating example.

13.3.3 Example. Let X be a (nontrivial) compact Hausdorff space. For a subset Y of X, set $I_Y := \{f \in C(X) : f(Y) = 0\}$. Then I_Y is easily seen to be a proper ideal of $C(X)$. We show that I_Y is maximal iff Y is a singleton.

To show that I_y is maximal, suppose that I_y is properly contained in an ideal I and let $f \in I \setminus I_y$, so that $f(y) \neq 0$. Define $g(x) = f(x) - f(y)$ $(x \in X)$. Then $g(y) = 0$, hence $g \in I_y \subseteq I$. It follows that the nonzero constant function $f - g = f(y)$ is in I, hence $I = C(X)$. Therefore, I_y is maximal.

Conversely, if Y has more than one element and $y \in Y$, then by Urysohn's lemma we can construct a function $f \in I_y \setminus I_Y$. Then I_Y is properly contained in I_y, so is not maximal. ◊

13.3.4 Proposition. *Every proper ideal $\mathscr{I}$ is contained in a maximal ideal and every maximal ideal is closed.*

Proof. Partially order the collection of proper ideals of $\mathscr{A}$ containing $\mathscr{I}$ by inclusion. The union $\mathscr{J}$ of a chain of proper ideals containing $\mathscr{I}$ is an ideal containing $\mathscr{I}$ and is proper since $e \notin \mathscr{J}$. Therefore, $\mathscr{J}$ is an upper bound for the chain. By Zorn's lemma, $\mathscr{I}$ is contained in a maximal ideal.

For the second part of the proposition, let $\mathscr{M}$ is a maximal ideal that is not closed. Then $\mathscr{M}$ is properly contained in $\mathrm{cl}(\mathscr{M})$. But $\mathrm{cl}(\mathscr{M})$ is a proper ideal (Ex. 13.8), contradicting the maximality of $\mathscr{M}$. □

Recall that the quotient space $\mathscr{A}/\mathscr{I}$ of $\mathscr{A}$ by a closed ideal $\mathscr{I}$ is a Banach algebra and the quotient map $Q : \mathscr{A} \to \mathscr{A}/\mathscr{I}$ is a continuous homomorphism (Ex. 13.7). The following theorem will be needed in the proof of 13.3.6 below.

13.3.5 Theorem. *A closed ideal $\mathscr{I}$ in $\mathscr{A}$ is maximal iff $\mathscr{A}/\mathscr{I}$ is a field.*

Proof. Assume that $\mathscr{I}$ is maximal. Let $Q_{\mathscr{I}} : \mathscr{A} \to \mathscr{A}/\mathscr{I}$ denote the quotient map. If $x \in \mathscr{A}$ and $Q(x)$ is not invertible, then the ideal $I := Q(\mathscr{A}x) = Q(\mathscr{A})Q(x)$ of $\mathscr{A}/\mathscr{I}$ cannot contain the identity and so is a proper ideal of $\mathscr{A}/\mathscr{I}$. It follows that $Q^{-1}(I)$ is a proper ideal of $\mathscr{A}$ containing $\mathscr{I} = Q^{-1}(0)$, so $Q^{-1}(I) = \mathscr{I}$ by maximality of $\mathscr{I}$. Since $x \in Q^{-1}(I)$, $Q(x) = 0$. Therefore, the only non-invertible member of $\mathscr{A}/\mathscr{I}$ is the zero element, hence $\mathscr{A}/\mathscr{I}$ is a field.

Conversely, suppose $\mathscr{A}/\mathscr{I}$ is a field and $\mathscr{I}$ is not maximal. Let $\mathscr{M}$ be a maximal ideal properly containing $\mathscr{I}$. Then there exists a continuous homomorphism $T : \mathscr{A}/\mathscr{I} \to \mathscr{A}/\mathscr{M}$ such that $TQ_{\mathscr{I}} = Q_{\mathscr{M}}$. Let $x \in \mathscr{M} \setminus \mathscr{I}$, so that $Q_{\mathscr{M}}x = 0$ and $Q_{\mathscr{I}}x \neq 0$. Then $(Q_{\mathscr{I}}x)^{-1}$ exists and so

$$Q_{\mathscr{M}}e = T(Q_{\mathscr{I}}x)T\big((Q_{\mathscr{I}}x)^{-1}\big) = (Q_{\mathscr{M}}x)T\big((Q_{\mathscr{I}}x)^{-1}\big) = 0,$$

impossible since $\mathscr{A}/\mathscr{M}$ is nontrivial. Therefore, $\mathscr{I}$ must be maximal. □

13.3.6 Theorem. *The mapping $\chi \to \ker\chi$ is a 1-1 correspondence between the spectrum of $\mathscr{A}$ and the collection of all maximal ideals of $\mathscr{A}$.*

Proof. Since $\chi(e) = 1$, $\ker\chi$ is a proper ideal of $\mathscr{A}$. Let $\mathscr{I}$ be an ideal properly containing $\ker\chi$ and let $x \in \mathscr{I} \setminus \ker\chi$. Then every member a of $\mathscr{A}$ may be written

$$a = \big[\chi(a)\chi(x)^{-1}x\big] + \big[a - \chi(a)\chi(x)^{-1}x\big].$$

The first term on the right is in $\mathscr{I}$ and the second is a member of $\ker\chi \subseteq \mathscr{I}$. Therefore, $a \in \mathscr{I}$, proving that $\mathscr{I} = \mathscr{A}$. Therefore, $\ker\chi$ is maximal.

Now let $\mathscr{M}$ be any maximal ideal. By 13.3.5, $\mathscr{A}/\mathscr{M}$ is a field, hence, by the Gelfand-Mazur theorem, $\mathscr{A}/\mathscr{M} = \{ze + \mathscr{M} : z \in \mathbb{C}\}$. Now define $\chi_0(ze + \mathscr{M}) = z$ and set $\chi = \chi_0 \circ Q_{\mathscr{M}}$. Then χ is a character with kernel $\mathscr{M}$.

Finally, if $\ker\chi_1 = \ker\chi_2$, then $c\chi_1 = \chi_2$ for some $c \in \mathbb{C}$ (0.2.3). Since $\chi_1(e) = \chi_2(e) = 1$, $\chi_1 = \chi_2$. □

Because of the 1-1 correspondence in 13.3.6, the spectrum $\sigma(\mathscr{A})$ of $\mathscr{A}$ is also called the **maximal ideal space** of $\mathscr{A}$.

Exercises

13.31 Show that $G^c_{\mathscr{A}}$ is the union of all maximal ideals in $\mathscr{A}$.

13.32 Let $\mathscr{A}$ and $\mathscr{B}$ be commutative, unital Banach algebras and $\Phi : \mathscr{A} \to \mathscr{B}$ a surjective homomorphism. Prove: if $\mathscr{M}$ is a maximal ideal in $\mathscr{A}$, then $\Phi(\mathscr{M})$ is a maximal ideal in $\mathscr{B}$.

13.33 [↑13.14] Show that the evaluation mapping $\widehat{z}$ is a homeomorphism from $\mathrm{cl}\,\mathbb{D}$ onto the spectrum $\Sigma := \sigma\big(A(\mathbb{D})\big)$ of the disk algebra. ⟦For surjectivity, let $\chi \in \Sigma$ and show that there exists a $z \in \mathbb{T}$ such that $\chi(P) = P(z)$ for every polynomial P on $\mathrm{cl}\,\mathbb{D}$.⟧

13.34 Let $\mathscr{A}$ be a unital Banach algebra, $x \in \mathscr{A}$ and let $\mathscr{B}$ be the closed subalgebra of $\mathscr{A}$ generated by x and e. Show that the map $F(\chi) = \chi(x)$ defines a homeomorphism from $\Sigma_{\mathscr{B}}$ onto $\sigma_{\mathscr{B}}(x)$.

13.35 Let χ be a linear functional on $\mathscr{A}$ with $\chi(e) \neq 0$. Prove that the following are equivalent.
(a) χ is a character of $\mathscr{A}$. (b) ker $\mathscr{A}$ is an ideal. (c) ker $\mathscr{A}$ is a subalgebra and $\chi(e) = 1$.

13.36 The **radical** of $\mathscr{A}$ is the intersection of all maximal ideals in $\mathscr{A}$. Prove that the radical consists of all $x \in \mathscr{A}$ such that $\lim_n (cx)^n = 0$ for all $c \in \mathbb{C}$.

13.4 Gelfand Theory

The Representation Theorem

Here is the main result of the chapter. Applications illustrating the broad utility of the theorem are given below and in §13.6.

13.4.1 Gelfand Representation Theorem. *Let $\mathscr{A}$ be a unital, commutative Banach algebra with spectrum $\Sigma := \sigma(\mathscr{A})$.*

(a) *Σ is a weak*-compact subset of the closed unit ball in $\mathscr{A}^*$.*

(b) *For each $x \in \mathscr{A}$, the function $\widehat{x}$ on Σ defined by $\widehat{x}(\chi) = \chi(x)$ is a continuous function such that $\widehat{x}(\Sigma) = \sigma(x)$, hence $\|\widehat{x}\| = r(x) \le \|x\|$.*

(c) *The map $x \to \widehat{x}$ is a continuous homomorphism from $\mathscr{A}$ into $C(\Sigma)$.*

(d) *A member x of $\mathscr{A}$ is invertible iff $\widehat{x}$ is invertible in $C(\Sigma)$.*

(e) *If $\mathscr{A}$ is a C^*-algebra, then $x \to \widehat{x}$ is an isometric $*$-isomorphism of $\mathscr{A}$ onto $C(\Sigma)$.*

Proof. (a) Since the pointwise limit of a net of characters is a character, Σ is w^*-closed subset of the closed unit ball of $\mathscr{A}'$ and so is w^*-compact by Alaoglu's theorem.

(b) Clearly, $\widehat{x}$ is w^*-continuous. If $z \in \sigma(x)$, then $e \notin (ze - x)\mathscr{A}$, hence $(ze - x)\mathscr{A}$ is a proper ideal and so is contained in a maximal ideal. Therefore, $\chi(ze - x) = 0$ for some character χ, hence $z = \widehat{x}(\chi) \in \widehat{x}(\Sigma)$. Conversely, if $z \notin \sigma(x)$, then $ze - x$ is not contained in any maximal ideal, hence $\chi(x) \neq z$ for all χ, that is, $z \notin \widehat{x}(\Sigma)$. Therefore, $\widehat{x}(\Sigma) = \sigma(x)$.

Part (c) is clear. For (d), if x is invertible, then $1 = \widehat{xx^{-1}} = \widehat{x}\widehat{x^{-1}}$, hence $(\widehat{x})^{-1} = \widehat{x^{-1}}$. Conversely, if $\widehat{x}$ is invertible in $C(\Sigma)$, then $0 \notin \widehat{x}(\Sigma) = \sigma(x)$, so x is invertible.

(e) By 13.2.9, $r(x) = \|x\|$, hence by (b) the map $x \mapsto \widehat{x}$ is an isometry. We prove next that $\widehat{x^*} = \overline{\widehat{x}}$. By 13.2.8, if x is self-adjoint, then its range $\sigma(x)$ is real, so the desired equation holds trivially. In the general case write $x = y + iz$, where $y = (x + x^*)/2$ and $z = (x - x^*)/2i$. Since these are self-adjoint, $x^* = y - iz$, hence $\widehat{x^*} = \widehat{y} - i\widehat{z} = \overline{\widehat{x}}$. We have shown that $x \to \widehat{x}$ is an isometric $*$-isomorphism of $\mathscr{A}$ into $C(\Sigma)$. It follows that $\widehat{\mathscr{A}}$ is a conjugate closed norm closed subalgebra of $C(\Sigma)$. Since $\widehat{\mathscr{A}}$ separates points of Σ and contains the constant functions, $\widehat{\mathscr{A}} = C(\Sigma)$ by the Stone-Weierstrass theorem. □

The function $\widehat{\boldsymbol{x}} \in C(\Sigma)$ in the representation theorem is called the **Gelfand transform** of $\boldsymbol{x}$. The map

$$\Gamma : \mathscr{A} \to C(\Sigma), \quad \Gamma(\boldsymbol{x}) = \widehat{\boldsymbol{x}},$$

is called the **Gelfand transform** of $\mathscr{A}$.

Application: The Stone-Čech Compactification

In this subsection, we apply the Gelfand representation theorem to unital C^*-subalgebras $\mathscr{F}$ of $C_b(S)$, where S is a topological space. We denote by $S^{\mathscr{F}}$ the spectrum of $\mathscr{F}$ with the Gelfand topology. For each $x \in S$, the mapping $\widehat{x}$ on $\mathscr{F}$ defined by

$$\widehat{x}(f) := f(x), \quad f \in \mathscr{F}$$

is clearly a member $S^{\mathscr{F}}$. The Gelfand transform $\widehat{f} \in C(S^{\mathscr{F}})$ of $f \in \mathscr{F}$ then satisfies $(\widehat{f})(\widehat{x}) = \widehat{x}(f) = f(x)$. The mapping

$$\iota_S : S \to S^{\mathscr{F}}, \ \iota_S(x) = \widehat{x},$$

is called the **canonical mapping** from S to $S^{\mathscr{F}}$. The Gelfand representation theorem yields a simple proof of the following generalization of the Stone-Čech compactification theorem:

13.4.2 Theorem. *Let S be a topological space and $\mathscr{F}$ a unital C^*-subalgebra of $C_b(S)$.*

(a) *$S^{\mathscr{F}}$ is compact Hausdorff topological space and ι_S is a continuous function from S onto a dense subset of $S^{\mathscr{F}}$.*

(b) *The adjoint map $\iota_S^* : C(S^{\mathscr{F}}) \to \mathscr{F}$ is a surjective isometric isomorphism.*

(c) *Let T be a topological space, $\mathscr{G}$ a unital C^*-subalgebra of $C_b(T)$, and $\varphi : S \to T$ a continuous function such that the dual map $\varphi^* : C_b(T) \to C_b(S)$ maps $\mathscr{G}$ into $\mathscr{F}$. Then there exists a continuous map $\widetilde{\varphi} : S^{\mathscr{F}} \to T^{\mathscr{G}}$ such that the following diagram commutes:*

$$\begin{array}{ccc} S^{\mathscr{F}} & \xrightarrow{\ \widetilde{\varphi}\ } & T^{\mathscr{G}} \\ {\scriptstyle \iota_S}\big\uparrow & & {\scriptstyle \iota_T}\big\uparrow \\ S & \xrightarrow{\ \varphi\ } & T \end{array}$$

In particular, $S^{\mathscr{F}}$ is unique up to homeomorphism.

Proof. The Gelfand transform $f \mapsto \widehat{f} : \mathscr{F} \to C(S^{\mathscr{F}})$ is an isometric isomorphism onto $C(S^{\mathscr{F}})$. To see that the mapping ι_S is continuous, let $x_\alpha \to x$ and note that $\widehat{f}(\widehat{x}_\alpha) = f(x_\alpha) \to f(x) = \widehat{f}(\widehat{x})$. Since $S^{\mathscr{F}}$ is compact and every function in $C(S^{\mathscr{F}})$ is of the form $\widehat{f}$, it follows that $\widehat{x}_\alpha \to \widehat{x}$ in $S^{\mathscr{F}}$. From the identity $\iota_S^*(\widehat{f})(x) = \widehat{f}(\widehat{x}) = f(x)$ we see that ι_S^* is the inverse of the Gelfand transform. Since every function $\widehat{f}$ that is zero on $\iota_S(S)$ is identically zero, $\iota_S(S)$ is dense in $S^{\mathscr{F}}$. This proves (a) and (b).

For (c), given $\chi \in S^{\mathscr{F}}$ define $\widetilde{\varphi}(\chi)$ to be the character $g \to \chi(\varphi^*(g))$ on $\mathscr{G}$. In particular, for $x \in S$ we have

$$\widetilde{\varphi}(\hat{x})(g) = \widehat{x}(g \circ \varphi) = g(\varphi(x)) = \widehat{\varphi(x)}(g), \quad g \in \mathscr{G},$$

hence $\widetilde{\varphi} \circ \iota_S = \iota_T \circ \varphi$. Clearly, $\widetilde{\varphi}$ is continuous in the Gelfand topology. □

The mapping $\iota = \iota_S$ need not be a homeomorphism or even 1-1. However, if S is completely regular and $\mathscr{F} = C_b(S)$, then ι *does* have these properties. Indeed, that ι is 1-1 in this case follows from the fact that $C_b(S)$ separates points of S. To see that ι is an open map and hence a homeomorphism, let $U \subseteq S$ and $x_1 \in U$. By the complete regularity of S, there exists $f \in C_b(S)$ such that $f(x_1) = 1$ and $f = 0$ on $S \setminus U$. The set $\{\widehat{x} : f(x) = \langle f, \widehat{x}\rangle > 0\}$ is then a weak* neighborhood of $\widehat{x}_1$ contained in $\iota(U)$. Thus $\iota(U)$ is open in $\iota(S)$. We have proved the following classical result:

13.4.3 Theorem (M. Stone, E. Čech). *If S is completely regular and $\mathscr{F} = C_b(S)$, then $\iota_S : S \to \iota_S(S)$ is a homeomorphism.*

The space $S^{C_b(S)}$ in the preceding theorem is called **Stone-Čech compactification** of S and is denoted by βS. We shall also use this notation for the spectrum of $C_b(S)$ for *any* topological space S.

Application: Wiener's Theorem

An **absolutely convergent trigonometric series** is an infinite series of the form

$$f_x(t) = \sum_{k=-\infty}^{\infty} x_k e^{ikt}, \ t \in \mathbb{R}, \quad \text{where } \|\boldsymbol{x}\|_1 = \sum_{k=-\infty}^{\infty} |x_k| < \infty.$$

In this subsection we use the Gelfand representation theorem to prove the following classical result:

13.4.4 Theorem (N. Wiener). *The reciprocal of an absolutely convergent, nonvanishing trigonometric series is an absolutely convergent trigonometric series.*

Proof. (Gelfand). We apply the representation theorem to the unital, commutative Banach algebra $\ell^1(\mathbb{Z})$ (13.1.1(f)). We claim that the characters of $\ell^1(\mathbb{Z})$ are the functions χ_z defined by

$$\chi_z(\boldsymbol{x}) = \sum_{n=-\infty}^{\infty} x_n z^n, \quad \boldsymbol{x} \in \ell^1(\mathbb{Z}), \ z \in \mathbb{T}.$$

Clearly $\chi_z(\boldsymbol{e}_0) = 1$. The calculation

$$\chi_z(\boldsymbol{x} * \boldsymbol{y}) = \sum_{n=-\infty}^{\infty} \sum_{k=-\infty}^{\infty} x_{n-k} y_k z^n = \sum_{k=-\infty}^{\infty} y_k z^k \sum_{n=-\infty}^{\infty} z^{n-k} x_{n-k} = \chi_z(\boldsymbol{x})\chi_z(\boldsymbol{y})$$

then shows that χ_z is a character. Now let χ be any character and define $\boldsymbol{e}_n$ $(n \in \mathbb{Z})$ by $\boldsymbol{e}_n(k) = \delta_{nk}$. Then $\boldsymbol{x} = \sum_{k=-\infty}^{\infty} x_n \boldsymbol{e}_n$ in the ℓ^1-norm. Moreover, from the identity $\boldsymbol{e}_n * \boldsymbol{e}_m = \boldsymbol{e}_{m+n}$ we see that $\boldsymbol{e}_1 * \cdots * \boldsymbol{e}_1 = \boldsymbol{e}_n$ and $\boldsymbol{e}_n * \boldsymbol{e}_{-n} = \boldsymbol{e}_0$, hence $\chi(\boldsymbol{e}_n) = \chi(\boldsymbol{e}_1)^n$ for all $n \in \mathbb{Z}$. Setting $z = \chi(\boldsymbol{e}_1) \in \mathbb{T}$ we have

$$\chi(\boldsymbol{x}) = \sum_{n=-\infty}^{\infty} x_n \chi(\boldsymbol{e}_n) = \sum_{k=-\infty}^{\infty} x_n z^n,$$

that is, $\chi = \chi_z$, verifying the claim.

The spectrum of $\ell^1(\mathbb{Z})$ may now be identified with $\mathbb{T}$ under the bijection $\chi_z \to z$. Consequently, the Gelfand transform $\widehat{\boldsymbol{x}}$ of $\boldsymbol{x} \in \ell^1(\mathbb{Z})$ may be written as

$$\widehat{\boldsymbol{x}}(z) = \widehat{\boldsymbol{x}}(\chi_z) = \sum_{n=-\infty}^{\infty} x_n z^n, \quad z \in \mathbb{T}.$$

Thus the absolutely convergent trigonometric series are precisely the characters $\widehat{\boldsymbol{x}}$. If $\widehat{\boldsymbol{x}}$ is never zero, then by (d) of the representation theorem the reciprocal $1/\widehat{\boldsymbol{x}}$ is the Gelfand transform of a member of $\ell^1(\mathbb{Z})$, proving the theorem. □

<u>**Exercises**</u>

13.37 Let $\mathscr{A}$ be a unital C^*-algebra and $x \in \mathscr{A}$ normal. Prove the following:

(a) x is unitary iff $\sigma(x) \subseteq \mathbb{T}$.

(b) x is self-adjoint iff $\sigma(x) \subseteq \mathbb{R}$.

(c) x is a projection iff $\sigma(x) \subseteq \{0, 1\}$.

(d) If $\mathscr{A}$ is commutative, then $x \geq 0$ iff $\sigma(x) \subseteq \mathbb{R}^+$. (See Ex. 13.51 for a strengthened version.)

⟦Consider $\mathscr{C}^*(x)$.⟧

13.38 Let $\mathscr{A}$ be a unital commutative C^* algebra and let $x \in \mathscr{A}$ be a projection such that $x \neq 0$ and $x \neq e$. Show that the spectrum of x is disconnected.

13.39 Show that the spectrum of $x \in \ell^1(\mathbb{Z})$ consists of all numbers $\sum_{n=-\infty}^{\infty} x_n z^n$ with $z \in \mathbb{T}$.

13.40 Let $\mathscr{A}$ be a unital Banach algebra and $x, y \in \mathscr{A}$ with $xy = yx$. Prove: $r(x+y) \leq r(x)+r(y)$.

13.41 Let $\mathscr{A}$ be the Banach algebra $C^1[0, 1]$ with the norm $\|f\| = \|f\|_\infty + \|f'\|_\infty$. One may argue exactly as in 13.3.2 that the mapping $x \to \widehat{x}$ is a homeomorphism from $[0, 1]$ onto the spectrum of $\mathscr{A}$, so the spectrum may be identified with $[0, 1]$. Show that the Gelfand transform $\Gamma : \mathscr{A} \to C[0, 1]$ is neither surjective nor an isometry.

---

*13.5 The Non-unital Case

The Unitization of a Banach Algebra

Let $\mathscr{A}$ be a non-unital, commutative Banach algebra. The spectrum $\sigma(\mathscr{A})$ of $\mathscr{A}$ is defined exactly as in the unital case, namely, the set of all non-identically zero homomorphisms $\chi : \mathscr{A} \to \mathbb{C}$. To make use of the Gelfand machinery, we adjoin an identity to $\mathscr{A}$ as follows: Let $\mathscr{A}_1 = \mathscr{A} \times \mathbb{C}$ be the product vector space and define multiplication in $\mathscr{A}_1$ by

$$(x, a)(y, b) = (xy + ay + bx, ab). \quad x, y \in \mathscr{A}, \ a \in \mathbb{C}.$$

Then $\mathscr{A}_1$ is an algebra with identity $(0, 1)$. Moreover, $(x, a) = (x, 0) + (0, a)(0, 1)$, so identifying $\mathscr{A} \times \{0\}$ with $\mathscr{A}$ and $\{0\} \times \mathbb{C}$ with $\mathbb{C}$ we may write $(x, a) = x + a$. With this algebraic identification, $\mathscr{A}$ is a maximal ideal in $\mathscr{A}_1$. Moreover, it is easy to check that $\mathscr{A}_1$ is a Banach algebra under the norm

$$\|(x, a)\| = \|x\| + |a|$$

and that $\mathscr{A}$ is isometrically isomorphic to $\mathscr{A} \times 0$. (In Ex. 13.43, the reader is asked to verify these assertions.) The algebra $\mathscr{A}_1$ is called the **unitization of** $\mathscr{A}$.

The spectrum of $\mathscr{A}_1$ is related to that of $\mathscr{A}$ as follows: For $\chi \in \sigma(\mathscr{A})$ define

$$\chi_1(x, a) = \chi(x) + a. \tag{13.3}$$

Then χ_1 is easily seen to be a character of $\mathscr{A}_1$ (Ex. 13.44). In particular, $|\chi(x)| = |\chi_1(x, 0)| \leq \|\chi_1\| \, \|(x, 0)\| = \|x\|$, so $\|\chi\| \leq 1$. Thus a character of $\mathscr{A}$ is a member of the closed unit ball of $\mathscr{A}'$. The spectrum of $\mathscr{A}$ may not be closed, but it *is* the case that $\sigma(\mathscr{A}) \cup \{0\}$ is closed and hence weak* compact. Indeed, if $\chi_\alpha \to \varphi$ in the weak* topology of $\sigma(\mathscr{A})$, then φ is easily seen to be a homomorphism, hence either $\varphi = 0$ or $\varphi \in \sigma(\mathscr{A})$. Now let φ be any character of $\sigma(\mathscr{A}_1)$. Then

$$\varphi(x, a) = \varphi(x, 0) + \varphi(0, a) = \varphi(x, 0) + a.$$

The map $\boldsymbol{x} \to \varphi(\boldsymbol{x}, 0)$ is either a character of $\mathscr{A}'$ or the zero homomorphism. In the former case, φ is of the form χ_1 as in (13.3), and in the latter case φ is the character $\varphi_0(\boldsymbol{x} + a) := a$. Thus we see that

$$\sigma(\mathscr{A}_1) = \{\chi_{_1} : \chi \in \sigma(\mathscr{A})\} \cup \{\varphi_0\} \text{ and } \sigma(\mathscr{A}_1)\big|_{\mathscr{A}} = \sigma(\mathscr{A}) \cup \{0\}. \tag{13.4}$$

In this way we may identify $\sigma(\mathscr{A}_1)$ with $\sigma(\mathscr{A}) \cup \{0\}$. From (13.3), the Gelfand transforms $\Gamma : \mathscr{A} \to \sigma(\mathscr{A})$ and $\Gamma_1 : \mathscr{A}_1 \to \sigma(\mathscr{A}_1)$ are related by

$$\Gamma_1(\boldsymbol{x}, a)(\chi_{_1}) = \Gamma(\boldsymbol{x})(\chi) + a.$$

The Non-unital Representation Theorem

13.5.1 Theorem. *Let $\mathscr{A}$ be a non-unital, commutative Banach algebra. Then*

(a) *The spectrum $\Sigma := \sigma(\mathscr{A})$ is locally compact in the w^*-topology.*

(b) *The Gelfand transform Γ is a continuous algebra homomorphism of $\mathscr{A}$ into $C_0(\Sigma)$.*

(c) *For all $\boldsymbol{x} \in \mathscr{A}$, $\|\widehat{\boldsymbol{x}}\|_\infty = \lim_n \|\boldsymbol{x}^n\|^{1/n}$.*

(d) *If $\mathscr{A}$ is a $*$-algebra and $\widehat{\boldsymbol{x}^*} = \overline{\widehat{\boldsymbol{x}}}$ for all $\boldsymbol{x}$, then $\Gamma(\mathscr{A})$ is dense in $C_0(\Sigma)$.*

Proof. Since $\sigma(\mathscr{A}) \cup \{0\}$ is weak* compact and since removing a point from a compact space produces a locally compact space, we see that $\sigma(\mathscr{A})$ is locally compact, proving (a). Thus if $\sigma(\mathscr{A})$ is not compact, then $\sigma(\mathscr{A}) \cup \{0\}$ is the one-point compactification of $\sigma(\mathscr{A})$.

Part (b) is clear. To prove (c) recall that, by the unital case,

$$\|\widehat{(\boldsymbol{x}, 0)}\|_\infty = \lim_n \|(\boldsymbol{x}, 0)^n\|^{1/n} = \lim_n \|\boldsymbol{x}^n\|^{1/n}.$$

Furthermore,

$$\widehat{(\boldsymbol{x}, 0)}(\chi_{_1}) = \Gamma((\boldsymbol{x}, 0))(\chi_{_1}) = \Gamma(\boldsymbol{x})(\chi) = \widehat{\boldsymbol{x}}(\chi) \text{ and } \widehat{(\boldsymbol{x}, 0)}(\varphi_0) = 0,$$

hence $\|\widehat{(\boldsymbol{x}, 0)}\|_\infty = \|\widehat{\boldsymbol{x}}\|_\infty$. Therefore, (b) holds.

Finally, the hypothesis in (d) implies that $\Gamma(\mathscr{A})$ is conjugate closed. Since $\Gamma(\mathscr{A})$ trivially separates points and characters are not identically zero, the locally compact version of the Stone-Weierstrass theorem (0.12.13) implies that $\Gamma(\mathscr{A})$ is dense in $C_0(X)$. □

The Spectrum of $C_0(X)$

Let X be a noncompact, locally compact Hausdorff space. We show that the mapping $F : x \mapsto \widehat{x}$ is a homeomorphism from X onto the spectrum Σ of $C_0(X)$, so that Σ may be identified with X. The argument uses Example 13.3.2 applied to the one-point compactification $X_\infty = X \cup \{\infty\}$ of X.

The mapping F is obviously a continuous injection into Σ. To show that F is surjective, let $\chi \in \Sigma$ and define χ_∞ on $C(X_\infty)$ by

$$\chi_\infty(f) = \chi(f_0) + f(\infty), \quad f_0 := f\big|_X - f(\infty), \quad f \in C(X_\infty).$$

Then χ_∞ agrees with χ on the space $C_0(X)$, which may be identified with the set of functions in $C(X_\infty)$ that are zero at ∞. We claim that χ_∞ is in the spectrum Σ_∞ of $C(X_\infty)$. Clearly, χ_∞ is linear and $\chi_\infty(1) = 1$. From

$$f_0 g_0 = \left[f\big|_X - f(\infty)\right]\left[g\big|_X - g(\infty)\right] = (fg)\big|_X - f(\infty)g_0 - g(\infty)f_0 - f(\infty)g(\infty)$$

we have

$$(fg)_0 = (fg)\big|_X - f(\infty)g(\infty) = f_0g_0 + f(\infty)g_0 + g(\infty)f_0.$$

Since $f_0, g_0 \in C_0(X)$,

$$\begin{aligned}\chi_\infty(fg) &= \chi\big((fg)_0\big) + (fg)(\infty) = \chi(f_0)\chi(g_0) + f(\infty)\chi(g_0) + g(\infty)\chi(f_0) + f(\infty)g(\infty)\\ &= \chi_\infty(f)\chi_\infty(g).\end{aligned}$$

Therefore, $\chi_\infty \in \Sigma_\infty$ and so $\chi_\infty = \widehat{\infty}$ or $\widehat{x}$ for some $x \in X$. But if the former, then for all $f \in C(X_\infty)$ we have $f(\infty) = \chi_\infty(f) = \chi(f_0) + f(\infty)$, which implies that $\chi(g) = 0$ for all $g \in C_0(X)$, contrary to the definition of character. Thus $\chi_\infty = \widehat{x}$ for some $x \in X$ and so $\chi = \widehat{x} = F(x)$, proving that F is surjective.

It remains to show that F^{-1} is continuous. This follows from the implications $\widehat{x}_\alpha \to \widehat{x}$ in $\Sigma \Rightarrow f(x_\alpha) \to f(x)$ for all $f \in C_0(X) \Rightarrow f(x_\alpha) \to f(x)$ for all $f \in C(X_\infty) \Rightarrow \widehat{x}_\alpha \to \widehat{x}$ in $\Sigma_\infty \Rightarrow x_\alpha \to x$ in $X_\infty \Rightarrow x_\alpha \to x$ in X. ◊

The Spectrum of $L^1(\mathbb{R}^d)$

We show that the spectrum Σ of the Banach algebra $L^1(\mathbb{R}^d)$ (13.1.1(g)) may be identified with $\mathbb{R}^d$ as follows: For $t \in \mathbb{R}^d$, define a function ϕ_t on $L^1(\mathbb{R}^d)$ by

$$\phi_t(f) := \int e^{i\,t\cdot x} f(x)\,dx. \tag{†}$$

Thus $\phi_t(f)$ is simply a variation of the Fourier transform of f evaluated at t and so ϕ_t is a member of Σ (see 6.2.1(a)). Now define $F : \mathbb{R}^d \to \Sigma,\ \ F(t) = \phi_t$. Then F is continuous by the dominated convergence theorem. We show that F is a homeomorphism, allowing us to identify the topological spaces $\mathbb{R}^d$ and Σ.

That F is 1-1 follows from (L^1, L^∞) duality. Indeed, if $\phi_{t_1} = \phi_{t_2}$, then $e^{i\,x\cdot t_1} = e^{i\,x\cdot t_2}$ for a.a. x, hence, by continuity, $t_1 = t_2$. To show that F is surjective, let $\phi \in \Sigma$ and use duality again to choose $h \in L^\infty(\mathbb{R}^d)$ such that

$$\phi(f) = \int f(x)h(x)\,dx \quad \text{for all } f \in L^1(\mathbb{R}^d).$$

For $f \in L^1$ and $y \in \mathbb{R}^d$, set $f_y(x) = f(x-y)$. If $f, g \in L^1(\mathbb{R}^d)$ we have

$$\phi(f * g) = \phi(f)\phi(g) = \int g(y)\phi(f)h(y)\,dy. \tag{†}$$

On the other hand, by Fubini's theorem,

$$\begin{aligned}\phi(f * g) &= \int (f * g)(x)h(x)\,dx = \iint f(x-y)g(y)h(x)\,dy\,dx\\ &= \int g(y) \int f(x-y)h(x)\,dx\,dy\\ &= \int g(y)\phi(f_y)\,dy.\end{aligned} \tag{‡}$$

Therefore, from (†) and (‡),

$$\int g(y)\phi(f)h(y)\,dy = \int g(y)\phi(f_y)\,dy \quad \forall\ g \in L^1(\mathbb{R}^d).$$

It follows that for each $f \in L^1$, $\phi(f)h(y) = \phi(f_y)$ for a.a. y. Choosing f so that $\phi(f) \neq 0$, we then have $h(y) = \phi(f)^{-1}\phi(f_y)$ for a.a. y. The right side of this equation is then a continuous version of h. Replace h by this version and note that h, which is uniquely determined by ϕ, does not depend of f. Thus for all $f \in L^1$ and $y \in \mathbb{R}^d$, $\phi(f)h(y) = \phi(f_y)$ and so

$$\phi(f)h(x+y) = \phi(f_{x+y}) = \phi((f_x)_y) = \phi(f_x)h(y) = \phi(f)h(x)h(y).$$

Therefore, $h(x+y) = h(x)h(y)$ for all x, $y \in \mathbb{R}^d$. Since h is continuous, $h(x) = \exp(i\,t \cdot x)$ for some $t \in \mathbb{R}^d$ (Ex. 13.42) and so $\phi = \phi_t$ from (†). Thus F is surjective.

It remains to show that F^{-1} is continuous. For ease of notation we do this for the case $d = 1$; the proof for general case is entirely similar. Let $t \in \mathbb{R}$ and let (t_α) be a net in $\mathbb{R}$ such that

$$F(t_\alpha) - F(t) = \int (e^{it_\alpha x} - e^{itx}) f(x)\,dx \to 0 \text{ for all } f \in L^1. \qquad (‡)$$

Replacing $f(x)$ by $e^{itx}f(x)$, we may take $t = 0$. Thus we must show that $t_\alpha \to 0$. Taking $f = \mathbf{1}_{[0,1]}$ in (‡) and integrating shows that the net (t_α) must be bounded. Let (t_β) be any convergent subnet, say $t_\beta \to s$. Then $e^{i\,t_\beta x} \to e^{i\,sx}$ uniformly in $x \in [0,1]$, and taking $f = (e^{-i\,t_\beta x} - 1)\mathbf{1}_{[0,1]}$ in (‡) we see that

$$\int_0^1 |e^{i\,sx} - 1|^2\,dx = 0.$$

Therefore, $e^{i\,sx} = 1$ for all $x \in [0,1]$, which is possible only if $s = 0$. This shows that $t_\alpha \to 0$, completing the argument. ◊

Exercises

13.42 Let $f \in C_b(\mathbb{R}^d)$ such that $f(x) \neq 0$ and $f(x+y) = f(x)f(y)$ for all x, $y \in \mathbb{R}^d$. Carry out the following steps to prove that there exists $t \in \mathbb{R}^d$ such that $f(x) = \exp(i\,t \cdot x)$ for all x.

(a) There exists $a > 0$ such that $\alpha := \int_0^a \cdots \int_0^a f(y_1, \ldots, y_d)\,dy_1 \ldots dy_d \neq 0$.

(b) $\alpha f(x) = \int_{x_1}^{a+x_1} \cdots \int_{x_d}^{a+x_d} f(y_1, \ldots, y_d)\,dy_1 \ldots dy_d$, hence f is continuously differentiable.

(c) There exists $z_j \in \mathbb{C}$ such that $\partial_j f(0, \ldots 0, x_j, 0, \ldots, 0) = z_j f(0, \ldots 0, x_j, 0, \ldots, 0)$.

(d) $f(0, \ldots 0, x_j, 0, \ldots, 0) = e^{iz_j x_j}$, $z_j = it_j$, t_j real. Draw the desired conclusion.

13.43 Let $\mathscr{A}$ be a Banach algebra and $\mathscr{A}_1$ the unitization of $\mathscr{A}$. Prove the following:.

(a) $\mathscr{A}_1$ is an algebra with identity $\mathbf{1} := (0, 1)$.

(b) $(x, a) = (x, 0) + (0, a)(0, 1)$, so that identifying $\mathscr{A} \times \{0\}$ with $\mathscr{A}$ with $\{0\} \times \mathbb{C}$ with $\mathbb{C}$ we may write $(x, a) = x + a$.

(c) $\mathscr{A}_1$ is commutative iff $\mathscr{A}$ is commutative.

(d) $\mathscr{A}_1$ is a Banach algebra with the norm $\|(x, a)\| = \|x\| + |a|$ and $\mathscr{A}$ is isometrically isomorphic to $\mathscr{A} \times 0$.

(e) $\mathscr{A}$ is a maximal ideal of $\mathscr{A}_1$.

13.44 Let $\mathscr{A}$ be a nonunital commutative Banach algebra with spectrum Σ and let $\mathscr{A}_1$ be the unitization of $\mathscr{A}$. Prove that χ_1 is a character of $\mathscr{A}_1$ and that the mapping $\chi \to \chi_1$ is an injection from Σ into the spectrum Σ_1 of $\mathscr{A}_1$.

13.6 Operator Calculus

In this section we use the Gelfand representation theorem to prove two general results leading to the construction of functions of a normal operator on a Hilbert space.

The Continuous Functional Calculus

13.6.1 Theorem. *Let $\mathscr{A}$ be a unital C^* algebra and let $\boldsymbol{x} \in \mathscr{A}$ be normal. Then there exists an isometric $*$-isomorphism $\Psi : f \mapsto f(\boldsymbol{x})$ from $C(\sigma(\boldsymbol{x}))$ onto $\mathscr{C}^*(\boldsymbol{x})$ such that the following hold:*

(a) *$\Psi(z) = \boldsymbol{x}$ and $\Psi(1) = \boldsymbol{e}$, where z denotes the identity function on $\sigma(\boldsymbol{x})$ and 1 denotes the constant function.*

(b) *For any polynomial $p(z, \overline{z})$ in z and $\overline{z}$ on $\sigma(\boldsymbol{x})$, $\Psi(p(z, \overline{z})) = p(\boldsymbol{x}, \boldsymbol{x}^*)$.*

(c) *$\sigma(f(\boldsymbol{x})) = f(\sigma(\boldsymbol{x}))$ for every $f \in C(\sigma(\boldsymbol{x}))$.*

(d) *The isometric $*$-isomorphism Ψ is unique with respect to property (a).*

(e) *If $\mathscr{B}$ is a unital C^*-algebra and $\Phi : \mathscr{A} \to \mathscr{B}$ is a $*$-homomorphism that maps identity onto the identity, then $\sigma(\Phi(\boldsymbol{x})) \subseteq \sigma(\boldsymbol{x})$ and $\Phi(f(\boldsymbol{x})) = f(\Phi(\boldsymbol{x}))$ for $f \in C(\sigma(\boldsymbol{x}))$.*

Proof. Set $\Sigma := \sigma(\mathscr{C}^*(\boldsymbol{x}))$. By 13.4.1, there exists an isometric $*$- isomorphism $\Gamma : \mathscr{C}^*(\boldsymbol{x}) \to C(\Sigma)$ such that $\Gamma(\boldsymbol{x}) = \widehat{\boldsymbol{x}}$ maps Σ continuously onto $\sigma(\boldsymbol{x})$, where $\widehat{\boldsymbol{x}}(\chi) = \chi(\boldsymbol{x})$. If $\chi_1(\boldsymbol{x}) = \chi_2(\boldsymbol{x})$, then, by taking complex conjugates, we have $\chi_1(\boldsymbol{x}^*) = \chi_2(\boldsymbol{x}^*)$. By considering limits of polynomials in $(\boldsymbol{x}, \boldsymbol{x}^*)$, we see that $\chi_1 = \chi_2$. Therefore, $\widehat{\boldsymbol{x}} : \Sigma \to \sigma(\boldsymbol{x})$ is a homeomorphism. Now define $\Psi : C(\sigma(\boldsymbol{x})) \to \mathscr{C}^*(\boldsymbol{x})$ by

$$\Psi(f) = f(\boldsymbol{x}) := \Gamma^{-1}(f \circ \widehat{\boldsymbol{x}}), \quad f \in C(\sigma(\boldsymbol{x})).$$

Then Ψ is an isometric $*$-isomorphism onto $\mathscr{C}^*(\boldsymbol{x})$. Moreover,

$$\Psi(z) = \Gamma^{-1}(\mathrm{id}_{\sigma(\boldsymbol{x})} \circ \widehat{\boldsymbol{x}}) = \Gamma^{-1}(\widehat{\boldsymbol{x}}) = \boldsymbol{x},$$

and since $\Gamma(\boldsymbol{e})$ is the constant function 1,

$$\Psi(1) = \Gamma^{-1}(1 \circ \widehat{\boldsymbol{x}}) = \Gamma^{-1}(1) = \boldsymbol{e}.$$

Also, since $\Gamma(\boldsymbol{x}^*) = \overline{\Gamma(\boldsymbol{x})} = \overline{\widehat{\boldsymbol{x}}} = \widehat{\boldsymbol{x}^*}$ we see that

$$\Psi(\overline{z}) = \Psi\left(\overline{\mathrm{id}_{\sigma(\boldsymbol{x})}}\right) = \Gamma^{-1}\left(\overline{\widehat{\boldsymbol{x}}}\right) = \Gamma^{-1}\left(\widehat{\boldsymbol{x}}^*\right) = \boldsymbol{x}^*.$$

Thus (b) holds.

For (c) we have $z \in \sigma(f(\boldsymbol{x}))$ iff $f(\boldsymbol{x}) - z\boldsymbol{e}$ is not invertible in $\mathscr{C}^*(\boldsymbol{x})$ iff $f - z \cdot 1$ is not invertible in $C(\sigma(\boldsymbol{x}))$ (by the isomorphism Ψ) iff $f(w) = z$ for some $w \in \sigma(\boldsymbol{x})$ iff $z \in f(\sigma(\boldsymbol{x}))$.

To verify (d), note that any two isometric $*$-star isomorphisms that satisfy (a) must agree on polynomials in $(z, \overline{z})$ and hence, by continuity and the Stone-Weierstrass theorem, must be equal.

The inclusion in (e) is Ex. 13.18. The equality in (e) clearly holds for polynomials f in z and $\overline{z}$. The general equality then follows from the Stone-Weierstrass theorem. □

The mapping $f \mapsto f(\boldsymbol{x})$ in the theorem is known as the *continuous functional calculus*.

Applications to Operators on Hilbert Space

The next few results use the functional calculus $f \mapsto f(T)$ to obtain several important properties of normal operators T on a complex Hilbert space $\mathscr{H}$, some of which were stated without proof in Chapter 11. The functional calculus provides a unified approach to the proofs.

13.6.2 Theorem (Fuglede-Putnam). *Let $S, T \in \mathscr{B}(\mathscr{H})$ with T normal. If $ST = TS$, then $ST^* = T^*S$.*

Proof. (Rosenblum [40]). For any $Q \in \mathscr{C}^*(T)$, the series $\exp(Q) := \sum_{n=0}^{\infty} Q^n/n!$ converges absolutely and is a member of $\mathscr{C}^*(T)$. For $z \in \mathbb{C}$ define

$$R(z) := \exp(-zT^*)S\exp(zT^*) \quad \text{and} \quad U(z) := \exp(zT^* - \overline{z}T).$$

Since $-i(zT^* - \overline{z}T)$ is self-adjoint, $U(z)$ is unitary (13.2.7). Moreover, by the functional calculus,

$$U(-z)\exp(-\overline{z}T) = \exp(-zT^* + \overline{z}T)\exp(-\overline{z}T) = \exp(-zT^*).$$

Therefore,

$$R(z) = U(-z)\exp(-\overline{z}T)SU(z)\exp(\overline{z}T). \tag{†}$$

Since S commutes with T it commutes with $\exp(-\overline{z}T)$. Furthermore, by the functional calculus, $\exp(-\overline{z}T)\exp(\overline{z}T) = I$ and $\exp(-\overline{z}T)$ commutes with $U(z)$. It follows from (†) that $R(z) = U(-z)SU(z)$. In particular, for fixed $\boldsymbol{x}, \boldsymbol{y} \in \mathscr{H}$, the function

$$f_{\boldsymbol{xy}}(z) := (R(z)\boldsymbol{x} \mid \boldsymbol{y}) = (SU(z)\boldsymbol{x} \mid U(z)\boldsymbol{y})$$

is bounded on $\mathbb{C}$. We claim that f is an entire function. Assuming this for the moment, we conclude from Liouville's theorem that $f_{\boldsymbol{xy}}$ is constant. Therefore,

$$(R(z)\boldsymbol{x} \mid \boldsymbol{y}) = (R(0)\boldsymbol{x} \mid \boldsymbol{y}) = (S\boldsymbol{x} \mid \boldsymbol{y})$$

for all $\boldsymbol{x}$, $\boldsymbol{y}$, that is, $R(z) = S$ for all z. Thus $S\exp(zT^*) = \exp(zT^*)S$ and so by induction

$$S\exp(zT^*)^n = \exp(zT^*)^n S \text{ for all } n \text{ and } z. \tag{‡}$$

Setting $c_n := (S(T^*)^n\boldsymbol{x} \mid \boldsymbol{y}) - ((T^*)^nS\boldsymbol{x} \mid \boldsymbol{y})$, we then have by (‡)

$$\sum_{n=0}^{\infty} \frac{z^n}{n!} c_n = 0 \text{ for all } z.$$

Since for some $M > 0$, $|c_n| \le M\,\|T^*\|^n$, the series converges uniformly on bounded sets and therefore defines an analytic function of z. Since the function is identically zero, the coefficients c_n are zero. In particular $c_1 = 0$, which implies the desired result.

To see that $f := f_{\boldsymbol{xy}}$ is entire, set $c_{m,n} := ((T^*)^nS(T^*)^m\boldsymbol{x} \mid \boldsymbol{y})$ and note that

$$f(z) = (\exp(-zT^*)S\exp(zT^*)\boldsymbol{x} \mid \boldsymbol{y}) = \sum_{m,n} \frac{(-1)^n}{n!m!} z^{n+m} c_{m,n}.$$

Since for some $C > 0$ $|c_{m,n}| \le C\,\|T^*\|^{m+n}$, the series converges uniformly on bounded sets. It follows that f is entire. □

13.6.3 Corollary. *Let $S, T \in \mathscr{B}(\mathscr{H})$ with T normal. If $ST = TS$, then $Sf(T) = f(T)S$ for all $f \in C(\sigma(T))$. That is, $f(T) \in \{T\}''$.*

Proof. By 13.6.2, S commutes with all polynomials in T, T^*. Since these generate $\mathscr{C}^*(T)$, the assertion follows. □

13.6.4 Theorem. *Let $T \in \mathscr{B}(\mathscr{H})$. Then T is a positive operator iff T is positive as an element of the C^*-algebra $\mathscr{B}(\mathscr{H})$. In this case, there exists a unique positive square root $T^{1/2}$, that is, a unique positive operator that satisfies $(T^{1/2})^2 = T$. Moreover, if T is invertible, then T^{-1} is positive, $T^{1/2}$ is invertible, and $(T^{1/2})^{-1} = (T^{-1})^{1/2}$.*

Proof. If T is positive in the C^*-algebra $\mathscr{B}(\mathscr{H})$, then $T = S^*S$ for some $S \in \mathscr{B}(\mathscr{H})$ and so $(T\boldsymbol{x} \mid \boldsymbol{x}) = (S\boldsymbol{x} \mid S\boldsymbol{x}) \geq 0$ for all $\boldsymbol{x}$. Therefore, T is a positive operator.

Conversely, assume that T is a positive operator. Then $\sigma(T) \subseteq \mathbb{R}^+$. Indeed, if $t < 0$ and $\boldsymbol{x} \in \mathscr{H}$, then

$$\|t\boldsymbol{x} - T\boldsymbol{x}\|^2 = \|T\boldsymbol{x}\|^2 - 2\,(T\boldsymbol{x} \mid \boldsymbol{x}) + t^2\|\boldsymbol{x}\| \geq -2t\,(T\boldsymbol{x} \mid \boldsymbol{x}) + t^2\|\boldsymbol{x}\| \geq t^2\|\boldsymbol{x}\|,$$

hence $tI - T$ is invertible by 12.3.7. Applying the functional calculus to $f(t) = \sqrt{t}\ (t \in \sigma(T))$ we have $f(T) = T^{1/2}$ and $T = f^2(T) = f(T)^2 = (T^{1/2})^2$. Therefore, T is positive and has a positive square root.

To show uniqueness, let A and B be positive operators with $A^2 = B^2 = T$. Let (p_n) be a sequence of polynomials converging uniformly on $\sigma(A) \cup \sigma(B)$ to the square root function. Since

$$\sigma(A^2) = \sigma(A)^2 = \{t^2 : t \in \sigma(A)\}$$

and $p_n(t^2) \to t$ on $\sigma(A)$, $p_n(T) = p_n(A^2) \to A$ by the functional calculus. Similarly, $p_n(T) \to B$. Therefore, $A = B$, verifying uniqueness.

If T is invertible, then T^{-1} is a limit of polynomials in T and is therefore a member of $\mathscr{C}^*(T)$. Moreover, $\sigma(T^{-1}) \subseteq (0, \infty)$ by 13.2.8(b), hence T^{-1} is positive. Therefore, $(T^{-1})^{1/2}$ is defined. The functional calculus with $f(t) = 1/\sqrt{t}$ proves the last part of the theorem. □

13.6.5 Theorem. *Let $T \in \mathscr{B}(\mathscr{H})$ be self-adjoint. Then there exists a unique pair of positive operators T^+ and T^- such that $T = T^+ - T^-$ and $T^+T^- = T^-T^+ = 0$. Moreover, $\|T\| = \max\{\|T^+\|, \|T^-\|\}$ and $|T| = T^+ + T^-$.*

Proof. Apply the functional calculus to the nonnegative functions x^+ and x^- on $\sigma(T) \subseteq \mathbb{R}$ using the relations

$$x = x^+ - x^-, \quad |x| = x^+ + x^-, \quad x^+x^- = 0, \text{ and } \sup_{x \in \sigma(T)} |x| = \max\Big\{\sup_{x \in \sigma(T)} x^+,\ \sup_{x \in \sigma(T)} x^-\Big\}.$$

This gives positive operators T^+ and T^- with the desired properties.

To show uniqueness, let $T = A - B$, where A and B are positive operators with $AB = BA = 0$. Note that $\sigma(T) \subseteq \mathbb{R}$ and $\sigma(A) \cup \sigma(B) \subseteq [0, \infty)$. Using the above relations, we see that $T^k = A^k + (-B)^k$ for all $k \in \mathbb{N}$ and therefore $p(T) = p(A) + p(-B)$ for all polynomials p with $p(0) = 0$. Let X be a compact, symmetric subinterval of $\mathbb{R}$ containing $\sigma(A) \cup \sigma(-B) \cup \sigma(T)$. By the Stone-Weierstrass theorem there exists a sequence of polynomials p_n with $p_n(0) = 0$ such that $p_n(x) \to x^+$ uniformly on X. In particular, $p_n(x) \to x$ on $\sigma(A)$ and $p_n(-x) \to 0$ on $\sigma(B)$. By the functional calculus, we then have $p_n(T) \to T^+$ and $p_n(T) = p_n(A) + p_n(-B) \to A$, hence $A = T^+$. Similarly, $B = T^-$. □

13.6.6 Theorem. *Every operator T on a Hilbert space $\mathscr{H}$ is a linear combination of unitary operators.*

Proof. Since T is a linear combination of self-adjoint operators (12.1.3), we may assume T is self-adjoint. Moreover, it is no loss of generality to assume that $\|T\| \leq 1$. Then $\sigma(T) \subseteq [-1, 1]$, hence the function $f(t) = t + i\sqrt{1 - t^2}$ is defined on $\sigma(T)$. Since $f\overline{f} = 1$ and $\mathrm{id}_{[-1,1]} = \frac{1}{2}(f + \overline{f})$, by the functional calculus $f(T)f(T)^* = I$ and $T = \frac{1}{2}(f(T) + f(T)^*)$, proving the theorem. □

The Borel Functional Calculus

Let $\mathscr{H}$ be a complex Hilbert space and $T \in \mathscr{B}(\mathscr{H})$ normal. In this subsection we extend the continuous functional calculus $f \mapsto f(T)$ to Borel functions f on $\sigma(T)$. For this we need the following lemma.

13.6.7 Lemma. *Let $K \subseteq \mathbb{C}$ be compact. Then the space $BL(K)$ of bounded complex-valued Borel functions on K is the smallest conjugate closed algebra of bounded functions that (a) contains all polynomials on K and (b) is closed under pointwise limits of uniformly bounded sequences of functions on K.*

Proof. Let $A(K)$ denote the intersection of all conjugate closed algebras of bounded functions on K with properties (a) and (b). Then $A(K)$ has properties (a) and (b), and since $BL(K)$ has properties (a) and (b), $A(K) \subseteq BL(K)$. Moreover, $A(K)$ is an algebra, and by the Stone-Weierstrass theorem and (b), $A(K)$ contains $C(K)$. Now let $\mathcal{A}$ denote the set of all $E \subseteq K$ such that $\mathbf{1}_E \in A(K)$. We claim that $\mathcal{A}$ is a σ-field that contains all the open sets. Assuming this for the moment, we see that $\mathcal{A}$ must then contain all Borel subsets of K. Therefore, $A(K)$ contains all simple functions, and since every bounded Borel function f on K is a pointwise limit of a uniformly bounded sequence of simple functions, $A(K) \supseteq BL(K)$, proving the lemma.

It remains to verify the claim. Let U be open in K and $K_n = \{x \in U : d(x, U^c) \geq 1/n\}$. Choosing $f_n \in C(K)$ such that $\mathbf{1}_{K_n} \leq f_n \leq \mathbf{1}_U$, we have $f_1 \vee \cdots \vee f_n \uparrow \mathbf{1}_U$, hence $\mathbf{1}_U \in A(K)$. Therefore, $A(K)$ contains all open sets. If E, $F \in \mathcal{A}$, then $\mathbf{1}_{E^c} = 1 - \mathbf{1}_E$ and $\mathbf{1}_{E \cap F} = \mathbf{1}_E \mathbf{1}_F$, hence $\mathcal{A}$ is closed under complements and finite intersections and therefore also under finite unions. If $E_n \in \mathcal{A}$ and $E_n \uparrow E$, then $\mathbf{1}_{E_n} \uparrow \mathbf{1}_E$, hence $\mathbf{1}_E \in A(K)$ and so $E \in \mathcal{A}$. Therefore, $\mathcal{A}$ is a σ-field. □

13.6.8 Theorem. *Let T be a normal operator on a Hilbert space $\mathscr{H}$ and set $K := \sigma(T)$. Then there exists a $*$-homomorphism $f \mapsto f(T)$ from $BL(K)$ into $\mathscr{B}(\mathscr{H})$ such that*

- **(a)** $1(T) = I$ *and* $\mathrm{id}_K(T) = T$.
- **(b)** *If (f_n) is a uniformly bounded sequence in $BL(K)$ that converges pointwise to f, then $f_n(T)x \overset{w}{\to} f(T)x$ for every $x \in \mathscr{H}$.*
- **(c)** $\left(\sum_{k=0}^n a_k z^k\right)(T) = \sum_{k=0}^n a_k T^k$.
- **(d)** $\|f(T)\| \leq \|f\|_\infty$.
- **(e)** *If $S \in \mathscr{B}(\mathscr{H})$ and $ST = TS$, then $Sf(T) = f(T)S$ for every $f \in BL(K)$.*

Moreover, the $*$-homomorphism $f \mapsto f(T)$ is unique with respect to properties (a) and (b).

Proof. (a) and (c) hold by the continuous functional calculus $f \mapsto f(T) : C(K) \to \mathscr{B}(\mathscr{H})$. We extend this to $BL(K)$ as follows: For each pair x, $y \in \mathscr{H}$, the mapping $f \to (f(T)x \,|\, y)$ is a bounded linear functional on $C(K)$, hence, by the Riesz representation theorem, there exists a complex measure $\mu(x, y)$ on K such that for each $f \in C(K)$

$$(f(T)x \,|\, y) = \int_K f \, d\mu(x, y). \tag{13.5}$$

We claim that

(i) $\mu(ax + by, z) = a\mu(x, z) + b\mu(y, z)$. (ii) $\mu(y, x) = \overline{\mu(x, y)}$.

(iii) $\mu(x, x) \geq 0$. (iv) $d\mu(g(T)x, y) = g\,d\mu(x, y)$, $g \in C(K)$.

Indeed, by integrating against a continuous function f and using (13.5), we see that (i) holds because $(f(T)\boldsymbol{x} \mid \boldsymbol{y})$ is sesquilinear in $(\boldsymbol{x}, \boldsymbol{y})$, and (ii) follows from the calculation

$$\int_K f\,d\mu(\boldsymbol{y}, \boldsymbol{x}) = (f(T)\boldsymbol{y} \mid \boldsymbol{x}) = (\boldsymbol{y} \mid f(T)^*\boldsymbol{x}) = \left(\boldsymbol{y} \mid \overline{f}(T)\boldsymbol{x}\right) = \overline{\left(\overline{f}(T)\boldsymbol{y} \mid \boldsymbol{x}\right)}$$
$$= \overline{\int_K \overline{f}\,d\mu(\boldsymbol{x}, \boldsymbol{y})} = \int_K f d\,\overline{\mu(\boldsymbol{x}, \boldsymbol{y})}.$$

For (iii), if $f \geq 0$ and $g = f^{1/2}$, then, by the continuous functional calculus, we have $f(T) = g^2(T) = g(T)g(T)$, hence

$$\int_K f\,d\mu(\boldsymbol{x}, \boldsymbol{x}) = (f(T)\boldsymbol{x} \mid \boldsymbol{x}) = (g(T)\boldsymbol{x} \mid g(T)\boldsymbol{x}) \geq 0.$$

Finally, (iv) follows from the calculation

$$\int_K fg\,d\mu(\boldsymbol{x}, \boldsymbol{y}) = ((fg)(T)\boldsymbol{x} \mid \boldsymbol{y}) = (f(T)g(T)\boldsymbol{x} \mid \boldsymbol{y}) = \int_K f d\mu(g(T)\boldsymbol{x}, \boldsymbol{y}).$$

Now, for $f \in BL(K)$ and $\boldsymbol{x}, \boldsymbol{y} \in \mathscr{H}$ we have $|\int_K f\,d\mu(\boldsymbol{x}, \boldsymbol{y})| \leq \|f\|_\infty \|\boldsymbol{x}\|\,\|\boldsymbol{y}\|$, hence by (i) and (ii) and 11.4.1 there exists a unique $f(T) \in \mathscr{B}(\mathscr{H})$ such that (13.5) holds. The mapping $f \mapsto f(T) : BL(K) \to \mathscr{B}(\mathscr{H})$ is clearly linear and satisfies $\|f(T)\| \leq \|f\|_\infty$ and $\overline{f}(T) = f(T)^*$, the latter by (ii). Moreover, if $g \in C(K)$, then from (iv) we have

$$\int_K gf\,d\mu(\boldsymbol{x}, \boldsymbol{y}) = \int_K f\,d\mu(g(T)\boldsymbol{x}, \boldsymbol{y}) = (f(T)g(T)\boldsymbol{x} \mid \boldsymbol{y}) = (g(T)\boldsymbol{x} \mid f(T)^*\boldsymbol{y})$$
$$= \int_K g\,d\mu(\boldsymbol{x}, f(T)^*\boldsymbol{y}),$$

hence $f\,d\mu(\boldsymbol{x}, \boldsymbol{y}) = d\mu(\boldsymbol{x}, f(T)^*\boldsymbol{y})$. Therefore, for all $f, g \in BL(K)$ and $\boldsymbol{x}, \boldsymbol{y} \in \mathscr{H}$,

$$(f(T)g(T)\boldsymbol{x} \mid \boldsymbol{y}) = (g(T)\boldsymbol{x} \mid f(T)^*\boldsymbol{y}) = \int_K g\,d\mu(\boldsymbol{x}, f(T)^*\boldsymbol{y}) = \int_K gf\,d\mu(\boldsymbol{x}, \boldsymbol{y})$$
$$= ((fg)(T)\boldsymbol{x} \mid \boldsymbol{y}),$$

which shows that $f(T)g(T) = (fg)(T)$. We have proved that the mapping $f \mapsto f(T)$ from $BL(K)$ into $\mathscr{B}(\mathscr{H})$ is a $*$-homomorphism satisfying (a), (c), and (d).

To verify (b), we apply the dominated convergence theorem to obtain

$$(f_n(T)\boldsymbol{x} \mid \boldsymbol{y}) = \int_K f_n\,d\mu(\boldsymbol{x}, \boldsymbol{y}) \to \int_K f\,d\mu(\boldsymbol{x}, \boldsymbol{y}) = (f(T)\boldsymbol{x} \mid \boldsymbol{y}).$$

For (e), if $ST = TS$ and $f \in C(K)$, then by the continuous functional calculus $Sf(T) = f(T)S$ and so for all $\boldsymbol{x}$, $\boldsymbol{y}$,

$$\int_K f\,d\mu(\boldsymbol{x}, S^*\boldsymbol{y}) = (f(T)\boldsymbol{x} \mid S^*\boldsymbol{y}) = (Sf(T)\boldsymbol{x} \mid \boldsymbol{y}) = (f(T)S\boldsymbol{x} \mid \boldsymbol{y}) = \int_K f\,d\mu(S\boldsymbol{x}, \boldsymbol{y}).$$

Therefore, $\mu(\boldsymbol{x}, S^*\boldsymbol{y}) = \mu(S\boldsymbol{x}, \boldsymbol{y})$, hence for $f \in BL(K)$

$$(Sf(T)\boldsymbol{x} \mid \boldsymbol{y}) = (f(T)\boldsymbol{x} \mid S^*\boldsymbol{y}) = \int_K f\,d\mu(\boldsymbol{x}, S^*\boldsymbol{y}) = \int_K f\,d\mu(S\boldsymbol{x}, \boldsymbol{y}) = (f(T)S\boldsymbol{x} \mid \boldsymbol{y}),$$

proving (e).

It remains to show uniqueness with respect to properties (a) and (b). Let $f \mapsto \widetilde{f}(T)$ be another $*$-homomorphism with these properties. Then the collection of all $f \in BL(K)$ for which $\widetilde{f}(T) = f(T)$ is a conjugate closed algebra containing all polynomials on K and is closed under pointwise limits of uniformly bounded sequences and so must coincide with $BL(K)$ by 13.6.7. □

The mapping $f \mapsto f(T)$ in the above theorem is known as the *Borel functional calculus.*

The Spectral Theorem for Normal Operators

It is possible to recast the Borel functional calculus in terms of *projection-valued measures,* as described in the following proposition.

13.6.9 Proposition. *Let $T \in \mathscr{B}(\mathscr{H})$ be normal with Borel functional calculus $f \mapsto f(T)$. For each Borel subset E of $K := \sigma(T)$, set $P(E) := \mathbf{1}_E(T)$. Then the mapping $P : \mathfrak{B}(K) \to \mathscr{B}(\mathscr{H})$ has the following properties:*

(a) *$P(E)$ is an orthogonal projection.*

(b) *$P(\emptyset) = 0$, $P(K) = I$.*

(c) *$P(E \cap F) = P(E)P(F)$.*

(d) *If $E_1, E_2, \ldots$ are disjoint, and $E = \bigcup_n E_n$, then the series $\sum_{n=1}^{\infty} P(E_n)\boldsymbol{x}$ converges in norm to $P(E)\boldsymbol{x}$ for every $\boldsymbol{x}$.*

Proof. Parts (a) – (c) follow immediately from the Borel functional calculus, as does (d) for finite sequences. For infinite sequences, set $F_n = \bigcup_{j=1}^{n} E_j$. Then $\mathbf{1}_{F_n} \to \mathbf{1}_E$ pointwise on K, hence $P(F_n)\boldsymbol{x} \overset{w}{\to} P(E)\boldsymbol{x}$ for all $\boldsymbol{x} \in \mathscr{X}$, by part (b) of 13.6.8. Set $T_n = P(E) - P(F_n) = P(E \setminus F_n)$. Then $T_n\boldsymbol{x} \overset{w}{\to} 0$ and

$$\|P(E)\boldsymbol{x} - P(F_n)\boldsymbol{x}\|^2 = (T_n\boldsymbol{x} \mid T_n\boldsymbol{x}) = (T_n^*T_n\boldsymbol{x} \mid \boldsymbol{x}) = (T_n\boldsymbol{x} \mid \boldsymbol{x}) \to 0,$$

proving (d). □

We may now formulate the functional calculus in terms of integrals. For each $\boldsymbol{x}$, $\boldsymbol{y} \in \mathscr{X}$, define $P_{(\boldsymbol{x},\boldsymbol{y})}(E) := (P(E)\boldsymbol{x} \mid \boldsymbol{y})$. Then

$$P_{\boldsymbol{x},\boldsymbol{y}}(E) = (\mathbf{1}_E(T)\boldsymbol{x} \mid \boldsymbol{y}) = \int_K \mathbf{1}_E \, d\mu_{(\boldsymbol{x},\boldsymbol{y})} = \mu(\boldsymbol{x}, \boldsymbol{y})(E),$$

so the set function $P_{(\boldsymbol{x},\boldsymbol{y})}$ is simply the measure $\mu_{(\boldsymbol{x},\boldsymbol{y})}$ of the Borel functional calculus, and (13.5) may be written

$$(f(T)\boldsymbol{x} \mid \boldsymbol{y}) = \int_{\sigma(T)} f(z) \, dP_{\boldsymbol{x},\boldsymbol{y}}(z)$$

or simply

$$f(T) := \int_{\sigma(T)} f(z) \, dP(z).$$

This expresses $f(T)$ as an integral with respect to the set function P, which is called the **spectral measure for** T. The special case

$$I = \int_{\sigma(T)} 1 \, dP(z)$$

is the motivation for the alternate terminology **spectral resolution of the identity**. The special case $f(z) = z$ results in the **spectral theorem for normal operators**:

13.6.10 Theorem. *If T is a normal operator and $P(z)$ is a spectral resolution of the identity, then*

$$T = \int_{\sigma(T)} z\, dP(z)$$

Note that if T is compact, then $\sigma(T)$ is a sequence $(\lambda_n) \in \mathfrak{c}_0$, hence the last integral reduces to an infinite series, giving the spectral theorem of §12.3.

Exercises

13.45 Let $\mathscr{A}$ be a unital C^* algebra. Show that if $x^* = -x$, then $\sigma(x) \subseteq i\,\mathbb{R}$.

13.46 Let $\mathscr{A}$ be a unital C^* algebra, $x \in \mathscr{A}$ unitary and $\sigma(x) \neq \mathbb{T}$. Show that $x = e^{iy}$ for some self-adjoint y .

13.47 Verify the following assertions to obtain an alternate proof that the operators $T^{\pm}$ are unique:

(a) Let $T = A - B$, for positive operators A and B with $AB = 0$. Then $AT = TA$ and $BT = TB$.

(b) A and B commute with $T^{\pm}$.

(c) If C and D are positive operators and $CD = DC$, then CD is positive.

(d) Set $S := T^+ - A = T^- - B$. Then $0 \le S^*S = S^2 = -(T^-A + T^+B) \le 0$.

(e) $S = 0$.

13.48 [↑13.18] Let $\mathscr{A}$ and $\mathscr{B}$ be unital C^*-algebras and $\Phi : \mathscr{A} \to \mathscr{B}$ an injective $*$-homomorphism that maps identity onto identity. Show that for $x \in \mathscr{A}$, $\sigma\big(\Phi(x)\big) = \sigma(x)$ and hence that Φ is an isometry. ⟦Assume first that x is self-adjoint and use the functional calculus on $C(\sigma(x))$. Apply this to x^*x for the general case.⟧

13.49 Let $\mathscr{A}$ be a unital C^* algebra, $x \in \mathscr{A}$ normal, and $c \in \mathbb{C}$. Show that

$$\|ce - x\| = \sup\{|c - z| : z \in \sigma(x)\}.$$

Use this to prove that if x is self-adjoint, then $\sigma(x) \subseteq [0, \infty)$ iff $\|ce - x\| \le c$ for some (for every) $c \ge \|x\|$ ⟦Use 13.2.5.⟧

13.50 Let $\mathscr{A}$ be a unital C^* algebra and x, $y \in \mathscr{A}$ positive. Use the preceding exercise to show that

$$\|(\|x\| + \|y\|)e - (x + y)\| \le \|x\| + \|y\|.$$

Conclude that $\sigma(x + y) \subseteq [0, \infty)$.

13.51 Let $\mathscr{A}$ be a unital C^* algebra and let $x \in \mathscr{A}$ be self-adjoint. Prove the following to conclude that $x \ge 0$ iff $\sigma(x) \subseteq \mathbb{R}^+$.

(a) If $\sigma(x) \subseteq \mathbb{R}^+$, then $x \ge 0$. ⟦Consider the functional calculus on $\mathscr{C}^*(x)$.⟧

(b) Let $x \ge 0$. Then $x = x^+ - x^-$, where $\sigma(x^{\pm}) \subseteq \mathbb{R}^+$. ⟦Use the functional calculus exactly as in 13.6.5.⟧

(c) Set $z = yx^-$. Then $\sigma(z^*z) \subset (-\infty, 0]$.

(d) $\sigma(zz^* + zz^*) \subseteq [0, \infty)$. ⟦Write $z = u + iv$, where u and v are self-adjoint and use Ex. 13.50.⟧

(e) $\sigma(zz^*) \subseteq [0, \infty)$. ⟦Use (c), (d), and Ex. 13.50.⟧

(f) $z^*z = 0$. ⟦Use (c), (e) and Ex. 13.26.⟧

(g) $\sigma(x) \subseteq \mathbb{R}^+$.

13.52 Show that the definition $x \le y$ iff $y - x \ge 0$ gives a partial order on the set of self-adjoint members of a unital C^*−algebra such that $x \le y \Rightarrow z^*xz \le z^*yz$.

13.53 Let $\mathscr{A}$ be a unital C^*-algebra. Show that if x, $y \in \mathscr{A}$ are positive and $xy = yx$, then xy is positive. ⟦First assume $\mathscr{A}$ is commutative.⟧

13.54 Prove that if $0 \leq S \leq T$ and S is invertible, then T is invertible and $0 \leq T^{-1} \leq S^{-1}$. ⟦Consider $S^{-1/2}$ and use Ex. 12.14.⟧

13.55 Let $T \in \mathscr{B}(\mathscr{H})$ be normal. Show that $(f \circ g)(T) = f(g(T))$, where g is a bounded Borel function on $\sigma(T)$ and f is a bounded Borel function on the closure K of $g(\sigma(T))$. ⟦Fix g and let B denote the set of Borel functions f on K for which the equality holds. Then B is a conjugate closed algebra with properties (a) and (b) of 13.6.7.⟧

Chapter 14

Miscellaneous Topics

In this chapter we consider some of the deeper aspects of functional analysis and give several important applications. Additional applications may be found in Chapters 15, 16, and 17

14.1 Weak Sequential Compactness

In this section we prove that weak compactness and weak sequential compactness in a Banach space are equivalent. We first prove the analogous result for the pointwise topology on the space of continuous functions on a compact Hausdorff space.

Pointwise Sequential Compactness

14.1.1 Theorem. *Let X be a compact Hausdorff topological space and let $A \subseteq C(X)$ be norm bounded. Let p denote the topology of pointwise convergence on $C(X)$. The following statements are equivalent:*

(a) *A is relatively p-sequentially compact, that is, every sequence (f_n) in A has subsequence that p-converges in $C(X)$.*

(b) *If (f_n) is a sequence in A and (x_m) is a sequence in X such that the iterated limits*

$$\lim_m \lim_n f_n(x_m) \quad \textit{and} \quad \lim_n \lim_m f_n(x_m)$$

exist, then the limits are equal.

(c) *A is relatively p-compact.*

Proof. (a) $\Rightarrow$ (b): Assume that the limits in (b) exist. By the hypothesis, (f_n) has a p sequential limit point, say $\lim_k f_{n_k} = f \in C(X)$. Let $x \in X$ be a limit point of (x_m), say $x_{m_\alpha} \to x$. Then

$$\lim_k \lim_\alpha f_{n_k}(x_{m_\alpha}) = \lim_k f_{n_k}(x) = f(x) = \lim_\alpha f(x_{m_\alpha}) = \lim_\alpha \lim_k f_{n_k}(x_{m_\alpha}).$$

It follows that the limits in (b) are equal.

(b) $\Rightarrow$ (c): Because A is norm bounded, $A \subseteq K^X$ for some compact $K \subseteq \mathbb{C}$. By Tychonoff's theorem, K^X is compact in the topology of pointwise convergence on X. To prove (c) it therefore suffices to show that the closure B of A in K^X consists entirely of continuous functions. Suppose, for a contradiction, that B contains a function $f : X \to K$ which is not continuous at some $x \in X$. Thus there exists $\varepsilon > 0$ such that if $D_\varepsilon := \{c \in \mathbb{K} : |c| < \varepsilon\}$, then

$$f(N_x) - f(x) \not\subseteq D_\varepsilon \text{ for every neighborhood } N_x \text{ of } x \text{ in } X. \tag{\dagger}$$

By induction there exist sequences (f_n) in A and (x_n) in X such that

(i) $|f_j(x) - f_j(x_n)| < 1/n, \quad j = 1, \dots, n,$

(ii) $|f(x) - f(x_n)| \geq \varepsilon$, and

(iii) $|f_1(x) - f(x)| < 1$ and $|f_{n+1}(y) - f(y)| < 1/(n+1), \quad y \in \{x, x_1, \dots, x_n\}$.

Indeed, the functions f_j in (i) define a neighborhood

$$N_x := \{y \in X : |f_j(x) - f_j(y)| < 1/n, \quad j = 1, \dots, n\}$$

of x which is used in (†) to obtain the point x_n in (i) and (ii); and (iii) uses the fact that f is in the pointwise closure of A. Now, since f is bounded, there exists a subsequence $(y_k := x_{m_k})$ such that $f(y_k) \to c$ for some $c \in \mathbb{K}$. Then, by (i) and (iii),

$$\lim_k \lim_n f_n(y_k) = \lim_k f(y_k) = c \quad \text{and} \quad \lim_n \lim_k f_n(y_k) = \lim_n f_n(x) = f(x).$$

But by (ii), $f(x) \neq c$, contradicting (b). Therefore, f must be continuous.

(c) $\Rightarrow$ (a): Let (f_n) be a sequence in A. Suppose first that X is metrizable. It then has a countable dense subset D. Since A is bounded, a standard diagonal argument shows that (f_n) has a subsequence (f_{n_k}) that converges pointwise on D. Since C is p-compact, a subnet of (f_{n_k}) converges pointwise on X to a member f of C. If another subnet of (f_{n_k}) converges pointwise on X to a member g of C, then $f = g$ on D and so, by continuity, $f = g$. It follows that (f_{n_k}) converges pointwise to f on X.

In the general case, define an equivalence relation on X by

$$x \sim x' \text{ iff } f_n(x) = f_n(x') \text{ for all } n.$$

Give $Z := X/\sim$ the quotient topology and let $Q : X \to Z$ denote the quotient map. Define $\widetilde{f}_n$ on Z by $\widetilde{f}_n \circ Q = f_n$. Since $\widetilde{f}_n$ is continuous, the initial topology τ defined by $(\widetilde{f}_n)$ is weaker than the quotient topology τ_q. Furthermore, τ is metrizable by

$$d(Q(x), Q(y)) = \sum_{n=1}^{\infty} \frac{1}{2^n} \frac{|f_n(x) - f_n(y)|}{1 + |f_n(x) - f_n(y)|}.$$

Since τ is Hausdorff and τ_q is compact, $\tau = \tau_q$. Now, by (a), (f_n) has a p-limit point f in $C(X)$, say $f_{n_\alpha} \to f$. Define $\widetilde{f}$ on Z so that $\widetilde{f} \circ Q = f$. Then $\widetilde{f}$ is well-defined, since $Q(x) = Q(y) \Rightarrow f(x) = \lim_\alpha f_{n_\alpha}(x) = \lim_n f_{n_\alpha}(y) = f(y)$. Since $\widetilde{f}$ is a p-limit point of $(\widetilde{f}_n)$, by the preceding paragraph $\widetilde{f}_{n_k} \xrightarrow{p} \widetilde{f}$ for some subsequence (f_{n_k}). Therefore $f_{n_k} \xrightarrow{p} f$, proving (a). □

Part (b) of the lemma is known as *Grothendieck's double limit property.*

The Eberlein-Šmulian Theorem

Here is the aforementioned result on the equivalence of weak and sequential weak compactness. The proof makes critical use of 14.1.1.

14.1.2 Theorem (Eberlein-Šmulian). *Let X be a Banach space and $A \subseteq \mathscr{X}$. Then A is relatively weakly compact iff every sequence $(\boldsymbol{x}_n)$ in A has a weakly convergent subsequence.*

Proof. For the necessity, it suffices to show that $(\boldsymbol{x}_n)$ has a subsequence $(\boldsymbol{x}_{n_k})$ such that for some $\boldsymbol{x} \in \mathscr{X}$, $\langle \boldsymbol{x}_{n_k}, f \rangle \to \langle \boldsymbol{x}, f \rangle$ for all f in the closed unit ball C_1' of $\mathscr{X}'$. We may assume that $\mathscr{X}$ is the closed linear span of $(\boldsymbol{x}_n)$; otherwise we could simply consider weak convergence in this space. It follows that C_1', which is weak* compact, is metrizable under the metric

$$d(f, g) = \sum_{n=1}^{\infty} \frac{1}{2^n} \frac{|f(x_n) - g(x_n)|}{1 + |f(x_n) - g(x_n)|}.$$

In particular, C_1' has a weak* dense sequence (f_m). By a diagonal argument, there exists a subsequence (y_n) of (x_n) such that $\alpha_m := \lim_n \langle y_n, f_m \rangle$ exists for each m. Since (y_n) is relatively weakly compact, there exists $y \in \mathcal{Y}$ and a subnet (y_α) of (y_n) such that $y_\alpha \xrightarrow{w} y$. Therefore, $\langle y, f_m \rangle = \lim_\alpha \langle y_\alpha, f_m \rangle = \alpha_m$ for all m. If z is another such limit point, then $\langle z, f_m \rangle = \alpha_m$ for all m, hence $y = z$ because (f_m) is weak* dense in C_1'. Therefore, (y_n) has a unique weak limit point and so must converge weakly.

For the sufficiency, note that the hypothesis and 10.1.2 imply that A is norm bounded. Let (x_α) be a net in A. Then $(\widehat{x}_\alpha)$ is a norm bounded net in $\widehat{A}$ and so has a subnet $(\widehat{x}_\beta)$ that weak* converges to some φ in $\mathcal{X}''$. It remains to show that $\varphi \in \widehat{\mathcal{X}}$, that is, φ is weak* continuous. By 10.2.9, it suffices to show that the restriction of φ to the closed unit ball C_1' in $\mathcal{X}'$ is w^*-continuous. But this topology is simply the topology of pointwise convergence on C_1'. Thus we have reduced the problem to showing that $\widehat{A}\big|_{C_1'}$ is relatively p-compact in the space of continuous functions on C_1'. But this follows from the hypothesis and 14.1.1, since $\widehat{A}\big|_{C_1'}$ is relatively p-sequentially compact. □

From Shur's theorem (10.1.4) we have

14.1.3 Corollary. *A subset of $\ell^1(\mathbb{N})$ is norm compact iff it is weakly compact.*

14.1.4 Corollary. *Let X be a compact Hausdorff space and A a norm-bounded subset of $C(X)$. Then A is relatively weakly compact iff A is relatively p-compact. In this case the topologies p and w coincide on A.*

Proof. The necessity is clear. For the sufficiency, it suffices to prove that A relatively weakly sequentially compact. Let (f_n) be a sequence be a sequence in A. By hypothesis, there exists a subsequence (f_{n_k}) that converges pointwise on X to some $f \in C(X)$. It follows from the Riesz representation theorem and the dominated convergence theorem that (f_{n_k}) converges weakly to f. □

14.1.5 Corollary. *Let X and Y be compact Hausdorff topological spaces and $F : X \times Y \to \mathbb{C}$ a bounded, separately continuous function. Then the function $y \to F(\cdot, y) : Y \to C(X)$ is continuous in the weak topology of $C(X)$.*

Proof. By separate continuity, the function $y \to F(\cdot, y) : Y \to C(X)$ is continuous in the p-topology of $C(X)$, hence $A := F(\cdot, Y)$ is p-compact. Since it is also norm bounded, the topologies p and w coincide on A. □

Part of the next result has already been proved (10.3.6). An application of the Eberlein-Šmulian theorem completes the proof.

14.1.6 Corollary. *Let $\mathcal{X}$ be a Banach space. Then the following are equivalent.*

(a) *$\mathcal{X}$ is reflexive.* **(b)** *C_1 is weakly compact.* **(c)** *C_1 is weakly sequentially compact.*

14.2 Weak Compactness in L^1

Let $(X, \mathcal{F}, \mu)$ be a σ-finite measure space. For $1 < p < \infty$, the space $L^p(X, \mathcal{F}, \mu)$ is reflexive, hence the weakly relatively compact sets of L^p are simply the bounded sets. For $p = 1$, the situation is more complicated. In this section we give a characterization of weak compactness in $L^1(\mu)$ for probability measures μ in terms of uniform integrability. Our treatment follows [14].

Weak Convergence and Uniform Integrability

14.2.1 Lemma. *Let $\mathcal{A}_0$ be a countable collection of subsets of X. Then the field $\varphi(\mathcal{A}_0)$ generated by $\mathcal{A}_0$ is countable.*

Proof. Define sequences $(\mathcal{A}_n)$ and $(\mathcal{B}_n)$ of subsets of X by the following scheme:

- $\mathcal{A}_1 :=$ the collection of all finite unions of members of $\mathcal{A}_0$,
- $\mathcal{B}_1 :=$ the collection of all complements of members of $\mathcal{A}_1$,
- $\mathcal{A}_2 :=$ the collection of all finite unions of members of $\mathcal{B}_1$,
- $\mathcal{B}_2 :=$ the collection of all complements of members of $\mathcal{A}_2$,
- etc.

Note that $A, B \in \mathcal{A}_n \Rightarrow A \cup B \in \mathcal{A}_n$, $A^c \in \mathcal{A}_{n+1}$, and $A \cap B = (A^c \cup B^c)^c \in \mathcal{A}_{n+2}$. Therefore, $\bigcup_n \mathcal{A}_n$ is a field and so coincides with $\varphi(\mathcal{A}_0)$. By induction, $\mathcal{A}_n$ is countable for every n, completing the proof. □

14.2.2 Lemma. *Let f be measurable. Then there exists a countable field $\mathcal{A} \subseteq \mathcal{F}$ such that f is measurable with respect to $\sigma(\mathcal{A})$.*

Proof. Choose a sequence of simple functions $g_n = \sum_{j=1}^{k_n} a_{n,j} \mathbf{1}_{A_{n,j}}$ converging pointwise to f. Let $\mathcal{A}_0$ denote the countable collection of the sets $A_{n,j}$. Then $\mathcal{A} := \varphi(\mathcal{A}_0)$ is countable (14.2.1) and f is measurable with respect to $\sigma(\mathcal{A})$. □

Here is the key step needed for the proof of the Dunford-Pettis theorem below.

14.2.3 Lemma. *Let $(X, \mathcal{F}, \mu)$ be a probability space. If (g_n) converges weakly to g in $L^1(\mu)$, then (g_n) is uniformly integrable.*

Proof. Let $d\mu_n = g_n \, d\mu$ and set

$$\eta(A) := \sum_{n=1}^{\infty} 2^{-n} \frac{|\mu_n|(A)}{1 + |\mu_n|(X)}, \quad A \in \mathcal{F}.$$

By the proof of the Vitali-Hahn-Saks theorem (5.2.4), given $\varepsilon > 0$ there exists $\delta > 0$ and $m \in \mathbb{N}$ such that

$$|\mu_n(A)| \le |\mu_m(A)| + 2\varepsilon \ \ \forall\ A \text{ with } \eta(A) \le \delta \text{ and } \forall\ n \ge m. \tag{†}$$

Since $|\mu_j(A)| \le 2^j \eta(A)\big(1 + |\mu_j|(X)\big)$, by taking a smaller δ if necessary we may assume that $\eta(A) \le \delta \Rightarrow |\mu_j(A)| < \varepsilon$ $(j = 1, \ldots, m)$. Combining this with (†), we have

$$|\mu_n(A)| \le 3\varepsilon \quad \forall\ n \text{ and } \forall\ A \text{ with } \eta(A) \le \delta. \tag{‡}$$

Now observe that $\eta \ll \mu$, hence we may choose $\delta_0 > 0$ so that $\mu(A) < \delta_0 \Rightarrow \eta(A) < \delta$. For such A, $\sup_n |\mu_n(A)| \le 3\varepsilon$ from (‡). Thus, by 4.4.2, (g_n) is uniformly integrable. □

The Dunford-Pettis Theorem

Here is the main result of the section.

14.2.4 Theorem (Dunford-Pettis). *Let $(X, \mathcal{F}, \mu)$ be a probability space. Then a subset $\mathcal{U}$ of $L^1(\mu)$ is relatively weakly compact iff $\mathcal{U}$ is uniformly integrable.*

Proof. First, recall that $\mathcal{U}$ is uniformly integrable iff

$$\lim_n \sup_{f \in \mathcal{U}} \int_{|f|>n} |f|\, d\mu = 0,$$

Suppose $\mathcal{U}$ is weakly relatively compact but not uniformly integrable. Then there exists $\delta > 0$ such that

$$\lim_{n\to\infty} \sup_{f \in \mathcal{U}} \int_{|f|>n} |f|\, d\mu \geq 2\delta.$$

For sufficiently large n, choose $f_n \in \mathcal{U}$ such that

$$\int_{|f_n|>n} |f_n|\, d\mu \geq \delta. \qquad (\alpha)$$

By the Eberlein-Šmulian theorem, (f_n) has a subsequence (g_n) that converges weakly to some g. But then (g_n) is uniformly integrable by 14.2.3, contradicting (α).

Conversely, suppose that $\mathcal{U}$ is uniformly integrable. We show that a sequence (f_n) in $\mathcal{U}$ has a weakly convergent subsequence. By considering real, imaginary, positive, and negative parts, we may assume that $f_n \geq 0$ for all n. Note that by 4.4.2

$$\lim_{\mu(E)\to 0} \sup\nolimits_n \|f_n \mathbf{1}_E\|_1 = 0. \qquad (\beta)$$

Now, by 14.2.2, there exists a countable field $\mathcal{A} \subseteq \mathcal{F}$ such that each f_n is $\mathcal{F}_0 := \sigma(\mathcal{A})$ measurable. Since $\left(\int_E f_n\right)$ is a bounded sequence and $\mathcal{A}$ is countable, we may use a diagonal argument to extract a subsequence (g_n) of (f_n) such that the limit

$$\eta(E) := \lim_n \mu_n(E), \quad \mu_n(E) := \int_E g_n\, d\mu,$$

exists for each $E \in \mathcal{A}$. Now let

- $\mathcal{G}$ = the collection of all $G \in \mathcal{F}$ such that $\eta(G \cap A)$ exists for every $A \in \mathcal{A}$, and
- $\mathcal{H}$ = the collection of all $H \in \mathcal{G}$ such that $\eta(H \cap G)$ exists for every $G \in \mathcal{G}$.

Clearly, $\mathcal{A} \subseteq \mathcal{H}$, and since $\mathcal{A}$ is closed under finite intersections, $\mathcal{A} \subseteq \mathcal{G}$ and so $\mathcal{H} \subseteq \mathcal{G}$. In particular, $\mathcal{H}$ is closed under finite intersections. Furthermore, since $\mu_n(H^c \cap G) = \mu_n(G) - \mu_n(H \cap G)$, $\mathcal{H}$ is closed under complements. Also, if $H_1, H_2 \in \mathcal{H}$ and $H_1 \cap H_2 = \emptyset$, then from $\mu_n\big((H_1 \cup H_2) \cap G\big) = \mu_n(H_1 \cap G) + \mu_n(H_2 \cap G)$ we see that $H_1 \cup H_2 \in \mathcal{H}$. Since an arbitrary union $H_1 \cup H_2$ may be written as a disjoint union of members of $\mathcal{H}$, the latter is a field.

We show next that $\mathcal{H}$ is closed under increasing unions. It will follow that $\mathcal{H}$ is a σ-field and hence contains $\mathcal{F}_0$. Let $H_k \in \mathcal{H}$, $H_k \uparrow H$, and $G \in \mathcal{G}$. To show that $\eta(H \cap G)$ exists, set $E_k = (H \cap G) \setminus (H_k \cap G)$ and note that $\mu(E_k) \downarrow 0$. By (β), $\mu_n(E_k) \downarrow_k 0$ uniformly in n, hence $\mu_n(H_k \cap G) \uparrow_k \mu_n(H \cap G)$ uniformly in n. Let $\varepsilon > 0$ and choose k such that

$$|\mu_n(H \cap G) - \mu_n(H_k \cap G)| < \varepsilon \ \ \forall\, n.$$

Then choose N so that $|\mu_n(H_k \cap G) - \mu_m(H_k \cap G)| < \varepsilon$ for all $m, n \geq N$. By the triangle inequality, $|\mu_n(H \cap G) - \mu_m(H \cap G)| < 3\varepsilon$ for all $n, m \geq N$. Therefore the sequence $(\mu_n(H \cap G))$ is Cauchy, hence $H \in \mathcal{H}$.

We have shown that the limit $\eta(E)$ exists for all $E \in \mathcal{F}_0$. By the Vitali-Hahn-Saks theorem, η is a measure, and clearly $\eta \ll \mu$. By the Radon-Nikodym theorem, there exists a $\mathcal{F}_0$-measurable function g such that $d\eta = g\,d\mu$. Thus

$$\int hg\,d\mu = \lim_n \int hg_n\,d\mu \qquad (\gamma)$$

holds for all $\mathcal{F}_0$-measurable indicator functions h, hence for all $\mathcal{F}_0$-simple functions. Since the simple functions are dense in $L^\infty(\mathcal{F}_0, \mu)$ (4.2.1), an approximation argument shows that (γ) holds for all $h \in L^\infty(\mathcal{F}_0, \mu)$. Therefore, $g_n \to g$ weakly in the subspace $L^1(\mathcal{F}_0, \mu)$, hence also in the ambient space $L^1(\mathcal{F}, \mu)$. □

14.3 Convexity and Compactness

Recall that the closed convex hull of a subset A of a topological vector space $\mathcal{X}_\tau$ is the intersection of all closed, convex sets containing A. It may also be characterized as the closure of co A (see Ex. 9.3). Furthermore, in a LCS the weak and norm closures of a convex set coincide (10.1.6). In this section we consider the closed convex hull of weakly compact subsets of Banach spaces and of τ-compact subsets of Fréchet spaces.

The Krein-Šmulian Theorem

14.3.1 Theorem (Krein-Šmulian). *The closed convex hull of a weakly compact set K in a Banach space $\mathcal{X}$ is weakly compact.*

Proof. We may assume that $K \subseteq C_1$. Let K have the weak topology and let C_1', the closed unit ball of $\mathcal{X}'$, have the weak* topology. Then K and C_1' are compact and Hausdorff in these topologies and the duality map $\langle \boldsymbol{x}, \boldsymbol{x}' \rangle$ restricted to $K \times C_1'$ is separately continuous. By 14.1.5, the mapping $C_1' \to C(K)$, $\boldsymbol{x}' \mapsto \boldsymbol{x}'\big|_K$, is continuous in the weak* topology of C_1' and the weak topology of $C(K)$. Thus for each $\mu \in C(K)'$, the linear functional $\boldsymbol{x}' \to \mu(\boldsymbol{x}'|_K)$ is weak* continuous on C_1' and hence equals $\widehat{\boldsymbol{x}_\mu}$ for a unique $\boldsymbol{x}_\mu \in \mathcal{X}$ (10.2.9). We now have

$$\mu(\boldsymbol{x}'|_K) = \langle \boldsymbol{x}_\mu, \boldsymbol{x}' \rangle \ \ \forall\, \mu \in C(K)' \text{ and } \forall\, \boldsymbol{x}' \in \mathcal{X}' \text{ with } \|\boldsymbol{x}'\| \leq 1. \qquad (\dagger)$$

It follows that the mapping $\mu \to \boldsymbol{x}_\mu$ is affine and weak* -weak continuous, hence the image A of the closed unit ball in $C(K)$ is weakly compact and convex. Since every $\boldsymbol{x} \in K$ is of the form $\boldsymbol{x}_\mu$ for $\mu = \widehat{\boldsymbol{x}}$ with $\|\widehat{\boldsymbol{x}}\| = \|\boldsymbol{x}\| \leq 1$, the closed convex hull of K is contained in A and so is weakly compact. □

14.3.2 Corollary. *The closure of the convex balanced hull of a weakly compact set K in a Banach space $\mathcal{X}$ is weakly compact.*

Proof. The convex balanced hull of K is the set B of all finite sums $\sum_j \alpha_j \boldsymbol{x}_j$, where $\boldsymbol{x}_j \in K$ and $\sum_j |\alpha_j| \leq 1$. Such a sum may be written

$$\sum_j t_j \beta_j \boldsymbol{x}_j, \quad t_j := |\alpha_j| \Big(\sum_i |\alpha_i| \Big)^{-1}, \quad \beta_j := \alpha_j |\alpha_j|^{-1} \sum_i |\alpha_i|.$$

Since $|\beta_j| \leq 1$, $\operatorname{cl} B$ is the closed convex hull of the weakly compact set $(\operatorname{cl} \mathbb{D}) \cdot K$ and hence is weakly compact. □

Mazur's Theorem

Here is an analog of the Krein-Šmulian theorem for Fréchet spaces, but in the original topology.

14.3.3 Theorem (Mazur). *Let K be a compact subset of a Fréchet space $\mathscr{X}$. Then* $\operatorname{cl}\operatorname{co} K$ *is compact.*

Proof. It suffices to show that $\operatorname{co} K$ is totally bounded. Let U be a neighborhood of zero in $\mathscr{X}$ and let V be a neighborhood of $\mathbf{0}$ such that $V + V \subseteq U$. Since K is totally bounded, $K \subseteq V + F$ for some finite set $F = \{\boldsymbol{x}_1, \ldots, \boldsymbol{x}_n\} \subseteq K$. Since the mapping

$$(t_1, \ldots, t_n) \mapsto \sum_{j=1}^{n} t_j \boldsymbol{x}_j : \Big\{(t_1, \ldots, t_n) : t_j \geq 0, \ \sum_{j=1}^{n} t_j = 1\Big\} \to \operatorname{co} F$$

is continuous, $\operatorname{co} F$ is compact. Therefore, there exists a finite set E such that $\operatorname{co} F \subseteq V + E$. It follows that $\operatorname{co} K \subseteq U + E$. Indeed, let $\boldsymbol{y} \in \operatorname{co} K$, say $\boldsymbol{y} := \sum_{j=1}^{m} t_j \boldsymbol{y}_j$, where $\boldsymbol{y}_j \in K$, $t_j \geq 0$, and $\sum_{j=1}^{m} t_j = 1$. By choice of F, there exist $\boldsymbol{z}_j \in F$ such that $\boldsymbol{y}_j - \boldsymbol{z}_j \in V$. By convexity of V,

$$\boldsymbol{y} = \sum_{j=1}^{m} t_j(\boldsymbol{y}_j - \boldsymbol{z}_j) + \sum_{j=1}^{m} t_j \boldsymbol{z}_j \in V + \operatorname{co} F \subseteq V + V + E \subseteq U + E.$$

Therefore, $\operatorname{co} K$ is totally bounded. □

The Finite Dimensional Case

For finite dimensional spaces we have the following stronger result:

14.3.4 Theorem. *Let K be a compact subset of $\mathbb{R}^d$. Then $\operatorname{co} K$ is compact.*

Proof. Let A be any nonempty subset of $\mathbb{R}^d$. We claim that for each $\boldsymbol{x} \in \operatorname{co} A$ there exists a subset B of A containing at most $d+1$ points such that $\boldsymbol{x} \in \operatorname{co} B$. To see this, let $\boldsymbol{x} = \sum_{j=1}^{m} t_j \boldsymbol{a}_j$, where $m > d+1$, $t_j \geq 0$ and $\sum_{j=1}^{m} t_j = 1$. We may assume that $t_j > 0$ for all j. Since $m - 1 > d$, the vectors $\boldsymbol{a}_j - \boldsymbol{a}_m$ $(1 \leq j \leq m-1)$ are linearly dependent, hence there exist c_j not all 0 such that $\sum_{j=1}^{m-1} c_j(\boldsymbol{a}_j - \boldsymbol{a}_m) = 0$. Setting $c_m = -\sum_{j=1}^{m-1} c_j$, we have

$$\sum_{j=1}^{m} c_j \boldsymbol{a}_j = 0 \ \text{ and } \ \sum_{j=1}^{m} c_j = 0. \tag{†}$$

We may assume each $c_j \neq 0$, otherwise reduce the above sums accordingly. Choose k so that $|c_j/t_j| \leq |c_k/t_k|$ for $j = 1, \ldots, m$. Then $t_j/t_k \geq |c_j/c_k| \geq c_j/c_k$, hence, using (†), we have

$$t_j - t_k\frac{c_j}{c_k} \geq 0, \quad \sum_{j=1}^{m}\left(t_j - t_k\frac{c_j}{c_k}\right) = 1, \quad \text{and} \quad \sum_{j=1}^{m}\left(t_j - t_k\frac{c_j}{c_k}\right)\boldsymbol{a}_j = \boldsymbol{x}.$$

Since the kth coefficient in the last sum is 0, $\boldsymbol{x}$ is now expressed as a convex combination of fewer than m vectors in A. Continuing this reduction process verifies the claim.

Now let $S = \{(t_1, \ldots, t_{d+1}) : t_j \geq 0, \ \sum_j t_j = 1\}$. By the result of the previous paragraph applied to $A = K$, we see that $\operatorname{co} K$ is the image of the compact set $S \times K \times \cdots \times K$ under the continuous map $(t_1, \ldots, t_{d+1}, \boldsymbol{x}_1, \ldots, \boldsymbol{x}_{d+1}) \mapsto \sum_{j=1}^{d+1} t_j \boldsymbol{x}_j$. Therefore, $\operatorname{co} K$ is compact. □

14.4 Extreme Points

Definitions and Examples

An **extreme point** of a nonempty, convex subset K of a vector space $\mathcal{X}$ is a point $z \in K$ with the property

$$x, y \in K,\ 0 < t < 1, \text{ and } z = tx + (1-t)y \Rightarrow x = y.$$

We denote the set of extreme points of K by $\operatorname{ex} K$. For example, if K is a triangle (inluding interior) in $\mathbb{R}^2$, then $\operatorname{ex} K$ consists of the vertices. The Krein-Milman theorem asserts that a compact, convex subset of a LCS is the closed convex hull of its extreme points. The theorem is among the most powerful tools in modern analysis with important applications in many diverse areas, some of which we explore in this and later sections.

Here are useful alternate characterizations of extreme point.

14.4.1 Proposition. *Let K be a nonempty convex subset of a vector space $\mathcal{X}$. The following are equivalent:*

(a) *z is an extreme point of K.*

(b) *If $z = \sum_{j=1}^n t_j x_j$, where $x_j \in K$, $0 < t_j < 1$, and $\sum_{j=1}^n t_j = 1$, then $z = x_j$ for some j.*

(c) *If $x, y \in K$ and $z = \frac{1}{2}x + \frac{1}{2}y$ then $x = y$.*

Proof. Assume that (a) holds. The verification of (b) is a simple induction argument. Indeed, the assertion is obviously true for $n = 2$, and if the assertion holds for n and $z = \sum_{j=1}^{n+1} t_j x_j$, then, setting $t = 1 - t_{n+1}$, we have

$$z = t\left[\sum_{j=1}^n \frac{t_j}{t} x_j\right] + (1-t)x_{n+1}, \quad \sum_{j=1}^n \frac{t_j}{t} = 1,$$

so the assertion holds for $n+1$.

Clearly, (b) $\Rightarrow$ (c). Now assume that (c) holds and let $x, y \in K$, $0 < t < 1$, and $z = tx + (1-t)y$ We may assume that $1/2 \le t < 1$. Then $s := 2t - 1 \in [0,1)$ and so $sx + (1-s)y \in K$. Since $z = \frac{1}{2}x + \frac{1}{2}(sx + (1-s)y)$, by hypothesis $x = sx + (1-s)y$. Therefore, $x = y$, hence z is extreme. □

The following proposition shows, as one might expect, that an extreme point of a convex set must lie on the boundary.

14.4.2 Proposition. *Let K be a convex subset of a TVS. Then no interior point of K can be an extreme point.*

Proof. Let $x \in \operatorname{int} K$ and choose $y \in \operatorname{int} K$ with $y \ne x$. Since the interior of a convex set is convex (9.1.1), the function $f(t) = y + t(x - y) = tx + (1-t)y$ maps $[0,1]$ into $\operatorname{int} K$. Since f is continuous, $f[0, 1+\varepsilon] \subseteq \operatorname{int} K$ for some $\varepsilon > 0$. Set $z = f(1+\varepsilon) = (1+\varepsilon)x - \varepsilon y$. Then z, $y \in K$, $z \ne y$, and $x = \varepsilon(1+\varepsilon)^{-1}y + (1+\varepsilon)^{-1}z$, hence x is not an extreme point. □

14.4.3 Examples.

(a) Let $\mathscr{X}$ be a strictly convex normed space. It follows from 8.1.8 that the extreme points of the closed unit ball in $\mathscr{X}$ are the points on the boundary S_1. In particular, this holds for Hilbert spaces and L^p spaces $(1 < p < \infty)$.

(b) The closed unit ball C_1 in $\mathfrak{c}_0$ has no extreme points. Indeed, if $\boldsymbol{x} = (x_n) \in S_1$ and n is chosen so that $|x_n| < 1/2$, then the equation

$$\boldsymbol{x} = \tfrac{1}{2}(x_1, \ldots, x_{n-1}, x_n - \tfrac{1}{2}, x_{n+1}, \ldots) + \tfrac{1}{2}(x_1, \ldots, x_{n-1}, x_n + \tfrac{1}{2}, x_{n+1}, \ldots)$$

expresses $\boldsymbol{x}$ as a nontrivial convex sum of members of C_1.

(c) The closed unit ball C_1 in $L^1(\mathbb{R}^d, \lambda^d)$ has no extreme points. To see this, let $\|f\|_1 = 1$ and choose a measurable set E such that $\int_E |f| = 1/2$ (Ex. 1.85). Set $f_1 := 2f\mathbf{1}_E$ and $f_2 := 2f\mathbf{1}_{E^c}$. Then f_1, $f_2 \in C_1$, $f = \frac{1}{2}f_1 + \frac{1}{2}f_2$, and $f_1 \neq f_2$, so f is not an extreme point.

(d) The extreme points of C_1 in $\ell^1(\mathbb{N})$ are the points $c\boldsymbol{e}_n = (0, \cdots, 0, c, 0, \cdots)$, $|c| = 1$. To see that such a point is extreme, assume that $c\boldsymbol{e}_n = t\boldsymbol{x} + (1-t)\boldsymbol{y}$, $\|\boldsymbol{x}\|_1 = \|\boldsymbol{y}\|_1 = 1$, so that $c = tx_n + (1-t)y_n$. If, say, $|x_n| < |y_n|$, then $|c| \leq t|x_n| + (1-t)|y_n| < |y_n| \leq 1$, impossible. Therefore, $|x_n| = |y_n|$, hence $c = x_n = y_n$, since c is an extreme point of the set $\{|z| \leq 1\}$. Since $\|\boldsymbol{x}\|_1 = \|\boldsymbol{y}\|_1 = 1$ and $|c| = 1$, it follows that $x_k = y_k = 0$ for all $k \neq n$.

No other point is extreme: Let $\boldsymbol{x} = (x_1, x_2, \ldots) \in S_1$ be an extreme point of C_1. It suffices to show that no two members of the sequence can be nonzero. Suppose, for example, that both x_1 and x_2 are nonzero. Set $t = \sum_{n=2}^{\infty} |x_n|$. Then $0 < t < 1$ and

$$\boldsymbol{x} = (1-t)(\operatorname{sgn}(x_1), 0, 0, \ldots) + t(0, x_2/t, x_3/t, \ldots),$$

which expresses $\boldsymbol{x}$ as a nontrivial convex combination of members of C_1.

(e) Let X be a compact Hausdorff topological space. The extreme points of the closed unit ball C_1 in $C(X)$ are the functions f such that $|f| = 1$. To see that such a function is extreme, let $f = tg + (1-t)h$, where $\|g\|_\infty = \|h\|_\infty = 1$ and $0 < t < 1$. Then for each $x \in X$, $f(x) = tg(x) + (1-t)h(x)$, hence, by the strict convexity of $\mathbb{C}$ = Euclidean space $\mathbb{R}^2$, $f(x) = g(x) = h(x)$.

Conversely, let $f \in C_1$ with $|f(x_0)| < 1$ for some x_0. The set

$$C := \left\{x \in X : |f(x) - f(x_0)| \geq \tfrac{1}{2}(1 - |f(x_0)|)\right\}$$

is closed and does not contain x_0, hence we may choose a nonzero $g \in C(X)$ such that $g = 0$ on C and $\|g\|_\infty < r$, $r > 0$ to be determined. Now, $f = \frac{1}{2}(f+g) + \frac{1}{2}(f-g)$ so if we can choose r so that $\|f \pm g\| \leq 1$ it will follow that f is not extreme. Thus it suffices to show that for suitable r, $|f(x)| + |g(x)| \leq 1$ for $x \in C^c$. But for such x,

$$|f(x)| \leq |f(x) - f(x_0)| + |f(x_0)| \leq \tfrac{1}{2}(1 - |f(x_0)|) + |f(x_0)| = \tfrac{1}{2}(1 + |f(x_0)|),$$

hence

$$|f(x)| + |g(x)| < \tfrac{1}{2}(1 + |f(x_0)|) + r.$$

Choosing $r = \frac{1}{2}(1 - |f(x_0)|)$ completes the argument.

A similar argument shows that the extreme points of the closed unit ball in L^∞ are the functions f with $|f(x)| = 1$ a.e. (Or one may use the fact that L^∞ is isometric and isomorphic to $C(X)$, where X is the spectrum of the C^*-algebra L^∞.)

(f) Let X be a compact Hausdorff space. Identify the dual of $C(X)$ with the space of all complex regular Borel measures μ on X with total variation norm $|\mu|(X)$. Let C_1' denote the closed unit ball in $C(X)'$ and $\mathscr{P}$ the convex subset of probability measures. We show:

(i) The extreme points of C_1' are the complex measures $c\delta_x$, $c \in \mathbb{C}$, $|c| = 1$.
(ii) The extreme points of $\mathscr{P}$ are the Dirac measures δ_x.

To see that $c\delta_x$ is extreme in C_1', let $c\delta_x = t\mu + (1-t)\nu$, where μ, $\nu \in C_1'$ and $0 < t < 1$. For any Borel set $E \ni x$,

$$c = c\delta_x(E) = t\mu(E) + (1-t)\nu(E), \quad \mu(E), \nu(E) \le 1,$$

hence $\mu(E) = \nu(E) = c$. In particular, $\mu(X) = \nu(X) = c$, hence $\mu(E^c) = \mu(X) - \mu(E) = 0 = \nu(E^c)$. Therefore, $\mu = \nu$ and so $c\delta_x$ is extreme in C_1'.

Conversely, suppose that $\|\mu\| = |\mu|(X) = 1$ and that the support K of $|\mu|$ contains at least two points x and y. Choose disjoint open sets $U \ni \boldsymbol{x}$ and $V \ni \boldsymbol{y}$. Then $|\mu|(U) > 0$ and $|\mu|(V) > 0$, hence also $|\mu|(U^c) > 0$. Define

$$\nu(E) = \frac{\mu(U \cap E)}{|\mu|(U)} \quad \text{and} \quad \eta = \frac{\mu(U^c \cap E)}{|\mu|(U^c)},$$

so that $|\nu|(X) = |\eta|(X) = 1$. Setting $t = |\mu|(U)$, we then have $\mu = t\nu + (1-t)\eta$, hence μ is not extreme. Therefore, $|\mu| = \delta_x$ for some $x \in X$. Thus $\mu(E) = 0$ if $x \notin E$ and $\mu(E) = \mu(\{x\})$ if $x \in E$, so $\mu = \mu(\{x\})\delta_x$ This verifies (i). Similar arguments establish (ii). ◊

For the proof of the Krein-Milman theorem, we need the following generalization of the notion of extreme point. Let A be a nonempty subset of a linear space $\mathscr{X}$. A nonempty subset E of A is an **extreme subset** of A if

$$\boldsymbol{x},\ \boldsymbol{y} \in A,\ 0 < t < 1, \text{ and } t\boldsymbol{x} + (1-t)\boldsymbol{y} \in E \Rightarrow \boldsymbol{x}, \boldsymbol{y} \in E.$$

For example, a side E of a triangle A in $\mathbb{R}^2$ is an extreme subset.

14.4.4 Lemma. *Let $\mathscr{X}$ be a LCS and $A \ne \emptyset$ a compact subset of $\mathscr{X}$. If f is a real continuous linear functional on $\mathscr{X}$ and $a := \min\{f(\boldsymbol{x}); \boldsymbol{x} \in A\}$, then $E := \{\boldsymbol{x} \in A : f(\boldsymbol{x}) = a\}$ is an extreme subset of A.*

Proof. Let $\boldsymbol{x}$, $\boldsymbol{y} \in A$, $0 < t < 1$, and $f(t\boldsymbol{x} + (1-t)\boldsymbol{y}) = a$. Then $f(\boldsymbol{x}), f(\boldsymbol{y}) \ge a$ and the assumption that one or the other is greater than a leads to the contradiction $a = f(t\boldsymbol{x} + (1-t)\boldsymbol{y}) = tf(\boldsymbol{x}) + (1-t)f(\boldsymbol{y}) > ta + (1-t)a = a$. Therefore, $\boldsymbol{x}$, $\boldsymbol{y} \in E$. □

The Krein-Milman Theorem

14.4.5 Theorem (Krein-Milman). *Let $\mathscr{X}$ be a LCS over $\mathbb{K}$ and let $K \subseteq \mathscr{X}$ be a nonempty compact convex subset of $\mathscr{X}$. Then K is the closed convex hull of its extreme points.*

Proof. We first show that $\operatorname{ex} K \ne \emptyset$. Let $\mathfrak{E}$ be the collection of all nonempty, closed extreme subsets of K. Since K is such a set, $\mathfrak{E} \ne \emptyset$. Partially order $\mathfrak{E}$ downward by inclusion, and let $\mathfrak{E}_0$ be a chain in $\mathfrak{E}$. We show that $E_0 := \bigcap_{E \in \mathfrak{E}_0} E \in \mathfrak{E}$. Now, $\mathfrak{E}_0$ has the finite intersection property since it is linearly ordered, hence by compactness $E_0 \ne \emptyset$. Also, it is easy to check that E_0 is an extreme set of K. Therefore, E_0 is a lower bound for $\mathfrak{E}_0$ in $\mathfrak{E}$. By Zorn's lemma, K has an extreme subset A that does not properly contain another extreme subset. We show that A consists of a single point. Suppose to the contrary that A contains distinct points $\boldsymbol{x}$ and $\boldsymbol{y}$. Then there exists a real continuous linear functional f such that $f(\boldsymbol{x}) \ne f(\boldsymbol{y})$. By the lemma, the set $E := \{\boldsymbol{z} \in A : f(\boldsymbol{z}) = a\}$ is an extreme subset of A. Since A is an extreme subset of K, it follows that E is an extreme subset of K. But E cannot contain both $\boldsymbol{x}$ and $\boldsymbol{y}$, hence E is a proper nonempty subset of A, contradicting the minimality of A. Therefore, A has just one point, necessarily an extreme point of K.

Now let B denote the closed convex hull of the set of extreme points of K. Since K is closed and convex, $B \subseteq K$. Suppose the containment is proper, and let $\boldsymbol{x} \in K \setminus B$. By 9.3.2, there exists a real continuous linear functional f such that $f(\boldsymbol{x}) < \inf f(B)$. Now, since the set $C := \{\boldsymbol{z} \in K : f(\boldsymbol{z}) = \inf f(K)\}$ is nonempty, compact, and convex, it has an extreme point $\boldsymbol{z}$, by the first paragraph. Since C is an extreme subset of K (by the lemma), $\boldsymbol{z}$ is an extreme point of K. In particular $\boldsymbol{z} \in B$, which is impossible, since $f(\boldsymbol{z}) = \inf f(K) \le f(\boldsymbol{x}) < f(\boldsymbol{y})$ for all $\boldsymbol{y} \in B$. Therefore, it must be the case that $B = K$. □

The following theorem describes a minimality property of $\operatorname{ex} K$. It asserts that the closure of any subset E of K that "generates" K must already contain the extreme points of K.

14.4.6 Theorem. *Let $\mathscr{X}$ be a LCS and let $K \subseteq \mathscr{X}$ be a nonempty, compact, convex, subset of $\mathscr{X}$. If $K = \operatorname{cl} \operatorname{co} E$, then $\operatorname{ex} K \subseteq \operatorname{cl} E$.*

Proof. We may assume that E is closed, hence compact. Suppose for a contradiction that $\boldsymbol{x}$ is an extreme point of K not contained in E. Let U be a closed, balanced, neighborhood of zero such that $(\boldsymbol{x} + U) \cap E = \emptyset$. By compactness, there exist $\boldsymbol{z}_1, \ldots, \boldsymbol{z}_n \in E$ such that the sets $\boldsymbol{z}_j + U$ cover E. Set $E_j := E \cap (\boldsymbol{z}_j + U)$, these sets being compact and contained in K. Now, the mapping

$$\Big\{(t_1, \ldots, t_n) : t_j \ge 0, \sum_{j=1}^n t_j = 1\Big\} \times E_1 \times \cdots \times E_n \to K : (t_1, \ldots, t_n, \boldsymbol{x}_1, \ldots, \boldsymbol{x}_n) \mapsto \sum_{j=1}^n t_j \boldsymbol{x}_j$$

is continuous and so has compact range $\operatorname{co}\Big(\bigcup_{j=1}^n E_j\Big)$. Since $E \subseteq \operatorname{co}\Big(\bigcup_{j=1}^n E_j\Big)$, we have

$$K = \operatorname{cl} \operatorname{co} E \subseteq \operatorname{cl} \operatorname{co}\Big(\bigcup_{j=1}^n E_j\Big) = \operatorname{co}\Big(\bigcup_{j=1}^n E_j\Big).$$

Thus $\boldsymbol{x}$ may be expressed as $\boldsymbol{x} = \sum_{j=1}^n t_j \boldsymbol{x}_j$, where $\boldsymbol{x}_j \in E_j$, $t_j \ge 0$, and $\sum_{j=1}^n t_j = 1$. Since $\boldsymbol{x}$ is extreme, $\boldsymbol{x} = \boldsymbol{x}_j$ for some j. Thus $\boldsymbol{x} \in E_j \subseteq \boldsymbol{z}_j + U \subseteq E + U$. But then $\boldsymbol{x} = \boldsymbol{e} + \boldsymbol{u}$ for some $\boldsymbol{e} \in E$ and $\boldsymbol{u} \in U$, producing the contradiction $\boldsymbol{x} - \boldsymbol{u} = \boldsymbol{e} \in (\boldsymbol{x} + U) \cap E = \emptyset$. □

14.4.7 Remarks.

(a) The set of extreme points of a compact convex set need not be closed, even in the finite dimensional case, as the figure illustrates.

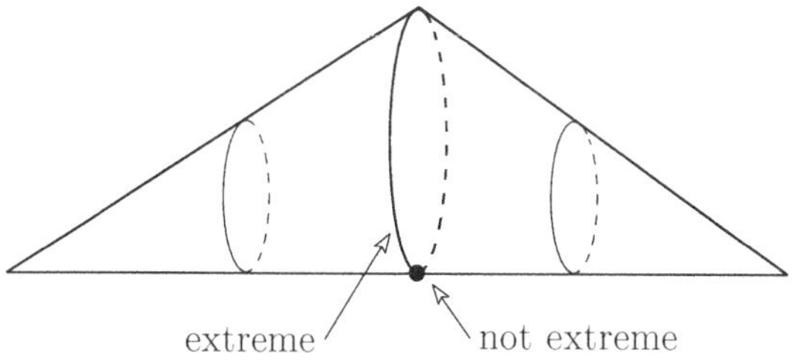

FIGURE 14.1: A nonclosed set of extreme points.

(b) If $\mathscr{X}$ is the dual of a normed space, then the closed unit ball C_1 is weak* compact and so C_1 is the closed convex hull of its extreme points. Thus, by (b) and (c) of 14.4.3, $L^1(\mathbb{R}^d, \lambda^d)$ and $\mathfrak{c}_0$ are not dual spaces.

(c) The space $C(X, \mathbb{R})$, where X is a nontrivial, compact, connected, Hausdorff topological space, is not a dual space. Indeed, the extreme points of C_1 are the functions f with $|f| = 1$. For such a function, $f^{-1}\{-1\}$ and $f^{-1}\{1\}$ are disjoint open sets whose union is X, hence one of these sets must equal X. Therefore, the extreme points of C_1 are the constant functions ± 1 and so $\operatorname{ex} C_1$ consists of constant functions. However, Urysohn's lemma implies the existence of nonconstant functions in C_1. Thus $\operatorname{cl}\operatorname{co}(\operatorname{ex} C_1) \neq C_1$, verifying the assertion. ◊

14.5 Applications of the Krein-Milman Theorem

The Existence of Ergodic Measures

Let X be a compact Hausdorff space and let $\mathscr{P} = \mathscr{P}(X)$ denote the convex, w^*-compact set of regular Borel probability measures on X. Let $\mathscr{T}$ be a family of homeomorphisms on X. A member μ of $\mathscr{P}$ is said to be $\mathscr{T}$**-invariant** if $T(\mu) = \mu$ for all $T \in \mathscr{T}$. We denote by $\mathscr{P}_{\mathscr{T}}$ the subset of $\mathscr{P}$ consisting of $\mathscr{T}$-invariant measures. A direct argument shows that if $\mathscr{P}_{\mathscr{T}}$ is nonempty, then it is convex and weak*-compact. A member μ of $\mathscr{P}_{\mathscr{T}}$ is said to be **ergodic** if $\mu(A) = 0$ or 1 for all $A \in \mathscr{F}$ with $T^{-1}(A) = A$ μ-a.e., that is, $\mu(A \triangle T^{-1}(A)) = 0$. We claim that if the mappings in $\mathscr{T}$ commute, then $\mathscr{P}_{\mathscr{T}} \neq \emptyset$ and the ergodic measures are the extreme points of $\mathscr{P}_{\mathscr{T}}$. In particular, the Krein-Milman theorem implies the existence of ergodic probability measures.

Suppose first that $\mathscr{T}$ consists of a single map T. Set $\mathscr{P}_T = \mathscr{P}_{\mathscr{T}}$. Abusing notation slightly, we let T also denote the weak* continuous affine, bijection $T: \mu \mapsto T(\mu)$: $\mathscr{P} \to \mathscr{P}$. To show that $\mathscr{P}_T$ is nonempty, fix $\mu \in \mathscr{P}$ and define $\mu_n \in \mathscr{P}$ by

$$\mu_n = \frac{1}{n}\sum_{j=0}^{n-1} T^j(\mu).$$

Let $f \in C(X)$. Since $T(\mu_n) = \frac{1}{n}\sum_{j=1}^{n} T^j(\mu)$ we have

$$|\mu_n(f) - T(\mu_n)(f)| \le n^{-1}|f - T^n(\mu)(f)| \le 2n^{-1}\|f\|_\infty .$$

Thus if ν is any weak* limit point of (μ_n), then $T(\nu)(f) = \nu(f)$ for all $f \in C(X)$, hence $\mathscr{P}_T$ is nonempty.

Now let $\mu \in \mathscr{P}_T$. The mapping $Uf := f \circ T$ maps $L^2(\mu)$ onto $L^2(\mu)$ and is unitary. By the mean ergodic theorem (12.1.16),

$$\lim_n \frac{1}{n}\sum_{j=0}^{n-1} U^j f = Pf \tag{†}$$

in $L^2(\mu)$ norm, where P is the projection onto the closed linear subspace of $L^2(\mu)$ consisting of those $g \in L^2$ with $Ug = g$. Applying the continuous linear functional $h \to \int h\,d\mu = \mu(h)$, we see by invariance of μ that $\mu(Pf) = \mu(f)$.

We claim that if μ is ergodic, then Pf must be constant and that constant must be $\mu(f)$. To see this, observe that U maps real functions onto real functions, hence so does P. By considering real and imaginary parts, we may take f to be real. Set $g = Pf$. Because $g = Ug = g \circ T$, the set $A_n := \{g \ge \mu(g) + 1/n\}$ satisfies $T^{-1}(A_n) = A_n$ μ-a.e. and so has measure zero or one. If the measure were one, then by integrating we would obtain the absurdity $\mu(g) \ge \mu(g) + 1/n$. Therefore, $\mu(A_n) = 0$ for all n and so $g \le \mu(g)$ μ-a.e. A similar argument shows that $g \ge \mu(g)$ μ-a.e. Thus $Pf = \mu(Pf) = \mu(f)$ for all $f \in C(X)$, verifying the claim.

To show that an ergodic measure μ is an extreme point, let $\mu = t\nu + (1-t)\eta$, where $\nu, \eta \in \mathscr{P}_T$ and $0 < t < 1$. Since $\nu \le \mu/t$ it follows that (†) also holds in the $L^2(\nu)$ norm. Applying the linear functional ν to (†) and using $Pf = \mu(f)$ and the invariance of ν we see that $\nu(f) = \nu(Pf) = \mu(f)$ for all $f \in C(X)$. Therefore, $\mu = \nu = \eta$, establishing that every ergodic measure is an extreme point. Conversely, suppose that $\mu \in \mathscr{P}_T$ is not ergodic. Then there exists $A \in \mathscr{F}$ such that $T(A) = A$ μ-a.e. and $0 < \mu(A) < 1$. The measure μ_A is then invariant:

$$\mu_A(T(E)) = \mu(A \cap T(E)) = \mu(T^{-1}(A \cap T(E))) = \mu((T^{-1}A) \cap E) = \mu_A(E).$$

Similarly, μ_{A^c} is invariant. Now define invariant probability measures

$$\nu := \frac{\mu_A}{\mu(A)} \quad \text{and} \quad \eta := \frac{\mu_{A^c}}{\mu(A^c)}.$$

Setting $t = \mu(A)$ we have $0 < t < 1$ and $\mu = t\nu + (1-t)\eta$, hence μ is not extreme. This verifies the claim for the case $\mathscr{T} = \{T\}$.

Now consider the general case. Let $S, T \in \mathscr{T}$. Because the maps commute, T maps $\mathscr{P}_S$ onto itself, hence we may restrict the mapping $\mu \to T(\mu)$ in the argument of the second paragraph to the set $\mathscr{P}_S$ to conclude that μ is both T- and S-invariant, that is, $\mathscr{P}_{\{S,T\}} \ne \emptyset$. More generally, $\mathscr{P}_F \ne \emptyset$ for every finite $F \subseteq \mathscr{T}$. Since these sets are compact, their intersection $\mathscr{P}_{\mathscr{T}}$ is nonempty. The entire argument in the preceding paragraphs then goes through if $\{T\}$ is replaced by $\mathscr{T}$.

The Stone-Weierstrass Theorem

In this section we use the Krein-Milman theorem to give a relatively short proof, due to de Branges, of Stone's generalization of the Weierstrass approximation theorem.

14.5.1 Theorem (Stone-Weierstrass). *Let X be a compact Hausdorff topological space and $\mathscr{A}$ a conjugate-closed subalgebra of $C(X)$ that contains the constant functions and separates points of X. Then $\mathscr{A}$ is dense in $C(X)$.*

Proof. It is clear that the closure of $\mathscr{A}$ is a conjugate-closed subalgebra of $C(X)$, hence we may assume that $\mathscr{A}$ is closed and then prove that $\mathscr{A} = C(X)$.

By the Hahn-Banach theorem, it suffices to show that $\mathscr{A}^\perp = \{0\}$. Suppose that this is not the case. By Alaoglu's theorem, the closed unit ball C_1' of $\mathscr{A}^\perp \subseteq C(X)'$ is weak* compact. By the Krein-Milman theorem, there exists a nontrivial extreme point μ of C_1', which we take to be a complex Radon measure on X with $K := \operatorname{supp}|\mu| \ne \emptyset$. We show that K consists of a single point x. It will then follow that $\mu = c\delta_x$, where $|c| - 1$. Since $1 \in \mathscr{A}$ and $\mu \in \mathscr{A}^\perp$, we will have the contradiction $c = \int 1\, d\mu = 0$, completing the proof.

Assume that K contains distinct points x and y. We show that this is incompatible with the fact that μ is an extreme point of C_1'. Note first that for any $h \in \mathscr{A}$ the complex measure $h\, d\mu$ is in $\mathscr{A}^\perp$. Indeed, since $\mathscr{A}$ is an algebra, if $g \in \mathscr{A}$, then $gh \in \mathscr{A}$ and so $\int gh\, d\mu = 0$, as required. Now, since $\mathscr{A}$ separates points, there exists a function $f \in \mathscr{A}$ such that $f(x) \ne f(y)$. Let $c = f(y)$. Then $g := f - c \in \mathscr{A}$ and $g(y) = 0 \ne g(x)$. Set

$$h = \frac{1}{1 + \|g\overline{g}\|} g\overline{g}, \ s := \int h\, d|\mu| = \|h\mu\| \text{ and } t := 1 - s = \int (1-h)\, d|\mu| = \|(1-h)\mu\|.$$

Then h, $1 - h \in \mathscr{A}$, hence $h\, d\mu$, $(1-h)\, d\mu \in \mathscr{A}^\perp$. Moreover, $h(y) = 0$, $0 < h(x) < 1$, and $0 \le h < 1$. Choose $0 < r < h(x)$ and an open neighborhood U of x such that $h \ge r$ on

U. Then $s \geq \int_U h\,d|\mu| \geq r|\mu|(U) > 0$, the last inequality because $U \cap K \neq \emptyset$. A similar calculation shows that $t > 0$. Since

$$\mu = s\,\frac{h\mu}{\|h\mu\|} + t\,\frac{(1-h)\mu}{\|(1-h)\mu\|}$$

and μ is an extreme point of C_1', $\mu = \|h\mu\|^{-1} h\mu$. By uniqueness of densities, $\|h\mu\|^{-1} h = 1$ $|\mu|$-a.e. But then, by continuity, $h = \|h\mu\| =$ constant on K, contradicting that $h(x) \neq h(y)$. $\square$

The Banach-Stone Isomorphism Theorem

Let X and Y be compact Hausdorff topological spaces. Given continuous mappings $\sigma : Y \to X$ and $\tau : Y \to \mathbb{C}$, define a mapping $T_{(\sigma,\tau)} : C(X) \to C(Y)$ by

$$(T_{(\tau,\sigma)} f)(y) = \tau(y)(f \circ \sigma)(y), \quad f \in C(X),\ y \in Y. \tag{14.1}$$

Clearly, $T_{(\tau,\sigma)}$ is linear and

$$\left\|T_{(\tau,\sigma)} f\right\|_\infty = \sup\{|\tau(y)(f \circ \sigma)(y)| : y \in Y\} \leq \|\tau\|_\infty \|f\|_\infty, \tag{14.2}$$

hence $\left\|T_{(\tau,\sigma)}\right\| \leq \|\tau\|_\infty$. In this subsection we prove that every isometric isomorphism of $C(X)$ onto $C(Y)$ is of the form $T_{(\tau,\sigma)}$, where σ is a homeomorphism and $\tau(Y) \subseteq \mathbb{T}$. We isolate part of the proof in the following lemma.

14.5.2 Lemma. *The linear map $T := T_{(\tau,\sigma)}$ is an isometry onto $C_b(Y)$ iff σ is a homeomorphism onto Y and $\tau(Y) \subseteq \mathbb{T}$.*

Proof. The sufficiency is clear. For the necessity, first take $f = 1$ to obtain $|\tau(y)| = |(T1)(y)| \leq \|T1\|_\infty \leq 1$. Next take $g = 1$ and set $f = T^{-1}g$. Since $\|g\|_\infty = 1$ and T is an isometry, $\|f\|_\infty = 1$, hence $1 = g(y) = |\tau(y)(f \circ \sigma)(y)| \leq |\tau(y)|$. Therefore, $\tau(Y) \subseteq \mathbb{T}$.

Now suppose $\sigma(Y) \neq X$ and let $x \in X \setminus \sigma(Y)$. Choose a continuous function f on X such that $f(x) = 1$ and $f\big(\sigma(Y)\big) = 0$. Then $Tf = 0$, yet $\|f\|_\infty \geq 1$. Therefore, σ must be surjective.

Finally, to show that σ is injective and hence is a homeomorphism, let $y_0 \neq y_1$ and let g be a continuous function on Y such that $0 \leq g \leq 1$, $g(y_0) = 0$, and $g(y_1) = 1$. Set $f = T^{-1}g$. Then

$$\tau(y_0)(f \circ \sigma)(y_0) = (Tf)(y_0) = 0 \quad \text{and} \quad \tau(y_1) f \circ \sigma(y_1) = (Tf)(y_1) = 1,$$

hence $f(\sigma(y_0)) = 0 \neq f(\sigma(y_1))$ and so $\sigma(y_0) \neq \sigma(y_1)$. $\square$

We may now prove

14.5.3 Theorem (Banach-Stone). *Let X and Y be compact Hausdorff topological spaces. If $T : C(X) \to C(Y)$ is a surjective, isometric isomorphism, then there exists a homeomorphism $\sigma : Y \to X$ and a continuous function $\tau : Y \to \mathbb{T}$ such that $T = T_{(\tau,\sigma)}$.*

Proof. The dual map $T' : M_{ra}(Y) \to M_{ra}(X)$ is an isometric isomorphism and hence maps the closed unit ball $C_{1,Y}$ of $M_{ra}(Y)$ onto the closed unit ball $C_{1,X}$ of $M_{ra}(X)$. Since T' is also a weak*-homeomorphism that preserves convexity, it maps the extreme points of $C_{1,Y}$ onto the extreme points of $C_{1,X}$. By 14.4.3(f), the extreme points of these balls are of the form $c\delta_z$, $|c| = 1$, where δ_z is the Dirac measure at z. Thus for each $y \in Y$ there exist unique $\tau(y) \in \mathbb{T}$ and $\sigma(y) \in X$ such that $T'\delta_y = \tau(y)\delta_{\sigma(y)}$. Therefore, for all $f \in C(X)$,

$$(Tf)(y) = \langle f, T'\delta_y\rangle = \big\langle f, \tau(y)\delta_{\sigma(y)}\big\rangle = \tau(y) f\big(\sigma(y)\big).$$

Taking $f = 1$, we have $(T1)(y) = \tau(y)$, hence τ is continuous. It follows that $f\big(\sigma(y)\big) = [\tau(y)]^{-1}(Tf)(y)$ is continuous in y for each f, which implies that σ is continuous. Since $T = T_{(\tau,\sigma)}$, σ is a homeomorphism by the lemma. $\square$

The Lyapunov Convexity Theorem

An $\mathbb{R}$-valued measure μ on a measurable space $(X, \mathcal{F})$ is said to be **non-atomic** if for each $A \in \mathcal{F}$ with $|\mu|(A) > 0$ there exists $B \in \mathcal{F}$ such that $B \subseteq A$ and $0 < |\mu|(B)| < |\mu|(A)$. For example, for any real-valued $g \in L^1(\lambda^d)$ the measure $g\, d\lambda^d$ on $\mathbb{R}^d$ is non-atomic (Ex. 1.85. On the other hand, counting measure on $\mathbb{N}$ is obviously atomic. The theorem in this subsection asserts that the range of a finite dimensional, non-atomic vector measure is compact and convex. For the proof we need

14.5.4 Lemma. *Let ν be a σ-finite, non-atomic measure on $(X, \mathcal{F})$. If $\nu(E) > 0$, then $L^\infty(E)$ is infinite dimensional.*

Proof. We identify the space $L^\infty(E)$ with the subspace of all $f \in L^\infty(X)$ such that $f = 0$ on E^c. Since ν is σ-finite, we may suppose that $\nu(E) < \infty$, otherwise consider a subset F of E with positive finite measure and work with the subspace $L^\infty(F)$. Set $E_0 = E$. Since ν is non-atomic, we may choose measurable sets E_n such that $E_n \subseteq E_{n-1}$ and $0 < \nu(E_n) < \nu(E_{n-1})$ for all n. Set $F_n = E_{n-1} \setminus E_n$. Then the sets F_n are disjoint and have positive measure implying that the indicator functions $\mathbf{1}_{F_n}$ are linearly independent. □

14.5.5 Theorem (Lyapunov). *Let $\mu_1, \ldots, \mu_d$ be real-valued non-atomic measures on $\mathcal{F}$. For $E \in \mathcal{F}$, define $\boldsymbol{\mu}(E) = \big(\mu_1(E), \ldots, \mu_d(E)\big)$. Then the set $\boldsymbol{\mu}(\mathcal{F}) := \{\boldsymbol{\mu}(E) : E \in \mathcal{F}\}$ is a compact convex subset of $\mathbb{R}^d$.*

Proof. Set $\nu = \sum_{j=1}^d |\mu_j|$ and note that ν is a non-atomic measure on $(X, \mathcal{F})$ with $|\mu_j| \ll \nu$ for each j. By the Radon-Nikodym theorem, there exists $g_j \in L^1(\nu)$ such that $d\mu_j = g_j\, d\nu$, so $\boldsymbol{\mu} = (g_1\nu, \ldots, g_d\nu)$. Define a linear map

$$T : L^\infty(\nu) \to \mathbb{R}^d, \quad Tf = \left(\int_X f\, d\mu_1, \ldots, \int_X f\, d\mu_d\right) = \left(\int_X f g_1\, d\nu, \ldots, \int_X f g_d\, d\nu\right).$$

Then T is continuous with respect to the weak* topology of $L^\infty(\nu) = (L^1)'$ and the norm topology of $\mathbb{R}^d$. Moreover, $T\mathbf{1}_E = \boldsymbol{\mu}(E)$ for all $E \in \mathcal{F}$. Now consider the convex set $C := \{f \in L^\infty(\nu) : 0 \le f \le 1\ \nu \text{ a.e.}\}$. If (f_α) is a net in C that w^*-converges to f, then $0 \le \int_E f\, d\nu \le \int_E 1\, d\nu$ for all E, hence $0 \le f \le 1$ ν-a.e. Therefore, C is w^*-closed and so is w^*-compact, by the Banach-Alaoglu theorem. Thus $T(C)$ is compact and convex in $\mathbb{R}^d$.

We claim that $\boldsymbol{\mu}(\mathcal{F}) = T(C)$, which will prove the theorem. By definition of C, we have $\boldsymbol{\mu}(E) = T(\mathbf{1}_E) \in T(C)$ for all E, that is, $\boldsymbol{\mu}(\mathcal{F}) \subseteq T(C)$. For the reverse inclusion, let $x \in T(C)$ and consider the convex, weak* compact set $K := \{f \in C : Tf = x\}$. By the Krein-Milman theorem, K has an extreme point g. We show that g is an indicator function. If not, then $\nu\{g(1-g) \ne 0\} > 0$ and so for some $\varepsilon > 0$ the set $E := \{\varepsilon \le g \le 1 - \varepsilon\}$ has positive ν measure. By the lemma, $L^\infty(E)$ is infinite dimensional. Since $T\big(L^\infty(E)\big)$ is finite dimensional, it follows that T cannot be 1-1, hence $Th = \mathbf{0}$ for some nonzero $h \in L^\infty(E)$ with $\|h\|_\infty \ne 0$. Multiplying by $\varepsilon\, \|h\|_\infty^{-1}$, we may assume that $-\varepsilon \le h \le \varepsilon$. But then $g \pm h$ are distinct members of K and $g = \frac{1}{2}(g+h) + \frac{1}{2}(g-h)$, contradicting that g is extreme in K. Therefore, g is an indicator function, completing the proof. □

The convexity theorem may be seen as the theoretical basis for the so-called "bang-bang" principle in optimal control. Control theory considers how systems (in physics, economics, etc.) that evolve in time are influenced by feedback. The bang-bang principle asserts that optimal change of system in minimal time may be achieved by the extreme values of the set of allowable controls. (A missile can find its target optimally by executing a sequence extreme left or extreme right rudder movements.) See, for example, [27].

The Ryll-Nardzewski Fixed Point Theorem

Let $\mathscr{X}$ be a LCS, C a nonempty, compact, convex subset of $\mathscr{X}$, and $\mathscr{T}$ a family of continuous affine mappings from C into itself. A point $x \in C$ is called a **fixed point** of $\mathscr{T}$ if $T(x) = x$ for all $T \in \mathscr{T}$. We denote the set of fixed points of a single mapping T by F_T. Then F_T is a compact, convex (possibly empty) subset of C, and $\mathscr{T}$ has a fixed point iff $\bigcap_{T \in \mathscr{T}} F_T \neq \emptyset$. In this chapter we establish an important fixed point theorem for mappings on C. The following lemma will be useful.

14.5.6 Lemma. *Suppose that $\bigcap_{T \in \mathscr{F}} F_T \neq \emptyset$ for each finite subset $\mathscr{F} \subseteq \mathscr{T}$. Then $\mathscr{T}$ has a fixed point.*

Proof. By hypothesis, the collection $\{F_T : T \in \mathscr{T}\}$ has the finite intersection property, hence, by compactness, has a nonempty intersection. □

For the theorems in this subsection, we require the following notions. A nonempty subset X of C is said to be **$\mathscr{T}$-invariant** if $TX \subseteq X$ for all $T \in \mathscr{T}$. A **minimal invariant set** is an invariant set not properly contained in an invariant set. If X is $\mathscr{T}$-invariant, then $\mathscr{T}$ is said to be **noncontracting on** X if

$$x, y \in X \text{ and } x \neq y \Rightarrow 0 \notin \mathrm{cl}\{Tx - Ty : T \in \mathscr{T}\};$$

equivalently, for any net (T_α) in $\mathscr{T}$,

$$T_\alpha x - T_\alpha y \to 0 \ \Rightarrow \ x = y.$$

Here is the key step needed to establish the Ryll-Nardzewski fixed point theorem. Its proof relies on the Krein-Milman theorem.

14.5.7 Lemma (Dugundji-Granas). *Let $\mathscr{T}$ be a semigroup of continuous affine maps from C into itself that is noncontracting on each minimal, closed $\mathscr{T}$-invariant subset of C. Then $\mathscr{T}$ has a fixed point.*

Proof. Let $\mathfrak{A}$ be the collection of all nonempty, compact, convex, $\mathscr{T}$-invariant subsets of C. In particular, $C \in \mathfrak{A}$, hence $\mathfrak{A} \neq \emptyset$. Partially order $\mathfrak{A}$ downward by inclusion. Clearly, a chain $\mathfrak{A}_0$ in $\mathfrak{A}$ has lower bound $\bigcap \mathfrak{A}_0$, hence, by Zorn's lemma, $\mathfrak{A}$ has a minimal element E. Now let $\mathfrak{B}$ be the collection of all nonempty, closed, invariant, subsets of E. Another application of Zorn's lemma shows that there is a closed, minimal, $\mathscr{T}$-invariant subset X of E. We show that X has a single member, completing the proof.

Assume that X has distinct members x and y. Since E is convex and invariant, the set $A := \{T\big((x+y)/2\big) : T \in \mathscr{T}\}$ is contained in E, hence so is the closure $\mathrm{cl}\, A$, which is also invariant. Since each T is affine, $\mathrm{cl}\,\mathrm{co}\, A$ is invariant. Thus, by minimality of E, $\mathrm{cl}\,\mathrm{co}\, A = E$. Now let z be an extreme point of E. By 14.4.6, $z \in \mathrm{cl}\, A$, so there exists a net $T_\alpha((x+y)/2) \to z$. Taking subnets if necessary, we may suppose that $T_\alpha(x) \to u$ and $T_\alpha(y) \to v$ for some $u, v \in E$. Thus $z = \frac{1}{2}u + \frac{1}{2}v$, and since z is extreme $u = v$. But then $T_\alpha x - T_\alpha y \to 0$, contradicting the noncontracting property of $\mathscr{T}$ on X. □

14.5.8 Corollary (F. Hahn). *Let $\mathscr{T}$ be a noncontracting semigroup of continuous affine maps from C into itself. Then $\mathscr{T}$ has a fixed point.*

14.5.9 Corollary (Kakutani). *Let $\mathscr{T}$ be a group of continuous, affine maps from C onto itself. Suppose that $\mathscr{T}$ is equicontinuous, that is, for each neighborhood V of 0 there exists a neighborhood U of 0 such that*

$$x - y \in U \Rightarrow Tx - Ty \in V \quad \forall\, T \in \mathscr{T}.$$

Then $\mathscr{T}$ has a fixed point.

Proof. By 14.5.8, is suffices to show that $\mathscr{T}$ is noncontracting. Let $x, y \in C$ and let (T_α) be a net in $\mathscr{T}$ such that $T_\alpha x - T_\alpha y \to 0$. Let V be an arbitrary neighborhood of 0 and choose U as in the theorem. Next, choose α_0 such that $T_\alpha(x) - T_\alpha(y) \in U$ for all $\alpha \geq \alpha_0$. For such α, $x - y = T_\alpha^{-1}T_\alpha x - T_\alpha^{-1}T_\alpha y \in V$. Since V was arbitrary, $x = y$. □

We are now in a position to prove the main result of the section. We give a nontrivial application in Chapter 16.

14.5.10 Theorem (Ryll-Nardzewski). *Let C be a nonempty, weakly compact, convex subset of a locally convex space X_τ and let $\mathscr{T}$ be a τ-noncontracting semigroup of weakly continuous affine maps from C into itself. Then $\mathscr{T}$ has a fixed point.*

Proof. (Dugundji-Granas) By 14.5.6, it suffices to prove that $\bigcap_{T\in\mathscr{F}} F_T \neq \emptyset$, where $\mathscr{F} \subseteq \mathscr{T}$ is finite. Let $\mathscr{S}$ denote the subsemigroup of $\mathscr{T}$ generated by $\mathscr{F}$. Then $\mathscr{S}$ consists of all products of members of $\mathscr{F}$. Choose any point $x_0 \in C$. Since $\mathscr{S}$ is countable, the $\mathscr{S}$-invariant convex set $K := \mathrm{cl}_\tau \mathrm{co}(\mathscr{S} x_0) \subseteq C$ is τ-separable. Moreover, by 10.1.6, K is weakly closed. Let X be a weakly closed, minimal, $\mathscr{S}$-invariant subset of K. We show that $\mathscr{S}$ is noncontracting on X in the weak topology. It will follow from 14.5.7 that $\mathscr{S}$ has a fixed point in K, proving the theorem.

Let x and y be distinct members of X. Since $\mathscr{S}$ is τ-noncontracting, there exists a τ-open, convex neighborhood U of 0 such that the neighborhood $V - V$ of zero is disjoint from $\{Sx - Sy : S \in \mathscr{S}\}$, where $V := \mathrm{cl}\, U$. Since $\{z + U : z \in X\}$ is a cover of X and X is τ-separable, there exist countably many sets $(z_n + V) \cap X$ that cover X. Since V is τ-closed and convex, it is weakly closed. Therefore, the weakly compact set X is a countable union of weakly closed sets $(z_n + V) \cap X$. By Baire's theorem (0.12.5), some set $(z_n + V) \cap X$ contains a nonempty, weakly open set W. Now, the collection $\{S^{-1}(W) : S \in \mathscr{S}\}$ of weakly open sets covers X; otherwise $X \setminus \bigcup_{S\in\mathscr{S}} S^{-1}(W)$ would be a weakly closed, nonempty, $\mathscr{S}$-invariant subset properly contained in the minimal set X. To show that $\mathscr{S}$ is noncontracting in the weak topology, suppose for a contradiction that there exists a net (S_α) in $\mathscr{S}$ such that $w\text{-}\lim_\alpha [S_\alpha x - \alpha S_\alpha y] = 0$. We may assume by the weak compactness of X that the limits $w\text{-}\lim_\alpha S_\alpha x$ and $w\text{-}\lim_\alpha S_\alpha y$ exist and hence are equal. Let z denote their common value. Since $z \in X \subseteq \bigcup_{S\in\mathscr{S}} S^{-1}(W)$, we may choose S so that $Sz \in W$, implying that $SS_\alpha x$ and $SS_\alpha y$ are eventually in $W \subseteq z_n + V$. But then $SS_\alpha x - SS_\alpha y$ is eventually in $V - V$, contradicting the choice of V. □

14.6 Vector-Valued Integrals

Let $\mathscr{X}$ be a LCS and $(X, \mathscr{F}, \mu)$ a measure space. A function $f : X \to \mathscr{X}$ is said to be **weakly measurable (weakly integrable)** if $x' \circ f$ is measurable (integrable) for each $x' \in \mathscr{X}'$. For a weakly integrable f we seek a member of $\mathscr{X}$, which we denote by $\int_X f(x)\, d\mu(x)$ and call the **weak integral of f with respect to μ**, that satisfies

$$\left\langle \int_X f\, d\mu, x' \right\rangle = \int_X \langle f, x' \rangle\, d\mu \;\left(= \int_X \langle f(x), x' \rangle\, d\mu(x) \right) \text{ for all } x' \in \mathscr{X}'. \tag{14.3}$$

If (14.3) holds, then, because $\mathscr{X}'$ separates points, the integral $\int_X f\, d\mu$ is unique, linear in f and, in the case of a normed space, satisfies the inequality

$$\left\| \int_X f\, d\mu \right\| \leq \int_X \|f(x)\|\, d\mu(x). \tag{14.4}$$

Moreover, if $\mathscr{X}$ is a Hilbert space and (14.3) holds for weakly integrable f and g, then

$$\left(\int_X f\,d\mu \,\Big|\, \int_X g\,d\mu\right) = \int_X (f(x)\,|\,g(y))\,d\mu(x)\,d\mu(y) \text{ for all } \boldsymbol{x} \in \mathscr{X}. \tag{14.5}$$

Indeed, taking $\boldsymbol{x}'(\cdot) = \left(\cdot \mid \int_X g\,d\mu\right)$ in (14.3), we have

$$\left(\int_X f\,d\mu \,\Big|\, \int_X g\,d\mu\right) = \int_X \left(f(x) \,\Big|\, \int_X g(y)\,d\mu(y)\right) d\mu(x),$$

and using (14.3) on the inner integral yields (14.5). Finally, if $T : \mathscr{X} \to \mathscr{Y}$ is continuous and linear and (14.3) holds for f and Tf, where $(Tf)(x) = T(f(x))$, then

$$\left\langle T\int_X f\,d\mu, \boldsymbol{y}'\right\rangle = \left\langle \int_X f\,d\mu, T'\boldsymbol{y}'\right\rangle = \int_X \langle f, T'\boldsymbol{y}'\rangle\,d\mu = \int_X \langle Tf, \boldsymbol{y}'\rangle\,d\mu = \left\langle \int_X Tf\,d\mu, \boldsymbol{y}'\right\rangle,$$

that is,

$$T\int_X f(x)\,d\mu(x) = \int_X T(f(x))\,d\mu(x). \tag{14.6}$$

For the construction of the weak integral, we consider first the case of a Banach space.

Weak Integrals in Banach Spaces

Let $\mathscr{X}$ be a Banach space, $(X, \mathfrak{F}, \mu)$ a σ-finite measure space, and $f : X \to \mathscr{X}$ weakly integrable. The mapping

$$T_f : \boldsymbol{x}' \to \boldsymbol{x}' \circ f : \mathscr{X}' \to L^1(\mu)$$

is clearly linear. Moreover, by the closed graph theorem, T_f is continuous. Indeed, if $\boldsymbol{x}'_n \to \boldsymbol{x}'$ in $\mathscr{X}'$ and $T_f\boldsymbol{x}'_n \to g$ in $L^1(\mu)$, then for some subsequence we have $T_f\boldsymbol{x}'_{n_k} = \boldsymbol{x}'_{n_k} \circ f \overset{a.e.}{\to} g$, hence $T_f\boldsymbol{x}' = g$ a.e., as required. Identifying $L^1(\mu)'$ with $L^\infty(\mu)$, we see that the dual map $T'_f : L^\infty(\mu) \to \mathscr{X}''$ satisfies

$$\langle \boldsymbol{x}', T'_f g\rangle = \langle T_f\boldsymbol{x}', g\rangle = \int_X \langle f(x), \boldsymbol{x}'\rangle g(x)\,d\mu(x), \quad g \in L^\infty(\mu),\ \boldsymbol{x}' \in \mathscr{X}'.$$

In particular,

$$\langle \boldsymbol{x}', T'_f\mathbf{1}_E\rangle - \int_E \langle f, \boldsymbol{x}'\rangle\,d\mu, \quad E \in \mathfrak{F},\ \boldsymbol{x}' \in \mathscr{X}'.$$

We denote $T'_f\mathbf{1}_E$ by $\int_E f\,d\mu$. Thus $\int_E f\,d\mu$ is the unique member of $\mathscr{X}''$ satisfying

$$\left\langle \boldsymbol{x}', \int_E f\,d\mu\right\rangle = \int_E \langle f(x), \boldsymbol{x}'\rangle\,d\mu(x), \quad \forall\ E \in \mathfrak{F} \text{ and } \boldsymbol{x}' \in \mathscr{X}'. \tag{14.7}$$

The vector $\int_E f\,d\mu$ is called the **Dunford integral of** f **over** E. If $\int_E f\,d\mu \in \mathscr{X}(=\widehat{\mathscr{X}})$ for all E, then f is said to be **Pettis integrable** and $\int_E f\,d\mu$ is called the **Pettis integral of** f **over** E. In this case, (14.7) may be written

$$\left\langle \int_E f\,d\mu, \boldsymbol{x}'\right\rangle = \int_E \langle f(x), \boldsymbol{x}'\rangle\,d\mu(x), \quad \forall\ E \in \mathfrak{F} \text{ and } \boldsymbol{x}' \in \mathscr{X}'. \tag{14.8}$$

If $\mathscr{X}$ is reflexive, then the Dunford and Pettis integrals clearly coincide. The following example shows this is not necessarily the case for nonreflexive spaces.

14.6.1 Example. Let μ be counting measure on $\mathbb{N}$ and let $\boldsymbol{a} = (a_n) \in \ell^\infty(\mathbb{N})$. Define $f : \mathbb{N} \to \mathfrak{c}_0$ by $f(n) = a_n \boldsymbol{e}_n$. For any $\boldsymbol{x} = (x_n) \in \ell^1(\mathbb{N}) = \mathfrak{c}_0'$,

$$\langle f(n), \boldsymbol{x} \rangle = \sum_{j=1}^{\infty} a_j x_j \boldsymbol{e}_n(j) = a_n x_n,$$

hence

$$\int_{\mathbb{N}} \langle f(n), \boldsymbol{x} \rangle \, d\mu(n) = \sum_{n=1}^{\infty} a_n x_n = \langle \boldsymbol{x}, \boldsymbol{a} \rangle.$$

Therefore, f is Dunford integrable with $\int_{\mathbb{N}} f \, d\mu = \boldsymbol{a} \in \ell^\infty(\mathbb{N}) = \mathfrak{c}_0''$, and f is Pettis integrable iff $\boldsymbol{a} \in \mathfrak{c}_0$. ◊

The next theorem gives a simple sufficient condition for Pettis integrability.

14.6.2 Theorem. *If $\mathcal{X}$ is a separable Banach space and $f : X \to \mathcal{X}$ is weakly measurable with $\|f(\cdot)\| \in L^1(\mu)$, then f is Pettis integrable.*

Proof. The condition $\|f(\cdot)\| \in L^1(\mu)$ implies that f is weakly integrable, hence (14.7) holds. It remains to show that $\boldsymbol{x}'' := \int_E f \, du \in \mathcal{X}$. For this it suffices by 10.2.9 to show that $\boldsymbol{x}''$ is weak* continuous on C_1'.

By the separability of $\mathcal{X}$, C_1', is metrizable in the weak* topology. Let $\boldsymbol{x}_n' \to \boldsymbol{x}'$ in the weak* topology of C_1'. Then $\langle f(x), \boldsymbol{x}_n' \rangle \to \langle f(x), \boldsymbol{x}' \rangle$ for each $x \in X$, and since $| \langle f(x), \boldsymbol{x}_n' \rangle | \le \|f(x)\|$ and $\|f(\cdot)\|$ is integrable, the dominated convergence theorem implies that

$$\langle \boldsymbol{x}_n', \boldsymbol{x}'' \rangle = \left\langle \boldsymbol{x}_n', \int_E f \, d\mu \right\rangle = \int_E \langle f(x), \boldsymbol{x}_n' \rangle \, d\mu(x) \to \int_E \langle f(x), \boldsymbol{x}' \rangle \, d\mu(x) = \langle \boldsymbol{x}', \boldsymbol{x}'' \rangle,$$

as required. □

The question of countable additivity of the Dunford and Pettis integrals is of critical importance in applications. The Dunford integral is countably additive in the weak* sense but not in the norm sense. To see the former, let (E_n) be a disjoint sequence in $\mathcal{F}$ and set $F_n = \bigcup_{j=1}^n E_n$ and $E = \bigcup_n E_n$. Then for all $\boldsymbol{x}'$,

$$\lim_n \left\langle \boldsymbol{x}', \int_{F_n} f \, d\mu \right\rangle = \lim_n \int_{F_n} \langle f(x), \boldsymbol{x}' \rangle \, d\mu(x) = \int_E \langle f(x), \boldsymbol{x}' \rangle \, d\mu(x) = \left\langle \boldsymbol{x}', \int_E f \, d\mu \right\rangle,$$

which implies that

$$\int_E f \, d\mu = w^*\text{-} \sum_{n=1}^{\infty} \int_{E_n} f \, d\mu.$$

On the other hand, taking $\boldsymbol{a} = (1, 1, \ldots)$ in Example 14.6.1 we have for any n

$$\left\langle \boldsymbol{e}_n, \int_{\mathbb{N}} f \, d\mu \right\rangle = \int_{\mathbb{N}} \langle \boldsymbol{e}_n, f(k) \rangle \, d\mu(k) = \int_{\mathbb{N}} \langle \boldsymbol{e}_n, \boldsymbol{e}_k \rangle \, d\mu(k) = \sum_k \langle \boldsymbol{e}_n, \boldsymbol{e}_k \rangle = 1,$$

and for $m < n$

$$\left\langle \boldsymbol{e}_n, \sum_{j=1}^{m} \int_{\{j\}} f \, d\mu \right\rangle = \sum_{j=1}^{m} \int_{\{j\}} \langle \boldsymbol{e}_n, \boldsymbol{e}_k \rangle \, d\mu(k) = \sum_{j=1}^{m} \langle \boldsymbol{e}_n, \boldsymbol{e}_j \rangle = 0,$$

hence for all m

$$\left\| \int_{\mathbb{N}} f\,d\mu - \sum_{j=1}^{m} \int_{\{j\}} f\,d\mu \right\| \geq \sup_n \left| \left\langle \boldsymbol{e}_n, \int_{\mathbb{N}} f\,d\mu - \sum_{j=1}^{m} \int_{\{j\}} f\,d\mu \right\rangle \right| \geq 1.$$

Therefore, the integral is not countably additive in the norm sense.

An argument similar to the above shows that the Pettis integral is countably additive in the weak sense:

$$\int_E f\,d\mu = w\text{-}\sum_{n=1}^{\infty} \int_{E_n} f\,d\mu.$$

It is a remarkable fact that, in contrast to the Dunford integral, the Pettis integral is also countably additive in the norm sense. This may be seen as a consequence of the Orlicz-Pettis theorem regarding weak subseries-convergence of sequences. For details the reader is referred to [12] or [45].

Weak Integrals in Locally Convex Spaces

Weak integrals may be defined for a LCS; however, for this we require a Radon measure.

14.6.3 Theorem. *Let $\mathscr{X}$ be a LCS, X a locally compact Hausdorff topological space, and μ a Radon measure on X. If $f : X \to \mathscr{X}$ is continuous with compact support K_f and $\operatorname{cl}\operatorname{co} f(X)$ is compact,*[1] *then there exists a unique member $\int_X f\,d\mu \in \operatorname{cl}\operatorname{co} f(X)$ such that* (14.3) *holds.*

Proof. By restricting the field of scalars to $\mathbb{R}$ and taking real parts of members of $\mathscr{X}'$, we may regard $\mathscr{X}$ as a real linear space. Moreover, by a scalar adjustment, we may suppose that $\mu(K_f) = 1$. Given a finite subset $F = \{\boldsymbol{x}'_1, \ldots, \boldsymbol{x}'_n\}$ of $\mathscr{X}'$, let Q_F denote the closed set of all $\boldsymbol{x} \in \operatorname{cl}\operatorname{co} f(X)$ with the property

$$\langle \boldsymbol{x}, \boldsymbol{x}' \rangle = \int_{K_f} \langle f, \boldsymbol{x}' \rangle \,d\mu \quad \text{for all} \quad \boldsymbol{x}' \in F. \tag{$\dagger$}$$

If each Q_F is nonempty, then by compactness $\bigcap_F Q_F \neq \emptyset$, hence the unique member of this set satisfies the requirements of the theorem. Set

$$\boldsymbol{a} := (a_1, \ldots, a_n), \quad \text{where} \quad a_j = \int_X \langle f, \boldsymbol{x}'_j \rangle \,d\mu.$$

Then Q_F will be nonempty if

$$\boldsymbol{a} \in \operatorname{cl}\operatorname{co} \left\{ \left(\langle f(x), \boldsymbol{x}'_1 \rangle, \ldots, \langle f(x), \boldsymbol{x}'_n \rangle \right) : x \in X \right\}. \tag{$\ddagger$}$$

Indeed, if (‡) holds, then $\boldsymbol{a}$ is the limit of a net of convex combinations of the form

$$\sum_{j=1}^{n} t_j \left(\langle f(x_j), \boldsymbol{x}'_1 \rangle, \ldots, \langle f(x_j), \boldsymbol{x}'_n \rangle \right) = \left(\langle \boldsymbol{y}, \boldsymbol{x}'_1 \rangle, \ldots, \langle \boldsymbol{y}, \boldsymbol{x}'_n \rangle \right), \quad \boldsymbol{y} := \sum_{j=1}^{n} t_j f(x_j),$$

hence the limit $\boldsymbol{x}$ of any convergent subnet of the corresponding $\boldsymbol{y}$'s $\left(\in \operatorname{co} f(X) \right)$ satisfies $\left(\langle \boldsymbol{x}, \boldsymbol{x}'_1 \rangle, \ldots, \langle \boldsymbol{x}, \boldsymbol{x}'_n \rangle \right) = \boldsymbol{a}$, which is (†). Now, if (‡) does not hold, then, by the separation theorem in $\mathbb{R}^n$, there exists a vector $\boldsymbol{b} = (b_1, \ldots, b_n) \in \mathbb{R}^n$ such that

$$\left(\langle f(x), \boldsymbol{x}'_1 \rangle, \ldots, \langle f(x), \boldsymbol{x}'_n \rangle \right) \cdot \boldsymbol{b} < \boldsymbol{a} \cdot \boldsymbol{b} \;\; \forall \;\; x \in X.$$

The integral of the left side over K_f is just $\boldsymbol{a} \cdot \boldsymbol{b}$. Recalling that $\mu(K_f) = 1$, this is also the integral of the right side. Since K_f is compact, integration preserves the strict inequality and produces the contradiction $\boldsymbol{a} \cdot \boldsymbol{b} < \boldsymbol{a} \cdot \boldsymbol{b}$. □

[1] This is automatically the case if $\mathscr{X}$ is a Fréchet space (14.3.4).

14.6.4 Corollary. *Let $\mathscr{X}$ be a LCS, X a compact Hausdorff topological space, and $\mathscr{P}(X)$ the set of all Radon probability measures μ on X. If $f : X \to \mathscr{X}$ is continuous and $\operatorname{cl}\operatorname{co} f(X)$ is compact, then the integral $\int_X f\,d\mu$ exists in $\mathscr{X}$. Moreover, the mapping $\mu \to \int_X f\,d\mu$ from $\mathscr{P}(X)$ to $\mathscr{X}$ is w^*-w continuous.*

The Bochner Integral

So far we have considered only weak integrals, that is, integrals defined in terms of continuous linear functionals. In a Banach space, it is possible to construct a stronger form of vector integral using limits of vector-valued simple functions in much the same way as scalar integrals are constructed.

Let $\mathscr{X}$ be a Banach space and $(X, \mathcal{F}, \mu)$ a measure space. An $\mathscr{X}$-**valued simple function** is a function $f : X \to \mathscr{X}$ with finite range such that the set $\{f = \boldsymbol{x}\}$ is measurable for all $\boldsymbol{x} \in \mathscr{X}$. If the distinct nonzero values of f are $\boldsymbol{x}_k$ $(1 \le k \le n)$, then f may be expressed in **standard form** as

$$f = \sum_{k=1}^{n} \mathbf{1}_{E_k} \boldsymbol{x}_k, \quad E_k := \{f = \boldsymbol{x}_k\}.$$

We say that f is **Bochner integrable** if $\mu(E_k) < \infty$ for all k. In this case we define the **Bochner integral of** f by

$$\int_X f\,d\mu = \sum_{k=1}^{n} \mu(E_k)\boldsymbol{x}_k.$$

Note that

$$\left\|\int_X f\,d\mu\right\| \le \sum_{k=1}^{n} \mu(E_k)\,\|\boldsymbol{x}_k\| = \int_X \|f\|\,d\mu. \tag{14.9}$$

An argument entirely similar to that of 3.1.1(b) shows that for Bochner integrable simple functions f and g and scalars c,

$$\int_X (f+g)\,d\mu = \int_X f\,d\mu + \int_X g\,d\mu \quad \text{and} \quad \int_X cf\,d\mu = c\int_X f\,d\mu.$$

Thus the integral is linear on the vector space of all Bochner integrable simple functions.

A function $f : X \to \mathscr{X}$ is **strongly measurable** if there exists a sequence of simple functions $f_n : X \to \mathscr{X}$ such that $\lim_n \|f_n(x) - f(x)\| = 0$ for μ-a.a. $x \in X$. In this case we write $f_n \overset{a.e.}{\to} f$. It is easy to check that the set of strongly measurable functions is a linear space under pointwise operations. Moreover, since $\|f_n\|$ is measurable and $\|f_n(x)\| \to \|f(x)\|$ a.e., we see that the norm of a strongly measurable function is measurable.

A strongly measurable function $f : X \to \mathscr{X}$ is said to be **Bochner integrable** if there exists a sequence of Bochner integrable simple functions f_n such that

$$f_n \overset{a.e.}{\to} f \quad \text{and} \quad \lim_n \int_X \|f_n - f\|\,d\mu = 0.$$

In this case the **Bochner integral of** f is defined by

$$\int_X f\,d\mu = \lim_n \int_X f_n\,d\mu. \tag{14.10}$$

We shall call the sequence of simple functions (f_n) in this definition a **defining sequence for the integral of** f. To see that the limit in (14.10) exists, note that by (14.9)

$$\left\|\int_X f_m\,d\mu - \int_X f_n\,d\mu\right\| \le \int \|f_m - f_n\|\,d\mu \le \int_X \|f_m - f\|\,d\mu + \int_X \|f_n - f\|\,d\mu,$$

hence $\left(\int f_n\,d\mu\right)$ is a Cauchy sequence and so converges. To see that the limit in (14.10) is independent of the defining sequence (f_n), let (g_n) be another such sequence. Then

$$\left\|\int_X f_n\,d\mu - \int_X g_n\,d\mu\right\| \le \int_X \|f_n - f\|\,d\mu + \int_X \|g_n - f\|\,d\mu \to 0.$$

We say that f is **Bochner integrable over** $E \in \mathcal{F}$, if $f\mathbf{1}_E$ is Bochner integrable.

14.6.5 Proposition. *The set of Bochner integrable functions is a linear space under pointwise operations, and the Bochner integral is linear. Moreover, if f is Bochner integrable, then it is integrable over every $E \in \mathcal{F}$.*

Proof. Let f and g be Bochner integrable with defining sequences (f_n) and (g_n), respectively. Then $f_n + g_n \overset{a.e.}{\to} f + g$ and $\lim_n \int_X \|f_n + g_n - (f+g)\|\,d\mu = 0$, hence $f+g$ is Bochner integrable. Since the integral is linear on simple functions,

$$\int_X (f+g)\,d\mu = \lim_n \int_X (f_n + g_n)\,d\mu = \lim_n \int_X f_n\,d\mu + \lim_n \int_X g_n\,d\mu = \int_X f\,d\mu + \int_X g\,d\mu,$$

and

$$\int_X cf\,d\mu = \lim_n \int_X cf_n = \lim_n c \int_X f_n\,d\mu = c\int_X f\,d\mu.$$

The last assertion of the proposition follows from the obvious fact $(f_n\mathbf{1}_E)$ is a defining sequence for $\int_E f$. □

Here is a useful characterization of Bochner integrability:

14.6.6 Theorem. *Let $f : X \to \mathcal{X}$ be strongly measurable. Then f is Bochner integrable iff $\|f(\cdot)\|$ is integrable. In this case, $\left\|\int_X f\,d\mu\right\| \le \int_X \|f\|\,d\mu$.*

Proof. If f is Bochner integrable and (f_n) is a defining sequence for $\int_X f$, then, by definition, $\|f_n - f\|$ is integrable. Since $\|f\| \le \|f_n - f\| + \|f_n\|$, $\|f\|$ is integrable.

Conversely, assume that $\|f\|$ is integrable. Choose a sequence of simple functions f_n with $\|f_n(x) - f(x)\| \to 0$ μ-a.e. Set $E_n = \{x : \|f_n(x)\| \le 2\|f(x)\|\}$ and $g_n := f_n\mathbf{1}_{E_n}$. Then g_n is simple, and because $\|f_n(x)\| \to \|f(x)\|$, $x \in E_n$ for all sufficiently large n and so $\|g_n(x) - f(x)\| \to 0$ a.e. Since $\|g_n(x) - f(x)\| \le 3\|f(x)\|$, $\int \|g_n - f\|\,d\mu \to 0$ by the dominated convergence theorem. Therefore, f is Bochner integrable, $\int g_n\,d\mu \to \int f\,d\mu$, and $\int \|g_n\|\,d\mu \to \int \|f\|\,d\mu$, the last limit by the dominated convergence theorem. Finally, taking limits in $\left\|\int_X g_n\,d\mu\right\| \le \int_X \|g_n\|\,d\mu$, we obtain the desired inequality. □

Next, we prove a dominated convergence theorem for the Bochner integral. For this we need the following lemma.

14.6.7 Lemma. *Suppose that $(X, \mathcal{F}, \mu)$ is σ-finite. If $f_n : X \to \mathcal{X}$ is strongly measurable for all n and $\lim_n \|f_n(x) - f(x)\| = 0$ a.e., then f is strongly measurable.*

Proof. Since $(X, \mathcal{F}, \mu)$ is σ-finite, there exists a positive, integrable function ψ on X (Ex. 3.25). Since $\|f_n - f_m\|$ is measurable, $\|f_n - f\| = \lim_m \|f_n - f_m\|$ is measurable. For measurable functions g and h, define

$$d(g,h) = \int \frac{\|g(x) - h(x)\|}{1 + \|g(x) - h(x)\|}\psi\,d\mu.$$

Since $\|f(x) - f_n(x)\| \to 0$ a.e., $d(f, f_n) \to 0$ by the dominated convergence theorem. For each n, let $(g_{n,k})_k$ be a sequence of simple functions converging a.e. in norm to f_n. Then $\lim_k d(f_n, g_{n,k}) = 0$. Since $d(f, g_{n,k}) \le d(f, f_n) + d(f_n, g_{n,k})$, we may choose a sequence of simple functions h_n such that $d(f, h_n) \to 0$. Passing to a subsequence if necessary, we may assume (since ψ is positive) that $\|h_n - f\| \to 0$ a.e. Therefore, f is strongly measurable. □

Here is the promised convergence theorem.

14.6.8 Theorem. *Let $(X, \mathcal{F}, \mu)$ be σ-finite and let $(f_n : X \to \mathscr{X})$ be a sequence of Bochner integrable functions such that $f_n \overset{a.e.}{\to} f$. Suppose there exists $g \in L^1(\mu)$ such that $\|f_n\| \le g$ a.e. for each n. Then f is Bochner integrable and*

$$\lim_n \int_X f_n\, d\mu = \int_X f\, d\mu.$$

Proof. By the lemma, f is strongly measurable. Moreover, $\|f\| \le g$ a.e., hence $\|f\|$ is integrable and so f is Bochner integrable (14.6.6). Moreover, since $\|f - f_n\| \to 0$ and $\|f - f_n\| \le 2g$, the dominated convergence theorem implies that $\int \|f - f_n\| \to 0$. The desired conclusion now follows from the inequality $\|\int_X (f - f_n)\, d\mu\| \le \int_X \|f - f_n\|\, d\mu$. □

14.6.9 Theorem. *Let $T \in \mathscr{B}(\mathscr{X})$. If f is Bochner integrable, then so is Tf and*

$$T\left(\int_X f\, d\mu\right) = \int_X Tf\, d\mu.$$

Proof. Let g be a Bochner integrable simple function, say $g = \sum_{k=1}^n \boldsymbol{x}_k \mathbf{1}_{E_k}$, $\mu(E_k) < \infty$. Then $Tg = \sum_{k=1}^n (T\boldsymbol{x}_k)\mathbf{1}_{E_k}$, hence Tg is a Bochner integrable simple function and

$$\int_X Tg\, d\mu = \sum_{k=1}^n (T\boldsymbol{x}_k)\mu(E_k) = T\sum_{k=1}^n \boldsymbol{x}_k \mu(E_k) = T\int_X g\, d\mu.$$

Now let (f_n) be a defining sequence of simple functions for the integral of f. From $\|Tf_n(x) - Tf(x)\| \le \|T\|\,\|f_n(x) - f(x)\|$, we see that (Tf_n) is a defining sequence of simple functions for the integral of Tf. Therefore,

$$\int_X Tf\, d\mu = \lim_n \int_X Tf_n\, d\mu = \lim_n T\int_X f_n\, d\mu = T\lim_n \int_X f_n\, d\mu = T\int_X f\, d\mu. \qquad \square$$

Here is the connection between the Pettis integral and the Bochner integral.

14.6.10 Proposition. *If $f : X \to \mathscr{X}$ is Bochner integrable, then f is Pettis integrable and the Bochner and Pettis integrals coincide.*

Proof. Let g be a Bochner integrable simple function, say $g = \sum_{k=1}^m \boldsymbol{x}_k \mathbf{1}_{E_k}$, $\mu(E_k) < \infty$. Then for any $\boldsymbol{x}' \in \mathscr{X}'$,

$$\begin{aligned}\left\langle \int_X g, \boldsymbol{x}' \right\rangle &= \left\langle \sum_{k=1}^m \mu(E_k)\boldsymbol{x}_k, \boldsymbol{x}' \right\rangle = \sum_{k=1}^m \mu(E_k)\langle \boldsymbol{x}_k, \boldsymbol{x}'\rangle = \int_X \sum_{k=1}^m \langle \boldsymbol{x}_k, \boldsymbol{x}'\rangle \mathbf{1}_{E_k}\, d\mu \\ &= \int_X \left\langle \sum_{k=1}^m \mathbf{1}_{E_k}\boldsymbol{x}_k, \boldsymbol{x}' \right\rangle d\mu = \int_X \langle g, \boldsymbol{x}' \rangle\, d\mu. \qquad (\dagger)\end{aligned}$$

Now let (g_n) be the defining sequence for the Bochner integral of f constructed in the proof of 14.6.6. For any $\boldsymbol{x}' \in \mathscr{X}'$, $(\boldsymbol{x}' \circ f)(x) = \lim_n (\boldsymbol{x}' \circ g_n)(x)$ a.e. and $|(\boldsymbol{x}' \circ g_n)(x)| \le \|\boldsymbol{x}'\|\,\|g_n(x)\| \le 2\,\|\boldsymbol{x}'\|\,\|f(x)\|$, hence by the dominated convergence theorem and $(\dagger)$

$$\left\langle \int_X f\, d\mu, \boldsymbol{x}' \right\rangle = \lim_n \left\langle \int_X g_n\, d\mu, \boldsymbol{x}' \right\rangle = \lim_n \int_X \langle g_n, \boldsymbol{x}' \rangle\, d\mu = \int_X \langle f, \boldsymbol{x}' \rangle\, d\mu.$$

Replacing f by $f\mathbf{1}_E$ shows that f is Pettis integrable and that the Bochner integral $\int_E f\, d\mu$ is the same as the Pettis integral. □

Note that the function f in Example 14.6.1 is Bochner integrable iff $\boldsymbol{a} \in \ell^1(\mathbb{N})$ (14.6.6). Choosing $\boldsymbol{a} \in \mathfrak{c}_0 \setminus \ell^1(\mathbb{N})$ produces an example of a Pettis integrable function that is not Bochner integrable.

14.7 Choquet's Theorem

Let K be a nonempty, compact, convex subset of a real LCS $\mathcal{X}$. Taking $X = \operatorname{cl}\operatorname{ex} K$ and $f(x) = x$ in 14.6.4 and noting by the Krein-Milman theorem that $\operatorname{cl}\operatorname{co} X = K$, we have

$$\int_X x\, d\mu(x) \in K, \quad \mu \in \mathscr{P}(X).$$

A vector $x_0 \in K$ is said to be **represented by** $\mu \in \mathscr{P}(X)$ if

$$x_0 = \int_{X=\operatorname{cl}\operatorname{ex} K} x\, d\mu(x). \tag{14.11}$$

Every point of K may be so represented. Indeed, since the mapping $F(\mu) := \int_X x\, d\mu(x)$ on $\mathscr{P}(X)$ is affine, $F(\operatorname{co}\delta_X) = \operatorname{co} X$, and since F is w^*-w continuous, it follows from 14.4.2(f) and the Krein-Milman theorem that $F(\mathscr{P}(X)) = K$. Choquet's theorem asserts that if K is metrizable, then the measure μ in 14.11 may be taken to have support in $\operatorname{ex} K$ rather than simply in $\operatorname{cl}\operatorname{ex} K$.

For the remainder of the section, K is assumed to be metrizable with metric d. For the proof of Choquet's theorem, we need the following lemmas.

14.7.1 Lemma. *$\operatorname{ex} K$ is a Borel subset of K.*

Proof. Let $x \in K$. Then x is not an extreme point of K iff there exist distinct points y and z in K such that $x = \frac{1}{2}(y + z)$ (14.4.1). Therefore, the complement of $\operatorname{ex} K$ in K is the Borel set $\bigcup_n F_n$, where $F_n = \{\frac{1}{2}(y + z) : y, z \in K, d(y, z) \geq 1/n\}$. □

14.7.2 Lemma. *Let $A(K, \mathbb{R})$ denote set real-valued, continuous, affine functions on K. For $f \in C(K, \mathbb{R})$, define a function $c(f) : K \to \mathbb{R}$ by*

$$c(f)(x) := \inf\{h(x) : h \in A(K, \mathbb{R}) \text{ and } f \leq h\}, \quad x \in K.$$

Then the following properties hold:

(a) $f \leq c(f) \leq \|f\|_\infty$.

(b) *$c(f)$ is concave and Borel measurable.*

(c) *If f is concave, then $f = c(f)$.*

(d) $c(f + g) \leq c(f) + c(g)$ *and* $c(tf) = tc(f)$ $(t > 0)$.

(e) $c(f + h) = c(f) + h$ *if* $h \in A(K, \mathbb{R})$.

(f) $c(f - g) \leq \|f - g\|_\infty$.

Proof. (a) The first inequality is clear, and the second follows from the fact that the constant function $h(x) := \|f\|_\infty$ is affine.

(b) Let $0 < t < 1$. If $h_j \in A(K, \mathbb{R})$ and $h_j \geq f_j$, then $th_1 + (1 - t)h_2 \in A(K, \mathbb{R})$ and $tf_1 + (1 - t)f_2 \leq th_1 + (1 - t)h_2$, hence

$$c\big(tf_1 + (1 - t)f_2\big)(x) \leq th_1(x) + (1 - t)h_2(x) \text{ for all } x \in K.$$

Taking infima over h_1 and h_2 yields

$$c\big(tf_1 + (1-t)f_2\big)(x) \le t\,c(f_1)(x) + (1-t)\,c(f_2)(x).$$

For the second part of (b), note that if $c(f)(x) < a$, then $h(x) < a$ for some $h \in A(K,\mathbb{R})$ with $h \ge f$. Since $h < a$ on some neighborhood U of x, the inequality $c(f) < a$ holds on U. Thus $\{c(f) < a\}$ is open, hence measurable.

(c) Assume for a contradiction that $f(y) < c(f)(y)$ for some y. Since f is continuous and concave, $C := \{(x,t) : t \le f(x)\}$ is a closed convex subset of the real LCS $\mathscr{X} \times \mathbb{R}$. Since $\big(y, c(f)(y)\big) \notin C$, by the separation theorem there exists an $a \in \mathbb{R}$ and a continuous linear functional F on $\mathscr{X} \times \mathbb{R}$ such that

$$F(x,t) \le a < F\big(y, c(f)(y)\big) \quad \forall\ (x,t) \in C.$$

In particular, $F(y, f(y)) < F\big(y, c(f)(y)\big)$, and by subtracting and normalizing we see that

$$F(0,1) = \frac{F\big(y, c(f)(y)\big) - F\big(y, f(y)\big)}{c(f)(y) - f(y)} > 0.$$

Now define

$$h(x) := \frac{a - F(x,0)}{F(0,1)}, \qquad x \in K.$$

Then

$$F\big(x, h(x)\big) = F\big(0, h(x)\big) + F\big(x, 0\big) = h(x)F(0,1) + F\big(x,0\big) = a.$$

If also $F\big(x,t\big) = a$, then $a - F(x,0) = tF\big(0,1\big)$, and dividing by $F\big(0,1\big)$ shows that $t = h(x)$. Thus $h(x)$ is the unique real number satisfying $F(x, h(x)) = a$. It follows that $h \in A(K,\mathbb{R})$. Since for all $x \in K$,

$$a \ge F\big(x, f(x)\big) = F(x,0) + f(x)F\big(0,1\big) = a - h(x)F(0,1) + f(x)F\big(0,1\big),$$

we see that $f \le h$. Therefore, $c(f)(y) \le h(y)$. On the other hand,

$$F(y,0) + h(y)F(0,1) = a < F\big(y, c(f)(y)\big) = F(y,0) + c(f)(y)F(0,1),$$

hence $h(y) < c(f)(y)$. With this contradiction we see that (c) holds.

(d) The first inequality is proved by considering affine functions h and k majorizing f and g, respectively, and noting that $h + k$ is affine. The second follows from the fact that h is affine iff th is affine.

(e) By (a) and (d) it suffices to show $c(f) + h \le c(f + h)$. But if k is affine and $k \ge f + h$ then $k - h$ is affine and majorizes f, so $k - h \ge c(f)$, or $k \ge c(f) + h$. Taking infima over k gives the desired inequality.

(f) By part (d), $c(f) = c(f-g+g) \le c(f-g) + c(g)$, hence $c(f) - c(g) \le c(f-g) \le \|f-g\|_\infty$, the last inequality by (a). $\square$

We may now prove

14.7.3 Theorem (Choquet). *Let $\mathscr{X}$ be a real LCS and K a nonempty, compact, convex, metrizable subset of $\mathscr{X}$. Then each $x_0 \in K$ is represented by a Radon probability measure μ on $\operatorname{cl}\operatorname{ex} K$ supported by $\operatorname{ex} K$.*

Proof. Since K is metrizable, $A(K,\mathbb{R})$ is separable. Let (h_n) be a dense sequence in the unit sphere of $A(K,\mathbb{R})$ and set $g = \sum_n h_n^2/2^n$. We claim that g is strictly convex on K. To see this, let $\boldsymbol{x} \neq \boldsymbol{y} \in K$ and $0 < t < 1$. Since $\mathscr{X}'$ separates points of $\mathscr{X}$, affine functions separate points of K, hence there exists an h_n such that $h_n(\boldsymbol{x}) \neq h_n(\boldsymbol{y})$. Since the function $x \mapsto x^2$ is strictly convex,

$$h_n^2\big(t\boldsymbol{x} + (1-t)\boldsymbol{y}\big) = \big[th_n(\boldsymbol{x}) + (1-t)h_n(\boldsymbol{y})\big]^2 < th_n^2(\boldsymbol{x}) + (1-t)h_n^2(\boldsymbol{y}).$$

Since weak inequality holds for the remaining functions h_k in the definition of g, it follows that $g(t\boldsymbol{x} + (1-t)cy) < tg(\boldsymbol{x}) + (1-t)g(cy)$, verifying the claim.

Fix $\boldsymbol{x}_0 \in K$ and define a functional p on $C(K,\mathbb{R})$ by $p(f) = c(f)(\boldsymbol{x}_0)$, where $c(f)$ is the function in 14.7.2. From 14.7.2(d), p is subadditive and positively homogeneous. Define a linear functional φ on the subspace $B := A(K,\mathbb{R})+\mathbb{R}\,g$ of $C(K,\mathbb{R})$ by $\varphi(h+rg) = (h+rg)(\boldsymbol{x}_0)$. In particular, $\varphi(h) = h(\boldsymbol{x}_0)$, $\varphi(g) = g(\boldsymbol{x}_0)$, and $\varphi(1) = 1$. We claim that $\varphi \leq p$ on B, that is,

$$h(\boldsymbol{x}_0) + rg(\boldsymbol{x}_0) \leq c(h+rg)(\boldsymbol{x}_0) \quad \forall \ \ r \in \mathbb{R} \ \text{ and } h \in A(K,\mathbb{R}).$$

Indeed, if $r \geq 0$, then $c(h+rg) = h + rc(g) \geq h + rg$ by 14.7.2(a,e), and if $r < 0$, then rg is concave, hence $c(h+rg) = h + rg$ by 14.7.2(c), verifying the claim. By the Hahn-Banach theorem, φ extends to a linear functional μ on $C(K,\mathbb{R})$ such that $\mu \leq p$ on $C(K,\mathbb{R})$. Noting that $f \leq 0 \Rightarrow \mu(f) \leq p(f) = c(f)(\boldsymbol{x}_0) \leq 0$, we see that μ is a positive linear functional on $C(K,\mathbb{R})$. Since $\varphi(1) = 1$, we may identify μ with a Radon probability measure on K. Since $g(\boldsymbol{x}_0) = \varphi(g) = \mu(g)$ for all $g \in A(F,\mathbb{R})$, μ represents $\boldsymbol{x}_0$. It remains to show that $\operatorname{supp}\mu \subseteq \operatorname{ex} K$.

We claim that $\mu(g) = \mu\big(c(g)\big)$ (recalling that $c(g)$ is a bounded Borel function). Indeed, by 14.7.2(a), $\mu(g) \leq \mu\big(c(g)\big)$. For the reverse inequality, let $h \in A(K,\mathbb{R})$ and $h \geq g$. Then $h \geq c(g)$, hence $h(\boldsymbol{x}_0) = \varphi(h) = \mu(h) \geq \mu\big(c(g)\big)$. Taking the infimum over all such h yields $\mu(g) = \varphi(g) = c(g)(\boldsymbol{x}_0) \geq \mu\big(c(g)\big)$.

We now have $\int [c(g) - g]\,d\mu = 0$. Since $c(g) - g \geq 0$ (14.7.2(a)), $\mu\big(g < c(g)\big) = 0$. To complete the proof it therefore suffices to show that $K \setminus \operatorname{ex} K \subseteq \{g < c(g)\}$. But if $\boldsymbol{x} \in K \setminus \operatorname{ex} K$ and $\boldsymbol{y} \neq \boldsymbol{z} \in K$ with $\boldsymbol{x} = \frac{1}{2}(\boldsymbol{y} + \boldsymbol{z})$, then, by the strict convexity of g and the concavity of $c(g)$, $g(\boldsymbol{x}) < \frac{1}{2}g(\boldsymbol{y}) + \frac{1}{2}g(\boldsymbol{z}) \leq \frac{1}{2}c(g)(\boldsymbol{y}) + \frac{1}{2}c(g)(\boldsymbol{z}) \leq c(g)(\boldsymbol{x})$, as required. □

The proof of Choquet's Theorem given above is due to Bonsall. This, as well as a proof of the more general Choquet-Bishop-deLeeuw theorem (where the metrizability hypothesis on K is removed), may be found in [37], which contains many related results. Here is one of particular interest:

14.7.4 Corollary (Rainwater). *Let $\mathscr{X}$ be a separable normed space, $\boldsymbol{x} \in \mathscr{X}$, and $(\boldsymbol{x}_n)$ a bounded sequence in $\mathscr{X}$. If $\lim_n \langle \boldsymbol{x}_n, \boldsymbol{x}' \rangle = \langle \boldsymbol{x}, \boldsymbol{x}' \rangle$ for every extreme point $\boldsymbol{x}'$ of the closed unit ball C_1' of $\mathscr{X}'$, then $\boldsymbol{x}_n \xrightarrow{w} \boldsymbol{x}$.*

Proof. We may assume that $\mathscr{X}$ is a real normed space. It suffices to show that $\lim_n \langle \boldsymbol{x}_n, \boldsymbol{y}' \rangle = \langle \boldsymbol{x}, \boldsymbol{y}' \rangle$ for $\boldsymbol{y}' \in C_1'$. Since C_1' is compact, convex, and metrizable in the weak* topology, $\boldsymbol{y}'$ may be represented by a probability measure μ on C_1' supported by the extreme points:

$$\boldsymbol{y}' = \int_{\operatorname{ex} C_1'} \boldsymbol{x}'\,d\mu(\boldsymbol{x}').$$

By the dominated convergence theorem,

$$\langle \boldsymbol{x}_n, \boldsymbol{y}' \rangle = \int_{\operatorname{ex} C_1'} \widehat{\boldsymbol{x}_n}(\boldsymbol{x}')\,d\mu(\boldsymbol{x}') \to \int_{\operatorname{ex} C_1'} \widehat{\boldsymbol{x}}(\boldsymbol{x}')\,d\mu(\boldsymbol{x}') = \langle \boldsymbol{x}, \boldsymbol{y}' \rangle.$$

□

Part III

Applications

Chapter 15

Distributions

Spaces of distributions are the duals of spaces of C^∞ functions on open subsets of $\mathbb{R}^d$. The operations of differentiation, convolution, and Fourier transform of functions may be extended by duality to distributions, opening up the possibility of finding non-differentiable solutions, so-called *weak solutions*, of differential equations that may not have smooth solutions. For example, consider the partial differential equation

$$\sum_{\alpha \in S} \psi_\alpha(x)\partial^\alpha f(x) = g(x)$$

on some open set $U \subseteq \mathbb{R}^d$, where S is a finite set of multi-indices, $\psi_\alpha \in C^\infty(U)$, and g is locally integrable. Multiplying the equation by a function ϕ in $C_c^\infty(U)$ and integrating by parts over $\mathbb{R}^d$ yields

$$\int_{\mathbb{R}^d} f \sum_{\alpha \in S} (-1)^{|\alpha|}\partial^\alpha(\psi_\alpha \phi)\, d\lambda^d = \int_{\mathbb{R}^d} g\phi\, d\lambda^d.$$

There are no constant terms here because ϕ has compact support. Functions f that satisfy the last equation for every $\phi \in C_c^\infty(U)$ are called weak solutions of the original PDE. There is no reason to assume that these solutions must be smooth.

It is beyond the scope of the text to delve into the distributional theory of PDEs. Our goal here is merely to define the main distribution spaces, describe their functional analytic properties, and discuss the standard operations on distributions. We do, however, give a simple application to PDEs in §15.6.

15.1 General Theory

The Fréchet Space $C_K^\infty(U)$

Let U be an open subset of $\mathbb{R}^d$. For a compact subset K of U, let

$$C_K^\infty(U) := \{\phi \in C^\infty(U) : \operatorname{supp}\phi \subseteq K\}.$$

Then $C_K^\infty(U)$ is a Fréchet space with respect to the countable family of norms $\phi \to \|\partial^\alpha \phi\|_\infty$, where $\alpha = (\alpha_1, \ldots, \alpha_d)$ is a multi-index. The verification that $C_K^\infty(U)$ is complete is entirely similar to the argument in 9.1.6(c) regarding the completeness of the Fréchet space $C^\infty(U)$. We denote the topology of $C_K^\infty(U)$ by τ_K. The following proposition is immediate from 9.2.4.

15.1.1 Proposition. *Let F be a linear functional on $C_K^\infty(U)$. Then F is τ_K-continuous iff there exists $m \in \mathbb{Z}^+$ and $M > 0$ such that $|\langle \phi, F\rangle| \le M p_m(\phi)$ for all $\phi \in C_K^\infty(U)$, where*

$$p_m(\phi) := \max\{\|\partial^\alpha \phi\|_\infty : |\alpha| \le m\}.$$

The Spaces $\mathcal{D}(U)$ and $\mathcal{D}'(U)$

Define an increasing sequence of compact subsets of U as in 9.1.6(a) by

$$K_n := \{x \in \mathbb{R}^d : |x| \le n,\ d(x, U^c) \ge 1/n\}.$$

Clearly, $K_n \uparrow U$ and $C_c^\infty(U)$ is the union of the spaces $C_{K_n}^\infty(U)$. Since every compact subset K of U is contained in some K_n, the following result is an immediate consequence of 9.4.6 and 9.4.7.[1]

15.1.2 Theorem. *There exists a LCS topology τ on $C_c^\infty(U)$ such that the following hold:*

(a) *A sequence (ϕ_n) τ-converges to ϕ in $C_c^\infty(U)$ iff there exists a compact set K such that $(\phi_n) \subseteq C_K^\infty(U)$ and $\partial^\alpha \phi_n \to \partial^\alpha \phi$ uniformly on U for all multi-indices α.*

(b) *If T is a linear mapping from $C_c^\infty(U)$ to a LCS $\mathcal{X}$, then T is τ-continuous iff for each compact subset K of U the restriction of T to $C_K^\infty(U)$ is τ_K continuous, that is, iff $T\phi_n \to T\phi$ in $\mathcal{X}$ whenever $(\phi_n) \subseteq C_K^\infty(U)$ and $\partial^\alpha \phi_n \to \partial^\alpha \phi$ uniformly on U for all multi-indices α.*

The space $C_c^\infty(U)$ with the topology τ is denoted by $\mathcal{D}(U)$ and is called the space of **test functions.** A member of the dual $\mathcal{D}'(U)$ of $\mathcal{D}(U)$ is called **a distribution** or **generalized function** on U. It is customary in this setting to reverse the duality notation $\langle \phi, F\rangle = F(\phi)$ and write instead

$$\langle F, \phi\rangle = F(\phi), \quad F \in \mathcal{D}'(U),\ \phi \in \mathcal{D}(U).$$

This convention frequently renders a distribution formula into a more readable and computationally convenient form. The following proposition follows directly from 15.1.1 and (b) of 15.1.2.

15.1.3 Proposition. *Let F be a linear functional on $\mathcal{D}(U)$. Then $F \in \mathcal{D}'(U)$ iff for each compact $K \subseteq U$ there exists $m \in \mathbb{Z}^+$ and $M > 0$ such that $|\langle F, \phi\rangle| \le M\, p_m(\phi)$ for all $\phi \in C_K^\infty(U)$.*

Examples of Distributions

(a) Let $f : U \to \mathbb{C}$ be *locally Lebesgue integrable*, that is, f is measurable and $f\big|_K$ is Lebesgue integrable for every compact subset K of U. Denote the space of all locally integrable functions by $L^1_{loc}(U)$. For each $f \in L^1_{loc}(U)$, the equation $\langle F_f, \phi\rangle := \int_U f\phi\, d\lambda^d$ defines a distribution, as may be seen by taking $m = 0$ in 15.1.3. Note that the mapping $f \to F_f : L^1_{loc}(U) \to \mathcal{D}'(U)$ is linear. Moreover, if we identity functions that are equal a.e., then the map is 1-1. Indeed, this amounts to the assertion that $\int_U f\phi = 0$ for all $\phi \in C_c^\infty(U)$ $\Rightarrow f = 0$ a.e., which is valid by a standard approximation argument, since $C_c^\infty(U)$ is dense in $L^1(U)$. In view of this correspondence and to simplify notation one frequently writes f for F_f, so that

$$\langle f, \phi\rangle = \langle F_f, \phi\rangle = \int_U f\phi\, d\lambda^d, \quad \phi \in \mathcal{D}(U). \tag{15.1}$$

(b) Let μ be a Radon measure on U. Then $\phi \to \int_U \phi\, d\mu$ defines a distribution, again by taking $m = 0$ in 15.1.3. More generally, for fixed α, the mapping $\phi \to \int_U \partial^\alpha \phi\, d\mu$ defines a distribution, this time by taking $m = |\alpha|$.

(c) A special case of (b) is obtained by taking μ to be the Dirac measure δ_x at $x \in U$. This gives the **Dirac delta distribution** $\phi \to \phi(x)$ **at** x.

[1] These refer to the existence and the basic properties of strict inductive limits. Since an understanding of the material in the current chapter does not depend on the abstract notion of inductive limit, the reader may simply accept the statement of Theorem 15.1.2.

15.1.4 Remarks. The Dirac delta distribution is not given by a function as in (a). To see this, take the special case $U = \mathbb{R}$ and $x = 0$. For $r > 0$ consider the test function

$$\phi_r(x) = \begin{cases} \exp\left(1 - [1 - (x/r)^2]^{-1}\right) & \text{if } |x| \le r \\ 0 & \text{otherwise.} \end{cases}$$

If δ_0 were given by a locally integrable function f we would have

$$1 = \phi_r(0) = \int_{|x|\le r} f(x)\phi_r(x)\,dx \le \int_{|x|\le r} |f(x)|\,dx \to 0 \text{ as } r \to 0.$$

Similarly, if $\alpha \neq 0$ then the distribution $\phi \mapsto \partial^\alpha \phi(x)$ is not given by a measure μ as in Example (b). Otherwise, for the test function $\psi_r(x) := x\phi_r(x)$, we would have

$$1 = \psi_r'(0) = \int_{\mathbb{R}} \psi_r(x)\,d\mu \le \int_{|x|\le r} r\,d\mu \to 0 \text{ as } r \to 0. \qquad \Diamond$$

15.2 Operations on Distributions

Derivative of a Locally Integrable Function

Let $f \in C^\infty(U)$, α a multi-index, and $\phi \in \mathcal{D}(U)$. The classical integration by parts formula gives

$$\langle \partial^\alpha f, \phi \rangle = \int_U \phi\,\partial^\alpha f = (-1)^{|\alpha|} \int_U f \partial^\alpha \phi = (-1)^{|\alpha|} \langle f, \partial^\alpha \phi \rangle .$$

Here we have used the convention described in (15.1), The right side of the equation makes sense for any locally integrable function f. Thus we define the **distributional** or **weak derivative** $\partial^\alpha f$ of f by

$$\langle \partial^\alpha f, \phi \rangle = (-1)^{|\alpha|} \langle f, \partial^\alpha \phi \rangle, \quad \phi \in \mathcal{D}(U), \quad f \in L^1_{loc}(U).$$

For example, the classical derivative of $f(x) = |x|$ does not exist on $\mathbb{R}$, but the distributional derivative of f exists and equals $\mathbf{1}_{(0,\infty)} - \mathbf{1}_{(-\infty,0)}$. Indeed, if $\phi \in \mathcal{D}(\mathbb{R})$, then integrating by parts we have

$$-\langle f, \phi' \rangle = \int_{-\infty}^0 x\phi'(x)\,dx - \int_0^\infty x\phi'(x)\,dx = -\int_{-\infty}^0 \phi + \int_0^\infty \phi = \langle \mathbf{1}_{(0,\infty)} - \mathbf{1}_{(-\infty,0)}, \phi \rangle.$$

Derivative of a Distribution

Generalizing the preceding, we define the **derivative** $\partial^\alpha F$ **of** $F \in \mathcal{D}'(U)$ by

$$\langle \partial^\alpha F, \phi \rangle := (-1)^{|\alpha|} \langle F, \partial^\alpha \phi \rangle, \quad \phi \in \mathcal{D}(U).$$

It follows directly from 15.1.3 that $\partial^\alpha F \in \mathcal{D}'(U)$. For an example, take $H = \mathbf{1}_{[0,\infty)}$, the so-called **Heaviside function** on $\mathbb{R}$. For any $\phi \in \mathcal{D}(\mathbb{R})$ we have

$$\left\langle \frac{d}{dx} F_H, \phi \right\rangle = -\int_0^\infty \phi' = \phi(0),$$

hence $F_H' = \delta_0(\phi)$.

Multiplication by a Smooth Function

Given $F \in \mathcal{D}'(U)$ and $f \in C^\infty(U)$, define fF by

$$\langle fF, \phi\rangle = \langle F, f\phi\rangle, \quad \phi \in \mathcal{D}(U).$$

By 15.1.3, $fF \in \mathcal{D}'(U)$. Note that if g is locally integrable, then

$$\langle fF_g, \phi\rangle = \langle F_g, f\phi\rangle = \int g(f\phi) = \langle F_{fg}, \phi\rangle,$$

that is, $fF_g = F_{fg}$. Furthermore, for any $1 \le k \le d$ and $\phi \in \mathcal{D}$, the classical product rule $\partial_k(f\phi) = f\partial_k\phi + \phi\partial_k f$ implies that

$$\begin{aligned}\langle \partial_k(fF), \phi\rangle &= -\langle fF, \partial_k\phi\rangle = -\langle F, f\partial_k\phi\rangle = -\langle F, \partial_k(f\phi)\rangle + \langle F, \phi\,\partial_k f\rangle \\ &= \langle \partial_k F, f\phi\rangle + \langle(\partial_k f)F, \phi\rangle = \langle f(\partial_k F), \phi\rangle + \langle(\partial_k f)F, \phi\rangle,\end{aligned}$$

that is,

$$\partial_k(fF) = f(\partial_k F) + (\partial_k f)F.$$

This is the **product rule for distributions**.

Composition with Linear Maps

Let $T \in \mathscr{B}(\mathbb{R}^d)$ be invertible and set $V := T(U)$, so that $T: U \to V$ is C^∞ with C^∞ inverse $T^{-1}: V \to U$. If $f \in C_c(V)$ and $\phi \in \mathcal{D}(U)$, then $F_f \in \mathcal{D}'(V)$, $\phi\circ T^{-1} \in \mathcal{D}(V)$, and by the change of variables theorem (3.2.18),

$$F_f(\phi\circ T^{-1}) = \int_V f\cdot(\phi\circ T^{-1}) = |\det T| \int_U (f\circ T)\phi = |\det T| F_{f\circ T}(\phi),$$

which we write as

$$F_{f\circ T}(\phi) = |\det T|^{-1} F_f(\phi\circ T^{-1}), \quad \phi \in \mathcal{D}(U), \quad f \in C_c(V).$$

The identification $f \leftrightarrow F_f$ then suggests the following definition of $F\circ T$ for an arbitrary distribution F:

$$F\circ T(\phi) = |\det T|^{-1} F(\phi\circ T^{-1}), \quad \phi \in \mathcal{D}(U), \quad F \in \mathcal{D}'(V).$$

One easily checks that $F\circ T \in \mathcal{D}'(U)$. In particular, for reflections $T(x) = -x$ we define the distribution $\widetilde{F}$ by

$$\widetilde{F}(\phi) := F\circ T(\phi) = F(\widetilde{\phi}), \quad \text{where } \widetilde{\phi}(x) := \phi(-x).$$

15.3 Distributions with Compact Support

Let V be an open subset of U. A distribution $F \in \mathcal{D}'(U)$ is said to be **zero on** V if $\langle F, \phi\rangle = 0$ for all $\phi \in C_c(U)$ with $\operatorname{supp}\phi \subseteq V$. The complement in U of the union of all open sets on which F is zero is called the **support of** F and is denoted by $\operatorname{supp} F$. Thus $\operatorname{supp} F$ is the closed set defined by

$$U \setminus \operatorname{supp} F = \bigcup\{V : V \subseteq U \text{ is open and } F = 0 \text{ on } V\}.$$

The support of a distribution is a generalization of the notion of support of a function in the following sense:

15.3.1 Proposition. *If $f \in C(U)$ is locally integrable, then* $\operatorname{supp}(F_f) = \operatorname{supp}(f)$.

Proof. Let $C := \operatorname{supp}(f)$. Since $\langle F_f, \phi \rangle = \int_C f\phi = 0$ for all $\phi \in C_c(U)$ with support contained in the open set $U \setminus C$, $\operatorname{supp}(F_f) \subseteq C$.

For the reverse inclusion, let V be any open set on which $F_f = 0$. Then

$$\int f\phi \, d\lambda^d = \langle F_f, \phi \rangle = 0 \;\forall\; \phi \in C_c(U) \text{ with } \operatorname{supp}(\phi) \subseteq V.$$

Let $K \subseteq V$ be compact. If $f \in C^\infty(U)$, then, replacing ϕ by $\overline{f}\phi$ in the above, where $\mathbf{1}_K \leq \phi \leq \mathbf{1}_U$, we see that $\int_K |f|^2 \, d\lambda^d = 0$ and so $f = 0$ on K. Since a continuous function is uniformly approximable on K by C^∞ functions, the same result holds for continuous functions f. Since K was arbitrary, $f = 0$ on V. Therefore, $\operatorname{supp}(f) \subseteq V^c$. Since V was arbitrary, $\operatorname{supp}(f) \subseteq \operatorname{supp}(F_f)$. □

15.3.2 Proposition. *$F = 0$ on $U \setminus \operatorname{supp} F$.*

Proof. Let $\{V_i : i \in \mathfrak{I}\}$ be the collection of all open subsets of U on which $F = 0$ and let $\phi \in C_c(U)$ with $K := \operatorname{supp}(\phi) \subseteq \bigcup_i V_i = U \setminus \operatorname{supp} F$. We show $F(\phi) = 0$, which will prove the proposition.

By the partition of unity theorem (0.14.6) applied to the open cover $\{V_i\}$ of K there exists a finite subcover $\{V_1, \ldots, V_p\}$ and nonnegative functions $\chi_j \in C_c^\infty(U)$ $(j = 1, \ldots, p)$ such that $\operatorname{supp}(\chi_j) \subseteq V_j$ and $\sum_{j=1}^p \chi_j = 1$ on K. Then, by definition of V_j, $F(\phi\chi_j) = 0$, and since $\phi = \sum_j \phi\chi_j$ we have $F(\phi) = \sum_j F(\phi\chi_j) = 0$. □

A distribution $F \in \mathcal{D}'(U)$ is said to have **compact support** if $\operatorname{supp}(F)$ is compact. For example, by 15.3.1, members of $C_c(U)$, considered as distributions, have compact support. We denote the space of all such distributions by $\mathcal{E}'(U)$:

$$\mathcal{E}'(U) := \{F \in \mathcal{D}'(U) : \operatorname{supp}(F) \text{ is compact}\}.$$

Recall from 9.1.6(c) that topology on the Fréchet space $C^\infty(U)$ is defined by the seminorms

$$p_{m,\alpha}(f) = \sup\{|\partial^\alpha f(x)| : x \in K_m\}, \tag{15.2}$$

where the K_m are compact, $K_m \subseteq \operatorname{int}(K_{m+1})$, and $K_m \uparrow U$, and that $C_c^\infty(U)$ is dense in $C^\infty(U)$. The next theorem asserts that the dual of $C^\infty(U)$ is $\mathcal{E}'(U)$. For the statement, we employ the following convenient notation: Let $\mathscr{X}$ be a LCS and $\mathscr{Y}$ a linear subspace of $\mathscr{X}$ with a locally convex topology with respect to which the inclusion mapping $\mathscr{Y} \hookrightarrow \mathscr{X}$ is continuous. This simply means that the given topology of $\mathscr{Y}$ is stronger than the relative topology from $\mathscr{X}$. It follows that the restriction to $\mathscr{Y}$ of every member of the dual $\mathscr{X}'$ of $\mathscr{X}$ is a member of $\mathscr{Y}'$. We express this by writing $\mathscr{X}'\big|_{\mathscr{Y}} \subseteq \mathscr{Y}'$.

15.3.3 Theorem. *The inclusion mapping $\mathcal{D}(U) \hookrightarrow C^\infty(U)$ is continuous, and the restriction to $\mathcal{D}(U)$ of a member G of the dual of $C^\infty(U)$ is a distribution F. Moreover, F has compact support, and every member of $\mathcal{D}(U)'$ with compact support arises in this manner, that is, extends (uniquely) to a member of the dual of $C^\infty(U)$. Thus*

$$C^\infty(U)'\Big|_{\mathcal{D}(U)} = \mathcal{E}(U)'.$$

Proof. Let (ϕ_n) $\mathcal{T}$-converge to 0 in $\mathcal{D}(U)$ as in (a) of 15.1.2. Thus there exists a compact $K \subseteq U$ such that $(\phi_n) \subset C_K^\infty(U)$ and $\partial^\alpha \phi_n \to 0$ uniformly on U for all multi-indices α. Then, trivially, $\partial^\alpha \phi_n \to 0$ uniformly on any compact subset of U. Thus by (b) of 15.1.2, $\mathcal{D}(U) \hookrightarrow C^\infty(U)$ is continuous. This shows that every continuous linear functional G on $C^\infty(U)$

restricts to a continuous linear functional F on $\mathcal{D}(U)$. To see that F has compact support, by continuity of G choose $C > 0$ and $m, N \geq 1$ such that $|G(\phi)| \leq C \max_{|\alpha| \leq N} p_{m,\alpha}(\phi)$ for all $\phi \in C^\infty(U)$. If $\phi \in C_c^\infty(U)$ and $\text{supp}(\phi) \subseteq K_m^c$, then $p_{m,\alpha}(\phi) = 0$ and so $F(\phi) = G(\phi) = 0$. Therefore, K_m^c is one of the open sets comprising $U \setminus \text{supp}(F)$, hence $\text{supp}\, F \subseteq K_m$.

Conversely, let $F \in \mathcal{D}'(U)$ have compact support. Choose $\psi \in C_c^\infty(U)$ such that $\psi = 1$ on $\text{supp}\, F$ and set $K := \text{supp}\, \psi \supseteq \text{supp}\, F$. Define G on $C^\infty(U)$ by $G(f) = F(f\psi)$. By continuity of F on $C_K^\infty(U)$, there exists $M > 0$ and $N \geq 1$ such that $|F(f\psi)| \leq M \max_{|\alpha| \leq N} \|\partial^\alpha (f\psi)\|_\infty$ for all $f \in C^\infty(U)$. Now, by the product rule, $\partial^\alpha(f\psi)$ is a sum of derivatives $(\partial^\beta f)(\partial^\gamma \psi)$ $(|\beta| + |\gamma| = |\alpha|)$, and each of the terms $\partial^\gamma \psi$ has support in K. Letting M' be a bound for the sum of the terms $|\partial^\gamma \psi|$, we then have for sufficiently large m

$$|G(f)| = |F(f\psi)| \leq MM' \max_{|\alpha| \leq N} \sup_K |\partial^\alpha(f)| = MM' \max_{|\alpha| \leq N} p_{m,\alpha}(f), \quad f \in C^\infty(U),$$

proving that G is continuous on $C^\infty(U)$.

Now note that $V := U \setminus K \subseteq U \setminus \text{supp}\, F$, hence $F = 0$ on V (15.3.2). Let $f \in C_c^\infty(U)$. Since $f\psi = f$ on K, $\text{supp}\,(f\psi - f) \subseteq V$ and so $F(f\psi - f) = 0$, that is, $G(f) = F(f)$. Therefore, G is an extension of F. Uniqueness of the extension follows from the fact that $C_c^\infty(U)$ is dense in $C^\infty(U)$ (9.1.6(b)). □

15.4 Convolution of Distributions

The convolution of functions f, $g \in L^1(\mathbb{R}^d)$ was defined in §6.1. The same construction is valid for $f \in L^1_{loc}(\mathbb{R}^d)$ and $\psi \in C_c^\infty(\mathbb{R}^d)$:

$$f * \psi(x) = \int f(x-y)\psi(y)\, dy = \int f(y)\psi(x-y)\, dy = \int f(y)\psi_x(y)\, dy = F_f(\psi_x),$$

where $\psi_x(y) := \psi(x - y)$. This suggests the definition

$$(F * \psi)(x) := F(\psi_x), \quad x \in \mathbb{R}^d,$$

for an arbitrary distribution F. The function $F * \psi$ is called the **convolution of F with ψ**. The basic properties of this convolution are given in the following proposition.

15.4.1 Proposition. *Let $F \in \mathcal{D}'(\mathbb{R}^d)$, $\psi \in \mathcal{D}(\mathbb{R}^d)$. Then*

(a) $F * \psi \in C^\infty(\mathbb{R}^d)$, **(b)** $\partial^\alpha(F * \psi)(x) = \langle F, (\partial^\alpha \psi)_x, \rangle$, *and* **(c)** $(\partial^\alpha F) * \psi = F * (\partial^\alpha \psi)$,

where $(\partial^\alpha \psi)_x(y) = (\partial^\alpha \psi)(x - y)$.

Proof. We show first that $F * \psi$ is continuous. Let $x_n \to x$ in $\mathbb{R}^d$, so that $(x_n) \subseteq C_r(x)$ for some r. Then the supports of ψ_{x_n} and ψ_x are contained in the compact set $C_r(x) - \text{supp}(\psi)$. Moreover, from $\partial^\alpha \psi_{x_n} = (-1)^{|\alpha|}(\partial^\alpha \psi)_{x_n}$ and $\partial^\alpha \psi_x = (-1)^{|\alpha|}(\partial^\alpha \psi)_x$ we see that $\partial^\alpha \psi_{x_n} \to \partial^\alpha \psi_x$ uniformly on $\mathbb{R}^d$. Therefore, by 15.1.2(a), $\psi_{x_n} \xrightarrow{T} \psi_x$ and so $F(\psi_{x_n}) \to F(\psi_x)$.

Now observe that if $\mathbf{t} := (t, 0, \ldots, 0)$, then

$$(\partial/\partial x_1)(F * \psi)(x) = \lim_{t \to 0} t^{-1}\big[F(\psi_{x+\mathbf{t}}) - F(\psi_x)\big] = \lim_{t \to 0} F\big(t^{-1}[\psi_{x+\mathbf{t}} - \psi_x]\big).$$

An argument similar to that of the preceding paragraph (using the mean value theorem) shows that if $t_n \to 0$, then

$$t_n^{-1}\big[\psi_{x+\mathbf{t}_n}(y) - \psi_x(y)\big] = t_n^{-1}\big[\psi(x + \mathbf{t}_n - y) - \psi(x-y)\big] \to [(\partial/\partial x_1)\psi](x-y)$$

in $\mathcal{D}(\mathbb{R}^d)$ and so $(\partial/\partial x_1)(F * \psi)(x) = F\big((\partial/\partial x_1 \psi)_x\big)$. Analogous arguments apply to the other variables. By induction we obtain $\partial^\alpha(F * \psi)(x) = F\big((\partial^\alpha\psi)_x\big)$, proving (a) and (b).

From the definitions of convolution and derivative,

$$[(\partial^\alpha F) * \psi](x) = (\partial^\alpha F)(\psi_x) = (-1)^{|\alpha|}F(\partial^\alpha \psi_x) = F((\partial^\alpha\psi)_x) = (F * \partial^\alpha\psi)(x),$$

verifying (c). □

15.4.2 Proposition. *If $F \in \mathcal{D}'(\mathbb{R}^d)$ and $\psi \in C_c^\infty(\mathbb{R}^d)$, then $\operatorname{supp} F * \psi \subseteq \operatorname{supp} F + \operatorname{supp}\psi$. In particular, the members of $\mathcal{E}'(\mathbb{R}^d) * C_c^\infty(\mathbb{R}^d)$ have compact support, that is, the inclusion $\mathcal{E}'(\mathbb{R}^d) * C_c^\infty(\mathbb{R}^d) \subseteq \mathcal{E}'(\mathbb{R}^d)$ holds.*

Proof. Since $\operatorname{supp} F$ is closed and $\operatorname{supp}\psi$ is compact, the set $C := \operatorname{supp} F + \operatorname{supp}\psi$ is closed. Let U be open with compact closure contained in C^c. Then $\operatorname{cl} U - \operatorname{supp}\psi$ is compact and does not meet the closed set $\operatorname{supp} F$, hence there exists $g \in C^\infty(\mathbb{R}^d)$ such that $g = 0$ on an open set $V \supseteq \operatorname{cl} U - \operatorname{supp}\psi$ and $g = 1$ on an open set $W \supseteq \operatorname{supp} F$. Then for all $\phi \in \mathcal{D}(\mathbb{R}^d)$, $\operatorname{supp}(g\phi - \phi) \subseteq W^c \subseteq (\operatorname{supp} F)^c$ and so $F(g\phi - \phi) = 0$. In particular, $F(g\psi_x - \psi_x) = 0$ that is, $F * \psi(x) = F(g\psi_x)$. But if $x \in U$, then $g\psi_x$ is identically equal to zero. Indeed, assume that $g(y)\psi_x(y) \neq 0$ for some y. Then $x - y \in \operatorname{supp}\psi$, hence $y \in V$. But $g = 0$ on V. Thus $F * \psi(x) = F(g\psi_x) = 0$ on U. Since U was arbitrary, $F * \psi = 0$ on the open set C^c and so $\operatorname{supp} F * \psi \subseteq C$. □

The following lemma will be used to prove the associative law for convolutions.

15.4.3 Lemma. *Let $F \in \mathcal{D}'(\mathbb{R}^d)$ and $\psi, \phi \in \mathcal{D}(\mathbb{R}^d)$. Then $\langle F * \psi, \phi\rangle = \langle F, \widetilde{\psi} * \phi\rangle$, where $\widetilde{\psi}(x) := \psi(-x)$.*

Proof. The left side of the desired equality may be written

$$\langle F * \psi, \phi\rangle = \int (F * \psi)(x)\phi(x)\,dx = \int \big\langle F, \phi(x)\,\psi_x\big\rangle\,dx.$$

To obtain a like expression for the right side, note that

$$(\widetilde{\psi} * \phi)(y) = \int \widetilde{\psi}(y - x)\phi(x)\,dx = \int \psi(x - y)\phi(x)\,dx = \int \phi(x)\psi_x(y)\,dx,$$

so that the right side of the desired equation may be written

$$\langle F, \widetilde{\psi} * \phi\rangle = \left\langle F, \int \phi(x)\psi_x(\cdot)\,dx\right\rangle,$$

where the integral may be taken to be a Bochner integral. Thus we must show that

$$\int \big\langle F, \phi(x)\psi_x\big\rangle\,dx = \left\langle F, \int \phi(x)\psi_x(\cdot)\,dx\right\rangle.$$

To this end, note first that the integrand on the right, as a function of y, is supported in the compact set $K := \operatorname{supp}(\phi) + \operatorname{supp}(\psi)$. Overlay K with a grid $\mathcal{Q}$ of cubes Q_j with volumes v_j and let $x_j \in Q_j$. Set

$$S(y, \mathcal{Q}) := \sum_j \phi(x_j)\psi_{x_j}(y)v_j.$$

Then

$$\int \phi(x)\psi_x(y)\,dx - S(y, \mathcal{Q}) = \sum_j \int_{Q_j} \big[\phi(x)\psi_x(y) - \phi(x_j)\psi_{x_j}(y)\big]\,dx.$$

By uniform continuity, the integrands on the right tend to zero uniformly in y as $\|\mathcal{Q}\| \to 0$. Therefore, the Riemann sums $S(y, \mathcal{Q})$ tend to $\int \phi(x)\psi_x(y)\, dx$ uniformly in y. A similar argument shows that

$$\partial^\alpha S(y, \mathcal{Q}) = \sum_j \phi(x_j)\partial_y^\alpha \psi_{x_j}(y)v_j \to \int \phi(x)\partial_y^\alpha \psi_x(y)\, dx = \partial_y^\alpha \int \phi(x)\psi_x(y)\, dx$$

uniformly in y as $\|\mathcal{Q}\| \to 0$. Thus $S(y, \mathcal{Q}) \to \int \phi(x)\psi_x\, dy$ in the topology of $\mathcal{D}(U)$ and so

$$\int \langle F, \phi(x)\psi_x \rangle\, dx = \lim_{\|\mathcal{Q}\|\to 0} \sum_j \langle F, \phi(x_j)\psi_{x_j} v_j \rangle = \lim_{\|\mathcal{Q}\|\to 0} \langle F, S(y, \mathcal{Q}) \rangle = \left\langle F, \int \phi(x)\psi_x\, dx \right\rangle$$

as required. □

Here is the aforementioned associative law for convolutions:

15.4.4 Theorem. *Let $F \in \mathcal{D}'(\mathbb{R}^d)$ and $\psi, \phi \in \mathcal{D}(\mathbb{R}^d)$. Then $F * (\psi * \phi) = (F * \psi) * \phi$.*

Proof. For all y,

$$(\psi * \phi)_x(y) = (\psi * \phi)(x - y) = \int \psi(z)\phi(x - y - z)\, dz = \int \widetilde{\psi}(z)\phi_x(y - z)\, dz = (\widetilde{\psi} * \phi_x)(y).$$

Therefore, by the lemma,

$$[F * (\psi * \phi)](x) = \langle F, (\psi * \phi)_x \rangle = \langle F, \widetilde{\psi} * \phi_x \rangle = \langle F * \psi, \phi_x \rangle = [(F * \psi) * \phi](x).$$ □

As in the classical case, convolution is continuous:

15.4.5 Theorem. *For $F \in \mathcal{D}'(\mathbb{R}^d)$ the linear mapping $T : \phi \mapsto F * \phi : \mathcal{D}(\mathbb{R}^d) \to C^\infty(\mathbb{R}^d)$ is continuous.*

Proof. We show that for any compact set K, the restriction of T to $C_K(\mathbb{R}^d)$ is continuous. The seminorms defining the topology of $C_K(\mathbb{R}^d)$ are of the form

$$p_{K,n}(\phi) = \sup\{|\partial^\alpha \phi(x)| : x \in \mathbb{R}^d,\ |\alpha| \le n\},$$

The seminorms defining the topology of $C^\infty(\mathbb{R}^d)$ are of the form

$$q_{H,k}(\phi) = \sup\{|\partial^\beta \phi(x)| : x \in H,\ |\beta| \le k\},\quad H \text{ compact},\ k \in \mathbb{Z}^+.$$

We show that given $q_{H,k}$ there exists $n \ge 0$ and $M > 0$ such that

$$q_{H,k}(F * \phi) \le M p_{K,n}(\phi)\ \ \forall\ \ \phi \in C_K(\mathbb{R}^d).$$

Now, F restricted to $C^\infty_{H-K}(\mathbb{R}^d)$ is continuous, hence there exists $M > 0$ and $m \in \mathbb{N}$ such that

$$|\langle F, \psi \rangle| \le M \sup\left\{|\partial^\alpha \psi(y)| : y \in \mathbb{R}^d,\ |\alpha| \le m\right\}\quad \forall\, \psi \in C^\infty_{H-K}(\mathbb{R}^d).$$

In particular, if $\phi \in C_K(\mathbb{R}^d)$ and $x \in H$, then $\phi_x \in C^\infty_{H-K}(\mathbb{R}^d)$, hence, recalling that $\partial^\beta(F * \phi)(x) = \langle F, (\partial^\beta \phi)_x \rangle$, we have

$$\begin{aligned} |\partial^\beta(F * \phi)(x)| &\le M \sup\left\{|\partial^\alpha(\partial^\beta \phi)_x(y)| : y \in H - K,\ |\alpha| \le m\right\} \\ &= M \sup\left\{|\partial_y^\alpha(\partial^\beta \phi)(x - y)| : y \in H - K,\ |\alpha| \le m\right\} \\ &= M \sup\left\{|(\partial^{\alpha+\beta} \phi)(x - y)| : y \in H - K,\ |\alpha| \le m\right\} \end{aligned}$$

and so

$$\begin{aligned} q_{H,k}(F*\phi) &= \sup\{|\partial^\beta(F*\phi)(x)| : x \in H,\ |\beta| \le k\} \\ &\le M \sup\{|\partial^{\alpha+\beta}\phi(x-y)| : x \in H,\ y \in H-K,\ |\alpha| \le m,\ |\beta| \le k\} \\ &\le M \sup\{|\partial^{\alpha+\beta}\phi(z)| : z \in K,\ |\alpha| \le m,\ |\beta| \le k\} \\ &= p_{K,m+k}(\phi). \end{aligned}$$

□

Recall that the space $C_c(\mathbb{R}^d)$ may be viewed as a subspace of $\mathcal{E}'(\mathbb{R}^d)$ via the identification $f \leftrightarrow F_f$. Since $F * \psi \in C_c^\infty(\mathbb{R}^d)$ for $F \in \mathcal{E}'(\mathbb{R}^d)$ and $\psi \in C_c^\infty(\mathbb{R}^d)$ (15.4.2), the following theorem implies that the space $C_c^\infty(\mathbb{R}^d)$ is weak* dense in $\mathcal{E}'(\mathbb{R}^d)$.

15.4.6 Theorem. *There exists a sequence (φ_n) in $C_c^\infty(\mathbb{R}^d)$ such that for every $F \in \mathcal{D}'(\mathbb{R}^d)$, $F * \varphi_n \xrightarrow{w^*} F$, that is,*

$$\langle F, \phi\rangle = \lim_n \langle F * \varphi_n, \phi\rangle = \lim_n \int (F * \varphi_n)\phi, \quad \phi \in \mathcal{D}(\mathbb{R}^d).$$

Proof. Let $(\varphi_n) \subseteq C_c^\infty(\mathbb{R}^d)$ be a sequence such that $f * \varphi_n \to f$ uniformly for all uniformly continuous and bounded functions f on $\mathbb{R}^d$, where $\operatorname{supp}(\varphi_n) \subseteq B_{1/n}(0)$ and each φ_n is an even function (6.1.3). By associativity,

$$\big\langle F * \varphi_n, \phi\big\rangle = \int (F*\varphi_n)(x)\phi(x)\,d\lambda^d(x) = [(F*\varphi_n)*\widetilde{\phi}](0) = [F*(\varphi_n*\widetilde{\phi})](0) = \langle F, \varphi_n * \phi\rangle,$$

the last equality from $(\varphi_n * \phi)_0 = \varphi_n * \phi$ (because φ_n is even). But the sequence $(\varphi_n * \phi)$ is supported in a compact set K and $\partial^\alpha(\varphi_n * \phi) = \varphi_n * \partial^\alpha\phi \to \partial^\alpha\phi$ uniformly on K for all α. Therefore, $\varphi_n * \phi \to \phi$ in $\mathcal{D}(\mathbb{R}^d)$ and so $\big\langle F * \varphi_n, \phi\big\rangle \to \big\langle F, \phi\big\rangle$. □

15.4.7 Remark. Lemma 15.4.3 suggests the following definition of convolution in $\mathcal{E}'(\mathbb{R}^d)$:

$$\langle F * G, \phi\rangle = \langle F, \widetilde{G} * \phi\rangle, \quad \phi \in C_c^\infty(\mathbb{R}^d), \quad F, G \in \mathcal{E}'(\mathbb{R}^d).$$

It may be shown that $F * G$ is a distribution with compact support and that convolution on $\mathcal{E}'(\mathbb{R}^d)$ is commutative, associative, and bilinear. (See, for example, [48]). ◊

15.5 Tempered Distributions

Recall that the Fréchet space $\mathcal{S}(\mathbb{R}^d)$ of rapidly decreasing functions on $\mathbb{R}^d$ is a subspace of $C^\infty(\mathbb{R}^d)$ whose topology is given by the equivalent families of seminorms

$$\Big\{q_{\alpha,m}(\phi) = \sup_{x\in\mathbb{R}^d} (1+|x|^2)^m|\partial^\alpha\phi(x)| : m, \alpha\Big\} \text{ and } \Big\{p_{\alpha,\beta}(\phi) = \sup_{x\in\mathbb{R}^d} |x^\alpha\partial^\beta\phi(x)| : \alpha, \beta\Big\}.$$

Recalling the definition

$$p_{m,\alpha}(\phi) = \sup_{x\in K_m} |\partial^\alpha\phi(x)|, \quad m \in \mathbb{N},\ \phi \in C^\infty(\mathbb{R}^d),$$

of a defining seminorm on the space $C^\infty(\mathbb{R}^d)$ (9.1.6(b)), we see that

$$p_{m,\alpha}(\phi) \le q_{\alpha,n}(\phi) \ \ \forall\ \ m, n \ \text{ and } \ \phi \in \mathcal{S}(\mathbb{R}^d). \tag{15.3}$$

Moreover, if $\text{supp}(\phi) \subseteq K_m$, then the supremum in the definition of $q_{\alpha,n}(\phi)$ may be taken over K_m, hence for a suitable $M > 0$ depending only on m,

$$q_{\alpha,n}(\phi) = \sup_{x \in K_m} (1 + |x|^2)^n |\partial^\alpha \phi(x)| \le M \sup_{x \in K_m} |\partial^\alpha \phi(x)| = M p_{m,\alpha}(\phi). \tag{15.4}$$

Using these relations we prove

15.5.1 Proposition. (a) $\mathcal{D}(\mathbb{R}^d) \subseteq \mathcal{S}(\mathbb{R}^d) \subseteq C^\infty(\mathbb{R}^d)$.

(b) *The inclusion mappings* $\mathcal{D}(\mathbb{R}^d) \hookrightarrow \mathcal{S}(\mathbb{R}^d) \hookrightarrow C^\infty(\mathbb{R}^d)$ *are continuous, hence*

$$C^\infty(\mathbb{R}^d)'\Big|_{\mathcal{S}(\mathbb{R}^d)} \subseteq \mathcal{S}'(\mathbb{R}^d) \quad \textit{and} \quad \mathcal{S}'(\mathbb{R}^d)\big|_{\mathcal{D}(\mathbb{R}^d)} \subseteq \mathcal{D}'(\mathbb{R}^d).$$

(c) $\mathcal{D}(\mathbb{R}^d)$ *is dense in* $\mathcal{S}(\mathbb{R}^d)$ *and* $\mathcal{S}(\mathbb{R}^d)$ *is dense in the Fréchet space* $C^\infty(U)$.

Proof. Part (a) is clear. For (b) let $\phi_n \to 0$ in $\mathcal{D}(\mathbb{R}^d)$. Then there exists m such that $\text{supp}(\phi_n) \subseteq K_m$ for all n, hence, by (15.4), $q_{\alpha,m}(\phi_n) \to 0$. This shows that $\mathcal{D}(\mathbb{R}^d) \hookrightarrow \mathcal{S}(\mathbb{R}^d)$ is continuous. A similar argument using (15.3) shows that $\mathcal{S}(\mathbb{R}^d) \hookrightarrow C^\infty(\mathbb{R}^d)$ is continuous.

(c) Since $C_c^\infty(\mathbb{R}^d)$ is contained in $\mathcal{S}(\mathbb{R}^d)$ and is dense in $C^\infty(\mathbb{R}^d)$ (9.1.6(b)), $\mathcal{S}(\mathbb{R}^d)$ must be dense in $C^\infty(\mathbb{R}^d)$. To show that $C_c^\infty(\mathbb{R}^d)$ is dense in $\mathcal{S}(\mathbb{R}^d)$, let $f \in \mathcal{S}(\mathbb{R}^d)$ and choose $\phi \in C_c^\infty(\mathbb{R}^d)$ such that $\phi(x) = 1$ for all $|x| \le 1$. The function $f_n(x) := f(x)\phi(x/n)$ is in $C_c^\infty(\mathbb{R}^d)$, hence the desired conclusion will follow if we show that $f_n \to f$ in the topology of $\mathcal{S}(\mathbb{R}^d)$, that is,

$$\left(1 + |x|^2\right)^k \left|\partial^\alpha \left[f(x)(1 - \phi(x/n))\right]\right| \to 0 \text{ uniformly on } \mathbb{R}^d.$$

Now, $\partial^\alpha f(x)(1 - \phi(x/n)$ is a sum of terms $\partial^\beta f(x) \cdot \partial^\gamma(1 - \phi(x/n))$. Moreover, for any compact set K, $\sup_{x \in K} |1 - \phi(x/n)| = 0$ for all large n. Thus the sequence $\partial^\alpha f(x)(1 - \phi(x/n)$ converges uniformly to zero on compact sets. Since $(1 + |x|^2)^k \partial^\beta f(x)$ is in $C_0(\mathbb{R}^d)$, it follows that $(1 + |x|^2)^k \partial^\beta f(x) \cdot \partial^\gamma(1 - \psi(x/n))$ converges uniformly to zero on $\mathbb{R}^d$, completing the proof. □

15.5.2 Proposition. *Let* β *be a multi-index,* $f \in \mathcal{S}(\mathbb{R}^d)$, *and* g *a polynomial on* $\mathbb{R}^d$. *Then the linear mappings* $\phi \mapsto \partial^\beta \phi$, $\phi \mapsto f\phi$, *and* $\phi \mapsto g\phi$ *on* $\mathcal{S}(\mathbb{R}^d)$ *are continuous.*

Proof. The equality

$$q_{\alpha,n}(\partial^\beta \phi) = \sup_{x \in \mathbb{R}^d} \left(1 + |x|^2\right)^n |\partial^{\alpha+\beta} \phi(x)| = q_{\alpha+\beta,n}(\phi)$$

implies that $\partial^\beta \phi \in \mathcal{S}(\mathbb{R}^d)$ and that the function $\phi \to \partial^\beta \phi$ is continuous. Now consider

$$q_{\alpha,n}(f\phi) = \sup_{x \in \mathbb{R}^d} \left(1 + |x|^2\right)^n |\partial^\alpha (f\phi)(x)|.$$

By the product rule, $\left(1 + |x|^2\right)^n \partial^\alpha (f\phi)(x)$ is a sum of products $\left(1 + |x|^2\right)^n \partial^\beta f(x) \cdot \partial^\gamma \phi(x)$, which are majorized by $q_{\beta,n}(f) \cdot q_{\gamma,0}(\phi)$. This shows that $f\phi \in \mathcal{S}(\mathbb{R}^d)$ and that $\phi \mapsto f\phi$ is continuous. A similar argument shows that $\phi \mapsto g\phi$ is continuous. □

The members of $\mathcal{S}'(\mathbb{R}^d)$ are called **tempered distributions**. By 15.5.1(b), they may be viewed as distributions that are continuous in a weaker topology and with an enlarged space of test functions. Their importance derives from connections with Fourier analysis, discussed in the next subsection.

15.5.3 Examples.

(a) A distribution F with compact support is tempered. To see this, let $K = \text{supp}(F)$ and for $\phi \in \mathcal{S}(\mathbb{R}^d)$ set $G(\phi) := F(\phi\psi)$ where $\psi \in C_c^\infty(\mathbb{R}^d)$ and $\psi = 1$ on K. For any $\phi \in C_c^\infty(\mathbb{R}^d)$, $\text{supp}(\phi(1-\psi)) \subseteq K^c$, hence $F\big(\phi(1-\psi)\big) = 0$, that is, $G = F$ on $C_c^\infty(\mathbb{R}^d)$. Therefore, G is a linear extension of F to $\mathcal{S}(\mathbb{R}^d)$. To see that G is continuous, let $\phi_n \to 0$ in $\mathcal{S}(\mathbb{R}^d)$. Then $\partial^\alpha \phi_n \to 0$ uniformly on $\mathbb{R}^d$, hence, by the product rule and the boundedness of the derivatives of ψ, $\partial^\alpha(\psi\phi_n) \to 0$ uniformly on $\mathbb{R}^d$. Therefore, $\psi\phi_n \to 0$ in $\mathcal{D}(\mathbb{R}^d)$ and so $G(\phi_n) = F(\psi\phi_n) \to 0$.

(b) A polynomial f on $\mathbb{R}^d$ is tempered. This is simply the assertion that the linear functional $\phi \to \int f\phi$ on $\mathcal{S}(\mathbb{R}^d)$ is continuous, that is, , for some continuous seminorm $q_{\alpha,m}$,

$$\Big| \int f\phi \Big| \le q_{\alpha,m}(\phi) \ \ \forall \ \ \phi \in \mathcal{S}(\mathbb{R}^d).$$

This is clear by taking $\alpha = 0$ and m sufficiently large.

(c) Every $f \in L^p$ is tempered. It must be shown as in (b) that for suitable m,

$$\Big| \int f\phi \Big| \le \sup_{x \in \mathbb{R}^d} \big(1 + |x|^2\big)^m |\phi(x)|.$$

For example, if $1 < p < \infty$ and q is the conjugate exponent, then

$$\int |\phi|^q = \int \frac{|\phi(x)|^q (1+|x|^2)^m}{(1+|x|^2)^m}\,dx \le \int \frac{dx}{1+|x|^2)^m} \sup_{y\in\mathbb{R}^d} \big(1+|y|^2\big)^m |\phi(y)|^q,$$

hence by Hölder's inequality

$$\int |f\phi| \le \|f\|_p \left(\int \frac{dx}{(1+|x|^2)^m} \right)^{1/q} \sup_{y\in\mathbb{R}^d} \big(1+|y|^2\big)^{m/q} |\phi(y)|.$$

It therefore suffices to choose m sufficiently large so that the term in parentheses is finite. (See 3.6.3.)

(d) If F is a tempered distribution, then so are $\partial^\alpha F$, fF ($f \in \mathcal{S}(\mathbb{R}^d)$) and gF (g a polynomial). This follows immediately from 15.5.2. ◊

The Fourier Transform of a Tempered Distribution

Recall that the Fourier transform of $f \in \mathcal{S} = \mathcal{S}(\mathbb{R}^d)$ is defined by

$$\widehat{f}(\xi) = \int e^{-2\pi i\,\xi\cdot x} f(x)\,dx, \quad \xi \in \mathbb{R}^d,$$

and the inverse by

$$\check{f}(x) = \int e^{2\pi i\,\xi\cdot x} f(\xi)\,d\xi, \quad x \in \mathbb{R}^d$$

Moreover, the mappings $f \mapsto \widehat{f}$ and $f \mapsto \check{f}$ are continuous in the topology of $\mathcal{S}$, and for $\phi \in \mathcal{S}$ we have

$$\langle \widehat{f}, \phi \rangle = \int \widehat{f} \cdot \phi = \int f \cdot \widehat{\phi} = \langle f, \widehat{\phi} \rangle$$

with a similar equation holding for $\check{f}$. (See the proof of 6.2.4.) This suggests the following definitions:

The **Fourier transform** $\widehat{F}$ and **inverse Fourier transform** $\check{F}$ of an arbitrary distribution F are defined, respectively, by

$$\langle \widehat{F}, \phi\rangle = \langle F, \widehat{\phi}\,\rangle \quad \text{and} \quad \langle \check{F}, \phi\rangle = \langle F, \check{\phi}\,\rangle, \qquad \phi \in \mathcal{S}.$$

Many of the standard results of Fourier analysis carry over to tempered distributions. For example, the formulas $\widehat{\check{\phi}} = \check{\widehat{\phi}} = \phi$ and $\widehat{\widehat{\phi}} = \widetilde{\phi}$ imply, by duality, their distribution counterparts

$$\widehat{\check{F}} = \check{\widehat{F}} = F \quad \text{and} \quad \widehat{\widehat{F}} = \widetilde{F},$$

as may be readily verified. As in the classical case, one sets $\mathfrak{F}(F) := \widehat{F}$, so that $\mathfrak{F}^{-1}(F) = \check{F}$.

15.6 Sobolev Theory

Sobolev Spaces

For an open subset U of $\mathbb{R}^d$ define linear spaces

$$L^p_k(U) := \big\{f \in L^p(U) : \partial^\alpha f \in L^p(U)\ \forall\ |\alpha| \le k\big\}, \quad 1 \le p < \infty,\ k \in \mathbb{N}.$$

The derivatives in the definition are assumed to be distributional derivatives. Thus $\partial^\alpha f \in L^p(U)$ satisfies

$$\int \phi\, \partial^\alpha f = (-1)^{|\alpha|} \int f \partial^\alpha \phi, \quad \phi \in C_c^\infty(U).$$

The Sobolev inequalities, proved below, imply that one actually obtains ordinary derivatives by taking $f \in L^p_m(U)$ for sufficiently large m.

Define a norm on $L^p_k(U)$ by

$$\|f\|_{k,p} := \Big(\sum_{|\alpha|\le k} \int_U |\partial^\alpha f|^p\Big)^{1/p}.$$

Thus a sequence (f_n) converges to f in $L^p_k(U)$ iff $\partial^\alpha f_n \to \partial^\alpha f$ in $L^p(U)$ for all α with $|\alpha| \le k$. For $p = 2$, the norm is given by an inner product

$$(f \,|\, g)_k := \sum_{|\alpha|\le k} \int_U (\partial^\alpha f)\overline{(\partial^\alpha g)}.$$

Sobolev spaces, being defined in terms of L^p norms, tend to be somewhat easier to manage than spaces of distributions. Moreover, they have an advantage over L^p spaces in that a derivative of a member of $L^p_k(U)$ is a member of $L^p_{k-1}(U)$. These features make Sobolev spaces important tools in the study of weak solutions of PDEs.

15.6.1 Theorem. *$L^p_k(U)$ is a Banach space and $L^2_k(U)$ is a Hilbert space.*

Proof. Let (f_n) be a Cauchy sequence in $L^p_k(U)$. Then for each α with $|\alpha| \le k$, $(\partial^\alpha f_n)$ is a Cauchy sequence in $L^p(U)$ and so converges to some $f_\alpha \in L^p(U)$. For any $\phi \in C_c^\infty(U)$ we then have

$$\langle f_\alpha, \phi\rangle = \int f_\alpha \phi = \lim_n \int (\partial^\alpha f_n)\phi = \lim_n (-1)^{|\alpha|} \int f_n(\partial^\alpha \phi) = (-1)^{|\alpha|} \int f(\partial^\alpha \phi) = \langle \partial^\alpha f, \phi\rangle,$$

hence $f_\alpha = \partial^\alpha f$. Thus $f_n \to f$ in $L^p_k(U)$, showing that $L^p_k(U)$ is complete and hence a Banach space. □

Application: Elliptic PDEs

The space $L_k^2(U)$ is of particular interest in Sobolev theory because of the availability of general Hilbert space techniques. We illustrate with the following application.

Define $H_k^2(U)$ to be the closure of $C_c^\infty(U)$ in $L_k^2(U)$. In general, $H_k^2(U)$ is a proper subset of $L_k^2(U)$. Of course, as a closed subspace of a Hilbert space, $H_k^2(U)$ is itself a Hilbert space. Assume that U is bounded and let $g_{ij} \in C^\infty(U_0)$, where U_0 is an open set containing the compact set $\operatorname{cl} U$. Consider the differential operator P on U defined by

$$Pf = -\sum_{i,j=1}^{d} \partial_i \left(g_{ij} \partial_j f\right), \tag{15.5}$$

where $\partial_i = \partial/\partial_{x_i}$. We assume all functions are real-valued. Further, we assume that the matrix $[g_{ij}]$ is strictly positive definite, that is,

$$\sum_{i,j=1}^{d} y_i y_j g_{ij}(x) > 0 \quad \text{for all } y_j \in \mathbb{R} \text{ and } x \in U_0. \tag{15.6}$$

Under these circumstances, the operator P in (15.5) is said to be **strongly elliptic**. We show the existence of a unique weak solution f of the PDE

$$Pf + af = h, \quad \text{where } a \geq 0 \text{ and } h \in H_1^2(U). \tag{15.7}$$

Since the sum in (15.6) is continuous in $(x, \boldsymbol{y})$, it has a minimum $m > 0$ on the compact set $\operatorname{cl} U \times S^1$. Therefore

$$\sum_{i,j=1}^{d} y_i y_j g_{ij}(x) \geq m|\boldsymbol{y}|^2 \quad \text{for all } y_j \in \mathbb{R} \text{ and } x \in \operatorname{cl} U. \tag{15.8}$$

For $\phi, \psi \in C_c^\infty(U)$, define a bilinear form B on $C_c^\infty(U)$ by

$$B(\phi,\psi) := \langle (P+aI)\phi, \psi\rangle = a\int \phi\psi - \sum_{i,j=1}^{d} \int \psi\partial_i(g_{ij}\partial_j\phi) = a\int \phi\psi + \sum_{i,j=1}^{d} \int g_{ij}(\partial_j\phi)(\partial_i\psi),$$

where the last equality comes from an integration by parts. Since the functions g_{ij} are bounded, it follows from the definition of inner product in $H_1^2(U)$ and the CBS inequality that for some constant $c > 0$

$$|B(\phi,\psi)| \leq c \left\|\phi\right\|_{1,2} \left\|\psi\right\|_{1,2}.$$

Therefore, B extends continuously to a sesquilinear form on $H_1^2(U)$. Furthermore, by (15.8),

$$B(\phi,\phi) = a\int \phi^2 + \sum_{i,j=1}^{d} \int g_{ij}(\partial_i\phi)(\partial_j\phi) \geq m \sum_{j=0}^{d} \int (\partial_j\phi)^2 = m \left\|\phi\right\|_{1,2}^2. \tag{15.9}$$

Now, since

$$\left|\int hg\right| \leq \left(\int |h|^2\right)^{1/2} \left(\int |g|^2\right)^{1/2} \leq \left(\int |h|^2\right)^{1/2} \|g\|_{1,2}^2,$$

the functional $g \to \int hg$ is continuous on $H_1^2(U)$. By the Lax-Milgram theorem (11.4.2), there exists a unique $f \in H_1^2(U)$ such that $B(f,g) = \int hg$. In particular, for all $\psi \in C_c^\infty(U)$,

$$\langle Pf + af, \psi\rangle = B(f,\psi) = \int h\psi = \langle h, \psi\rangle,$$

which shows that f is the desired weak solution of (15.7). ◊

Sobolev Inequalities

These inequalities are important tools in determining existence and uniqueness of solutions of a variety of PDEs, as well as in the study of regularity properties of these solutions. In this subsection we give the reader a flavor of the subject by proving two such inequalities.

15.6.2 Theorem. *If $f \in L_d^1(\mathbb{R}^d)$, then $\|f\|_\infty \le c\|f\|_{1,d}$ and there exists $g \in C_b(\mathbb{R}^d)$ such that $f = g$ a.e. Moreover, if $f \in L_{d+k}^1(\mathbb{R}^d)$ $(k \ge 1)$, then one may take $g \in C_b^k(\mathbb{R}^d)$.*

Proof. Consider first the case $f \in C^\infty(\mathbb{R}^d)$ and $d = 2$. For any $\psi \in C_c^\infty(\mathbb{R}^2)$,

$$\psi(x, y) = \int_{-\infty}^{y} \int_{-\infty}^{x} \frac{\partial^2 \psi}{\partial x\, \partial y}(s, t)\, ds\, dt,$$

hence

$$|\psi(x, y)| \le \int_{-\infty}^{y} \int_{-\infty}^{x} \left|\frac{\partial^2 \psi}{\partial x\, \partial y}(s, t)\right| ds\, dt \le \left\|\frac{\partial^2 \psi}{\partial x\, \partial y}\right\|_1.$$

Replacing ψ by $f\psi$, we have

$$|f\psi(x, y)| \le \left\|\frac{\partial^2 (f\psi)}{\partial x\, \partial y}\right\|_1, \quad x,\, y \in \mathbb{R}.$$

Since

$$\frac{\partial^2 (f\psi)}{\partial x\, \partial y} = \psi \frac{\partial^2 f}{\partial x\, \partial y} + \frac{\partial f}{\partial x}\frac{\partial \psi}{\partial y} + \frac{\partial f}{\partial y}\frac{\partial \psi}{\partial x} + f\frac{\partial^2 \psi}{\partial x\, \partial y},$$

we see that

$$|f\psi(x, y)| \le \|\psi\|_\infty \left\|\frac{\partial^2 f}{\partial x\, \partial y}\right\|_1 + \left\|\frac{\partial \psi}{\partial y}\right\|_\infty \left\|\frac{\partial f}{\partial x}\right\|_1 + \left\|\frac{\partial \psi}{\partial x}\right\|_\infty \left\|\frac{\partial f}{\partial y}\right\|_1 + \|f\|_1 \left\|\frac{\partial^2 \psi}{\partial x\, \partial y}\right\|_\infty.$$

Now let $0 \le \psi \le 1$ such that $\psi = 1$ on $[-1, 1] \times [-1, 1]$ and $\psi = 0$ outside $[-2, 2] \times [-2, 2]$. Set $\psi_n(x) = \psi(x/n)$. Since the partial derivatives of ψ are bounded, there exists a constant c depending only on ψ such that for all (x, y) and n,

$$|f\psi_n(x, y)| \le c\left(\left\|\frac{\partial^2 f}{\partial x\, \partial y}\right\|_1 + \left\|\frac{\partial f}{\partial x}\right\|_1 + \left\|\frac{\partial f}{\partial y}\right\|_1 + \|f\|_1\right).$$

Since $\psi_n(x, y) \to 1$, we obtain the Sobolev inequality for $d = 2$:

$$\|f\|_\infty \le c\left(\left\|\frac{\partial^2 f}{\partial x\, \partial y}\right\|_1 + \left\|\frac{\partial f}{\partial x}\right\|_1 + \left\|\frac{\partial f}{\partial y}\right\|_1 + \|f\|_1\right).$$

For an arbitrary d, start off the argument with

$$\psi(x_1, \ldots, x_d) = \int_{-\infty}^{x_d} \cdots \int_{-\infty}^{x_1} \frac{\partial^d}{\partial x_1 \ldots \partial x_d} \psi(t_1, \ldots, t_d)\, dt_1 \ldots dt_2.$$

and proceed as above.

Now consider the general case $f \in L_d^1(\mathbb{R}^d)$. Choose an approximate identity (ϕ_n) in $C_c^\infty(\mathbb{R}^d)$ for $L^1(\mathbb{R}^d)$. Since $f * \phi_n$ is C^∞, by the preceding paragraph

$$\|f * \phi_n\|_\infty \le c\|f * \phi_n\|_{1,d} \quad \text{for all } n. \tag{†}$$

The norm on the right is a sum of terms that are L^1 norms of the derivatives

$$\begin{aligned}\partial^\alpha(f*\phi_n)(x) &= \int f(y)\partial_x^\alpha\phi_n(x-y)\,dy = (-1)^{|\alpha|}\int f(y)\partial_y^\alpha\phi_n(x-y)\,dy\\ &= \int(\partial^\alpha f)(y)\phi_n(x-y)\,dy = (\partial^\alpha f*\phi_n)(x),\end{aligned}$$

where $\partial^\alpha f$ is the distributional derivative. Taking absolute values and integrating with respect to x, recalling that $\|\phi_n\|_1 = 1$, we see that $\|\partial^\alpha(f*\phi_n)\|_1 \le \|\partial^\alpha f\|_1$. Taking the sum over all $|\alpha| \le d$ and using (†) we obtain

$$\|f*\phi_n\|_\infty \le c\|f\|_{1,d} \quad \text{for all } n.$$

Since $f*\phi_n \xrightarrow{L^1} f$, there exists a subsequence such that $f*\phi_{n_k} \to f$ a.e. Thus $\|f\|_\infty \le c\|f\|_{1,d}$. Since $f - f*\phi_n \in L^1_d$, we may replace f in the last inequality by $f - f*\phi_n$ to conclude that

$$\|f - f*\phi_n\|_\infty \le c\|f - f*\phi_n\|_{1,d}.$$

The norm on the right is a sum of terms $\|\partial^\alpha f - \partial^\alpha(f*\phi_n)\|_1 = \|\partial^\alpha f - (\partial^\alpha f)*\phi_n\|_1$ which tend to zero by the approximate identity property. Therefore, $\|f - f*\phi_n\|_\infty \to 0$, hence f has a version that is continuous. This proves the theorem for the case $k = 0$. For $k \ge 1$ one need only replace f in this argument by $\partial^\alpha f$, $|\alpha| \le k$. □

Theorem 15.6.2 has general versions where L^1_d is replaced by L^p_m, $1 \le p < \infty$, $m \ge d$. We illustrate with the case $p = 2$, which has an elegant proof using Fourier transforms.

15.6.3 Theorem. *Let $m > d/2$. If $f \in L^2_m(\mathbb{R}^d)$, then $\|f\|_\infty \le c\|f\|_{m,2}$ and there exists $g \in C_0(\mathbb{R}^d)$ such that $f = g$ a.e. Moreover, ff $f \in L^2_{m+k}(\mathbb{R}^d)$ $(k \ge 1)$, then g may be taken to be a C^k function.*

Proof. In the following we employ the convention of describing a function by displaying the independent variable. We show first that $\widehat{f} \in L^1(\mathbb{R}^d)$. From the formula

$$\partial^\alpha\mathfrak{F}(\phi) = \mathfrak{F}((-2\pi\, i\, x)^\alpha\phi(x)), \quad \phi \in \mathcal{S}$$

and the definition of the Fourier transform of a distribution we have for $|\alpha| \le m$

$$\begin{aligned}\langle\mathfrak{F}(\partial^\alpha f), \phi\rangle &= \langle\partial^\alpha f, \mathfrak{F}(\phi)\rangle = \langle f, \partial^\alpha\mathfrak{F}(\phi)\rangle = \langle f, \mathfrak{F}((-2\pi\, i\, x)^\alpha\phi(x))\rangle\\ &= \langle\mathfrak{F}(f), (-2\pi\, i\, x)^\alpha\phi(x)\rangle = \langle(-2\pi\, i\, \xi)^\alpha\mathfrak{F}(f), \phi\rangle.\end{aligned}$$

Since $\mathcal{S}(\mathbb{R}^d)$ is dense in $L^2(\mathbb{R}^d)$, $\mathfrak{F}(\partial^\alpha f) = (-2\pi\, i\, \xi)^\alpha\mathfrak{F}(f)$. Taking L^2 norms of the last equation and using the Plancherel theorem $\|\mathfrak{F}(\partial^\alpha f)\|_2 = \|\partial^\alpha f\|_2$, we have for a suitable constant M_1

$$\int|\xi^\alpha|^2|\widehat{f}(\xi)|^2\,d\xi = M_1\|\partial^\alpha f\|_2, \quad |\alpha| \le m.$$

Summing over α we obtain

$$\int h(\xi)|\widehat{f}(\xi)|^2\,d\xi = M_1\sum_{|\alpha|\le m}\|\partial^\alpha f\|_2, \quad \text{where } h(\xi) := \sum_{|\alpha|\le m}|\xi^\alpha|^2.$$

We claim that $\|\widehat{f}\|_1 < \infty$. By the CBS inequality,

$$\|\widehat{f}\|_1 = \int h(\xi)^{1/2}|\widehat{f}(\xi)|h(\xi)^{-1/2}\,d\xi \le \left(\int h(\xi)|\widehat{f}(\xi)|^2\,d\xi\right)^{1/2}\left(\int h(\xi)^{-1}\,d\xi\right)^{1/2},$$

hence it suffices to show that the second factor on the right is finite. Now, by taking α's of the form $(0,\ldots,0)$ and $(0,\ldots,0,m,0\ldots,0)$, we have $h(\boldsymbol{\xi}) \geq 1+\sum_{j=1}^{d}|\xi_j|^{2m}$. The inequality

$$|\boldsymbol{\xi}|^{2m} = \left(\sum_{j=1}^{d}|\xi_j|^2\right)^m \leq \left(d \max_{1\leq j\leq d}|\xi_j|^2\right)^m = d^m \max_{1\leq j\leq d}|\xi_j|^{2m} \leq d^m \sum_{j=1}^{d}|\xi_j|^{2m}$$

then shows that $h(\boldsymbol{\xi}) \geq 1 + c|\boldsymbol{\xi}|^{2m}$. By 3.6.3 (with $s = 2m$, $t = 1$, and $p = 2$) we see that $\int h^{-1} < \infty$ and so for a suitable $M_2 > 0$

$$\|\widehat{f}\|_1 \leq M_2 \sum_{|\alpha|\leq m} \|\partial^\alpha f\|_2 < \infty.$$

It now follows from the Fourier inversion formula and the Riemann-Lebesgue lemma that $f \in C_0(\mathbb{R}^d)$. Finally, since

$$f(x) = \int \widehat{f}(\boldsymbol{\xi})e^{2\pi i\, x\cdot\boldsymbol{\xi}}\, d\boldsymbol{\xi}$$

we have $\|f\|_\infty \leq \|\widehat{f}\|_1 \leq C_2\sum_{|\alpha|\leq m}\|\partial^\alpha f\|_2$. This completes the proof of the first part of theorem. The second part may be proved in a similar manner by replacing f throughout by $\partial^\beta f$, $|\beta| \leq k$. □

Chapter 16

Analysis on Locally Compact Groups

Lebesgue measure on $\mathbb{R}$ and counting measure on $\mathbb{Z}$ are examples of measures μ that are *translation invariant*, that is, $\mu(B + x) = \mu(B)$ for all Borel sets B. These are special cases of a general construct called *Haar measure*. As we shall see, the existence Haar measure leads to a unification and generalization of Fourier analysis, the basic aspects of which are presented in this chapter.

16.1 Topological Groups

Definitions and Basic Properties

A **topological group** is a group G with a topology relative to which the group operations

$$(s,t) \to st : G \times G \to G \quad \text{and} \quad t \to t^{-1} : G \to G$$

are continuous. For example, a TVS, and in particular $\mathbb{K}^d$, is an abelian topological group under addition. The set of nonzero members of $\mathbb{K}$ is an abelian topological group under multiplication. The set of $n \times n$ matrices over $\mathbb{K}$ with determinant one is a nonabelian topological group under matrix multiplication.

Here are useful alternate characterizations of a topological group that will be needed in the chapter.

16.1.1 Proposition. *Let G be a group with a topology. The following are equivalent:*

(a) *G is a topological group.*

(b) *The map $(s,t) \mapsto s^{-1}t : G \times G \to G$ is continuous.*

(c) *The map $(s,t) \mapsto s^{-1}t : G \times G \to G$ is continuous at (e,e), and for each $a \in G$ the translation mappings $x \to ax$ and $x \to xa$ are continuous.*

Proof. (a) $\Rightarrow$ (b): The map is a composition of the continuous mapping $(s,t) \mapsto (s^{-1},t)$ and the multiplication map, hence is continuous.

(b) $\Rightarrow$ (c): The first statement is clear. If $x_\alpha \to x$, then $(a^{-1}, x_\alpha) \to (a^{-1}, x)$, hence $ax_\alpha \to ax$. Therefore, $x \mapsto ax$ is continuous. Similarly, $x \mapsto xa$ is continuous.

(c) $\Rightarrow$ (a): If $s_\alpha \to s$, then, by the second part of the hypothesis, $(s^{-1}s_\alpha, s^{-1}s) \to (e,e)$. Applying the first part of the hypothesis, we have $s_\alpha^{-1}s = (s^{-1}s_\alpha)^{-1}(s^{-1}s) \to e$, hence $s_\alpha^{-1} \to s^{-1}$, which shows that inversion is continuous. Since multiplication is the composition of the continuous maps $(s,t) \mapsto (s^{-1},t)$ and $(s^{-1},t) \mapsto st$, multiplication is continuous at (e,e). Now let $s_\alpha \to s$ and $t_\beta \to t$. Then $s^{-1}s_\alpha \to e$ and $t_\beta t^{-1} \to e$, hence $s^{-1}s_\alpha t_\beta t^{-1} \to e$ and so $s_\alpha t_\beta \to st$. Therefore, multiplication is continuous. □

The basic properties of topological groups are given in the next proposition.

16.1.2 Proposition. *Let G be a topological group and H a subgroup of G.*

(a) *For fixed $a \in G$, the mappings $t \to at$, $t \to ta$, and $t \to t^{-1}$ are homeomorphisms.*

(b) *Each neighborhood U of e contains a* **symmetric** *neighborhood of e, that is, a neighborhood V of e such that $V = V^{-1}$ $(= \{x^{-1} : x \in V\})$.*

(c) *Each neighborhood U of e contains a neighborhood V of e such that $VV \subseteq U$.*

(d) *The closure of H is a subgroup.*

(e) *If H is open, then it is also closed.*

(f) *If G is Hausdorff and H is locally compact, then H is closed.*

Proof. Part (a) follows from 16.1.1. For part (b), take $V = U \cap U^{-1}$. Part (c) follows from the continuity of the mapping $(s,t) \to st$ at (e,e) and part (d) from the continuity of the group operations. For (e), let $x \in \text{cl}(H)$. Since xH is a neighborhood of x, $xH \cap H \neq \emptyset$. Then $xy \in H$ for some $y \in H$ and so $x = (xy)y^{-1} \in H$.

To prove (f), let $x \in \text{cl}_G(H)$, $(x_\alpha) \subseteq H$, and $x_\alpha \to x$. Since H is locally compact, there exists an open neighborhood V of e in G such that $\text{cl}_H(V \cap H)$ is compact in H. Therefore, $\text{cl}_H(V \cap H)$ is compact in G, hence also closed. From $x_\alpha^{-1} \to x^{-1}$ and $x_\alpha^{-1} \in H$ we have $x^{-1} \in \text{cl}_G(H)$. Thus since Vx^{-1} is a neighborhood of x^{-1}, $H \cap Vx^{-1} \neq \emptyset$. Choose $y \in H \cap Vx^{-1}$. Then $yx \in V$ so yx_α is eventually in $V \cap H$. Thus yx is in the closed set $\text{cl}_H(V \cap H) \subseteq H$ and so $x = y^{-1}yx \in H$. □

Translation and Uniform Continuity

The **left** and **right translates** of a function f on a topological group G are defined by

$$L_t f(s) = f(ts)^{\,1} \quad \text{and} \quad R_t f(s) = f(st), \quad s, t \in G. \tag{16.1}$$

The set of all left (respectively, right) translates of f is denoted by $L_G f$ (respectively, $R_G f$). A subset $\mathscr{F}$ of functions on G is said to be **left translation invariant** if $L_G f \subseteq \mathscr{F}$ for all $f \in \mathscr{F}$. **Right translation invariance** is defined analogously. A subset that is both left and right translation invariant is said to be **translation invariant**. For example, by continuity of multiplication in G, the spaces $C_b(G)$, $C_0(G)$, and $C_c(G)$ are translation invariant. Note that the translation operators L_t and R_t satisfy

$$L_{st} = L_t L_s \quad \text{and} \quad R_{st} = R_s R_t.$$

A function $f : G \to \mathbb{C}$ is said to be **right uniformly continuous** on G if the function $t \mapsto R_t f$ is norm continuous at e. Thus for each $\varepsilon > 0$ there exists a neighborhood V of e such that $\|R_t f - f\|_\infty < \varepsilon$ for all $t \in V$ or, equivalently,

$$|f(st) - f(s)| < \varepsilon \quad \text{for all } t \in V \text{ and } s \in G.$$

Setting $x = st$ and $y = s$ we may write this as

$$|f(x) - f(y)| < \varepsilon \quad \text{for all } x, y \in G \text{ with } x \in yV.$$

Left uniform continuity is defined by replacing R_t by L_t, and for this the following equivalent formulation holds:

$$|f(x) - f(y)| < \varepsilon \quad \text{for all } x, y \in G \text{ with } x \in Vy.$$

A function is **uniformly continuous** if it is both left and right uniformly continuous.

[1]Some authors give the alternate definition $L_t f(s) = f(t^{-1}s)$, which results in the pleasing relation $L_{st} = L_s L_t$. Our choice is dictated by the desire for a uniform notation for groups and semigroups. (See Chapter 17.)

16.1.3 Proposition. *Let G be a topological group. If $f \in C_c(G)$, then f is uniformly continuous.*

Proof. It suffices to prove right continuity. Let $K := \operatorname{supp}(f)$ and $\varepsilon > 0$. By continuity, for each $s \in K$ there exists a neighborhood U_s of e such that $|f(st) - f(s)| < \varepsilon/2$ for all $t \in U_s$. By 16.1.2, there exists a symmetric neighborhood V_s of e such that $V_sV_s \subseteq U_s$. Since $\{sV_s : s \in K\}$ is an open cover of K, by compactness there exist $s_1, \ldots, s_n \in K$ such that $K \subseteq \bigcup_{j=1}^n s_jV_{s_j}$. Set $V := \bigcap_{j=1}^n V_{s_j}$. We claim that if $t \in V$, then $|f(st) - f(s)| < \varepsilon$ for all $s \in G$. This obviously holds if both s, $st \notin K$. Now consider the remaining cases:

(a) $s \in K$. Then $s \in s_jV_{s_j}$ for some j, hence $x := s_j^{-1}s \in V_{s_j} \subseteq U_{s_j}$ and $y := (s_j^{-1}s)t \in V_{s_j}V_{s_j} \subseteq U_{s_j}$, so by definition of U_{s_j}

$$|f(st) - f(s)| \le |f(st) - f(s_j)| + |f(s_j) - f(s)| = |f(s_jy) - f(s_j)| + |f(s_j) - f(s_jx)| < \varepsilon.$$

(b) $s \notin K$ and $st \in K$. Then $st = s_jt_j$ for some $t_j \in V_{s_j} \subseteq U_{s_j}$, hence

$$|f(st) - f(s_j)| = |f(s_jt_j) - f(s_j)| < \varepsilon/2.$$

Moreover, since V_{s_j} is symmetric, $x := s_j^{-1}s = (s_j^{-1}st)t^{-1} = t_jt^{-1} \in V_{s_j}V_{s_j} \subseteq U_{s_j}$, hence

$$|f(s_j)| = |f(s) - f(s_j)| = |f(s_jx) - f(s_j)| < \varepsilon/2.$$

Therefore, $|f(st) - f(s)| \le |f(st) - f(s_j)| + |f(s_j)| < \varepsilon$, as required. □

For a function f on G, define $\widetilde{f}$ by

$$\widetilde{f}(x) = f(x^{-1}), \quad x \in G.$$

Then f is right uniformly continuous iff $\widetilde{f}$ is left uniformly continuous. A function f is **symmetric** if $\widetilde{f} = f$. For example, for any g the functions $g + \widetilde{g}$ and $g \cdot \widetilde{g}$ are symmetric.

16.2 Haar Measure

Definition and Basic Properties

For the remainder of the chapter we restrict our attention to locally compact, Hausdorff topological groups. Such a group is traditionally referred to as simply a **locally compact group**. The fundamental property of a locally compact group that allows such a rich supply of analytical techniques is the existence of translation invariant measures.

A Borel measure μ on locally compact group G is said to be **left invariant** if

$$\mu(sB) = \mu(B) \text{ for all } s \in G \text{ and all Borel sets } B. \tag{16.2}$$

Right invariance is defined by replacing sB by Bs. A nontrivial (that is, not identically zero) left (right) invariant Radon measure on G is called a **left (right) Haar measure**. A measure that is both a left Haar measure and a right Haar measure is called a **Haar measure**. Lebesgue measure on $\mathbb{R}^d$ and counting measure on $\mathbb{Z}^d$ are Haar measures. One may show directly that the set function $B \mapsto \int_B x^{-1}\,dx$ defines a Haar measure on the multiplicative group of nonzero real numbers. We shall see other examples later.

Now define a Borel measure $\widetilde{\mu}$ by

$$\widetilde{\mu}(B) = \mu(B^{-1}), \quad B \in \mathcal{B}(G). \tag{16.3}$$

Then $\widetilde{\mu}(Bs) = \mu((Bs)^{-1}) = \mu(s^{-1}B^{-1})$, hence μ is left invariant iff $\widetilde{\mu}$ is right invariant. Moreover, since inversion is a homeomorphism, $\widetilde{\mu}$ is regular iff μ is regular. Thus μ is a left Haar measure iff $\widetilde{\mu}$ is a right Haar measure. In view of this duality, we shall frequently state results only for left Haar measures.

The next proposition summarizes the elementary properties of left Haar measure. For the statement we need the following notation, which will be used throughout the chapter.

$$C_c^+ = C_c^+(G) := \{f \in C_c(G) : f \geq 0 \text{ and } \|f\|_\infty > 0\}.$$

16.2.1 Proposition. *Let μ be a Radon measure on G. The following statements are equivalent:*

(a) *μ is a left Haar measure.*

(b) *$\mu(sK) = \mu(K)$ for all compact $K \subseteq G$ and $s \in G$.*

(c) *$\mu(sU) = \mu(U)$ for all open $U \subseteq G$ and $s \in G$.*

(d) *$\int L_s f\, d\mu = \int f\, d\mu$ for all $f \in L^1(\mu)$ and $s \in G$.*

(e) *$\int L_s f\, d\mu = \int f\, d\mu$ for all $f \in C_c^+$ and $s \in G$.*

Proof. That (a), (b), and (c) are equivalent follows easily from the regularity properties of Radon measures (7.1). Clearly, (d) implies (b) and (e).

Now suppose that (a) holds. Then

$$\int L_s \mathbf{1}_B\, d\mu = \int \mathbf{1}_B(st)\, d\mu(t) = \mu(s^{-1}B) = \mu(B),$$

hence (d) holds for indicator functions. The usual arguments then show that (d) holds for all $f \in L^1$. That (e) $\Rightarrow$ (d) follows by approximation (7.1.2). □

16.2.2 Proposition. *Let μ be a left Haar measure on G. Then*

(a) *$\mu(U) > 0$ for all for all open $U \subseteq G$.*

(b) *$\mu(G) < \infty$ iff G is compact.*

(c) *If E is a Borel set and $\mu(E) = 0$, then E^c is dense in G.*

Proof. (a) Suppose $\mu(U) = 0$ for some nonempty open set. Since any compact set K may be covered by finitely many translates sU and since $\mu(sU) = \mu(U) = 0$, $\mu(K) = 0$. By regularity, $\mu(B) = 0$ for all Borel sets B, contradicting the definition of Haar measure.

(b) The sufficiency follows from the definition of Radon measure. For the necessity, assume that G is not compact. Choose any open neighborhood U of e with compact closure. Then G cannot be covered by finitely many left translates of U. Letting s_1 be arbitrary, we may construct a sequence (s_n) such that $s_{n+1} \notin \bigcup_{k=1}^n s_n U$. Now let V be a symmetric open neighborhood of e with $VV \subset U$. The sets $s_n V$ are disjoint. Indeed, if $m > n$ and $(s_n V) \cap (s_m V) \neq \emptyset$, then $s_n v_n = s_m v_m$ for some $v_n, v_m \in V$ and we have the contradiction $s_m = s_n v_n v_m^{-1} \in s_n U$. Now set $B = \bigcup_n s_n V$. By left invariance, $\mu(B) = \sum_n \mu(V)$. But since $\mu(B) < \infty$, $\mu(V) = 0$, contradicting (a).

Part (c) follows from (a). □

Existence of Haar Measure

16.2.3 Theorem. *Every locally compact group G has a left Haar measure and a right Haar measure.*

Proof. (Weil). It suffices to construct a left Haar measure on G. This is accomplished by first constructing a left invariant positive linear functional I on $C_c(G)$. An application of the Riesz representation theorem then yields the desired measure.

To construct I, let f, $\phi \in C_c^+$ and define the *Haar covering number of f with respect to ϕ* by

$$(f:\phi) = \inf\left\{\sum_{i=1}^n c_i : f \le \sum_{i=1}^n c_i L_{s_i}\phi \quad n\in\mathbb{N},\ s_j\in G, \text{ and } c_j \ge 0\right\}.$$

To see that such constants c_i exist and hence that $(f:\phi)<\infty$, consider the nonempty open set $U := \{t : \phi(t) > \frac{1}{2}\|\phi\|_\infty\}$ and let s_jU be finitely many left translates of U that cover the compact set $\text{supp}(f)$. If $x \in \text{supp}\, f$, then $2\phi(s_j^{-1}x) > \|\phi\|_\infty$ for some j, and setting $c := 2\|f\|_\infty \|\phi\|_\infty^{-1}$ we have $f(x) \le \|f\|_\infty \le \sum_i c\phi(s_i^{-1}x)$.

The basic idea in the construction of I is to show that a suitably normalized version of $(f:\phi)$ tends to $I(f)$ as $\text{supp}(\phi)$ tends to $\{e\}$. This is accomplished in the following steps, the first of which suggests that (f,ϕ) is an appropriate precursor of I. (All functions considered in these steps are assumed to be in C_c^+.)

(1) *$(f:\phi)$ has the following properties:*

- **(a)** $(f_1+f_2:\phi) \le (f_1:\phi)+(f_2:\phi)$.
- **(b)** $(cf:\phi) = c(f:\phi)\ \forall\, c>0$.
- **(c)** $f\le g \Rightarrow (f:\phi)\le(g:\phi)$.
- **(d)** $(L_sf:\phi) = (f:\phi)\ \forall\, s\in G$.
- **(e)** $(f:\phi)\le(f:g)(g:\phi)$.
- **(f)** $(f:g)\ge \|f\|_\infty \|g\|_\infty^{-1}$.

⟦Parts (a) – (d) follow directly from the definition of $(f:\phi)$. For (e), let

$$f \le \sum_{i=1}^m a_i L_{s_i} g \quad\text{and}\quad g \le \sum_{j=1}^n b_j L_{t_j}\phi.$$

Then $L_{s_i}g \le \sum_{j=1}^n b_j L_{t_j s_i}\phi$, hence $f \le \sum_{i,j} a_i b_j L_{t_j s_i}\phi$ and so

$$(f:\phi) \le \sum_{i,j} a_i b_j = \left(\sum_{i=1}^m a_i\right)\left(\sum_{j=1}^n b_j\right).$$

Taking infima over all sums $\sum_{i=1}^m a_i$ and $\sum_{j=1}^n b_j$ gives (e).

Now let $f \le \sum_{i=1}^m c_i L_{s_i} g$. Then $f(x) \le \|g\|_\infty \sum_{i=1}^m c_i$ for all x and so we have $\|f\|_\infty \le \|g\|_\infty \sum_{i=1}^m c_i$. Taking infima over the sums $\sum_{i=1}^m c_i$ yields (f).⟧

(2) *Let f_0 be an arbitrary member of C_c^+ and define $I_\phi(f) := \dfrac{(f:\phi)}{(f_0:\phi)}$. Then I_ϕ has the following properties:*

- **(a)** $I_\phi(f_1+f_2) \le I_\phi(f_1)+I_\phi(f_2)$.
- **(b)** $I_\phi(cf) = cI_\phi(f)\ \forall\, c>0$.
- **(c)** $f\le g \Rightarrow I_\phi(f)\le I_\phi(g)$.
- **(d)** $I_\phi(L_sf) = I_\phi(f)\ \forall\ s\in G$.
- **(e)** $(f_0:f)^{-1} \le I_\phi(f) \le (f:f_0)$.

⟦By (f) of (1), $(f_0:\phi)>0$, hence I is well-defined. Properties (a) – (e) then follow immediately from the corresponding parts (a) – (e) of (1).⟧

(3) *For each $\varepsilon > 0$, there exists a neighborhood V of e such that*

$$I_\phi(f_1) + I_\phi(f_2) \le I_\phi(f_1 + f_2) + \varepsilon \;\text{ for all } \phi \text{ with } \operatorname{supp}(\phi) \subseteq V.$$

⟦Let $g \in C_c^+$ such that $g = 1$ on $\operatorname{supp}(f_1 + f_2)$, and let $\delta > 0$ be arbitrary. Set $h := f_1 + f_2 + \delta g$ and $h_k := f_k/h$. Note that if $h(x) = 0$, then $f_k(x) = 0$, in which case the value of $h_k(x)$ is taken to be zero. With this definition, one easily checks that $h_k \in C_c^+$. By 16.1.3, there exists a neighborhood V of e such that $|h_k(x) - h_k(y)| < \delta$ whenever $y^{-1}x \in V$. If $K := \operatorname{supp}(\phi) \subseteq V$ and $h \le \sum_i c_i L_{s_i}\phi$, then for $k = 1, 2$ we have

$$f_k(x) = h(x)h_k(x) \le \sum_i c_i\phi(s_i x)h_k(x).$$

Since the only contribution to the sum on the right comes from terms for which $s_i x \in K$, and since for these $|h_k(x) - h_k(s_i^{-1})| < \delta$, we see that

$$f_k(x) \le \sum_i c_i\phi(s_i x)\big[h_k(s_i^{-1}) + \delta\big].$$

Therefore, $(f_k : \phi) \le \sum_i c_i[h_k(s_i^{-1}) + \delta]$, hence

$$(f_1 : \phi) + (f_2 : \phi) \le \sum_i c_i\big[h_1(s_i^{-1}) + h_2(s_i^{-1}) + 2\delta\big] \le (1 + 2\delta)\sum_i c_i,$$

the last inequality because $h_1 + h_2 \le 1$. Taking the infimum over all such sums $\sum_i c_i$ and dividing by $(f_0 : \phi)$ we have

$$\begin{aligned} I_\phi(f_1) + I_\phi(f_2) &\le (1 + 2\delta)I_\phi(h) \le (1 + 2\delta)\big[I_\phi(f_1 + f_2) + \delta I_\phi(g)\big] \\ &= I_\phi(f_1 + f_2) + \big[2\delta I_\phi(f_1 + f_2) + \delta(1 + 2\delta)I_\phi(g)\big], \end{aligned}$$

the second inequality by (a) and (b) of (2) applied to $h = f_1 + f_2 + \delta g$. By (e) of (2), the term in square brackets is $\le 2\delta(f_1 + f_2 : f_0) + \delta(1 + 2\delta)(g : f_0)$. Choosing δ so that this expression is less than ε completes the proof of (3).⟧

(4) *There exists a positive linear functional I on $C_c(G)$ such that $I(L_s f) = I(f)$ for all $s \in G$.*

⟦The aforementioned limiting process $I_\phi \to I$ is provided by Tychonoff's theorem, using part (e) of (2): For each $f \in C_c^+$, let J_f denote the interval $[(f_0 : f)^{-1}, (f : f_0)]$ and let $X := \prod_{f \in C_c^+} J_f$. Then X is compact in the product topology, that is, the topology with basic open neighborhoods

$$N_F := \{G \in X : |G(f_i) - F(f_i)| < \varepsilon,\ i = 1, \ldots, n\},\ \varepsilon > 0,\ n \in \mathbb{N},\ f_i \in C_c^+. \quad (\dagger)$$

Moreover, $I_\phi \in X$ for each $\phi \in C_c^+$. For each neighborhood V of e set

$$C_V := \operatorname{cl}_X\{I_\phi : \phi \in C_c^+,\ \operatorname{supp}(\phi) \subseteq V\}.$$

Then C_V is compact and has the finite intersection property, since $C_{V_1} \cap \cdots \cap C_{V_n} \supseteq C_{V_1 \cap \cdots \cap V_n}$. By compactness of X, $\bigcap_V C_V \ne \emptyset$. If I is a member of this intersection, then, from $(\dagger)$, for each V, $\varepsilon > 0$, and $f_i \in C_c^+$ there exists ϕ with $\operatorname{supp}(\phi) \subseteq V$ such that

$$|I_\phi(f_i) - I(f_i)| < \varepsilon, \quad i = 1, \ldots, n.$$

It follows from (3) that I is additive on C_c^+ and has properties (b) – (e) of (2). Extending I to $C_c(G)$ by defining $I(f) := I(f^+) - I(f^-)$ produces the desired functional.⟧ □

Essential Uniqueness of Haar Measure

Haar measure μ is not unique, since multiplying μ by a positive constant obviously produces another Haar measure. However, Haar measure is *essentially unique* in the sense that there is no other way of producing new Haar measures.

16.2.4 Theorem. *If μ and ν are left Haar measures on a locally compact group G, then $\mu = c\nu$ for some $c > 0$.*

Proof. (Loomis). Let $f_1, f_2 \in C_c^+$ and let U be a fixed compact, symmetric neighborhood of e. For $i = 1, 2$, the set $K_i := [\text{supp}(f_i)U] \cup [U\text{supp}(f_i)]$ is compact. Moreover, for each $y \in U$ the functions $x \mapsto f_i(xy)$ and $x \mapsto f_i(yx)$ have support contained in K_i. We show that

$$\frac{\int f_1\, d\mu}{\int f_1\, d\nu} = \frac{\int f_2\, d\mu}{\int f_2\, d\nu}. \tag{†}$$

Assuming (†) holds and setting the expression on the right equal to c, we then have

$$\int f_1\, d\mu = c \int f_1\, d\nu \quad \text{for all } f_1 \in C_c^+.$$

It follows from the uniqueness part of 7.2.1 that $\mu = c\nu$, proving the theorem.

To verify (†), for a given $\varepsilon > 0$ choose a compact symmetric neighborhood V of e contained in U such that

$$|f_i(xy) - f_i(yx)| < \varepsilon \text{ for all } y \in V \text{ and } x \in G, \quad i = 1, 2.$$

This is possible by the uniform continuity of f_i. Next, choose $g \in C_c^+$ such that $g(x) = g(x^{-1})$ and $\text{supp}(g) \subseteq V$. (For example, one could choose $h \in C_c^+$ such that $\mathbf{1}_{\{e\}} \le h \le \mathbf{1}_V$ and then take $g(x) = h(x) + h(x^{-1})$.) By left invariance of μ,

$$\int g\, d\nu \int f_i\, d\mu = \iint g(y) f_i(x)\, d\mu(x)\, d\nu(y) = \iint g(y) f_i(yx)\, d\mu(x)\, d\nu(y),$$

and by left invariance of μ and ν, the symmetry property of g, and Fubini's theorem for Radon measures (7.3.2),

$$\begin{aligned}\int g\, d\mu \int f_i\, d\nu &= \iint g(x) f_i(y)\, d\mu(x)\, d\nu(y) = \iint g(y^{-1}x) f_i(y)\, d\mu(x)\, d\nu(y)\\ &= \iint g(x^{-1}y) f_i(y)\, d\nu(y)\, d\mu(x) = \iint g(y) f_i(xy)\, d\nu(y)\, d\mu(x)\\ &= \iint g(y) f_i(xy)\, d\mu(x)\, d\nu(y).\end{aligned}$$

Thus

$$\begin{aligned}\left|\int g\, d\nu \int f_i\, d\mu - \int g\, d\mu \int f_i\, d\nu\right| &\le \int_V \int_G g(y) |f_i(yx) - f_i(xy)|\, d\mu(x)\, d\nu(y)\\ &\le \varepsilon \mu(K_i) \int g\, d\nu\end{aligned}$$

and so

$$\left|\frac{\int f_i\, d\mu}{\int f_i\, d\nu} - \frac{\int g\, d\mu}{\int g\, d\nu}\right| \le \varepsilon \frac{\mu(K_i)}{\int f_i\, d\mu}, \quad i = 1,\ 2.$$

Therefore,

$$\left|\frac{\int f_1\, d\mu}{\int f_1\, d\nu} - \frac{\int f_2\, d\mu}{\int f_2\, d\nu}\right| \le \varepsilon \left(\frac{\mu(K_1)}{\int f_1\, d\mu} + \frac{\mu(K_2)}{\int f_2\, d\mu}\right).$$

Letting $\varepsilon \to 0$ shows that the ratios on the left are equal. □

The Modular Function

The modular function makes an important connection between left and right Haar measures. Let μ be a left Haar measure. Then for each x the set function $B \mapsto \mu(Bx)$, $B \in \mathcal{B}(G)$, is again a left Haar measure, hence, by essential uniqueness, there exists a unique positive real number $\Delta(x)$ such that

$$\mu(Bx) = \Delta(x)\mu(B) \quad \text{for all } x \in G \text{ and all } B \in \mathcal{B}(G).$$

Since this obviously holds for μ replaced by $c\mu$, $c > 0$ and since all left Haar measures are of this form, we see that Δ is independent of the measure μ. The function Δ is called the **modular function of** G. It is an intrinsic feature of G.

Theorem 16.2.6 below gives the key properties of the modular function. For the proof we need the following lemma, a generalization of which is given later.

16.2.5 Lemma. *Let $f \in C_c(G)$ and $1 \le p < \infty$. Then the mapping $x \mapsto R_x f$ is continuous at e in the L^p norm.*

Proof. Let U be a compact, symmetric neighborhood of e and set $K := \operatorname{supp} f$, so that KU is compact and $\operatorname{supp}(R_x f) \subseteq KU$ for $x \in U$. By uniform continuity of f (16.1.3), given $\varepsilon > 0$ we may choose a neighborhood V of e contained in U such that $|f(yx) - f(y)| < \varepsilon(\mu(KU))^{-1/p}$ for all $x \in V$ and $y \in G$. For such x we then have

$$\|R_x f - f\|_p^p = \int_{KU} |f(yx) - f(y)|^p \, d\mu(y) \le \varepsilon^p. \qquad \square$$

16.2.6 Theorem. *The modular function Δ is a continuous homomorphism from G into the group of positive reals under multiplication. Moreover,*

$$\int R_x f \, d\mu = \Delta\left(x^{-1}\right) \int f \, d\mu, \quad f \in L^1(\mu). \tag{16.4}$$

Proof. The calculation

$$\Delta(xy)\mu(B) = \mu(Bxy) = \Delta(y)\mu(Bx) = \Delta(x)\Delta(y)\mu(B)$$

shows that Δ is a homomorphism. To prove (16.4), note that

$$\int R_x \mathbf{1}_B(y) \, d\mu(y) = \int \mathbf{1}_{Bx^{-1}}(y) \, d\mu(y) = \mu(Bx^{-1}) = \Delta(x^{-1}) \int \mathbf{1}_B(y) \, d\mu(y),$$

hence (16.4) holds for measurable indicator functions f. The usual arguments then show that the equation holds for all $f \in L^1(\mu)$.

Now take $f \in C_c(G)$ such that $\int f \, d\mu \neq 0$. By the lemma, the left side of (16.4) is continuous in x at e. It follows that Δ is continuous at e, and since Δ is a homomorphism, it is continuous on G. $\square$

It follows directly from the definition that a left Haar measure is right invariant iff $\Delta(x) \equiv 1$. In this case, G is said to be **unimodular**. Abelian groups are obviously unimodular. Here is another important class of unimodular groups.

16.2.7 Proposition. *Every compact group is unimodular.*

Proof. If G is compact, then $0 < \mu(G) < \infty$ (16.2.2). Since $Gx = G$, we have $\mu(G) = \mu(Gx) = \Delta(x)\mu(G)$, hence $\Delta(x) = 1$. $\square$

For a compact group G, the unique Haar measure μ for which $\mu(G) = 1$ is called **normalized Haar measure.** For a finite group $G = \{x_1, \ldots, x_n\}$ normalized Haar measure is given by

$$\mu(B) = \frac{1}{n}\sum_{j=1}^{n} \mathbf{1}_B(x_j), \quad B \subseteq G.$$

We conclude this section with a result that relates a left Haar measure μ to the right Haar measure $\widetilde{\mu}$ (see (16.3)).

16.2.8 Proposition. *Let μ be a left Haar measure on G. If one side of the following equation exists, then so does the other and the equality is then valid.*

$$\int f(y^{-1})\Delta(y^{-1})\, d\mu(y) = \int f(y)\, d\mu(y).$$

In particular, μ is inverse invariant, that is, $\widetilde{\mu} = \mu$, iff G is unimodular.

Proof. Replacing f by $\widetilde{f}$ shows that the assertion is equivalent to

$$\int f(y)\Delta(y^{-1})\, d\mu(y) = \int f(y^{-1})\, d\mu(y) \quad \left(= \int f(y)\, d\widetilde{\mu}(y)\right).$$

For $f \in C_c(G)$, let $I(f)$ denote the left side of this equation and $\widetilde{I}(f)$ the right side. For $x \in G$,

$$\begin{aligned} I(R_x f) &= \int f(yx)\Delta(y^{-1})\, d\mu(y) = \Delta(x)\int f(yx)\Delta\big((yx)^{-1}\big)\, d\mu(y) \\ &= \Delta(x)\Delta(x^{-1})\int f(y)\Delta(y^{-1})\, d\mu(y) \\ &= I(f), \end{aligned}$$

the third equality by (16.4). Therefore, I is a right Haar integral. Since $\widetilde{I}$ is also a right Haar integral, there exists $c > 0$ such that $I = c\widetilde{I}$, that is,

$$\int f(y)\Delta(y^{-1})\, d\mu(y) = c\int f(y^{-1})\, d\mu(y),$$

in the sense that if one side is finite then so is the other, in which case equality holds. In particular, if f is symmetric, then

$$(1-c)\int f\, d\mu = \int [1 - \Delta(y^{-1})] f(y)\, d\mu(y).$$

Let $\varepsilon > 0$ and let U be a compact symmetric neighborhood of e on which $|\Delta - 1| < \varepsilon$. Taking $f = \mathbf{1}_U$ in the last equation we have $|1 - c|\mu(U) \leq \varepsilon\mu(U)$. Since ε was arbitrary, $c = 1$, hence $\widetilde{I} = I$, completing the proof. $\square$

Note that the conclusion of the proposition may be written

$$\int f(y)\Delta(y)\, d\widetilde{\mu}(y) = \int f(y)\, d\mu(y),$$

which shows that $\Delta d\widetilde{\mu} = d\mu$.

16.3 Some Constructions

Haar Measure on Direct Products

The **direct product** of groups G and H is the group $G \times H$ with multiplication

$$(a, b)(x, y) = (ax, by), \quad a, x \in G, \quad b, y \in H.$$

If G and H are topological groups, then $G \times H$ is easily seen to be a topological group in the product topology.

Now let G and H be locally compact groups and let μ and ν be left Haar measures on G and H, respectively. As noted in 7.3.4, $\mu \otimes \nu$ on $\mathfrak{B}(G) \otimes \mathfrak{B}(H)$ extends to a Radon measure $\mu \overline{\otimes} \nu$ on $\mathfrak{B}(G \times H)$ such that for all $f \in C_c(G \times H)$

$$\int_{G\times H} f(x, y)\, d(\mu \overline{\otimes} \nu)(x, y) = \int_G \int_H f(x, y)\, d\nu(y)\, d\mu(x) = \int_H \int_G f(x, y)\, d\mu(x)\, d\nu(y).$$

From the left invariance of μ and ν we then have

$$\int_{G\times H} L_{(a,b)} f\, d(\mu \overline{\otimes} \nu) = \int_{G\times H} f\, d(\mu \overline{\otimes} \nu),$$

hence $\mu \overline{\otimes} \nu$ is left Haar measure on $G \times H$. Furthermore,

$$\begin{aligned}\int_{G\times H} R_{(a,b)} f\, d(\mu \overline{\otimes} \nu) &= \int_H \int_G f(xa, yb)\, d\mu(x)\, d\nu(y)\\ &= \Delta_G(a^{-1})\Delta_H(b^{-1}) \int_H \int_G f(x, y)\, d\mu(x)\, d\nu(y),\end{aligned}$$

hence

$$\Delta_{G\times H}(a, b) = \Delta_G(a)\Delta_H(b).$$

It follows that if $G \times H$ is unimodular iff both G and H are unimodular.

Haar Measure on Semidirect Products

Let G and H be groups and let $\sigma : H \times G \to H$ be a function such that for each $x \in G$ the map $\sigma_x := \sigma(\cdot, x)$ is a member of the automorphism group $\mathrm{Aut}(H)$ of H, that is, the group of isomorphisms of H onto H, and such that the mapping $x \mapsto \sigma_x : G \to \mathrm{Aut}(H)$ is a homomorphism into $\mathrm{Aut}(H)$. Thus we require that

$$\sigma(y_1 y_2, x) = \sigma(y_1, x)\sigma(y_2, x), \ \sigma(e, x) = e, \ \sigma(y, x_1 x_2) = \sigma\big(\sigma(y, x_1), x_2\big), \ \sigma(\cdot, e) = \mathrm{id}_H.$$

Define multiplication on $G \times H$ by

$$(a, b)(x, y) = \big(ax, \sigma(b, x)y\big) = \big(ax, \sigma_x(b)y\big), \quad a,\ x \in G, \quad b,\ y \in H.$$

It is straightforward to check that under this multiplication $G \times H$ is a group with identity (e, e) and inverse

$$(x, y)^{-1} = \big(x^{-1}, \sigma(y^{-1}, x^{-1})\big).$$

The group $G \times H$ is called a **semidirect product** of G and H and is denoted by $G \circledS H$. The semidirect product reduces to the direct product by taking $\sigma_x = \mathrm{id}_H$ for all x. If G and

H are topological groups we require additionally that σ be continuous, in which case $G \textcircled{\sigma} H$ is easily seen to be a topological group in the product topology.

Now let G and H be locally compact groups and let μ and ν be left Haar measures on G and H, respectively. For $f \in C_c(G \times H)$,

$$\begin{aligned}\int_{G\times H} L_{(a,b)} f(x,y)\, d(\mu\overline{\otimes}\nu)(x,y) &= \int_G \int_H f\big(ax, \sigma_x(b)y\big)\, d\nu(y)\, d\mu(x) \\ &= \int_G \int_H f(x,y)\, d\nu(y)\, d\mu(x),\end{aligned}$$

hence $\mu\overline{\otimes}\nu$ is a left Haar measure on $G \textcircled{\sigma} H$. To find the modular function, let $\sigma_a(\nu)$ denote the image measure on $\mathcal{B}(H)$:

$$\sigma_a(\nu)(B) = \nu\big(\sigma_a^{-1}(B)\big) = \nu\big(\sigma_{a^{-1}}(B)\big), \quad B \in \mathcal{B}(H).$$

Then, by left invariance of ν,

$$\sigma_a(\nu)(yB) = \nu\big(\sigma_{a^{-1}}(y)\sigma_{a^{-1}}(B)\big) = \sigma_a(\nu)(B),$$

so $\sigma_a(\nu)$ is a left Haar measure. By essential uniqueness, $\sigma_a(\nu) = \delta(a)\nu$ for some $\delta(a) > 0$. From

$$\delta(ax)\nu = \sigma_{ax}(\nu) = (\sigma_a \circ \sigma_x)(\nu) = \sigma_a\big(\delta(x)(\nu)\big) = \delta(a)\delta(x)\nu$$

we see that $\delta : G \to (0, \infty)$ is a homomorphism. Moreover, from

$$\delta(a) \int_H f\, d\nu = \int_H f\, d\sigma_a(\nu) = \int_H f\big(\sigma_a(y)\big)\, d\nu(y), \quad f \in C_c(H),$$

and the continuity of σ it follows that $\delta(a)$ is continuous. We now have

$$\begin{aligned}&\int_{G\times H} R_{(a,b)} f(x,y)\, \mu\overline{\otimes}\nu(x,y) = \int_H \int_G f\big(xa, \sigma_a(y)b\big)\, d\mu(x)\, d\nu(y) \\ &= \Delta_G(a^{-1})\Delta_H(b^{-1})\delta(a) \int_H \int_G f(x,y)\, d\mu(x)\, d\nu(y),\end{aligned}$$

hence

$$\Delta_{G \textcircled{\sigma} H}(a,b) = \Delta_G(a)\Delta_H(b)\delta(a).$$

It follows that if $G \textcircled{\sigma} H$ is unimodular iff both G and H are unimodular and $\delta \equiv 1$.

16.3.1 Example. Let G be the group of nonzero real numbers under multiplication, let $H = (\mathbb{R}, +)$, and take $\sigma(y,x) = \sigma_x(y) = xy$, $x \in G$, $y \in H$. Then multiplication in $G \textcircled{\sigma} H$ has the form

$$(a,b)(x,y) = \big(ax, \sigma_x(b) + y\big) = (ax, bx + y).$$

Since

$$\begin{bmatrix} x & y \\ 0 & 1 \end{bmatrix} \begin{bmatrix} a & b \\ 0 & 1 \end{bmatrix} = \begin{bmatrix} ax & bx + y \\ 0 & 1 \end{bmatrix},$$

$G \textcircled{\sigma} H$ may be realized concretely as a matrix group. ◊

Haar Measure on Quotient Groups

Let G be a group and H a subgroup of G. Recall that $x \sim y$ iff $x^{-1}y \in H$ defines an equivalence relation on G with equivalence classes xH (see §0.2). Moreover, if G is a topological group, then the quotient map $Q : G \to G/H$ is continuous. The following theorem summarizes the basic properties of quotients of topological groups.

16.3.2 Theorem. *Let G be a topological group and H a subgroup of G.*

(a) *The quotient topology is the unique topology on G/H relative to which Q is open and continuous.*

(b) *If H is closed, then G/H is Hausdorff.*

(c) *If H is normal in G, then G/H is a topological group.*

(d) *If G is locally compact, then so is G/H.*

Proof. (a) If U is open in G, then $Q^{-1}(Q(U)) = UH = \bigcup_{y \in H} Uy$, which is open in G. Uniqueness follows easily from the fact that Q is open and continuous.

(b) If $Q(x) \neq Q(y)$, then $x^{-1}y$ is in the open set H^c. Since $(s,t) \to sx^{-1}yt$ is continuous, there exists a symmetric neighborhood U of e such that $Ux^{-1}yU \subseteq H^c$. It follows that $Q(xU)$ and $Q(yU)$ are disjoint neighborhoods of $Q(x)$ and $Q(y)$.

(c) By 16.1.1, it suffices to show that the map $(Q(x), Q(y)) \mapsto Q(x)^{-1}Q(y)$ is continuous at $(Q(e), Q(e))$, and that for each $a \in G$ the translation mappings $Q(x) \mapsto Q(ax)$ and $Q(x) \mapsto Q(xa)$ are continuous. We establish the former and leave the latter as an exercise. An arbitrary open neighborhood of $Q(e)$ is of the form $Q(U)$, where U is an open neighborhood of e. Choose an open neighborhood V of e such that $v, w \in V \Rightarrow v^{-1}w \in U$ and let $Q(x)$, $Q(y) \in Q(V)$, say $Q(x) = Q(v)$ and $Q(y) = Q(w)$, $(v, w \in V)$. Then $x \in Hv$ and $y \in wH$, hence $x^{-1}y \in v^{-1}HwH = v^{-1}wH$, where the last equality uses the normality of H. Therefore, $Q(x)^{-1}Q(y) = Q(v^{-1}w) \in Q(U)$.

(d) This follows from the continuity and openness of Q. □

16.3.3 Theorem. *Let G be a locally compact group, H a closed normal subgroup of G, ν a left Haar measure on H, and η a left Haar measure on G/H. Then there exists a Haar measure μ on G such that*

$$\int f(x)\, d\mu(x) = \int_{G/H} \int_H f(xy)\, d\nu(y)\, d\eta(xH), \quad f \in C_c(G).$$

Moreover, $\Delta_G = \Delta_H$ on H.

Proof. We show first that the right side of the equation, which we denote by $I(f)$, is well-defined. Let $F(x)$ denote the inner integral:

$$F(x) = \int_H f(xy)\, d\nu(y) = \int_{(x^{-1}K)\cap H} f(xy)\, d\nu(y), \quad x \in G,\ K := \operatorname{supp}(f). \qquad (\dagger)$$

If $aH = bH$, then $b^{-1}a \in H$ and so by left invariance of ν,

$$F(b) = \int_H f(by)\, d\nu(y) = \int_H f(b(b^{-1}ay))\, d\nu(y) = F(a).$$

It follows that $F = f' \circ Q$ for a unique function $f' \in C_b(G/H)$ depending only on f. Now, $y \in (x^{-1}K) \cap H \Rightarrow Q(x) = Q(xy) \in Q(K)$, so $(x^{-1}K) \cap H = \emptyset$ for all x for which

$Q(x) \in Q(K)^c$. It follows from (†) that $F(x) = 0$ for such x and so $\mathrm{supp}(f') \subseteq Q(K)$. Therefore, $f' \in C_c(G/H)$ and

$$I(f) = \int_{G/H} F(x)\, d\eta(xH) = \int_{G/H} f'(xH)\, d\eta(xH), \tag{‡}$$

which shows that I is well-defined.

Now, I is clearly a positive linear functional on $f \in C_c(G)$. We show that I is left invariant. Let $a \in G$. In the notation of the first paragraph,

$$(L_aF)(x) = F(ax) = \int_H f(axy)\, d\nu(y) = \int_H (L_af)(xy)\, d\nu(y).$$

Comparing with (†) we see that $(L_af)' \circ Q = L_aF$, that is,

$$(L_af)'(Q(x)) = F(ax) = f'(Q(ax)) = f'(Q(a)Q(x)).$$

Thus by (‡) and the left invariance of η,

$$I(L_af) = \int_{G/H} (L_af)'(Q(x))\, d\eta(xH) = \int_{G/H} f'(aHxH)\, d\eta(xH) = I(f),$$

verifying the claim.

The first part of the theorem now follows from the Riesz representation theorem. For the second part, if $b \in H$ and $f \in C_c(G)$, then

$$\Delta_G(b^{-1})I(f) = I(R_bf) = \int_{G/H}\int_H f(xyb)\, d\nu(y)\, d\eta(xH) = \Delta_H(b^{-1})I(f). \qquad \square$$

16.4 The L^1-Group Algebra

For this and the remaining sections of the chapter, G is a locally compact group with left Haar measure, written as dx, and modular function Δ. Integrals with respect to dx will frequently be denoted by $\int f$. The left Haar measure of a Borel set E is denoted by $|E|$.

Convolution and Involution

The **convolution** $f * g$ and **involution** f^* of functions $f, g : G \to \mathbb{C}$ are defined by

$$f * g(x) = \int f(y)g(y^{-1}x)\, dy \quad \text{and} \quad f^*(x) = \Delta(x^{-1})\overline{f(x^{-1})}, \qquad x \in G.$$

If $A, B \subseteq L^1(G)$, we set $A * B := \{f * g : f \in A,\ g \in B\}$. The next two propositions summarize the basic properties of these operations.

16.4.1 Proposition. *Let f, g, $h \in L^1(G)$ and $c \in \mathbb{C}$.*

(a) $f * g \in L^1(G)$ *and* $\|f * g\|_1 \le \|f\|_1\|g\|_1$.

(b) *If G is abelian, then* $f * g = g * f$.

(c) $(f * g) * h = f * (g * h)$.

(d) $f * (cg) = (cf) * g = c(f * g)$.

(e) $(f * (g + h) = f * g + f * h$.

(f) $C_c(G) * C_c(G) \subseteq C_c(G)$.

Proof. The proofs of (a)–(e) are entirely similar to the corresponding parts of 6.1.1, except that care must be taken to allow for the fact that the group is not necessarily abelian and dx is not necessarily right invariant. For example, to prove (c) use left invariance and Fubini's theorem[2] to obtain

$$\begin{aligned} f*(g*h)(x) &= \int f(z)(g*h)(z^{-1}x)\,dz = \iint f(z)g(y)h(y^{-1}z^{-1}x)\,dy\,dz \\ &= \iint f(z)g(z^{-1}y)h(y^{-1}x)\,dy\,dz = \int (f*g)(y)h(y^{-1}x)\,dy \\ &= (f*g)*h(x). \end{aligned}$$

To prove (f), let $\varphi_i \in C_c(G)$ and set $K_i = \operatorname{supp}\varphi_i$. From $\varphi_1 * \varphi_2(x) = \int_{K_1} \varphi_1(y)\varphi_2(y^{-1}x)\,dy$ we see that if $y \in K_1$ and $x \notin yK_2$, then the integrand is zero. Therefore, $\operatorname{supp}\varphi_1 * \varphi_2 \subseteq K_1K_2$, which is compact. □

16.4.2 Proposition. *Let f, g, $h \in L^1(G)$ and $c \in \mathbb{C}$.*

(a) $(f+g)^* = f^* + g^*$. **(b)** $(cf)^* = \bar{c}f^*$.
(c) $f^{**} = f$. **(d)** $(f*g)^* = g^* * f^*$.
(e) $\|f^*\|_1 = \|f\|_1$. **(f)** $f, g \in L^1 \cap L^2 \Rightarrow (f^* * g)(x) = (R_x g \,|\, f)$.

Proof. Parts (a) – (c) are clear. For (d), we have

$$\begin{aligned} (f^* * g^*)(x) &= \int f^*(y)g^*(y^{-1}x)\,dy = \int \Delta(y^{-1})\Delta((y^{-1}x)^{-1})\overline{f(y^{-1})g(x^{-1}y)}\,dy \\ &= \Delta(x^{-1}) \int \overline{f(y^{-1})g(x^{-1}y)}\,dy = \Delta(x^{-1}) \int \overline{g(y)f(y^{-1}x^{-1})}\,dy \\ &= (g*f)^*(x). \end{aligned}$$

Parts (e) and (f) follow from 16.2.8:

$$\|f^*\|_1 = \int \Delta(x^{-1})|f(x^{-1})|\,dx = \int |f(x)|\,dx = \|f\|_1$$

and

$$(f^* * g)(x) = \int \Delta(y^{-1})\overline{f(y^{-1})}g(y^{-1}x)\,dy = \int \overline{f(y)}g(yx)\,dy = (R_x g \,|\, f). \qquad \square$$

From the preceding propositions we have

16.4.3 Theorem. *$L^1(G)$ is a Banach $*$-algebra with respect to convolution and involution, and is commutative if G is abelian.*

The following result complements 16.4.1(f):

16.4.4 Proposition. *Let G be unimodular and let p and q be conjugate exponents with $1 < p$, $q < \infty$. If $f \in L^p(G)$ and $g \in L^q(G)$, then $f * g \in C_0(G)$ and $\|f*g\|_\infty \le \|f\|_p \|g\|_q$.*

[2] Fubini's theorem requires the spaces to be σ-finite, which is not necessarily the case in the present setting. However, the theorem is valid for functions $f \in L^p$, $1 \le p < \infty$, since it may be shown that such functions are zero outside a σ-compact set. We may therefore invoke Fubini's theorem and shall do so without further comment. For the technical details, the reader is referred to [21] or [34].

Proof. By Hölder's inequality and translation and inversion invariance,

$$|f * g(x)| \le \int |f(y)g(y^{-1}x)|\,dy \le \left(\int |f(y)|^p\,dy\right)^{1/p}\left(\int |g(y^{-1}x)|^q\,dy\right)^{1/q} = \|f\|_p\,\|g\|_q,$$

establishing the inequality.

Now let $f_n, g_n \in C_c(G)$ such that $\|f_n - f\|_p \to 0$ and $\|g_n - g\|_q \to 0$ (7.1.2). By the inequality just established, $\|f_n * g_n - f * g\|_\infty \to 0$. But by 16.4.1(f), $f_n * g_n \in C_c(G)$. Therefore, $f * g \in C_0(G)$. □

Approximate Identities

In this subsection we generalize to arbitrary locally compact groups the existence of an approximate identity, established for the group $\mathbb{R}^d$ in 6.1.2. [3] The proof uses the following lemma, which expresses an important continuity property of left and right translations, extending 16.2.5.

16.4.5 Lemma. *Let $f \in L^p(G)$, $1 \le p < \infty$. Then the mappings $x \to L_x f$ and $x \to R_x f$ are continuous in the L^p norm.*

Proof. We prove the right translation version. Let $\varepsilon > 0$, $g \in C_c(G)$, U a compact neighborhood of U of e, and $x \in U$. Then

$$\begin{aligned}\|R_x f - f\|_p &\le \|R_x(f-g)\|_p + \|R_x g - g\|_p + \|f - g\|_p\\ &\le \sup_{x\in U}\Delta(x^{-1})^{1/p}\,\|f-g\|_p + \|f-g\|_p + \|R_x g - g\|_p.\end{aligned}$$

Since $C_c(G)$ is dense in L^p we may choose $g \in C_c(G)$ so that the sum of first two terms in the last expression is $< \varepsilon/2$. By 16.2.5, there exists a neighborhood V of e contained in U such that the third term is $< \varepsilon/2$ for $x \in V$. For such x, $\|R_x f - f\|_p < \varepsilon$, which shows that $x \mapsto R_x f$ is L^p continuous at e. Continuity at arbitrary x_0 follows from

$$\|R_x f - R_{x_0} f\| = \Delta(x_0^{-1})\|R_{x_0^{-1}x} f - f\|.$$ □

16.4.6 Theorem. *Let $f \in L^p$, $1 \le p < \infty$, and $\varepsilon > 0$. Then there exists a neighborhood V of the identity such that $\|f * \psi - f\|_p < \varepsilon$ and $\|\psi * f - f\|_p < \varepsilon$ for all symmetric $\psi \in C_c^+(G)$ with $\operatorname{supp}(\psi) \subseteq V$ and $\int \psi = 1$. Moreover, if $p = \infty$, then the first inequality holds if f is right uniformly continuous, and the second holds if f is left uniformly continuous.*

Proof. We prove only the part concerning $f * \psi$. Given $\varepsilon > 0$, by the preceding lemma we may choose a neighborhood V of e such that $\|R_y f - f\|_p < \varepsilon$ for all $y \in V$. If f is right uniformly continuous, then we may choose V so that $\|R_y f - f\|_\infty < \varepsilon$. Now let $\psi \in C_c^+(G)$ be symmetric with $\operatorname{supp}(\psi) \subseteq V$ and $\int \psi = 1$. Then, by left invariance and symmetry of ψ,

$$f * \psi(x) - f(x) = \int f(y)\psi(y^{-1}x)\,dy - f(x)\int \psi(y)\,dy = \int [f(xy) - f(x)]\psi(y)\,dy.$$

By Minkowski's integral inequality (4.1.5),

$$\begin{aligned}\|f * \psi - f\|_p &\le \left[\int\left|\int |f(xy) - f(x)|\psi(y)\,dy\right|^p dx\right]^{1/p} \le \int\left[\int |f(xy) - f(x)|^p\,dx\right]^{1/p}\psi(y)\,dy\\ &= \int_V \|R_y f - f\|_p\,\psi(y)\,dy < \varepsilon,\end{aligned}$$

[3]For a discrete group G, $L^1(G)$ actually has an identity, namely the indicator function $\mathbf{1}_{\{e\}}$.

verifying the desired inequality for $p < \infty$. If f is right uniformly continuous, then

$$\|f * \psi - f\|_\infty \le \int \|R_y f - f\|_\infty \psi(y)\, dy < \varepsilon. \qquad \square$$

Theorem 16.4.6 is typically used as follows: Since the set of all neighborhoods V of the identity is directed downward by inclusion, we may form a net $(\psi_V)_V$, where ψ_V has the properties in the theorem. We then have

$$\lim_V f * \psi_V = f \text{ in } L^p,\ 1 \le p \le \infty.$$

(For $p = \infty$, f must be uniformly continuous.) The net (ψ_V) is called an **approximate identity** for $L^p(G)$. More generally, this term is applied to any net (ψ_α) in $C_c^+(G)$ for which $f * \psi_\alpha \to f$ in $L^p(G)$ and $\int \psi_\alpha = 1$. Here is an application:

16.4.7 Theorem. *A closed subspace $\mathscr{I}$ of $L^1(G)$ is a right (left) ideal iff it is closed under right (left) translations.*

Proof. We prove the right version. Suppose $\mathscr{I}$ is a right ideal. For $f \in \mathscr{I}$ and (ψ_α) an approximate identity, we have

$$R_x(f * \psi_\alpha)(y) = \int f(z)\psi_\alpha(z^{-1}yx)\, dz = \int f(z)(R_x\psi_\alpha)(z^{-1}y)\, dz = (f * R_x\psi_\alpha)(y).$$

Since R_x is continuous on L^1, $R_x f = \lim_\alpha R_x(f * \psi_\alpha) = \lim_\alpha(f * R_x\psi_\alpha) \in \mathscr{I}$, proving that $\mathscr{I}$ is closed under right translation.

Conversely, assume that $\mathscr{I}$ is right translation invariant. For $f \in \mathscr{I}$ and $g \in C_c(G)$, we have

$$f * g(x) = \int f(y)g(y^{-1}x)\, dy = \int f(xy)g(y^{-1})\, dy = \int g(y^{-1})(R_y f)(x)\, dy.$$

The function $y \mapsto g(y^{-1})R_y f$ is L^1-continuous and has compact support, hence may be weakly integrated as in 14.6.3. The result is a unique function $F = \int g(y^{-1})R_y f\, dy$ in the closed linear span of the right translations of f, and therefore in $\mathscr{I}$, such that

$$\int h(x)F(x)\, dx = \iint h(x)g(y^{-1})R_y f(x)\, dy\, dx = \int h(x)(f * g)(x)\, dx, \quad h \in L^\infty(G).$$

It follows that $f * g = F \in \mathscr{I}$. $\square$

The Measure Algebra

Let μ and ν be complex Radon measures on $\mathcal{B}(G)$. For $\phi \in C_c(G)$ define

$$I(\phi) := \iint \phi(xy)\, d\mu(x)\, d\nu(y).$$

Then I is a continuous linear functional on $C_c(G)$ and so is given by a complex Radon measure $\mu * \nu$ on $\mathcal{B}(G)$. Thus

$$\int \phi\, d(\mu * \nu) = \iint \phi(xy)\, d\mu(x)\, d\nu(y), \quad \phi \in C_c(G).$$

The measure $\mu * \nu$ is called the **convolution of** μ **and** ν. By 7.3.2 and 7.3.3, $\mu * \nu$ may also be seen as the image measure $m(\mu \overline{\otimes} \nu)$ of $\mu \overline{\otimes} \nu$ under the multiplication mapping $m(x, y) = xy$. Therefore, we have

$$\int h(z)\, d(\mu * \nu)(z) = \int h(xy)\, d(\mu \overline{\otimes} \nu)(x, y)$$

in the usual sense that whenever one side exists then so does the other and equality holds.

It is easy to check that the collection $M_{ra}(G)$ of Radon measures on G is a Banach algebra under the operation of convolution. The proof is the same as for the special case $M_{ra}(\mathbb{R}^d)$ (see 6.4.1). Moreover, $M_{ra}(G)$ is a $*$-algebra under involution $\mu \to \mu^*$ defined by

$$\mu^*(E) = \overline{\mu(E^{-1})}, \quad E \in \mathcal{B}(G)$$

or, equivalently,

$$\int \phi(x)\, d\mu^*(x) = \int \phi(x^{-1})\, d\overline{\mu}(x) = \left(\int \overline{\phi}(x^{-1})\, d\mu(x)\right)^{-}, \quad \phi \in C_c(G).$$

For example, the requirement $(\mu * \nu)^* = \nu^* * \mu^*$ follows from the calculations

$$\int \phi\, d(\mu * \nu)^* = \left(\int \overline{\phi}(x^{-1})\, d(\mu * \nu)(x)\right)^{-} = \left(\iint \overline{\phi}((xy)^{-1})\, d\mu(x)\, d\nu(y)\right)^{-} \quad \text{and}$$

$$\begin{aligned}\int \phi\, d(\nu^* * \mu^*) &= \iint \phi(yx)\, d\nu^*(y)\, d\mu^*(x) = \int \left(\int \overline{\phi}(y^{-1}x)\, d\nu\right)^{-} d\mu^*(x)\\ &= \left(\iint \overline{\phi}(y^{-1}x^{-1})\, d\nu(y)\, d\nu(x)\right)^{-}.\end{aligned}$$

Finally, the Dirac measure δ_e is an identity for $M_{ra}(G)$ as is seen, for example, from

$$\int \phi\, d(\mu * \delta_e) = \iint \phi(xy)\, d\mu(x) d\delta_e(y) = \int \phi(xe)\, d\mu(x) = \int \phi(x)\, d\mu(x).$$

16.5 Representations

Positive-Definite Functions

A function $\phi : G \to \mathbb{C}$ is said to be **positive definite** if

$$\sum_{j,k=1}^{n} c_j \overline{c}_k \phi(x_k^{-1} x_j) \geq 0 \quad \text{for all } c_j \in \mathbb{C},\ x_j \in G, \text{ and } n \in \mathbb{N}. \tag{16.5}$$

Using the Euclidean inner product, we may write this condition as $(A\boldsymbol{c} \mid \boldsymbol{c}) \geq 0$, where $\boldsymbol{c} = (c_1, \ldots, c_n)$ and $A = [a_{jk}]_{n \times n}$, $a_{jk} := \phi(x_k^{-1} x_j)$. Thus ϕ is a positive definite function iff A is a positive definite matrix.

16.5.1 Proposition. *Let ϕ be positive definite and $x, y \in G$. Then*

(a) $\phi(x^{-1}) = \overline{\phi(x)}$.

(b) $|\phi(x)| \leq \phi(e)$, *hence* $\|\phi\|_\infty = \phi(e)$.

(c) $|\phi(x) - \phi(y)| \leq 2\phi(e) \operatorname{Re}[\phi(e) - \phi(xy^{-1})]$.

Proof. For the parameters $n = 1$, $x_1 = e$, and $c_1 = 1$, we have $\phi(e) \geq 0$. Now take $n = 2$, $x_1 = x$, $x_2 = e$. Then for $|c| = 1$,

$$0 \leq (A(1,c) \mid (1,c)) = \begin{bmatrix} 1 & \overline{c} \end{bmatrix} \begin{bmatrix} \phi(e) & \phi(x) \\ \phi(x^{-1}) & \phi(e) \end{bmatrix} \begin{bmatrix} 1 \\ c \end{bmatrix} = 2\phi(e) + c\phi(x) + \overline{c}\phi(x^{-1}). \tag{$\dagger$}$$

Therefore, $c\phi(x) + \overline{c}\phi(x^{-1})$ is real. Taking $c = 1$ and $c = i$ shows that $\phi(x) + \phi(x^{-1})$ and $i[\phi(x) - \phi(x^{-1})]$ are real, which implies (a). Choosing c in (†) so that $c\phi(x) = -|\phi(x)|$ and using (a), we have $0 \le 2\phi(e) - |\phi(x)| + \overline{c}\phi(x^{-1}) = 2\phi(e) - 2|\phi(x)|$, proving (b).

For (c), take $n = 3$, $x_1 = e$, $x_2 = x$, $x_3 = y$. For $|c| = 1$ and t real,

$$0 \le (A(1, tc, -tc) \,|\, (1, tc, -tc)) = \begin{bmatrix} 1 & t\overline{c} & -t\overline{c} \end{bmatrix} \begin{bmatrix} \phi(e) & \phi(x^{-1}) & \phi(y^{-1}) \\ \phi(x) & \phi(e) & \phi(y^{-1}x) \\ \phi(y) & \phi(x^{-1}y) & \phi(e) \end{bmatrix} \begin{bmatrix} 1 \\ tc \\ -tc \end{bmatrix}$$

$$= \phi(e)(1 + 2t^2) + \overline{c}t\big[\phi(x) - \phi(y)\big] + ct\big[\phi(x^{-1}) - \phi(y^{-1})\big] - t^2\big[\phi(y^{-1}x) + \phi(x^{-1}y)\big]$$

$$= \phi(e)(1 + 2t^2) + 2t\mathrm{Re}\big\{\overline{c}\big[\phi(x) - \phi(y)\big]\big\} - 2t^2 \mathrm{Re}\, \phi(y^{-1}x),$$

the last equality by (a). Taking $\overline{c} = |\phi(x) - \phi(y)|[\phi(x) - \phi(y)]^{-1}$ we have for all real t

$$0 \le 2[\phi(e) - \mathrm{Re}\, \phi(y^{-1}x)]t^2 + 2\,\big|\phi(x) - \phi(y)\big|t + \phi(e) =: at^2 + bt + c.$$

The discriminant $b^2 - 4ac$ is therefore ≤ 0, implying (c). □

Functions of Positive Type

A function $\phi \in L^\infty(G)$ is said to be of **positive type** if

$$\int (f^* * f)\phi \ge 0 \quad \text{for all} \quad f \in L^1(G).$$

Since $C_c(G)$ is dense in $L^1(G)$, to test for this property it suffices to take $f \in C_c(G)$. Indeed, if $f_n \in C_c(G)$ satisfies the preceding inequality for all n and if $\|f_n - f\|_1 \to 0$, then $0 \le \int (f_n^* * f_n)\phi \to \int (f^* * f)\phi$ by L^1 continuity of convolution.

For future reference we note that

$$\int (g^* * f)\phi = \iint \Delta(y^{-1})\overline{g(y^{-1})} f(y^{-1}x)\phi(x)\, dy\, dx = \iint \overline{g(y)} f(x)\phi(y^{-1}x)\, dx\, dy, \quad (16.6)$$

where we have used 16.2.8 and the left invariance of dx. Taking $g = f$ and considering the conjugate of the last integral, we see that ϕ is of positive type iff $\overline{\phi}$ is of positive type.

We denote the set of all continuous functions of positive-type by $\mathscr{P}(G)$:

$$\mathscr{P}(G) := \left\{\phi \in C_b(G) : \int (f^* * f)\phi \ge 0 \text{ for all } f \in L^1(G)\right\}.$$

Note that $\mathscr{P}(G)$ is convex.

The following proposition shows that for bounded continuous functions, the notions of positive-definite function and function of positive-type coincide.

16.5.2 Proposition. *Let $\phi : G \to \mathbb{C}$ be bounded and continuous. Then ϕ is of positive type iff ϕ is positive definite.*

Proof. Let ϕ be positive definite and let $f \in C_c(G)$, $K := \mathrm{supp}(f)$. Then the function $g(x, y) := f(x)\overline{f(y)}\phi(y^{-1}x)$, which is the integrand in (16.6), is continuous and has support contained in $K \times K$ and so is uniformly continuous. Given $\varepsilon > 0$, choose a neighborhood V of e such that

$$|g(x, y) - g(a, b)| < \varepsilon \quad \text{whenever } x \in aV \text{ and } y \in bV.$$

By compactness, K is a finite union of sets of the form $aV \cap K$ $(a \in K)$, hence there exists a measurable partition $E_1, \ldots, E_n$ of K and points $x_j \in E_j$ such that $|g(x,y) - g(x_j, x_k)| < \varepsilon$ for all $(x,y) \in E_j \times E_k$. We then have

$$\begin{aligned} I &:= \int (f^* * f)\phi = \sum_{j,k} \iint \mathbf{1}_{E_j \times E_k}(x,y) g(x,y)\,dx\,dy, \\ S_\varepsilon &:= \sum_{j,k} |E_j|\,|E_k|\,g(x_j,x_k) = \sum_{j,k} |E_j|\,f(x_j)\overline{|E_k|\,f(x_k)}\phi(x_k^{-1}x_j) \geq 0, \text{ and} \\ I - S_\varepsilon &= \sum_{j,k} \iint \mathbf{1}_{E_j \times E_k}(x,y)\big[g(x,y) - g(x_j,x_k)\big]\,dx\,dy. \end{aligned}$$

Since

$$|I - S_\varepsilon| \leq \sum_{j,k} \iint \mathbf{1}_{E_j \times E_k}(x,y)|g(x,y) - g(x_j,x_k)|\,dx\,dy \leq \varepsilon |K|^2,$$

$I = I - S_\varepsilon + S_\varepsilon \geq I - S_\varepsilon \to 0$, hence $I \geq 0$. Therefore, ϕ is of positive type.

Conversely, let ϕ be of positive type and let $c_j \in \mathbb{C}$, $x_j \in G$. For an approximate identity (ψ_U), set $f_U(x) := \sum_{j=1}^n c_j \psi_U(x_j^{-1}x)$ and $I_U := \int (f_U^* * f_U)\phi$. By (16.6) and left invariance,

$$I_U = \int_U \int_U \overline{f_U(y)} f_U(x) \phi(y^{-1}x)\,dx\,dy = \sum_{j,k} c_j \bar{c}_k \int_U \int_U \psi_U(x)\psi_U(y)\phi(y^{-1}x_k^{-1}x_j x)\,dx\,dy.$$

Since $\int \psi_U = 1$ we see that

$$I_U - \sum_{j,k} c_j \bar{c}_k \phi(x_k^{-1}x_j) = \sum_{j,k} c_j \bar{c}_k \int_U \int_U \psi_U(x)\psi_U(y)\big[\phi(y^{-1}x_k^{-1}x_j x) - \phi(x_k^{-1}x_j)\big]\,dx\,dy.$$

Given $\varepsilon > 0$, choose U_0 so that $|\phi(y^{-1}x_k^{-1}x_j x) - \phi(x_k^{-1}x_j)| < \varepsilon$ for all $x, y \in U_0$ and all j, k. Then for any $U \subseteq U_0$,

$$\left| I_U - \sum_{j,k} c_j \bar{c}_k \phi(x_k^{-1}x_j) \right| \leq \varepsilon \sum_{j,k} |c_j \bar{c}_k|,$$

which shows that $\lim_U I_U = \sum_{j,k} c_j \bar{c}_k \phi(x_k^{-1}x_j)$. Since $I_U \geq 0$, the limit is nonnegative. □

Unitary Representations

Let $\mathscr{X}$ be a normed space. The **strong operator topology** of $\mathscr{B}(\mathscr{X})$ is the locally convex topology defined by the seminorms

$$p(T) = \max\{\|T\boldsymbol{x}_j\| : \boldsymbol{x}_j \in \mathscr{X},\ 1 \leq j \leq n\}.$$

The **weak operator topology** of $\mathscr{B}(\mathscr{X})$ is the locally convex topology defined by the seminorms

$$p(T) = \max\{|\langle T\boldsymbol{x}_j, \boldsymbol{x}'_j\rangle| : \boldsymbol{x}_j \in \mathscr{X}, \boldsymbol{x}'_j \in \mathscr{X}',\ 1 \leq j \leq n\}.$$

Thus a net (T_α) in $\mathscr{B}(\mathscr{X})$ converges to T in the strong operator topology (resp., weak operator topology) iff $T_\alpha \boldsymbol{x} \stackrel{s}{\to} T\boldsymbol{x}$ (resp., $T_\alpha \boldsymbol{x} \stackrel{w}{\to} T\boldsymbol{x}$) for each $\boldsymbol{x} \in \mathscr{X}$.

A **representation of** G on $\mathscr{X}$ is a mapping π from G into $\mathscr{B}(\mathscr{X})$ such that

$$\pi(xy) = \pi(x)\pi(y), \quad x,\, y \in G.$$

If $\mathscr{X} = \mathscr{H}$ is a Hilbert space and each $\pi(x)$ is unitary, then π is called a **unitary representation of** G. In this case we shall require that π be continuous in the strong operator topology. Thus a unitary representation $\pi : G \to \mathscr{B}(\mathscr{H})$ satisfies

$$\pi(xy) = \pi(x)\pi(y),\ \pi(x^{-1}) = \pi(x)^{-1} = \pi(x)^*, \text{ and } x \mapsto \pi(x)\boldsymbol{x} \text{ is continuous } \forall\ \boldsymbol{x} \in \mathscr{H}.$$

It is notable that continuity of π in the strong operator topology is equivalent to continuity in the weak operator topology. Indeed, if (U_α) is a net of unitary operators converging in the weak operator topology to a unitary operator U, then

$$\begin{aligned}\|U_\alpha\boldsymbol{x} - U\boldsymbol{x}\|^2 &= \|U_\alpha\boldsymbol{x}\|^2 - 2\mathrm{Re}\,(U_\alpha\boldsymbol{x}\,|\,U\boldsymbol{x}) + \|U\boldsymbol{x}\|^2 = 2\,\|\boldsymbol{x}\|^2 - 2\mathrm{Re}\,(U_\alpha\boldsymbol{x}\,|\,U\boldsymbol{x})\\ &\to 2\,\|\boldsymbol{x}\|^2 - 2\,(U\boldsymbol{x}\,|\,U\boldsymbol{x}) = 0.\end{aligned}$$

An important example of a unitary representation of G is the **left regular representation** π_L on $L^2(G)$ defined by

$$\pi_L(x)f = L_{x^{-1}}f.$$

Indeed, the unitary property follows immediately from the left invariance of dx:

$$\big(\pi_L(x)f \mid \pi_L(x)g\big) = \int f(x^{-1}y)\overline{g(x^{-1}y)}\,dx = \int f(y)\overline{g(y)}\,dx = (f \mid g).$$

Since $\big(\pi_L(x)f \mid g\big) = \int f(x^{-1}y)\overline{g(y)}\,dy = \int (L_{x^{-1}}f(y))\overline{g(y)}\,dy$, continuity of π follows from 16.4.5.

There is a close connection between unitary representations and functions of positive-type. Here is one part of the connection. A converse is given by 16.5.7.

16.5.3 Proposition. *Let π be a unitary representation of G on $\mathscr{H}$. For a vector $\boldsymbol{x} \in \mathscr{H}$ define*

$$\phi(x) = (\pi(x)\boldsymbol{x}\,|\,\boldsymbol{x}), \quad x \in G, \tag{16.7}$$

Then $\phi \in \mathscr{P}(G)$.

Proof. By (16.6), for any $f \in C_c(G)$

$$\begin{aligned}\int (f^* * f)\phi &= \iint \overline{f(y)}f(x)\phi(y^{-1}x)\,dx\,dy = \iint \overline{f(y)}f(x)\left(\pi(y^{-1}x)\boldsymbol{x}\,|\,\boldsymbol{x}\right)\,dx\,dy\\ &= \iint \left(f(x)\pi(x)\boldsymbol{x}\,|\,f(y)\pi(y)\boldsymbol{x}\right)\,dx\,dy = (\boldsymbol{u}\,|\,\boldsymbol{u}) \ge 0,\end{aligned}$$

where $\boldsymbol{u}$ is the weak vector integral $\int f(x)\pi(x)\boldsymbol{x}\,dx$ (14.6.3). □

16.5.4 Corollary. *Let $f \in L^2(G)$ and $\widetilde{f}(x) := \overline{f(x^{-1})}$ $(= \Delta(x)f^*(x))$. Then $f * \widetilde{f} \in \mathscr{P}(G)$.*

Proof. $\overline{f * \widetilde{f}(x)} = \int f(x^{-1}y)\overline{f(y)}\,dy = \big(\pi_L(x)f \mid f\big)$. □

16.5.5 Corollary. *Let $\mathscr{P}_c(G) := C_c(G) \cap \mathscr{P}(G)$. Then $C_c(G) * C_c(G) \subseteq \operatorname{span}\mathscr{P}_c(G))$. Moreover, $\operatorname{span}\mathscr{P}_c(G)$ is dense in $C_c(G)$ in the uniform norm and is dense in $L^p(G)$ in the L^p norm for $1 \le p < \infty$.*

Proof. Let $f \in C_c(G)$ and $K := \operatorname{supp} f$. By 16.5.4, $f * \widetilde{f} \in \mathscr{P}(G)$. Also, from $f * \widetilde{f}(x) = \int_K f(y)\overline{f(x^{-1}y)}\,dy$ we see that $\operatorname{supp}(f * \widetilde{f}) \subseteq KK^{-1}$. Therefore, $f * \widetilde{f} \in \mathscr{P}_c(G)$. Since the mapping $(g, h) \mapsto g * \widetilde{h}$ on $C_c(G) \times C_c(G)$ is sesquilinear, by the polarization identity we have $g * \widetilde{h} = \frac{1}{4}\sum_{k=1}^4 i^k (g + i^k h) * (g + i^k h)\widetilde{\ }$. Replacing h by $\widetilde{h}$ we see that $g * h \in \operatorname{span}\mathscr{P}_c(G)$. Taking h to be an approximate identity, we conclude that $\operatorname{span}\mathscr{P}_c(G)$ is dense in $C_c(G)$ in the uniform and L^p norms and hence is dense in L^p. □

The converse of 16.5.3 is considerably deeper. We shall the following lemma.

16.5.6 Lemma. *Let ϕ be a bounded Borel function on G such that $\int f\phi = 0$ for all $f \in L^1(G)$. Then $\phi = 0$ a.e. on each Borel set E with $|E| < \infty$. If, additionally, ϕ is continuous, then $\phi = 0$ everywhere.*

Proof. Let $K \subseteq E$ be compact and let $g \in C_c(G)$ be nonnegative with $g = 1$ on K. Taking $f = g\overline{\phi}$ in the hypothesis, we have $\int_K |\phi|^2 = 0$, hence $\phi = 0$ a.e. on K. By regularity, there exist compact K_n with $K_n \uparrow B \subseteq E$ and $|K_n| \uparrow |E|$. Then $\phi = 0$ a.e. on B and $|E \setminus B| = 0$, hence $\phi = 0$ a.e. on E. If ϕ is continuous, then ϕ is identically equal to zero on any compact set K by 16.2.2(c), so ϕ is the zero function. □

A **cyclic vector** for a unitary representation $\pi : G \to \mathscr{B}(\mathscr{H})$ is a member $\boldsymbol{x}$ of $\mathscr{H}$ such that the linear span of $\pi_\phi(G)\boldsymbol{x}$ is dense in $\mathscr{H}$. Here is the aforementioned converse of 16.5.3.

16.5.7 Theorem. *Let ϕ be of positive type on G. Then there exists a Hilbert space $\mathscr{H}_\phi$, a unitary representation $\pi_\phi : G \to \mathscr{B}(\mathscr{H}_\phi)$, and a cyclic vector $\boldsymbol{x} \in \mathscr{H}_\phi$ such that $\phi(\cdot) = (\pi(\cdot)\boldsymbol{x} \mid \boldsymbol{x})_\phi$ a.e. on every Borel set E with $|E| < \infty$. Thus if $\phi \in \mathscr{P}(G)$, then $\phi(\cdot) = (\pi(\cdot)\boldsymbol{x} \mid \boldsymbol{x})_\phi$ everywhere.*

Proof. Assume that $\phi \neq 0$. Define

$$(f \mid g)_\phi = \int (g^* * f)\phi = \iint \overline{g(y)} f(x)\phi(y^{-1}x)\, dx\, dy, \quad f, g \in L^1(G), \tag{16.8}$$

where the second equality is from (16.6). Then $(f \mid g)_\phi$ is a positive sesquilinear form on $L^1(G)$ and by the CBS inequality

$$(f \mid g)_\phi \le (f \mid f)_\phi (g \mid g)_\phi \le \|f\|_1 \|g\|_1 \|\phi\|_\infty . \tag{16.9}$$

Now define $\mathscr{N} := \{f \in L^1 : (f \mid f)_\phi = 0\}$. Then $\mathscr{N}$ is a closed linear subspace of $L^1(G)$. Let $f \mapsto \breve{f}$ denote the quotient map $L^1(G) \to L^1(G)/\mathscr{N}$ and define

$$\left(\breve{f} \mid \breve{g}\right)_\phi = (f \mid g)_\phi, \quad f,\ g \in L^1(G).$$

If $\breve{f}_1 = \breve{f}_2$ and $\breve{g}_1 = \breve{g}_2$, then $(f_1 - f_2 \mid f_1 - f_2)_\phi = (g_1 - g_2 \mid g_1 - g_2)_\phi = 0$ and so by the CBS inequality

$$|(f_1 \mid g_1)_\phi - (f_2 \mid g_2)_\phi| \le |(f_1 - f_2 \mid g_1)_\phi| + |(f_2 \mid g_2 - g_2)_\phi| = 0.$$

Therefore, $(\breve{f} \mid \breve{g})_\phi$ is well-defined. It is readily established that $(\breve{f} \mid \breve{g})_\phi$ is an inner product on $L^1(G)/\mathscr{N}$. Denote the Hilbert space completion of $L^1(G)/\mathscr{N}$ by $\mathscr{H}_\phi$ (11.1.7). From (16.9),

$$\left|\left(\breve{f} \mid \breve{g}\right)_\phi\right| \le \|g\|_1 \|f\|_1 \|\phi\|_\infty .$$

Next, for $x \in G$ define $\breve{L}_x$ on $L^1(G)/\mathscr{N}$ by $\breve{L}_x \breve{f} = (L_x f)\breve{}$. By left invariance,

$$(L_x f \mid L_x g)_\phi = \iint \overline{g(xz)} f(xy)\phi(z^{-1}y)\, dy\, dz = \iint \overline{g(z)} f(y)\phi(z^{-1}y)\, dy\, dz = (f \mid g)_\phi ,$$

hence $\breve{L}_x$ is well-defined, preserves the inner products, and therefore extends to a unitary operator on $\mathscr{H}_\phi$. Now define a mapping $\pi_\phi : G \to \mathscr{B}(\mathscr{H}_\phi)$ by $\pi_\phi(x) = \breve{L}_{x^{-1}}$. Then

$$\pi_\phi(x)\pi_\phi(y)f = \breve{L}_{x^{-1}}(\breve{L}_{y^{-1}} f) = \breve{L}_{x^{-1}}(L_{y^{-1}} f)\breve{} = (L_{x^{-1}} L_{y^{-1}} f)\breve{} = (L_{y^{-1}x^{-1}} f)\breve{} = \pi_\phi(xy)f,$$

hence π_ϕ is a unitary representation of G on $\mathscr{H}_\phi$.

It remains to find a cyclic vector $\boldsymbol{x} \in \mathscr{H}_\phi$ such that (16.7) holds. To this end, let (ψ_α) be an approximate identity in $L^1(G)$. Then for $f \in L^1(G)$, $(\check{f} \mid \check{\psi}_\alpha)_\phi = \int (\psi_\alpha^* * f)\phi \to \int f\phi$. Since the net $(\check{\psi}_\alpha)$ is bounded in the norm of $\mathscr{H}_\phi$, there exists a subnet $(\check{\psi}_\beta)$ that converges weakly to some $\boldsymbol{x} \in \mathscr{H}_\phi$. Thus

$$(\check{f} \mid \boldsymbol{x})_\phi = \int f\phi, \quad f \in L^1(G), \tag{$\dagger$}$$

and so

$$(\check{f} \mid \pi_\phi(y)\boldsymbol{x})_\phi = (\pi_\phi(y)^{-1}\check{f} \mid \boldsymbol{x})_\phi = \int f(yx)\phi(x)\,dx = \int f(x)\phi(y^{-1}x)\,dx.$$

Therefore, for any $f, g \in L^1(G)$,

$$(\check{f} \mid \check{g})_\phi = \iint \overline{g(y)} f(x)\phi(y^{-1}x)\,dx\,dy = \int \overline{g(y)} \big(\check{f} \mid \pi_\phi(y)\boldsymbol{x}\big)_\phi\,dy. \tag{$\ddagger$}$$

It follows that if $\big(\check{f} \mid \pi_\phi(y)\boldsymbol{x}\big) = 0$ for all y, then $\check{f} = 0$, which shows that the linear span of $\pi_\phi(G)\boldsymbol{x}$ is dense in $\mathscr{H}_\phi$. Moreover, if $g \in C_c(G)$, then the vector integral $I(g) := \int g(y)\pi_\phi(y)\boldsymbol{x}\,dy$ exists, and from ($\ddagger$) we have $(\check{f} \mid \check{g})_\phi = \big(\check{f} \mid I(g)\big)_\phi$ for all f. Therefore,

$$\int \big(\boldsymbol{x} \mid \pi_\phi(y)\boldsymbol{x}\big)_\phi \overline{g(y)}\,dy = \big(\boldsymbol{x} \mid I(g)\big)_\phi = \lim_\beta \big(\check{\psi}_\beta \mid I(g)\big)_\phi = (\boldsymbol{x} \mid \check{g})_\phi = \int \overline{g\phi},$$

the last equality from ($\dagger$). The desired conclusion now follows from the preceding lemma, since $C_c(G)$ is dense in L^1. □

It is not necessarily the case that $\phi(\cdot) = (\pi(\cdot)\boldsymbol{x} \mid \boldsymbol{x})$ a.e. on G. Indeed, as the proof shows, such a conclusion would depend on (L^1, L^∞) duality, which holds generally only in the σ-finite case.

Irreducible Representations

Let π be a unitary representation of G on a Hilbert space $\mathscr{H}$. An **invariant subspace** for π is a subspace $\mathscr{M}$ of $\mathscr{H}$ such that $\pi(x)\mathscr{M} \subseteq \mathscr{M}$ for all $x \in G$. If the only invariant subspaces for π are the trivial subspaces $\{0\}$ and $\mathscr{H}$, then π is said to be **irreducible**; otherwise π is **reducible**. Also, call an operator in $\mathscr{B}(\mathscr{H})$ **nontrivial** if it is not a multiple of the identity operator I. The following result is a fundamental tool in the study of representations.

16.5.8 Schur's Lemma. *A unitary representation π is reducible iff there exists a nontrivial $T \in \mathscr{B}(\mathscr{H})$ that commutes with every $\pi(x)$.*

Proof. Assume that π is reducible and let $\mathscr{M}$ be a nontrivial closed subspace of $\mathscr{H}$ such that $\pi(x)\mathscr{M} \subseteq \mathscr{M}$ for all $x \in G$. For $\boldsymbol{x} \in \mathscr{M}$ and $\boldsymbol{x}^\perp \in \mathscr{M}^\perp$, $\big(\boldsymbol{x} \mid \pi(x)\boldsymbol{x}^\perp\big) = \big(\pi(x^{-1})\boldsymbol{x} \mid \boldsymbol{x}^\perp\big) = 0$, hence $\pi\big(\mathscr{M}^\perp\big) \subseteq \mathscr{M}^\perp$. If P denotes the orthogonal projection onto $\mathscr{M}$, then

$$\pi(x)P(\boldsymbol{x} + \boldsymbol{x}^\perp) = \pi(x)\boldsymbol{x} = P\pi(x)\boldsymbol{x} = P\pi(x)(\boldsymbol{x} + \boldsymbol{x}^\perp),$$

hence P is a nontrivial operator commuting with every $\pi(x)$.

Conversely, let T be a nontrivial operator commuting with every $\pi(x)$. Then

$$\pi(x)T^* = \big(T\pi(x^{-1})\big)^* = \big(\pi(x^{-1})T\big)^* = T^*\pi(x),$$

hence T^* commutes with each $\pi(x)$. Therefore, the self-adjoint operators $T_r := (T+T^*)/2$ and $T_i := (T-T^*)/(2i)$ commute with $\pi(x)$. Since $T = T_r + iT_i$, at least one of the operators is nontrivial. Thus we may as well assume that the original operator T is self-adjoint. Now consider the Borel functional calculus $f \mapsto f(T)$. Since $\pi(x)$ commutes with T it commutes with the projections $P_E := \mathbf{1}_E(T)$, where E is a nontrivial Borel subset of $\sigma(T)$. Then ran P_E is a nontrivial subspace of $\mathscr{H}$ invariant under every $\pi(x)$, hence π is reducible. □

16.5.9 Corollary. *If G is abelian and π is irreducible, then* $\dim(\mathscr{H}) = 1$.

Proof. For each $y \in G$, $\pi(y)\pi(x) = \pi(x)\pi(y)$ for all $x \in G$, hence there exists a constant $c(y)$ such that $\pi(y) = c(y)I$. Thus all subspaces of $\mathscr{H}$ are invariant and so, by irreducibility, $\dim(\mathscr{H}) = 1$. □

The connection between irreducible unitary representations and functions of positive-type is given in the next theorem. For this, let C_1 and S_1 denote, respectively, the closed unit ball and the unit sphere in $L^\infty(G)$.

16.5.10 Theorem. *Let $\phi \in \mathscr{P}(G) \cap S_1$ and let π_ϕ be as in 16.5.7. Then π_ϕ is irreducible iff ϕ is an extreme point of $\mathscr{P} \cap S_1$.*

Proof. Let π_ϕ be reducible, so that $\mathscr{H}_\phi = \mathscr{M} \oplus \mathscr{M}^\perp$ for some nontrivial closed linear subspace $\mathscr{M}$ of $\mathscr{H}_\phi$ invariant under π_ϕ. Let $\boldsymbol{x} \in \mathscr{H}_\phi$ be a cyclic vector for π_ϕ. Since the spaces $\mathscr{M}$ and $\mathscr{M}^\perp$ are invariant and nontrivial, $\boldsymbol{x}$ is a member of neither, hence $\boldsymbol{x} = \boldsymbol{m} + \boldsymbol{m}^\perp$, where $\boldsymbol{m}, \boldsymbol{m}^\perp \neq 0$. Setting $\boldsymbol{y} = \boldsymbol{m}/\|\boldsymbol{m}\|$ and $\boldsymbol{z} = \boldsymbol{m}^\perp/\|\boldsymbol{m}^\perp\|$, we then have

$$\begin{aligned}\phi(x) &= (\pi_\phi(x)\boldsymbol{x} \mid \boldsymbol{x})_\phi = (\pi_\phi(x)\boldsymbol{m} \mid \boldsymbol{m})_\phi + \left(\pi_\phi(x)\boldsymbol{m}^\perp \mid \boldsymbol{m}^\perp\right)_\phi \\ &= \|\boldsymbol{m}\|^2 (\pi_\phi(x)\boldsymbol{y} \mid \boldsymbol{y})_\phi + \|\boldsymbol{m}^\perp\|^2 (\pi_\phi(x)\boldsymbol{z} \mid \boldsymbol{z})_\phi . \qquad (\dagger)\end{aligned}$$

Since

$$\|\boldsymbol{m}\|^2 + \left\|\boldsymbol{m}^\perp\right\|^2 = \|\boldsymbol{x}\|^2 = (\pi_\phi(e)\boldsymbol{x} \mid \boldsymbol{x})_\phi = \phi(e) = 1,$$

equation (†) exhibits ϕ as a proper convex combination of members of $\mathscr{P} \cap S_1$. Therefore, ϕ is not extreme.

Now assume that π_ϕ is irreducible and let $\phi = \theta + \psi$, $\theta, \psi \in \mathscr{P}(G)$. Then, by (16.8), $(f \mid g)_\phi = (f \mid g)_\theta + (f \mid g)_\psi$, which implies that $(f \mid f)_\theta \le (f \mid f)_\phi$ and so

$$|(f \mid g)_\theta|^2 \le (f \mid f)_\theta (g \mid g)_\theta \le (f \mid f)_\phi (g \mid g)_\phi .$$

It follows that $B(\breve{f}, \breve{g}) := (f \mid g)_\theta$ is a well-defined bounded Hermitian sesquilinear form on $\mathscr{H}_\phi$. By 11.4.1 there exists $T \in \mathscr{B}(\mathscr{H}_\phi)$ such that $(T\breve{f} \mid \breve{g})_\phi = (f \mid g)_\theta$ for all $f, g \in L^1(G)$. Recalling that

$$(\pi_\phi(x)\breve{f} \mid \breve{g})_\phi = (\breve{L}_{x^{-1}}\breve{f} \mid \breve{g})_\phi = (L_{x^{-1}}f \mid g)_\phi,$$

with the analogous equations holding for θ, we have

$$\begin{aligned}(T\pi_\phi(x)\breve{f} \mid \breve{g})_\phi &= (T(L_{x^{-1}}f)^\vee \mid \breve{g})_\phi = (L_{x^{-1}}f \mid g)_\theta = (f \mid L_x g)_\theta = (T\breve{f} \mid \breve{L}_x\breve{g})_\phi \\ &= (\pi_\phi(x)T\breve{f} \mid \breve{g})_\phi.\end{aligned}$$

Thus T commutes with $\pi_\phi(x)$ for all x and so $T = cI$ for some $c \in \mathbb{C}$ by Schur's lemma. Therefore,

$$\int (g^* * f)\theta = (f \mid g)_\theta = (T\breve{f} \mid \breve{g})_\phi = (c\breve{f} \mid \breve{g})_\phi = (cf \mid g)_\phi = \int (g^* * f)c\phi$$

for all $f, g \in L^1$. Since the functions $g^* * f$ form a dense subset of L^1, $\theta = c\phi$.

Now let $\phi = t\theta + (1-t)\psi$, where $\theta, \psi \in \mathscr{P}(G) \cap S_1$ and $0 < t < 1$. By the preceding paragraph, $t\theta = a\phi$ for some $a \in \mathbb{C}$, hence $(a-t)\phi = (1-t)\psi$. Taking norms, we have $|a-t| = 1-t$. Therefore, $\phi = b\psi$ for some b with $|b| = 1$ and so $b\int(f^* * f)\psi = \int(f^* * f)\phi$ for all $f \in L^1$. Since the integrals are nonnegative, $b > 0$, hence $b = 1$ and $\phi = \psi$. Therefore, ϕ is an extreme point of $\mathscr{P}(G) \cap S_1$. □

The next theorem, a fundamental result in representation theory, asserts that the irreducible unitary representations of G separate points of G. For the proof we need the following.

16.5.11 Lemma. *Let $f \in C_c(G)$, $f \neq 0$. Then there exists a $\psi \in \mathscr{P}(G)$ with $\int(f^* * f)\psi > 0$.*

Proof. Observe that $f^* * f$ is continuous and $f^* * f(e) = \|f\|_2^2 > 0$. Thus there exists a compact, symmetric neighborhood V of e on which $f^* * f \geq c > 0$. Take $g \in C_c(G)$ with $g \geq 0$ and $g = 1$ on V and set $\psi := g * g$. By left invariance,

$$\begin{aligned}\int (f^* * f)\psi &\geq \iint (f^* * f)(x)\mathbf{1}_V(x)\mathbf{1}_V(y^{-1}x)\,dy\,dx = \iint (f^* * f)(x)\mathbf{1}_V(x)\mathbf{1}_V(y^{-1})\,dy\,dx \\ &\geq c|V|^2 > 0.\end{aligned}$$ □

16.5.12 Theorem (Gelfand-Raikov). *Given distinct points $x, y \in G$, there exists an irreducible unitary representation π of G such that $\pi(x) \neq \pi(y)$.*

Proof. Let $a := x^{-1}y$ and choose $g \in C_c(G)$ such that $L_a g \neq g$. Set $f := L_a g - g \in C_c(G)$ and choose $\psi \in \mathscr{P}(G)$ as in the lemma. Normalizing, we may assume $\psi \in \mathscr{P}(G) \cap S_1$. By the Krein-Milman theorem, ψ is a weak* limit of convex combinations of extreme points of $\mathscr{P}(G) \cap S_1$, hence there must exist an extreme point ϕ such that $\int(f^* * f)\phi > 0$. Thus, in the notation of 16.5.7, $(f \mid f)_\phi > 0$. Since $\left\|\pi_\phi(a^{-1})\breve{g} - \breve{g}\right\|_\phi^2 = (f \mid f)_\phi > 0$, $\pi_\phi(x)\breve{g} \neq \pi_\phi(y)\breve{g}$. Finally, by 16.5.10, π_ϕ is irreducible. □

Unitary Representations of Compact Groups

In this subsection G is assumed to be compact and π denotes a generic unitary representation of G on a Hilbert space $\mathscr{H}$.

16.5.13 Lemma. *For a fixed unit vector $\boldsymbol{u} \in \mathscr{H}$, define $T \in \mathscr{B}(\mathscr{H})$ by the vector integral*

$$T\boldsymbol{x} = \int (\boldsymbol{x} \mid \pi(x)\boldsymbol{u})\,\pi(x)\boldsymbol{u}\,dx.$$

Then T is a compact, positive, nonzero operator and $T\pi(x) = \pi(x)T$ for all $x \in G$.

Proof. For any $\boldsymbol{x}, \boldsymbol{y} \in \mathscr{H}$,

$$(T\boldsymbol{x} \mid \boldsymbol{y}) = \int (\boldsymbol{x} \mid \pi(x)\boldsymbol{u})\,(\pi(x)\boldsymbol{u} \mid \boldsymbol{y})\,dx.$$

In particular, $(T\boldsymbol{x} \mid \boldsymbol{x}) = \int |(\boldsymbol{x} \mid \pi(x)\boldsymbol{u})|^2\,dx \geq 0$, and because $|(\boldsymbol{u} \mid \pi(x)\boldsymbol{u})|^2$ is continuous in x and positive at $x = e$, $(T\boldsymbol{u} \mid \boldsymbol{u}) > 0$. Therefore, T is a nonzero, positive operator. Furthermore, by translation invariance,

$$\begin{aligned}(T\pi(y)\boldsymbol{x} \mid \boldsymbol{y}) &= \int (\pi(y)\boldsymbol{x} \mid \pi(x)\boldsymbol{u})\,(\pi(x)\boldsymbol{u} \mid \boldsymbol{y})\,dx = \int (\boldsymbol{x} \mid \pi(y^{-1}x)\boldsymbol{u})\,(\pi(x)\boldsymbol{u} \mid \boldsymbol{y})\,dx \\ &= \int (\boldsymbol{x} \mid \pi(x)\boldsymbol{u})\,(\pi(yx)\boldsymbol{u} \mid \boldsymbol{y})\,dx = \int (\boldsymbol{x} \mid \pi(x)\boldsymbol{u})\,(\pi(x)\boldsymbol{u} \mid \pi(y^{-1})\boldsymbol{y})\,dx \\ &= (T\boldsymbol{x} \mid \pi(y^{-1})\boldsymbol{y}) = (\pi(y)T\boldsymbol{x} \mid \boldsymbol{y}).\end{aligned}$$

Therefore, $T\pi(x) = \pi(x)T$.

Now, because G is compact, the function $\pi(\cdot)\boldsymbol{u}$ is uniformly continuous, hence given $\varepsilon > 0$, there exists a measurable partition $E_1, \ldots, E_n$ of G and $x_j \in E_j$ such that $x \in E_j \Rightarrow \|\pi(x)\boldsymbol{u} - \pi(x_j)\boldsymbol{u}\| < \varepsilon$. Now set

$$T_j\boldsymbol{x} := \int_{E_j} (\boldsymbol{x} \mid \pi(x)\boldsymbol{u})\, \pi(x)\boldsymbol{u}\, dx \quad \text{and} \quad F_j\boldsymbol{x} := |E_j|\, (\boldsymbol{x} \mid \pi(x_j)\boldsymbol{u})\, \pi(x_j)\boldsymbol{u}$$

and note that

$$\|T_j\boldsymbol{x} - F_j\boldsymbol{x}\| \leq \int_{E_j} \|\, (\boldsymbol{x} \mid \pi(x)\boldsymbol{u})\, \pi(x)\boldsymbol{u} - (\boldsymbol{x} \mid \pi(x_j)\boldsymbol{u})\, \pi(x_j)\boldsymbol{u}\|\, dx.$$

For $x \in E_j$, the integrand is less than or equal to

$$\left\| (\boldsymbol{x} \mid [\pi(x) - \pi(x_j)]\boldsymbol{u})\, \pi(x)\boldsymbol{u} \right\| + \left\| (\boldsymbol{x} \mid \pi(x_j)\boldsymbol{u})\, [\pi(x) - \pi(x_j)]\boldsymbol{u} \right\| \leq 2\varepsilon \, \|\boldsymbol{x}\|,$$

hence $\|T_j\boldsymbol{x} - F_j\boldsymbol{x}\| < 2\varepsilon\, \|\boldsymbol{x}\|\, |E_j|$ and so $\left\|T - \sum_{j=1}^n F_j\right\| < 2\varepsilon$. This shows that T may be approximated in norm by operators of finite rank and hence is compact. □

16.5.14 Corollary. *If π is irreducible, then $\mathscr{H}$ is finite dimensional.*

Proof. The compact operator T of the lemma commutes with every $\pi(x)$, hence is a nonzero multiple of the identity. Thus the identity operator on $\mathscr{H}$ is compact, which implies that $\mathscr{H}$ is finite dimensional. □

The following lemma will allow us to express a representation in terms of irreducible representations.

16.5.15 Lemma. *If $\mathscr{H}$ is finite dimensional, then π is a direct sum of irreducible representations. That is, $\mathscr{H} = \mathscr{M}_1 \oplus \cdots \oplus \mathscr{M}_n$, where π is irreducible on $\mathscr{M}_j$.*

Proof. If π is reducible, then it has a nontrivial invariant subspace $\mathscr{M}$. Since $\mathscr{M}^\perp$ is also invariant, we may assume by induction that $\mathscr{M}$ and $\mathscr{M}^\perp$ are each direct sums of irreducible subspaces, hence so is $\mathscr{H}$. □

16.5.16 Theorem. *Every representation π of G is a direct sum of irreducible (hence finite dimensional) representations. That is, there exists a family of mutually orthogonal invariant subspaces of $\mathscr{H}$ with linear span dense in $\mathscr{H}$ such that the restriction of π to each subspace is irreducible.*

Proof. Since the operator T of 16.5.13 is compact and positive, it has a nonzero eigenvalue (12.3.8) and hence a finite dimensional eigenspace $\mathscr{M}$. Since T commutes with π, $\mathscr{M}$ is π-invariant. By 16.5.15, $\mathscr{M}$ is a direct sum of irreducible representations. In particular, there exist irreducible subrepresentations.

Now consider families of mutually orthogonal irreducible π-invariant subspaces. Ordering these families by inclusion and applying Zorn's lemma yields a maximal family $\{\mathscr{M}_i : i \in \mathfrak{I}\}$. It then follows that $\mathscr{H}$ is the closed linear span S of all the $\mathscr{M}_i$, otherwise $S^\perp$ would contain a π-invariant irreducible subspace, contradicting maximality. □

A **coefficient** of π is a function on G of the form $(\pi(\cdot)\boldsymbol{x} \mid \boldsymbol{y})$, $\boldsymbol{x}, \boldsymbol{y} \in \mathscr{H}$. Let $\mathscr{C}(G)$ denote the linear span of all coefficients of finite dimensional representations of G. By 16.5.15, $\mathscr{C}(G)$ is also the linear span of coefficients of all finite dimensional *irreducible* representations of G. The final and main theorem of this subsection asserts that $\mathscr{C}(G)$ is dense in $C(G)$. For this we need the following lemma.

16.5.17 Lemma. *Let $\mathscr{K}$ be a finite dimensional complex Hilbert space and let $\mathscr{V}$ be a group of operators on $\mathscr{K}$ (under composition) whose identity is the identity operator. If $\mathscr{V}$ is compact in $\mathscr{B}(\mathscr{K})$, then there exists an inner product on $\mathscr{K}$ relative to which each member of $\mathscr{V}$ is unitary.*

Proof. Clearly, $\mathscr{V}$ is a topological group under composition. If dV denotes normalized Haar measure on $\mathscr{V}$ and $(\boldsymbol{x} \,|\, \boldsymbol{y})$ is the given inner product on $\mathscr{K}$, then

$$\langle \boldsymbol{x} \mid \boldsymbol{y} \rangle := \int_{\mathrm{cl}\,\mathscr{V}} (V\boldsymbol{x} \,|\, V\boldsymbol{y})\, dV$$

is the required new inner product on $\mathscr{K}$. For example, the calculation

$$\langle V_0\boldsymbol{x} \mid V_0\boldsymbol{y} \rangle = \int_{\mathscr{V}} (VV_0\boldsymbol{x} \,|\, VV_0\boldsymbol{y})\, dV = \langle \boldsymbol{x} \mid \boldsymbol{y} \rangle$$

shows that $V_0 \in \mathscr{V}$ is unitary. □

We may now prove

16.5.18 Theorem (Peter-Weyl). *Let G be a compact topological group. Then $\mathscr{C}(G)$ is dense in $C(G)$.*

Proof. By the Gelfand-Raikov theorem, $\mathscr{C} = \mathscr{C}(G)$ separates points of G. We show that $\mathscr{C}$ is closed under multiplication and complex conjugation. The desired conclusion will then follow from the Stone-Weierstrass theorem.

The product of typical members of $\mathscr{C}$ is of the form

$$\left[\sum_{j=1}^{n} (\pi_j(x)\boldsymbol{x}_j \,|\, \boldsymbol{y}_j)\right]\left[\sum_{k=1}^{m} (\widetilde{\pi}_k(x)\widetilde{\boldsymbol{x}}_k \,|\, \widetilde{\boldsymbol{y}}_k)\right] = \sum_{j,k} (\pi_j(x)\boldsymbol{x}_j \,|\, \boldsymbol{y}_j)\, (\widetilde{\pi}_k(x)\boldsymbol{x}_k \,|\, \boldsymbol{y}_k).$$

To prove closure under multiplication, it therefore suffices to show that if $\pi : G \to \mathscr{B}(\mathscr{H})$ and $\widetilde{\pi} : G \to \mathscr{B}(\widetilde{\mathscr{H}})$ are finite dimensional unitary representations of G, then the function $x \mapsto (\pi(x)\boldsymbol{x} \,|\, \boldsymbol{y})\, (\widetilde{\pi}(x)\widetilde{\boldsymbol{x}} \,|\, \widetilde{\boldsymbol{y}})$ is a member of $\mathscr{C}$. For this we use 12.4.6, which implies that for each $x \in G$ there exists a unique bounded linear operator $\pi(x) \otimes \widetilde{\pi}(x)$ on the finite dimensional Hilbert space $\mathscr{B}_2(\mathscr{H}, \widetilde{\mathscr{H}})$ such that

$$\big((\pi(x) \otimes \widetilde{\pi}(x))\boldsymbol{x} \otimes \widetilde{\boldsymbol{x}} \,|\, \boldsymbol{y} \otimes \widetilde{\boldsymbol{y}}\big) = (\pi(x)\boldsymbol{x} \,|\, \boldsymbol{y})\, (\widetilde{\pi}(x)\widetilde{\boldsymbol{x}} \,|\, \widetilde{\boldsymbol{y}}).$$

This defines a unitary representation $\pi \otimes \widetilde{\pi}$ on $\mathscr{B}_2(\mathscr{H}, \widetilde{\mathscr{H}})$.

To show that $\mathscr{C}(G)$ is closed under complex conjugation, it suffices the show that if $\pi : G \to \mathscr{B}(\mathscr{H})$ is a finite dimensional unitary representations of G, then $(\boldsymbol{y}_0 \,|\, \pi(\cdot)\boldsymbol{x}_0)$ is a coefficient of a finite dimensional unitary representation on G. To this end, let $\mathscr{F}$ denote the finite dimensional subspace of $C(G)$ consisting of all functions $g_{\boldsymbol{x}}$ defined by

$$g_{\boldsymbol{x}}(t) = (\boldsymbol{y}_0 \,|\, \pi(t)\boldsymbol{x}), \quad t \in G,\ \boldsymbol{x} \in \mathscr{H}.$$

Since $\mathscr{F}$ is right translation invariant, $t \mapsto R_t$ is a continuous representation of G on the space $\mathscr{F}$. By the lemma, there exists an inner product $\langle \cdot \mid \cdot \rangle$ on $\mathscr{F}$ relative to which the operators R_t are unitary. Since the evaluation map $\widehat{e}$ is a continuous linear functional on $\mathscr{F}$, by the Riesz representation theorem there exists a member $g_{\boldsymbol{x}_1}$ of $\mathscr{F}$ such that

$$g_{\boldsymbol{x}}(e) = \langle g_{\boldsymbol{x}} \mid g_{\boldsymbol{x}_1} \rangle, \quad \boldsymbol{x} \in \mathscr{H}.$$

It follows that

$$(\boldsymbol{y}_0 \,|\, \pi(s)\boldsymbol{x}_0) = g_{\boldsymbol{x}_0}(s) = g_{\pi(s)\boldsymbol{x}_0}(e) = \big\langle g_{\pi(s)\boldsymbol{x}_0} \mid g_{\boldsymbol{x}_1} \big\rangle = \langle R_s g_{\boldsymbol{x}_0} \mid g_{\boldsymbol{x}_1} \rangle,$$

which shows that $(\boldsymbol{y}_0 \,|\, \pi(\cdot)\boldsymbol{x}_0)$ is a coefficient of the unitary representation R, completing the proof. □

16.6 Locally Compact Abelian Groups

In this section, G is assumed to be abelian.

The Dual Group

A **character** of G is a continuous homomorphism ξ from G into the circle group $\mathbb{T}$. The set of all characters is denoted by $\widehat{G}$. Two characters ξ_1 and ξ_2 may be multiplied together to form another character:

$$(\xi_1\xi_2)(xy) = \big[\xi_1(xy)\big]\big[\xi_2(xy)\big] = \big[\xi_1(x)\xi_1(y)\big]\big[\xi_2(x)\xi_2(y)\big] = (\xi_1\xi_2)(x)\cdot(\xi_1\xi_2)(x)$$

Furthermore, the map $x \mapsto \xi^{-1}(x) = \overline{\xi}(x)$ is easily seen to define a character. Thus $\widehat{G}$ is an abelian group with identity the constant function 1. We show in this subsection that $\widehat{G}$ is locally compact under a natural topology. We use the standard notation

$$\langle x, \xi\rangle = \xi(x), \quad x \in G, \quad \xi \in \widehat{G}.$$

The **Fourier transform** $\widehat{f} : \widehat{G} \to \mathbb{C}$ of $f \in L^1(G)$ is defined by

$$\widehat{f}(\xi) := \int \overline{\langle x, \xi\rangle} f(x)\, dx = \int \langle x, \xi^{-1}\rangle\, f(x)\, dx.$$

As in the case $G = \mathbb{R}^d$ (see 6.2.1),

$$\widehat{f * g} = \widehat{f}\cdot\widehat{g} \text{ and } \widehat{f^*} = \overline{\widehat{f}}. \tag{16.10}$$

Now define $\Phi_\xi(f) = \widehat{f}(\xi)$. Clearly $\|\Phi_\xi(f)\| = \|\widehat{f}\|_\infty \le \|f\|_1$. Moreover, the mapping $\xi \mapsto \Phi_\xi : \widehat{G} \to L^1(G)$ is the restriction to $\widehat{G} \subseteq L^\infty(G)$ of the isometric isomorphism that identifies $L^\infty(G)$ with the dual of $L^1(G)$. [4] More can be said:

16.6.1 Theorem. *The function $\xi \mapsto \Phi_\xi$ is a bijection from $\widehat{G}$ onto the spectrum $\Sigma = \sigma(L^1(G))$ of the commutative Banach algebra $L^1(G)$.*

Proof. Recall that Σ is the set of continuous, nontrivial homomorphisms from $L^1(G)$ into $\mathbb{C}$, these being members of the dual of $L^1(G)$. The calculation

$$\begin{aligned}\Phi_\xi(f*g) &= \iint \overline{\langle x, \xi\rangle} f(y^{-1}x)g(y)\, dy\, dx = \iint \overline{\langle x, \xi\rangle} f(y^{-1}x)g(y)\, dx\, dy \\ &= \iint \overline{\langle xy, \xi\rangle} f(x)g(y)\, dx\, dy = \Phi_\xi(f)\Phi_\xi(g)\end{aligned}$$

shows that $\Phi_{\widehat{G}} \subseteq \Sigma$. For the reverse inclusion, let $\Phi \in \Sigma\ \big(\subseteq L^1(G)'\big)$ and choose $\phi \in L^\infty(G)$ (see footnote) such that

$$\Phi(f) = \int \phi(y) f(y)\, dy, \quad f \in L^1(G).$$

[4] In the non-σ-finite case, the assertion that the dual of $L^1(G)$ is $L^\infty(G)$ requires a modification of the definition of $L^\infty(G)$ using the notion of *local measurability*. We shall assume that $L^\infty(G)$ has been so modified. (see [21]). Alternatively, the reader may simply assume in what follows that G is σ-finite.

Fix $g \in L^1(G)$ such $\Phi(g) \neq 0$. For any $f \in L^1(G)$,

$$\begin{aligned}\int \phi(y)f(y)\,dy = \Phi(f) = \frac{\Phi(f*g)}{\Phi(g)} &= \frac{1}{\Phi(g)}\iint \phi(x)g(y^{-1}x)f(y)\,dy\,dx \\ &= \frac{1}{\Phi(g)}\iint \phi(x)g(y^{-1}x)f(y)\,dx\,dy = \int \frac{1}{\Phi(g)}\Phi\big(L_{y^{-1}}g\big)f(y)\,dy.\end{aligned}$$

Therefore, ϕ may be identified with, and hence replaced by, the continuous function $y \mapsto \Phi(g)^{-1}\Phi\big(L_{y^{-1}}g\big)$, which is a nonzero continuous homomorphism from G into $\mathbb{C}$. Since $\phi(y^n) = \phi(y)^n$ for every $n \in \mathbb{Z}$ and ϕ is bounded, we see that $|\phi(y)| = 1$, hence $\phi \in \widehat{G}$. $\square$

Recall that Σ is locally compact in the weak* (Gelfand) topology of the dual of $L^1(G)$. Let $\widehat{G}$ have the unique topology that makes the mapping $\xi \to \Phi_\xi : \widehat{G} \to \Sigma$ a homeomorphism. Then $\widehat{G}$ is locally compact, and a basic neighborhood of $\xi_0 \in \widehat{G}$ is of the form

$$V(\xi_0; f_1, \ldots, f_n; \varepsilon) = \Big\{\xi \in \widehat{G} : \Big|\widehat{f_j}(\xi) - \widehat{f_j}(\xi_0)\Big| < \varepsilon,\ j = 1, \ldots, n\Big\}, \tag{16.11}$$

where $f_j \in L^1$ and $\varepsilon > 0$. Thus a net (ξ_α) in $\widehat{G}$ converges to $\xi_0 \in \widehat{G}$ iff $\widehat{f}(\xi_\alpha) \to \widehat{f}(\xi_0)$ for all $f \in L^1(G)$. Note that, by virtue of the homeomorphism $\xi \to \Phi_\xi$, the two meanings of $\widehat{f}$, one as the Gelfand transform of f and the other as the Fourier transform of f, coincide:

$$\widehat{f}\big(\Phi_\xi\big) = \Phi_\xi(f) = \widehat{f}(\xi), \quad \xi \in \widehat{G}.$$

From this identification and 13.5.1 we have

16.6.2 Proposition. *The space of Fourier transforms of members of $L^1(G)$ is a conjugate closed subalgebra of $C_0(\widehat{G})$ that is dense in $C_0(\widehat{G})$.*

We now show that $\widehat{G}$ is a topological group under the topology described in the preceding paragraph. For this it is helpful to introduce an equivalent neighborhood system on $\widehat{G}$. The following lemmas accomplish this.

16.6.3 Lemma. *Every $\xi \in \widehat{G}$ is uniformly continuous. Moreover, $\langle x, \xi\rangle$ is jointly continuous in $(x, \xi) \in G \times \widehat{G}$.*

Proof. For $f \in L^1(G)$,

$$\widehat{L_x f}(\xi) = \int f(xy)\overline{\xi(y)}\,dy = \int f(y)\overline{\xi(x^{-1}y)}\,dy = \xi(x)\int f(y)\overline{\xi(y)}\,dy = \xi(x)\widehat{f}(\xi),$$

hence if $\widehat{f}(\xi) \neq 0$, then $\xi(x) = \widehat{f}(\xi)^{-1}\widehat{L_x f}(\xi)$. Since

$$\Big|\widehat{L_x f}(\xi) - \widehat{L_y f}(\xi)\Big| \leq \|L_x f - L_y f\|_1,$$

ξ is uniformly continuous. The calculation

$$\Big|\widehat{L_x f}(\xi) - \widehat{L_{x_0} f}(\xi_0)\Big| \leq \|L_x f - L_{x_0} f\|_1 + \Big|\widehat{L_{x_0} f}(\xi) - \widehat{L_{x_0} f}(\xi_0)\Big|$$

shows that the map $(x, \xi) \mapsto \langle x, \xi\rangle$ is jointly continuous. $\square$

16.6.4 Corollary. *If $x_\alpha \to x_0$ in G, then $\langle x_\alpha, \xi\rangle \to \langle x_0, \xi\rangle$ uniformly in ξ on compact subsets of $\widehat{G}$.*

16.6.5 Lemma. *The sets*

$$W(\xi_0, K, \varepsilon) := \Big\{\xi \in \widehat{G} : \sup_{x\in K} |\langle x, \xi\rangle - \langle x, \xi_0\rangle| < \varepsilon\Big\},$$

where $K \subseteq G$ is compact and $\varepsilon > 0$, form a basis of open neighborhoods of $\xi_0 \in \widehat{G}$. Thus $\xi_\alpha \to \xi_0$ in $\widehat{G}$ iff $\langle x, \xi_\alpha\rangle \to \langle x, \xi_0\rangle$ uniformly in x on each compact subset of G.

Proof. Let $F(x, \xi) = |\langle x, \xi\rangle - \langle x, \xi_0\rangle|$. By 16.6.3, F is continuous on $G \times \widehat{G}$. To show that $W := W(\xi_0, K, \varepsilon)$ is open in $\widehat{G}$, fix $\xi \in W$ and let $x \in K$. Then $F(x, \xi) < \varepsilon$, hence there exists an open set $U_x \subseteq G$ containing x and an open set $V_x \subseteq \widehat{G}$ containing ξ such that $F < \varepsilon$ on $U_x \times V_x$. Since K is compact, there exist $x_1, \ldots, x_n \in K$ such that $K \subset \bigcup_j U_{x_j}$. Then $V := \bigcap_j V_{x_j}$ is a neighborhood of ξ in $\widehat{G}$ and $F < \varepsilon$ on $K \times V$, that is, $V \subseteq W(\xi_0, K, \varepsilon)$. Therefore, $W(\xi_0, K, \varepsilon)$ is open in $\widehat{G}$.

It remains to show that every neighborhood $V(\xi_0; f_1, \ldots, f_n; \delta)$ in (16.11) contains $W(\xi_0, K, \varepsilon)$ for suitable K and $\varepsilon > 0$. Since

$$W(\xi_0, K_1 \cup K_2, \varepsilon_1 \wedge \varepsilon_2) \subseteq W(\xi_0, K_1, \varepsilon_1) \cap W(\xi_0, K_2, \varepsilon_2),$$

it suffices to show that, given $f \in L^1(G)$ and $\delta > 0$, $W(\xi_0, K, \varepsilon) \subseteq V(\xi_0; f; \delta)$ for some K and ε, that is,

$$|\langle x, \xi\rangle - \langle x, \xi_0\rangle| < \varepsilon \ \forall\, x \in K \Rightarrow \left|\widehat{f}(\xi) - \widehat{f}(\xi_0)\right| < \delta. \tag{†}$$

But for any compact $K \subseteq G$,

$$\left|\widehat{f}(\xi) - \widehat{f}(\xi_0)\right| \le \int_K |(\xi - \xi_0)\cdot f| + \int_{K^c} |(\xi - \xi_0)\cdot f| \le \int_K |(\xi - \xi_0)\cdot f| + 2\int_{K^c} |f|,$$

and choosing K so that the second term in the last inequality is $< \delta/2$ and taking $\varepsilon = \delta/(2\,\|f\|_1)$ we see that (†) holds. □

We may now prove the main result of the subsection:

16.6.6 Theorem. *$\widehat{G}$ is a locally compact abelian topological group in the Gelfand topology.*

Proof. All that needs to be proved is that $\widehat{G}$ is a topological group. This follows easily from the characterization of convergence given in 16.6.5: Let $\xi_\alpha \to \xi$ and $\zeta_\alpha \to \zeta$ uniformly on compact sets K. Then $\xi_\alpha^{-1} = \overline{\xi_\alpha} \to \overline{\xi} = \xi^{-1}$ uniformly on K, and from the inequality

$$|\xi_\alpha\zeta_\alpha - \xi\zeta| \le |\xi_\alpha\zeta_\alpha - \xi\zeta_\alpha| + |\xi\zeta_\alpha - \xi\zeta| = |\xi_\alpha - \xi| + |\zeta_\alpha - \zeta|$$

we see that $\xi_\alpha\zeta_\alpha \to \xi\zeta$ uniformly on K. □

The topological group $\widehat{G}$ is called the **dual group** of G. The following examples give concrete representations of various dual groups.

16.6.7 Examples.

(a) *The dual of $\mathbb{R}$ is $\mathbb{R}$*: Every character of $(\mathbb{R}, +)$ is of the form $\xi_y(x) := e^{iyx}$, where $y \in \mathbb{R}$. Indeed, if ξ is a character of $\mathbb{R}$, then for any $a, x \in \mathbb{R}$,

$$\int_x^{a+x} \xi(t)\,dt = \int_0^a \xi(x+t)\,dt = \xi(x)\int_0^a \xi(t)\,dt.$$

Choosing a such that $\alpha := \int_0^a \xi(t)\,dt \neq 0$ (possible because $\xi(0) = 1$), we have

$$\xi(x) = \frac{1}{\alpha}\int_x^{a+x} \xi(t)\,dt,$$

which shows that ξ is differentiable with derivative

$$\xi'(x) = \alpha^{-1}\big[\xi(a+x) - \xi(x)\big] = \beta\,\xi(x), \quad \beta := \alpha^{-1}[\xi(a) - 1].$$

Therefore, $\xi = \xi_y$, where $y = \beta/i$, verifying the assertion. The mapping $y \to \xi_y : \mathbb{R} \to \widehat{\mathbb{R}}$ is easily seen to be a homeomorphism and a group isomorphism. Thus $(\mathbb{R}, +)$ is its own dual.

(b) *The dual of* $\mathbb{T}$ *is* $\mathbb{Z}$: Every character ξ of $(\mathbb{T}, \cdot)$ is of the form $\xi_n(z) = z^n$, where $n \in \mathbb{Z}$. Indeed, the mapping $x \mapsto \xi(e^{ix})$ is a character of $\mathbb{R}$, hence $\xi(e^{ix}) = e^{ixy}$ for some $y \in \mathbb{R}$ and all x. Since $e^{2\pi iy} = \xi(e^{2\pi i}) = \xi(1) = 1$, y must be an integer n. Therefore, $\xi(e^{ix}) = (e^{ix})^n = \xi_n(e^{ix})$ for all x, verifying the assertion. Using 16.6.5, one easily sees that the topology on $\widehat{\mathbb{T}}$ is discrete and that the mapping $n \to \xi_n : \mathbb{Z} \to \widehat{\mathbb{T}}$ group isomorphism.

(c) *The dual of* $\mathbb{Z}$ *is* $\mathbb{T}$: Every character ξ of $(\mathbb{Z}, +)$ is of the form $\xi_z(n) = z^n$, where $|z| = 1$; simply take $z = \xi(1)$. The mapping $z \to \xi_z : \mathbb{T} \to \widehat{\mathbb{Z}}$ is clearly a homeomorphism and a group isomorphism. ♢

Higher dimensional versions of the above examples may be obtained with the aid of the following.

16.6.8 Proposition. *Let G_j be a locally compact abelian topological group $(1 \le j \le d)$ and let G denote the product group $\prod_j G_j$. Then the product group $\prod_j \widehat{G}_j$ is isomorphic and homeomorphic to $\widehat{G}$ under the mapping $(\xi_1, \ldots, \xi_d) \to \xi_1 \otimes \cdots \otimes \xi_d$, where*

$$\langle (x_1, \ldots, x_d), \xi_1 \otimes \cdots \otimes \xi_d \rangle := \prod_j \langle x_j, \xi_j \rangle .$$

Proof. $\xi_1 \otimes \cdots \otimes \xi_d$ is clearly a character, and an arbitrary character ξ is of this form, where $\xi_j(x) := \xi(e, \ldots, e, \overset{j}{x}, e \ldots, e)$. □

From the proposition and the above examples, we have the identifications

$$\widehat{\mathbb{R}^d} \cong \mathbb{R}^d, \quad \widehat{\mathbb{T}^d} \cong \mathbb{Z}^d, \text{ and } \widehat{\mathbb{Z}^d} \cong \mathbb{T}^d.$$

Note that in each case $\widehat{\widehat{G}} \cong G$. That this holds in general is the content of the Pontrjagin Duality Theorem, proved later.

We conclude this subsection with the following characterization of the dual of a quotient group.

16.6.9 Theorem. *Let H be a closed subgroup of G, $Q : G \to G/H$ the quotient map, and set $H^\perp := \{\xi \in \widehat{G} : \xi(H) = \{1\}\}$. Then $\Psi(\zeta) := \zeta \circ Q$ defines a topological isomorphism of $\widehat{G/H}$ onto $H^\perp$.*

Proof. First, $\zeta \circ Q$ is a continuous homomorphism and $(\zeta \circ Q)(H) = \{\zeta(Q(e))\} = \{1\}$, hence Ψ maps into $H^\perp$. Since $\Psi(\zeta_1\zeta_2) := (\zeta_1\zeta_2) \circ Q = (\zeta_1 \circ Q)(\zeta_2 \circ Q)$, Ψ is a homomorphism. Now let $\xi \in H^\perp$ and define ζ on G/H by $\zeta \circ Q = \xi$. Then ζ is well-defined, since $Q(x) = Q(y)$ $\Rightarrow Q(xy^{-1}) = Q(e) \Rightarrow xy^{-1} \in H \Rightarrow \xi(x) = \xi(y)$. Also, ζ is a homomorphism into $\mathbb{T}$ with $\zeta(Q(e)) = 1$, and since Q is an open map, ζ is continuous. Thus Ψ maps $\widehat{G/H}$ onto $H^\perp$. Since $\Psi(\zeta) = 1$ implies $\zeta = 1$, Ψ is a group isomorphism.

It remains to show that Ψ is a homeomorphism, that is, $\zeta_\alpha \to 1$ uniformly on compact subsets of G/H iff $\zeta_\alpha \circ Q \to 1$ on uniformly compact subsets of G. The necessity is clear, since if K is compact in G, then $Q(K)$ is compact in G/H. For the sufficiency, it suffices to show that if C is compact in G/H, then there exists a compact $K \subseteq G$ such that $Q(K) = C$. To construct K, let U be an open neighborhood of e in G with compact closure V. Since the open sets $Q(xU)$ $(x \in G)$ cover C, there exist $x_1, \ldots, x_n \in G$ such that $C \subseteq \bigcup_j Q(x_jU)$. Then $K := Q^{-1}(C) \cap \bigcup_j x_jV$ satisfies the requirements. □

For example, by the theorem the dual of $\mathbb{R}/\mathbb{Z}$ consists of all characters on $\mathbb{R}$ of the form $x \mapsto e^{2\pi i n x}$ ($n \in \mathbb{Z}$), which implies that $\widehat{\mathbb{R}/\mathbb{Z}}$ is isomorphic to $\mathbb{Z}$. The latter can also be seen from the fact that $\mathbb{R}/\mathbb{Z}$ is topologically isomorphic to $\mathbb{T}$ under the map $x + \mathbb{Z} \mapsto e^{ix}$ and that the dual of $\mathbb{T}$ is $\mathbb{Z}$.

16.6.10 Corollary. *If $x \in G \setminus H$, then there exists $\xi \in H^{\perp}$ such that $\langle x, \xi \rangle \neq 1$.*

Proof. By 16.5.9 and the Gelfand-Raikov theorem, the characters of a locally compact abelian group separate points. Thus we may choose $\zeta \in \widehat{G/H}$ such that $\zeta(xH) \neq 1$. Then $\xi := \zeta \circ Q$ has the desired properties. □

Bochner's Theorem

A function ϕ on G is said to be **represented by** $\mu \in M_{ra}(\widehat{G})$ if

$$\phi(x) = \int \langle x, \xi \rangle \, d\mu(\xi), \quad x \in G. \tag{16.12}$$

The theorem proved in this subsection gives necessary and sufficient conditions on ϕ for such a representation to exist. We shall need the following lemma.

16.6.11 Lemma. *Let μ and ν be complex Radon measures on $\widehat{G}$ such that*

$$\int \langle x, \xi \rangle \, d\mu(\xi) = \int \langle x, \xi \rangle \, d\nu(\xi) \quad \textit{for all} \quad x \in G.$$

Then $\mu = \nu$.

Proof. First, note that for $f \in L^1(G)$,

$$\iint f(x) \langle x, \xi \rangle \, d\mu(\xi) \, dx = \iint f(x) \langle x, \xi \rangle \, dx \, d\mu(\xi) = \int \widehat{f}(\xi^{-1}) \, d\mu(\xi),$$

and similarly for ν. Thus $\int \widehat{f}(\xi^{-1}) \, d\nu(\xi) = \int \widehat{f}(\xi^{-1}) \, d\mu(\xi)$ for all $f \in L^1(G)$. Since the space of Fourier transforms is dense in $C_0(\widehat{G})$ (16.6.2), the measures μ and ν are equal. □

16.6.12 Theorem (Bochner). *A function ϕ on G is represented by $\mu \in M_{ra}(\widehat{G})$ iff $\phi \in \mathscr{P}(G)$, in which case μ is unique. Moreover, if $\|\phi\|_\infty = 1$, then μ is a probability measure.*

Proof. Uniqueness follows from 16.6.11. If (16.12) holds, then for any $f \in L^1(G)$,

$$\begin{aligned}\int (f^* * f)\phi &= \iint f(x)\overline{f(y)}\phi(y^{-1}x) \, dx \, dy = \iiint f(x)\overline{f(y)} \langle y^{-1}x, \xi \rangle \, d\mu(\xi) \, dx \, dy \\ &= \iiint f(x) \langle x, \xi \rangle \overline{f(y) \langle y, \xi \rangle} \, dx \, dy \, d\mu(\xi) = \int |\widehat{f}(\xi^{-1})|^2 \, d\mu(\xi) \geq 0,\end{aligned}$$

hence ϕ is of positive type. That ϕ is continuous follows from inner regularity of μ and 16.6.4. Therefore, $\phi \in \mathscr{P}(G)$.

Conversely, let $\phi \in \mathscr{P}(G)$. We may assume that $\|\phi\|_\infty = 1$. By the CBS inequality (see proof of 16.5.7),

$$\left| \int (g^* * f)\phi \right|^2 \leq \left[\int (f^* * f)\phi \right] \left[\int (g^* * g)\phi \right], \quad f, g \in L^1(G). \tag{†}$$

Now let (ψ_V) be an approximate identity in $L^1(G)$ and take $g = \psi_V$ in (†). Since $\|\phi\|_\infty = 1$ and $\int \psi_V^* * \psi_V = \left|\int \psi_V\right|^2 = 1$, we have

$$\left|\int (\psi_V^* * f)\phi\right|^2 \le \int (f^* * f)\phi.$$

Letting $V \to e$ we obtain

$$\left|\int f\phi\right| \le \left|\int (f^* * f)\phi\right|^{1/2}, \quad f \in L^1(G).$$

Now set $h = f^* * f$ and $h_n = h * \cdots * h$ (n factors). Iterating the preceding inequality, noting that $h^* = h$, we have

$$\left|\int f\phi\right| \le \left|\int h\phi\right|^{1/2} \le \left|\int h_2\phi\right|^{1/4} \le \cdots \le \left|\int h_{2n}\phi\right|^{1/2^{n+1}} \le \|h_{2n}\|_1^{1/2^{n+1}}.$$

By 13.5.1 $\|h_{2n}\|_1^{1/2^{n+1}} \to \|\widehat{h}\|_\infty^{1/2} = \|\,|\widehat{f}|^2\|_\infty^{1/2} = \|\widehat{f}\|_\infty$, hence

$$\left|\int f\phi\right| \le \|\widehat{f}\|_\infty, \quad f \in L^1(G).$$

Define a linear functional F on $L^1(G)\widehat{\ } \subseteq C_0(\widehat{G})$ by $F(\widehat{f}) = \int f\phi$. The preceding inequality shows that F is well-defined and $\|F(\widehat{f})\|_\infty \le \|\widehat{f}\|_\infty$. Since $L^1(G)\widehat{\ }$ is dense in $C_0(\widehat{G})$ (16.6.2), F has a continuous extension to $C_0(\widehat{G})$ with $\|F\|_\infty \le 1$. By the Riesz representation theorem, there exists a $\nu \in M_{ra}(\widehat{G})$ such that for all $f \in L^1(G)$,

$$\int f\phi = F(\widehat{f}) = \int \widehat{f}\, d\nu = \iint \langle x, \xi^{-1}\rangle f(x)\, dx\, d\nu(\xi) = \int f(x) \int \langle x, \xi^{-1}\rangle\, d\nu(\xi)\, dx,$$

hence

$$\phi(x) = \int \langle x, \xi^{-1}\rangle\, d\nu(\xi) = \int \langle x, \xi\rangle\, d\mu(\xi),$$

where $d\mu(\xi) := d\nu(\xi^{-1})$. Finally, from $\mu(\widehat{G}) = \phi(e) = \|\phi\|_\infty = 1$ we see that $\mu \ge 0$. □

The Inversion Theorem

In this subsection we show that, for a suitable class of functions f, the Fourier transform $f \mapsto \widehat{f}$ may be inverted. The proof for the special case $G = \mathbb{R}^d$ given in Chapter 6 relied on rapidly decreasing functions. As these are not available here, the proof for the general case is based instead on functions of positive type. We begin with

16.6.13 Lemma. *Let $K \subseteq \widehat{G}$ be compact. Then there exists $f \in \mathscr{P}_c(G) := \mathscr{P}(G) \cap C_c(G)$ such that $\widehat{f} \ge 0$ and $\widehat{f} > 0$ on K.*

Proof. Let $g = \psi^* * \psi$, where $\psi \in C_c(G)$ and $\int \psi = 1$. By translation invariance and unimodularity,

$$\widehat{g}(\xi) = \int \overline{\langle x, \xi\rangle} g(x)\, dx = \iint \overline{\langle x, \xi\rangle \psi(y^{-1})}\psi(y^{-1}x)\, dy\, dx = \iint \overline{\langle x, \xi\rangle \psi(y)}\psi(yx)\, dy\, dx$$
$$= \iint \overline{\langle x, \xi\rangle \psi(y)}\psi(yx)\, dx\, dy = \iint \overline{\langle y^{-1}x, \xi\rangle \psi(y)}\psi(x)\, dy\, dx = |\widehat{\psi}(\xi)|^2.$$

In particular, $\widehat{g} \geq 0$ and $\widehat{g}(e) = 1$. By continuity, there exists an open neighborhood U of e in $\widehat{G}$ such that $\widehat{g} > 0$ on U. Since K is compact, there exist $\xi_j \in K$ such that $K \subseteq \bigcup_{j=1}^n \xi_j U$. Set $f = g \sum_{j=1}^n \xi_j$. Then $f \in C_c(G)$ and

$$\widehat{f}(\xi) = \sum_{j=1}^n \int \overline{\langle x, \xi \rangle}\, \langle x, \xi_j \rangle\, g(x)\, dx = \sum_{j=1}^n \int \overline{\langle x, \xi\xi_j^{-1} \rangle} g(x)\, dx = \sum_{j=1}^n \widehat{g}(\xi\xi_j^{-1}),$$

hence $\widehat{f} \geq 0$ on $\widehat{G}$ and $\widehat{f} > 0$ on K. Finally, for any $\xi \in \widehat{G}$ and $h \in L^1(G)$,

$$\begin{aligned}\int (h^* * h)(\xi g) &= \iint \overline{h(y^{-1})} h(y^{-1}x)\xi(x)g(x)\, dy\, dx = \iint \overline{(\xi h)(y^{-1})}(\xi h)(y^{-1}x)g(x)\, dy\, dx \\ &= \int [(\xi h)^* * (\xi h)]g \geq 0,\end{aligned}$$

hence $\int (h^* * h)f \geq 0$ and so $f \in \mathscr{P}(G)$. □

Here is the promised inversion theorem. For convenience, we indicate the property of the function f described in the conclusion of the last lemma by writing $f \sim K$.

16.6.14 Theorem. *If $f \in \mathscr{S} := L^1(G) \cap \operatorname{span} \mathscr{P}(G)$, then $\widehat{f} \in L^1(\widehat{G})$ and*

$$f(x) = \int \langle x, \xi \rangle\, \widehat{f}(\xi)\, d\xi, \quad x \in G, \tag{16.13}$$

where $d\xi$ is a suitably normalized Haar measure on $\widehat{G}$.

Proof. We give the proof in several steps:

(1) *For each $f \in \mathscr{S}$ there exists $\mu_f \in M_{ra}\widehat{G}$ such that $f(x) = \int \langle x, \xi \rangle\, d\mu_f(\xi)$. Moreover, $\widehat{f}\, d\mu_g = \widehat{g}\, d\mu_f$.*

⟦The first assertion follows from Bochner's theorem. For the second, let $h \in L^1(G)$. Then

$$\begin{aligned}\int \widehat{h}\,\widehat{g}\, d\mu_f &= \int \widehat{h * g}\, d\mu_f = \int \overline{\langle x, \xi \rangle}(h * g)(x)\, dx\, d\mu_f(\xi) = \int (h * g)(x) f(x^{-1})\, dx \\ &= [(h * g) * f](e).\end{aligned}$$

Similarly $\int \widehat{h}\,\widehat{f}\, d\mu_g = [(h * f) * g](e)$. Since $[(h * g) * f] = [(h * f) * g]$ we have

$$\int \widehat{h}\,\widehat{g}\, d\mu_f = \int \widehat{h}\,\widehat{f}\, d\mu_g \quad \text{for all} \quad h \in L^1(G).$$

Since the Fourier transforms $\widehat{h}$ are dense in $C_0(\widehat{G})$, $\widehat{g}\, d\mu_f = \widehat{f}\, d\mu_g$.⟧

(2) *For $\varphi \in C_c(\widehat{G})$, by 16.6.13 choose $f \in \mathscr{P}_c(G)$ such that $f \sim \operatorname{supp} \varphi$. Define*

$$I(\varphi) = \int \frac{\varphi}{\widehat{f}}\, d\mu_f = \int_{\operatorname{supp} \varphi} \frac{\varphi}{\widehat{f}}\, d\mu_f.$$

Then I is independent of the choice of f and is a positive linear functional on $C_c(\widehat{G})$.

⟦If also $g \sim \operatorname{supp} \varphi$, then by step (1),

$$\int \frac{\varphi}{\widehat{f}}\, d\mu_f = \int \frac{\varphi}{\widehat{f}\,\widehat{g}} \widehat{g}\, d\mu_f = \int \frac{\varphi}{\widehat{f}\,\widehat{g}} \widehat{f}\, d\mu_g = \int \frac{\varphi}{\widehat{g}}\, d\mu_g.$$

Clearly, I is positive and $I(cf) = cI(f)$. To verify additivity, let $f \sim \operatorname{supp}\varphi_1 \cup \operatorname{supp}\varphi_2$. Then $f \sim \operatorname{supp}\varphi_j$, hence

$$I(\varphi_1 + \varphi_2) = \int \frac{\varphi_1 + \varphi_2}{\widehat{f}}\, d\mu_f = \int \frac{\varphi_1}{\widehat{f}}\, d\mu_f + \int \frac{\varphi_2}{\widehat{f}}\, d\mu_f = I(\varphi_1) + I(\varphi_2).]\!]$$

(3) *For $g \in \mathcal{S}$ and $\varphi \in C_c(\widehat{G})$, $I(\varphi\widehat{g}) = \int \varphi\, d\mu_g$. In particular, I is nontrivial.*

[[By step (1),

$$I(\varphi\widehat{g}) = \int \frac{\varphi\widehat{g}}{\widehat{f}}\, d\mu_f = \int \frac{\varphi}{\widehat{f}} \widehat{f}\, d\mu_g = \int \varphi\, d\mu_g.$$

Now choose g and φ so that $\int \varphi\, d\mu_g \neq 0$.]]

(4) *I is translation invariant.*

[[Fix $\zeta \in \widehat{G}$ and set $\tau(\xi) := \xi\zeta$. For the image measure $\tau(\mu_f)$ we have

$$\int \langle x, \xi\rangle\, d\mu_{\zeta f}(\xi) = (\zeta f)(x) = \int \langle x, \zeta\xi\rangle\, d\mu_f(\xi) = \int \langle x, \xi\rangle\, d\tau(\mu_f)(\xi),$$

hence $d\mu_{\zeta f} = d\tau(\mu_f)$ by 16.6.11. Therefore, if $f > 0$ on $\operatorname{supp}\varphi \cup \operatorname{supp} R_\zeta\varphi$, then

$$I(R_\zeta \varphi) = \int \frac{\varphi(\xi\zeta)}{\widehat{f}(\xi)}\, d\mu_f = \int \frac{\varphi(\xi)}{\widehat{f}(\xi\zeta^{-1})}\, d\mu_{\zeta f}(\xi) = \int \frac{\varphi(\xi)}{\widehat{(\zeta f)}(\xi)}\, d\mu_{\zeta f}(\xi) = I(\varphi).]\!]$$

To complete the proof of the theorem, let $d\xi$ denote the Haar measure corresponding to the linear functional I, and let $f \in \mathcal{S}$. By step (3),

$$\int \varphi(\xi)\widehat{f}(\xi)\, d\xi = \int \varphi(\xi)\, d\mu_f(\xi) \quad \text{for all} \quad \varphi \in C_c(\widehat{G}).$$

It follows that $\widehat{f}(\xi)\, d\xi = d\mu_f(\xi)$ and $\widehat{f} \in L^1(\widehat{G})$. Recalling the defining property of μ_f in step (1), we see that (16.13) holds. □

The following is a special case of the Gelfand-Raikov theorem. We give a simple independent proof based on the preceding theorem.

16.6.15 Corollary. *The characters of G separate points of G.*

Proof. Let $x, y \in G$ with $x \neq y$. Choose $f \in C_c(G)$ such that $f(x) \neq f(y)$. Since $\operatorname{span}\mathscr{P}_c(G)$ is dense in $C_c(G)$, $g(x) \neq g(y)$ for some $g \in \operatorname{span}\mathscr{P}_c(G)$. Since

$$\int \langle x, \xi\rangle\, \widehat{g}(\xi)\, d\xi = g(x) \neq g(y) = \int \langle y, \xi\rangle\, \widehat{g}(\xi)\, d\xi,$$

$\langle x, \xi\rangle \neq \langle y, \xi\rangle$ for some $\xi \in \widehat{G}$. □

For a given Haar measure dx on G, the measure $d\xi$ for which the conclusion of the theorem holds is called the **dual measure of** dx. For example, in 6.2.4 we had the formulas

$$\widehat{f}(\xi) = \int f(x)\overline{\langle x, \xi\rangle}\, dx \quad \text{and} \quad f(x) = \int \widehat{f}(\xi)\, \langle x, \xi\rangle\, d\xi, \quad \langle x, \xi\rangle := e^{2\pi i \xi x}.$$

The map $\langle x, \xi\rangle$ identifies $\mathbb{R}$ with its dual, and under this identification the dual of Lebesgue measure is itself.

16.6.16 Proposition. *If G is compact, then $\widehat{G}$ has the discrete topology. Moreover, if Haar measure on G is normalized so that $|G| = 1$, then the characters form an orthonormal set in $L^2(G) \subseteq L^1(G)$ and the dual measure is counting measure.*

Proof. If $|G| = 1$ and $\xi \in \widehat{G}$, then $\xi \in L^1(G)$ and for all y

$$\int \langle x, \xi \rangle \, dx = \int \langle xy, \xi \rangle \, dx = \langle y, \xi \rangle \int \langle x, \xi \rangle \, dx.$$

Thus if $\xi \neq 1$, then $\int_G \xi = 0$. It follows that $\int_G \xi\overline{\zeta} = 1$ or 0 according as $\xi = \zeta$ or $\xi \neq \zeta$, that is, the characters form an orthonormal set in $L^2(G)$.

Since the function $\phi \mapsto \int \phi$ is weak* continuous on $C(G)$, $U := \{\zeta \in \widehat{G} : |\int \xi - 1| < 1/2\}$ is open in $\widehat{G}$. But $\int \xi = 0$ or 1, hence $U = \{1\}$. Therefore, $\{1\}$ is open, which implies that $\widehat{G}$ is discrete.

Now, if $g = 1$ on G, then $\widehat{g}(\xi) = \int \overline{\xi} = \mathbf{1}_{\{1\}}(\xi)$. Therefore, if $d\mu(\xi)$ denotes the dual measure on $\widehat{G}$, then, by the inversion theorem,

$$1 = g(e) = \int \langle e, \xi \rangle \, \widehat{g}(\xi) \, d\mu(\xi) = \mu\{e\}.$$

By translation invariance, $\mu\{x\} = 1$ for all $x \in G$. Thus μ is counting measure. □

Here is the dual of the preceding proposition.

16.6.17 Proposition. *If G is discrete, then $\widehat{G}$ is compact. Moreover, if Haar measure on G is counting measure, then the dual measure on $\widehat{G}$ satisfies $|\widehat{G}| = 1$.*

Proof. If G is discrete, then the Dirac function δ_e is an identity for $L^1(G)$, hence the spectrum $\widehat{G}$ of $L^1(G)$ is compact. If Haar measure dx on G is counting measure and $f = \mathbf{1}_{\{e\}}$, then

$$\widehat{f}(\xi) = \int_G \overline{\langle x, \xi \rangle} f(x) \, dx = \overline{\langle e, \xi \rangle} = 1, \quad \xi \in \widehat{G},$$

hence

$$1 = f(e) = \int_{\widehat{G}} \langle e, \xi \rangle \widehat{f}(\xi) \, d\xi = \int_{\widehat{G}} 1 \, d\xi = |\widehat{G}|.$$ □

For example, consider the compact group $\mathbb{T}$ with Haar measure $d\theta/2\pi$ and dual group $\mathbb{Z}$ with counting measure. The characters are, respectively, $\xi_n(z) = z^n$ and $\xi_\theta(n) = e^{in\theta}$ hence, the inversion theorem in this setting is

$$\widehat{f}(n) = \int_0^{2\pi} f(\theta) e^{-in\theta} \frac{d\theta}{2\pi}, \quad f(\theta) = \sum_{n=-\infty}^{\infty} \widehat{f}(n) e^{in\theta}.$$

The Plancherel Theorem

The L^2 properties of the Fourier transform on G are given in the following result:

16.6.18 Theorem (Plancherel). *The Fourier transform $f \mapsto \widehat{f}$ on $L^1(G) \cap L^2(G)$ extends uniquely to a unitary transformation from $L^2(G)$ onto $L^2(\widehat{G})$.*

Proof. Let $f \in L^1(G) \cap L^2(G)$. By 16.5.4, $f * f^* \in \mathscr{P}(G)$, hence we may apply the inversion formula to $f * f^*$ to obtain

$$\int |f(x)|^2 \, dx = f * f^*(e) = \int \langle e, \xi \rangle \, (\widehat{f * f^*})(\xi) \, d\xi = \int |\widehat{f}(\xi)|^2 \, d\xi,$$

the last equality by (16.10). This shows that the Fourier transform is an L^2-isometry from $L^1(G)\cap L^2(G)$ to $L^2(\widehat{G})$. Since $L^1(G)\cap L^2(G)$ contains $C_c(G)$, which is dense in $L^2(G)$, the transform has a unique extension to an isometry T from $L^2(G)$ into $L^2(\widehat{G})$. It remains to show that T is surjective. For this it suffices to show that the image of $L^1(G)\cap L^2(G)$ under T has a trivial orthogonal complement. To this end, let $\varphi \in L^2(\widehat{G})$ with $\varphi \perp \big(L^1(G)\cap L^2(G)\big)\widehat{\ }$. For any $x\in G$, $\xi\in\widehat{G}$, and $f\in L^1(G)\cap L^2(G)$,

$$\widehat{(R_x f)}(\xi) = \int \overline{\langle y,\xi\rangle} f(yx)\,dy = \int \overline{\langle yx^{-1},\xi\rangle} f(y)\,dy = \langle x,\xi\rangle \int \overline{\langle y,\xi\rangle} f(y)\,dy = \langle x,\xi\rangle\,\widehat{f}(\xi),$$

hence,

$$\int \varphi(\xi)\,\langle x,\xi\rangle\,\widehat{f}(\xi)\,d\xi = \int \varphi(\xi)\widehat{(R_x f)}(\xi)\,d\xi = 0.$$

Since $\varphi\widehat{f}\in L^1(\widehat{G})$, $d\mu(\xi) := \varphi(\xi)\widehat{f}(\xi)\,d\xi$ is a complex Radon measure on $\widehat{G}$. We now have $\int \langle x,\xi\rangle\,d\mu = 0$ for all x, so by 16.6.11, μ is the zero measure. This implies that $\varphi\widehat{f}=0$ a.e. Since this holds for all $f\in L^1(G)\cap L^2(G)$, it follows from 16.6.13 that $\varphi=0$ a.e. on each compact subset of $\widehat{G}$. Therefore,

$$(\varphi\,|\,g) = \int \varphi\overline{g} = 0 \quad \text{for all } g\in C_c(\widehat{G}).$$

Since $C_c(\widehat{G})$ is dense in $L^2(\widehat{G})$ (7.1.2), $\varphi=0$ a.e. on $\widehat{G}$. Therefore, T is surjective. □

We shall use the notation $\widehat{f}$ to indicate the image of $f\in L^2(G)$ under the unitary transformation of the theorem. By the unitary property we have **Parseval's formula**

$$\int f(x)\overline{g(x)}\,dx = \int \widehat{f}(\xi)\overline{\widehat{g}(\xi)}\,d\xi \quad f,\, g\in L^2(G). \tag{16.14}$$

16.6.19 Corollary. *If G is compact with $|G|=1$, then $\widehat{G}$ is an orthonormal basis for $L^2(G)$.*

Proof. By 16.6.16, $\widehat{G}$ is orthonormal. To show that $\widehat{G}$ is complete, let $f\in L^2(G)$ such that $f\perp\widehat{G}$. Then $\widehat{f}(\xi)=\int f(x)\overline{\langle x,\xi\rangle}\,dx = 0$ for all ξ. Since $f\mapsto\widehat{f}$ is an isometry, $f=0$. □

The Pontrjagin Duality Theorem

We showed earlier that the mapping $(x,\xi)\mapsto\langle x,\xi\rangle : G\times\widehat{G}\to\mathbb{T}$ is jointly continuous. Moreover, by definition of multiplication on $\widehat{G}$, for each $x\in G$ $\langle x,\xi\zeta\rangle = \langle x,\xi\rangle\langle x,\zeta\rangle$. Thus the function $\widehat{x} := \langle x,\cdot\rangle$ is a character on $\widehat{G}$, that is, $\widehat{x}\in\widehat{\widehat{G}}$. Moreover, the mapping

$$\Phi : G\to\widehat{\widehat{G}},\quad \Phi(x)=\widehat{x}.$$

is a group homomorphism, since $\langle\xi,\widehat{xy}\rangle = \langle xy,\xi\rangle = \langle x,\xi\rangle\langle y,\xi\rangle = \langle\xi,\widehat{x}\rangle\langle\xi,\widehat{y}\rangle$. Since Φ is 1-1 (16.6.15), Φ a group isomorphism. In this section we show that $\Phi(G)=\widehat{\widehat{G}}$ and that Φ is a homeomorphism. For this we need the following:

16.6.20 Lemma. *If K is a proper closed subset of G, then there exist $\phi_i\in C_c(G)$ such that $\varphi_1*\varphi_2\ge 0$, $\varphi_1*\varphi_2\ne 0$, and $\varphi_1*\varphi_2=0$ on K.*

Proof. Choose $x\in K^c$ and a symmetric neighborhood U of e such that $xUU\subseteq K^c$. Let $\varphi_i\in C_c^+(G)$ such that $\operatorname{supp}\varphi_1\subseteq xU$ and $\operatorname{supp}\varphi_2\subseteq U$. Then $\varphi_1*\varphi_2$ has the required properties. For example, from $\varphi_1*\varphi_2(y)=\int_{xU}\varphi_1(z)\varphi_2(z^{-1}y)\,dz$ we see that if $y\in K$, then $y\notin xUU$, hence the integrand is zero over xU. □

16.6.21 Lemma. *If $\phi_i \in C_c(\widehat{G})$, then $\phi_1 * \phi_2 = \widehat{f}$ for some $f \in L^1(G)$.*

Proof. Set $\phi := \phi_1 * \phi_2$. For $x \in G$, define

$$f_j(x) = \int \langle x, \xi \rangle \, \phi_j(\xi) \, d\xi \quad \text{and} \quad f(x) = \int \langle x, \xi \rangle \, \phi(\xi) \, d\xi.$$

Since $\phi_j(\xi) \, d\xi$, $\phi(\xi) \, d\xi \in M_{ra}(\widehat{G})$, f_j, f are in the linear span of $\mathscr{P}(G)$ (16.6.12). Moreover, for any $g \in L^1(G) \cap L^2(G)$ we have

$$\int_G f_j \overline{g} = \iint \langle x, \xi \rangle \, \phi_j(\xi) \overline{g(x)} \, d\xi \, dx = \int \phi_j \overline{\widehat{g}}.$$

By the CBS inequality, the absolute value of the last term is $\le \|\phi_j\|_2 \, \|g\|_2$. Since g was arbitrary and $L^1(G) \cap L^2(G)$ is dense in $L^2(G)$, $f_j \in L^2(G)$. Since

$$f(x) = \iint \langle x, \xi \rangle \, \phi_1(\xi\zeta^{-1})\phi_2(\zeta) \, d\zeta \, d\xi = \iint \langle x, \xi\zeta \rangle \, \phi_1(\xi)\phi_2(\zeta) \, d\xi \, d\zeta = f_1(x) f_2(x),$$

$f \in L^1(G)$. Since also $f \in \operatorname{span} \mathscr{P}(G)$, by the inversion theorem $f(x) = \int \langle x, \xi \rangle \, \widehat{f}(\xi) \, d\xi$. Thus

$$\int \langle x, \xi \rangle \, \phi(\xi) \, d\xi = f(x) = \int \langle x, \xi \rangle \, \widehat{f}(\xi) \, d\xi \quad \text{for all } x \in G$$

and so by 16.6.11, $\widehat{f} = \phi$. □

16.6.22 Theorem (Pontrjagin). *The mapping is $\Phi : x \mapsto \langle x, \cdot \rangle$ a homeomorphism and group isomorphism from G onto $\widehat{\widehat{G}}$.*

Proof. Let $x_\alpha \to x$ in G. Since $\langle x_\alpha, \xi \rangle \to \langle x, \xi \rangle$ uniformly in ξ on compact subsets of $\widehat{G}$ (16.6.4), $\Phi(x_\alpha) \to \Phi(x)$ in $\widehat{\widehat{G}}$. Conversely, let $\Phi(x_\alpha) \to \Phi(x)$ in $\widehat{\widehat{G}}$. By 16.6.5,

$$\langle x_\alpha, \xi \rangle = \langle \xi, \widehat{x}_\alpha \rangle \to \langle \xi, \widehat{x} \rangle = \langle x, \xi \rangle$$

uniformly in ξ on compact subsets of $\widehat{G}$. Thus for all $f \in \operatorname{span} \mathscr{P}_c(G)$,

$$f(x_\alpha) = \int \langle x_\alpha, \xi \rangle \, \widehat{f}(\xi) \, d\xi \to \int \langle x, \xi \rangle \, \widehat{f}(\xi) \, d\xi = f(x).$$

Since $\operatorname{span} \mathscr{P}_c(G)$ is dense in $C_c(G)$ (16.5.5), $f(x_\alpha) \to f(x)$ for all $f \in C_c(G)$. This implies that $x_\alpha \to x$. Otherwise, there would exist a compact neighborhood U of x such that x_α is frequently in U^c, and we obtain a contradiction by choosing $f \in C_c(G)$ such that $f(x) = 1$ and $f = 0$ on U^c. Therefore, Φ a homeomorphism of G onto $\Phi(G)$.

It remains to show that $\Phi(G) = \widehat{\widehat{G}}$. Now, because Φ is a homeomorphism and group isomorphism, $\Phi(G)$ is a locally compact subgroup of $\widehat{\widehat{G}}$ and hence is closed (16.1.2). Suppose for a contradiction that $\Phi(G) \subsetneqq \widehat{\widehat{G}}$. By 16.6.20, there exists a nonzero convolution $\varphi := \varphi_1 * \varphi_2$ of functions $\varphi_j \in C_c(\widehat{\widehat{G}})$ that vanishes identically on $\Phi(G)$. By 16.6.21, $\varphi = \widehat{f}$ for some $f \in L^1(\widehat{G})$. In particular, for all $x \in G$,

$$0 = \varphi(\widehat{x}) = \int_{\widehat{G}} \langle \xi, \widehat{x} \rangle \, f(\xi) \, d\xi = \int_{\widehat{G}} \langle x, \xi \rangle \, f(\xi) \, d\xi.$$

But then by 16.6.11, $f = 0$, producing the contradiction $\varphi \equiv 0$. Therefore, $\Phi(G) = \widehat{\widehat{G}}$, completing the proof. □

Here is an application, the analog of 16.6.9.

16.6.23 Theorem. *If H is a closed subgroup of G, then $\widehat{H}$ is topologically isomorphic to $\widehat{G}/H^{\perp}$ and $\widehat{H} = \widehat{G}\big|_H$.*

Proof. Let $Q : \widehat{G} \to \widehat{G}/H^{\perp}$ denote the quotient map. By 16.6.9 applied to $\widehat{G}/H^{\perp}$, $\big(\widehat{G}/H^{\perp}\big)\widehat{\ }$ is topologically isomorphic to $H^{\perp\perp}$ under the mapping $F \mapsto F \circ Q$. By 16.6.10, H is topologically isomorphic to $H^{\perp\perp}$ under duality. Therefore, H is topologically isomorphic to $\big(\widehat{G}/H^{\perp}\big)\widehat{\ }$under a mapping $x \mapsto F_x$, where $F_x \circ Q = \widehat{x}$. It follows that $\widehat{H}$ is topologically isomorphic to $\big(\widehat{G}/H^{\perp}\big)\widehat{\ }\widehat{\ }$ under a mapping $\zeta \mapsto T_\zeta$, where $\zeta(x) = T_\zeta(F_x)$. By the duality theorem, $T_\zeta = \widehat{Q(\xi)}$ for some $\xi \in \widehat{G}$. Thus $\zeta(x) = \widehat{Q(\xi)}(F_x) = F_x(Q(\xi)) = \xi(x)$ $(x \in H)$, that is, $\zeta = \xi\big|_H$. □

Chapter 17

Analysis on Semigroups

In this chapter we study representations of semigroups with a topology. Some of the results here rely on, and indeed may be be seen as extensions of, results in the Fourier analysis of groups discussed in the last chapter. In particular, compact topological groups and unitary representations play a central role.

Much of the material in this chapter is based on the papers [10], [11] and [16]. Generalizations and additional material, as well as detailed references, may be found in [4] and [41].

17.1 Semigroups with Topology

The underlying object of study in this chapter is the **semitopological semigroup**, defined as a semigroup S with a topology relative to which multiplication $(s,t) \to st$ is separately continuous. As the structure of a semitopological semigroup is not as rich as that of a locally compact topological group, one must rely on extrinsic techniques, notably functional analytic. In particular, the Gelfand theory applied to various C^*-algebras of functions on S is an important tool. For this we shall initially rely on the left and right translation operators on $C_b(S)$. These are defined exactly as in the group case, as are the notions of left translation invariance, right translation invariance, and translation invariance of subsets of $C_b(S)$. Note that the hypothesis of separate continuity of multiplication in S implies that $C_b(S)$ is itself translation invariant. The left and right translates $L_S f$ and $R_S f$ of f are left and right translation invariant, respectively, as may be seen from the inclusions

$$L_t L_S f = L_{St} f \subseteq L_S f \quad \text{and} \quad R_t R_S f = R_{tS} f \subseteq R_S f.$$

Much of the material in the chapter depends on the following result obtained in 13.4.2. Given a unital C^*-subalgebra $\mathscr{F}$ of $C_b(S)$ with spectrum $S^{\mathscr{F}}$, the evaluation mapping $\widehat{s}$ on $\mathscr{F}$ defined by

$$\widehat{s}(f) := f(s), \quad f \in \mathscr{F}$$

is member $S^{\mathscr{F}}$, and the canonical mapping

$$\iota = \iota_{\mathscr{F}} : S \to S^{\mathscr{F}}, \quad \iota(s) = \widehat{s},$$

is continuous with dense range and satisfying $\iota^* C\big(S^{\mathscr{F}}\big) = \mathscr{F}$.

In the sequel we shall also need the notions of **semitopological group** and **topological semigroup**. The former is a group with separately continuous multiplication (inversion is *not* assumed to to be continuous) and the latter is a semigroup with a topology relative to which multiplication is jointly continuous.

For the remainder of the chapter, unless otherwise stated,
S denotes an arbitrary semitopological semigroup.

17.2 Weakly Almost Periodic Functions

Definition and Basic Properties

A function $f \in C_b(S)$ is said to be **weakly almost periodic** if $R_S f$ is relatively weakly compact in $C_b(S)$. The set of all weakly almost periodic functions on S is denoted by $WAP(S)$. For example, if S is a compact semitopological semigroup, then $WAP(S) = C(S)$, as may be seen from the pointwise continuity of the mapping $s \to R_s f$ and 14.1.4.

17.2.1 Theorem. *$WAP(S)$ is a translation invariant unital C^*-algebra of $C_b(S)$.*

Proof. The relations

$$R_S(cf+g) \subseteq cR_S f + R_S g,\ R_s\overline{f} = \overline{R_s f},\ R_s R_S f = R_{sS} f \subseteq R_S f,\ R_S L_s f = L_s R_S f, \quad (17.1)$$

show that $WAP(S)$ is a conjugate closed translation invariant subspace of $C_b(S)$. Since $R_S(fg) \subseteq (R_S f)(R_S g)$, to show that $WAP(S)$ is an algebra it suffices to prove that the product $AB = \{fg : f \in A,\ g \in B\}$ of weakly compact subsets of $C_b(S)$ it weakly compact. Now, by 13.4.2, $C_b(S)$ is (canonically) isometrically isomorphic to $C(\beta S)$, where βS is the spectrum of $C_b(S)$. The images of A and B in $C(\beta S)$ are then weakly compact in $C(\beta S)$, and the assertion follows easily from the equivalence of pointwise and weak compactness in $C(\beta S)$ (14.1.1).

To show that $WAP(S)$ is closed, let $f_n \in WAP(S)$ and $f_n \to f \in C_b(S)$. We show that an arbitrary sequence $(R_{s_k} f)$ of right translates of f has a weakly convergent subsequence. It will follow from the Eberlein-Šmulian theorem that $R_S f$ is relatively weakly compact, as required. Now, since each $R_S f_n$ relatively weakly compact, a standard diagonal argument produces a subsequence (t_k) of (s_k) and a sequence (g_n) in $C_b(S)$ such that $g_n = w\text{-}\lim_k R_{t_k} f_n$ for all n. For any $\varphi \in C_b(S)'$ with $\|\varphi\| \le 1$ we then have

$$|\varphi(g_n) - \varphi(g_m)| = \lim_k |\varphi\big(R_{t_k}(f_n - f_m)\big)| \le \|f_n - f_m\|_\infty,$$

hence $\|g_n - g_m\|_\infty \le \|f_n - f_m\|_\infty$. Thus (g_n) converges in norm to some $g \in C_b(S)$. The inequality

$$\begin{aligned} |\varphi(R_{s_k} f - g)| &\le |\varphi(R_{s_k} f - R_{s_k} f_n)| + |\varphi(R_{s_k} f_n - g_n)| + |\varphi(g_n - g)| \\ &\le \|f_n - f\| + |\varphi(R_{s_k} f_n - g_n)| + \|g_n - g\| \end{aligned}$$

then shows that $\varphi(R_{s_k} f) \to \varphi(g)$. Therefore, $R_{s_k} f \xrightarrow{w} g$, as required. □

17.2.2 Proposition. *Let T be a semitopological semigroup and $\theta : S \to T$ a continuous homomorphism. Then $\theta^*\big(WAP(T)\big) \subseteq WAP(S)$, where $\theta^* : C(T) \to C(S)$ is the dual map. In particular, if S is a subsemigroup of T, then $WAP(T)|_S \subseteq WAP(S)$.*

Proof. The first assertion follows from the weak continuity of θ^* and the identity $R_s\theta^*(g) = \theta^*\big(R_{\theta(s)} g\big)$. For the second, take θ to be the inclusion map. □

The Dual of the Space of Weakly Almost Periodic Functions

We now give the dual space $WAP(S)'$ a multiplication that makes it a Banach algebra. For this we introduce the following operators on $WAP(S)$: Given $\varphi \in WAP(S)'$ define $R_\varphi : WAP(S) \to B(S)$ by

$$(R_\varphi f)(s) = \varphi(L_s f), \quad f \in WAP(S), \quad s \in S.$$

Then R_φ is obviously linear and satisfies

$$\|R_\varphi\| \le \|\varphi\|, \quad R_{\widehat{t}} = R_t, \quad \text{and} \quad L_t R_\varphi = R_\varphi L_t.$$

Additional properties of R_φ are given in the following lemma.

17.2.3 Lemma. *Let C_1' denote the closed unit ball in $WAP(S)'$, $M := M(WAP(S))$ the set of means on $WAP(S)$, and $X := S^{WAP}$ the spectrum of $WAP(S)$. Let $f \in WAP(S)$.*

(a) *$R_X f$ is the weak closure of $R_S f$ and is weakly compact.*

(b) *$R_M f$ is the weakly closed convex hull of $R_S f$ and is weakly compact.*

(c) *$R_{C_1'} f$ is the weakly closed convex balanced hull of $R_S f$ and is weakly compact.*

(d) *$R_\varphi WAP(S) \subseteq WAP(S)$ for all $\varphi \in WAP(S)'$.*

(e) *The mapping $\varphi \mapsto R_\varphi f$ from $WAP(S)'$ into $WAP(S)$ is weak* -weak continuous.*

Proof. For part (a), recall that X is the weak* closure of the set of mappings $\widehat{s} : f \to f(s)$ (13.4.2). Now let $\varphi \in X$ and let $\widehat{t_\alpha} \to \varphi$ in the weak* topology of $WAP(S)'$. Then for each $s \in S$,

$$R_{t_\alpha} f(s) = \widehat{t_\alpha}(L_s f) \to \varphi(L_s f) = R_\varphi f(s),$$

that is, $R_{t_\alpha} f \to R_\varphi f$, pointwise. Since $R_S f$ is relatively weakly compact, the convergence is also in the weak topology, proving (a).

For (b), recall from 10.2.7 that the convex hull of the set $\{\widehat{s} : s \in S\}$ is weak* -dense in M. Thus for each $\varphi \in M$ there exists a net of convex sums $\sum_j c_j^\alpha \widehat{s_j^\alpha}$ such that $\sum_j c_j^\alpha g(s_j^\alpha) \to \varphi(g)$ for all $g \in WAP(S)$. Taking $g = L_s f$ we have the pointwise convergence

$$\sum_j c_j^\alpha R_{s_j^\alpha} f(s) = \sum_j c_j^\alpha f(s s_j^\alpha) \to \varphi(L_s f) = R_\varphi f(s), \quad s \in S.$$

By the Krein-Šmulian theorem, the convex hull of R_S is relatively weakly compact, hence the convergence of $\sum_j c_j^\alpha R_{s_j^\alpha} f$ is also in the weak topology. Thus $R_M f$ is the weak closure of the convex hull of R_S and hence is weakly compact, proving (b). Similar arguments prove (c) (see 10.30).

Part (d) follows from (c) and right translation invariance of $WAP(S)$.

For (e), it suffices by 10.2.10 to prove that the restriction of the mapping to C_1' is weak* -weak continuous. But this follows because the mapping $\varphi \mapsto R_\varphi f : C_1' \to WAP(S)$ is w^*-pointwise continuous and the range $R_{C_1'} f$ is weakly compact by (c). □

By (c) of the lemma, we have

17.2.4 Corollary. *Let $\mathscr{F}$ be a translation invariant, norm closed, conjugate closed linear subspace of $WAP(S)$. Then $R_\varphi \mathscr{F} \subseteq \mathscr{F}$ for all $\varphi \in WAP(S)'$.*

Now define multiplication $(\varphi_1, \varphi_2) \mapsto \varphi_1 \cdot \varphi_2$ in $WAP(S)'$ by

$$(\varphi_1 \cdot \varphi_2)(f) = \varphi_1(R_{\varphi_2} f), \quad f \in WAP(S). \tag{17.2}$$

The calculation $R_{\varphi_1 \cdot \varphi_2} f(s) = \varphi_1 \cdot \varphi_2(L_s f) = \varphi_1(R_{\varphi_2} L_s f) = \varphi_1(L_s R_{\varphi_2} f) = R_{\varphi_1}(R_{\varphi_2} f)(s)$ shows that

$$R_{\varphi_1 \cdot \varphi_2} = R_{\varphi_1} R_{\varphi_2}. \tag{17.3}$$

17.2.5 Theorem. *Under multiplication defined in* (17.2), $WAP(S)'$ *is a Banach algebra and a semitopological semigroup in the weak* topology. Moreover, in the notation of 17.2.3,* C_1', M, *and* S^{WAP} *are weak* compact subsemigroups of* $WAP(S)'$.

Proof. Associativity follows from (17.3):

$$(\varphi_1 \cdot \varphi_2) \cdot \varphi_3(f) = (\varphi_1 \cdot \varphi_2)(R_{\varphi_3} f) = \varphi_1(R_{\varphi_2} R_{\varphi_3} f) = \varphi_1(R_{\varphi_2 \cdot \varphi_3} f) = \varphi_1 \cdot (\varphi_2 \cdot \varphi_3)(f).$$

It is immediate from the definition of multiplication that $(\varphi_1 + \varphi_2) \cdot \varphi_3 = \varphi_1 \cdot \varphi_3 + \varphi_2 \cdot \varphi_3$. Since $R_{\varphi_2 + \varphi_3} = R_{\varphi_2} + R_{\varphi_3}$, we also have $\varphi_1 \cdot (\varphi_2 + \varphi_3) = \varphi_1 \cdot \varphi_2 + \varphi_1 \cdot \varphi_3$. The remaining properties of normed algebra multiplication are clear.

That $WAP(S)'$ is a semitopological semigroup in the weak* topology may be seen directly from the definition of multiplication by applying 17.2.3(e). The last assertions of the theorem follow from the fact that sets M, C_1', and S^{WAP} are closed under multiplication $\varphi_1 \cdot \varphi_2$. Indeed, this is obviously the case for $\varphi_j = \widehat{s}_j$, and taking various iterated weak* limits proves the general case. □

The Weakly Almost Periodic Compactification

The following theorem summarizes the general properties of the spectrum S^{WAP} of $WAP(S)$ and the canonical map $\iota = \iota_{WAP} : S \to S^{WAP}$, $\iota(s) = \widehat{s}$.

17.2.6 Theorem. S^{WAP} *is a compact semitopological semigroup and* ι *is a continuous homomorphism onto a dense subset of* S^{WAP} *such that* $\iota^*\big(C(S^{WAP})\big) = WAP(S)$.

Proof. By 13.4.2, ι is a continuous mapping onto a dense subset of S^{WAP} such that the asserted equality holds. Moreover, for $f \in WAP(S)$ and $s_1, s_2 \in S$ we have

$$\iota(s_1) \cdot \iota(s_2)(f) = \iota(s_1)(R_{\iota(s_2)} f) = (R_{\iota(s_2)} f)(s_1) = \iota(s_2)(L_{s_1} f) = f(s_1 s_2) = \iota(s_1 s_2)(f),$$

hence ι is a homomorphism. □

The pair (ι, S^{WAP}) is called the **weakly almost periodic compactification** of S. A key feature of this compactification is the following *extension property*:

17.2.7 Theorem. *Given a continuous homomorphism* θ *from* S *into a semitopological semigroup* T, *there exists a continuous homomorphism* $\widetilde{\theta} : S^{WAP} \to T^{WAP}$ *such that the following diagram commutes:*

$$\begin{array}{ccc} S^{WAP} & \xrightarrow{\ \widetilde{\theta}\ } & T^{WAP} \\ {\scriptstyle \iota}\uparrow & & {\scriptstyle \iota}\uparrow \\ S & \xrightarrow{\ \theta\ } & T \end{array}$$

Proof. By 17.2.2, θ^* maps $WAP(T)$ into $WAP(S)$. Thus the assertion is an immediate consequence of 13.4.2. □

17.2.8 Corollary. *A function* $f \in C_b(S)$ *is weakly almost periodic iff* $L_S f$ *is relatively weakly compact.*

Proof. Let $f \in WAP(S)$. Then the mapping $x \mapsto L_x \widehat{f}$ on $X := S^{WAP}$ is pointwise continuous hence weakly continuous by 14.1.4. Therefore $L_X \widehat{f}$ is weakly compact in $C(X)$ and so $L_S f$ is relatively weakly compact in $WAP(S)$. The converse may be proved by considering the reverse semigroup obtained from S by reversing the order of multiplication. □

Since a compact semitopological semigroup is its own WAP compactification, we have

17.2.9 Corollary. *If T is a compact semitopological semigroup and $\theta : S \to T$ is a continuous homomorphism, then there exists a continuous homomorphism $\widetilde{\theta} : S^{WAP} \to T$ such that $\widetilde{\theta} \circ \iota_{WAP} = \theta$.*

17.2.10 Corollary (Eberlein). *If G is a locally compact group, then every member of $WAP(G)$ is uniformly continuous.*

Proof. Let $X = G^{WAP}$ and $f \in WAP(G)$. We show that the maps $s \mapsto L_s f$ and $s \mapsto R_s f$ are norm continuous at e.

Since the function $F(x, y) = \widehat{f}(xy)$ is separately continuous, there exists a point $x_0 \in X$ such that F is jointly continuous at all points of $\{x_0\} \times X$ (B.0.14). By compactness of X, $x \mapsto L_x\widehat{f}$ is norm continuous at x_0 (B.0.9). Therefore, the set $U := \{x \in X : \|L_x\widehat{f} - L_{x_0}\widehat{f}\|_\infty < \varepsilon/2\}$ is open in X. Since $\iota(G)$ is dense in X, $V := \iota^{-1}(U)$ is nonempty. Choose any $t \in V$. Then Vt^{-1} is a neighborhood of e in G, and for $s \in Vt^{-1}$ we have

$$\|L_s f - f\|_\infty = \|L_{st}f - L_t f\|_\infty \le \|L_{\widehat{st}}\widehat{f} - L_{x_0}\widehat{f}\|_\infty + \|L_{x_0}\widehat{f} - L_{\widehat{t}}\widehat{f}\|_\infty < \varepsilon.$$

Thus $s \mapsto L_s f$ is norm continuous at e. Similarly $s \mapsto R_s f$ is norm continuous at e. □

The converse of 17.2.10 is false. (See example below.)

17.2.11 Corollary. *If G is a noncompact, locally compact group, then $C_0(G) \subseteq WAP(G)$. In particular, if G is abelian, then the Fourier transform $\widehat{f}$ of $f \in L^1(G)$ is weakly almost periodic.*

Proof. Let $X = G \cup \{\infty\}$ denote the one-point compactification of G. Extend multiplication in G to X by defining $x \cdot \infty = \infty \cdot x = \infty$ for all x. If s_α, $s \in G$, and $s_\alpha \to \infty$, then for any compact $K \subseteq G$, eventually $ss_\alpha \notin K$, hence $ss_\alpha \to \infty$. Similarly, $s_\alpha s \to \infty$. Thus X is a compact semitopological semigroup. By 17.2.9, the inclusion map $\theta : G \hookrightarrow X$ extends to a continuous homomorphism $\widetilde{\theta} : G^{WAP} \to X$ with $\theta = \widetilde{\theta} \circ \iota_{WAP}$. Considering dual maps we then have

$$C_0(G) = \theta^*(C(X)) = (\iota^*_{WAP} \circ \widetilde{\theta}^*)(C(X)) \subseteq \iota^*_{WAP}(C(G^{WAP})) = WAP(G). \qquad \square$$

17.2.12 Corollary. *The coefficients of a unitary representation $\pi : G \to \mathscr{B}(\mathscr{H})$ of a locally compact group G are weakly almost periodic. In particular, every continuous function of positive type is weakly almost periodic.*

Proof. Let $\pi : G \to \mathscr{B}(\mathscr{H})$ be a unitary representation. Now, $\mathscr{B}(\mathscr{H})$ is a semitopological semigroup in the weak operator topology, that is, the operation $(T, S) \mapsto TS$ of composition in continuous in this topology. Indeed, if $T_\alpha \to T$ in that topology, then for all $\boldsymbol{x}$ and $\boldsymbol{x}'$,

$$\langle T_\alpha S\boldsymbol{x}, \boldsymbol{x}'\rangle \to \langle TS\boldsymbol{x}, \boldsymbol{x}'\rangle \quad \text{and} \quad \langle ST_\alpha\boldsymbol{x}, \boldsymbol{x}'\rangle = \langle T_\alpha\boldsymbol{x}, S'\boldsymbol{x}'\rangle \to \langle T\boldsymbol{x}, S'\boldsymbol{x}'\rangle = \langle ST\boldsymbol{x}, \boldsymbol{x}'\rangle,$$

hence $T_\alpha S \to TS$ and $ST_\alpha \to ST$ in the weak operator topology. Thus, by the extension theorem, there exists a continuous homomorphism $\widetilde{\pi} : G^{WAP} \to \mathscr{B}(\mathscr{H})$. Then any coefficient $(\pi(\cdot)\boldsymbol{x} \mid \boldsymbol{y})$ extends to a continuous function $(\widetilde{\pi}(\cdot)\boldsymbol{x} \mid \boldsymbol{y})$ on G^{WAP} and so is weakly almost periodic. The last assertion follows from 16.5.7. □

From 17.2.6 we see that if $f \in WAP(S)$ then $|f| \in WAP(S)$. The converse is false:

17.2.13 Example. Let $f(x) = \tan^{-1}(x)$. By 17.2.11, $|f|$ is weakly almost periodic on $S := (\mathbb{R}, +)$. On the other hand, while f is uniformly continuous, it is not weakly almost periodic. To see the latter, choose a subnet (n_α) of the sequence $(1, 2, \ldots)$ such that the limits $x = \lim_\alpha \iota(n_\alpha)$ and $y = \lim_\alpha \iota(-n_\alpha)$ exist in S^{WAP}. If $f \in WAP(S)$ we then have the contradiction

$$-\frac{\pi}{2} = \lim_\alpha \lim_\beta f(n_\alpha - n_\beta) = \lim_\alpha \lim_\beta \widehat{f}(\iota(n_\alpha)\iota(-n_\beta)) = \lim_\alpha \widehat{f}(\iota(n_\alpha)y) = f(xy), \text{ and}$$
$$\frac{\pi}{2} = \lim_\beta \lim_\alpha f(n_\alpha - n_\beta) = \lim_\beta \lim_\alpha \widehat{f}(\iota(n_\alpha)\iota(-n_\beta)) = \lim_\beta \widehat{f}(x\iota(-n_\alpha)y) = f(xy). \qquad \Diamond$$

Invariant Means on Weakly Almost Periodic Functions

A mean m on a translation invariant, conjugate closed, unital subspace $\mathscr{F}$ of $C_b(S)$ is said to be **left invariant** if $m(L_s f) = m(f)$ for all $f \in \mathscr{F}$ and $s \in S$. The notion of **right invariant mean** is defined analogously. The set of left (respectively, right) invariant means is denoted by $M_\ell(\mathscr{F})$ (respectively, $M_r(\mathscr{F})$). A member of the intersection $M_\ell(\mathscr{F}) \cap M_r(\mathscr{F})$ is called an **invariant mean**. It is easy to check that, if nonempty, these sets are convex and weak* closed. For example, if S is commutative, then $M\left(C_b(S)\right) \neq \emptyset$ (8.6.2), and the same holds if S is a compact Hausdorff topological group, since then the integral with respect to normalized Haar measure is an invariant mean.

Note that a mean m on $WAP(S)$ induces a mean on $C(S^{WAP})$, which in turn induces a probability measure μ on S^{WAP}:

$$m(f) = \int_{S^{WAP}} \widehat{f}(x)\, d\mu(x), \quad f \in WAP(S).$$

Thus for $y \in S^{WAP}$,

$$\begin{aligned} m(R_y f) &= \int_{S^{WAP}} \widehat{(R_y f)}(x)\, d\mu(x) = \int_{S^{WAP}} x(R_y f)\, d\mu(x) = \int_{S^{WAP}} (xy)(f)\, d\mu(x) \\ &= \int_{S^{WAP}} \widehat{f}(xy)\, d\mu(x), \end{aligned}$$

from we conclude that $m \in M_r$ iff μ is a right invariant measure on S^{WAP}. The left version follows by considering the reverse semigroup of S.

The following result makes a connection between the invariance of means on $WAP(S)$ and multiplication in the Banach algebra $WAP(S)'$.

17.2.14 Proposition. *Let m be a mean on $WAP(S)$. Then $m \in M_\ell$ ($m \in M_r$) iff $\varphi \cdot m = m$ ($m \cdot \varphi = m$) for all $\varphi \in WAP(S)'$.*

Proof. For any $f \in WAP(S)$ and $s \in S$,

$$(\widehat{s} \cdot m)(f) = \widehat{s}(R_m f) = (R_m f)(s) = m(L_s f) \quad \text{and} \quad (m \cdot \widehat{s})(f) = m(R_{\widehat{s}} f) = m(R_s f).$$

Therefore, $m \in M_\ell$ ($m \in M_r$) iff $\widehat{s} \cdot m = m$ ($m \cdot \widehat{s} = m$) for all $s \in S$. The desired equivalence then follows by taking suitable limits, noting that C_1' is the weak*-closed convex balanced hull of δ_S (Ex. 10.30). □

17.2.15 Corollary. *If $WAP(S)$ has a left invariant mean and a right invariant mean, then it has an invariant mean.*

Proof. Let m_ℓ be a left invariant mean, m_r a right invariant mean, and set $m := m_r \cdot m_\ell$. By 17.2.14,

$$\varphi \cdot m = (\varphi \cdot m_r) \cdot m_\ell = m_\ell = m_r \cdot m_\ell = m$$

and similarly

$$m \cdot \varphi = m_r \cdot (m_\ell \cdot \varphi) = m_r = m_r \cdot m_\ell = m,$$

hence m is an invariant mean. □

17.2.16 Theorem. *$WAP(S)$ has a left invariant mean iff for each $f \in WAP(S)$ the set $C(f) := \mathrm{cl}_w \, \mathrm{co}(R_S f)$ contains a constant function. The analogous assertion holds for right invariant means.*

Proof. By 17.2.3(b), $C(f) = R_M f$. If m is a left invariant mean for $WAP(S)$, then $R_m(f)(s) = m(L_s) = m(f)$ for all s, hence $R_m f$ is the required constant function.

Conversely, assume that $R_M f$ contains a constant function $R_{\mu_f} f$ for each $f \in WAP(S)$. Then for each $s \in S$, $R_{\mu_f}(L_s f) = L_s R_{\mu_f} f = R_{\mu_f} f$, hence the set

$$M(f, s) := \{\mu \in M : R_\mu(f - L_s f) = 0\}$$

is nonempty. Furthermore, $M(f, s)$ is weak* compact and $M(WAP(S)) \cdot M(f, s) \subseteq M(f, s)$, as may be seen from $R_{m \cdot \mu} = R_m R_\mu$. It follows by induction that

$$\bigcap_{j=1}^{n} M(f_j, s_j) \neq \emptyset, \quad f_j \in WAP(S), \quad s_j \in S.$$

Indeed, if $\mu \in \bigcap_{j=1}^{n-1} M(f_j, s_j)$ and $\nu \in M(R_\mu f_n, s_n)$, then $\nu \cdot \mu \in \bigcap_{j=1}^{n} M(f_j, s_j)$. Thus the sets $M(f, s)$ have the finite intersection property, so by compactness their intersection contains a point η. Then $\eta^2(L_s f) = \eta(R_\eta L_s f) = \eta(R_\eta f) = \eta^2(f)$ for all $f \in WAP(S)$ and $s \in S$, hence η^2 is a left invariant mean. □

17.2.17 Corollary. *If S is a semitopological group, then $WAP(S)$ has an invariant mean.*

Proof. R_S restricted to $C(f)$ is a group of weakly continuous, noncontracting affine maps from $C(f)$ into itself. By the Ryll-Nardzewski fixed point theorem, $C(f)$ has a fixed point g. Thus $g(st) = g(s)$ for all s and t. Taking s to be the identity of S shows that g is a constant function. By the theorem, $WAP(S)$ has a left invariant mean. A similar argument shows that $WAP(S)$ has a right invariant mean. By 17.2.15, $WAP(S)$ has an invariant mean. □

17.3 Almost Periodic Functions

Definition and Basic Properties

A member f of $C_b(S)$ is said to be **almost periodic** if $R_S f$ is relatively compact in the norm topology of $C_b(S)$. The set of all almost periodic functions is denoted by $AP(S)$. For example, the characters of a locally compact abelian group are almost periodic, as is easily established. In particular, the function $n \mapsto e^{in}$ is almost periodic on $(\mathbb{Z}, +)$. On the other hand, the function $f(n) = e^{in^2}$ is *not* almost periodic on $(\mathbb{Z}, +)$. Indeed, by a result of Dirichlet, [1] $e^{i\mathbb{N}}$ is dense in $\mathbb{T}$, hence we may choose a sequence (n_k) in $\mathbb{N}$ with $e^{in_k} \to 1$ and $e^{in_k^2} \to c \in \mathbb{T}$. Then the right translations $e^{i(n+n_k)^2} = e^{i(n^2+2nn_k+n_k^2)}$ tend pointwise to ce^{in^2}. If the convergence were uniform in $n \in \mathbb{Z}$, then e^{2inn_k} would converge uniformly to 1 and so z^{2n_k} would converge uniformly to 1 in $z \in \mathbb{T}$, or, equivalently, $e^{2n_k ti}$ would converge uniformly to 1 in $t \in \mathbb{R}$. But this is impossible, since $e^{2n_k t_k i} \to -1$, where $t_k = \pi/(2n_k)$.

Clearly $AP(S) \subseteq WAP(S)$. If S is a compact topological semigroup, then joint continuity of multiplication in S implies that $AP(S) = C(S)$, hence $AP(S) = WAP(S)$. On the other hand, we have

[1] See, for example [], Example 8.3.9.

17.3.1 Proposition. *If G is a locally compact group, then $C_0(G)\setminus\{0\} \subseteq WAP(G)\setminus AP(G)$, hence $AP(G) \subsetneqq WAP(G)$.*

Proof. We have already seen that $C_0(G) \subseteq WAP(G)$ (17.2.11). Let $f \in C_0(G) \cap AP(G)$ and choose a net $s_\alpha \to \infty$. By almost periodicity, we may suppose that $\|R_{s_\alpha} f - g\|_\infty \to 0$ for some $g \in C_b(G)$. Given $\varepsilon > 0$, choose α_0 so that $|f(ss_\alpha) - g(s)| < \varepsilon$ for all $\alpha \geq \alpha_0$ and $s \in G$. Taking limits shows that $|g(s)| < \varepsilon$ for all s and ε, hence $g = 0$. Thus $|f(ss_\alpha)| < \varepsilon$ for all $\alpha \geq \alpha_0$ and s. Replacing s by ss_α^{-1}, we have $|f(s)| < \varepsilon$ for all s and so $f = 0$. □

17.3.2 Theorem. *$AP(S)$ is a translation invariant, unital $C*$-subalgebra of $C_b(S)$.*

Proof. The relations in (17.1) show that $AP(S)$ is a translation invariant, conjugate closed subspace of $C_b(S)$. Since multiplication in $C_b(S)$ is a norm continuous operation, the product of norm compact sets is norm compact, hence $AP(S)$ is an algebra. Moreover, if $f_n \in AP(S)$ and $f_n \to f$ in $C_b(S)$, then a straightforward total boundedness argument shows that $f \in AP(S)$. Therefore, $AP(S)$ is closed in $C_b(S)$. □

17.3.3 Proposition. *Let T be a semitopological semigroup and $\theta : S \to T$ a continuous homomorphism. Then $\theta^*\big(AP(T)\big) \subseteq AP(S)$, where $\theta^* : C(T) \to C(S)$ is the dual map. In particular, if S is a subsemigroup of T, then $AP(T)|_S \subseteq AP(S)$.*

Proof. This follows from the norm continuity of θ^* and $R_s\theta^*(g) = \theta^*\big(R_{\theta(s)}g\big)$. □

The Almost Periodic Compactification

The following theorem summarizes the general properties of the spectrum S^{AP} of $AP(S)$ and the canonical map $\iota = \iota_{AP} : S \to S^{AP}$, $\iota(s) = \widehat{s}$.

17.3.4 Theorem. *S^{AP} is a compact topological semigroup and $\iota : S \to S^{AP}$ is a continuous homomorphism onto a dense subsemigroup such that $\iota^*\big(C(S^{AP})\big) = AP(S)$.*

Proof. By 17.2.4, $R_\varphi AP(S) \subseteq AP(S)$, hence multiplication $\varphi_1 \cdot \varphi_2$ is defined on $AP(S)'$. Thus, as in the WAP case, $AP(S)'$ is a semitopological semigroup in the weak* topology, S^{AP} is a compact semitopological semigroup, and ι is a continuous homomorphism onto a dense subset of S^{AP}. It remains only to show that multiplication in S^{AP} is jointly continuous. Let $f \in AP(S)$. By the relative norm compactness of $R_S f$, the map $\varphi \to R_\varphi f$ on C_1' is w^*-norm continuous. It follows that $\varphi_1 \cdot \varphi_2(f) = \varphi_1(R_{\varphi_2} f)$ is jointly continuous in (φ_1, φ_2) on C_1' and the conclusion follows. □

Note that the theorem implies that $L_S f$ is relatively compact for all $f \in AP(S)$. Thus the notions of *right almost periodicity* and *left almost periodicity* coincide.

The pair (ι, S^{AP}) is called the **almost periodic compactification** of S. Analogous to the weakly almost periodic case we have the following *extension property*, which is immediate from 17.3.3.

17.3.5 Theorem. *For each continuous homomorphism θ from S into a semitopological semigroup T, there exists a continuous homomorphism $\widetilde{\theta} : S^{AP} \to T^{AP}$ such that the following diagram commutes:*

$$\begin{array}{ccc} S^{AP} & \xrightarrow{\ \widetilde{\theta}\ } & T^{AP} \\ \iota\big\uparrow & & \iota\big\uparrow \\ S & \xrightarrow{\ \theta\ } & T \end{array}$$

Since a compact topological semigroup is its own AP compactification, we have

17.3.6 Corollary. *For each continuous homomorphism θ from S into a compact topological semigroup T, there exists a continuous homomorphism $\widetilde{\theta} : S^{AP} \to T$ such that $\widetilde{\theta} \circ \iota_{AP} = \theta$.*

17.4 The Structure of Compact Semigroups

For deeper results we need to determine the algebraic structure of compact semitopological semigroups. This structure is based largely on the existence of idempotents and properties of closed ideals. We begin with an important result on joint continuity of multiplication in a compact semitopological group.

Ellis's Theorem

17.4.1 Theorem (Ellis) *A compact Hausdorff semitopological group G is a topological group.*

Proof. To establish joint continuity of multiplication, it is enough to show that multiplication is continuous at each point of $\{e\} \times G$. Indeed, if $x_\alpha \to x$ and $y_\alpha \to y$, then, by separate continuity, $x^{-1}x_\alpha \to e$, hence if multiplication is continuous at (e, y), then $(x^{-1}x_\alpha)y_\alpha \to y$ and so $x_\alpha y_\alpha \to xy$.

Fix $y \in G$. To verify continuity of multiplication at (e, y), we show first that for each $x \in G$ with $x \neq y$ there are neighborhoods N_x of e, U_x of x, and V_x of y such that $(N_x V_x) \cap U_x = \emptyset$. To see this, let $g \in C(G)$ with $\operatorname{ran} g \subseteq [-1, 1]$, $g(y) = 0$ and $g(x) \neq g(y)$. Define $f : G \times G \to [-1, 1]$ by $f(s, z) = g(sz)$. By B.0.8, there exists a dense subset A of G such that f is jointly continuous at every point of $A \times Y$. Since $\{s \in G : f(s, x) \neq f(s, y)\}$ is open and nonempty, it contains a member s of A. Set

$$\varepsilon := |f(s, x) - f(s, y)| \ \ (> 0).$$

By joint continuity of f at (s, y), there exist neighborhoods N_x of e in G and V_x of y in K such that

$$|f(t, v) - f(s, y)| < \varepsilon/2 \ \ \forall\, (t, v) \in sN_x \times V_x. \tag{†}$$

Set

$$U_x := \{u \in K : |f(s, x) - f(s, u)| < \varepsilon/2\}. \tag{‡}$$

Then U_x is a neighborhood of x in K. Now suppose that $(N_x V_x) \cap U_x \neq \emptyset$. Then there exists $t \in N_x$ and $v \in V_x$ such that $u := tv \in U_x$. From (†) and (‡),

$$\begin{aligned} |f(s, x) - f(s, y)| &\leq |f(s, x) - f(s, tv)| + |f(s, tv) - f(s, y)| \\ &= |f(s, x) - f(s, u)| + |f(st, v) - f(s, y)| \\ &< \varepsilon/2 + \varepsilon/2 = \varepsilon, \end{aligned}$$

contradicting the definition of ε. Therefore, $(N_x V_x) \cap U_x = \emptyset$, as claimed.

Now let W be a neighborhood of y. Since $G \setminus W$ is compact, there is a finite set $F \subseteq G$ such that $G \setminus W \subseteq \bigcup_{x \in F} U_x$. Set $N := \bigcap_{x \in F} N_x$ and $V := \bigcap_{x \in F} V_x$. Then N and V are neighborhoods of e and y, respectively, and $NV \subset W$. This completes the proof of joint continuity of multiplication in G.

To verify continuity of inversion, Let $x_\alpha \to x \in G$. By compactness, we may assume that $x_\alpha^{-1} \to y$ for some $y \in G$. Then $xy = \lim_\alpha x_\alpha x_\alpha^{-1} = e$, hence $y = x^{-1}$ and so $x_\alpha^{-1} \to x^{-1}$. $\square$

Ellis has shown that the conclusion of the above theorem holds if the topology of G is merely locally compact (and Hausdorff) [17], [18].

Existence of Idempotents

An **idempotent** in a semigroup is an element e satisfying $e^2 = e$. A semigroup need not have an idempotent, as is the case, for example, for $(1, \infty)$ under multiplication or addition. However, in the compact case one always has idempotents:

17.4.2 Lemma. *A compact Hausdorff semitopological semigroup X has an idempotent.*

Proof. Order the collection of closed subsemigroups of X downward by inclusion. If $\mathscr{C}$ is a chain of such semigroups, then $\bigcap \mathscr{C} \neq \emptyset$ by compactness. By Zorn's lemma, X has a minimal closed subsemigroup Y. Let $e \in Y$. Then eY is a closed subsemigroup of Y, hence $eY = Y$ by minimality. Choose $y \in Y$ such that $e = ey$. The set $Z = \{z \in Y : ez = e\}$ is then a nonempty closed subsemigroup of Y and so $Z = Y$. In particular, $e \in Z$ and so $e^2 = e$. □

Ideal Structure

A nonempty subset Y of a semigroup X is a **left ideal** if $XY \subseteq Y$. A left ideal is a **minimal** if it properly contains no left ideal. Right ideals and minimal right ideals are defined similarly. An **ideal** is a subset of X that is both a left ideal and a right ideal. An ideal is a **minimal** if properly contains no ideal.

The left and right minimal ideal structures are given in the following theorems.

17.4.3 Theorem. *Let X be a compact, Hausdorff semitopological semigroup.*

(a) *Minimal left (resp., right) ideals exist and are of the form Xe (resp., eX), where e is an idempotent.*

(b) *Distinct minimal left (right) ideals are disjoint.*

(c) *If R is a minimal right ideal and L is a minimal left ideal, then RL is a topological group. If e is the identity of RL, then $RL = eXe$.*

Proof. (a) We prove the left case. A Zorn's lemma argument in the spirit of the proof of 17.4.2 shows that minimal *closed* left ideals L exist. If L' is left ideal contained in L and $x \in L'$, then Xx is a closed left ideal contained in L', which forces $Xx = L' = L$. Therefore, all minimal left ideals are closed. Taking x to be an idempotent completes the proof of (a).

(b) Let L_1 and L_2 be distinct minimal left ideals. Then L_1 and L_2 are disjoint; otherwise, by minimality, $L_1 = L_1 \cap L_2 = L_2$.

(c) Clearly $RL \subseteq R \cap L$. Since $LRL \subseteq L$, $(RL)(RL) = R(LRL) \subseteq RL$, hence RL is a semigroup. We show next that RL is a group. Let $t \in RL$. Then $Lt \subseteq L$, hence, by minimality, $Lt = L$. Therefore, $RLt = RL$ for all $t \in RL$. Similarly, $tRL = RL$ for all $t \in RL$. Let $e \in RL$ such that $et = t$. If $s \in RL$, there exist $x \in RL$ such that $s = tx$, hence $es = etx = tx = s$. Similarly there exists $e' \in RL$ such that $se' = s$ for all $s \in RL$. Then $e = ee' = e'$, so e is an identity for RL. To see that every $t \in RL$ has an inverse, choose $y, z \in RL$ such that $yt = e = tz$. Then $z = ez = ytz = ye = y$. Therefore, RL is a group. Since $e \in L$, $Xe \subseteq L$ and so $Xe = L$ by minimality. Similarly $eX = R$. Therefore, $RL = eXXe \subseteq eXe \subseteq RL$, so $RL = eXe$. Finally, by 17.4.1, eXe is a compact topological group. □

17.4.4 Theorem. *Let X be a compact Hausdorff semitopological semigroup and let $K = K(X)$ be the union of all minimal left ideals. Then K is also the union of all minimal right ideals and is an ideal contained in every other ideal.*

Proof. K is obviously a left ideal. Let Xe be a minimal left ideal and $s \in X$. We claim that the left ideal Xes is minimal. To see this, let L be a left ideal contained in Xes. Every member of L is of the form ys for some $y \in Xe$, hence the set $\{y \in Xe : ys \in L\}$ is nonempty. Since it is a left ideal it must equal Xe. Thus $y \in Xe \Rightarrow ys \in L$, that is, $Xes \subseteq L$. Therefore, $Xes \subseteq K$, so K is a right ideal and hence is an ideal.

Now, if I is any ideal in X and Xe is a minimal left ideal, then IXe is a left ideal contained in Xe and so $IXe = Xe$. Since also $IXe \subseteq I$, $Xe \subseteq I$. Therefore, $K \subseteq I$, so K is contained in every ideal of X. Similar arguments show that the union K' of all minimal right ideals is an ideal contained in every ideal of X. Therefore $K = K'$. □

17.4.5 Corollary. *K is the union of disjoint, compact topological groups eXe, where $e^2 = e \in K$.*

Proof. By minimality $K^2 = K$. But K^2 is the union of disjoint topological groups RL. □

17.4.6 Corollary. *X is a topological group iff it satisfies the left and right cancellation laws*

$$xy = xz \Rightarrow y = z \quad \text{and} \quad yx = zx \Rightarrow y = z.$$

Proof. For the sufficiency, let $e^2 = e \in K$. Then for any $x \in X$, $eex = ex$, hence $ex = x$ and so $X = eX$. Similarly $X = Xe$. Thus $X = eXe$, so X is a group. □

17.4.7 Corollary. *$WAP(S)$ has an invariant mean iff $K(S^{WAP})$ is a compact topological group.*

Proof. By the preceding, $K(S^{WAP})$ is a compact topological group iff has S^{WAP} has a unique minimal right ideal and a unique minimal left ideal.

Let m be an invariant mean on $WAP(S)$. If L_1 and L_2 are minimal left ideals of S^{WAP}, then, choosing any $\eta_j \in L_j$, we have $m = m \cdot \eta_j \in L_1 \cap L_2$ (17.2.14), so $L_1 = L_2$ by (b) of 17.4.3. Therefore, X has a unique minimal left ideal. Similarly, X has a unique minimal right ideal. Thus $K(S^{WAP})$ is a compact topological group.

Conversely, assume $K = K(S^{WAP})$ is a compact topological group. Define a mean on $WAP(S)$ by

$$m(f) = \int_K \widehat{f}(x)\, d\mu(x),$$

where μ is normalized Haar measure on K. Then m is invariant. □

17.5 Strongly Almost Periodic Functions

Definition and Basic Properties

A unitary representation π of S on a Hilbert space $\mathscr{H}$ is defined exactly as in the case of a locally compact group, the difference being that, while the operator $\pi(s)^{-1}$ is defined for all s, it may not be in the range of π. The space $SAP(S)$ of **strongly almost periodic functions** on S is defined as the closed linear span of the set of coefficients of finite dimensional unitary representations of S. Since the unitary group in a finite dimensional space is compact, $SAP(S) \subseteq AP(S)$. Indeed, if π is a finite dimensional unitary representation of S on $\mathscr{H}$ and if (s_α) is a net in S, then a subnet $\pi(s_\beta)$ converges to some unitary operator U and so for $\boldsymbol{x}, \boldsymbol{y} \in \mathscr{H}$,

$$R_{s_\beta}\left(\pi(s)\boldsymbol{x} \mid \boldsymbol{y}\right) = \left(\pi(s_\beta)\boldsymbol{x} \mid \pi(s)^{-1}\boldsymbol{y}\right) \to \left(U\boldsymbol{x} \mid \pi(s)^{-1}\boldsymbol{y}\right)$$

uniformly in s.

For a compact Hausdorff topological group G, the Peter-Weyl theorem (16.5.18) implies that $SAP(G) = C(G)$, hence $SAP(G) = AP(G)$. We show in 17.5.9 that

$$C_0(\mathbb{R}^+) \setminus \{0\} \subseteq AP(\mathbb{R}^+, +) \setminus SAP(\mathbb{R}^+, +), \tag{17.4}$$

hence $SAP(\mathbb{R}^+, +) \subsetneq AP(\mathbb{R}^+, +)$.

Our immediate goal is to show that $SAP(S)$ is a unital C^*-subalgebra of $C_b(S)$. For this we need the following lemma.

17.5.1 Lemma. *Let T be a compact topological semigroup and H a subgroup of T. Then $G := \mathrm{cl}\, H$ is a topological group.*

Proof. We show first that inversion may be extended to G. Given $x \in G$, let $x_\alpha \in H$ with $x_\alpha \to x$. By compactness, we may assume that $x_\alpha^{-1} \to y$ for some $y \in G$. Then $xy = \lim_\alpha x_\alpha x_\alpha^{-1} = e$. Similarly, $zx = e$ for some $z \in G$. Therefore, G is a group. That inversion in G is continuous is proved as in 17.4.1. □

We may now prove

17.5.2 Theorem. *$SAP(S)$ is a translation invariant unital C^*-subalgebra of $C_b(S)$.*

Proof. Let π be a finite dimensional unitary representation of S on $\mathscr{H}$. The relations $R_s\,(\pi(t)\boldsymbol{x} \mid \boldsymbol{y}) = (\pi(t)\pi(s)\boldsymbol{x} \mid \boldsymbol{y})$ and $L_s\,(\pi(t)\boldsymbol{x} \mid \boldsymbol{y}) = (\pi(t)\boldsymbol{x} \mid \pi(s)^*\boldsymbol{y})$ show that $SAP(S)$ is translation invariant. Furthermore, the proof that $SAP(S)$ is closed under multiplication is the same as in the proof of the Peter-Weyl theorem (16.5.18).

It remains to show that $SAP(S)$ is conjugate closed. For this it suffices to show that if $\boldsymbol{x}_0, \boldsymbol{y}_0 \in \mathscr{H}$, then $f(s) := (\boldsymbol{y}_0 \mid \pi(s)\boldsymbol{x}_0)$ is a coefficient of some unitary representation. The proof of this is similar to but somewhat more involved than the corresponding part of the proof of the Peter-Weyl theorem. As in the latter, let $\mathscr{F}$ denote the finite dimensional subspace of $C_b(S)$ consisting of all functions g_x defined by

$$g_x(s) = (\boldsymbol{y}_0 \mid \pi(s)\boldsymbol{x}), \quad s \in S, \ \boldsymbol{x} \in \mathscr{H}.$$

Since $\mathscr{F}$ is right translation invariant, the mapping $s \mapsto R_s$ is a continuous representation of S on the space $\mathscr{F}$. Since $R_s g_x = g_{U_s x}$, R_s is surjective, hence invertible. Thus R_S is contained in a bounded group of operators on $\mathscr{H}$ and hence, by 17.5.1, is contained in a compact group of such operators. Thus, by 16.5.17, there exists an inner product $\langle \cdot \mid \cdot \rangle$ on $\mathscr{F}$ relative to which the operators R_s are unitary. Since the closure of $\pi(S)$ is a group, there exist a sequence (s_n) in S such that $\pi(s_n) \to I$. We may assume that the evaluation maps $\widehat{s}_n$ on $\mathscr{F}$ converge to a member of the dual space $\mathscr{F}'$, which, by the Riesz representation theorem, is given by a member g_{x_1} of $\mathscr{F}$. Thus

$$\lim_n g_x(s_n) = \langle g_x \mid g_{x_1} \rangle, \quad \boldsymbol{x} \in \mathscr{H}.$$

Since $\mathscr{F}$ is translation invariant, the limit relation holds for $R_S g_x$ as well. It follows that

$$f(s) = (\boldsymbol{y}_0 \mid \pi(s)\boldsymbol{x}_0) = \lim_n\,(\boldsymbol{y}_0 \mid \pi(s_n)\pi(s)\boldsymbol{x}_0) = \lim_n R_s g_{x_0}(s_n) = \langle R_s g_{x_0} \mid g_{x_1} \rangle,$$

which shows that f is a coefficient of the unitary representation R, completing the proof. □

17.5.3 Proposition. *Let T be a semitopological semigroup and $\theta : S \to T$ a continuous homomorphism. Then $\theta^*\big(SAP(T)\big) \subseteq SAP(S)$, where $\theta^* : C(T) \to C(S)$ is the dual map. In particular, if S is a subsemigroup of T, then $SAP(T)|_S \subseteq SAP(S)$.*

Proof. This follows essentially from the fact that if π is a continuous, finite dimensional unitary representation of T, then $\pi \circ \theta$ is a continuous finite dimensional unitary representation of S. □

The Strongly Almost Periodic Compactification

The following theorem summarizes the general properties of the spectrum S^{SAP} of $SAP(S)$ and the canonical map $\iota = \iota_{SAP} : S \to S^{SAP}$, $\iota(s) = \widehat{s}$.

17.5.4 Theorem. *S^{SAP} is a compact topological group and ι is a continuous homomorphism onto a dense subsemigroup such that $\iota^*\big(C(S^{SAP})\big) = SAP(S)$.*

Proof. Since $SAP(S) \subseteq AP(S)$, S^{SAP} is a topological semigroup. It remains to show that S^{SAP} is a group. For this we show that S^{SAP} has the cancellation properties in 17.4.6. We show that if $yx = zx$ in S^{SAP}, then $\widehat{f}(y) = \widehat{f}(z)$ for all $f \in SAP(S)$, where $\iota^*(\widehat{f}) = f$. It suffices to show this for $f(s) = (U_s\boldsymbol{x} \mid \boldsymbol{y})$, where U is a continuous, finite dimensional, unitary representation U of S. Let $\iota(s_\alpha) \to x$. We may assume that $U_{s_\alpha} \to V$ for some unitary operator V. Let g be the coefficient $g(s) = \big(U_sV^{-1}\boldsymbol{x} \mid \boldsymbol{y}\big)$. Then for all s

$$R_x\widehat{g}\big(\iota(s)\big) = \widehat{g}\big(\iota(s)x\big) = \lim_\alpha \widehat{g}\big(\iota(ss_\alpha)\big) = \lim_\alpha \big(U_sU_{s_\alpha}V^{-1}\boldsymbol{x} \mid \boldsymbol{y}\big) = (U_s\boldsymbol{x} \mid \boldsymbol{y}) = f(s),$$

hence $R_x\widehat{g} = \widehat{f}$ and so $\widehat{f}(y) = \widehat{g}(yx) = \widehat{g}(zx) = \widehat{f}(z)$. □

The pair (ι, S^{SAP}) is called the **strongly almost periodic compactification of** S. As in the WAP and AP cases, we have the following extension property, which may be proved using 17.5.3.

17.5.5 Theorem. *For each continuous homomorphism θ from S into a semitopological semigroup T, there exists a continuous homomorphism $\widetilde{\theta} : S^{SAP} \to T^{SAP}$ such that the following diagram commutes:*

$$\begin{array}{ccc} S^{SAP} & \xrightarrow{\ \widetilde{\theta}\ } & T^{SAP} \\ \iota\big\uparrow & & \iota\big\uparrow \\ S & \xrightarrow{\ \theta\ } & T \end{array}$$

Since a compact topological group is its own SAP compactification, we have

17.5.6 Corollary. *For each continuous homomorphism θ from S into a compact topological group T, there exists a continuous homomorphism $\widetilde{\theta} : S^{SAP} \to T$ such that $\widetilde{\theta} \circ \iota_{SAP} = \theta$.*

17.5.7 Corollary. *If S is a group, then $AP(S) = SAP(S)$.*

Proof. By 17.5.1, S^{AP} is a topological group. Applying 17.5.6 to $T = S^{AP}$ and $\theta = \iota_{AP}$, we obtain a continuous homomorphism $\widetilde{\theta} : S^{SAP} \to S^{AP}$ such that $\widetilde{\theta} \circ \iota_{SAP} = \iota_{AP}$. Thus

$$AP(S) = \iota_{AP}^*\big(C(S^{AP})\big) = \iota_{SAP}^* \circ \widetilde{\theta}^*\big(C(S^{AP}\big) \subseteq SAP(S).$$ □

17.5.8 Corollary. *Let $WAP(S)$ have an invariant mean m. Then*

$$WAP(S) = SAP(S) \oplus WAP(S)_0,$$

where

$$WAP(S)_0 := \{f \in WAP(S) : m(|f|) = 0\}.$$

Moreover, $WAP(S)_0$ is an ideal of the C^*-algebra $WAP(S)$. In particular, these assertions hold of S is a group or is commutative.

Proof. By 17.4.7, the minimal ideal $K = K(S^{WAP})$ is a compact topological group. We denote the identity in S^{SAP} by 1 and the identity of K by e, so that $K = S^{WAP}e = eS^{WAP}e$. The map $\theta(s) = \iota_{_{WAP}}(s)e$ from S into K is a continuous homomorphism, hence, by 17.5.6, there exists a continuous homomorphism $\bar{\theta} : S^{SAP} \to K$ such that $\theta = \bar{\theta} \circ \iota_{_{SAP}}$. Therefore,

$$\theta^*(C(K)) = (\iota^*_{_{SAP}} \circ \bar{\theta}^*)(C(K)) \subseteq SAP(S).$$

In particular, if $f \in WAP(S)$, then the function $R_e f(s) = \widehat{f}(\iota_{_{WAP}}(s)e) = \widehat{f}(\theta(s))$ is strongly almost periodic. Therefore, $R_e WAP(S) \subseteq SAP(S)$. Now let $g \in SAP(S)$ and choose $\widehat{g} \in C(S^{SAP})$ such that $g = \iota^*_{_{SAP}}(\widehat{g})$. If $(\iota_{_{WAP}}(t_\alpha)) \to e$, then $\iota_{_{SAP}}(t_\alpha) = \widetilde{\theta} \circ \iota_{_{WAP}}(t_\alpha) \to \widetilde{\theta}(e) = 1$, so

$$R_e g(t) = \widehat{g}(t\iota_{_{SAP}}(t_\alpha)) \to \widehat{g}(t \cdot 1) = g(t).$$

We have proved that R_e is a projection from $WAP(S)$ onto $SAP(S)$. It remains to show that $\ker R_e = WAP_0(S)$ and that $WAP(S)_0$ is an ideal of $WAP(S)$. Now,

$$m(|f|) = \int_K |\widehat{f}(x)|\, dx, \quad f \in WAP(S),$$

where dx is Haar measure on K. Thus $m(|f|) = 0$ iff $\widehat{f}(x) = 0$ for all $x \in K$ iff $\widehat{f}(xe) = 0$ for all $x \in S^{WAP}$ iff $R_e f = 0$. Therefore, $\ker R_e = WAP_0(S)$. That $WAP(S)_0$ is an ideal follows from the inequality $m(|fg|) \leq \|g\|_\infty m(|f|)$. □

17.5.9 Corollary. $AP(\mathbb{R}^+, +) = AP(\mathbb{R}, +)\big|_{\mathbb{R}^+} \oplus C_0(\mathbb{R}^+)$.

Proof. Set $S = \mathbb{R}^+$. Since S is commutative, $C_b(S)$ has an invariant mean m. By an obvious modification of the preceding corollary,

$$AP(S) = SAP(S) \oplus AP(S)_0, \quad \text{where} \quad AP(S)_0 = \{f \in AP(S) : m(|f|) = 0\} = \ker R_e.$$

Here $m(|f|) = \int_K |\widehat{f}(x)|\, dx$, where $\widehat{f} \circ \iota_{_{AP(S)}} = f$ and dx is Haar measure on the compact group $K = S^{AP}e$. Thus we need to show that $SAP(S) = AP(\mathbb{R})|_S$ and $AP(S)_0 = C_0(S)$.

For the first equality, note that by 17.5.7 and 17.5.3, $AP(\mathbb{R})|_{\mathbb{R}^+} \subseteq SAP(S)$. Now let $f \in SAP(S)$, $f = \iota^*_{_{SAP}}(\widehat{f})$. We show that f may be extended to a function $g \in AP(\mathbb{R})$. To this end, define $\varphi : \mathbb{R} \to S^{SAP}$ by

$$\varphi(t) = \begin{cases} \iota_{_{SAP(S)}}(t) & \text{if } t \geq 0, \\ \iota_{_{SAP(S)}}(-t)^{-1} & \text{if } t < 0. \end{cases}$$

By considering cases, φ is easily seen to be a continuous homomorphism. By the extension property, there exists a continuous homomorphism $\widetilde{\varphi} : \mathbb{R}^{AP} \to S^{SAP}$ such that $\widetilde{\varphi} \circ \iota_{_{AP}} = \varphi$. Then $g := \varphi^*(\widehat{f}) \in AP(\mathbb{R})$, and for $t \geq 0$, $g(t) = \widehat{f}(\varphi(t)) = \widehat{f}(\iota_{_{SAP}}(t)) = f(t)$. Therefore, $SAP(S) = AP(\mathbb{R})|_S$.

Now, trivially, $C_0(S) \subseteq AP(S)$. We claim that $e \notin \iota_{_{AP(S)}}(S)$. Assuming this we may then choose a net $\iota_{_{AP(S)}}(s_\alpha) \to e$ with $s_\alpha \to \infty$, and so $R_e f = 0$ iff $f(s + s_\alpha) \to 0$ iff $f \in C_0(S)$. To verify the claim, assume $\iota_{_{AP(S)}}(s_0) = e$ for some $s_0 \in S$. Then $\iota_{_{AP(S)}}(s_0 + s_0) = e^2 = e = \iota_{_{AP(S)}}(s_0)$, and since $C_0(S)$ separates points of $\mathbb{R}^+$, $s_0 + s_0 = s_0$ and so $s_0 = 0$. Thus $\iota_{_{AP(S)}}(0) = e$, which implies that $S^{AP}e = S^{AP}$. Therefore, S^{AP} is a topological group with identity e. Now, the one-point compactification $[0, \infty]$ of $[0, \infty)$ is easily seen to be a topological semigroup under $s + \infty := \infty$, hence there exists a continuous surjective homomorphism $\varphi : S^{AP} \to [0, \infty]$ such that $\iota_{_{AP(S)}}(s) = s$ for all s (17.3.6). In particular, $[0, \infty]$ is a group with identity 0, impossible. □

17.6 Semigroups of Operators

In this section we extend previous results based on the semigroup of operators R_S to an arbitrary semigroup of bounded linear operators on a Banach space $\mathscr{X}$.

Definitions and Basic Properties

Let $\mathscr{U}$ be semigroup of operators on $\mathscr{X}$, that is, a nonempty subset of $\mathscr{B}(\mathscr{X})$ closed under operator composition. A point $\boldsymbol{x} \in \mathscr{X}$ is said to be **almost periodic** (**weakly almost periodic**) if the set $\mathscr{U}\boldsymbol{x} := \{U\boldsymbol{x} : U \in \mathscr{U}\}$ is relatively compact in the norm (weak) topology of $\mathscr{X}$.

17.6.1 Proposition. *Let $\mathscr{U}$ be uniformly bounded. The sets $\mathscr{X}_a$ and $\mathscr{X}_w$ of almost periodic and weakly almost periodic vectors in $\mathscr{X}$ are closed, $\mathscr{U}$-invariant linear subspaces of $\mathscr{X}$.*

Proof. We prove only the weakly almost periodic part. Clearly, $0 \in \mathscr{X}_w$. The relations

$$\mathscr{U}(\boldsymbol{x}+\boldsymbol{y}) \subseteq \mathscr{U}\boldsymbol{x} + \mathscr{U}\boldsymbol{y}, \quad \mathscr{U}(c\boldsymbol{x}) = c\mathscr{U}\boldsymbol{x}, \quad \text{and} \quad \mathscr{U}(U\boldsymbol{x}) \subseteq \mathscr{U}(\boldsymbol{x}), \quad U \in \mathscr{U},$$

show that $\mathscr{X}_w$ is an invariant linear subspace of $\mathscr{X}$. To show that $\mathscr{X}_w$ is closed in $\mathscr{X}$, let $(\boldsymbol{x}_n)$ be a sequence in $\mathscr{X}_w$ converging in norm to $\boldsymbol{x}$ in $\mathscr{X}$. By the Eberlein-Šmulian theorem, it suffices to show that $\mathscr{U}\boldsymbol{x}$ is weakly relatively sequentially compact. Let $(U_n\boldsymbol{x})$ be a sequence in $\mathscr{U}\boldsymbol{x}$. Since each set $\mathscr{U}\boldsymbol{x}_n$ is relatively weakly sequentially compact, a standard diagonal argument shows that there exists a subsequence (U_k) of (U_n) and a sequence $(\boldsymbol{y}_n) \in \mathscr{X}$ such that $U_k\boldsymbol{x}_n \stackrel{w}{\to} \boldsymbol{y}_n$ for each n. For any $\boldsymbol{x}' \in \mathscr{X}'$ with $\|\boldsymbol{x}'\| \leq 1$ we then have

$$|\langle \boldsymbol{y}_n, \boldsymbol{x}'\rangle - \langle \boldsymbol{y}_m, \boldsymbol{x}'\rangle| = \lim_k |\langle U_k\boldsymbol{x}_n, \boldsymbol{x}'\rangle - \langle U_k\boldsymbol{x}_m, \boldsymbol{x}'\rangle| \leq C\|\boldsymbol{x}_n - \boldsymbol{x}_m\|,$$

where $C = \sup_{U\in\mathscr{U}}\|U\|$. Therefore, $\|\boldsymbol{y}_n - \boldsymbol{y}_m\| \leq C\|\boldsymbol{x}_n - \boldsymbol{x}_m\|$, which shows that $(\boldsymbol{y}_n)$ is a Cauchy sequence. Let $\boldsymbol{y} := \lim_n \boldsymbol{y}_n$. Given $\varepsilon > 0$, choose n such that $\|\boldsymbol{x} - \boldsymbol{x}_n\| < \varepsilon$ and $\|\boldsymbol{y} - \boldsymbol{y}_n\| < \varepsilon$. For all k,

$$\begin{aligned}|\langle U_k\boldsymbol{x}, \boldsymbol{x}'\rangle - \langle \boldsymbol{y}, \boldsymbol{x}'\rangle| &\leq |\langle U_k(\boldsymbol{x}-\boldsymbol{x}_n), \boldsymbol{x}'\rangle| + |\langle U_k\boldsymbol{x}_n, \boldsymbol{x}'\rangle - \langle \boldsymbol{y}_n, \boldsymbol{x}'\rangle| + |\langle \boldsymbol{y}_n - \boldsymbol{y}, \boldsymbol{x}'\rangle| \\ &\leq (C+1)\varepsilon + |\langle U_k\boldsymbol{x}_n, \boldsymbol{x}'\rangle - \langle \boldsymbol{y}_n, \boldsymbol{x}'\rangle|,\end{aligned}$$

hence

$$\overline{\lim_k}|\langle U_k\boldsymbol{x}, \boldsymbol{x}'\rangle - \langle \boldsymbol{y}, \boldsymbol{x}'\rangle| \leq (C+1)\varepsilon + \overline{\lim_k}||\langle U_k\boldsymbol{x}_n, \boldsymbol{x}'\rangle - \langle \boldsymbol{y}_n, \boldsymbol{x}'\rangle| = (C+1)\varepsilon.$$

Therefore, $\langle U_k\boldsymbol{x}, \boldsymbol{x}'\rangle \to \langle \boldsymbol{y}, \boldsymbol{x}'\rangle$ and so $\boldsymbol{x} \in \mathscr{X}_w$. □

A semigroup of operators $\mathscr{U}$ on $\mathscr{X}$ is said to be **almost periodic** (resp., **weakly almost periodic**) if $\mathscr{X} = \mathscr{X}_a$ (resp., $\mathscr{X} = \mathscr{X}_w$). For example, if $\mathscr{X}$ is reflexive and $\mathscr{U}$ is uniformly bounded, then $\mathscr{U}$ is weakly almost periodic. Here is an example for the nonreflexive case.

17.6.2 Example. Let $(X, \mathfrak{F}, \mu)$ be a probability space and S a semigroup under composition of measurable transformations $s : X \to X$. For each $s \in S$, let μ_s denote the image measure of μ, so

$$\mu_s(E) = \mu(s^{-1}(E)), \quad E \in \mathfrak{F}.$$

We assume that $\mu_s \ll \mu$ for each $s \in S$ and that $c := \sup_{s\in S}\left\|\frac{d\mu_s}{d\mu}\right\|_\infty < \infty$. This is obviously the case if the members of S are measure-preserving, i.e., $\mu_s = \mu$ for all $s \in S$. Define $U_s f = f \circ s$, $f \in L^1$. Then

$$\|U_s f\|_1 = \int |f \circ s|\, d\mu = \int |f|\, d\mu_s = \int |f| \frac{d\mu_s}{d\mu}\, d\mu \le c \|f\|_1,$$

hence U_S is uniformly bounded in L^1. Since $\|U_s \mathbf{1}_A\|_\infty \le 1$, $U_S \mathbf{1}_A$ uniformly integrable and so is relatively weakly compact, by the Dunford-Pettis theorem (14.2.4). Therefore, $U_S f$ weakly relatively compact for every simple function f. Since these are dense in L^1, the proposition shows that U_S is weakly almost periodic on L^1. ♢

17.6.3 Theorem. *Let $\mathscr{U}$ be a semigroup of operators on a Banach space $\mathscr{X}$.*

(a) *If $\mathscr{U}$ is weakly almost periodic, then in the weak operator topology of $\mathscr{B}(\mathscr{X})$ the closure $\mathscr{U}^w$ of $\mathscr{U}$ is a compact semitopological semigroup of uniformly bounded operators.*

(b) *If $\mathscr{U}$ is almost periodic, then in the strong operator topology of $\mathscr{B}(\mathscr{X})$ the closure $\mathscr{U}^a$ of $\mathscr{U}$ is a compact topological semigroup of uniformly bounded operators.*

Proof. The uniform boundedness principle shows that $\mathscr{U}^w$ and $\mathscr{U}^a$ are uniformly bounded. For each $\boldsymbol{x} \in \mathscr{X}$, let $K_{\boldsymbol{x}}$ denote the closure of $\mathscr{U}\boldsymbol{x}$ in the weak topology of $\mathscr{X}$. The product space $K := \prod_{\boldsymbol{x} \in \mathscr{X}} K_{\boldsymbol{x}}$ contains $\mathscr{U}$ and is compact by Tychonoff's theorem. Therefore, the closure $\mathrm{cl}(\mathscr{U})$ of $\mathscr{U}$ in C is compact. But $\mathrm{cl}(\mathscr{U}) \subseteq \mathscr{B}(\mathscr{X})$. To see this, let (T_α) be a net in $\mathscr{U}$ such that $T_\alpha \to T$ in the product topology. Thus for all $\boldsymbol{x}, \boldsymbol{y} \in \mathscr{X}$,

$$T_\alpha(\boldsymbol{x} + \boldsymbol{y}) \xrightarrow{w} T_\alpha(\boldsymbol{x} + \boldsymbol{y}), \quad T_\alpha(\boldsymbol{x}) \xrightarrow{w} T(\boldsymbol{x}) \text{ and } T_\alpha(\boldsymbol{y}) \xrightarrow{w} T(\boldsymbol{y}).$$

It follows that T is linear, and an application of the uniform boundedness principle shows that T is bounded. Therefore, $\mathscr{U}^w = \mathrm{cl}(\mathscr{U})$, proving that $\mathscr{U}^w$ is compact in the weak operator topology. A similar argument shows that $\mathscr{U}^a$ is compact in the strong operator topology.

We have already seen in the proof of 17.2.12 that operator composition in $\mathscr{B}(\mathscr{X})$ is weak operator continuous. It follows that $\mathscr{U}^w$ is closed under operator composition and so is a semitopological semigroup. It remains to show that operator composition in $\mathscr{U}^a$ is continuous in the strong operator topology. But if $T_\alpha \to T$ and $S_\alpha \to S$ in that topology, then for all $\boldsymbol{x} \in \mathscr{X}$

$$\begin{aligned}\|T_\alpha S_\alpha \boldsymbol{x} - TS\boldsymbol{x}\| &\le \|T_\alpha S_\alpha \boldsymbol{x} - T_\alpha S\boldsymbol{x}\| + \|T_\alpha S\boldsymbol{x} - TS\boldsymbol{x}\| \\ &\le M \|S_\alpha \boldsymbol{x} - S\boldsymbol{x}\| + \|T_\alpha S\boldsymbol{x} - TS\boldsymbol{x}\| \\ &\to 0,\end{aligned}$$

where $M = \sup_{U \in \mathscr{U}} \|U\| < \infty$. □

Dynamical Properties of Semigroups of Operators

A representation of the semitopological semigroup S by operators on a Banach space $\mathscr{X}$ is defined as in the group case, namely as a homomorphism $U : s \mapsto U_s$ from S into $\mathscr{B}(\mathscr{X})$. A representation U is said to be **almost periodic** (respectively, **weakly almost periodic**) if it is continuous in the strong operator (respectively, weak operator) topology and the semigroup $\mathscr{U} := U_S$ is strongly (resp.,weakly) almost periodic. For example, the representation $s \mapsto R_s$ is weakly almost periodic on $WAP(S)$ and almost periodic on $AP(S)$. If $\mathscr{X}$ is reflexive, $s \mapsto U_s$ is weak operator continuous, and U_S is uniformly bounded, then U_S is weakly almost periodic.

Let $U : s \mapsto U_s$ be a weakly almost periodic representation of S on $\mathscr{X}$. The **coefficient algebra** of the representation is the unital C^*-subalgebra $\mathscr{A}_U$ of $C_b(S)$ generated by the coefficients $s \to \langle U_s \boldsymbol{x}, \boldsymbol{x}' \rangle$.

17.6.4 Proposition. (a) *$\mathscr{A}_U$ is a translation invariant subalgebra of $WAP(S)$.*

(b) *The map $\psi : C(\mathscr{U}^w) \to C_b(S)$ defined by $\psi(g)(s) = g(U_s)$ is a C^*-algebra isomorphism onto $\mathscr{A}_U$ that commutes with translations.*

(c) *$\mathscr{A}_U$ has an invariant mean m iff $K := K(\mathscr{U}^w)$ is a compact topological group. In this case,*

$$m(\psi(g)) = \int_K g(V)\,dV, \quad g \in C(\mathscr{U}^w),$$

where dV is normalized Haar measure on K.

Proof (a) Since right and left translations of coefficients are coefficients, $\mathscr{A}_U$ is translation invariant. Moreover, since $\mathscr{U}$ is weakly almost periodic, it follows exactly as in the proof of 17.2.12 that a coefficient is weakly almost periodic. Therefore, $\mathscr{A}_U \subseteq WAP(S)$.

(b) Clearly, ψ is a C^* isomorphism into $C_b(S)$. Let $\boldsymbol{x}$, $\boldsymbol{y} \in \mathscr{X}$ and define $g \in C(\mathscr{U}^w)$ by $g(T) = \langle T\boldsymbol{x}, \boldsymbol{y}\rangle$. Then $\psi(g)$ is a coefficient, hence $\operatorname{ran}\psi$ contains $\mathscr{A}_U$. Since $\psi^{-1}(\mathscr{A}_U)$ is a conjugate closed unital subalgebra of $C(\mathscr{U}^w)$ that separates points of $\mathscr{U}^w$, it must coincide with $C(\mathscr{U}^w)$.

(c) For any mean m on $\mathscr{A}_U$, $\psi^*(m)$ is a mean on $C(\mathscr{U}^w)$, and conversely. Since $\psi(R_{U_t}g)(s) = g(U_sU_t) = g(U_{st}) = R_t\psi(g)(s)$ for all s, we have $\psi^*(m)(R_{U_t}g) = m(R_t\psi(g))$. Similarly, $\psi^*(m)(L_{U_t}g) = m(L_t\psi(g))$. Thus m is an invariant mean iff $\psi^*(m)$ is an invariant mean. The first part of (c) now follows from 17.4.7 applied to the semigroup $\mathscr{U}^w$. If m is an invariant mean, then $\psi^*(m)$ defines a normalized Haar measure dV, verifying the last assertion. □

A $\mathscr{U}$-invariant finite dimensional subspace $\mathscr{Y}$ of $\mathscr{X}$ is said to be **unitary** if $\mathscr{U}|_{\mathscr{Y}}$ is contained in a uniformly bounded group of operators on $\mathscr{Y}$ whose identity is the identity operator. The space of **strongly almost periodic vectors** in $\mathscr{X}$ is the closed linear subspace $\mathscr{X}_p$ of $\mathscr{X}$ generated by the unitary subspaces of $\mathscr{X}$. The set of **dissipative vectors** in $\mathscr{X}$ is defined by

$$\mathscr{X}_0 = \{\boldsymbol{x} \in \mathscr{X} : \boldsymbol{0} \in \operatorname{cl}_w U_S\boldsymbol{x}\}.$$

The following theorem asserts that every member $\boldsymbol{x}$ of $\mathscr{X}$ is a unique sum of a strongly almost periodic vector and a dissipative vector. Thus for some net (s_α), $U_{s_\alpha}\boldsymbol{x}$ converges to a vector $\boldsymbol{x}_p$ with a "stable group orbit" $\mathscr{U}^w\boldsymbol{x}_p$.

17.6.5 Theorem (deLeeuw-Glicksberg). *Let $\mathscr{U}$ be weakly almost periodic and let $\mathscr{A}_U$ have an invariant mean m. Then*

(a) *$\mathscr{X}_0$ is a closed, $\mathscr{U}$-invariant subspace of $\mathscr{X}$ and $\mathscr{X} = \mathscr{X}_p \oplus \mathscr{X}_0$.*

(b) *$\mathscr{X}_p$ is the largest closed, $\mathscr{U}$-invariant subspace of $\mathscr{X}$ on which $\mathscr{U}^w$ acts as a group with identity the identity operator,*

(c) *$\mathscr{X}_0 = \big\{\boldsymbol{x} \in \mathscr{X} : m\big(|\langle U_{(\cdot)}\boldsymbol{x}, \boldsymbol{x}'\rangle| \ \forall\ \boldsymbol{x}' \in \mathscr{X}'\big) = 0\big\}$.*

Proof. By 17.6.4, there exists an idempotent $E \in K := K(\mathscr{U}^w)$ such that $K = \mathscr{U}^wE = E\mathscr{U}^w = E\mathscr{U}^wE$ is a compact topological group with identity E. Thus $\mathscr{X} = E\mathscr{X} \oplus (I - E)\mathscr{X}$. We show that $E\mathscr{X} = \mathscr{X}_p$ and $(I - E)\mathscr{X} = \mathscr{X}_0$.

For the inclusion $\mathscr{X}_p \subseteq E\mathscr{X}$, it suffices to show that every unitary subspace $\mathscr{Y}$ of $\mathscr{X}$ is contained in $E\mathscr{X}$. But $\mathscr{U}|_{\mathscr{Y}}$ is contained in a group with identity the identity operator I, and since $E^2 = E$ we have $E|_{\mathscr{Y}} = I$ and so $\mathscr{Y} = E\mathscr{Y} \subseteq E\mathscr{X}$.

Next, we show that $\mathscr{Z} := E\mathscr{X} \subseteq \mathscr{X}_p$. Since $EV = VE = EVE$ for all $V \in \mathscr{U}^w$, $\mathscr{Z}$ is $\mathscr{U}^w$-invariant and $V = EVE$ on $\mathscr{Z}$. Thus $G := \mathscr{U}^w|_{\mathscr{Z}}$ is a compact topological group. Let

dV be normalized Haar measure on G and let (ϕ_α) be a symmetric approximate identity in $C(G)$. Then for $f \in C(G)$,

$$\int_G \phi_\alpha(V)f(V)\,dV = (\phi_\alpha * f)(E) \to f(E).$$

It follows that for fixed $\mathscr{z} \in \mathscr{Z}$, the vector integrals $V_{\phi_\alpha}\mathscr{z} := \int_G \phi_\alpha(V)V\mathscr{z}\,dV$ converge weakly to $\mathscr{z}$:

$$\langle V_{\phi_\alpha}\mathscr{z}, \mathscr{x}'\rangle = \lim_\alpha \int_G \phi_\alpha(V)\,\langle V\mathscr{z}, \mathscr{x}'\rangle\,dV \to \langle E\mathscr{z}, \mathscr{x}'\rangle = \langle \mathscr{z}, \mathscr{x}'\rangle.$$

It therefore suffices to show that $V_{\phi_\alpha}\mathscr{z} \in \mathscr{X}_p$. Now, $C(G) = SAP(G)$ is generated by finite dimensional, translation invariant subspaces, hence every ϕ_α is uniformly approximable by functions ϕ from such spaces $\mathscr{G}$. Since $V_{\phi_\alpha}\mathscr{z}$ is norm approximable by $V_\phi\mathscr{z}$, it now suffices to show that the finite dimensional space $\{V_\phi\mathscr{z} : \phi \in \mathscr{G}\}$ is $\mathscr{U}$ invariant (hence unitary). But this follows from

$$WV_\phi\mathscr{z} = \int_G \phi(V)WV\mathscr{z}\,dV = \int_G \phi(W^{-1}V)V\mathscr{z}\,dV = \int_G L_{W^{-1}}\phi(V)V\mathscr{z}\,dV = V_{W^{-1}\phi}.$$

This completes the proof that $E\mathscr{X} = \mathscr{X}_p$, which implies that $\mathscr{U}^w$ restricted to $\mathscr{X}_p$ is a group of operators on $\mathscr{X}_p$ with identity the identity operator. Now let $\mathscr{Y}$ be any $\mathscr{U}$-invariant subspace on which $\mathscr{U}^w$ acts as a group with identity the identity operator. Since $E^2 = E$, $E|_{\mathscr{Y}} = I$ and so $\mathscr{Y} = E\mathscr{Y} \subseteq E\mathscr{X} = \mathscr{X}_p$. Therefore, $\mathscr{X}_p$ is the largest such space.

Next, we show that $(I - E)\mathscr{X}$ $(= \ker E) = \mathscr{X}_0$. Since $\mathrm{cl}_w\, U_S\mathscr{x} = \mathscr{U}^w\mathscr{x}$ it follows that $\mathscr{x} \in \mathscr{X}_0$ iff $V\mathscr{x} = 0$ for some $V \in \mathscr{U}^w$. Thus if $\mathscr{x} \in \mathscr{X}_0$, then $\{V \in \mathscr{U}^w : V\mathscr{x} = 0\}$ is nonempty, hence is a closed left ideal and so must contain the idempotent E. Therefore, $\mathscr{X}_0 = \ker E$.

Finally, let m be an invariant mean on $\mathscr{A}_U$ and let $g(V) := |\langle V\mathscr{x}, \mathscr{x}'\rangle|$, so that $\psi(g)(s) = |\langle U_s\mathscr{x}, \mathscr{x}'\rangle|$. By (c) of 17.6.4, $m\big(|\langle U_{(\cdot)}\mathscr{x}, \mathscr{x}'\rangle|\big) = \int_K |\langle V\mathscr{x}, \mathscr{x}'\rangle|\,dV$. It follows that $m\big(|\langle U_{(\cdot)}\mathscr{x}, \mathscr{x}'\rangle|\big) = 0$ for all $\mathscr{x}'$ iff $V\mathscr{x} = 0$ for all $V \in K$ iff $E\mathscr{x} = 0$ (since $K = KE$) iff $\mathscr{x} \in (E - I)\mathscr{X} = \mathscr{X}_0$. □

The conclusions of the theorem hold if either S is commutative or a group, since in each case, $WAP(S)$ has an invariant mean. One also has

17.6.6 Corollary (deLeeuw-Glicksberg). *If $\|U_s\| \le 1$ for all s and if both $\mathscr{X}$ and $\mathscr{X}'$ are strictly convex, then the conclusions of the theorem hold.*

Proof. We show that $E_1 = E_1E_2 = E_2$ for all idempotents in $K(\mathscr{U}^w)$. It will follow that $K(\mathscr{U}^w)$ is a compact topological group, and we can then apply the theorem.

By minimality, $\mathscr{U}^wE_1E_2 = \mathscr{U}^wE_2$, hence we may choose V so that $VE_1E_2 = E_2$. Then

$$\|E_2\mathscr{x}\| = \|VE_1E_2\mathscr{x}\| \le \|E_1E_2\mathscr{x}\| \le \|E_2\mathscr{x}\|$$

so $\|E_2\mathscr{x}\| = \|E_1E_2\mathscr{x}\|$. It follows that $E_2\mathscr{x} = E_1E_2\mathscr{x}$; otherwise, by strict convexity of $\mathscr{X}$,

$$\|E_2\mathscr{x}\| = \left\|\tfrac{1}{2}E_1(E_1E_2\mathscr{x} + E_2\mathscr{x})\right\| \le \left\|\tfrac{1}{2}(E_1E_2\mathscr{x} + E_2\mathscr{x})\right\| < \|E_2\mathscr{x}\|.$$

To show that $E_1 = E_1E_2$ we use minimality again to choose V so that $E_1E_2V = E_1$. Then $V^*E_2^*E_1^* = E_1^*$, and since E_2^* is a projection the argument of the preceding paragraph shows that $E_2^*E_1^* = E_1^*$ and so $E_1E_2 = E_1$. □

Ergodic Properties of Semigroups of Operators

Let $U : s \mapsto U_s$ be a weakly almost periodic representation of S on a Banach space $\mathscr{X}$. Then $\mathscr{V} := \operatorname{co}\mathscr{U}$ is a semigroup of operators on $\mathscr{X}$, and by the Krein-Šmulian theorem, $\mathscr{V}\boldsymbol{x} = \operatorname{co}\mathscr{U}\boldsymbol{x}$ is relatively weakly compact. Therefore, $\mathscr{V}^w$ is a weakly almost periodic semigroup of operators on $\mathscr{X}$. The results of the preceding subsection may then be applied to $\mathscr{V}$.

The **coefficient space** of the representation U is the closed linear subspace $\mathscr{F}_U$ of $C_b(S)$ generated by the coefficients $s \mapsto \langle U_s\boldsymbol{x}, \boldsymbol{x}'\rangle$, their conjugates, and the constant functions. Here is the appropriate analog of 17.6.4 in this setting.

17.6.7 Proposition. *Let $A(\mathscr{V}^w)$ denote the space of continuous affine functions on $\mathscr{V}^w$*

(a) *$\mathscr{F}_U$ is a translation invariant subspace of $WAP(S)$.*

(b) *The map $\psi : A(\mathscr{V}^w) \to C_b(S)$ defined by $\psi(g)(s) = g(U_s)$ is an isometry onto $\mathscr{F}_U$ that commutes with translations.*

(c) *$\mathscr{F}_U$ has an invariant mean iff there exists an idempotent E in $\mathscr{V}^w$ such that $EV = VE = E$ for all $V \in \mathscr{V}^w$.*

Proof. The proof of (a) is essentially the same as that of part (a) of 17.6.4. The details are left to the reader.

(b) That ψ is an isometry into $C_b(S)$ is clear. Given a coefficient $h(s) = \langle U_s\boldsymbol{x}, \boldsymbol{y}\rangle$, define $g \in A(\mathscr{V}^w)$ by $g(V) = \langle V\boldsymbol{x}, \boldsymbol{y}\rangle$. Then $\psi(g) = h$, which shows that $\mathscr{F}_U \subseteq \operatorname{ran}\psi$ and so $\psi^{-1}(\mathscr{F}_U) \subseteq A(\mathscr{V}^w)$. To show equality, let $\mu \in C(\mathscr{V}^w)'$ such that $\mu = 0$ on $\psi^{-1}(\mathscr{F}_U)$. We show that $\mu = 0$ on $A(\mathscr{V}^w)$; it will follow from the Hahn-Banach theorem that $\psi^{-1}(\mathscr{F}_U) = A(\mathscr{V}^w)$ and hence that $\operatorname{ran}\psi = \mathscr{F}_U$.

Now, μ may be identified with a complex measure on $\mathscr{V}^w$ and hence may be written as a linear combination of probability measures μ_j on $\mathscr{V}^w$, say

$$\mu = a_1\mu_1 - a_2\mu_2 + i(a_3\mu_3 - a_4\mu_4), \quad a_j \geq 0.$$

Since $1 \in \psi^{-1}(\mathscr{F}_U)$, we have $0 = a_1 - a_2 + i(a_3 - a_4)$, hence $a_1 = a_2$ and $a_3 = a_4$. Therefore, we may assume that $\mu_1 = \mu_2$ and $\mu_3 = \mu_4$ on $\psi^{-1}(\mathscr{F}_U)$. By 10.31, each μ_j is in the weak* closed convex hull C of the set $\delta_{\mathscr{V}^w}$ of all Dirac measures on $\mathscr{V}^w$. But since $\mathscr{V}^w$ closed and convex, the restriction of C to $A(\mathscr{V}^w)$ is simply $\delta_{\mathscr{V}^w}$. Thus each μ_j restricted to $A(\mathscr{V}^w)$ is a Dirac measure δ_{V_j}. Therefore, $f(V_1) = f(V_2)$ and $f(V_3) = f(V_4)$ for all $f \in \psi^{-1}(\mathscr{F}_U)$. But $\psi^{-1}(\mathscr{F}_U)$ contains all functions $V \to \langle V\boldsymbol{x}, \boldsymbol{x}'\rangle$ and so separates points of $\mathscr{V}^w$. Thus $V_1 = V_2$ and $V_3 = V_4$, proving that $\mu = 0$ on $A(\mathscr{V}^w)$.

(c) For any mean m on $\mathscr{F}_U$, $\psi^*(m)$ is a mean on $A(\mathscr{V}^w)$, and conversely. By the argument in (b), $\psi^*(m) = \delta_E$ for some $E \in \mathscr{V}^w$. By the argument in (c) of 17.6.4, m is an invariant mean on $\mathscr{F}$ iff $\psi^*(m)$ is an invariant mean on $A(\mathscr{V}^w)$, that is, iff $f(VE) = f(EV) = f(E)$ for all $V \in \mathscr{V}^w$. Since $A(\mathscr{V}^w)$ separates points of $\mathscr{V}^w$, (c) follows. □

Here is the main result of the subsection.

17.6.8 Theorem (deLeeuw-Glicksberg). *If $\mathscr{F}_U$ has an invariant mean, then*

$$\mathscr{X} = \{\boldsymbol{x} : U_s\boldsymbol{x} = \boldsymbol{x} \ \forall\ s \in S\} \oplus \operatorname{cl}\operatorname{span}\{U_s\boldsymbol{x} - \boldsymbol{x} : \boldsymbol{x} \in \mathscr{X}, s \in S\}.$$

Proof. By 17.6.4, there exists $E \in \mathscr{V}^w$ such that $EV = VE = E$ for all V. In particular, $E^2 = E$, so $\mathscr{X} = E\mathscr{X} \oplus (I - E)\mathscr{X}$. Now, $U_s\boldsymbol{x} = \boldsymbol{x}$ for all $s \in S$ iff $V\boldsymbol{x} = \boldsymbol{x}$ for all $V \in \mathscr{V}^w$. Since $VE = E$, the latter is equivalent to $E\boldsymbol{x} = \boldsymbol{x}$. This shows that $E\mathscr{X}$ is the first space in the above direct sum. Since the second space is the span of the vectors $V\boldsymbol{x} - \boldsymbol{x}$ $(V \in \mathscr{V}^w)$ and since $E(V\boldsymbol{x} - \boldsymbol{x}) = 0$, the second space is $\ker E = (I - E)\mathscr{X}$, completing the proof. □

The preceding theorem allows a simple proof of the following generalization of the mean ergodic theorem of von Neumann.

17.6.9 Corollary. *Let* $U \in \mathscr{B}(\mathcal{X})$ *such that the semigroup* $\{U^n : n \in \mathbb{N}\}$ *is weakly almost periodic. Then* $A_n = n^{-1}\sum_{j=0}^{n-1} U^j$ *converges in the strong operator topology to a projection* $E \in \mathscr{B}(\mathcal{X})$ *satisfying* $EU = UE = E$.

Proof. Let E be the projection in the proof of the theorem for the representation $n \to U^n$. We need only show that for fixed k, $A_n(U^k\boldsymbol{x} - \boldsymbol{x}) \to \mathbf{0}$. This follows from the identity $A_n(U\boldsymbol{x} - \boldsymbol{x}) = \frac{1}{n}(U^n\boldsymbol{x} - \boldsymbol{x})$ and the uniform boundedness of $U^{\mathbb{N}}$. □

The preceding corollary holds for an operator U of norm ≤ 1 on a reflexive Banach space. For a nonreflexive example, let $(X, \mathscr{F}, \mu)$ be a probability space and $\varphi : X \to X$ measurable such that $\mu(\varphi^{-1}(E)) \leq \mu(E)$ for all $E \in \mathfrak{F}$. Define U in L^1 by $Uf = f \circ \varphi$. By 17.6.2, $U^{\mathbb{N}}$ is weakly almost periodic, hence the corollary is applicable and we have L^1 convergence $n^{-1}\sum_{j=0}^{n-1} U^j f = Ef$ A more refined version of this result in the special case of a measure preserving φ is proved in 18.5.

Chapter 18

Probability Theory

Probability theory has long been a subject of great interest, its roots dating back to the analysis of games of chance in the sixteenth century. The development of modern probability theory as a branch of measure theory was initiated by Kolmogorov in the early twentieth century.

Intuitively, a probability is a number between 0 and 1 that expresses the likelihood of an outcome in an **experiment**. In this context, the term experiment simply refers to a repeatable procedure that has a well-defined set of outcomes; something as simple as tossing a die or as complex as noting the first time a stock dips below a prescribed level. In practice, the determination of probabilities may be based on logical deduction, analytical methods, or statistical analysis (as in polling). For our purposes, we shall take as given a particular assignment of probabilities and not be concerned with their origin. More precisely, our development of the subject begins in the modern tradition with a given probability space $(\Omega, \mathcal{F}, P)$.[1]

18.1 Random Variables

Many terms in modern probability theory reflect the classical origins of the subject as well as its use in analysing real data. For example, the set Ω of a probability space $(\Omega, \mathcal{F}, P)$ is called the **sample space** (in practice, the set of outcomes of an experiment), and members of $\mathcal{F}$ are called **events** (sets of outcomes). Properties holding almost everywhere are said to hold **almost surely** (a.s.). A real-valued (Borel) measurable function is called a **random variable** and may be viewed as a numerical description of an outcome of an experiment. A measurable function that takes values in $\mathbb{R}^d$ is called a **d-dimensional random variable**.

Random variables are typically denoted by letters X, Y, etc. A d-dimensional random variable may be written $(X_1, \ldots, X_d)$, where each X_j is a (1-dimensional) random variable. If $\mathcal{X}$ is a family of d-dimensional random variables on Ω, we denote by $\sigma(\mathcal{X})$ the σ-sub-algebra of $\mathcal{F}$ generated by the sets $\{X \in B\}$, where $X \in \mathcal{X}$ and $B \in \mathcal{B}(\mathbb{R}^d)$.

Expectation and Variance

The integral with respect to P of a random variable X is called the **mean** or **expectation** of X and is denoted by $E(X)$ (or, simply, $E\,X$):

$$E\,X = \int_\Omega X\,dP.$$

[1]Here, in keeping with standard conventions, we write Ω instead of X and use the symbol P for a probability measure. Other changes of notation to accommodate convention, as well as changes in terminology, are given in §18.1.

The **variance** of an L^2 random variable X is defined by

$$V(X) := E[X - E(X)]^2 = E(X^2) - [E(X)]^2. \tag{18.1}$$

Variance may be seen as a measure of the dispersion of the data X from the mean. The quantity

$$\sigma(X) := \sqrt{V(X)}$$

is called the **standard deviation of** X. The **covariance** of L^2 random variables X and Y is the quantity

$$\operatorname{cov}(X, Y) := E[(X - E(X))(Y - E(Y))] = E(XY) - E(X)E(Y). \tag{18.2}$$

Covariance measures the degree of correlation between X and Y. For example, independent random variables have covariance zero (see 18.2.3).

The **characteristic function** ϕ_X of a d-dimensional random variable $X = (X_1, \ldots, X_d)$ is defined by

$$\phi_X(t) = E(e^{it\cdot X}).$$

Note that this is simply a variation of the Fourier transform of the image measure $X(P)$ (see next subsection).

Probability Distributions

A probability measure on $\mathcal{B}(\mathbb{R}^d)$ is called a d-**dimensional probability distribution**. An important example is the image measure $P_X := X(P)$ on $\mathcal{B}(\mathbb{R}^d)$ of a d-dimensional random variable $X = (X_1, \ldots, X_d)$. The measure P_X is called the **distribution of** X or the **joint distribution of** $(X_1, \ldots, X_d)$. By the image measure theorem 3.2.15,

$$E\, g(X_1, \ldots, X_n) = \int g(x_1, \ldots, x_n)\, dP_X(x_1, \ldots, x_n)$$

for any Borel function g for which one side or the other of the equation is defined. Every probability distribution Q on $\mathcal{B}(\mathbb{R}^d)$ arises in this manner, that is, as the distribution of a random variable X on a probability space $(\Omega, \mathscr{F}, P)$: simply take $\Omega = \mathbb{R}^d$, $\mathcal{F} = \mathcal{B}(\mathbb{R}^d)$, $P = Q$ and X the identity mapping on $\mathbb{R}^d$. A family $\mathcal{X}$ of d-dimensional random variables is said to be **identically distributed** if $P_X = P_Y$ for all $X, Y \in \mathcal{X}$.

For $d = 1$ the function

$$F_X(x) = P(X \le x) = P_X\big((-\infty, x]\big)$$

is called the **cumulative distribution function** (cdf) of X. In many cases of interest, the cdf is given by a **probability density** f_X, so that

$$F_X(x) = \int_{-\infty}^{x} f_X(t)\, dt \quad \text{and} \quad E\big(g(X)\big) = \int_{\mathbb{R}} g(t) f_X(t)\, dt.$$

If $\operatorname{ran} X$ is countable, then the cdf is given by the **probability mass function** (pmf)

$$p_X(x) := P(X = x).$$

In this case

$$F_X(x) = \sum_{t \le x} p_X(x) \quad \text{and} \quad E\big(g(X)\big) = \sum_{x} g(x) p_X(x).$$

The following are standard distributions given in terms of the probability mass function or density. In each case X denotes a random variable with the given distribution.

18.1.1 Examples.

- **Bernoulli distribution with parameter** $p \in (0,1)$:

$$p_X(1) = 1 - p_X(0) = p.$$

For example, the number of heads (0 or 1) that appear on a single toss of a fair coin has a Bernoulli distribution with parameter 1/2. By an easy calculation,

$$E(X) = p, \quad V(X) = pq, \quad \text{and} \quad \phi_X(t) = e^{it}p + q, \quad \text{where } q := 1 - p.$$

- **Binomial distribution with parameters** (n, p), $n \in \mathbb{N}$, $0 < p < 1$:

$$p_X(k) = \binom{n}{k} p^k q^{n-k}, \quad 0 \le k \le n, \quad q := 1 - p.$$

For example, the number of heads that appear in n tosses of a coin has a binomial distribution, where p is the probability of a head on a single toss. One may check that

$$E(X) = np, \quad V(X) = npq, \quad \text{and} \quad \phi_X(t) = (e^{it}p + q)^n.$$

- **Geometric distribution with parameter** $p \in (0,1)$:

$$p_X(k) = q^k p \quad k \in \mathbb{N}, \quad q := 1 - p.$$

For example, in repeated tosses of a coin, the number of tails that appear before the first head is geometrically distributed. One easily calculates

$$E(X) = \frac{q}{p}, \quad V(X) = \frac{q}{p^2}, \quad \text{and} \quad \phi_X(y) = \frac{p}{1 - qe^{it}}.$$

- **Uniform distribution on the interval** (a, b):

$$f_X = \frac{1}{b-a} \mathbf{1}_{(a,b)}.$$

For example, a number drawn randomly from the interval (a, b) is uniformly distributed. For such a random variable,

$$E(X) = \frac{a+b}{2}, \quad V(X) = \frac{(b-a)^2}{12}, \quad \text{and} \quad \phi_X = \frac{e^{itb} - e^{itb}}{it(b-a)}.$$

- **Exponential distribution with parameter** $\lambda > 0$:

$$f_X = \lambda e^{-\lambda x} \mathbf{1}_{[0,\infty)}.$$

The exponential distribution is often used to model the life of a biological, electrical, or mechanical system. It is easily seen that

$$E(X) = \frac{1}{\lambda}, \quad V(X) = \frac{1}{\lambda^2}, \quad \text{and} \quad \phi_X = \frac{\lambda}{\lambda - it}.$$

- **Normal distribution with mean** $m \in \mathbb{R}$ **and standard deviation** $\sigma > 0$:

$$f_X = \frac{1}{\sigma\sqrt{2\pi}} \exp\left(-\frac{1}{2}\left(\frac{x-m}{\sigma}\right)^2\right).$$

For example, samples drawn randomly from a large population of independent data have a nearly normal distribution. One calculates that

$$E(X) = m, \quad V(X) = \sigma^2, \quad \text{and} \quad \phi_X = e^{itm - \sigma^2 t^2/2}.$$

The first two equations may be established using $\int_{-\infty}^{\infty} e^{-x^2/2}\,dx = \sqrt{2\pi}$ and a substitution. The third equation follows from 6.2.3. ◊

18.2 Independence

The notion of independence is specific to probability theory and may be seen as one of several major points of departure of the subject from general measure theory.

Independent Events

Let $(\Omega, \mathcal{F}, P)$ be a probability space. A family $\{A_i : i \in \mathfrak{I}\}$ of events in $\mathcal{F}$ is said to be **independent** if

$$P(A_{i_1} \cap \cdots \cap A_{i_n}) = P(A_{i_1}) \cdots P(A_{i_n})$$

for all choices of distinct indices i_k in $\mathfrak{I}$. A family $\{\mathcal{A}_i : i \in \mathfrak{I}\}$ of subcollections $\mathcal{A}_i$ of $\mathcal{F}$ is **independent** if the collection $\{A_i : i \in \mathfrak{I}\}$ is independent for all choices $A_i \in \mathcal{A}_i$, $i \in \mathfrak{I}$.

For example, if $(\Omega, \mathcal{F}, P) = (\Omega_1 \times \Omega_2, \mathcal{F}_1 \otimes \mathcal{F}_2, P_1 \times P_2)$, then, by definition of the product measure, the σ-fields $\mathcal{F}_1 \times \Omega_2$ and $\Omega_2 \times \mathcal{F}_2$ are independent families. This is the basis of the notion of **independent trials**. Indeed, if (ω_1, ω_2) represents the outcome of a two stage experiment, then in this model the events $A_1 \times \Omega_2$ and $\Omega_1 \times A_2$, occurring in stages one and two, respectively, are independent. This idea generalizes to arbitrary finite sequences of trials and even to infinite sequences (see §18.4).

18.2.1 Proposition. *Let $(\Omega, \mathcal{F}, P)$ be a probability space and $\{\mathcal{A}_i : i \in \mathfrak{I}\}$ an independent family of π-systems contained in $\mathcal{F}$. Then the family $\{\sigma(\mathcal{A}_i) : i \in \mathfrak{I}\}$ is independent.*

Proof. We may suppose that $\Omega \in \mathcal{A}_i$ for every i, since adjoining Ω does not alter the independence property. Since the notion of independence involves only finitely many sets at a time, we may also assume that $\mathfrak{I}$ is finite, say, $\mathfrak{I} = \{1, \ldots, n\}$. The property of independence may now be expressed as

$$P(A_1 \cap \cdots \cap A_n) = P(A_1) \cdots P(A_n), \tag{†}$$

for all $A_j \in \sigma(\mathcal{A}_j)$, $j = 1, \ldots, n$. Fix $A_j \in \mathcal{A}_j$, $j = 2, \ldots, n$. By hypothesis, (†) holds for all $A_1 \in \mathcal{A}_1$. Since each side is a finite measure in $A_1 \in \sigma(\mathcal{A}_1)$, it follows from the measure uniqueness theorem (1.6.8) that (†) holds for all $A_1 \in \sigma(\mathcal{A}_1)$. Now fix $A_1 \in \sigma(\mathcal{A}_1)$ and $A_j \in \mathcal{A}_j$, $j > 2$. Arguing as before, this time on A_2, we conclude that (†) holds for all $A_1 \in \sigma(\mathcal{A}_1)$, $A_2 \in \sigma(\mathcal{A}_2)$, and $A_j \in \mathcal{A}_j$, $j = 3, \ldots, n$. Continuing in this manner, we see that (†) holds for all $A_j \in \sigma(\mathcal{A}_j)$ and all j. $\square$

Independent Random Variables

A collection $\{X_i : i \in \mathfrak{I}\}$ of random variables is said to be **independent** if the family $\{\mathcal{A}_i := X_i^{-1}(\mathcal{B}(\mathbb{R})) : i \in \mathfrak{I}\}$ is independent. In particular, finitely many random variables $X_1, \ldots, X_n$ are independent iff

$$P(X_1 \in B_1, \ldots, X_n \in B_n) = P(X_1 \in B_1) \cdots P(X_n \in B_n)$$

for all Borel sets B_j. Note that by 18.2.1, to test for independence it suffices to take B_j in a generating π-system. The preceding equation may be written in terms of probability distributions as

$$P_{(X_1, \ldots, X_n)}(B_1 \times \cdots \times B_n) = P_{X_1}(B_1) \cdots P_{X_n}(B_n) = \big(P_{X_1} \otimes \cdots \otimes P_{X_n}\big)(B_1 \times \cdots \times B_n).$$

Thus we have

18.2.2 Proposition. *Random variables $X_1, \ldots, X_n$ are independent iff*

$$P_{(X_1,\ldots,X_n)} = P_{X_1} \otimes \cdots \otimes P_{X_n}.$$

Proposition 18.2.2 gives a precise characterization of the notion of independent trials. If the random variable X_j is the numerical outcome of the jth stage of an experiment, then in a model described by the law $P_{X_1} \otimes \cdots \otimes P_{X_n}$ no trial will have influence on future trials.

18.2.3 Theorem. *If $X_1, \ldots, X_n$ are independent and either $X_j \geq 0$ for all j or X_j is integrable for all j, then*

$$E(X_1 \cdots X_n) = E(X_1) \cdots E(X_n).$$

Proof. We prove the second case. By 18.2.2,

$$\begin{aligned} E(|X_1 \cdots X_n|) &= \int |x_1 \cdots x_n| \, dP_{(X_1,\ldots,X_n)}(x_1, \ldots, x_n) \\ &= \int \cdots \int |x_1| \cdots |x_n| \, dP_{X_1}(x_1) \cdots dP_{X_n}(x_n) \\ &= E|X_1| \cdots E|X_n| < \infty \end{aligned}$$

By Fubini's theorem, the absolute value signs in this equation may be removed, proving the theorem. □

18.2.4 Proposition. *Let $X_1, \ldots, X_n$ be independent and $X_j \in L^2(P)$. Then*

$$V(X_1 + \cdots + X_n) = V(X_1) + \cdots + V(X_n).$$

Proof. Set $S := X_1 + \cdots + X_n$, $m_j := E(X_j)$, and $m := m_1 + \cdots + m_n$. Then

$$(S - m)^2 = \left[\sum_{j=1}^{n}(X_j - m_j)\right]^2 = \sum_{j=1}^{n}(X_j - m_j)^2 + \sum_{i \neq j}(X_i - m_i)(X_j - m_j),$$

and the conclusion follows by taking expectations, noting that the expectation of the second sum on the right is zero, by independence. □

18.2.5 Proposition. *Let $X_1, \ldots, X_n$ be independent random variables. Then*

$$P_{X_1 + \cdots + X_n} = P_{X_1} * \cdots * P_{X_n}.$$

Proof. For $A \in \mathcal{B}(\mathbb{R})$,

$$\begin{aligned} P_{X_1+\cdots+X_n}(A) &= \int \mathbf{1}_A(X_1 + \cdots + X_n) \, dP \\ &= \int \mathbf{1}_A(x_1 + \cdots + x_n) \, dP_{(X_1,\ldots,X_n)} \\ &= \int \cdots \int \mathbf{1}_A(x_1 + \cdots + x_n) \, dP_{X_1} \cdots dP_{X_n} \\ &= (P_{X_1} * \cdots * P_{X_n})(A), \end{aligned}$$

the third equality from 18.2.2. □

18.3 Conditional Expectation

Let $(\Omega, \mathcal{F}, P)$ be a probability space, $\mathcal{G}$ a sub-σ-field of $\mathcal{F}$, and X an L^1 random variable. Recall from §5.3 that the conditional expectation of X given $\mathcal{G}$ is a $\mathcal{G}$-random variable Y such that $\int_A Y\,dP = \int_A X\,dP$ for all $A \in \mathcal{G}$ (see 5.3.6). The standard notation for Y is $E(X \mid \mathcal{G})$. Thus, by definition, $E(X \mid \mathcal{G})$ is the unique (up to a set of $\mathcal{G}$ measure zero) $\mathcal{G}$ random variable with the property

$$\int_A E(X \mid \mathcal{G})\,dP = \int_A X\,dP \quad \text{for all} \quad A \in \mathcal{G}. \tag{18.3}$$

Note that by uniqueness of measures (1.6.8), the equation holds for all $A \in \mathcal{G}$ iff it holds for A in a generating π-system for $\mathcal{G}$. In the special case $\mathcal{G} = \sigma(X_1, X_2, \ldots)$, $E(X \mid \mathcal{G})$ is called the **conditional expectation of X given $X_1, X_2, \ldots$** and is denoted by $E(X \mid X_1, X_2, \ldots)$. To test whether 18.3 holds in this case, it suffices to restrict consideration to events A of the form $\{X_1 \in B_1, \ldots, X_n \in B_n\}$.

A sub-σ-field $\mathcal{G}$ of $\mathcal{F}$ may be viewed as information regarding the location of an outcome. For example, in the case of a repeated coin toss, the σ-field generated by all events of the form $\{H\} \times A_2 \times A_3 \times \cdots$ tells us with certainty that the first toss came up heads. Conditional expectation generalizes the notion of standard expectation by incorporating such information into its definition. It may be viewed as the best prediction of X given the information $\mathcal{G}$. The two extreme cases are $E\big(X \mid \{\emptyset, \Omega\}\big) = E(X)$ and $E(X \mid \mathcal{P}(X)) = X$. In the first case, the σ-field $\{\emptyset, \Omega\}$ provides no information, and one merely obtains the mean of X. In the second case, the best prediction of X given all possible information is X itself.

The following theorem summarizes the main properties of conditional expectation. The reader will note that several of these properties are analogs of those of ordinary expectation.

18.3.1 Theorem. *Let X, $Y \in L^1(\Omega, \mathcal{F}, P)$ and let $\mathcal{G}$ and $\mathcal{H}$ be σ-fields with $\mathcal{H} \subseteq \mathcal{G} \subseteq \mathcal{F}$.*

(a) $E(1 \mid \mathcal{G}) = 1$.

(b) $E(aX + bY \mid \mathcal{G}) = aE(X \mid \mathcal{G}) + bE(Y \mid \mathcal{G})$, a, $b \in \mathbb{R}$.

(c) $X \le Y \Rightarrow E(X \mid \mathcal{G}) \le E(Y \mid \mathcal{G})$.

(d) $|E(X \mid \mathcal{G})| \le E(|X| \mid \mathcal{G})$.

(e) (Conditional Jensen's Inequality). *If $\phi : \mathbb{R} \to \mathbb{R}$ is convex and $\phi(X) \in L^1$, then $\phi(E(X \mid \mathcal{G})) \le E(\phi(X) \mid \mathcal{G})$. In particular, if $X \in L^p$ $(1 \le p < \infty)$, then $\|E(X \mid \mathcal{G})\|_p \le \|X\|_p$.*

(f) (Factor Property). *If X is a $\mathcal{G}$-random variable, then $E(XY|\mathcal{G}) = XE(Y|\mathcal{G})$. In particular, $E(X \mid \mathcal{G}) = X$.*

(g) (Independence Property). *If $\sigma(X)$ and $\mathcal{G}$ are independent, then $E(X \mid \mathcal{G}) = E(X)$.*

(h) (Tower Property). $E\left[E(X \mid \mathcal{G}) \mid \mathcal{H}\right] = E(X \mid \mathcal{H})$.

(i) (Monotone convergence theorem). *Let (X_n) be a sequence of nonnegative random variables with $X_n \uparrow X$. If X is integrable, then $E(X_n \mid \mathcal{G}) \uparrow E(X \mid \mathcal{G})$.*

Proof. Properties (a)–(c) follow by taking integrals over sets $A \in \mathcal{G}$. Part (d) follows from part (c) and the inequalities $\pm X \le |X|$. Part (e) is proved the same way as the standard Jensen's inequality (4.5.4).

For (f), note first that the random variable $XE(Y|\mathcal{G})$ is $\mathcal{G}$-measurable. Now let $A \in \mathcal{G}$. To establish the required property that $E[\mathbf{1}_A XE(Y|\mathcal{G})] = E(\mathbf{1}_A XY)$, we may assume that $X, Y \geq 0$. Now, for $\mathcal{G}$-simple functions $X = \sum_{j=1}^n a_j \mathbf{1}_{A_j}$, we have, by definition of $E(Y \mid \mathcal{G})$ and linearity,

$$E[\mathbf{1}_A XE(Y \mid \mathcal{G})] = \sum_{j=1}^n a_j E[\mathbf{1}_{A \cap A_j} E(Y \mid \mathcal{G})] = \sum_{j=1}^n a_j E(\mathbf{1}_{A \cap A_j} Y) = E(\mathbf{1}_A XY).$$

The desired equality now follows by considering an increasing sequence of simple functions X_n and applying the monotone convergence theorem.

For (g), simply note that by independence of $\mathcal{G}$ and $\sigma(X)$ and by 18.2.3 we have

$$E(\mathbf{1}_A X) = (E\,\mathbf{1}_A)(E\,X) = E[\mathbf{1}_A E(X)], \ A \in \mathcal{G}.$$

Property (h) follows from

$$\int_A E(E(X \mid \mathcal{G})|\mathcal{H})\,dP = \int_A E(X \mid \mathcal{G})\,dP = \int_A X\,dP = \int_A E(X \mid \mathcal{H})\,dP, \ A \in \mathcal{H}.$$

Finally, for (i) we apply (c) to conclude that $E(X_n \mid \mathcal{G}) \uparrow Y$ for some $\mathcal{G}$-random variable Y. By the monotone convergence theorem, for any $A \in \mathcal{G}$,

$$\int_A Y\,dP = \lim_n \int_A E(X_n \mid \mathcal{G})\,dP = \lim_n \int_A X_n\,dP = \int_A X\,dP.$$

Therefore, $Y = E(X \mid \mathcal{G})$. □

18.4 Sequences of Independent Random Variables

A sequence of independent random variables may be viewed as the numerical outcomes of of independent trials of an experiment. In this chapter, we consider the main theorems regarding such sequences, including zero-one laws, laws of large numbers, and the central limit theorem. The first step is to construct the infinite product of a sequence of probability spaces. The construction is motivated by the following example.

18.4.1 Example. Suppose a fair coin is tossed repeatedly. The outcomes of the experiment may be identified with infinite sequences of heads (H) and tails (T). We seek to assign a probability to the event E_n that the first head appears on the nth toss. For this we argue as follows: Since on the first toss the outcomes H or T are equally likely, we should set $P(E_1) = 1/2$. Similarly, the outcomes HH, HT, TH, TT of the first two tosses are equally likely, hence $P(E_2)$, the probability that TH occurs, should be $1/4$. In general, by this argument we should set $P(E_n) = 2^{-n}$, $n \geq 1$. If $\Omega_n = \{H, T\}$ denotes the sample space of outcomes on the nth toss and if $P_n\{H\} = P_n\{T\} = 1/2$, then E_n is of the form $E_n = A_1 \times \cdots \times A_n \times \Omega_{n+1} \times \cdots$, where $A_j \subseteq \Omega_j$, and $P(E_n) = P_1(A_1) \cdots P_n(A_n)$. The last equation describes a general way of assigning probabilities $P(E_n)$. Theorem 18.4.4 below shows that such assignments of probabilities may be extended to obtain a probability measure P on the σ-field generated by all sets E_n. Thus, in the example at hand, we conclude that the probability of a head eventually appearing is $P(\bigcup_n E_n) = \sum_{n=1}^{\infty} 2^{-n} = 1$. ◊

Infinite Product Measures

Let $(\Omega_n, \mathcal{F}_n, P_n)$ $(n = 1, 2, \ldots)$ be probability spaces and $\Omega := \prod_{n=1}^\infty \Omega_n$. A **cylinder set** in Ω is a subset of the form

$$B \times \Omega_{n+1} \times \Omega_{n+2} \times \cdots, \quad B \in \mathcal{F}_1 \otimes \cdots \otimes \mathcal{F}_n.$$

A special case is the **rectangular cylinder set**

$$A_1 \times \cdots \times A_n \times \Omega_{n+1} \times \Omega_{n+2} \times \cdots, \quad A_k \in \mathcal{F}_k. \tag{$\dagger$}$$

Interpreting A_k as an event that occurs at "time k", cylinder sets may be seen as events occurring in finite time. The σ-field generated by all the cylinder sets (hence also by the rectangular cylinder sets) is called the **product σ-field** and is denoted by $\bigotimes_{n=1}^\infty \mathcal{F}_n$. The following analog of 2.1.5 is readily established.

18.4.2 Proposition. *Let $\mathcal{F} = \bigotimes_{n=1}^\infty \mathcal{F}_n$ and let $\pi_n : \Omega \to \Omega_n$ be the nth projection map $\pi_n(\omega_1, \omega_2 \ldots) = \omega_n$. Then π_n is $\mathcal{F}/\mathcal{F}_n$-measurable. Moreover, if $(\Omega_0, \mathcal{F}_0)$ is a measurable space, then a mapping $T : \Omega_0 \to \Omega$ is $\mathcal{F}_0/\mathcal{F}$-measurable iff $\pi_n \circ T$ is $\mathcal{F}_0/\mathcal{F}_n$-measurable for every n.*

For the construction of a suitable probability measure on $\bigotimes_{n=1}^\infty \mathcal{F}_n$, we follow the elegant argument of Saeki [42], which begins with the following lemma.

18.4.3 Lemma. *Let Ω be a nonempty set and $\mathcal{A}$ a semiring of subsets of Ω containing Ω. Let P be a set function on $\mathcal{A}$ such that $P(\emptyset) = 0$ and $\sum_{n=1}^\infty P(A_n) = 1$ whenever (A_n) is a disjoint sequence in $\mathcal{A}$ with union Ω. Then P extends to a probability measure on $\sigma(\mathcal{A})$.*

Proof. Let $\mathcal{A}_u$ denote the set of all finite disjoint unions of members of $\mathcal{A}$. By the proof of 1.6.4, $\mathcal{A}_u$ is a field. Moreover, since $\emptyset \in \mathcal{A}$, every member A of $\mathcal{A}_u$ can be written (non-uniquely) as an *infinite* disjoint union of members A_n of $\mathcal{A}$. We shall call (A_n) a *representing sequence for A*. Now extend P to $\mathcal{A}_u$ by defining $P(A) = \sum_{n=1}^\infty P(A_n)$, where (A_n) is any representing sequence for A. To see that the extension is well-defined, write $A^c \in \mathcal{A}_u$ as a disjoint union $B_1 \cup \cdots \cup B_m$, $B_j \in \mathcal{A}$. By hypothesis,

$$\sum_{n=1}^{\infty} P(A_n) = 1 - \sum_{j=1}^{m} P(B_j).$$

As the right side is independent of the representing sequence for A, the extension P is well-defined. Since, by definition, P is countably additive on $\mathcal{A}_u$, Theorem 1.6.4 guarantees the existence of an extension of P to $\sigma(\mathcal{A})$. □

We may now prove

18.4.4 Theorem. *There exists a unique probability measure P on $\bigotimes_{n=1}^\infty \mathcal{F}_n$ such that*

$$P\big(A_1 \times \cdots \times A_n \times \Omega_{n+1} \times \Omega_{n+2} \times \cdots\big) = P_1(A_1) \cdots P_n(A_n)$$

for all n and all $A_k \in \mathcal{F}_k$.

Proof. Uniqueness follows from 1.6.8. To establish existence, for each $n \in \mathbb{N}$ let $\mathcal{A}_n$ denote the collection of all rectangular cylinder sets of the form $A_1 \times \cdots \times A_n \times \Omega_{n+1} \times \cdots$, and let $\mathcal{A}$ denote the collection of all rectangular cylinder sets. Since $\mathcal{A}_n \uparrow \mathcal{A}$ and each $\mathcal{A}_n$ is a semiring, $\mathcal{A}$ is a semiring. Now set $Q_n = P_1 \otimes \cdots \otimes P_n$ and define a set function P on $\mathcal{A}$ by

$$P\left(A_1 \times \cdots \times A_n \times \Omega_{n+1} \times \Omega_{n+2} \times \cdots\right) := Q_n(A_1 \times \cdots \times A_n).$$

In particular,

$$P(\Omega_1 \times \cdots \times \Omega_{n-1} \times A_n \times \Omega_{n+1} \times \Omega_{n+2} \times \cdots) = P_n(A_n).$$

To see that P is well-defined, suppose that $n < m$ and

$$A \times \Omega_{n+1} \times \Omega_{n+2} \times \cdots = B \times \Omega_{m+1} \times \Omega_{m+2} \times \cdots,$$

where

$$A = A_1 \times \cdots \times A_n \quad \text{and} \quad B = B_1 \times \cdots \times B_m, \quad A_j, B_j \in \mathfrak{F}_j.$$

Then $B = A \times \Omega_{n+1} \times \cdots \times \Omega_m$ and so

$$P(A \times \Omega_{n+1} \times \Omega_{n+2} \times \cdots) = Q_n(A) = Q_m(B) = P(B \times \Omega_{m+1} \times \Omega_{m+2} \times \cdots).$$

Next, we show that P has the property of the lemma. Let (A_n) be a disjoint sequence in $\mathcal{A}$ with union Ω. Then

$$A_n = \prod_{j=1}^{\infty} A_{nj}, \quad A_{nj} \in \mathfrak{F}_j, \quad A_{nj} = \Omega_j, \ j > j_n, \quad \text{and} \quad P(A_n) = \prod_{j=1}^{\infty} P(A_{nj}).$$

Suppose, for a contradiction, that $\sum_{n=1}^{\infty} P(A_n) \neq 1$. Then there must exist an $\omega_1 \in \Omega_1$ such that

$$\sum_{n=1}^{\infty} \mathbf{1}_{A_{n1}}(\omega_1) \prod_{j=2}^{\infty} P(A_{nj}) \neq 1;$$

otherwise, integrating over $\omega_1 \in \Omega_1$ would produce $\sum_{n=1}^{\infty} P(A_n) = 1$. It follows by similar reasoning that there exists $\omega_2 \in \Omega_2$ such that

$$\sum_{n=1}^{\infty} \mathbf{1}_{A_{n1}}(\omega_1)\mathbf{1}_{A_{n2}}(\omega_2) \prod_{j=3}^{\infty} P(A_{nj}) \neq 1.$$

By induction, we obtain a point $\omega = (\omega_1, \omega_2, \ldots) \in \Omega$ such that for all m,

$$\sum_{n=1}^{\infty} \prod_{j=1}^{m} \mathbf{1}_{A_{nj}}(\omega_j) \prod_{j=m+1}^{\infty} P(A_{nj}) \neq 1. \tag{a}$$

Now, ω is in A_p for some p. Recalling that

$$A_p = A_{p1} \times \cdots \times A_{pj_p} \times \Omega_{j_p+1} \times \Omega_{j_p+2} \cdots,$$

we see that

$$\prod_{j=1}^{j_p} \mathbf{1}_{A_{pj}}(\omega_j) \prod_{j=j_p+1}^{\infty} P(A_{pj}) = 1. \tag{b}$$

Next, we show that

$$\prod_{j=1}^{j_p} \mathbf{1}_{A_{nj}}(\omega_j) \prod_{j=j_p+1}^{\infty} P(A_{nj}) = 0, \quad n \neq p. \tag{c}$$

Let $\omega_j' \in \Omega_j$, $j > j_p$. Then $\mathbf{1}_{A_p}(\omega_1, \ldots, \omega_{j_p}, \omega_{j_p+1}', \ldots) = 1$, and since $\sum_{n=1}^{\infty} \mathbf{1}_{A_n} = 1$ it follows that for $n \neq p$,

$$\prod_{j=1}^{j_p} \mathbf{1}_{A_{nj}}(\omega_j) \prod_{j=j_p+1}^{\infty} \mathbf{1}_{A_{nj}}(\omega_j') = \mathbf{1}_{A_n}(\omega_1, \ldots, \omega_{j_p}, \omega_{j_p+1}', \ldots) = 0. \tag{d}$$

Let $N \geq j_p + 1$ and restrict ω_j' to lie in A_{nj} for $j > N$. Then $\prod_{j>N} \mathbf{1}_{A_{nj}}(\omega_j') = 1$, hence from (d)

$$\prod_{j=1}^{j_p} \mathbf{1}_{A_{nj}}(\omega_j) \prod_{j=j_p+1}^{N} \mathbf{1}_{A_{nj}}(\omega_j') = \mathbf{1}_{A_n}(\omega_1, \ldots, \omega_{j_p}, \omega_{j_p+1}', \ldots) = 0.$$

Integrating with respect to P_j over all $\omega_j' \in \Omega_j$ $(j = j_p + 1, \ldots, N)$, we have

$$\prod_{j=1}^{j_p} \mathbf{1}_{A_{nj}}(\omega_j) \prod_{j=j_p+1}^{N} P(A_{nj}) = 0.$$

Since N was arbitrary, (c) holds. From (b) and (c) we have

$$\sum_{n=1}^{\infty} \prod_{j=1}^{j_p} \mathbf{1}_{A_{nj}}(\omega_j) \prod_{j=j_p+1}^{\infty} P(A_{nj}) = 1.$$

But this contradicts (a) with $m = j_p$. □

The probability measure P constructed in the theorem is called the **product of the probability measures** (P_n) and is denoted by $\bigotimes_{n=1}^{\infty} P_n$. The probability space

$$\left(\prod_{n=1}^{\infty} \Omega_n, \bigotimes_{n=1}^{\infty} \mathcal{F}_n, \bigotimes_{n=1}^{\infty} P_n\right)$$

is called the **product of the probability spaces** $(\Omega_n, \mathcal{F}_n, P_n)$. An important special case is the countable product of probability spaces of the form $(\mathbb{R}, \mathcal{B}(\mathbb{R}), P_n)$. Note that in this case, $\bigotimes_{n=1}^{\infty} \mathcal{B}(\mathbb{R}) = \mathcal{B}(\mathbb{R}^{\infty})$, where $\mathbb{R}^{\infty}$ is the topological Cartesian product of countably many copies of $\mathbb{R}$. This follows from the fact that a basis for the product topology consists of countable unions of sets of the form $U_1 \times \cdots \times U_n \times \mathbb{R} \times \mathbb{R} \times \cdots$, where U_j is in a countable basis for $\mathbb{R}$. Similar remarks apply to a countable product of probability spaces $(\mathbb{R}^d, \mathcal{B}(\mathbb{R}^d), P_n)$.

The Distribution of a Sequence of Random Variables

Using 18.4.4, we may extend 18.2.2 to the infinite case as follows: Let $X := (X_n)$ be a infinite sequence of random variables on a probability space $(\Omega, \mathcal{F}, P)$. The **distribution of** X is the probability measure P_X on $\bigotimes_{n=1}^{\infty} \mathcal{B}(\mathbb{R})$ defined by

$$P_X(B) = P\big((X_1, X_2, \cdots) \in B\big).$$

By definition, the random variables X_j are independent iff for each n and $B_j \in \mathcal{B}(\mathbb{R})$,

$$P\big((X_1, \ldots, X_n) \in B_1 \times \cdots B_n\big) = P(X_1 \in B_1) \cdots P(X_n \in B_n),$$

which may be written

$$P_X\big(B_1 \times \cdots B_n \times \Omega_{n+1} \times \times \Omega_{n+2} \times \cdots\big) = \left(\bigotimes_{j=1}^{\infty} P_{X_j}\right)(B_1 \times \cdots B_n \times \Omega_{n+1} \times \Omega_{n+2} \times \cdots).$$

Thus, by the uniqueness of measures theorem, we have

18.4.5 Proposition. *A sequence $X = (X_n)$ of random variables is independent iff $P_X = \bigotimes_{j=1}^{\infty} P_{X_j}$.*

The question still remains as to whether there exist sequences of independent random variables. Theorem 18.4.4 neatly settles that question: Consider the sequence of probability spaces $(\Omega_n, \mathcal{F}_n, P_n)$, where $\Omega_n = \mathbb{R}^d$, $\mathcal{F}_n = \mathcal{B}(\mathbb{R}^d)$, and P_n is an arbitrary d-dimensional probability distribution on $\mathbb{R}^d$, and let $(\Omega, \mathcal{F}, P)$ denote the product space. The projection maps $X_n : \Omega \to \mathbb{R}^d$ are then d-dimensional random variables such that

$$\{(X_1, \cdots, X_n) \in B_1 \times \cdots \times B_n\} = \{\omega : \omega_j \in B_j,\ 1 \le j \le n\} = B_1 \times \cdots \times B_n \times \Omega_{n+1} \cdots$$

In particular, $P_{X_j} = P_j$, and the X_j are independent since

$$\begin{aligned} P\{(X_1, \cdots, X_n) \in B_1 \times \cdots \times B_n\} &= P(B_1 \times \cdots \times B_n \times \Omega_{n+1} \cdots) = \prod_{j=1}^{n} P_j(B_j) \\ &= \prod_{j=1}^{n} P(X_j \in B_j) \end{aligned}$$

We have proved

18.4.6 Proposition. *Given a sequence (P_n) of d-dimensional probability distributions, there exists a probability space $(\Omega, \mathcal{F}, P)$ and a sequence of independent d-dimensional random variables X_n on $(\Omega, \mathcal{F}, P)$ such that $P_{X_n} = P_n$ for all n.*

We return to the coin toss experiment:

18.4.7 Example. By the proposition, there exists a probability space and a sequence of independent random variables X_n such that $P(X_n = 1) = p = 1 - P(X_n = 0)$, where $0 < p < 1$. This may be taken as a model for an infinite sequence of coin tosses, where $X_n = 1$ if the nth toss is heads, $X_n = 0$ if the nth toss is tails, and p is the probability of heads on a single toss. Using this model, we may determine probabilities of various interesting events. For example, the probability that a head occurs on an even toss is

$$P(X_2 = 1) + P(X_2 = 0, X_4 = 1) + P(X_2 = X_4 = 0, X_6 = 1) + \cdots = p(1 + q + q^2 + \cdots) = 1,$$

where $q := 1 - p$. The probability that the *first* head occurs on an even toss is

$$P(X_1 = 0, X_2 = 1) + P(X_1 = X_2 = X_3 = 0, X_4 = 1) + \cdots = p(q + q^3 + q^5 + \cdots) = \frac{q}{1+q}.$$

For a fair coin, the latter probability is 1/3. ◊

Zero-One Laws

The **tail σ-field** of a sequence of random variables X_n is the σ-field

$$\mathcal{T} = \bigcap_{n=1}^{\infty} \sigma(X_n, X_{n+1}, \cdots).$$

Members of $\mathcal{T}$ are called **tail events**. Thus tail events are unaffected by changes that occur in finite time. For example, the events

$$\left\{\omega : \frac{1}{n}\sum_{k=1}^{n} X_k(\omega) \to \frac{1}{2}\right\} \quad \text{and} \quad \{\omega : X_n(\omega) \to 0\}$$

are tail events, but

$$\left\{\omega : \sum_{n=1}^{\infty} X_n(\omega) = 0\right\}$$

is not. Of particular interest are tail events for independent random variables. These have considerably restricted probabilities, as the following theorem shows.

18.4.8 Kolmogorov's Zero-One Law. *If (X_n) is a sequence of independent random variables and $A \in \mathfrak{T}$, then $P(A) = 0$ or 1.*

Proof. Since $\sigma(X_1, X_2, \ldots)$ is generated by the field $\bigcup_{n=1}^{\infty} \sigma(X_1, \ldots, X_n)$, there exists by 1.6.5 a strictly increasing sequence of integers n_k and sets $A_k \in \sigma(X_i : 1 \le i \le n_k)$ such that $P(A \triangle A_k) < 1/k$. Now

$$|P(A) - P(A_k)| = \left| \int (\mathbf{1}_A - \mathbf{1}_{A_k})\, dP \right| \le \int |\mathbf{1}_A - \mathbf{1}_{A_k}|\, dP = P(A \triangle A_k) \quad \text{and}$$

$$|P(A) - P(A_k \cap A)| \le \int |\mathbf{1}_A - \mathbf{1}_{A_k \cap A}|\, dP = \int \mathbf{1}_A |\mathbf{1}_A - \mathbf{1}_{A_k}|\, dP \le P(A \triangle A_k),$$

hence $P(A_k) \to P(A)$ and $P(A_k \cap A) \to P(A)$. But $A \in \sigma(X_i : i > n_k)$, so by independence $P(A_k \cap A) = P(A_k)P(A)$. Therefore, $P^2(A) = P(A)$. □

For a simple application, consider an infinite series $\sum_{n=1}^{\infty} X_n$ of independent random variables. Since $\sum_{n=1}^{\infty} X_n(\omega)$ converges iff $\sum_{n=m}^{\infty} X_n(\omega)$ converges, the event

$$A = \left\{ \omega : \sum_{n=1}^{\infty} X_n(\omega) \ \text{converges} \right\}$$

is a tail event and so has probability 0 or 1. In 18.4.14 we give sufficient conditions for which $P(A) = 1$, that is, for which the series converges almost surely.

The next result concerns a particularly important tail event and gives sufficient conditions that determine the probability of the event.

18.4.9 Borel-Cantelli Lemma. *Let (A_n) be a sequence of events and let $A = \limsup_n A_n$, the event that A_n occurs infinitely often (i.o.).*

(a) *If $\sum_{n=1}^{\infty} P(A_n) < \infty$, then $P(A) = 0$.*

(b) *If the events A_n are independent and $\sum_{n=1}^{\infty} P(A_n) = \infty$, then $P(A) = 1$.*

Proof. Part (a) follows from $P(A) \le \sum_{k=n}^{\infty} P(A_k)$ for all n. For (b) we have

$$1 - P(A) = P\left(\bigcup_{n=1}^{\infty} \bigcap_{k=n}^{\infty} {A_k}^c \right) = \lim_n \lim_m P\left(\bigcap_{k=n}^{m} {A_k}^c \right) = \lim_n \lim_m \prod_{k=n}^{m} P({A_k}^c),$$

the last equality by independence of (A_n). By the inequality $\ln(1 - x) \le -x$ on $[0, 1)$ we have for each n

$$\lim_m \ln \prod_{k=n}^{m} P(A_k^c) = \lim_m \ln \prod_{k=n}^{m} (1 - P(A_k)) \le -\lim_m \sum_{k=n}^{m} P(A_k) = -\infty,$$

which implies that

$$\lim_m \prod_{k=n}^{m} P({A_k}^c) = \lim_m \prod_{k=n}^{m} (1 - P(A_k)) = 0.$$

Therefore, $P(A) = 1$. □

Note that the independence hypothesis in (b) is crucial. For example, if P is Lebesgue measure on $[0, 1]$ and $A_n = [0, 1/n]$, then $\sum_{n=1}^{\infty} P(A_n) = \infty$ but $P(A) = 0$.

18.4.10 Example. Let (X_n) be a sequence of independent random variables such that $X_n \to 0$ a.s. Then $\sum_{n=1}^{\infty} P(X_n \ge \varepsilon) < \infty$. Otherwise, we would have $P(X_n \ge \varepsilon \text{ i.o.}) = 1$ by 18.4.9(b) and so $P(X_n \to 0) = 0$. ◊

Laws of Large Numbers

Consider the coin toss example 18.4.7, where X_n is the number of heads (0 or 1) appearing on the nth toss. The average number of heads in n tosses is then $(X_1 + \cdots + X_n)/n$, and for a fair coin we would expect this to be close to $1/2$ for large n. Thus we should have, in some sense,

$$\frac{X_1 + \cdots + X_n}{n} \to \frac{1}{2}.$$

In this subsection we derive several results regarding the convergence of such averages, these known generally as *laws of large numbers*. For the first law we need

18.4.11 Chebyshev's Inequality. *Let X be an L^2 random variable with mean zero and variance σ^2. Then*

$$P(|X| \geq r) \leq \frac{\sigma^2}{r^2}.$$

Proof. $\sigma^2 = \int X^2\,dP \geq \int_{\{|X|\geq r\}} X^2\,dP \geq r^2 P(|X| \geq r)$. □

18.4.12 Weak Law of Large Numbers. *Let (X_n) be a sequence of independent L^2 random variables. If $v_n := n^{-2}\sum_{k=1}^{n} V(X_n) \to 0$, then*

$$Y_n := \frac{1}{n}\sum_{k=1}^{n} \left[X_k - E(X_k)\right] \xrightarrow{P} 0.$$

Proof. Since Y_n has mean zero and variance v_n (18.2.4), $P(|Y_n| \geq \varepsilon) \leq v_n/\varepsilon^2 \to 0$. □

By strengthening the hypothesis of the weak law of large numbers, we obtain more powerful conclusions in the form of strong laws. For these laws we need the following generalization of Chebyshev's inequality.

18.4.13 Kolmogorov's Inequality. *Let $X_1, \ldots, X_n$ be independent L^2 random variables with mean 0 and set $S_j := X_1 + \cdots + X_j$, $j = 1, \ldots, n$. Then*

$$P\left(\max_{1\leq j\leq n} |S_j| \geq \varepsilon\right) \leq \frac{1}{\varepsilon^2}\sum_{j=1}^{n} V(X_j).$$

Proof. By replacing X_j by $X_j - E(X_j)$ and noting that $V(X + c) = V(X)$, we may assume that $E(X_j) = 0$. Set

$$A = \left\{\max_{1\leq j\leq n} |S_j| \geq \varepsilon\right\}, \quad B_1 = \{|S_1| \geq \varepsilon\}, \quad \text{and} \quad B_k = \{|S_k| \geq \varepsilon\} \cap \bigcap_{j=1}^{k-1}\{|S_j| < \varepsilon\} \quad (k \geq 2).$$

Clearly, $A = \bigcup_{k=1}^{n} B_k$ (disjoint). Now, for $k < n$ the random variables $S_n - S_k$ and $S_k \mathbf{1}_{B_k}$ are independent, being Borel functions of $(X_{k+1}, \cdots, X_n)$ and $(X_1, \cdots, X_k)$, respectively. Therefore, by 18.2.3,

$$E\big((S_n - S_k)S_k\mathbf{1}_{B_k}\big) = E(S_n - S_k)E(S_k\mathbf{1}_{B_k}) = 0.$$

Writing

$$S_n^2 = (S_n - S_k + S_k)^2 = (S_n - S_k)^2 + 2(S_n - S_k)S_k + S_k^2,$$

we then have

$$E(S_n^2\mathbf{1}_{B_k}) = E\big[(S_n - S_k)^2\mathbf{1}_{B_k}\big] + 2E\big[(S_n - S_k)S_k\mathbf{1}_{B_k}\big] + E(S_k^2\mathbf{1}_{B_k}) \geq E(S_k^2\mathbf{1}_{B_k}) \geq \varepsilon^2 P(B_k),$$

which implies

$$\sum_{k=1}^{n} V(X_k) = V(S_n) \geq \int_A S_n^2\,dP = \sum_k \int_{B_k} S_n^2\,dP \geq \sum_k \varepsilon^2 P(B_k) = \varepsilon^2 P(A). \qquad \square$$

The following is a precursor to the strong laws.

18.4.14 Theorem. *Let (X_n) be a sequence of independent L^2 random variables. If $\sum_{n=1}^\infty V(X_n) < \infty$, then $\sum_{n=1}^\infty (X_n - E(X_n))$ converges a.s. Hence if $\sum_{n=1}^\infty E(X_n)$ also converges, then $\sum_{n=1}^\infty X_n$ converges a.s.*

Proof. Let $S_n := \sum_{j=1}^n (X_j - E(X_j))$. We claim that $\overline{\lim}_n S_n$ and $\underline{\lim}_n S_n$ are finite a.s. To see this, note first that

$$|\overline{\lim_n} S_n| \leq \overline{\lim_n} |S_n|, \quad |\underline{\lim_n} S_n| = |\overline{\lim_n}(-S_n)| \leq \overline{\lim_n} |S_n|,$$

and

$$\overline{\lim_n} |S_n| \leq \overline{\lim_n} |S_n - S_1| + |S_1| \leq \sup_{n\geq 2} |S_n - S_1| + |S_1|.$$

Therefore, the claim will follow if we show that $\sup_{n\geq 2} |S_n - S_1| < \infty$ a.s. Now, for $r > 0$,

$$\begin{aligned} P\left(\sup_{n\geq 2} |S_n - S_1| \geq 2r\right) &\leq P\left(\bigcup_{n=2}^\infty \{|S_n - S_1| \geq r\}\right) = \lim_{N\to\infty} P\left(\bigcup_{n=2}^N \{|S_n - S_1| \geq r\}\right) \\ &\leq \lim_{N\to\infty} P\left(\max_{2\leq n\leq N} |S_n - S_1| \geq r\right) \\ &\leq \frac{1}{r^2}\sum_{j=2}^\infty V(X_j), \end{aligned} \qquad (\dagger)$$

the last inequality by Kolmogorov's inequality applied to $X_n - E(X_n)$. Thus

$$P\left(\sup_{n\geq 2} |S_n - S_1| = \infty\right) \leq P\left(\sup_{n\geq 2} |S_n - S_1| \geq 2r\right) \leq \frac{1}{r^2}\sum_{j=2}^\infty V(X_j)$$

for all r and so $P\left(\sup_{n\geq 2} |S_n - S_1| = \infty\right) = 0$, verifying the claim.

Now note that for any real sequence (a_n), $a \in \mathbb{R}$, and $m \in \mathbb{N}$, if $a - \varepsilon < a_n < a + \varepsilon$ for all $n \geq m$, then $a - \varepsilon \leq \underline{\lim}_n a_n \leq \overline{\lim}_n a_n \leq a + \varepsilon$, hence $\overline{\lim}_n a_n - \underline{\lim}_n a_n \leq 2\varepsilon$. Put another way,

$$\overline{\lim_n} a_n - \underline{\lim_n} a_n > 2\varepsilon \Rightarrow \sup_{n\geq m} |a_n - a| \geq \varepsilon.$$

Thus for all m, by an obvious extension of $(\dagger)$,

$$P\left(\overline{\lim_n} S_n - \underline{\lim_n} S_n > 2\varepsilon\right) \leq P\left(\sup_{n\geq m} |S_n - S_m| \geq \varepsilon\right) \leq \frac{4}{\varepsilon^2}\sum_{j\geq m} V(X_j).$$

Letting $m \to \infty$ shows that $P\left(\overline{\lim}_n S_n - \underline{\lim}_n S_n > 2\varepsilon\right) = 0$ for all ε. It follows that $\overline{\lim}_n S_n = \underline{\lim}_n S_n$ almost surely. $\square$

18.4.15 Kronecker's Lemma. *If the series $\sum_{n=1}^\infty a_n$ converges, then*

$$\lim_n \frac{1}{n}\sum_{k=1}^n k a_k = 0.$$

Proof. Set $s := \sum_{n=1}^{\infty} a_n$ and $s_n := \sum_{k=1}^{n} a_k$. Since $s_n \to s$, it follows that the averages $t_n := (s_1 + \cdots + s_n)/n$ also tend to s. Then $(n-1)t_{n-1} = s_1 + \cdots + s_{n-1} = \sum_{k=1}^{n-1}(n-k)a_k$ and so

$$\frac{1}{n}\sum_{k=1}^{n} ka_k = \frac{1}{n}\sum_{k=1}^{n}[n-(n-k)]a_k = s_n - \frac{1}{n}\sum_{k=1}^{n-1}(n-k)a_k = s_n - \frac{n-1}{n}t_{n-1} \to 0. \qquad \square$$

We may now prove

18.4.16 L^2-Strong Law of Large Numbers. *Let (X_n) be a sequence of independent L^2 random variables. If $\sum_{n=1}^{\infty} n^{-2}V(X_n) < \infty$, then*

$$\lim_n \frac{1}{n}\sum_{k=1}^{n}\left(X_k - E(X_k)\right) = 0 \quad a.s.$$

Proof. By 18.4.14 applied to the sequence $(n^{-1}X_n)$, the series $\sum_{k=1}^{\infty} k^{-1}\big(X_k - E(X_k)\big)$ converges almost surely. Kronecker's lemma completes the proof. $\square$

In the case of identically distributed random variables, the requirement of square summability may be weakened. To prove this we need

18.4.17 Lemma. *For $n \in \mathbb{Z}^+$, let $A_n = \{|X| \geq n\}$ and $B_n = \{n \leq |X| < n+1\}$. Then*

$$\sum_{n=1}^{\infty} P(A_n) \leq \sum_{n=1}^{\infty} nP(B_n) \leq E\,|X| \leq \sum_{n=0}^{\infty}(n+1)P(B_n) = \sum_{n=0}^{\infty} P(A_n).$$

Proof. Since $P(A_m) = \sum_{n \geq m} P(B_n)$,

$$1 + \sum_{m=1}^{\infty} P(A_m) = \sum_{m=0}^{\infty} P(A_m) = \sum_{n=0}^{\infty}(n+1)P(B_n) \leq \sum_{n=1}^{\infty} nP(B_n) + 1,$$

which proves the first inequality. For the remaining inequalities, note that $n\mathbf{1}_{B_n} \leq |X|\mathbf{1}_{B_n} \leq (n+1)\mathbf{1}_{B_n}$, hence

$$\sum_{n=1}^{\infty} nP(B_n) \leq E\,|X| = \sum_{n=0}^{\infty}\int_{B_n} |X|\,dP \leq \sum_{n=0}^{\infty}(n+1)P(B_n). \qquad \square$$

18.4.18 L^1-Strong Law of Large Numbers. *Let (X_n) be a sequence of independent and identically distributed L^1 random variables. Then*

$$\lim_n \frac{1}{n}\sum_{k=1}^{n} X_k = E(X_1) \quad a.s.$$

Proof. Set $Y_n := X_n\mathbf{1}_{\{|X_n|<n\}}$. Then $P(X_n \neq Y_n) = P(|X_n| \geq n) = P(|X_1| \geq n)$, hence, by the lemma,

$$\sum_{n=1}^{\infty} P(X_n \neq Y_n) \leq E|X_1| < \infty.$$

By the Borel-Cantelli lemma, $X_n = Y_n$ eventually with probability one, hence it suffices to prove that $n^{-1}\sum_{k=1}^{n} Y_k \to E(X_1)$ a.s. Now $E(Y_n) = E(X_1\mathbf{1}_{\{|X_1|<n\}}) \to E(X_1)$, hence also

$n^{-1}\sum_{k=1}^{n} E(Y_k) \to E(X_1)$. We must therefore show that $n^{-1}\sum_{k=1}^{n}\big(Y_k - E(Y_k)\big) \to 0$ a.s. For this it is sufficient by the L^2 strong law to prove that $\sum_{k=1}^{\infty} V(Y_k)/k^2 < \infty$. Now,

$$V(Y_k) = E(Y_k^2) - E^2(Y_k) \le E(Y_k^2) = E\big(X_k^2 \mathbf{1}_{\{|X_k|<k\}}\big) = E\big(X_1^2 \mathbf{1}_{\{|X_1|<k\}}\big),$$

hence, setting $B_n = \{n-1 \le |X_1| < n\}$, we have

$$\sum_{k=1}^{\infty} \frac{1}{k^2} V(Y_k) \le \int \sum_{k=1}^{\infty} \frac{1}{k^2} X_1^2 \mathbf{1}_{\{|X_1|<k\}}\, dP = \sum_{n=1}^{\infty} \int_{B_n} \sum_{k=1}^{\infty} \frac{1}{k^2} X_1^2 \mathbf{1}_{\{|X_1|<k\}}\, dP.$$

Since $\mathbf{1}_{\{|X_1|<k\}}\mathbf{1}_{B_n} = 0$ $(k \le n-1)$, we have

$$\int_{B_n} \sum_{k=1}^{\infty} \frac{1}{k^2} X_1^2 \mathbf{1}_{\{|X_1|<k\}}\, dP \le \int_{B_n} \sum_{k=n}^{\infty} \frac{n^2}{k^2} \mathbf{1}_{\{|X_1|<k\}}\, dP \le \int_{B_n} \sum_{k=n}^{\infty} \frac{n^2}{k^2}\, dP \le 2nP(B_n),$$

where, for the last inequality, we have used the estimate $\sum_{k \ge n} k^{-2} \le 2/n$, as may be verified by a simple induction argument. Thus

$$\sum_{k=1}^{\infty} \frac{V(Y_k)}{k^2} \le 2 \sum_{n=1}^{\infty} nP(B_n),$$

which, by the lemma, is finite. □

The Central Limit Theorem

The central limit theorem is one of the most important results in probability theory, underlying much of statistical analysis. The theorem asserts (remarkably) that (suitably adjusted) sample means of independent data are approximately normally distributed, *regardless* of the actual distribution of the data. Here is the precise statement:

18.4.19 Theorem. *Let (X_n) be a sequence of independent, identically distributed L^2 random variables with mean μ and variance σ^2. Set $S_n := X_1 + \cdots + X_n$. Then for all x*

$$\lim_n P\left(\frac{S_n - n\mu}{\sigma\sqrt{n}} \le x\right) = \frac{1}{\sqrt{2\pi}} \int_{-\infty}^{x} e^{-t^2/2}\, dt. \tag{18.4}$$

Proof. Since

$$\frac{S_n - n\mu}{\sigma\sqrt{n}} = \frac{1}{\sqrt{n}} \sum_{k=1}^{n} \frac{X_n - \mu}{\sigma}, \quad E\left(\frac{X_n - \mu}{\sigma}\right) = 0, \quad \text{and} \quad V\left(\frac{X_n - \mu}{\sigma}\right) = 1,$$

the general result follows from the special case $\mu = 0$ and $\sigma = 1$, which we assume. Let $F_n(x)$ denote the distribution function of $n^{-1/2}S_n$ and F the distribution function of a standard normally distributed random variable. The assertion of the theorem is that $F_n(x) \to F(x)$ for all x. By 7.4.3, this is equivalent to the vague convergence $Q_n \to Q$, where

$$Q_n(B) = P\left(S_n \in \sqrt{n}B\right) \quad \text{and} \quad Q(B) = \frac{1}{\sqrt{2\pi}} \int_B e^{-t^2/2}\, dt.$$

To establish vague convergence, it suffices by 7.4.2 to show that $\widehat{Q}_n(\xi) \to \widehat{Q}(\xi)$ for every $\xi \in \mathbb{R}$. Now, differentiating under the integral in

$$\widehat{Q_1}(\xi) = \int e^{-2\pi i\,\xi x}\, dQ_1(x) = \int e^{-2\pi i\,\xi x}\, dP_{X_1}(x)$$

we have $\widehat{Q}_1{}'(0) = -2\pi i E(X_1) = 0$ and $\widehat{Q}_1{}''(0) = -4\pi^2 V(X_1) = -4\pi^2$. By Taylor's theorem,

$$\widehat{Q}_1(\xi) = 1 + \tfrac{1}{2}\widehat{Q}_1''(\theta(\xi))\xi^2, \quad \text{where} \quad |\theta(\xi)| \le |\xi|.$$

By independence, $Q_n(B) = P_{X_1} * \cdots * P_{X_n}(\sqrt{n}\, B)$, hence

$$\begin{aligned}\widehat{Q}_n(\xi) &= \int \cdots \int e^{-2\pi i n^{-1/2}\xi(x_1+\cdots+x_n)}\, dP_{X_1}(x_1)\cdots dP_{X_n}(x_n) = \left[\widehat{Q}_1(n^{-1/2}\xi)\right]^n \\ &= \left[1 + \widehat{Q}_1''\left(\theta(n^{-1/2}\xi)\right)\frac{\xi^2}{2n}\right]^n = \left(1 + \frac{a_n}{n}\right)^n,\end{aligned}$$

where

$$a_n := Q_1''\left(\theta(n^{-1/2}\xi)\right)\xi^2/2 \to \widehat{Q}_1''(0)\,\xi^2/2 = -2\pi^2\xi^2.$$

Also, by 6.2.3 with $a = 1/2$ and $b = -2\pi i$,

$$\widehat{Q}(\xi) = \frac{1}{\sqrt{2\pi}}\int_{-\infty}^{\infty} e^{-x^2/2 - 2\pi i \xi x}\, dx = e^{-2\pi^2\xi^2}.$$

By a Taylor series expansion, $\ln(1+z) = z + z^2\, O(z)$. Taking $z = a_n/n$ we see that

$$\ln \widehat{Q}_n(\xi) = n\left[n^{-1}a_n + (n^{-1}a_n)^2 O(n^{-1}a_n)\right] = a_n + a_n^2 O(n^{-1}a_n)/n \to -2\pi^2\xi^2 = \ln \widehat{Q}(\xi).$$

Therefore, $\widehat{Q}_n(\xi) \to \widehat{Q}(\xi)$, completing the proof. □

18.4.20 Remark. In the special case that the random variables X_j are Bernoulli with parameter $p \in (0,1)$, (18.4) becomes

$$\lim_{n\to\infty} P\left(\frac{S_n - np}{\sqrt{npq}} \le x\right) = \frac{1}{\sqrt{2\pi}}\int_{-\infty}^{x} e^{-t^2/2}\, dt, \quad q := 1 - p,$$

a result is known as the **DeMoivre-Laplace theorem**. One can use this equation to obtain the following approximation for the pmf of the binomial random variable S_n:

$$\begin{aligned}P(S_n = k) &= P(k - .5 < S_n < k + .5) = P\left(\frac{k - .5 - np}{\sqrt{npq}} < \frac{S_n - np}{\sqrt{npq}} < \frac{k + .5 - np}{\sqrt{npq}}\right) \\ &\approx \Phi\left(\frac{k + .5 - np}{\sqrt{npq}}\right) - \Phi\left(\frac{k - .5 - np}{\sqrt{npq}}\right), \quad k = 0, 1, \ldots, n,\end{aligned}$$

where

$$\Phi(x) := \frac{1}{\sqrt{2\pi}}\int_{-\infty}^{x} e^{-t^2/2}\, dt.$$

In particular, for $p = .5$ we have the approximation

$$P(S_n = k) \approx \Phi\left(\frac{2k + 1 - n}{\sqrt{n}}\right) - \Phi\left(\frac{2k - 1 - n}{\sqrt{n}}\right). \qquad \diamond$$

The Individual Ergodic Theorem

A measurable transformation $T : \Omega \to \Omega$ on a probability space $(\Omega, \mathcal{F}, P)$ is said to be **measure preserving** if $T(P) = P$, that is, $P(T^{-1}(A)) = P(A)$ for all $A \in \mathcal{F}$. For the purposes of this section, it is useful to think of $T^n := T \circ T \cdots \circ T$ (n factors) as representing the state of an evolving system at time n. If X is a random variable that measures some aspect of the system at time 0, then $X_n := X \circ T^n$ is that measurement at time n.

An event $A \in \mathcal{F}$ is said to be **invariant** under a measure preserving map T if $T^{-1}(A) = A$. It is easy to see that the collection $\mathcal{I}$ of all invariant events is a σ-field. The transformation T is said to be **ergodic** if $P(A) = 0$ or 1 for all $A \in \mathcal{I}$. Thinking again of $X \circ T^n$ as some measurement of an evolving system, we see that the quantity $\frac{1}{n}\sum_{k=1}^{n} X\big(T^{k-1}(\omega)\big)$ is then a **time average** of the measurement over the states $\omega, T\omega, \ldots, T^{n-1}\omega$. Josiah Gibbs argued that for large n the time average should approximate the **space average** $E(X)$, an approximation that may be viewed as a "uniform mixing" criterion.[2] The fact that mixing does not compress or dilate sets (that is, $P(T^{-1}(A)) = P(A)$) is then a natural condition in this context. Moreover, for uniform mixing to occur, it is reasonable to expect the absence of nontrivial invariant sets A ($0 < P(A) < 1$), these representing "unmixed pockets." Thus ergodicity is another natural requirement. The ergodic theorem of Birkhoff makes these ideas precise. For the proof we need the following result.

18.4.21 Maximal Ergodic Theorem. *Let $T : \Omega \to \Omega$ be measure preserving and let X be an L^1 random variable. Define random variables*

$$S_n := \sum_{k=0}^{n-1} X \circ T^k \quad \textit{and} \quad M_n := \max\{0, S_1, S_2, \ldots, S_n\}.$$

Then for all n, $\int_{M_n>0} X\,dP \ge 0$.

Proof. (Garsia). For $1 \le k \le n$, $M_n(T\omega) \ge S_k(T\omega) = S_{k+1}(\omega) - X(\omega)$, hence

$$X(\omega) \ge S_{k+1}(\omega) - M_n(T\omega).$$

Since $S_1(\omega) = X(\omega)$ and $M_n(T\omega) \ge 0$, the inequality also holds for $k = 0$. Therefore,

$$X(\omega) \ge \max_{1\le j\le n} S_j(\omega) - M_n(T\omega).$$

Noting that $M_n(\omega) = \max_{1\le j\le n} S_j(\omega)$ on the set $\{M_n > 0\}$, we have

$$\begin{aligned}\int_{\{M_n>0\}} X\,dP &\ge \int_{\{M_n>0\}} M_n(\omega)\,dP(\omega) - \int_{\{M_n>0\}} M_n(T\omega)\,dP(\omega)\\ &\ge \int_{\{M_n>0\}} M_n(\omega)\,dP(\omega) - \int M_n(T\omega)\,dP(\omega).\end{aligned}$$

Since T is measure preserving, the last difference reduces to $-\int_{\{M_n\le 0\}} M_n\,dP \ge 0$. □

18.4.22 Birkhoff Ergodic Theorem. *Let $T : \Omega \to \Omega$ be measure preserving and let $X \in L^1(P)$. Set $S_n := \sum_{k=0}^{n-1} X \circ T^k$. Then*

$$\lim_n \frac{S_n}{n} = E(X \mid \mathcal{I}) \quad a.s., \tag{18.5}$$

where $\mathcal{I}$ is the σ-field of all invariant events. In particular, if T is ergodic, then $E(X \mid \mathcal{I}) = E(X)$.

Proof. First, note that $E(X \mid \mathcal{I}) \circ T = E(X \mid \mathcal{I})$. Indeed, since T is $\mathcal{I}$-measurable so is $E(X \mid \mathcal{I}) \circ T$, and, by the measure preserving property of T,

$$\int_A E(X \mid \mathcal{I}) \circ T\,dP = \int_A E(X \mid \mathcal{I})\,dP \quad \text{for all } A \in \mathcal{I}.$$

[2]The cocktail example of Halmos [22] comes to mind.

We may now express (18.5) as $\lim_n n^{-1}\sum_{k=1}^{n}\big(X - E(X \mid \mathfrak{I})\big)\circ T^k = 0$ a.s.. Thus we may assume in the statement of the theorem that $E(X \mid \mathfrak{I}) = 0$.

We show first that $Y := \overline{\lim}_n S_n/n \le 0$. Given $\varepsilon > 0$, set $A = \{Y > \varepsilon\}$. Noting that

$$Y(T\omega) = \overline{\lim_n}\,\frac{S_n(T\omega)}{n} = \overline{\lim_n}\,\frac{S_{n+1}(\omega) - X(\omega)}{n} = \overline{\lim_n}\,\frac{S_{n+1}(\omega)}{n+1} = Y(\omega),$$

we see that $A \in \mathfrak{I}$. Now apply the maximal ergodic theorem to $\widetilde{X} := [X - \varepsilon]\mathbf{1}_A$ with corresponding $\widetilde{M}_n$ and $\widetilde{S}_n$ to obtain

$$\int_{\{\widetilde{M}_n>0\}} \widetilde{X}\,dP \ge 0. \qquad (\alpha)$$

By the invariance of A,

$$\frac{1}{k}\widetilde{S}_k(\omega) = \frac{1}{k}\sum_{j=0}^{k-1}[X(T^j\omega) - \varepsilon]\mathbf{1}_A(T^j\omega) = \left[\frac{1}{k}S_k(\omega) - \varepsilon\right]\mathbf{1}_A(\omega),$$

and since $\sup_k \frac{1}{k}S_k \ge Y > \varepsilon > 0$ on A we have

$$\left\{\widetilde{M}_n > 0\right\} = \left\{\max_{1\le k\le n}\widetilde{S}_k > 0\right\} \uparrow \left\{\sup_k \widetilde{S}_k > 0\right\} = \left\{\sup_k \frac{1}{k}\widetilde{S}_k > 0\right\} = A. \qquad (\beta)$$

Since $\widetilde{X} \in L^1$, (α) and (β) imply that

$$0 \le \int_A \widetilde{X}\,dP = \int_A X\,dP - \varepsilon P(A) = \int_A E(X \mid \mathfrak{I})\,dP - \varepsilon P(A) = -\varepsilon P(A).$$

Therefore, $P(A) = 0$, that is, $Y \le \varepsilon$ a.s. Since ε was arbitrary, $\overline{\lim}_k S_k/k \le 0$ a.s. Since the argument holds for $-X$ as well, we also have $\overline{\lim}_k(-S_k/k) \le 0$, that is, $\underline{\lim}_k(S_k/k) \ge 0$ a.s. Therefore, $\lim_k S_k/k = 0$ a.s., as required.

Finally, if T is ergodic, then $\mathfrak{I}$ consists only of sets with measure zero or one, hence $E(X \mid \mathfrak{I}) = E(X)$. □

The following result is an analog of a result proved in the general setting of weakly almost periodic semigroups of operators on Banach spaces. (See 17.6.9 and the paragraph following.) We give a probabilistic proof here.

18.4.23 Corollary. *The convergence in* (18.5) *is also in* L^1.

Proof. Set $A_n(X) = n^{-1}\sum_{j=1}^{n} X \circ T^{j-1}$. If X is bounded, then so is the L^1 random variable $A_n(X) - E(X \mid \mathfrak{I})$, and since $A_n(X) - E(X \mid \mathfrak{I}) \to 0$ a.s., the convergence is in the L^1 norm by the dominated convergence theorem.

In the general case, given $\varepsilon > 0$ choose a bounded random variable Y such that $\|X - Y\|_1 < \varepsilon$. Then

$$\|A_n(X) - E(X \mid \mathfrak{I})\|_1 \le \|A_n(X - Y)\|_1 + \|A_n(Y) - E(Y \mid \mathfrak{I})\|_1 + \|E(X - Y \mid \mathfrak{I})\|_1. \quad (\dagger)$$

We show that each of the terms on the right may be made arbitrarily small.

By the result of the first paragraph, for sufficiently large n,

$$\|A_n(Y) - E(Y \mid \mathfrak{I})\|_1 < \varepsilon.$$

Also, by the measure preserving property of P,

$$\|A_n(X-Y)\|_1 \le \frac{1}{n}\sum_{j=1}^{n} \|(X-Y)\circ T^{j-1}\|_1 = \frac{1}{n}\sum_{j=1}^{n} \|X-Y\|_1 < \varepsilon.$$

Finally, by Fatou's lemma and (18.5),

$$\|E(X \mid \mathfrak{I}) - E(Y \mid \mathfrak{I})\|_1 = \int \lim_n |A_n(X-Y)|\, dP \le \liminf_n \int |A_n(X-Y)| \le \varepsilon.$$

Therefore, by (†), $\|A_n(X) - E(X \mid \mathfrak{I})\|_1 < 3\varepsilon$ for all sufficiently large n. □

Stationary Processes

A sequence of random variables X_n on a probability space $(\Omega, \mathfrak{F}, P)$ is called a **stationary process** if

$$P\big((X_n, X_{n+1}, \ldots) \in B\big) = P\big((X_1, X_2, \ldots) \in B\big) \ \ \forall\, B \in \mathfrak{B}(\mathbb{R}^\infty) \text{ and } n \in \mathbb{N}. \tag{18.6}$$

In particular, taking $B = B_1 \times \mathbb{R} \times \cdots$, we see that $P(X_n \in B_1) = P(X_1 \in B_1)$, so that the random variables X_n are identically distributed. In this section we prove an ergodic theorem for stationary processes.

For an example of a stationary process, let $T : \Omega \to \Omega$ be a measure preserving transformation and X_1 a random variable. Then the sequence $(X_n := X_1 \circ T^{n-1})$ is a stationary process. To see this, let

$$A_n := \{(X_n, X_{n+1}, \ldots) \in B\}$$

and note that because $X_n = X_{n-1} \circ T$ we have $T^{-1}(A_{n-1}) = A_n$ and so $P(A_n) = P(T^{-1}(A_{n-1})) = P(A_{n-1})$. Iterating, we obtain (18.6).

Now let $X = (X_n)$ be an arbitrary stationary process on $(\Omega, \mathfrak{F}, P)$ and let $T : \mathbb{R}^\infty \to \mathbb{R}^\infty$ denote the left shift operator $T(x_1, x_2, \ldots) = (x_2, x_3, \ldots)$. Thus for all n

$$(X_{n+1}, X_{n+2}, \ldots) = T^n(X_1, X_2, \ldots) =: T^n \circ X.$$

Clearly, T is a measurable transformation on $(\mathbb{R}^\infty, \mathfrak{B}(\mathbb{R}^\infty), P_X)$. Moreover, T is measuring preserving. Indeed, from

$$T^{-1}(B_1 \times \cdots \times B_n \times \mathbb{R} \times \cdots) = \mathbb{R} \times B_1 \times \cdots \times B_n \times \mathbb{R} \times \cdots$$

we have, by stationarity,

$$\begin{aligned} T\big(P_X\big)(B_1 \times \cdots \times B_n \times \mathbb{R} \times \cdots) &= P\left((X_1, X_2, \ldots) \in \mathbb{R} \times B_1 \times \cdots \times B_n \times \mathbb{R} \times \cdots\right) \\ &= P\left((X_2, X_3, \ldots) \in B_1 \times \cdots \times B_n \times \mathbb{R} \times \cdots\right) \\ &= P\left((X_1, X_2, \ldots) \in B_1 \times \cdots \times B_n \times \mathbb{R} \times \cdots\right) \\ &= P_X(B_1 \times \cdots \times B_n \times \mathbb{R} \times \cdots). \end{aligned}$$

Thus the measures $T(P_X)$ and P_X agree on the sets $B_1 \times \cdots \times B_n \times \mathbb{R} \times \cdots$ and so are equal, by the uniqueness theorem for measures.

Now call $A \in \mathfrak{F}$ **invariant** if there exists a $B \in \mathfrak{B}(\mathbb{R}^\infty)$ such that

$$A = \{(X_{n+1}, X_{n+2}, \ldots) \in B\} \ \text{ for all } n \ge 0.$$

The set $\mathfrak{I}$ of all invariant sets is easily seen to be a σ-subfield of $\mathfrak{F}$. Since the preceding relationship between A and B may be written as

$$\mathbf{1}_A = \mathbf{1}_B \circ T^n \circ X \ \text{ for all } n \ge 0,$$

the usual arguments show that a function f on Ω is $\mathcal{I}$-measurable iff there exists a $\mathcal{B}(\mathbb{R}^\infty)$-measurable function g such that

$$f = g \circ T^n \circ X \quad \text{for all } n \geq 0.$$

With this background we may now prove

18.4.24 Ergodic Theorem for Stationary Processes. *Let (X_n) be a stationary process on $(\Omega, \mathcal{F}, P)$. If $X_1 \in L^1(P)$ then*

$$\lim_n \frac{1}{n} \sum_{j=1}^n X_j = E(X_1 \mid \mathcal{I}) \quad \textit{in } L^1 \textit{ and a.s.}$$

Proof. Let $\pi_n : \mathbb{R}^\infty \to \mathbb{R}$ denote the nth coordinate projection. Then π_n is a random variable on $(\mathbb{R}^\infty, \mathcal{B}(\mathbb{R}^\infty), P_X)$ and $\pi_n = \pi_1 \circ T^{n-1}$. Since T is measure preserving, by the Birkhoff ergodic theorem the averages $n^{-1}\sum_{j=1}^n \pi_j$ converge a.s. and in L^1 on $(\mathbb{R}^\infty, \mathcal{B}(\mathbb{R}^\infty), P_X)$. Since (π_n) has distribution P_X, $n^{-1}\sum_{j=1}^n X_n$ converges a.s. and in L^1 to some random variable $Y \in L^1(P)$. To see that $Y = E(X_1 \mid \mathcal{I})$, note first that since $Y = g \circ T^n \circ X$ for all n, where g is the measurable function $g(x_1, x_2, \ldots) = \overline{\lim}_k (x_1 + \cdots + x_k/k)$, the random variable Y is $\mathcal{I}$-measurable. It remains to show that

$$\int_A Y\, dP = \int_A X_1\, dP \quad \text{for all } A \in \mathcal{I}. \tag{$\dagger$}$$

But if A is invariant, say $A = \{(X_k, X_{k+1}, \ldots) \in B\}$ for all k, then, by the stationary property of X,

$$\begin{aligned}\int_A X_k\, dP &= \int_{\{(x_k, x_{k+1}, \ldots) \in B\}} x_k\, dP_X(x_1, x_2, \ldots) = \int_{\{(x_1, x_2, \ldots) \in B\}} x_1\, dP_X(x_1, x_2, \ldots)\\ &= \int_A X_1\, dP,\end{aligned}$$

hence

$$\int_A \frac{1}{n} \sum_{k=1}^n X_k\, dP = \int_A X_1\, dP.$$

Taking limits yields $(\dagger)$. $\square$

18.5 Discrete-Time Martingales

A sequence (X_n) of random variables on a probability space $(\Omega, \mathcal{F}, P)$ is called a **discrete time stochastic process** or simply a **process**. If $X_n \in L^p$ for all n, then (X_n) is called an L^p**-process**. For example, a stationary process is an L^1 process, and L^1 and L^2 processes arose in the context of the laws of large numbers. If we view a process as a model for the numerical outcomes of an ongoing experiment, then a mathematical model of the history of the experiment becomes important. Such a model in known as a *filtration*.

Filtrations

A **(discrete-time) filtration** on $(\Omega, \mathcal{F}, P)$ is a sequence of σ-fields $\mathcal{F}_n$ such that

$$\mathcal{F}_n \subseteq \mathcal{F}_{n+1} \subseteq \mathcal{F} \text{ for all } n \in \mathbb{N}.$$

A probability space with a filtration is called a **filtered probability space** and is denoted by $(\Omega, \mathcal{F}, (\mathcal{F}_n), P)$. It is sometimes useful to view a filtration as a mathematical description of the information produced by an experiment consisting of repeated trials. At the completion of the nth trial, $\mathcal{F}_n$ encapsulates the information revealed by the outcome of this and previous trials.

A stochastic process (X_n) is said to be **adapted to a filtration** $(\mathcal{F}_n)$ if for all n the random variable X_n is $\mathcal{F}_n$-measurable. For example, (X_n) is clearly adapted to the filtration

$$\mathcal{F}^X = (\mathcal{F}_n^X), \quad \mathcal{F}_n^X := \big(\sigma(X_1, \ldots, X_n)\big),$$

which is called the **natural filtration of** (X_n). As noted above, a filtration models the evolution of information. Thus if (X_n) is adapted to a filtration $(\mathcal{F}_n)$, then the σ-field $\mathcal{F}_n$ includes all knowable information about the process up to time n. The natural filtration includes all knowable information about the process up to time n but nothing more.

Definition and General Properties of Martingales

Let $(\Omega, \mathcal{F}, (\mathcal{F}_n), P)$ be a filtered probability space and let (X_n) be an L^1-process adapted to $(\mathcal{F}_n)$. Then the pair $(X_n, \mathcal{F}_n)$ is said to be a

- **submartingale** if $X_n \leq E(X_{n+1} \mid \mathcal{F}_n)$ for all $n \in \mathbb{N}$.
- **supermartingale** if $X_n \geq E(X_{n+1} \mid \mathcal{F}_n)$ for all $n \in \mathbb{N}$.
- **martingale** if $X_n = E(X_{n+1} \mid \mathcal{F}_n)$ for all $n \in \mathbb{N}$.

We omit $\mathcal{F}_n$ from the notation $(X_n, \mathcal{F}_n)$ if the filtration is understood or is the natural filtration,

A submartingale has the **multistep property**

$$X_n \leq E(X_m \mid \mathcal{F}_n), \quad \text{for all } \ m \geq n, \tag{18.7}$$

that is,

$$\int_A X_n \, dP \leq \int_A X_m \, dP \ \text{ for all } A \in \mathcal{F}_n \text{ and } m \geq n.$$

Indeed, from $X_{m-1} \leq E(X_m \mid \mathcal{F}_{m-1})$ and the tower property we have

$$X_{m-2} \leq E(X_{m-1} \mid \mathcal{F}_{m-2}) \leq E\big(E(X_m \mid \mathcal{F}_{m-1}) \mid \mathcal{F}_{m-2}\big) = E(X_m \mid \mathcal{F}_{m-2}).$$

Iterating we obtain (18.7). Submartingales and martingales have analogous multistep properties. Note that $(X_n, \mathcal{F}_n)$ is a submartingale iff $(-X_n, \mathcal{F}_n)$ is a supermartingale.

We may think of a martingale as the accumulated winnings of a gambler in a sequence of fair games. The martingale condition, which may be written $E(X_{n+1} - X_n \mid \mathcal{F}_n) = 0$, then asserts that the best prediction of the gain $X_{n+1} - X_n$ on the next play, based on the information $\mathcal{F}_n$ obtained during the first n plays, is zero, the hallmark of a fair game. The games favor the house (respectively, the player), if the winnings constitute a supermartingale (respectively, a submartingale).

18.5.1 Examples of Martingales.

(a) Let $Y_1, Y_2, \ldots$ be a sequence of independent L^1 random variables on $(\Omega, \mathscr{F}, \mathbb{P})$ with mean one. Set $X_n := Y_1 \cdots Y_n$. By the factor and independence properties

$$E(X_{n+1} - X_n \mid X_1, \ldots, X_n) = X_n E(Y_{n+1} - 1 \mid X_1, \ldots, X_n) = X_n E(Y_{n+1} - 1) = 0,$$

which is the desired martingale property.

(b) Let (Y_n) be a sequence of independent L^1 random variables on $(\Omega, \mathscr{F}, \mathbb{P})$ with mean p and set $X_n = Y_1 + Y_2 + \cdots + Y_n - np$. Then

$$E(X_{n+1} - X_n \mid X_1, \ldots, X_n) = E(Y_{n+1} - p \mid X_1, \ldots, X_n) = E(Y_{n+1} - p) = 0.$$

(c) Let $(\Omega, \mathcal{F}, (\mathcal{F}_n), P)$ be a filtered probability space and $X \in L^1(\Omega, \mathscr{F}, \mathbb{P})$. Define $X_n = E(X|\mathcal{F}_n)$. By the tower property $E(X_{n+1} \mid \mathcal{F}_n) = E(X \mid \mathcal{F}_{n+1} \mid \mathcal{F}_n) = E(X \mid \mathcal{F}_n) = X_n$. Note that (X_n) is uniformly integrable, that is, $\lim_{t\to\infty} \sup_n \int_{\{|X_n|\geq t\}} |X_n|\, dP = 0$. Indeed, since $|X_n| = |E(X \mid \mathcal{F}_n)| \leq E(|X| \mid \mathcal{F}_n)$ and $\{|X_n| \geq t\} \in \mathcal{F}_n$,

$$\int_{\{|X_n|\geq t\}} |X_n|\, dP \leq \int_{\{|X_n|\geq t\}} E(|X| \mid \mathcal{F}_n)\, dP = \int_{\{|X_n|\geq t\}} |X|\, dP,$$

and by the Markov inequality,

$$P\{|X_n| \geq t\} \leq \frac{1}{t} \int |X_n|\, dP \leq \frac{1}{t} \int E(|X| \mid \mathcal{F}_n)\, dP = \frac{1}{t} \int |X|\, dP \to 0.$$

(d) Consider a sequence of finite partitions $\mathcal{P}_n := \{A_{n,1}, A_{n,2}, \ldots, A_{n,m_n}\}$ of Ω such that each member of $\mathcal{P}_n$ is a union of members of $\mathcal{P}_{n+1}$. Let Q be a probability measure such that $Q \ll P$. Define

$$X_n = \sum_{j=1}^{m_n} a_{n,j} \mathbf{1}_{A_{n,j}}, \quad a_{n,j} := \frac{Q(A_{n,j})}{P(A_{n,j})},$$

where $a_{n,j}$ is defined to be 0 if $P(A_{n,j}) = 0$. Clearly, (X_n) is adapted to the filtration $(\sigma(\mathcal{P}_n))$. If $A_{n,k} = \bigcup_{j\in F_k} A_{n+1,j}$, then $A_{n,k} \cap A_{n+1,j} = \emptyset$ unless $j \in F_k$, in which case the intersection is $A_{n+1,j}$. Therefore, for all k

$$\int_{A_{n,k}} X_{n+1}\, dP = \sum_{j=1}^{m_{n+1}} a_{n+1,j} \int_{A_{n,k}} \mathbf{1}_{A_{n+1,j}}\, dP = \sum_{j\in F_k} Q(A_{n+1,j}) = Q(A_{n,k}) = \int_{A_{n,k}} X_n\, dP,$$

which implies the martingale property. ◊

The next result is a direct consequence of the linearity and order properties of conditional expectation.

18.5.2 Proposition. *Let (X_n) and (Y_n) be processes on a filtered probability space $(\Omega, \mathcal{F}, (\mathcal{F}_n), P)$ and set $Z_n := aX_n + bY_n$, $a, b \in \mathbb{R}$.*

(a) *If $(X_n, \mathcal{F}_n)$, $(Y_n, \mathcal{F}_n)$ are martingales, then $(Z_n, \mathcal{F}_n)$ is a martingale.*

(b) *If $(X_n, \mathcal{F}_n)$ and $(Y_n, \mathcal{F}_n)$ are sub (super) martingales and $a, b \geq 0$, then $(Z_n, \mathcal{F}_n)$ is a sub (super) martingale.*

For the remainder of the subsection, we focus mainly on submartingales. Corresponding results for supermartingales may be obtained by considering $(-X_n)$. The next result describes several ways of generating submartingales.

18.5.3 Theorem. *Let (X_n) and (Y_n) be processes on a filtered probability space $(\Omega, \mathcal{F}, (\mathcal{F}_n), P)$.*

(a) *If $(X_n, \mathcal{F}_n)$ and $(Y_n, \mathcal{F}_n)$ are submartingales, then $(X_n \vee Y_n, \mathcal{F}_n)$ is a submartingale. In particular, $(X_n^+, \mathcal{F}_n)$ is a submartingale.*

(b) *If $(X_n, \mathcal{F}_n)$ is a submartingale, ϕ is convex and increasing, and $\phi(X_n) \in L^1$ for all n, then $\big(\phi(X_n), \mathcal{F}_n\big)$ is a submartingale.*

(c) *If $(X_n, \mathcal{F}_n)$ is a martingale, ϕ is convex, and $\phi(X_n) \in L^1$ for all n, then $\big(\phi(X_n), \mathcal{F}_n\big)$ is a submartingale. In particular, $(|X_n|, \mathcal{F}_n)$ is a submartingale.*

Proof. For (a) we have $E(X_{n+1} \vee Y_{n+1} \mid \mathcal{F}_n) \geq E(X_{n+1} \mid \mathcal{F}_n) \geq X_n$, with a similar inequality for Y. Therefore, $E(X_{n+1} \vee Y_{n+1} \mid \mathcal{F}_n) \geq X_n \vee Y_n$. Part (b) follows from the conditional form of Jensen's inequality: $\phi(X_n) \leq \phi\big(E(X_{n+1} \mid \mathcal{F}_n)\big) \leq E\big(\phi(X_{n+1}) \mid \mathcal{F}_n\big)$. The proof of part (c) is similar. □

The following theorem asserts that reducing the amount of information provided by a filtration preserves the submartingale property. (The same is not necessarily true if information is *increased*.)

18.5.4 Theorem. *Let $(\mathcal{G}_n)$ and $(\mathcal{F}_n)$ be filtrations with $\mathcal{G}_n \subseteq \mathcal{F}_n \subseteq \mathcal{F}$. If (X_n) is adapted to $(\mathcal{G}_n)$ and is a submartingale with respect to $(\mathcal{F}_n)$, then it is also a submartingale with respect to $(\mathcal{G}_n)$.*

Proof. $E(X_{n+1} \mid \mathcal{G}_n) = E\big(E(X_{n+1} \mid \mathcal{F}_n) \mid \mathcal{G}_n\big) \geq E(X_n \mid \mathcal{G}_n) = X_n$. □

Stopping Times. Optional Sampling

Let $\big(\Omega, \mathcal{F}, (\mathcal{F}_n), P\big)$ be a filtered probability space and let $\mathcal{F}_\infty$ denote the σ-field generated by $\bigcup_n \mathcal{F}_n$. A **stopping time** (relative to the given filtration) is a function $\tau : \Omega \to \mathbb{N} \cup \{\infty\}$ such that

$$\{\tau = n\} \in \mathcal{F}_n \quad \text{for all } n \in \mathbb{N} \cup \{\infty\}.$$

Note that if τ is a stopping time, then the set $\{\tau \leq n\}$ is a union of sets $\{\tau = j\} \in \mathcal{F}_j$ and so is a member of $\mathcal{F}_n$. It follows that $\{\tau > n\} = \{\tau \leq n\}^c$ also lies in $\mathcal{F}_n$.

The constant function $\tau = m$, where m is a fixed positive integer, is trivially a stopping time. Also, if τ and σ are stopping times, then so is $\tau \wedge \sigma$, as may be seen from

$$\{\tau \wedge \sigma = n\} = \{\tau = n, \sigma \geq n\} \cup \{\sigma = n, \tau \geq n\}.$$

In particular, $\tau \wedge m$ is a stopping time.

18.5.5 Example. Consider a stochastic process (X_n). Let τ denote the first time $X_n < 0$. Thus

$$\tau(\omega) = \begin{cases} \min\{n : X_n(\omega) < 0\} & \text{if } \{n : X_n(\omega) < 0\} \neq \emptyset \\ \infty & \text{otherwise,} \end{cases}$$

The calculations

$$\{\tau = n\} = \{X_n < 0\} \cap \bigcap_{j=0}^{n-1} \{X_j \geq 0\} \quad \text{and} \quad \{\tau = \infty\} = \bigcap_{j=1}^{\infty} \{X_j \geq 0\},$$

show that τ is a stopping time relative to the natural filtration of (X_n). In this connection, note that the function

$$\sigma(\omega) = \begin{cases} \max\{n : X_n(\omega) < 0\} & \text{if } \{n : X_n(\omega) < 0\} \neq \emptyset \\ \infty & \text{otherwise} \end{cases}$$

is *not* a stopping time. This is a mathematical formulation of the self-evident fact that one cannot predict the future: By knowing merely the past history of the process, one cannot expect (in the absence of prescience) to know when the process will be negative for the last time. ◊

One of the most important facts regarding stopping times is that they may be combined with submartingales to produce submartingales indexed by random times, a construct useful in contexts where one may wish to stop a process when a certain goal is achieved. (Think of a gambler who resolves to stop playing as soon as he has amassed sufficient winnings.) We shall need these so-called *stopped processes* in the proof of Doob's martingale convergence theorem below.

The main result of the current subsection depends on the following notions: Let $(\Omega, \mathcal{F}, (\mathcal{F}_n), P)$ be a filtered probability space and let (X_n) be a process adapted to $(\mathcal{F}_n)$. If τ is a stopping time taking values in $\mathbb{N} \cup \{\infty\}$, then the **stopped random variable** X_τ is defined by

$$X_\tau(\omega) := X_{\tau(\omega)}(\omega)\mathbf{1}_{\{\tau<\infty\}}(\omega) = \sum_{j=1}^{\infty} \mathbf{1}_{\{\tau(\omega)=j\}} X_j(\omega).$$

The σ-field $\mathcal{F}_\tau$ of **events up to time** τ is defined by

$$\mathcal{F}_\tau = \{A \in \mathcal{F} : A \cap \{\tau \le n\} \in \mathcal{F}_n \ \forall\, n \in \mathbb{N}\}.$$

That $\mathcal{F}_\tau$ is indeed a σ-field is a straightforward calculation. For example, to see that $\mathcal{F}_\tau$ is closed under complements, note that if $A \in \mathcal{F}_\tau$, then $A^c \cup \{\tau > n\} = (A \cap \{\tau \le n\})^c \in \mathcal{F}_n$, hence $A^c \cap \{\tau \le n\} = (A^c \cup \{\tau > n\}) \cap \{\tau \le n\} \in \mathcal{F}_n$.

18.5.6 Optional Sampling. *If (X_n) is a submartingale and τ and σ are stopping times with $\sigma \le \tau$, then $\mathcal{F}_\sigma \subseteq \mathcal{F}_\tau$. If τ is bounded, then X_τ and X_σ are integrable and $X_\sigma \le E(X_\tau \mid \mathcal{F}_\sigma)$.*

Proof. Let $A \in \mathcal{F}_\sigma$. Then $A \cap \{\tau \le n\} = A \cap \{\sigma \le n\} \cap \{\tau \le n\} \in \mathcal{F}_n$, hence $A \in \mathcal{F}_\tau$. Therefore, $\mathcal{F}_\sigma \subseteq \mathcal{F}_\tau$.

For the second assertion, assume that τ (hence also σ) takes values in $\{1, 2, \ldots, n\}$. Suppose first that $\tau - \sigma \le 1$. Then for $A \in \mathcal{F}_\sigma$,

$$\int_A (X_\tau - X_\sigma)\, dP = \sum_{j=1}^{n-1} \int_{A\cap\{\sigma=j,\tau=j+1\}} (X_{j+1} - X_j)\, dP.$$

Since

$$A \cap \{\sigma = j, \tau = j+1\} = A \cap \{\sigma = j\} \cap \{\tau > j\} \in \mathcal{F}_j,$$

the terms in the above sum are nonnegative by the submartingale property. Therefore, $X_\sigma \le E(X_\tau \mid \mathcal{F}_\sigma)$. For the general case, define stopping times $\rho_i = \tau \wedge (\sigma + i)$ $(0 \le i \le n)$. Then $\sigma = \rho_0 \le \rho_1 \le \cdots \le \rho_n = \tau$ and $\rho_{i+1} - \rho_i \le 1$, hence, by the special case,

$$\int_A X_\sigma\, dP \le \int_A X_{\rho_1}\, dP \le \cdots \le \int_A X_\tau\, dP, \quad A \in \mathcal{F}_\sigma. \qquad \square$$

18.5.7 Corollary. *Let (X_n) be a submartingale and let τ be a stopping time. Then $(X_{n\wedge\tau})$ is a submartingale.*

Proof. This follows immediately from 18.5.6 and the inequality $n \wedge \tau \le (n+1) \wedge \tau$. $\square$

The process $(X_{n\wedge\tau})$ in the corollary is called the **stopped process** relative to (X_n) and τ.

Upcrossings

The martingale convergence theorem, proved in the next subsection, is one of the key results in martingale theory. The proof is based on Doob's notion of **upcrossings**, which we now describe.

Let (x_n) be any sequence in $\mathbb{R}$. Given real numbers $a < b$, define a sequence (τ_n) with values in $\mathbb{N} \cup \{\infty\}$ by

$$\begin{aligned} \tau_1 &:= \inf\{j \geq 1 : x_j \leq a\}, & \tau_2 &:= \inf\{j > \tau_1 : x_j \geq b\}, \\ \tau_{2n-1} &:= \inf\{j > \tau_{2n-2} : x_j \leq a\}, & \tau_{2n} &:= \inf\{j > \tau_{2n-1} : x_j \geq b\}. \end{aligned} \tag{18.8}$$

Here, as usual, we set $\inf \emptyset = \infty$. Clearly, the sequence (τ_n) is increasing, $\tau_n \geq n$, and

$$x_{\tau_{2n-1}} \leq a < b \leq x_{\tau_{2n}} \quad \text{if} \quad \tau_{2n-1} < \infty.$$

From the definition we see that τ_1 is the first time the sequence is below a, τ_2 the first time after τ_1 that the sequence is above b, etc. It follows that τ_2 is the time of the first *upcrossing* of the interval $[a, b]$, τ_4 the time of the second upcrossing, and in general τ_{2k} is the time of the kth upcrossing. The number $U^n_{[a,b]}$ of upcrossings of the interval $[a, b]$ up to time n by the sequence (x_n) is the largest k for which $\tau_{2k} \leq n$:

$$U^n_{[a,b]} := \sup\{k : \tau_{2k} \leq n\}.$$

If the set in the definition is empty, we define $U^n_{[a,b]} = 0$. Obviously,

$$U^n_{[a,b]} \leq n, \quad \text{and } k > U^n_{[a,b]} \text{ iff } \tau_{2k} > n.$$

The total number of upcrossings is defined as

$$U_{[a,b]} = \sup_n U^n_{[a,b]} = \sup\{k : \tau_{2k} < \infty\}.$$

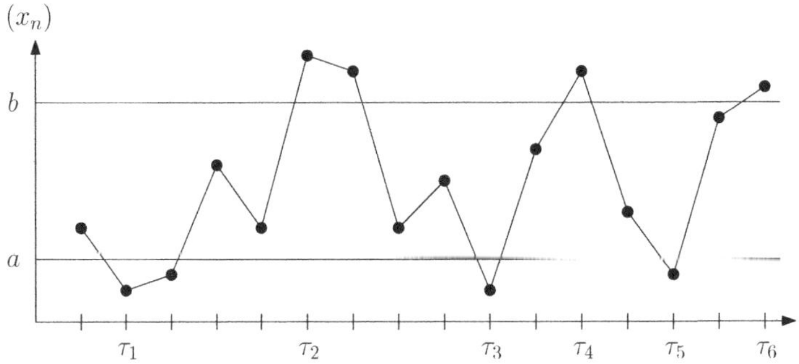

FIGURE 18.1: $U^{16}_{[a,b]} = 3$.

The connection between upcrossings and convergence of the sequence (x_n) is given in the following lemma.

18.5.8 Lemma. *A sequence (x_n) of real numbers converges in $\overline{\mathbb{R}}$ iff $U_{[a,b]} < \infty$ for all $a, b \in \mathbb{Q}$ with $a < b$.*

Proof. Set $\alpha := \underline{\lim}_n x_n$ and $\beta := \overline{\lim}_n x_n$. If $U_{[a,b]} = \infty$ for some $a < b$, then $x_n \leq a$ for infinitely many n and $x_n \geq b$ for infinitely many n, hence $\alpha \leq a < b \leq \beta$ and so (x_n) cannot converge in $\overline{\mathbb{R}}$. Conversely, if (x_n) does not converge in $\overline{\mathbb{R}}$, then there exist rationals a and b such that $\alpha < a < b < \beta$. It follows that $x_n < a$ infinitely often and $x_n > b$ infinitely often, hence $U_{[a,b]} = \infty$. $\square$

Now consider a process $X = (X_n)$. For each ω and pair of real numbers a, b, we may apply the above construction to the sequence $(X_n(\omega))$ to obtain $\mathbb{N} \cup \{\infty\}$-valued the functions τ_n, $U^n_{[a,b]}$, and $U_{[a,b]}$ on Ω. It is easily established by induction that τ_n is a stopping time. For example,

$$\{\tau_1 = k\} = \{X_1 > a\} \cap \cdots \cap \{X_{k-1} > a\} \cap \{X_k \le a\} \in \mathcal{F}_k \text{ and}$$

$$\{\tau_2 = k\} = \bigcup_{j=1}^{k-1} \{\tau_1 = j\} \cap \{X_{j+1} < b\} \cap \cdots \cap \{X_{k-1} < b\} \cap \{X_k \ge b\} \in \mathcal{F}_k.$$

In particular, $U^n_{[a,b]}$ and $U_{[a,b]}$ are $\mathcal{F}_\infty$-measurable. Here is the key result regarding upcrossings.

18.5.9 Upcrossing Inequality (Doob). *Let (X_n) be a submartingale on a filtered probability space $(\Omega, \mathcal{F}, (\mathcal{F}_n), P)$. Then, for any $a < b$,*

$$E\big(U^n_{[a,b]}\big) \le \frac{1}{b-a} E\big((X_n - a)^+ + |a|\big) \le \frac{1}{b-a} E\big(|X_n| + 2|a|\big), \quad n \in \mathbb{N}.$$

Proof. Set $X_0 = 0$ and $\widetilde{X}_n = (X_n - a)^+$ $(n \ge 0)$. By 18.5.3, $(\widetilde{X}_n)$ is a submartingale. Let $(\tau_n(\omega))$ be defined as in (18.8) for the sequence $(\widetilde{X}_n(\omega))$ but with a replaced by 0 and b replaced by $c := b - a$, and set $\tau_0 = 0$. Let $\widetilde{U}^n_{[0,c]}$ be defined by (τ_n). Clearly, $\widetilde{U}^n_{[0,c]} = U^n_{[a,b]}$. Let $2k > n$. Then $\tau_{2k} > n$, hence

$$\widetilde{X}_n - \widetilde{X}_0 = \sum_{j=1}^{2k} \big[\widetilde{X}_{\tau_j \wedge n} - \widetilde{X}_{\tau_{j-1} \wedge n}\big] = \sum_{j=1}^{k} \big[\widetilde{X}_{\tau_{2j} \wedge n} - \widetilde{X}_{\tau_{2j-1} \wedge n}\big] + \sum_{j=0}^{k-1} \big[\widetilde{X}_{\tau_{2j+1} \wedge n} - \widetilde{X}_{\tau_{2j} \wedge n}\big].$$

Denote the first sum on the right by S_1 and the second by S_2. Now, if $\tau_{2j-1} \ge n$, then $\widetilde{X}_{\tau_{2j} \wedge n} - \widetilde{X}_{\tau_{2j-1} \wedge n} = 0$, and if $\tau_{2j-1} < n$ then $X_{\tau_{2j-1}} \le a$ so $\widetilde{X}_{\tau_{2j-1}} = 0$. Therefore, S_1 includes all differences corresponding to the upcrossings of $[0, c]$ up to time n, hence $S_1 \ge c\widetilde{U}^n_{[0,c]}$ and so $E\,S_1 \ge cE\,\widetilde{U}^n_{[0,c]}$. Moreover, by optional stopping, $E\,S_2 \ge 0$. Therefore, $E(\widetilde{X}_n - \widetilde{X}_0) \ge E\,S_1 + E\,S_2 \ge cE\,\widetilde{U}^n_{[0,c]}$. Since $-\widetilde{X}_0 = -(0-a)^+ \le |a|$, the desired inequalities follow. □

Convergence of Martingales

Throughout this subsection, (X_n) is an adapted process on a filtered probability space $(\Omega, \mathcal{F}, (\mathcal{F}_n), P)$. There are several important results on the convergence of martingales. One of the most basic is the following:

18.5.10 Martingale Convergence Theorem (Doob). *Let $(X_n, \mathcal{F}_n)$ be a submartingale such that $\sup_n \|X_n\|_1 < \infty$. Then (X_n) converges almost surely to an L^1 random variable X_∞. If, additionally, (X_n) is uniformly integrable, then the convergence is in L^1.*

Proof. By 18.5.9, for each n

$$E\big(U^n_{[a,b]}\big) \le \frac{1}{b-a} E\big(|X_n| + 2|a|\big) \le \frac{1}{b-a} \sup_n \|X_n\|_1 + \frac{2|a|}{b-a}.$$

Since and $U^n_{[a,b]} \uparrow U_{[a,b]}$, by the monotone convergence theorem

$$E\big(U_{[a,b]}\big) \le \frac{1}{b-a} \sup_n \|X_n\|_1 + \frac{2|a|}{b-a} < \infty.$$

Therefore, $U_{[a,b]}$ is finite a.s. By 18.5.8, (X_n) converges a.s. to a measurable function $X_\infty : \Omega \to \overline{\mathbb{R}}$. By Fatou's lemma, $\int |X_\infty|\, dP \le \underline{\lim}_n \int |X_n|\, dP < \infty$, hence $X_\infty \in L^1$. This proves the first part of the theorem. The last part follows from 4.4.5. □

18.5.11 Corollary. *Let (X_n) be a submartingale such that $X_n \le 0$ for all n. Then (X_n) converges almost surely to an L^1 random variable X_∞.*

Proof. By the submartingale property, $E\, X_1 \le E\, X_n$, hence $E\,|X_n| = -E\, X_n \le -E\, X_1$ and so $\sup_n \|X_n\|_1 < \infty$. □

18.5.12 Corollary. *Let $1 \le p < \infty$ and let (X_n) be a submartingale such that $(|X_n|^p)$ is uniformly integrable. Then (X_n) converges almost surely and in L^p to an L^p random variable X_∞.*

Proof. By uniform integrability, $\sup_n \|X_n\|_p < \infty$ (4.4.2). Since $\|X_n\|_1 \le \|X_n\|_p$, (X_n) converges almost surely to an L^1 random variable X_∞. By 4.4.5, the convergence is in L^p norm. □

18.5.13 Corollary. *Let (X_n) be a martingale such that (X_n) is uniformly integrable. Then (X_n) converges almost surely and in L^1 to a random variable X_∞ with the property that $X_n = E(X_\infty \mid \mathfrak{F}_n)$ a.s. for all n.*

Proof. All but the last assertion follows from the preceding corollary. For the desired equality, let $A \in \mathfrak{F}_n$ and note that for all $m \ge n$,

$$\int_A X_n\, dP = \int_A X_m\, dP \overset{m\to\infty}{\to} \int_A X_\infty\, dP = \int_A E(X_\infty| \mid \mathfrak{F}_n)\, dP$$

so $X_n = E(X_\infty \mid \mathfrak{F}_n)$ a.s. □

18.5.14 Corollary. *Let $X \in L^1$ and denote by $\mathfrak{F}_\infty$ the σ-field generated by $\bigcup \mathfrak{F}_n$. Then*

$$E(X \mid \mathfrak{F}_n) \to E\big(X \mid \mathfrak{F}_\infty\big) \quad \textit{a.s. and in } L^1.$$

Proof. Let $X_n = E\big(X \mid \mathfrak{F}_n\big)$. Then (X_n) is a uniformly integrable martingale (18.5.1(c)) and $\sup_n \|X_n\|_1 \le \|X\|_1 < \infty$, hence (X_n) converges a.s. and in L^1 to some $\mathfrak{F}_\infty$-random variable X_∞. If $A \in \mathfrak{F}_m$ and $n > m$, then

$$\int_A E\big(X \mid \mathfrak{F}_\infty\big)\, dP = \int_A X\, dP = \int_A E\big(X \mid \mathfrak{F}_n\big)\, dP \to \int_A X_\infty\, dP.$$

Therefore $\int_A E\big(X \mid \mathfrak{F}_\infty\big)\, dP = \int_A X_\infty\, dP$ for all $A \in \mathfrak{F}_\infty$ and so $X_\infty = E\big(X \mid \mathfrak{F}_\infty\big)$ a.s. □

Recall that the L^1 strong law of large numbers asserts that for independent and identically distributed (iid) L^1 random variables X_n, the sample averages S_n/n tend to $E(X_1)$ a.s. or, equivalently, $\frac{1}{n}\sum_{k=1}^n (X_n - E(X_n)) \to 0$ a.s. The following generalization removes the iid requirement.

18.5.15 Corollary. *Let $\sup_n \|X_n\|_1 < \infty$. Set $\mathfrak{F}_0 = \{\emptyset, \Omega\}$. Then*

$$\frac{1}{n}\sum_{j=1}^n \big(X_j - E(X_j \mid \mathfrak{F}_{j-1})\big) \to 0 \quad \textit{a.s.}$$

Proof. Set $Y_n = \sum_{j=1}^{n} j^{-1}[X_j - E(X_j \mid \mathcal{F}_{j-1})]$. Then

$$Y_{n+1} - Y_n = \frac{1}{n+1}\big[X_{n+1} - E(X_{n+1} \mid \mathcal{F}_n)\big],$$

hence $E(Y_{n+1} - Y_n \mid \mathcal{F}_n) = 0$, that is, (Y_n) is a martingale. Since $\sup_n \|Y_n\|_1 < \infty$, (Y_n) converges a.s. to a random variable Y_∞. The conclusion now follows from Kronecker's lemma. □

Reversed Martingales

A **reversed filtration** on a probability space $(\Omega, \mathcal{F}, P)$ is a sequence of sub σ-fields $\mathcal{F}_n$ of $\mathcal{F}$ such that

$$\cdots \subseteq \mathcal{F}_{n+1} \subseteq \mathcal{F}_n \cdots \subseteq \mathcal{F}_1.$$

For example, if (X_n) is a sequence of random variables, then $\mathcal{F}_n := \sigma(X_k : k \geq n)$ defines a reversed filtration.

Now let (X_n) be an L^1 process such that X_n is $\mathcal{F}_n$-measurable for each n. Then $(X_n, \mathcal{F}_n)$ is a **reversed martingale** if

$$E(X_n \mid \mathcal{F}_{n+1}) = X_{n+1}, \quad n \geq 1.$$

Iterating, we obtain

$$E(X_n \mid \mathcal{F}_{n+p}) = X_{n+p}. \tag{18.9}$$

One may also formulate in an analogous way the notions of reversed submartingales and reversed supermartingales. We consider only the martingale case.

Here is the reversed martingale analog of Doob's convergence theorem. Note that the hypothesis $\sup_n \|X_n\|_1 < \infty$ is not needed in this setting.

18.5.16 Theorem. *Let $(X_n, \mathcal{F}_n)$ be a reversed martingale. Then there exists a random variable X_∞ such that*

$$\lim_n X_n = X_\infty \quad \text{a.s. and in } L^1.$$

Moreover, $X_\infty = E(X_1 \mid \mathcal{F}_\infty)$ a.s., where $\mathcal{F}_\infty = \bigcap_n \mathcal{F}_n$.

Proof. We apply Doob's upcrossing inequality to the number $U^n[a,b]$ of upcrossings of $[a,b]$ of the sequence $X_n, X_{n-1}, \ldots, X_1$. This gives

$$E\big(U^n_{[a,b]}\big) \leq \frac{1}{b-a} E\big(|X_1| + 2|a|\big).$$

Since $U^n_{[a,b]} \uparrow U_{[a,b]}$, which is the number of upcrossings of the infinite sequence $\cdots X_n, X_{n-1}, \cdots X_1$, we see that $U_{[a,b]} < \infty$ a.s. and so $X_n \to X_\infty$ a.s., as before. Since $X_n = E(X_1 \mid \mathcal{F}_n)$, (X_n) is uniformly integrable (18.5.1(c)), hence the convergence is also L^1.

Since X_m is $\mathcal{F}_n$-measurable for all $m \geq n$, X_∞ is $\mathcal{F}_n$-measurable for all n, that is, X_∞ is $\mathcal{F}_\infty$-measurable. Also, from $E(X_1 \mid \mathcal{F}_n) = X_n$ we have

$$E(X_1 \mid \mathcal{F}_\infty) = E\big(E(X_1 \mid \mathcal{F}_n) \mid \mathcal{F}_\infty\big) = E(X_n \mid \mathcal{F}_\infty),$$

hence

$$\int_A E(X_1 \mid \mathcal{F}_\infty)\, dP = \int_A X_n \to \int_A X_\infty\, dP, \quad \forall\, A \in \mathcal{F}_\infty.$$

Therefore, $X_\infty = E(X_1 \mid \mathcal{F}_\infty)$ a.s. □

For an application, consider an iid process (X_n) with $|E(X_1)| < \infty$ and set $S_n = X_1 + \cdots + X_n$ and $Y_n = n^{-1}S_n - E(X_1)$. Then (Y_n) is a reversed martingale with respect to the reversed filtration $\mathfrak{F}_n := \sigma(X_k : k \ge n)$. Indeed, from

$$n(n+1)(Y_{n+1} - Y_n) = nX_{n+1} - (X_1 + \cdots + X_n),$$

and we see by applying (18.9) that

$$n(n+1)E(Y_{n+1} - Y_n \mid \mathfrak{F}_{n+1}) = nX_{n+1} - nX_{n+1} = 0$$

and so $Y_{n+1} = E(Y_n \mid \mathfrak{F}_{n+1})$. Since $E(Y_1) = 0$, the theorem implies that $Y_n \to 0$ a.s. and in L^1, which is the law of large numbers with the added feature of L^1 convergence (which could also have been established originally.)

18.6 General Stochastic Processes

A d-**dimensional stochastic process** is an indexed family $\{X_i : i \in \mathfrak{I}\}$ of d-dimensional random variables X_i on some probability space $(\Omega, \mathfrak{F}, P)$. We shall also use the notation $X(i)$ or $X(i,\omega)$. In applications, $\mathfrak{I}$ is typically one of the sets $\mathbb{N}$ or $[0,\infty)$. We have already considered processes of the former type, so-called *discrete-time processes.* In this chapter we consider general processes with particular emphasis on the important special case $\mathfrak{I} = [0,\infty)$, so-called *continuous-time processes.*

The set $\mathbb{R}^d$ in a d-dimensional process $\{X_i : i \in \mathfrak{I}\}$ is called the **state space** of the process. For each ω, the function $i \to X_i(\omega)$ from $\mathfrak{I}$ to $\mathbb{R}^d$ is called a **path of the process.** For example, if X_t is the position at time t of a molecule of a gas, then a path $t \to X_t(\omega)$ represents one possible trajectory of the molecule. If X_t is the price of a stock at time t, then a path symbolizes a particular market scenario.

The Consistency Conditions

Let $\{X_i : i \in \mathfrak{I}\}$ be a d-dimensional process and set $S = \mathbb{R}^d$ and $\mathfrak{B} = \mathfrak{B}(\mathbb{R}^d)$. For each n-tuple $(i_1, \ldots, i_n)$ of distinct members of $\mathfrak{I}$, let $P_{i_1 \cdots i_n}$ denote the distribution of $(X_{i_1}, \ldots, X_{i_n})$, that is,

$$P_{i_1 \cdots i_n}(B) = P\big((X_{i_1}, \ldots, X_{i_n}) \in B\big), \quad B \in \mathfrak{B}(S^n := S \times \cdots \times S). \tag{18.10}$$

The probability measures $P_{(i_1 \ldots, i_n)}$ are called the **finite dimensional distributions** of the process X. Note that these distributions satisfy the following **consistency conditions**: For all n and $B_j \in \mathfrak{B}$,

C1. $P_{i_{\tau 1} \cdots i_{\tau n}}\big(B_{i_{\tau 1}} \times \cdots \times B_{i_{\tau n}}\big) = P_{i_1 \cdots i_n}\big(B_{i_1} \times \cdots \times B_{i_n}\big)$ $\forall$ permutation τ of $(1, \ldots, n)$.

C2. $P_{i_1 \cdots i_{n+1}}\big(B_{i_1} \times \cdots \times B_{i_n} \times S\big) = P_{i_1 \cdots i_n}\big(B_{i_1} \times \cdots \times B_{i_n}\big)$.

The problem we consider in this section is the converse: Given an index set $\mathfrak{I}$ and a family $\mathscr{D}(\mathfrak{I})$ of finite dimensional distributions satisfying the above consistency conditions, find a probability space $(\Omega, \mathfrak{F}, P)$ and a process such that (18.10) holds for all members of $\mathscr{D}(\mathfrak{I})$. To construct such a process, we must first define the product measurable space $S^{\mathfrak{I}}$ for a general measurable space $(S, \mathfrak{F})$.

The Product of Measurable Spaces

Let $\mathfrak{I}$ be an arbitrary index set, $(S, \mathcal{F})$ an arbitrary measurable space, and $S^{\mathfrak{I}}$ the collection of all functions $f : \mathfrak{I} \to S$. For $n \in \mathbb{N}$, let $(S^n, \mathcal{F}^n) = (S \times \cdots \times S, \mathcal{F} \otimes \cdots \otimes \mathcal{F})$ denote the n-fold product σ-field. In what follows, we consider finite sequences $(i_1, \ldots, i_n)$ of *distinct* members in $\mathfrak{I}$. These will be called **index sequences**. For such sequences, we write $(i_1, \ldots, i_n) \subseteq (j_1, \ldots, j_p)$ if $\{i_1, \ldots, i_n\} \subseteq \{j_1, \ldots, j_p\}$. Define the **projection map** corresponding to the index sequence $(i_1, \ldots, i_n)$ by

$$\pi_{i_1 \cdots i_n} : S^{\mathfrak{I}} \to S^n, \quad \pi_{i_1 \cdots i_n}(f) = \big(f(i_1), \ldots, f(i_n)\big).$$

A **cylinder set** over $A \in \mathcal{F}^n$ is a set of the form

$$\pi_{i_1 \cdots i_n}^{-1}(A) = \{f \in S^{\mathfrak{I}} : \big(f(i_1), \ldots, f(i_n)\big) \in A\}.$$

18.6.1 Lemma. *If $(i_1, \ldots, i_n) \subseteq (j_1, \ldots, j_p)$ and $A \in \mathcal{F}^n$, then there exists $A' \in \mathcal{F}^p$ such that*

$$\pi_{i_1 \cdots i_n}^{-1}(A) = \pi_{j_1 \cdots j_p}^{-1}(A').$$

Proof. Let τ be a permutation of $\{1, \ldots, p\}$ such that the first n coordinates of $(j_{\tau 1}, \ldots, j_{\tau p})$ are $i_1, \ldots, i_n$. Define

$$A' = \{(x_1, \ldots, x_p) : (x_{\tau 1}, \ldots, x_{\tau n}) \in A\}.$$

Then A' is the preimage of A under a measurable mapping $S^p \to S^n$, hence $A' \in \mathcal{F}^p$. Moreover, $\big(f(i_1), \ldots, f(i_n)\big) = \big(f(j_{\tau 1}), \ldots, f(j_{\tau n})\big) \in A$ iff $\big(f(j_1), \ldots, f(j_p)\big) \in A'$, so the desired equation holds. □

18.6.2 Corollary. *Given cylinder sets*

$$\pi_{i_1 \cdots i_n}^{-1}(A) \quad \textit{and} \quad \pi_{j_1 \cdots j_m}^{-1}(B), \quad A \in \mathcal{F}^n, \; B \in \mathcal{F}^m, \tag{18.11}$$

and $(k_1, \ldots, k_p) \supseteq (i_1, \ldots, i_n) \cup (j_1, \ldots, j_m)$, there exist A', $B' \in \mathcal{F}^p$ such that

$$\pi_{i_1 \cdots i_n}^{-1}(A) = \pi_{k_1 \cdots k_p}^{-1}(A') \quad \textit{and} \quad \pi_{j_1 \cdots j_m}^{-1}(B) = \pi_{k_1 \cdots k_p}^{-1}(B'). \tag{18.12}$$

We denote the collection of all cylinder sets by $\mathcal{C}(\mathfrak{I})$. Except in trivial cases, $\mathcal{C}(\mathfrak{I})$ is not a σ-field. However,

18.6.3 Proposition. *$\mathcal{C}(\mathfrak{I})$ is a field.*

Proof. Consider the cylinder sets in (18.11) represented as in (18.12). Then

$$\pi_{i_1 \cdots i_n}^{-1}(A) \cup \pi_{j_1 \cdots j_m}^{-1}(B) = \pi_{k_1 \cdots k_p}^{-1}(A' \cup B'),$$

which shows that $\mathcal{C}(\mathfrak{I})$ is closed under finite unions. Since the complement of $\pi_{i_1 \cdots i_n}^{-1}(A)$ is $\pi_{i_1 \cdots i_n}^{-1}(A^c)$, $\mathcal{C}(\mathfrak{I})$ is also closed under complementation and hence is a field. □

The σ-field generated by $\mathcal{C}(\mathfrak{I})$ is called the **product σ-field** and is denoted by $\mathcal{F}^{\mathfrak{I}}$. The equality

$$\pi_{i_1 \cdots i_n}^{-1}(B_1 \times \cdots \times B_n) = \bigcap_{j=1}^{n} \pi_{i_j}^{-1}(B_j)$$

shows that $\mathcal{F}^{\mathfrak{I}}$ is also the σ-field generated by all the projection mappings π_i. As a consequence, we have

18.6.4 Proposition. *The projection mappings are measurable transformations. A function T from a measurable space $(\Omega, \mathcal{G})$ to $S^{\mathfrak{I}}$ is measurable iff $\pi_i \circ T : \Omega \to S$ is measurable for every $i \in \mathfrak{I}$.*

While the σ-field $\mathcal{F}^{\mathfrak{I}}$ is adequate for many purposes, some important sets may not be members of $\mathcal{F}^{\mathfrak{I}}$. The reason for this shortcoming is the following result, which shows that members of $\mathcal{F}^{\mathfrak{I}}$ are determined by countable subsets of $\mathfrak{I}$.

18.6.5 Proposition. *For every member A of $\mathcal{F}^{\mathfrak{I}}$ there is a countable subset $\mathfrak{I}_A$ of $\mathfrak{I}$ depending on A such that*

$$f \in A,\ g \in S^{\mathfrak{I}},\ \text{and}\ f(j) = g(j)\ \forall\ j \in \mathfrak{I}_A \Rightarrow g \in A.$$

Proof. Let $\mathcal{G}$ denote the collection of all subsets A of $S^{\mathfrak{I}}$ with the stated property. We show that $\mathcal{F}^{\mathfrak{I}} \subseteq \mathcal{G}$. To this end, note first that $\mathcal{G}$ contains all sets of the form $A := \pi_i^{-1}(B)$, $B \in \mathcal{F}$; indeed, one need only take $\mathfrak{I}_A = \{i\}$. Since these are generators for $\mathcal{F}^{\mathfrak{I}}$, the desired inclusion will follow if we show that $\mathcal{G}$ is a σ-field.

Suppose $A \in \mathcal{G}$ and let $f \in A^c$, $g \in S^{\mathfrak{I}}$. If $f(j) = g(j)$ for all $j \in \mathfrak{I}_A$, then $g \in A^c$; otherwise, g would lie in A forcing f to lie in A. Therefore, we may take $\mathfrak{I}_{A^c} = \mathfrak{I}_A$, showing that $\mathcal{G}$ is closed under complementation. Now let (A_n) be a sequence in $\mathcal{G}$ and set $A := \bigcup_n A_n$. Let $f \in A$, $g \in S^{\mathfrak{I}}$, such that $f(j) = g(j)$ for all $j \in \bigcup_n \mathfrak{I}_{A_n}$. Since $f \in A_m$ for some m, $g \in A_m \subseteq A$. Therefore we may take $\mathfrak{I}_A = \bigcup_n \mathfrak{I}_{A_n}$. Since this is countable, $A \in \mathcal{G}$. □

18.6.6 Corollary. *$C[0, \infty)$ is not a member of $\mathcal{B}^{[0,\infty)}$.*

Proof. Suppose, for a contradiction, that $C[0, \infty) \in \mathcal{B}^{[0,\infty)}$. By the proposition there exists a countable subset D of $[0, \infty)$ with the property

$$f \in C[0, \infty),\ g \in \mathbb{R}^{\mathfrak{I}},\ \text{and}\ f(t) = g(t)\ \forall\ t \in D \Rightarrow g \in C\big[0, \infty)\big).$$

Now take $f \equiv 0$ and $g = \mathbf{1}_{\{s\}}$, where $s \notin D$. Then $f = g$ on D, yet g is not continuous. □

The Kolmogorov Extension Theorem

Set $S := \mathbb{R}^d$ and $\mathcal{B} := \mathcal{B}(\mathbb{R}^d)$. Here is the main result of the section.

18.6.7 Theorem (Kolmogorov). *Let $\mathfrak{I}$ be an arbitrary nonempty index set and let $\mathscr{D}(\mathfrak{I})$ be a collection of finite dimensional probability distributions satisfying the consistency conditions C1 and C2. Then there exists a unique probability measure P on the product space $(S^{\mathfrak{I}}, \mathcal{B}^{\mathfrak{I}})$ such that for every index sequence $(i_1, \ldots, i_n)$,*

$$P\big(\pi_{i_1 \cdots i_n}^{-1}(A)\big) = P_{i_1 \cdots i_n}(A), \quad A \in \mathcal{B}^n. \tag{18.13}$$

Proof. Define P on $\mathcal{C}(\mathfrak{I})$ by (18.13). To see that P is well-defined, suppose that $\pi_{i_1 \cdots i_n}^{-1}(A) = \pi_{j_1 \cdots j_m}^{-1}(B)$. Represent these cylinder sets as in (18.12). Then $A' = B'$, and it follows from the consistency conditions that

$$P_{i_1 \ldots i_n}(A) = P_{k_1 \ldots k_p}(A') = P_{k_1 \ldots k_p}(B') = P_{j_1 \ldots j_m}(B).$$

We show next that P is a probability measure on $\mathcal{C}(\mathfrak{I})$. The conclusion of the theorem will then follow from the measure extension theorem (1.6.4). Clearly $P(S^{\mathfrak{I}}) = 1$. To see that P is finitely additive on $\mathcal{C}(\mathfrak{I})$, represent disjoint cylinder sets as in (18.12). Then A' and B' must be disjoint and

$$\pi_{i_1 \cdots i_n}^{-1}(A) \cup \pi_{j_1 \cdots j_m}^{-1}(B) = \pi_{k_1 \cdots k_p}^{-1}(A' \cup B'),$$

hence

$$P\big(\pi^{-1}_{i_1\cdots i_n}(A)\cup\pi^{-1}_{j_1\cdots j_m}(B)\big) = P_{k_1\cdots k_p}(A'\cup B') = P_{k_1\cdots k_p}(A') + P_{k_1\cdots k_p}(B')$$
$$= P\big(\pi^{-1}_{i_1\cdots i_n}(A)\big) + P\big(\pi^{-1}_{j_1\cdots j_m}(B)\big).$$

Since $\mathcal{C}(\mathfrak{I})$ is a field, and P is finitely additive, P is monotone. It remains to show that if (A_n) is a sequence in $\mathcal{C}(\mathfrak{I})$ and $A_n \downarrow \emptyset$, then $P(A_n) \to 0$. Let $r := \lim_n P(A_n)$. We show that the assumption $r > 0$ implies the contradiction $\bigcap_n A_n \neq \emptyset$. Now, by 18.6.1, it is possible, without affecting monotonicity or changing the intersection $\bigcap_n A_n$, to precede the sequence (A_n) by terms $S^{\mathfrak{I}}$ and to insert duplicate terms A_j. Thus we may assume there exists an infinite sequence of distinct indices i_n such that

$$A_n = \pi^{-1}_{I_n}(B_n), \quad \text{where } I_n = (i_1,\ldots,i_n) \text{ and } B_n \in \mathcal{B}^n.$$

By regularity, choose a compact set $C_n \subseteq B_n$ with $P_{I_n}(B_n \setminus C_n) < r/2^{n+1}$ and set

$$D_n = \pi^{-1}_{I_n}(C_n) \text{ and } E_n = \bigcap_{j=1}^{n} D_j.$$

Then by monotonicity

$$P(A_n \setminus E_n) = P\Big(\bigcup_{j=1}^{n}(A_n\setminus D_j)\Big) \le \sum_{j=1}^{n} P(A_j\setminus D_j) = \sum_{j=1}^{n} P_{I_j}(B_j\setminus C_j) \le r/2,$$

and since $E_n \subseteq A_n$ we see that $P(E_n) \ge P(A_n) - r/2 \ge r/2 > 0$. Therefore, $E_n \neq \emptyset$. Choosing $f_n \in E_n$ we have

$$\big(f_n(i_1),\ldots,f_n(i_n)\big) \in C_j, \quad 1 \le j \le n.$$

In particular, $f_n(i_1) \in C_1$ for all $n \ge 1$, and since C_1 is compact there exists a subsequence $(f_n^{(1)})$ of (f_n) such that $f_n^{(1)}(i_1) \xrightarrow{n} x_1 \in C_1$. Likewise, since $(f_n^{(1)}(i_1), f_n^{(1)}(i_2)) \in C_2$ for all $n \ge 2$, exists a subsequence $(f_n^{(2)})$ of $(f_n^{(1)})$ such that $(f_n^{(2)}(i_1), f_n^{(2)}(i_2)) \xrightarrow{n} (x_1, x_2) \in C_2$. By induction we may construct successive subsequences $(f_n^{(k)})$ such that for all k,

$$\big(f_n^{(k)}(i_1), f_n^{(k)}(i_2),\ldots,f_n^{(k)}(i_k)\big) \xrightarrow{n} (x_1,x_2,\ldots,x_k) \in C_k. \qquad (\dagger)$$

For each k, the diagonal sequence $(f_n^{(n)}(i_k))_n$ then converges to x_k. Now choose any f such that $f(i_j) = x_j$ for all j. Then by $(\dagger)$, $f \in D_k \subseteq A_k$ for all k, which is the desired contradiction. This proves that P is a probability measure on $\mathcal{C}(\mathfrak{I})$. □

Taking X_i to be the projection map π_i, we now have the following resolution to the problem stated at the beginning of the section.

18.6.8 Corollary. *Given $\mathcal{D}(\mathfrak{I})$ as above, there exists a probability space $(\Omega, \mathcal{F}, P)$ and a family of $\mathbb{R}^d$-valued random variables such that 18.10 holds for every finite sequence $(i_1,\ldots,i_n)$.*

The following version of the theorem is useful in the important special case $\mathfrak{I} = (0,\infty)$ and $d = 1$.

18.6.9 Corollary. *Suppose that for each finite ordered sequence $t_1 < t_2 < \cdots < t_n$ in $(0,\infty)$*

there exists a probability distribution $P_{t_1\cdots t_n}$ *with* cdf $F_{t_1\cdots t_n}$ *such that for all* n *and* k *with* $1 \le k \le n$,

$$\lim_{x_k \to \infty} F_{t_1\cdots t_n}(x_1, \ldots, x_n) - F_{t_1\cdots t_{k-1}\,t_{k+1}\cdots t_n}(x_1, \ldots, x_{k-1}, x_{k+1}, \ldots, x_n). \tag{18.14}$$

Then there exists a unique probability measure P *on the product space* $(\mathbb{R}^{(0,\infty)}, \mathcal{B}(\mathbb{R})^{(0,\infty)})$ *such that for every sequence* $t_1 < t_2 < \cdots < t_n$,

$$P\big(\pi^{-1}_{t_1\cdots t_n}(A)\big) = P_{t_1\cdots t_n}(A) \;\; \forall \;\; A \in \mathcal{B}(\mathbb{R}^n).$$

Proof. The hypothesis implies that for all $B_j \in \mathcal{B}(\mathbb{R})$ and $t_1 < t_2 < \cdots < t_n$,

$$\begin{aligned} &P_{t_1\cdots t_{k-1}\,t_{k+1},\cdots t_n}(B_1 \times \cdots \times B_{k-1} \times B_{k+1} \cdots \times B_n) \\ &\qquad = P_{t_1\cdots t_{k-1}\,t_k\,t_{k+1},\ldots,t_n)}(B_1 \times \cdots \times B_{k-1} \times \mathbb{R} \times B_{k+1} \times \cdots \times B_n). \qquad (\dagger) \end{aligned}$$

The idea is to enlarge the collection of probability distributions to include *all* index sequences $(s_1, \ldots, s_n)$ and then apply the extension theorem. This is accomplished as follows: Given an arbitrary sequence $(s_1, \ldots, s_n)$ of distinct s_j, define

$$P_{s_1\cdots s_n}(B_1 \times \cdots \times B_n) = P_{s_{\tau 1}\cdots s_{\tau n}}(B_{\tau 1} \times \cdots \times B_{\tau n}),$$

where τ is the unique permutation of $(1, \ldots, n)$ that *orders* $s_1, \ldots, s_n$, that is, that produces the natural ordering $s_{\tau 1} < \cdots < s_{\tau n}$. If σ is any permutation of $(1, \ldots, n)$ and τ is the permutation that orders $s_{\sigma 1}, \ldots, s_{\sigma n}$, then $\tau\sigma$ is the permutation that orders $s_1, \ldots, s_n$, hence

$$P_{s_{\sigma 1}\cdots s_{\sigma n}}(B_{\sigma 1} \times \cdots \times B_{\sigma n}) = P_{s_{\tau\sigma 1}\cdots s_{\tau\sigma n}}(B_{\tau\sigma 1} \times \cdots \times B_{\tau\sigma n}) = P_{s_1\cdots s_n}(B_1 \times \cdots \times B_n).$$

This shows that consistency condition C1 holds. To verify C2, we must show that

$$P_{s_1\cdots s_n}(B_1 \times \cdots \times B_{n-1} \times \mathbb{R}) = P_{s_1\cdots s_{n-1}}(B_1 \times \cdots \times B_{n-1}).$$

But a permutation that orders $s_1, \ldots, s_n$ then places $\mathbb{R}$ in some position k, and an application of $(\dagger)$ yields the desired equation. □

18.7 Brownian Motion

The phrase *Brownian motion*, in the classical sense, refers to a phenomenon discovered in 1827 by the Scottish botanist Robert Brown, who observed that microscopic particles suspended in a fluid (liquid or gas) exhibited highly irregular motion characterized by seemingly independent random movements. Later it was determined that this motion resulted from collisions of the particles with molecules in the ambient fluid. In 1905, Albert Einstein gave a physical interpretation of Brownian motion. A rigorous mathematical model of Brownian motion was developed in the 1920s by Norbert Wiener. The model, known as a *Brownian motion process* or a *Wiener process*, has come to play an indispensable role in many areas of pure and applied mathematics. For example, in pure mathematics the process has spawned the study of continuous time martingales and stochastic calculus. In applied mathematics the Wiener process is used as mathematical model of "white noise." In mathematical finance, geometric Brownian motion is the fundamental component in the

Black-Scholes model for option pricing (discussed in §18.9). In the current section we consider a one-dimensional version of (mathematical) Brownian motion, which may be viewed as a model for the motion of a Brownian particle projected onto a vertical axis.

For a mathematical description of Brownian motion, we need to extend some earlier terminology. A **(continuous-time) filtration** on a probability space $(\Omega, \mathcal{F}, P)$ is a family of σ-fields $\mathcal{F}_t$ indexed by $t \in [0, \infty)$ such that $\mathcal{F}_s \subseteq \mathcal{F}_t \subseteq \mathcal{F}$ for all $0 \le s \le t$. A probability space with a filtration is called a **filtered probability space** and is denoted by $\big(\Omega, \mathcal{F}, (\mathcal{F}_t)_{t\ge 0}, P\big)$. As in the case of discrete time, a filtration $(\mathcal{F}_t)$ may be viewed as a mathematical model for ever more precise information produced by an experiment evolving in time. An important example is the **natural filtration** $\mathcal{F}^X = (\mathcal{F}_t^X)$ of a process $X = (X_t)$, where $\mathcal{F}_t^X$ is the σ-field $\sigma(X_s : 0 \le s \le t)$, which consists precisely of the information revealed by the process up to time t. A stochastic process (X_t) is said to be **adapted to a filtration** $(\mathcal{F}_t)$ if for all t the random variable X_t is $\mathcal{F}_t$-measurable. For example, a process is always adapted to its natural filtration, but there may be reason to consider larger filtrations.

A (one-dimensional) **Brownian motion** or **Wiener process** on a filtered probability space $(\Omega, \mathcal{F}, (\mathcal{F}_t)_{t\ge 0}, \mathbb{P})$ is a stochastic process $W = (W(t))_{t\ge 0}$ adapted to $(\mathcal{F}_t)_{t\ge 0}$ such that the following conditions hold:

(a) $W_0 = 0$;

(b) For $0 \le s < t$, the increment $W(t) - W(s)$ is normal with mean zero and variance $t - s$, that is,

$$P\big(W(t) - W(s) \in B\big) = \frac{1}{\sqrt{2\pi(t-s)}} \int_B \exp\left[-\frac{x^2}{t-s}\right] dx, \quad B \in \mathcal{B}(\mathbb{R}).$$

(c) For $0 \le s < t$, $W(t) - W(s)$ is independent of $\mathcal{F}_s$.

(d) The paths $t \to W_t(\omega)$ of W are continuous.

Note that $W(t)$ has **independent increments**, that is, if $0 < t_1 < t_2 < \cdots < t_n$ then the random variables $W(t_1), W(t_2) - W(t_1), \ldots, W(t_n) - W(t_{n-1})$ are independent. This follows by induction from (c) and the fact that (W_t) is adapted to $(\mathcal{F}_t)$.

Construction of Brownian Motion

The existence of a process with properties (a)–(c) is a consequence of the Kolmogorov extension theorem: For $0 < t_1 < \cdots < t_n$ set

$$f_{t_1 \cdots t_n}(x_1, \cdots, x_n) = \prod_{k=1}^{n} \frac{1}{\sqrt{2\pi \Delta t_k}} \exp\left[-\frac{[\Delta x_k]^2}{2\Delta t_k}\right],$$

where $\Delta x_k = x_k - x_{k-1}$ and $\Delta t_k = t_k - t_{k-1}$ $(x_0 = t_0 = 0)$. Then $f_{t_1 \cdots t_n}$ is a density for an n-dimensional cdf $F_{t_1 \cdots t_n}$ that satisfies the consistency condition (18.14). By 18.6.9 there exists a probability space $(\Omega, \mathcal{F}, P)$ and a process (X_t) such that

$$P\big((X_{t_1}, \ldots, X_{t_n}) \in B\big) = \int_B f_{t_1 \cdots t_n}(x)\, dx, \quad B \in \mathcal{B}(\mathbb{R}^n).$$

Taking $B = B_1 \times \cdots \times B_n$ and setting $x_j = y_1 + \cdots + y_j$ we have, by a substitution,

$$\begin{aligned} P\big((X_{t_1}, X_{t_2} - X_{t_1}, \cdots X_{t_n} - X_{t_{n-1}}) \in B\big) &= \int \mathbf{1}_B(x_1, x_2 - x_1, \ldots, x_n - x_{n-1}) f_{t_1 \cdots t_n}(x)\, dx \\ &= \int \mathbf{1}_B(y_1, y_2, \ldots, y_n) \prod_{k=1}^n \frac{1}{\sqrt{2\pi \Delta t_k}} \exp\left[-\frac{y_k^2}{2\Delta t_k}\right] dy \\ &= \prod_{k=1}^n \int \mathbf{1}_{B_k}(y_k) \frac{1}{\sqrt{2\pi\Delta t_k}} \exp\left[-\frac{y_k^2}{2\Delta t_k}\right] dy_k \\ &= \prod_{k=1}^n P\big(X_{t_k} - X_{t_{k-1}} \in B_k\big). \end{aligned}$$

This shows that (X_t) satisfies (b) and (c) of the definition of Brownian motion. Setting $X_0 = 0$ completes the construction.

It remains to show that there exists a *continuous* process satisfying (a)–(c). The idea is to modify the process (X_t) obtained in the preceding paragraph on a set of probability zero to produce the desired continuous process. This is accomplished by the following general theorem. The proof depends on the density of the dyadic rationals $j/2^n$ $(j \geq 0,\ n \geq 1)$ in $[0, \infty)$.

18.7.1 Theorem. *Let $g(t)$ and $h(t)$ be nonnegative even functions on some interval $(-a, a)$ that are increasing on $(0, a)$ such that g is continuous at zero and the series*

$$\sum_{n=1}^\infty g(2^{-n}) \quad \textit{and} \quad \sum_{n=1}^\infty n 2^n h(2^{-n})$$

converge. Let (X_t) be a stochastic process on a probability space $(\Omega, \mathcal{F}, P)$ that satisfies

$$P\Big(|X_t - X_s| \geq g(t-s)\Big) \leq h(t-s) \quad \textit{whenever } t, s \geq 0 \textit{ and } |t - s| < a.$$

Then there exists a process (Y_t) on $(\Omega, \mathcal{F}, P)$ with continuous paths such that for any t, $Y_t = X_t$ a.s. (the exceptional set depending on t). In particular, Y and X have the same finite-dimensional distributions.

Proof. Set $t_{n,j} = j2^{-n}$ $(n = 1, 2, \ldots,\ j = 0, 1, \ldots)$. For each n and ω, imbed in the path $t \to X_t(\omega)$ a polygonal line $X_n(\cdot, \omega)$ with vertices $(t_{n,j}, X(t_{n,j}, \omega))$:

$$X_n(t, \omega) := X(t_{n,j}, \omega) + 2^n(t - t_{n,j}) \left[X\big(t_{n,j+1}, \omega\big) - X(t_{n,j}, \omega)\right], \quad t_{n,j} \leq t \leq t_{n,j+1}.$$

This defines a sequence of processes $X_n(\cdot)$. Note that since $t_{n+1,2j} = t_{n,j}$,

$$X_n(t, \omega) = X(t_{n+1,2j}, \omega) + 2^n(t - t_{n+1,2j}) \left[X\big(t_{n+1,2j+2}, \omega\big) - X(t_{n+1,2j}, \omega)\right].$$

A direct calculation shows that if $t_{n,j} \leq t \leq t_{n,j+1}$, then

$$\Big|X_{n+1}(t, \omega) - X_n(t, \omega)\Big| \leq \Big|X\big(t_{n+1,2j+1}, \omega\big) - \tfrac{1}{2}\big[X(t_{n+1,2j}, \omega) + X(t_{n+1,2j+2}, \omega)\big]\Big|. \qquad (\alpha)$$

The figure below illustrates the idea. Here, A and B are consecutive points of the polygon for the process $X_n(\cdot)$, and C is the interpolation point required to pass to the polygon for the process $X_{n+1}(\cdot)$.

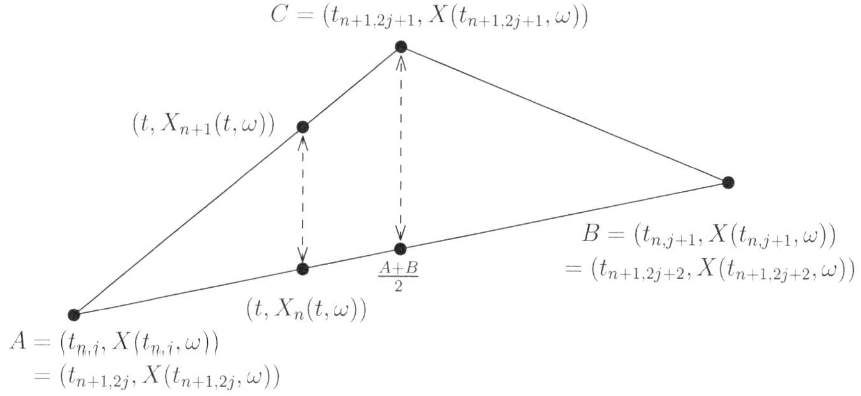

FIGURE 18.2:

From (α) we have

$$\begin{aligned}M_{n,j}(\omega) &:= \sup\{\left|X_{n+1}(t,\omega) - X_n(t,\omega)\right| : t_{n,j} \le t \le t_{n,j+1}\}\\ &\le \tfrac{1}{2}\Big|X\big(t_{n+1,2j+1},\omega\big) - X\big(t_{n+1,2j},\omega\big)\Big| + \tfrac{1}{2}\Big|X\big(t_{n+1,2j+1},\omega\big) - X\big(t_{n+1,2j+2},\omega\big)\Big]\Big|. \quad (\beta)\end{aligned}$$

Since the processes X_n have continuous paths, $M_{n,j}(\omega)$ may be calculated as the supremum over the rational interval $[t_{n,j}, t_{n,j+1}] \cap \mathbb{Q}$, hence $M_{n,j}$ is measurable. Moreover, from (β),

$$\begin{aligned}P\Big(M_{n,j} \ge g(2^{n+1})\Big) &\le P\left(\left|X\big(t_{n+1,2j+1},\omega\big) - X\big(t_{n+1,2j},\omega\big)\right| \ge g(2^{n+1})\right)\\ &\quad + P\left(\left|X\big(t_{n+1,2j+1},\omega\big) - X\big(t_{n+1,2j+2},\omega\big)\right| \ge g(2^{(n+1)})\right)\\ &\le 2h\big(1/2^{n+1}\big), \qquad (\gamma)\end{aligned}$$

the last inequality holding for all n with $1/2^{n+1} < a$. Now set

$$M_n(\omega) := \sup\{\left|X_{n+1}(t,\omega) - X_n(t,\omega)\right| : 0 \le t \le n\}.$$

Since $[0,n] = \bigcup_{j=0}^{n2^n-1} [t_{n_j}, t_{n_{j+1}}]$,

$$\{M_n \ge g\big(1/2^{n+1}\big)\} \subseteq \bigcup_{j=0}^{n2^n-1} \{M_{n,j} \ge g\big(1/2^{n+1}\big)\},$$

hence from (c)

$$P\{M_n \ge g\big(1/2^{n+1}\big)\} \le \sum_{j=0}^{n2^n-1} P\Big\{M_{n,j} \ge g\big(1/2^{n+1}\big)\Big\} \le (n+1)2^{n+1}h\big(1/2^{n+1}\big).$$

It follows from the hypothesis that the series $\sum_{n=1}^{\infty} P\left(M_n \ge g\big(1/2^{n+1}\big)\right)$ converges. By the Borel-Cantelli lemma we then have

$$P(A) = 0, \quad \text{where } A := \overline{\lim_n}\Big\{M_n \ge g\big(1/2^{n+1}\big)\Big\}.$$

Now, let $\omega \in A^c$ and $b > 0$. For any $p \in \mathbb{N}$, $t \in [0, b]$, and all sufficiently large n we have

$$\begin{aligned}\left|X_{n+p}(t,\omega) - X_n(t,\omega)\right| &\le \sum_{k=1}^{p} \left|X_{n+k}(t,\omega) - X_{n+k-1}(t,\omega)\right| \le \sum_{k=1}^{p} M_{n+k-1}(\omega)\\ &\le \sum_{k=1}^{p} g\big(1/2^{n+k}\big).\end{aligned}$$

Since the series $\sum_{k=1}^{\infty} g\big(1/2^k\big)$ converges, the preceding inequality implies that the sequence $(X_n(t,\omega))$ is uniformly Cauchy on $[0,b]$ and therefore converges uniformly on $[0,b]$. Now define

$$Y(t,\omega) = \begin{cases} \lim_n X_n(t,\omega) & \omega \notin A,\\ 0 & \omega \in A.\end{cases}$$

Then (Y_t) has continuous paths and $Y(j/2^n, \omega) = X(j/2^n, \omega)$ for all $\omega \in A^c$, $n \in \mathbb{N}$, and $j \in \mathbb{Z}^+$.

It remains to show that $Y_t = X_t$ a.s. This is clear if t is a dyadic rational. For arbitrary t, choose a sequence of dyadic rationals s_n so that $0 \le t - s_n < 2^{-n}$. Since g and h are increasing,

$$\begin{aligned}P\big(|X(s_n,\omega) - X(t,\omega)| \ge g(1/2^n)\big) &\le P\big(|X(s_n,\omega) - X(t,\omega)| \ge g(t - s_n)\big)\\ &\le h(t-s_n) \le h(1/2^n).\end{aligned}$$

Since the series $\sum_n h(1/2^n)$ converges, by the Borel-Cantelli lemma again we have

$$P(B) = 0, \quad \text{where } B := \overline{\lim_n}\,\big\{|X(s_n,\omega) - X(t,\omega)| \ge g(1/2^n)\big\}.$$

If $\omega \notin B$, then, eventually, $|X(s_n,\omega) - X(t,\omega)| \le g(1/2^n) \to 0$, hence $X(s_n,\omega) \to X(t,\omega)$. Therefore, $Y(t,\omega) = X(t,\omega)$ for $\omega \in (A \cup B)^c$. □

18.7.2 Theorem. *Brownian motion exists.*

Proof. Take X to be any process satisfying (a)–(c) of the definition of Brownian motion. Define

$$g(t) = |t|^{1/4} \quad \text{and} \quad h(t) = \sqrt{\frac{2}{\pi}}\,|t|^{1/4} \exp\left(\frac{-1}{2\sqrt{|t|}}\right).$$

We show that these functions satisfy the hypotheses of 18.7.1. We may then apply that theorem to obtain a continuous version of X, which is the desired Brownian motion.

The functions g and h are clearly even and increasing in $t > 0$. Moreover,

$$\sum_{n=1}^{\infty} g(1/2^n) = \sum_{n=1}^{\infty} \frac{1}{2^{n/4}} < \infty \quad \text{and} \quad \sum_{n=1}^{\infty} n2^n h(1/2^n) = \sqrt{\frac{2}{\pi}} \sum_{n=1}^{\infty} n2^{3n/4} \exp\left(\frac{-2^{n/2}}{2}\right) < \infty.$$

We claim that for $t \ne s$,

$$P\big(|X_t - X_s| \ge g(|t-s|)\big) \le h(|t-s|),$$

which may be written

$$\frac{1}{2|t-s|^{1/2}} \int_{|x| \ge |t-s|^{1/4}} \exp\left(-\frac{x^2}{2|t-s|}\right) dx \le |t-s|^{1/4} \exp\left(\frac{-1}{2|t-s|^{1/2}}\right). \qquad (\dagger)$$

Making the substitution $y = x|t-s|^{-1/2}$, setting $z := |t-s|^{-1/4}$, and integrating by parts, the left side of the inequality becomes

$$\int_z^\infty \exp(-y^2/2)\,dy = -\int_z^\infty \frac{1}{y}\frac{d}{dy}\exp(-y^2/2)\,dy = \frac{1}{z}\exp(-z^2/2) - \int_z^\infty \frac{1}{y^2}\exp(-y^2/2)\,dy$$
$$\le \frac{1}{z}\exp(-z^2/2),$$

which is (†). □

Non-Differentiability of Brownian Paths

The following result models some aspects of the behavior observed by Brown in his experiments with suspended particles.

18.7.3 Theorem. *Let W_t be a Brownian motion. Then there exists a set B of probability one such that, for each $\omega \in B$, the path $t \to W_t(\omega)$ is nowhere differentiable.*

Proof. We follow the argument in [15]. Let $\omega \in \Omega$. If $t \to W(t,\omega)$ has a derivative at some point s and if $a > |W'(s,\omega)|$, then for all $n \ge$ some $n(\omega, s, a)$

$$|W(t,\omega) - W(s,\omega)| \le a|s-t| \quad \text{whenever} \quad |t-s| < 1/2^{n-1}. \tag{†}$$

For $a > 0$, let E_a denote the set of all ω such that (†) holds for some s and for all $n \ge n(\omega, s, a)$. Thus if $\omega \notin E := \bigcup_{a\in\mathbb{Q}^+} E_a$, then $W(\cdot,\omega)$ is nowhere differentiable. We show that there exists a set A_a of probability zero such that $E_a \subseteq A_a$. Setting $A := \bigcup_{a\in\mathbb{Q}^+} A_a$ we then have $E \subseteq A$, $P(A) = 0$, and $W(\cdot,\omega)$ is nowhere differentiable for every $\omega \in A^c \subseteq E^c$, completing the proof.

Set $t_{j,n} := j/2^n$ and

$$M_{n,k} := \max_{1\le j\le 3} \big|W(t_{k+j,n}) - W(t_{k+j-1,n})\big|.$$

The increments in the definition are independent and have the distribution of $W(1/2^n)$, which is the same as that of $2^{-n/2}W(1)$, these being normally distributed with mean zero and variance $1/2^n$. Thus if g denotes the standard normal density, then

$$P(M_{n,k} \le \varepsilon) = P^3\big(|W(1)| \le 2^{n/2}\varepsilon\big) \le \left(\int_{-2^{n/2}\varepsilon}^{2^{n/2}\varepsilon} g(t)\,dt\right)^3 \le (2^{1+n/2}\varepsilon)^3.$$

Setting $M_n = \min_{0\le k<n2^n} M_{n,k}$ we have

$$P(M_n \le \varepsilon) \le \sum_{k=0}^{n2^n-1} P(M_{n,k} \le \varepsilon) \le n2^n(2^{1+n/2}\varepsilon)^3. \tag{‡}$$

Now let $\omega \in E_a$, so that (†) holds for some s and for all $n > n(\omega, s, a)$. We assume that $s > 0$. (A separate, one-sided argument may be given for the case $s = 0$.) Then, for each sufficiently large n, there exists $k > 0$ such that $t_{k+1,n} \le s < t_{k+2,n}$. It follows that, for $0 \le j \le 3$, $|t_{k+j,n} - s| < 1/2^{n-1}$, hence, by (†), $|W(t_{k+j,n},\omega) - W(s,\omega)| \le a/2^{n-1}$. By the triangle inequality, $M_{n,k}(\omega) \le a/2^{n-2}$. Taking $n > s$ we have $k < k+1 \le 2^n s \le n2^n$, hence $M_n(\omega) \le a/2^{n-2}$. We have shown that

$$E_a \subseteq A_a := \varlimsup_n A_n, \quad \text{where} \quad A_n := \{M_n \le a/2^{n-2}\}.$$

By (‡), $P(A_n) \le n2^n(2^{1+n/2}a/2^{n-2})^3 \to 0$. It follows that $P(A_a) = 0$, completing the proof. □

Variation of Brownian Paths

A useful way to quantify the volatile behavior of Brownian motion is by the *variation* of its paths, defined as follows: Let $[a,b]$ be a fixed interval and let

$$\mathcal{P} = \{a = t_0 < t_1 < \cdots < t_n = b\}$$

be a partition of $[a,b]$. For $p > 0$, define the pth **variation of** W **over** $\mathcal{P}$ as the random variable

$$V_{\mathcal{P}}^{(p)} = \sum_{j=1}^{n} |\Delta W(t_j)|^p, \quad \Delta W(t_j) := W(t_j) - W(t_{j-1}).$$

A path $t \mapsto W_t(\omega)$ is said to have **bounded (unbounded)** p**th variation on** $[a,b]$ if the quantities $V_{\mathcal{P}}^{(p)}(\omega)$, taken over all partitions $\mathcal{P}$, form a bounded (unbounded) set of real numbers. By 18.7.3 and 5.5.8 we have

18.7.4 Proposition. *With probability one, the paths of Brownian motion have unbounded first variation on every interval* $[a,b]$.

It may be shown that the paths of Brownian motion have unbounded p variation for all $p \le 2$ [47]. This state of affairs is partially redeemed by the following important result.

18.7.5 Theorem. $\lim_{||\mathcal{P}|| \to 0} V_{\mathcal{P}}^{(2)} = b - a$ *in* $L^2(P)$.

Proof. Given a partition $\mathcal{P}$ as above, define

$$A_{\mathcal{P}} = V_{\mathcal{P}}^{(2)}(W) - (b-a) = \sum_{j=1}^{n} D_j, \quad D_j := (\Delta W_{t_j})^2 - \Delta t_j.$$

By independent increments, for $j \ne k$ we have $E(D_j D_k) = (E\,D_j)(E\,D_k)$, which equals zero since ΔW_k has variance Δt_k. Therefore,

$$E(A_{\mathcal{P}}^2) = \sum_{j,k} E(D_j D_k) = \sum_{j=0}^{n-1} E\,D_j^2 = \sum_{j=0}^{n-1} E(Z_j^2 - 1)^2 (\Delta t_j)^2, \quad Z_j := \frac{\Delta W_{t_j}}{\sqrt{\Delta t_j}}. \tag{18.15}$$

Since Z_j is normal with mean zero and variance one, the quantity $c := E(Z_j^2 - 1)^2$ is finite, as may be verified by expressing c as an integral, using the standard normal density. We now have

$$E(A_{\mathcal{P}}^2) \le c||\mathcal{P}|| \sum_{j-0}^{n-1} \Delta t_j = c||\mathcal{P}||(b-a).$$

Letting $||\mathcal{P}|| \to 0$ forces $E(A_{\mathcal{P}}^2) \to 0$, which is the assertion of the theorem. □

18.7.6 Corollary. *For* $p > 2$, $\lim_{||\mathcal{P}|| \to 0} V_{\mathcal{P}}^{(p)} = 0$ *a.s.*

Proof. By the theorem, for any sequence of partitions with mesh tending to zero there exists a subsequence $(\mathcal{P}_n)$ such that $V_{\mathcal{P}_n}^{(2)} \stackrel{a.s.}{\to} b - a$. The inequality

$$V_{\mathcal{P}_n}^{(p)} = \sum_{j=0}^{n-1} |\Delta W_j|^{p-2} |\Delta W_j|^2 \le \max_j |W_{t_j} - W_{t_{j-1}}|^{p-2} V_{\mathcal{P}_n}^{(2)}$$

and the uniform continuity of the paths of W on $[a,b]$ imply that $\lim_n V_{\mathcal{P}_n}^{(2)} = 0$ a.s. □

The L^2 limit $\lim_{||\mathcal{P}|| \to 0} V_{\mathcal{P}}^{(2)}$ is called the *quadratic variation* of Brownian motion on the interval $[a,b]$. That Brownian motion has nonzero quadratic variation on any interval is a key property of Brownian motion that accounts for some of the differences between stochastic calculus, discussed below, and classical calculus.

Brownian Motion as a Martingale

The definitions of discrete-time martingales carry over in a natural way to the continuous case: Let $(\Omega, \mathcal{F}, (\mathcal{F}_t), P)$ be a filtered probability space and let (X_t) be an L^1 process adapted to $(\mathcal{F}_t)$. Then (X_t) is said to be a

- **supermartingale** if $X_s \ge E(X_t \mid \mathcal{F}_s)$ for all $0 \le s < t$,
- **submartingale** if $X_s \le E(X_t \mid \mathcal{F}_s)$ for all $0 \le s < t$,
- **martingale** if $X_s = E(X_t \mid \mathcal{F}_t)$ for all $0 \le s < t$.

The continuous-time analogs of 18.5.2, 18.5.3, and 18.5.4 hold and are proved as before. A martingale convergence theorem for continuous time is established below.

The following examples are taken relative to the natural (Brownian) filtration $\mathcal{F}^W$.

18.7.7 Examples. **(a)** Brownian motion (W_t) is a martingale. Indeed, since $W_t - W_s$ is independent of $\mathcal{F}_s^W$ for all $s \le t$, $E(W_t - W_s \mid \mathcal{F}_s^W) = E(W_t - W_s) = 0$.

(b) The process $(W_t^2 - t)_{t \ge 0}$ is a martingale: For $0 \le s \le t$ write

$$W_t^2 = [(W_t - W_s) + W_s]^2 = (W_t - W_s)^2 + 2W_s(W_t - W_s) + W_s^2.$$

Taking conditional expectations and using linearity and the factor and independence properties yields

$$E(W_t^2 \mid \mathcal{F}_s^W) = E(W_t - W_s)^2 + 2W_s E(W_t - W_s) + W_s^2 = t - s + W_s^2.$$

(c) The exponential process $\big(\exp(aW_t - a^2 t/2)\big)_{t \ge 0}$ is a martingale. This follows from the calculation (for $t > s$)

$$E(e^{aW_t} \mid \mathcal{F}_s^W) = e^{aW_s} E(e^{aW_t - aW_s} \mid \mathcal{F}_s^W) = e^{aW_s} E(e^{a(W_t - W_s)}) = e^{aW_s + a^2(t-s)/2}.$$

The last equality is seen as follows: Set $\sigma = \sqrt{t-s}$. Then

$$\begin{aligned} E(e^{a(W_t - W_s)}) &= \frac{1}{\sigma\sqrt{2\pi}} \int_{-\infty}^{\infty} \exp\left(ax - \frac{x^2}{2\sigma^2}\right) dx \\ &= \frac{\exp(a^2\sigma^2/2)}{\sigma\sqrt{2\pi}} \int_{-\infty}^{\infty} \exp\left(-\frac{1}{2}\left(\frac{x-a}{\sigma}\right)^2\right) dx \\ &= \exp(a^2\sigma^2/2). \end{aligned}$$ ◊

Here is a continuous time analog of the martingale convergence theorem (18.5.10). As with the latter, Doob's notion of upcrossing figures prominently in the proof.

18.7.8 Martingale Convergence Theorem (Doob). *Let $(X_t)_{t\ge 0}$ be a submartingale on a filtered probability space $(\Omega, \mathcal{F}, (\mathcal{F}_t)_{t>0}, P)$ such that $\sup_{t>0} \|X_t\|_1 < \infty$. Then X_t converges almost surely to an L^1 random variable X_∞ as $t \to \infty$. If, additionally, (X_t) is uniformly integrable, then the convergence is in L^1.*

Proof. For fixed m, let $R =: \{r_1 < \cdots < r_n\}$ be a finite sequence of rationals contained in $[0, m]$ and let $U_{[a,b]}^R$ be the number of upcrossings of $X_{r_1}, \ldots, X_{r_n}$ of $[a, b]$. By the upcrossing lemma,

$$E\,U_{[a,b]}^R \le \frac{1}{b-a} E\big(|X_m| + 2|a|\big).$$

Let $U_{[a,b]}^m$ denote the supremum of $U_{[a,b]}^R$ over all sets R. A sequence (R_k) of such sets increases to $[0,m]\cap\mathbb{Q}$, hence $U_{[a,b]}^{R_k}\uparrow U_{[a,b]}^m$. By the monotone convergence theorem,

$$E\,U_{[a,b]}^m \le \frac{1}{b-a}E\big(|X_m|+2|a|\big) \le \frac{1}{b-a}\Big(\sup_t \|X_t\|_1 + 2|a|\Big) < \infty.$$

Now let $U_{[a,b]} = \sup_m U_{[a,b]}^m$. By the monotone convergence theorem again,

$$E\,U_{[a,b]} \le \frac{1}{b-a}\Big(\sup_t \|X_t\|_1 + 2|a|\Big).$$

In particular, $U_{[a,b]}$ is finite a.s. Set

$$S_{a,b} := \{\omega : \varliminf_{t\to\infty} X_t(\omega) < a < b < \varlimsup_{t\to\infty} X_t(\omega)\} \quad\text{and}\quad S := \bigcup_{a,b\in\mathbb{Q},a<b} S_{a,b}$$

For $\omega\in S_{a,b}$, there exists a strictly increasing sequence (r_n) in $\mathbb{Q}^+$ tending to ∞ such that

$$X_{r_{2n-1}}(\omega) < a < b < X_{r_{2n}}(\omega),$$

implying that $U_{[a,b]}(\omega)=\infty$. Since $U_{[a,b]}$ is finite a.s., $P(S_{a,b})=0$ and so $P(S)=0$. One argues as in the discrete case that $X_\infty(\omega) := \lim_{t\to\infty} X_t(\omega)$ exists in $\overline{\mathbb{R}}$ for each $\omega\in S^c$. Setting $X_\infty = 0$ on S and letting $t\to\infty$ through $\mathbb{Q}$, we may apply Fatou's lemma as in the discrete case to obtain $X_\infty\in L^1(\Omega,\mathcal{F}_\infty,P)$. □

The continuous time analogs of 18.5.11–18.5.14 are valid. The proofs follow from 18.7.8 in much the same way as before. The continuous time notion of reversed martingale may be formulated as in the discrete case, and a martingale convergence theorem may be proved in this setting.

18.8 Stochastic Integration

A subject that has attracted a great deal of interest over recent years is an extension of classical calculus called *stochastic calculus*, loosely described as the calculus of continuous time processes. The basis for this calculus is the Ito integral, which we construct in this section. We assume throughout that W is a Brownian motion on a filtered probability space $(\Omega,\mathcal{F},(\mathcal{F}_t),P)$.

The Ito Integral of a Step Process

An **Ito step process** on a subinterval $[a,b]$ of $[0,\infty)$ is a process of the form

$$f_t(\omega) = f(t,\omega) = \sum_{j=1}^n \xi_{j-1}(\omega)\mathbf{1}_{[t_{j-1},t_j)}(t), \quad t_0 = a < t_1 < \cdots < t_n = b, \tag{18.16}$$

where ξ_{j-1} is $\mathcal{F}_{t_{j-1}}$-measurable and $\xi_{j-1}\in L^2(\Omega)$. The **Ito integral** of f on $[a,b]$ is defined by

$$I_a^b(f) = \int_a^b f(t)\,dW(t) := \sum_{j=1}^n \xi_{j-1}\Delta W(t_j), \quad\text{where } \Delta W(t_j) := W(t_j) - W(t_{j-1}).$$

Note that by refining the partition in (18.16) one still has an Ito step process, and the Ito integral is unchanged. For example, if a point s is inserted into (a, t_1), then

$$f(t,\omega) = \xi_0(\omega)\mathbf{1}_{[a,s)}(t) + \xi_0(\omega)\mathbf{1}_{[s,t_1)}(t) + \sum_{j=2}^{n} \xi_{j-1}(\omega)\mathbf{1}_{[t_{j-1},t_j)}(t).$$

In particular, if $g(t,\omega)$ is another Ito step process, then, by taking the common refinement of the partitions, one may assume that f and g are defined on the same partition. It follows that the collection $\mathscr{S}[a,b]$ of all Ito step processes on $[a,b]$ is a linear space and I_a^b is a linear map on $\mathscr{S}[a,b]$. Moreover, by Fubini's theorem,

$$\|f\|^2_{L^2([a,b]\times\Omega)} = \int_{[a,b]} E(f_t^2)\,dt = \sum_{j=1}^{n}(t_j - t_{j-1})E(\xi_{j-1}^2) < \infty, \tag{18.17}$$

hence $\mathscr{S}[a,b] \subseteq L^2([a,b]\times\Omega)$.

The following proposition shows that $I_a^b : \mathscr{S}[a,b] \to L^2(\Omega)$ is an isometry.

18.8.1 Proposition. *Let f be as in* (18.16). *Then $I_a^b(f)$ has mean zero and variance*

$$\left\|I_a^b(f)\right\|^2_{L^2(\Omega)} = \|f\|^2_{L^2([a,b]\times\Omega)}\,. \tag{18.18}$$

Proof. For the first assertion, we have

$$E(\xi_{j-1}\Delta W(t_j) \mid \mathscr{F}_{t_j-1}) = \xi_{j-1}E(\Delta W(t_j \mid \mathscr{F}_{t_j-1}) = \xi_{j-1}E(\Delta W(t_j)) = 0.$$

Here we have used the independence and factor properties of conditional expectation and the fact that Brownian increments have mean zero. Taking expectations yields

$$E\, I_a^b(f) = \sum_{j=1}^{n} E(\xi_{j-1}\Delta W(t_j)) = 0.$$

To verify (18.18), note that

$$I_a^b(f)^2 = \sum_{i\neq j} \xi_{i-1}\xi_{j-1}\Delta W(t_i)\Delta W(t_j) + \sum_{j=1}^{n} \xi_{j-1}^2[\Delta W(t_j)]^2,$$

hence

$$E[I_a^b(f)]^2 = \sum_{i\neq j} E\big(\xi_{i-1}\xi_{j-1}\Delta W(t_i)\Delta W(t_j)\big) + \sum_{j=1}^{n} E\big(\xi_{j-1}^2[\Delta W(t_j)]^2\big). \tag{α}$$

If $i < j$, then, by conditioning and using the factor and independence properties again, we have

$$\begin{aligned} E\big(\xi_{i-1}\xi_{j-1}\Delta W(t_i)\Delta W(t_j)\big) &= E\big(E(\xi_{i-1}\xi_{j-1}\Delta W(t_i)\Delta W(t_j \mid \mathscr{F}_{t_{j-1}})\big) \\ &= E\big(\xi_{i-1}\xi_{j-1}\Delta W(t_i)E(\Delta W(t_j) \mid \mathscr{F}_{t_{j-1}})\big) \\ &= E\big(\xi_{i-1}\xi_{j-1}\Delta W(t_i)\big)E\big(\Delta W(t_j)\big) \\ &= 0. \end{aligned} \tag{β}$$

Similarly,

$$\begin{aligned} E\big(\xi_{j-1}^2[\Delta W(t_j)]^2\big) &= E\big(E(\xi_{j-1}^2[\Delta W(t_j)]^2 \mid \mathscr{F}_{t_{j-1}})\big) = E\big(\xi_{j-1}^2 E([\Delta W(t_j)]^2 \mid \mathscr{F}_{t_{j-1}})\big) \\ &= E(\xi_{j-1}^2)(t_j - t_{j-1}). \end{aligned} \tag{γ}$$

Thus from (α), (β), (γ)

$$E[I_a^b(f)]^2 = \sum_{j=1}^{n} E(\xi_{j-1}^2)(t_j - t_{j-1}) = \|f\|^2_{L^2([a,b]\times\Omega)},$$

the last equality by (18.17). □

The General Ito Integral

By (18.18), the mapping $I_a^b : \mathscr{S}[a,b] \to L^2(\Omega)$ extends to an isometry on the closure $\operatorname{cl}\mathscr{S}[a,b]$ of $\mathscr{S}[a,b]$ in $L^2([a,b]\times\Omega)$. This defines the **Ito integral** for functions $f \in \operatorname{cl}\mathscr{S}[a,b]$:

$$I_a^b(f) = \int_a^b f\,dW = \int_a^b f(t)\,dW(t) := \lim_n \int_a^b f_n(t)\,dW(t), \quad f \in \operatorname{cl}\mathscr{S}[a,b],$$

where (f_n) is a sequence in $\mathscr{S}[a,b]$ such that $f_n \to f$ in $L^2([a,b]\times\Omega)$, that is,

$$\int_{[a,b]} E[f_n(t) - f(t)]^2\,dt \to 0.$$

By the isometric property,

$$E\left(\int_a^b f\,dW\right)^2 = E[I_a^b(f)]^2 = \int_a^b E[f^2(t)]\,dt, \tag{18.19}$$

hence, by the polarization identity,

$$E\left(\int_a^b f\,dW\right)\left(\int_a^b g\,dW\right) = \int_a^b E[f(t)g(t)]\,dt. \tag{18.20}$$

18.8.2 Proposition. *Let $f \in \operatorname{cl}\mathscr{S}[a,b]$ and $a < c < b$. Then*

$$\int_a^b f(t)\,dW(t) = \int_a^c f(t)\,dW(t) + \int_c^b f(t)\,dW(t).$$

Proof. Assume first that $f \in \mathscr{S}[a,b]$, as given in 18.16. If $c \in [t_k - 1, t_k)$, then

$$\begin{aligned}\int_a^b f(t)\,dW(t) &= \sum_{j<k}^{n} \xi_{j-1}\mathbf{1}_{[t_{j-1},t_j)}(t)\xi_{j-1}\Delta W(t_j) + \xi_{k-1}[W(c) - W(t_{k-1})] \\ &\quad + \xi_{k-1}[W(t_k) - W(c))] + \sum_{j>k}^{n} \xi_{j-1}\mathbf{1}_{[t_{j-1},t_j)}(t)\xi_{j-1}\Delta W(t_j) \\ &= \int_a^c f(t)\,dW(t) + \int_c^b f(t)\,dW(t).\end{aligned}$$

In the general case, let $f_n \in \mathscr{S}[a,b]$ such that $\int_{[a,b]} E(f_n(t) - f(t))^2\,dt \to 0$. Then clearly $f_n\big|_{[a,c]}$ and $f_n\big|_{[c,b]}$ are Ito step functions, and both $\int_{[a,c]} E(f_n(t) - f(t))^2\,dt \to 0$ and $\int_{[c,b]} E(f_n(t) - f(t))^2\,dt \to 0$, hence

$$\int_a^b f\,dW = \lim_n \int_a^b f_n\,dW = \lim_n \int_a^c f_n\,dW + \lim_n \int_c^b f_n\,dW = \int_a^c f\,dW + \int_c^b f\,dW. \quad \square$$

The following proposition shows that in certain circumstances the Ito integral is a limit of Riemann-Stieltjes sums.

18.8.3 Proposition. *Let* $f \in L^2([a,b] \times \Omega)$ *such that* f_t *is* $\mathcal{F}_t$*-measurable and the mapping* $(s,t) \to E(f_s f_t)$ *is continuous. Then* $f \in \operatorname{cl} \mathcal{S}[a,b]$ *and*

$$\int_a^b f(t)\,dW(t) = \lim_{\|\mathcal{P}\| \to 0} \sum_{j=1}^n f(t_{j-1})\Delta W(t_j), \quad \textit{where } \mathcal{P} := \{a = t_0 < t_1 < \cdots < t_n\}.$$

Proof. Define an Ito step process $f_{\mathcal{P}}$ by $f_{\mathcal{P}}(t,\omega) = f(t_{j-1},\omega)$ $(t_{j-1} < t \le t_j)$, where the $(t_{j-1}, t_j]$ are the intervals of the partition $\mathcal{P}$. Let $(\mathcal{P}_n)$ be any sequence of partitions with $\|\mathcal{P}_n\| \to 0$ and set $f_n := f_{\mathcal{P}_n}$. From the calculation

$$E[f(t) - f(s)]^2 = E\,f(t)^2 - 2E[f(s)f(t)] + E\,f(s)^2$$

we see that $\lim_{s \to t} E[f(t) - f(s)]^2 = 0$. Since $\|\mathcal{P}_n\| \to 0$ it follows that

$$\lim_n E[f(t) - f_n(t)]^2 = 0.$$

From the inequality

$$|f(t) - f_n(t)|^2 \le 2|f(t)|^2 + 2|f_n(t)|^2$$

we have

$$E\,|f(t) - f_n(t)|^2 \le 2E\,|f(t)|^2 + 2E\,|f_n(t)|^2 \le 4 \sup_{a \le s \le b} E\,|f(s)|^2.$$

By continuity the supremum is finite, so we may apply the dominated convergence theorem to conclude that

$$\lim_n \int_a^b E\,|f(t) - f_n(t)|^2\,dt = 0,$$

that is, $f_n \to f$ in $L^2([a,b] \times \Omega)$. Therefore, $f \in \operatorname{cl} \mathcal{S}[a,b]$ and $\int_a^b f_n\,dW \to \int_a^b f\,dW$. Since the sequence $(\mathcal{P}_n)$ was arbitrary, the conclusion follows. □

18.8.4 Example. Let $\mathcal{P} = \{a = t_0 < t_1 < \cdots < t_n = b\}$ be an arbitrary partition of $[a,b]$. By direct expansion

$$\sum_{j=0}^{n-1} W(t_{j-1})\Delta W_{t_j} = \tfrac{1}{2}\big[W^2(b) - W^2(a) - V_{\mathcal{P}}^{(2)}(W)\big].$$

It follows from 18.8.3 and 18.7.5 that

$$\int_a^b W(t)\,dW(t) = \frac{W^2(b) - W^2(a)}{2} - \frac{b-a}{2}. \qquad \Diamond$$

The Ito Integral as a Martingale

18.8.5 Theorem. *Let* $f \in \operatorname{cl} \mathcal{S}[a,b]$. *Then the process* $X_t = \int_a^t f\,dW$ *is a martingale with respect to the filtration* $(\mathcal{F}_t)$.

Proof. Let $b \ge t > s \ge a$. Since $X_t - X_s = \int_s^t f\,dW$, we need to show that

$$E\left(\int_s^t f\,dW \,\middle|\, \mathcal{F}_s\right) = 0.$$

Assume first that f is an Ito step process, say

$$f_u(\omega) = \sum_{j=1}^{n} \xi_{j-1}(\omega)\mathbf{1}_{[t_{j-1},t_j)}(u), \quad s = t_0 < t_1 < \cdots < t_n = t.$$

Then

$$\begin{aligned} E\left(\int_s^t f\,dW \,\middle|\, \mathcal{F}_s\right) &= \sum_{j=1}^{n} E(\xi_{j-1}\Delta W(t_j) \mid \mathcal{F}_s) \\ &= \sum_{j=1}^{n} E\big(E(\xi_{j-1}\Delta W(t_j) \mid \mathcal{F}_{t_j-1}) \mid \mathcal{F}_s\big) \\ &= \sum_{j=1}^{n} E\big(\xi_{j-1}E(\Delta W(t_j) \mid \mathcal{F}_{t_j-1}) \mid \mathcal{F}_s\big). \end{aligned}$$

The last sum is zero since, as noted earlier, $E(\Delta W(t_j) \mid \mathcal{F}_{t_j-1}) = 0$.

For a general f, let $f_n \in \mathcal{S}[s,t]$ such that $\int_s^t E\,|f_n(u) - f(u)|^2\,du \to 0$. By the first paragraph,

$$E\left(\int_s^t f\,dW \,\middle|\, \mathcal{F}_s\right) = E\left(\int_s^t (f - f_n)\,dW \,\middle|\, \mathcal{F}_s\right). \tag{$\dagger$}$$

Now, by Jensen's inequality,

$$\left|E\left(\int_s^t (f - f_n)\,dW \,\middle|\, \mathcal{F}_s\right)\right|^2 \le E\left(\left|\int_s^t (f - f_n)\,dW\right|^2 \,\middle|\, \mathcal{F}_s\right).$$

Taking expectations we have

$$E\left|E\left(\int_s^t (f - f_n)\,dW \,\middle|\, \mathcal{F}_s\right)\right|^2 \le E\left|\int_s^t (f - f_n)\,dW\right|^2 = \int_s^t E|f - f_n|^2\,dW,$$

the last equality by (18.19). Thus

$$E\left(\int_s^t (f - f_n)\,dW \,\middle|\, \mathcal{F}_s\right) \to 0 \text{ in } L^2,$$

hence a subsequence converges a.s. to zero. It follows from (†) that $E\left(\int_s^t f\,dW \,\middle|\, \mathcal{F}_s\right) = 0$ as required. □

It may be shown that almost all paths of the integral process X are continuous. (See, for example, [29].)

18.9 An Application to Finance

In this section we outline the argument that leads to the Black-Scholes formula for the price of a call option. For details the reader is referred to [24] or [43].

The Stock Price Process

Let W be a Brownian motion on a filtered probability space $(\Omega, \mathcal{F}, (\mathcal{F}_t), P)$, where we take $(\mathcal{F}_t)$ to be the natural filtration for W. In the Black-Scholes-Merton model, the price (in dollars) of a single share of a stock at time t is assumed to be a random variable S_t satisfying the stochastic integral equation

$$S_t = S_0 + \sigma \int_0^t S(s)\,dW(s) + \mu \int_0^t S(s)\,ds. \tag{18.21}$$

Here σ and μ are constants called, respectively, the *volatility* and *drift* of the stock. The integral equation is frequently written as a **stochastic differential equation**

$$dS = \sigma S\,dW + \mu S\,dt \quad \text{or} \quad \frac{dS}{S} = \sigma\,dW + \mu\,dt. \tag{18.22}$$

The latter form expresses the fact that the relative change in the stock price has a deterministic part $\mu\,dt$, which accounts for the general trend of the stock, and a component $\sigma\,dW$, which reflects the random nature of the stock.

The solution of (18.21) may be shown to be the **geometric Brownian motion process**

$$S_t = S_0 \exp\left[\sigma W_t + \left(\mu - \tfrac{1}{2}\sigma^2\right)t\right]. \tag{18.23}$$

Note that because of the relationship between S_t and W_t, $\mathcal{F}_t = \sigma(S_s : 0 \leq s \leq t)$. Thus the Brownian filtration $(\mathcal{F}_t)$ reveals stock price information. We show how these facts lead to a formula for the price of an option.

Self-Financing Portfolios

The key to determining the value of an option is the construction of a self-financing portfolio based on the stock and a risk-free bond. Assuming that the bond earns interest at a continuously compounded annual rate r and that the initial value of the bond is one dollar, the value of the bond at time t is seen to be $B_t := e^{rt}$. Now let ϕ and θ be stochastic process adapted to the filtration $(\mathcal{F}_t)$, these representing, respectively, the number of dollar bonds and number of shares of the stock held at time t. The value of the portfolio at time t is the random variable

$$V_t = \phi_t B_t + \theta_t S_t, \quad 0 \leq t \leq T,$$

where V_0 is the initial investment in the portfolio, assumed to be a constant. The portfolio is said to be **self-financing** if

$$dV = \phi\,dB + \theta\,dS, \tag{18.24}$$

where the differentials represent small changes. The equation may be best understood by considering a discrete version at times $t_0 = 0 < t_1 < t_2 < \cdots < t_n = T$. At time t_j, the value of the portfolio before the price S_j is known is

$$\phi_j B_{j-1} + \theta_j S_{j-1},$$

where we write S_j for S_{t_j}, etc. After S_j becomes known and the new bond value B_j is noted, the portfolio has value

$$V_j = \phi_j B_j + \theta_j S_j.$$

At this time, stocks and bonds may be bought and sold (based on the information provided by $\mathcal{F}_{t_j}$). For the portfolio to be self-financing, this rebalancing must not change the current value of the portfolio. Thus the new values ϕ_{j+1} and θ_{j+1} must satisfy

$$\phi_{j+1} B_j + \theta_{j+1} S_j = \phi_j B_j + \theta_j S_j.$$

It follows that

$$\begin{aligned}\Delta V_j &= \phi_{j+1}B_j + \theta_{j+1}S_j - (\phi_j B_j + \theta_j S_j)\\ &= \phi_{j+1}B_{j+1} + \theta_{j+1}S_{j+1} - (\phi_{j+1}B_j + \theta_{j+1}S_j)\\ &= \phi_{j+1}\Delta B_j + \theta_{j+1}\Delta S_j,\end{aligned}$$

which is the discrete version of (18.24).

Call Options

A **call option** based on a stock is a contract made between two parties, the buyer (holder) of the option and the seller (writer) of the option. The contract requires the writer to offer to sell the stock to the holder at a future time T for a predetermined amount K. At this time, the holder may or may not decide to exercise the option. Thus the payoff for the holder is $(S_T - K)^+$. A self-financing portfolio may be used by the writer as a hedging strategy, that is, an investment in shares of the stock and units of the bond devised to exactly cover the writer's obligation at maturity T. In this case, the portfolio is said to *replicate* the option. The writer initiates the portfolio with an amount V_0, the price of the option (cost to the holder). Here, V_0 is chosen so that $V_T = (S_T - K)^+$, which is the cost to the writer of the transaction. The law of one price (in an arbitrage-free market) then asserts that V_0 is the fair price of the option.

The Black-Scholes Option Price

To determine the fair price V_0 of the option, one introduces a new probability measure P^* on $(\Omega, \mathscr{F})$, called the **risk-neutral probability measure**, defined by

$$dP^* = Z_T\, dP, \quad \text{where } Z_T := \exp\left(-\alpha W_T - \tfrac{1}{2}\alpha^2 T\right) \text{ and } \alpha := \frac{\mu - r}{\sigma}.$$

The corresponding expectation operator is denoted by E^*. It may be shown that the process

$$W_t^* := W_t + \alpha t, \quad 0 \le t \le T, \quad \alpha := \frac{\mu - r}{\sigma},$$

is a Brownian motion under P^* on the interval $[0, T]$. By (18.23),

$$S_t = S_0 \exp\left(\sigma W_t^* + \left(r - \tfrac{1}{2}\sigma^2\right) t\right), \quad 0 \le t \le T. \tag{18.25}$$

Now form the **discounted price process** $\widetilde{S}$, given by

$$\widetilde{S}_t := e^{-rt} S_t = S_0 \exp\left(\sigma W_t^* - \tfrac{1}{2}\sigma^2 t\right), \quad 0 \le t \le T.$$

By 18.7.7(c), $\widetilde{S}_t$ is a P^*-martingale. One may show, as a consequence, that the **discounted value process** $\widetilde{V}$, given by

$$\widetilde{V}_t := e^{-rt} V_t,$$

is also a P^*-martingale. This implies the key fact $E^* \widetilde{V}_t$ is constant in t. In particular,

$$V_0 = E^* V_0 = E^* \widetilde{V}_T = e^{-rT} E^* V_T.$$

Since the portfolio value V_T is assumed to be the payoff to the holder of the option,

$$V_0 = e^{-rT} E^* (S_T - K)^+. \tag{18.26}$$

Now use (18.25) to write

$$S_T = S_0 \exp\left(\sigma\sqrt{T}\,Y + (r - \tfrac{1}{2}\sigma^2)T\right), \quad Y := T^{-1/2}W_T^*.$$

Since Y is a standard normal random variable under P^*,

$$E^*(S_T - K)^+ = \int_{-\infty}^{\infty} \left(S_0 \exp\left\{\sigma\sqrt{T}\,y + (r - \tfrac{1}{2}\sigma^2)T\right\} - K\right)^+ \varphi(y)\,dy, \tag{18.27}$$

where φ is the standard normal density. From (18.26) and (18.27) we see that the price of the option is given by the formula

$$V_0 = e^{-rT} \int_{-\infty}^{\infty} \left(S_0 \exp\left\{\sigma\sqrt{T}\,y + (r - \tfrac{1}{2}\sigma^2)T\right\} - K\right)^+ \varphi(y)\,dy. \tag{18.28}$$

A more succinct formula for the option price may be obtained as follows. Define

$$d_1 := \frac{\ln(S_0/K) + (r + \frac{1}{2}\sigma^2)T}{\sigma\sqrt{T}} \quad \text{and} \quad d_2 := \frac{\ln(S_0/K) + (r - \frac{1}{2}\sigma^2)T}{\sigma\sqrt{T}} = d_1 - \sigma\sqrt{T}.$$

Since the integrand in (18.28) is zero when $y < -d_2$, we may write the integral as

$$\begin{aligned} S_0 &\int_{-d_2}^{\infty} \exp\left\{\sigma\sqrt{T}\,y + \left(r - \tfrac{1}{2}\sigma^2\right)T\right\}\varphi(y)\,dy - K\int_{-d_2}^{\infty} \varphi(y)\,dy \\ &= S_0 \frac{e^{(r-\sigma^2/2)T}}{\sqrt{2\pi}} \int_{-d_2}^{\infty} \exp\left\{-\tfrac{1}{2}y^2 + \sigma\sqrt{T}\,y\right\} dy - K\left[1 - \Phi(-d_2)\right] \\ &= \quad S_0 e^{rT}\Phi(d_1) - K\Phi(d_2), \end{aligned}$$

where $\Phi(x) = \int_{-\infty}^{x} \varphi(y)\,dy$. Using (18.28), we finally arrive at the celebrated *Black-Scholes option pricing formula*

$$V_0 = S_0\Phi(d_1) - Ke^{-rT}\Phi(d_2).$$

Part IV

Appendices

Appendix A

Change of Variables Theorem

The goal of this appendix is prove the following result.

A.0.1 Change of Variables Theorem. *Let $U, V \subseteq \mathbb{R}^d$ be open and let $\varphi : U \to V$ be C^1 with C^1 inverse $\varphi^{-1} : V \to U$. If $f : V \to \mathbb{R}$ is Lebesgue measurable and either $f \geq 0$ or f is integrable, then*

$$\int_V f(y)\,dy = \int_U (f \circ \varphi)(x)|J_\varphi(x)|\,dx, \tag{A.1}$$

where J_φ is the Jacobian of φ on U.

We prove the theorem first for Borel functions. By the usual considerations, we may assume that $f \geq 0$. It then suffices to prove that

$$\int_V f\,d\lambda^d \leq \int_U (f \circ \varphi)|J_\varphi|\,d\lambda^d \tag{A.2}$$

for all Borel measurable functions $f : V :\to [0, +\infty]$. Indeed, if this inequality holds for all f and φ, then switching the roles of U and V we also have

$$\int_U g\,d\lambda^d \leq \int_V (g \circ \varphi^{-1})|J_{\varphi^{-1}}|\,d\lambda^d$$

for all Borel measurable $g : U :\to [0, +\infty]$. Taking $g = (f \circ \varphi)|J_\varphi|$ and recalling that $J_\varphi J_{\varphi^{-1}} = 1$, we obtain the reverse of inequality (A.2). Finally, by the standard arguments, it suffices to verify (A.2) for indicator functions $f = \mathbf{1}_B$, where $B \in \mathfrak{B}(V)$. Then (A.2) reduces to

$$\lambda^d(B) \leq \int_{\varphi^{-1}(B)} |J_\varphi|\,d\lambda^d, \quad B \in \mathfrak{B}(V).$$

Taking $B = \varphi(E)$ we obtain the equivalent statement

$$\lambda^d\big(\varphi(E)\big) \leq \int_E |J_\varphi|\,d\lambda^d, \quad E \in \mathfrak{B}(U). \tag{A.3}$$

The proof of (A.3) is accomplished by a sequence of lemmas. The first treats the case of a linear change of variable.

A.0.2 Lemma. *If $T : \mathbb{R}^d \to \mathbb{R}^d$ is linear and nonsingular, then*

$$\lambda^d(T(E)) = |\det T|\lambda^d(E), \quad E \in \mathfrak{B}(\mathbb{R}^d). \tag{A.4}$$

Proof. Since T is a homeomorphism, $T(E) \in \mathfrak{B}(\mathbb{R}^d)$, so the left side of (A.4) is defined. Furthermore, if (A.4) holds for T_1 and T_2, then

$$\lambda^d\big(T_1T_2(E)\big) = |\det T_1|\lambda^d\big(T_2(E)\big) = |\det T_1|\,|\det T_2|\lambda^d(E) = |\det(T_1T_2)|\lambda^d(E).$$

Since T is a product of elementary linear transformations, we may therefore assume that T is such a transformation. Thus we assume that the matrix of T is obtained from the identity matrix by one of the following operations:

(a) Interchange of two rows.

(b) Multiplication of a row by a nonzero constant.

(c) Addition of one row to another.

To prove (A.4) in this setting, suppose first that $E = I_1 \times \cdots \times I_d$ is a bounded d-dimensional interval. In case (a), $\det T = -1$ and $T(E)$ is the interval obtained from E by interchanging a pair of intervals I_i and I_j, hence (A.4) holds in this case. In (b), $T(E)$ is the interval obtained from E by multiplying one of the I_j by a nonzero constant a, hence $\lambda^d(T(E)) = |a|\lambda^d(E)$. Since $|\det T| = |a|$, (A.4) holds in this case as well. For case (c), suppose, for example, that the matrix of T is obtained by adding row two of the identity matrix to row one. Then

$$T(x_1, x_2, x_3, \ldots, x_n) = (x_1 + x_2, x_2, x_3, \ldots, x_n),$$

hence, by Fubini's theorem and translation invariance,

$$\begin{aligned}
\lambda^d(T(E)) &= \int \mathbf{1}_{T(E)}(x)\,dx = \int \mathbf{1}_E(x_1 - x_2, x_2, \ldots, x_n)\,dx \\
&= \iint \cdots \int \mathbf{1}_{I_1}(x_1 - x_2)\mathbf{1}_{I_2}(x_2)\cdots \mathbf{1}_{I_n}(x_n)\,dx_n \cdots dx_2\,dx_1 \\
&= |I_n|\cdots|I_3| \int \mathbf{1}_{I_2}(x_2) \int \mathbf{1}_{I_1}(x_1 - x_2)\,dx_1\,dx_2 \\
&= |I_n|\cdots|I_3|\,|I_2|\,|I_1| \\
&= \lambda^d(E).
\end{aligned}$$

Since $\det T = 1$, (A.4) holds in case (c). Therefore (A.4) holds for all nonsingular T and all bounded intervals E.

Now let I be a fixed bounded interval and let $\mathcal{G}_I$ denote the collection of all $E \in \mathcal{B}(\mathbb{R}^d)$ for which

$$\lambda^d(T(E \cap I)) = |\det T|\lambda^d(E \cap I). \qquad (\dagger)$$

By the first part of the proof, $\mathcal{G}_I$ contains the collection $\mathcal{I}$ all intervals of $\mathbb{R}^d$. We show that $\mathcal{G}_I$ is a λ-system (see 1.5). Let A, $B \in \mathcal{G}_I$ with $A \subseteq B$, and set $C = A \cap I$ and $D = B \cap I$. Then $(B \setminus A) \cap I = D \setminus C$ and

$$\lambda^d(T(D \setminus C)) = \lambda^d(T(D)) - \lambda^d(T(C)) = |\det T|(\lambda^d(D) - \lambda^d(C)) = |\det T|\lambda^d(D \setminus C),$$

hence $B \setminus A \in \mathcal{G}_I$. Now let $A_k \in \mathcal{G}_I$, $A_k \uparrow A$. Letting $k \to +\infty$ in

$$\lambda^d(T(A_k \cap I)) = |\det T|\lambda^d(A_k \cap I)$$

we see that $A \in \mathcal{G}_I$. Therefore, $\mathcal{G}_I$ is a λ-system. By Dynkin's theorem (1.2.6), $\mathcal{G}_I$ contains $\sigma(\mathcal{I}) = \mathcal{B}(\mathbb{R}^d)$. Thus ($\dagger$) holds for every $E \in \mathcal{B}(\mathbb{R}^d)$. Taking a sequence of bounded intervals I in ($\dagger$) increasing to $\mathbb{R}^d$ we obtain (A.4). □

For the next lemma, recall that $df_x : \mathbb{R}^d \to \mathbb{R}^d$ denotes the differential of a function $f : U \to \mathbb{R}^d$ at x, that is, the linear operator whose matrix is the Jacobian matrix of f evaluated at x.

A.0.3 Lemma. *Let $f : U \to \mathbb{R}^d$ be C^1 and let $K \subseteq U$ be compact and convex. Then $M := \sup_{z \in K} \|df_z\| < \infty$ and $\|f(x) - f(y)\| \leq M\|x - y\|$ for all x, $y \in K$.*

Proof. Since $z \mapsto df_z$ is continuous and K is compact, $M < \infty$. Let $x, y \in K$ and $u \in \mathbb{R}^d$ and set $\alpha(t) := tx + (1-t)y$. By the mean value theorem applied to the scalar function $g := u \cdot f \circ \alpha$, there exists a point $z = \phi(t) \in [x : y] \subseteq K$ such that

$$u \cdot \big[f(x) - f(y)\big] = g(1) - g(0) = g'(t) = u \cdot df_z(x-y).$$

Taking $u = f(x) - f(y)$ and using the CBS and operator norm inequalities, we have

$$|f(x) - f(y)|^2 = \big[f(x) - f(y)\big] \cdot \big[df_c(x-y)\big] \le M|f(x) - f(y)|\,|x-y|. \qquad \square$$

For the remaining lemmas, we use the following terminology and notation: The *cube with center $y \in \mathbb{R}^d$ and edge $r > 0$* is the half-closed interval

$$Q = Q_r(y) := \{x \in \mathbb{R}^d : y_j - r/2 < x_j \le y_j + r/2,\ j = 1, \ldots, d\}.$$

Note that $\lambda(Q) = r^d$ and that the diameter of Q is $r\sqrt{d}$. Thus

$$B_{r/2}(y) \subseteq Q_r(y) \subseteq C_{r\sqrt{d}/2}(y). \tag{A.5}$$

A.0.4 Lemma. *Let ψ be C^1 on U, Q a cube contained in U, and I_d the identity operator on $\mathbb{R}^d$. If $\|d\psi_x - I_d\| \le c$ for all $x \in Q$, then $\lambda^d\big(\psi(Q)\big) \le [(1+c)d]^d \lambda^d(Q)$.*

Proof. Let $\widetilde{\psi}(x) = \psi(x) - x$, so that $d\widetilde{\psi}_x = d\psi_x - I_d$. By A.0.3, for suitable $c > 0$,

$$\|\widetilde{\psi}(x) - \widetilde{\psi}(y)\| \le c\|x - y\| \quad \text{for all } x,\ y \in Q.$$

Thus, if Q has center x_0 and edge r, then recalling (A.5) we have for all $x \in Q$,

$$\|\psi(x) - \psi(x_0)\| \le \|\widetilde{\psi}(x) - \widetilde{\psi}(x_0)\| + \|x - x_0\| \le (c+1)\|x - x_0\| \le \tfrac{1}{2}(c+1)r\sqrt{d}.$$

Thus $\psi(Q)$ is contained in the closed ball C with center $\psi(x_0)$ and radius $\frac{1}{2}(c+1)r\sqrt{d}$. Since C is contained in the cube with center $\psi(x_0)$ and edge $(c+1)dr$, we have

$$\lambda^d\big(\psi(Q)\big) \le [(c+1)dr]^d = [(c+1)d]^d \lambda^d(Q). \qquad \square$$

We call a finite collection $\mathcal{Q}_r$ of pairwise disjoint cubes with edge r that covers a subset A of $\mathbb{R}^d$ a **paving** of A. Pavings $\mathcal{Q}_r = \{Q_r(x_j) : 1 \le j \le m\}$ and $\mathcal{Q}_s = \{Q_s(x_j) : 1 \le j \le m\}$ with the same centers are said to be **concentric**. Clearly, any bounded set has a paving $\mathcal{Q}_r$ with arbitrarily small r.

A.0.5 Lemma. *Let $K \subseteq U$ be compact. Then, for all sufficiently small δ and each $0 < r < \delta$, there exists a compact set K_δ and a paving $\mathcal{Q}_r$ of K with $K \subseteq \bigcup \mathcal{Q}_r \subseteq K_\delta \subseteq U$.*

Proof. Since K is compact and U^c is closed, $d(U^c, K) > 0$. For $0 < \delta < d(U^c, K)/\sqrt{d}$, let

$$K_\delta = \{x : d(x, K) \le \delta\sqrt{d}\}.$$

Then K_δ is compact and $K \subseteq K_\delta \subseteq U$. Let $0 < r < \delta$ and let Q be a cube with edge r. If $x \in Q \cap K$ and $y \in Q \cap K_\delta^c$, then

$$\delta\sqrt{d} < d(y, K) \le |x - y| \le r\sqrt{d}.$$

Therefore, if $r < \delta$ and $Q \cap K \ne \emptyset$, then $Q \cap K_\delta^c = \emptyset$, that is, $Q \subseteq K_\delta$. Since K is bounded, there exists a paving $\mathcal{Q}_r$ of K. Removing those members of $\mathcal{Q}_r$ that do not meet K produces a paving of K contained in K_δ. $\square$

A.0.6 Lemma. *Let $\varphi : U \to \mathbb{R}^d$ be C^1 on U and let $K \subseteq U$ be compact. Then for each $\varepsilon > 0$ there exists $\delta > 0$, a compact set K_δ with $K \subseteq K_\delta \subseteq U$, and concentric pavings $\mathcal{Q}_r$, $\mathcal{Q}_{dr}$ of K contained in K_δ with arbitrarily small r such that for any $Q_r(y) \in \mathcal{Q}_r$,*

$$\lambda^d\Big(\varphi\big(Q_r(y)\big)\Big) \le (1+\varepsilon)^d |J_\varphi(y)| \lambda^d\big(Q_{dr}(y)\big). \tag{A.6}$$

Moreover, δ may be chosen so that

$$\int_{K_\delta} |J_\varphi(x)|\,dx < \int_K |J_\varphi(x)|\,dx + \varepsilon. \tag{A.7}$$

Proof. Let $M = \sup\{\|(d\varphi_y)^{-1}\| : y \in K_\delta\}$, where K_δ is chosen as in A.0.5. For $x,\ y \in U$ define

$$\psi^y(x) = \big(d\varphi_y\big)^{-1}\big(\varphi(x) - \varphi(y)\big) = \big(d\varphi_y\big)^{-1}\big(\varphi(x)\big) - \big(d\varphi_y\big)^{-1}\big(\varphi(y)\big). \tag{a}$$

Since $\big(d\varphi_y\big)^{-1}$ is linear, by the chain rule

$$d(\psi^y)_x = (d\varphi_y)^{-1} \circ d\varphi_x.$$

Thus for all $x \in U$, $y \in K_\delta$, and $\mathbf{z} \in \mathbb{R}^d$,

$$\|d(\psi^y)_x(\mathbf{z}) - \mathbf{z}\| = \Big\|\big(d\varphi_y\big)^{-1}\big(d\varphi_x(\mathbf{z}) - d\varphi_y(\mathbf{z})\big)\Big\| \le M\|d\varphi_x - d\varphi_y\|\,\|\mathbf{z}\|$$

and so, by definition of the operator norm,

$$\|d(\psi^y)_x - I_d\| \le M\|d\varphi_x - d\varphi_y\|. \tag{b}$$

By the uniform continuity of $d\varphi$ on K_δ we may choose $0 < \delta_1 < \delta$ such that

$$\|d\varphi_x - d\varphi_y\| \le \varepsilon/M \quad \forall\ x, y \in K_\delta \text{ with } \|x - y\| < \delta_1\sqrt{d}. \tag{c}$$

Let $r < \delta_1/d$, and by A.0.5 let $\mathcal{Q}_r$, $\mathcal{Q}_{dr}$ be concentric pavings of K contained in K_δ. If $x \in Q := Q_r(y) \in \mathcal{Q}_r$, then $\|x-y\| < r\sqrt{d} < \delta_1\sqrt{d}$, hence, by (b) and (c), $\|d(\psi^y)_x - I_d\| < \varepsilon$. Applying A.0.4 we have

$$\lambda^d\big(\psi^y(Q)\big) \le [(1+\varepsilon)d]^d\lambda^d(Q) = (1+\varepsilon)^d\lambda^d\big(Q_{dr}(y)\big). \tag{d}$$

But by (a), $\psi^y(Q) = \big(d\varphi_y)\big)^{-1}\big(\varphi(Q)\big) - \big(d\varphi_y\big)^{-1}\big(\varphi(y)\big)$, hence, by translation invariance and A.0.2,

$$\lambda^d\big(\psi^y(Q)\big) = \lambda^d\big[\big(d\varphi_y\big)^{-1}\big(\varphi(Q)\big)\big] = |J_\varphi(y)|^{-1}\lambda^d\big(\varphi(Q)\big). \tag{e}$$

Inequality (A.6) now follows from (d) and (e).

For (A.7), note that since $K_{1/n} \downarrow K$ and $\mu(A) := \int_A |J_\varphi|\,d\lambda^d$ is a measure on the Borel sets, $\mu\big(K_{1/n}\big) \downarrow \mu(K)$. Thus there exists k such that $\mu\big(K_{1/n}\big) < \mu(K) + \varepsilon$. Taking $\delta < 1/n$ completes the proof. □

A.0.7 Lemma. *If $K \subseteq U$ is compact, then*

$$\lambda^d\big(\varphi(K)\big) \le \int_K |J_\varphi(y)|\,dy.$$

Proof. Let $\varepsilon > 0$ and choose $\delta > 0$ as in A.0.6. By uniform continuity of $J_\varphi(x)$ on K_δ, there exists $\delta_1 < \delta$ such that

$$|J_\varphi(x) - J_\varphi(y)| < \varepsilon \quad \forall\ x,\ y \in K_\delta \text{ with } \|x - y\| < \delta_1.$$

Choose pavings $\mathcal{Q}_r = \{Q_r(y)\}_y$ and $\mathcal{Q}_{dr} = \{Q_{dr}(y)\}_y$ as in A.0.6. For $x \in Q_{dr}(y)$ we have $|J_\varphi(y)| \le |J_\varphi(x) - J_\varphi(y)| + |J_\varphi(x)| < \varepsilon + |J_\varphi(x)|$, hence, applying (A.6),

$$(1+\varepsilon)^{-d}\lambda^d\big(\varphi(Q_r(y))\big) \le |J_\varphi(y)|\lambda^d(Q_{dr}(y)) \le \int_{Q_{dr}(y)} \big(|J_\varphi(x)| + \varepsilon\big)\, dx.$$

Therefore,

$$\begin{aligned}(1+\varepsilon)^{-d}\lambda^d\big(\varphi(K)\big) &\le \sum_y (1+\varepsilon)^{-d}\lambda^d\big(\varphi(Q_r(y))\big) \le \int_{K_\delta} \big(|J_\varphi(x)| + \varepsilon\big)\, dx \\ &\le \int_K |J_\varphi(x)|\, dx + \varepsilon\big(1 + \lambda^d(K_\delta)\big),\end{aligned}$$

the last inequality by (A.7). Letting $\varepsilon \to 0$ gives the desired inequality. □

To prove (A.3), use regularity to obtain an increasing sequence of compact sets $K_n \subseteq E$ such that $\lambda^d(K_n) \uparrow \lambda^d(E)$. Then $\lambda^d\big(\varphi(K_n)\big) \uparrow \lambda^d\big(\varphi(E)\big)$, hence by A.0.7 we obtain

$$\lambda^d\big(\varphi(E)\big) = \lim_n \lambda^d\big(\varphi(K_n)\big) \le \varliminf_n \int_{K_n} |J_\varphi(y)|\, dy \le \int_E |J_\varphi(y)|\, dy,$$

as required. This completes the proof of the change of variables theorem for the case f Borel.

Now let $f \ge 0$ be Lebesgue measurable on V. Then $f = g$ on $V \setminus E$, where $g \ge 0$ is Borel measurable, $E \subseteq V$, and $\lambda^d(E) = 0$. By the first part of the proof,

$$\int_V g(y)\, dy = \int_U (g \circ \varphi)(x)|J_\varphi(x)|\, dx.$$

But the left side equals $\int_V f(y)\, dy$, and since $f \circ \varphi = g \circ \varphi$ on $U \setminus \varphi^{-1}(E)$ the right side equals $\int_U (f \circ \varphi)(x)|J_\varphi(x)|\, dx$ provided we can show that

$$\lambda^d(\varphi^{-1}(E)) = 0.$$

To verify this, suppose first that E is bounded. Then $E \subseteq K$ for a compact interval K with $\lambda^d(K)$ arbitrarily small. Applying A.0.1 "in reverse," we have

$$\int_U h\, d\lambda^d = \int_V (h \circ \varphi^{-1})|J_{\varphi^{-1}}|\, d\lambda^d$$

for Borel functions $h \ge 0$ on U. Taking $h = \mathbf{1}_{\varphi^{-1}(K)}$ yields

$$\lambda^d(\varphi^{-1}(E)) \le \lambda^d(\varphi^{-1}(K)) = \int_V \mathbf{1}_{\varphi^{-1}(K)} \circ \varphi^{-1})|J_{\varphi^{-1}}|\, d\lambda^d = \int_K |J_{\varphi^{-1}}|\, d\lambda^d.$$

Since the right side may be made arbitrarily small, $\lambda^d(\varphi^{-1}(E)) = 0$. If E is unbounded, take a sequence of bounded set E_n of measure zero with $E_n \uparrow E$. □

Appendix B

Separate and Joint Continuity

In this appendix we prove the following theorem, which is used in Chapter 17 to establish joint continuity of multiplication in certain algebraic structures.

B.0.8 Theorem. *Let X and Y be topological spaces with X locally compact or a complete metric space and Y compact Hausdorff. If $f : X \times Y \to \mathbb{C}$ is bounded and separately continuous, then there exists a dense G_δ subset A of X such that f is jointly continuous at every point of $A \times Y$.*

The proof is based on the following lemmas. For these, we assume the hypotheses of the theorem, except we allow X to be an arbitrary topological space. We shall need the functions $F : X \to C(Y)$ and $G : X \to \mathbb{R}^+$ defined by

$$F(x) = f(x, \cdot) \text{ and } G(x) = \inf_U \sup\{\|F(x') - F(x'')\|_\infty : x' x'' \in U\},$$

where the infimum is taken over all neighborhoods U of x.

B.0.9 Lemma. *The function f is jointly continuous at every point of $\{x\} \times Y$ iff F is norm continuous at x.*

Proof. If F is not norm continuous at x, then there exists an $\varepsilon > 0$ and nets (y_α) and (x_α) with $x_\alpha \to x$ such that $|f(x_\alpha, y_\alpha) - f(x, y_\alpha)| \geq 2\varepsilon$ for all α. By compactness of Y, we may assume that $y_\alpha \to y \in Y$. Then, eventually, $|f(x, y_\alpha) - f(x, y| < \varepsilon$ and so

$$|f(x_\alpha, y_\alpha) - f(x, y)| \geq |f(x_\alpha, y_\alpha) - f(x, y_\alpha)| - |f(x, y_\alpha) - f(x, y| > \varepsilon.$$

Therefore, F is not jointly continuous at (x, y).

Now assume that F is norm continuous at x and let $x_\alpha \to x$ and $y_\alpha \to y$. Then from

$$\begin{aligned}|f(x_\alpha, y_\alpha) - f(x, y)| &\leq |f(x_\alpha, y_\alpha) - f(x, y_\alpha)| + |f(x, y_\alpha) - f(x, y)| \\ &\leq \|F(x_\alpha) - F(x)\|_\infty + |f(x, y_\alpha) - f(x, y)|\end{aligned}$$

we see that $f(x_\alpha, y_\alpha) \to f(x, y)$. □

The next lemma follows from 0.6.6 and 0.6.7

B.0.10 Lemma. *Define sets*

$$A_\varepsilon := \{x \in X : G(x) < \varepsilon\} \text{ and } A := \{x \in X : G(x) = 0\}.$$

Then A_ε is open and A is a G_δ. Moreover, $G(x) = 0$ iff F is continuous at x.

B.0.11 Lemma. *If $K \subseteq C(Y)$ is norm-compact and $r > 0$, then the set*

$$K_r := \{x \in X : d(F(x), K) \leq r\}$$

is closed in X. Moreover, if X is a Baire space and $G \geq \varepsilon > 0$ on X, then $\operatorname{int} K_r = \emptyset$ for $r = \varepsilon/12$.

Proof. Let $x_0 \in K_r^c$ and $r < s < t < d(F(x_0), K)$. For any $g \in K$, because Y is compact there exists $y_0 \in Y$ such that

$$d(F(x_0), K) = \min\{\|F(x_0) - h\|_\infty : h \in K\} \le \|F(x_0) - g\|_\infty = |f(x_0, y_0) - g(y_0)|.$$

Therefore, $|f(x_0, y_0) - g(y_0)| > t$, hence, by separate continuity of f, there exists a neighborhood U_g of x_0 such that

$$|f(x, y_0) - g(y_0)| > t \ \ \forall \ \ x \in U_g.$$

Thus if h is in the ball $B_{t-s}(g)$ in $C(Y)$ and $x \in U_g$, then

$$\|F(x) - h\|_\infty \ge \|F(x) - g\|_\infty - \|g - h\|_\infty \ge |f(x, y_0) - g(y_0)| - \|g - h\|_\infty > s. \qquad (\dagger)$$

Now, by compactness of K, there exist $g_1, \ldots, g_n \in K$ such that $K \subseteq \bigcup_j B_{t-s}(g_j)$. Therefore, by (†)

$$\|F(x) - h\|_\infty > s > r \ \ \forall \ \ h \in K \ \text{ and } \ x \in U := \bigcap_j U_{g_j}.$$

Taking the infimum of all h shows that the neighborhood U of x_0 is contained in K_r^c. Therefore, K_r is closed.

Now assume that X is a Baire space and $G \ge \varepsilon$ on X. Since K is compact, we may cover K with closed balls $C_s(g_1), \ldots, C_s(g_k)$, where $g_j \in K$ and $s = \varepsilon/4$. It follows that for $r = \varepsilon/12$, $\{h \in C(Y) : d(h, K) \le r\} \subseteq \bigcup_{j=1}^k C_s(g_j)$ and so

$$K_r \subseteq F^{-1}(\{h \in C(Y) : d(h, K) \le r\}) \subseteq \bigcup_{j=1}^{k} F^{-1}(C_s(g_j)).$$

By the first paragraph, K_r is closed, as are the sets $F^{-1}(C_b(g_j))$ (take $K = \{g_j\}$). Since X is a Baire space, if $\operatorname{int} K_r \neq \emptyset$, then $U := \operatorname{int} F^{-1}(C_s(g_j)) \neq \emptyset$ for some j. It follows that $\|F(x') - F(x'')\|_\infty \le 2s$ (x', $x'' \in U$) and so $G(x) \le 2s = \varepsilon/2$ ($x \in U$), contradicting the hypothesis. □

B.0.12 Lemma. *Let (x_n) be a sequence in X such that every subsequence has a cluster point in X. If x' is a cluster point of (x_n), then $F(x')$ is in the norm-closed convex hull of the set $\{F(x_n) : n \in \mathbb{N}\}$.*

Proof. We show first that the set $S := \{F(x_n) : n \in \mathbb{N}\}$ is relatively sequentially compact in the topology p of pointwise convergence in $C(Y)$. To see this, (g_k) be a sequence in S. If (g_k) has infinitely many distinct terms, then it has a subsequence that is in fact a subsequence of $(F(x_n))$. Since F is clearly p-continuous, the hypothesis on (x_n) implies that (g_k) has a subsequence that p-converges to some $g \in C(Y)$. On the other hand, if (g_k) has only finitely many distinct terms, then it has a constant subsequence, and the same conclusion holds.

By 14.1.4, S is relatively w-compact in $C(Y)$, hence the weak and pointwise closures of S coincide. Since $F(x')$ is in the pointwise closure and since the norm and weak closures of $\operatorname{co} S$ are the same, the conclusion of the lemma follows. □

The proof of B.0.8 is based on the following "game" on a topological space X. The game has two players, α and β. Player β starts the game by choosing a nonempty open set U_1. Player α then chooses a nonempty open set $V_1 \subseteq U_1$ and a point $x_1 \in V_1$. Next, player β chooses a nonempty open set $U_2 \subseteq V_1$. In general, move n of β is the choice of an open set $U_n \subseteq V_{n-1}$, and α's subsequent move n is the choice (V_n, x_n), where V_n is open and $x_n \in V_n \subseteq U_n$. In this way we obtain two decreasing sequences (U_n) and (V_n) of open sets and a sequence (x_n) of points in X. Player α wins the game (and defeats β) if every subsequence of (x_n) has a cluster point in the common intersection $\bigcap_{n=1}^{\infty} U_n = \bigcap_{n=1}^{\infty} V_n$. A **strategy** for α is a rule that governs each of α's moves based only on the immediately preceding move of β. A **winning strategy** for α is a strategy that results in the defeat of β no matter how β moves. A topological space X for which a winning strategy for α exists is called **α-favorable.**

B.0.13 Proposition. **(a)** *A complete metric space is α-favorable.*

(b) *A locally compact Hausdorff space is α-favorable.*

(c) *A nonempty open subset X' of an α-favorable space X is α-favorable.*

(d) *An α-favorable space X is a Baire space.*

Proof. (a) If β makes the move U_n, α counters it with the move $(V_n := B_{r_n}(x_n), x_n)$, where $C_{r_n}(x_n) \subseteq U_n$ and $r_n \downarrow 0$. By Cantor's intersection theorem, $\bigcap_{n=1}^{\infty} V_n = \bigcap_{n=1}^{\infty} C_{r_n}(x_n)$ contains a point x and $x_n \to x$. Therefore, X is α-favorable.

(b) If β makes the move U_n, α counters it with the move (V_n, x_n), where $\operatorname{cl} V_n \subseteq U_n$ is compact and $x_n \in V_n$ is arbitrary (0.12.3). The compactness of $\operatorname{cl} V_1$ implies that every subsequence of (x_n) has a cluster point in the common intersection $\bigcap_{n=1}^{\infty} V_n = \bigcap_{n=1}^{\infty} \operatorname{cl}, V_n$. Therefore, X is α-favorable.

(c) On the nth move in the game on X', player β chooses an open subset U_n of X'. Since U_n is open in X, player α may (and does) counter by a move (V_n, x_n) from the winning strategy.

(d) If X is not a Baire space, then there exists a sequence of open, dense subsets W_n of X such that $W := \bigcap_n W_n$ is not dense in X. Let U be a nonempty open set that does not meet W. The initial move $U_1 = U$ of β then defeats α. Indeed, no matter how α moves at stage $n-1$, β need only choose $U_n = V_{n-1} \cap W_{n-1}$, which results in $\bigcap_n U_n \subseteq U \cap W = \emptyset$. □

We may now prove the following generalization of B.0.8.

B.0.14 Theorem. *Let X and Y be topological spaces with X α-favorable and Y compact Hausdorff. If $f : X \times Y \to \mathbb{C}$ is bounded and separately continuous, then there exists a dense G_δ subset A of X such that f is jointly continuous at every point of $A \times Y$.*

Proof. Let A and A_ε be the sets in B.0.10. By B.0.9 and B.0.10, it suffices to show that A is dense in X. We assume that is not the case and seek a contradiction.

Since X is a Baire space, some A_ε is not dense in X. Thus $G \geq \varepsilon$ on the nonempty open set $X' := (\operatorname{cl} A_\varepsilon)^c$. Since X' is α-favorable, (B.0.13(c)) we may as well assume that $G \geq \varepsilon$ on X. To deduce the desired contradiction, we start the game.

Assume that α plays according to the winning strategy. For the first move, β chooses $U_1 = X$ and α makes a move (V_1, x_1) $(x_1 \in V_1 \subseteq U_1)$ from the winning strategy. Now consider the game immediately after the moves $U_1, \ldots, U_{n-1}$ and $(V_1, x_1), \ldots, (V_{n-1}, x_{n-1})$. For the nth move, β applies B.0.11 to the compact set $K_n := \operatorname{co}\{F(x_1), \ldots, F(x_{n-1})\}$ to obtain a set $K_{n,r} = \{x \in X : d(F(x), K_n) \leq r\}$ with void interior and then chooses the nonempty open set $U_n = V_{n-1} \cap K_{n,r}^c$. Thus, for all n, $d(F(x), K_n) \geq r$ on U_n. Now, since α plays with a winning strategy, (x_n) has a cluster point $x' \in \bigcap_n U_n$. By B.0.12, a convex combination of members of $\{F(x_n) : n \in \mathbb{N}\}$ is within $r/2$ of $F(x')$. But then for some n, $d(F(x'), K_n) < r$, the desired contradiction. □

B.0.15 Remark. Joint continuity results like those considered here go back at least to Baire [3], who proved B.0.8 for $X = Y = [0, 1]$. More refined results were obtained much later by Ellis [18], Namioka [35], and Lawson [31], [32]. For additional references see [41]. The treatment in this appendix via the notion of topological game follows Christensen [7], where f is allowed more generally to take values in a pseudo-metric space. ◊

References

[1] R. Ash and C. Doleans-Dade, *Probability and Measure Theory*, 2nd Ed., Academic Press, San Diego, 2000.

[2] G. Bachman, and L. Narici, *Functional Analysis*, Academic Press, New York, 1966.

[3] R. Baire, *Sur les fonctions de variables réelles*, Ann. di Mat. **3**, 1–123, 1899.

[4] J. Berglund, H. Junghenn, and P. Milnes, *Analysis on Semigroups: Function Spaces, Compactifications, Representations*, Wiley, New York, 1988.

[5] P. Billingsly, *Probability and Measure*, Wiley, New York, 1979.

[6] H. Brezis, *Functional Analysis, Sobolev Spaces, and Partial Differential Equations*, Springer-Verlag, New York, 2011.

[7] J.P.R. Christensen, *Joint continuity of separately continuous functions*, Proc. Amer. Math. Soc. **82**, 455–461, 1981.

[8] J. Clarkson, *Uniformly convex spaces*, Trans. Amer. Math. Soc. **40**, 415–420, 1936.

[9] J. Conway, *A Course in Functional Analysis*, Springer-Verlag, New York, 1990.

[10] K. deLeeuw, and I. Glicksberg, *Applications of almost periodic compactifications*, Acta Math. **105**, 63–97, 1961.

[11] K. deLeeuw, and I. Glicksberg, *Almost periodic functions on semigroups*, Acta Math. **105**, 99–140, 1961.

[12] J. Diestel and J.J. Uhl, *Vector Measures*, Mathematical Surveys, 15, Amer. Math, Soc. Providence, 1977.

[13] J. Dugundji and A. Granas, *Fixed Point Theory*, Springer-Verlag, New York, 2003.

[14] N. Dunford and J.T. Schwartz, *Linear Operators, Vol. I*, Wiley-Interscience, New York, 1958.

[15] A. Dvoretski, P. Erdos, and S. Kakutani, *Nonincrease everywhere of the Brownian motion process*, Proc. 4th Berkeley Symposium on Math Stat. and Prob., Vol. II, 103–116, 1961.

[16] W. Eberlein, *Abstract ergodic theorems and weak almost periodic functions*, Trans. Amer. Math. Soc. **94**, 217–240, 1949.

[17] R. Ellis, *Locally compact transformation groups*, Duke Math Journal, Vol. 24, Number 2, 119–125, 1957.

[18] R. Ellis, *A note on the continuity of the inverse*, Proc. Amer. Math. Soc. 8, 372–373, 1957.

[19] L. Fejer, *Beispiele stetiger Funktionen mit divegenter Fourierreihe*, J. Reine Angew. Math. **137**, 1–5, 1910.

[20] G. Folland, *Real Analysis. Modern Techniques and Their Applications*, 2nd Ed. John Wiley & Sons, New York, 1999.

[21] G. Folland, *A Course in Abstract Harmonic Analysis*, CRC Press, Boca Raton, 1995.

[22] P. Halmos, *Lectures on Ergodic Theory*, Chelsea, New York, 1956.

[23] P. Halmos, *Naive Set Theory*, Springer-Verlag, New York, 1994.

[24] H. Junghenn, *Option Valuation: A First Course in Financial Mathemtics*, CRC Press, Boca Raton, 2012.

[25] H. Junghenn, *A Course in Real Analysis*, CRC Press, Boca Raton, 2015.

[26] J. Kindler, *A simple proof of the Daniell-Stone representation theorem*, Amer. Math. Monthly **90**, 396–397, 1983.

[27] I. Kluvnek and G. Knowles, *Vector measures and control systems*, North-Holland Mathematics Studies 20, North-Holland, New York, 1976.

[28] E. Kreyszig, *Introduction to Functional Analysis with Applications*, John Wiley & Sons, New York, 1978.

[29] H. Kuo, *Introduction to Stochastic Integration*, Springer-Verlag, New York, 2006.

[30] S. Lang, *Real and Functional Analysis*, 3rd Ed., Springer-Verlag, New York, 1993.

[31] J. D. Lawson, *Joint continuity in semitopological semigroups*, Illinois J. Math **18**, 275–285, 1974.

[32] J. D. Lawson, *Additional notes on continuity in semitopological semigroups*, Semigroup Forum **12**, 265–280, 1976.

[33] P. Lax, *Functional Analysis*, 3rd Ed., Wiley Interscience, John Wiley & Sons, 2002.

[34] L. Loomis, *An Introduction to Abstract Harmonic Analysis*, D. Van Nostrand, Princeton, 1953.

[35] I. Namioka, *Separate continuity and joint continuity*, Pacific J. Math, **51**, 515–531, 1974.

[36] G. Pedersen, *Analysis Now*, Springer-Verlag, New York, 1995.

[37] R. Phelps, *Lectures on Choquet's Theorem*, 2nd Ed., Lecture Notes in Mathematics 1757, Springer-Verlag, New York, 2001.

[38] I. Rana, *An Introduction to Measure and Integration*, 2nd Ed., Graduate Studies in Mathematics Vol. 45, AMS, Providence, 2002.

[39] J. Ringrose, *A note on uniformly convex spaces*, J. London Math. Soc. **34**, p.92, 1959.

[40] M. Rosenblum, *On a theorem of Fuglede and Putnam*, J. London Math. Soc. **33**, 376–377, 1958.

[41] W. Ruppert, *Compact Semitopological Semigroups: An Intrinsic Theory*, Lecture Notes in Mathematics 1079, Springer-Verlag, New York, 1984.

[42] S. Saeki, *A proof of the existence of infinite product probability measures*, Amer. Math. Monthly Vol. 103, No. 8, 682-683, Oct. 1996.

[43] S. Shreve, *Stochastic Calculus for Finance*, Springer-Verlag, New York, 2004.

[44] I. Singer, *Bases in Banach Spaces I*, Springer-Verlag, Heidelberg, 1970.

[45] C. Swartz, *An Introduction to Functional Analysis*, Marcel Dekker, New York, 1992.

[46] M. Taylor, *Measure Theory and Integration*, Graduate Studies in Mathematics Vol. 76, American Mathematical Society, Providence, 2006.

[47] S. Taylor, *Exact asymptotic estimates of Brownian path variation*, Duke Math. J. Vol. 39, No. 2, 219–241, 1972.

[48] F. Treves, *Topological Vector Spaces, Distributions, and Kernels*, Academic Press, New York, 1967.

List of Symbols

Analysis on Groups and Semigroups

$L_t f$, $R_t f$, 386; $f * g$, 169,397; f^*, 397; $\widetilde{f}$, 397; $\mathfrak{F}(f)$, $\widehat{f}$, 171,411; $\check{f}$, 172; $\widehat{F}$, $\check{F}$, 380; $\mathscr{P}(G)$, 402; $\widehat{G}$, 411; $\mu * \nu$, 178,400; $\widehat{\mu}$, 179; $WAP(S)$, 424; $AP(S)$, 429; $SAP(S)$, 433; S^{WAP}, 426; S^{AP}, 430; S^{SAP}, 435.

Convergence

$\mathcal{T}\text{-}\lim_\alpha x_\alpha = x$, $x_\alpha \xrightarrow{\mathcal{T}} x$, 21; $f_n \xrightarrow{\text{a.e.}} f$, $f_n \xrightarrow{\mu} f$, $f_n \xrightarrow{\text{a.u.}} f$, 85; $f_n \xrightarrow{L^p} f$, 131; $\mu_n \xrightarrow{v} \mu$, 191; $x_\alpha \xrightarrow{w} x$, 257; $x_\alpha \xrightarrow{w^*} x$, 262.

Functions

id_X, $\iota_A : A \hookrightarrow X$, $\delta_{i,j}$, $\mathbf{1}_A$, x^+, x^-, 5; $\operatorname{Re} z$, $\operatorname{Im} z$, $\overline{z}$, $|z|$, $\operatorname{sgn}(z)$, $\widehat{x}$, δ_x, 6; f^+, f^-, $f_1 \vee \cdots \vee f_n$, $f_1 \wedge \cdots \wedge f_n$, $\sup_n f_n$, $\inf_n f_n$, $\overline{\lim}_n f_n$, $\underline{\lim}_n f_n$, $\operatorname{Re} f$, $\operatorname{Im} f$, $\overline{f}$ and $|f|$, 6; x^α, 172; $\Delta(x)$, 392; $f(\mathcal{A})$, $f^{-1}(\mathcal{B})$, 5.

Function Spaces

$B(X)$, 16; $C_b(X)$, $C(X)$, 25; $C_c(X)$, 34; $C_0(X)$, 35; $C^k(U)$, $C^\infty(U)$, $C_c^k(U)$, $C_c^\infty(U)$, 36; $L^p(X, \mathcal{F}, \mu)$, 123; $L^\infty(X, \mathcal{F}, \mu)$, 126; $L^0(X, \mathcal{F}, \mu)$, 245; $BV(I)$, 160; $AC(I)$, 164; $\mathcal{S} = \mathcal{S}(\mathbb{R}^d)$, 175; $A(\mathbb{D})$, 201; $\ell^p(\mathbb{N})$, 127; $\ell^p(\mathbb{Z})$, $\mathfrak{c}_{00}$, $\mathfrak{c}_0$, $\mathfrak{c}$, 200; $C_K^\infty(U)$, 369; $\mathcal{D}(U)$, 370; $L_k^p(U)$, 380; $\mathscr{P}(G)$, 402; $\widehat{G}$, 411; $WAP(S)$, 424; $AP(S)$, 429; $SAP(S)$, 433.

Measure

$\sigma(\mathcal{A})$, $\varphi(\mathcal{A})$, $\mathcal{B}(X)$, 45; $\mathcal{O}_I$, $\mathcal{C}_I$, $\mathcal{H}_I$, 46; $\mathcal{B}(\overline{\mathbb{R}})$, 46; $\mathcal{A}_1 \times \cdots \times \mathcal{A}_d$, $\mathcal{F}_1 \otimes \cdots \otimes \mathcal{F}_d$, 46; $(X, \mathcal{F}, \mu)$ 51; δ_x, μ_E, 52; $(X, \mathcal{F}_\mu, \overline{\mu})$, $\mathcal{M}(\mu^*)$, 54; μ_E, 52; μ^*, 56; $\mathcal{M}(\mu^*)$, 56; λ, λ^d, 63; $\mathcal{F}/\mathcal{G}$, 75; $T(\mu)$, $h\mu$, $h\,d\mu$, 96; $\mu \otimes \nu$, 112,188; $\mu \overline{\otimes} \nu$, 189; $\mu_1 \otimes \cdots \otimes \mu_d$, 114; $\mu_{i_1} \overline{\otimes} \cdots \overline{\otimes} \mu_{i_n}$, 189; $\nu \perp \eta$, 140; μ^+, μ^-, 140; $|\mu|$, 140,144; μ_r, μ_i, 143; $M(X, \mathcal{F})$, 146; $\nu \ll \mu$, 148; $\frac{d\nu}{d\mu}$, 149; $\overline{D}(\mu; x, r)$, $\underline{D}(\mu; x, r)$, $\overline{D}\,\mu$, $\underline{D}\,\mu$, $D\,\mu$, 154; $V_{I,\mathcal{P}}(f)$, $V_I(f)$, 159; T_f, 161; $M_{ra}(X)$, 182.

Metric Spaces

$d(x, y)$, (X, d), 10; $B_r(x)$, $C_r(x)$, $S_r(x)$, 11; $\operatorname{int}(E)$, $\operatorname{cl}(E)$, $\operatorname{bd}(E)$, 12,19; $d(A, B)$, $d(x, A)$, $d(E)$, 14.

Integration

$\int f$, $\int f\,d\mu$, $\int_E f(x)\,d\mu(x)$, $\int_E f(x)\mu(dx)$, $\int f\,dF$, 89; $\int_E f\,d\mu$, 91; $\underline{S}(f, \mathcal{P})$, $\overline{S}(f, \mathcal{P})$, $\underline{\int_a^b} f$, $\overline{\int_a^b} f$, $\int_a^b f$, 101; $\|\mathcal{P}\|$, 102; $S(f, \mathcal{P}, \xi)$, 103; $\int_{X \times Y} f(x, y)\,d(\mu \otimes \nu)(x, y)$, $\int_Y \int_X f(x, y)\,d\mu(x)\,d\nu(y)$, 112; $\int\!\int f(x_1, \ldots, x_d)\,d\mu_1(x_1) \ldots d\mu_d(x_d)$, 114.

Normed and Locally Convex Spaces

Norms, Seminorms, and Related Concepts

Probability

Sets

Spaces of Linear Mappings

Topological Spaces

Index

Printed and bound by CPI Group (UK) Ltd, Croydon, CR0 4YY
17/10/2024
01775672-0006